EMERGING DEVELOPMENTS IN THE POWER AND ENERGY INDUSTRY

T0122073

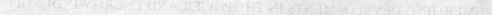

Emerging Developments in the Power and Energy Industry

Editors

Rodolfo Dufo-López
University of Zaragoza, Spain

Jaroslaw Krzywanski
Jan Dlugosz University in Czestochowa, Poland

Jai Singh
Dr. Harisingh Gour Central University, India

CRC Press
Taylor & Francis Group
Boca Raton London New York

CRC Press is an imprint of the
Taylor & Francis Group, an **informa** business

A BALKEMA BOOK

Published by:
CRC Press/Balkema
P.O. Box 447, 2300 AK Leiden, The Netherlands
e-mail: Pub.NL@taylorandfrancis.com
www.crcpress.com – www.taylorandfrancis.com

First issued in paperback 2021

ISBN 13: 978-1-03-223629-2 (pbk)
ISBN 13: 978-0-367-27169-5 (hbk)

Typeset by Integra Software Services Pvt. Ltd., Pondicherry, India

Table of contents

Preface

It is my great pleasure to present the proceedings of 11th Asia-Pacific Power and Energy Engineering Conference (APPEEC2019), held in Xiamen, China from April 19-21, 2019. I would like to take this opportunity to express my sincere thanks to all the authors and participants for their support to our conference.

The continuous researches on Power and Energy Engineering are now of critical significance to the sustainable development of science and the society. Power and Energy Engineering is a broad field of engineering dealing with renewable energy, energy efficiency, smart grid, electric power system operation and planning. It combines knowledge from the fields of physics, math, and chemistry with economic and environmental engineering practices. More and more scholars and practitioners, both within China and abroad, are committed themselves to the cause of this area.

With the development of society and technology, a great variety of research results are emerging. Here, APPEEC provides a platform for academic professionals and industry players to exchange the most updated information and achievements in those exciting research areas.

On behalf of the organizing committee, we appreciate the contribution from all the paper reviewers and the committee members. It is impossible to organize such a conference without their help. The papers in the proceedings of APPEEC provide details beyond what is possible to be included in an oral presentation and constitute a concise but timely medium for the dissemination of recent research results. I hope that you will find these proceedings interesting, exciting and informative.

Prof. Tao Lin
APPEEC Committee Chair

Emerging Developments in the Power and Energy Industry – Dufo-López, Krzywanski & Singh (eds)
© 2020 Taylor & Francis Group, London, ISBN 978-0-367-27169-5

Scientific Committee

General Chair
• **Prof. Tao Lin**
Wuhan University, China

General Co-Chair
• **Prof. Hossein Ganjidoust**
Tarbiat Modares University, Iran

TPC Co-Chair
• **Dr. Zhouyang Ren**
Chongqing University, China

Technical Program Committee
• **Dr. Hedayat Omidvar**
National Iranian Gas Company, Iran
• **Prof. Jaroslaw Krzywanski**
Jan Dlugosz University in Czestochowa, Poland
• **Prof. Jai Singh**
Dr. Harisingh Gour Central University, India
• **Prof. Rodolfo Dufo López**
University of Zaragoza, Spain
• **Dr. Sudhir Sharma**
DAV Institute of Engineering and Technology, India
• **Dr. Dan Li**
China Three Gorges University, China
• **Dr. Hejun Yang**
Hefei University of Technology, China
• **Dr. Qianggang Wang**
Chongqing University, China
• **Dr. Yunfeng Wen**
Hunan University, China
• **Dr. Ming Yang**
Shandong University, China
• **Dr. Shan Cheng**
China Three Gorges University, China
• **Dr. Longfei Liao**
Chongqing Electric Power Company, China
• **Prof. Zunhua Zhang**
Wuhan University of Technology, China
• **Prof. Yongtu Liang**
University of Petroleum-Beijing, China

Emerging Developments in the Power and Energy Industry – Dufo-López, Krzywanski & Singh (eds)
© 2020 Taylor & Francis Group, London, ISBN 978-0-367-27169-5

A safety analysis strategy for transmission lines considering power system oscillation

Jian Hu, Xiaofu Xiong & Jian Wang
State Key Laboratory of Power Transmission Equipment & System Security and New Technology, Chongqing University, Chongqing, P.R. China

ABSTRACT: The safety of transmission lines during power system oscillation has been a long-neglected problem. The conductor temperature exceeds the upper limit of safety during long-term oscillation, which leads to conductor damage or faults. Different operating parameters of transmission lines are closely related to the conductor temperature. Thus, the safety of transmission lines during power system oscillation can be analyzed through the conductor temperature. The time-domain expression of the oscillation current, which is regarded as the cutting point, was initially deduced in this work. Then, the transient thermal balance equation was adopted to calculate the process of conductor temperature increase for the safety analysis of transmission lines during oscillation. Finally, the effectiveness of the proposed method was validated with a case study of typical scenarios. Results show that the response to an increase in conductor temperature is affected mainly by the amplitude of the oscillation current, the oscillation duration, and the meteorological condition of the transmission line. Under severe weather conditions, the conductor temperature exceeds the short-term maximum allowable temperature in a relatively short time. Therefore, a reasonable block time of distance relay or protection against overheating is necessary to ensure the safe operation of transmission lines and to avoid conductor breakdown or cascading faults during power system oscillation.

Keywords: oscillation current, power system oscillation, safety analysis, temperature calculation, transient thermal balance, transmission line

1 INTRODUCTION

Oscillation is defined as the out-of-sync running of a power system or plant in parallel operation (Xu 2002). As a severe system accident, oscillation seriously degrades the stability of the system and the safety of units and restricts the power transfer capacity of a power grid. The wide spread of oscillation can easily cause a system cascading trip and even trigger a large-scale blackout, which would lead to huge losses for the national economy and severe negative effects on the public.

Power system oscillations have occurred many times in various countries and regions. A power system separation was directly caused by the accident-induced oscillation in the US WSCC system in 1996 (Kosterev et al. 1999). In 2008, the Colombian power system experienced large low-frequency oscillations that lasted nearly 90 min (Arango et al. 2010). Several oscillations occurred in central China's power grid during the massive winter storms of early 2008, which brought heavy ice, snows, and below freezing temperature. The

---

*Corresponding author: exphujian@foxmail.com

0loscillations affected the power grids of Hunan, Hubei, and Henan and are thus far the widest-reaching oscillations in central China's power grid (Dong et al. 2008). An extra-low-frequency oscillation existed in the asynchronous interconnection system of the Yunnan power grid in China (Lu et al. 2017). Thus, oscillation is a common occurrence in large-scale interconnected power systems. With the increase in power grid scale, oscillation has become one of the significant factors that threaten the security and stability of power systems. Accordingly, the safety analysis of a power system during oscillation is of great practical significance.

Many scholars and engineers have conducted extensive and in-depth research on the dynamic characteristics, induced mechanism (P. Zhang & T. Bi. 2016), analytical approaches (Chen et al. 2014), and suppression methods (Huang et al. 2015) of oscillation since it was proposed. However, existing research focuses on the effect of oscillation on power systems. A method for the small signal stability analysis of a power system based on online recursive stochastic subspace identification was previously proposed (Yang et al. 2017). The method increases the rapidity of security and stability assessment of modern power systems. Xu (2004) proposed an integrated analysis method and control strategy for oscillation research, which can be applied to enhance the security of a power supply. Li (2016) designed a low-frequency oscillation monitoring and damping system from the steam turbine side, which increases the maximum stability of power systems and guarantees the safe and stable operation of power grids. The aforementioned studies have analyzed the safety of power systems during oscillation from the system level, but few studies have analyzed the safety of power systems from the component level.

As an important and indispensable component of power grids, the transmission line is responsible for power transmission, and its safe operation is essential to the development of the national economy. The voltage and current of a transmission line change remarkably during power system oscillation. However, existing research rarely highlights the effect of oscillation on transmission lines.

Temperature, sag, and stress are important indexes for the safe operation of transmission lines, and they are mutually influenced by one another. The safe operation of transmission lines can be impaired when the sag is extremely large or small. When the sag is extremely large, the stress subject to the transmission line will decrease, which cannot guarantee the safe distance to the ground or conductor of the other phase, leading to short-circuit faults with wind swings. When the sag is extremely small, the stress will be extremely large, possibly leading to wire breakage accidents. Therefore, the stress is bound to the sag, whereas the sag is bound to the conductor temperature of the transmission line. The greater the temperature is, the longer the conductor length and the larger the sag will be. By contrast, the lower the temperature becomes, the less the conductor shrinks, thereby decreasing the stress. The different operation parameters of the transmission line are directly or indirectly related to the conductor temperature. Thus, the safety of transmission lines can be characterized by the conductor temperature. Conductor temperature is applied in this work to judge the safety of transmission lines during oscillation.

To summarize, the time-domain expression of the oscillation current is initially derived on the basis of the current flowing through the transmission line, and the variation of the changes in oscillation current is analyzed. Then, the transient thermal balance equation is used to calculate the conductor temperature increase for the safety analysis of the transmission line during oscillation. Finally, the validity of the proposed method is verified with a case study. Results show that the transmission line will continue to operate in an unsafe condition due to the thermal effect generated by the oscillation current, which will cause conductor damage. The proposed method can be used to improve the oscillating block of distance protection and the criteria for establishing protection against overheating to avoid damage to the transmission line resulting from overheating during long-term oscillation and provide an effective safety analysis method for transmission lines during oscillation.

This article is organized as follows. Section 2 analyzes oscillation current and its characteristics. Section 3 introduces the safety analysis method for transmission lines during oscillation. Section 4 summarizes the case study and result analysis. Section 5 concludes this article.

2 OSCILLATION CURRENT AND ITS CHARACTERISTICS

Power system oscillation may be triggered when the power transmission capacity is extremely large and exceeds the limitation of static stability; the reactive power is deficient, which leads to a voltage drop; the short-circuit faults are removed slowly; or the nonsynchronous automatic reclosing is unsuccessful. Many scholars and engineers have conducted work on changes in electrical quantities during oscillation. Overall, the voltage, current amplitude, and phase change periodically during power system oscillation. Significant attention has been paid to the change in the electrical field caused by the voltage, which is irrelevant to our study. The present work aims to characterize the safety of transmission lines during oscillation through temperature increase response analysis; thus, the heating condition of the conductor must be considered. Therefore, the oscillation current is used as the breakthrough point of this study.

2.1 *Analytic expression of the oscillation current*

The change in current during power system oscillation is illustrated by a two-source power system, as shown in Figure 1. When the oscillation occurs in an all-phase operation mode, we can analyze the change of the oscillation current in a single-phase system mode because of the symmetry of the three phases.

Figure 1 presents the parameters of the power system and transmission line and the assumed directions of electromotive force and current. f_M and f_N are the frequencies of power sources M and N, respectively. E_M is the electromotive force of source M, whose amplitude is equal to E_N, which is the electromotive force of source N with E_M as a reference. E_N is perceived to revolve or swing around E_M, and the oscillation angle of E_N behind E_M varies from 0° to 360°.

The phase expression of the oscillation current can be obtained from Equation (1):

$$I_{sw} = \frac{\Delta E}{Z_\Sigma} = \frac{E_M}{Z_\Sigma}\sqrt{1 + h^2 - 2h\cos\delta}\,e^{j(\theta-\varphi)}. \tag{1}$$

where ΔE is the electromotive force difference between sides, Z_Σ is the system impedance, h is the amplitude ratio of the electromotive force on two sides, θ is the angle ΔE ahead of E_M, and φ is the system impedance angle.

Equation (1) indicates that the amplitude and phase of the oscillation current are related with the oscillation angle δ. Only when δ is constant are oscillation current I_{sw} and θ constant, and the oscillation current will be a pure sine function. When δ is an even multiple of the circumference ratio π, I_{sw} is the minimum. Conversely, when δ is an odd multiple of π, I_{sw} is the maximum.

To obtain the time-domain expression of the oscillation current, the initial phase angle of the current generated by E_M is assumed to be α_0 at any angle based on the premise that E_N is the reference. Then, Equation (1) can be written as a time-domain expression:

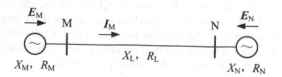

Figure 1. Single-line diagram of the two-source power system.

$$i_{sw}(t) = \frac{\sqrt{2}E_M\sin(2\pi f_M t + \alpha_o) - \sqrt{2}E_N\sin(2\pi f_N t)}{|Z_\Sigma|}. \tag{2}$$

The presumption of the equality of electromotive forces on both sides, that is, $E = |E_M| = |E_N|$, in Equation (2) can be described as

$$i_{sw}(t) = I_m 2\cos\left(\frac{\omega_M + \omega_M}{2}t + \frac{\alpha_o}{2}\right)\sin\left(\frac{\omega_M - \omega_M}{2}t + \frac{\alpha_o}{2}\right), \tag{3}$$

where I_m is the maximum amplitude of the oscillation current, $I_m = |E_{max}/Z_\Sigma|$; ω_M is the angular frequency of source M, $\omega_M = 2\pi f_M$; and ω_N is the angular frequency of source N, $\omega_N = 2\pi f_N$. The frequency of the oscillation current is $(f_M + f_N)/2$ from the cosine function in Equation (3), and the instantaneous values of the sine function change periodically at a period of $2/(f_M - f_N)$. Thus, the oscillation period is $T = 1/(f_M - f_N)$.

When $\omega_o = (\omega_M + \omega_M)/2$, $\omega_s = \omega_M - \omega_N$, Equation (3) can be expressed as follows:

$$i_{sw}(t) = 2I_m\cos\left(\omega_o t + \frac{\alpha_o}{2}\right) \times \sin\left(\frac{\omega_s}{2}t + \frac{\alpha_o}{2}\right). \tag{4}$$

Evidently, $\delta = \omega_s t$, and Equation (4) can be rewritten as

$$i_{sw}(t) = 2I_m\cos\left(\omega_o t + \frac{\alpha_o}{2}\right) \times \sin\left(\frac{\delta}{2} + \frac{\alpha_o}{2}\right). \tag{5}$$

2.2 Characteristics of the oscillation current

On the basis of Equation (3), the oscillation period is $T = 1/(f_M - f_N)$. However, in actual situations, the oscillation period is not always consistent in terms of the oscillation duration due to the coherent, strong nonlinearity of the power system itself. Hence, the oscillation current is an aperiodic quantity. The oscillation period is changing in the entire process of oscillation. Statistics (Cui et al. 1993) show that the first oscillation period is the longest after stability is destroyed. If the oscillation continues to develop, then the oscillation period will be increasingly short. Meanwhile, the last period is long before the stability of power system recovery, and the oscillation period ranges from 0.1 s to 3 s. Furthermore, the oscillation duration, which changes from seconds to minutes, rests with the measurements to restore stability. The amplitude and phase of the oscillation current are constantly changing in the oscillation duration, and both are about the oscillation angle. In an extremely large sustained oscillation current, the temperature increase as a consequence of the oscillation current can break the conductor, which might cause the transmission line to operate continuously in an unsafe condition.

3 SAFETY ANALYSIS STRATEGY OF THE TRANSMISSION LINE

3.1 Transient thermal balance of the transmission line

The thermal balance of transmission lines is affected not only by the conductor current but also by the prevailing ambient weather parameters, such as wind speed and direction, ambient temperature, and solar radiation (Wang 2007). The change in conductor temperature of the transmission line follows a transient heat balance during oscillation:

$$m \cdot c \cdot \frac{dTe}{dt} = q_s + q_J - q_c - q_r, \tag{6}$$

where m is the mass per unit length of the conductor (kg/m), c is the equivalent specific heat capacity of the conductor ($J \cdot kg^{-1} {}^\circ C^{-1}$), T_e is the conductor temperature ($^\circ C$), and q_s, q_J, q_c, and q_r are the solar heat gain, Joule heat gain, convective heat loss, and radiative heat loss rates per unit length, respectively (W/m).

The formulas for q_s, q_J, q_c, and q_r in Equation (6) are briefly described as follows (CIGRE Working Group B2.43 2014):

3.1.1 Solar heat gain

The solar heat gain rate per unit length by conductor q_s (W/m) is directly proportional to the solar absorptivity of the conductor α, the outer diameter of the conductor D_0 (m), the solar radiation intensity at the altitude of the conductor Q_s (W/m^2), and the angle between the incident direction of sunlight and the conductor θ_s ($^\circ$), and it can be expressed by the following general equation:

$$q_s = \alpha D_\circ Q_s \sin \theta_s. \tag{7}$$

3.1.2 Joule heat gain

Joule heat depends on the current flowing through the conductor. The Joule heat of the oscillation current in conductor changing with time is expressed as

$$q_J = \left(\frac{1}{T} \int_{t_1}^{t_2} i_{sw}^2(t)dt \right) R(T_c), \tag{8}$$

where $R(T_c)$ is the alternating current resistance per unit length (Ω/m) when the conductor temperature is T_c, which can be obtained by linear interpolation.

3.1.3 Convective heat loss

Convective heat loss can be expressed as a function of the dimensionless Nusselt number as follows:

$$q_c = \pi \lambda_f (T_c - T_a) N_u, \tag{9}$$

where λ_f is the thermal conductivity coefficient of air, T_c is the surface temperature of the conductor, and T_a is the ambient temperature.

Convection is an important factor for cooling conductors. Two types of convection are considered, natural and forced, which correspond to windless and windy conditions. Forced convection depends on the wind speed and direction relative to the transmission line. At moderate to high wind speeds, forced convection dominates and natural convection can be ignored. At low wind speeds, natural convection has a significant effect, that is, it becomes the dominant mechanism at very low wind speeds.

The forced convective heat loss can be calculated as follows:

$$q_{cf} = \pi \lambda_f (T_c - T_a) Nu_\delta, \tag{10}$$

where $Nu_{\delta w}$ is the Nusselt number when the wind direction is δ_w.

The natural convective heat loss can be calculated as follows:

$$q_{cn} = \pi \lambda_f (T_c - T_a) Nu_{nat}, \tag{11}$$

where Nu_{nat} is the Nusselt number when the wind direction is zero.

When calculating the convective heat loss of the conductor, the maximum values of q_{cn} and q_{cf} are taken as the convective heat loss:

$$q_c = \max\{q_{cn}, q_{cf}\}. \tag{12}$$

3.1.4 *Radiative heat loss*

Radiative heat loss is significantly correlated with the temperature difference between the conductor and ambient air. On the basis of the Stefan–Boltzmann law, the heat loss from the conductor due to radiation can be expressed as follows:

$$q_r = \pi D_0 \sigma_B \varepsilon \left[(T_c + 237)^4 - (T_a + 237)^4 \right], \tag{13}$$

where ε is the emissivity of the conductor and σ_B is the Stefan–Boltzmann constant, which is 5.6697×10^{-8}.

3.2 *Temperature increase response and safety analysis*

Under short-circuit faults, the Joule heat gain by the short-circuit current can be approximately considered to contribute to the conductor temperature increase, and no dissipation occurs due to the short duration of the short circuit, which is an adiabatic process (Miao & Zhu 2015). However, the oscillation period varies in the entire process of oscillation, and its maximum value is 3 s. In addition, the oscillation duration ranges from 10 s to 20 s based on statistics, and the duration can be longer in actual situations. For example, a 4- to 5-min oscillation of a 110-kV transmission line occurred in the southwestern power grid of China on June 27, 1987, due to the malfunction of distance protection, leading to tie-line melting in succession. Therefore, the temperature increase by the oscillation current cannot be regarded as an adiabatic process during power system oscillation.

The oscillation current flowing through the conductor and the ambient weather lead to the change in conductor temperature, which is a transient thermal balance as expressed by Equation (6).

The oscillation period is considerably small and generally smaller than the thermal time constant (Miao & Zhu 2015). Thus, for each interval of oscillation period, the weather parameters remain constant, and the oscillation current is a periodic function related to time t. Consequently, the temperature change (ΔT_e) is easy to calculate by using the oscillation period as the time interval (Δt). Equation (6) is applied to each interval of the oscillation period as follows:

$$\Delta T_e = \frac{q_s + q_J - q_c - q_r}{mc} \Delta t \tag{14}$$

On the basis of Equation (14), the temperature change can be calculated easily. The final temperature for each interval is the initial temperature for the next one. Then, the curve of the conductor temperature increase response is obtained during oscillation. Therefore, many results can be obtained, such as the conductor temperature at the specific time and the time to reach the specific temperature, which can provide important support to the safety analysis of the transmission line and power system.

From the perspective of safe conductor operation, a maximum temperature value will be set for the transmission line, namely, the maximum allowable temperature. In general, T_N is used to indicate the maximum allowable temperature when the conductor is in normal or long-term operation. Nevertheless, for the temperature increase response caused by the increase in current, the short-time maximum allowable temperature (T_E) can be the maximum allowable temperature of the transmission line, which varies with the type, model, and operating

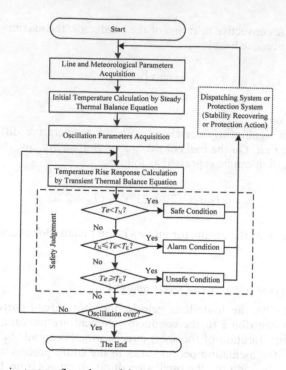

Figure 2. Safety analysis strategy flow chart of the transmission line during system oscillation.

condition of the conductor. For example, the short-time maximum allowable temperature (T_E) of the ACSR is 100°C. Therefore, the safety of the transmission line in oscillation condition can be judged in accordance with the temperature. When the conductor temperature is lower than the maximum allowable temperature (T_N) of the conductor, the transmission line is in a safe condition. When the conductor temperature is higher than the maximum allowable temperature (T_N) and lower than the short-time maximum allowable temperature (T_E), the transmission line is in a safety-alarm condition. When the conductor temperature is higher than the short-time maximum allowable temperature (T_E), the transmission line is in an unsafe condition and urgently requires a transmission line trip-out.

On the basis of the foregoing analysis, the safety analysis flow chart of the transmission line during oscillation can be obtained, as shown in Figure 2.

4 CASE STUDY

To demonstrate the effectiveness of the proposed method, a modified version of the IEEE 39-bus benchmark system is adopted as the test example, as shown in Figure 3. The length of transmission lines L_1–L_3 is 340 km, the conductor type is LGJ 400/35, and the voltage level is 500 kV. Table 1 lists the parameters.

The thermal balance of the system is steady before oscillation (i.e., the differential term on the left side of Equation (6) is zero), and the original temperature of the transmission line can be calculated by the conductor and corresponding meteorological parameters in Tables 1 and 2.

In this work, we assume that line L_3 is in the state of interruption maintenance and failure occurred in line L_2, resulting in relay action and breaker-tripping, which will cause system oscillations.

On the basis of the analysis in the second section, line L_1 is equal to the MN dual-power system. The frequency of the M side is increased to 51 Hz and the frequency of the N side is reduced to 48.5 Hz, leading to system oscillation due to a fault. Then, the calculated

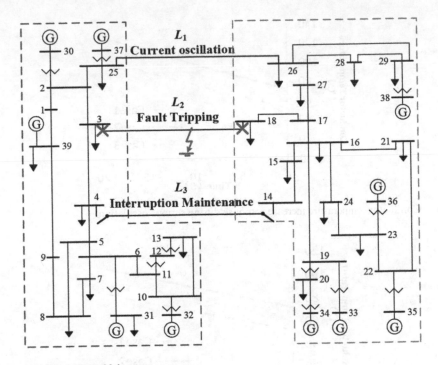

Figure 3. Modified IEEE 39-bus test system.

Table 1. Parameters of LGJ 400/35.

Conductor Parameters	Value
Conductor outside diameter D_0 (mm)	26.82
Steel core diameter (mm)	7.5
Aluminum outer strand diameter (mm)	3.22
Sectional area (mm^2)	425.24
Emissivity ε	0.9
Steel specific heat capacity at 20°C (J/kg • °C)	481
Steel temperature coefficient of specific heat capacity at 20°C	1×10^{-4}
Aluminum specific heat capacity at 20°C (J/kg • °C)	897
Aluminum temperature coefficient of specific heat capacity at 20°C	3.8×10^{-4}
Mass per unit length of steel (kg/m)	0.4059

Table 2. Parameters of ambient weather conditions.

Cases[a]	Ambient Weather(°C)	Wind speed (m/s)	Radiation Intensity (W/m^2)
Case 1	20	0	1000
Case 2	40	0	1000
Case 3	40	2	1000

[a]Case 1, mild weather in spring and autumn; Case 2, high temperature in summer with no wind; Case 3, high temperature in summer with wind.

oscillation period is 0.4 s (i.e., $T = 1 / (f_M - f_N) = 0.4$ s) in the ideal case. The oscillation period is ideally coincident with Equation (5).

The maximum amplitude of the transmission line oscillation current is 5242.88 A from Equation (3) (i.e., the maximum amplitude of the oscillation current of a single conductor is 1310.72 A).

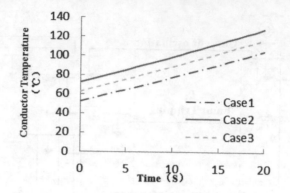

Figure 4.   Conductor temperature increase response during serious oscillation.

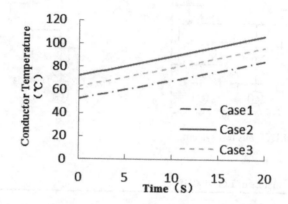

Figure 5.   Conductor temperature increase response during weak oscillation.

In the synthesis of factors, the temperature increase response caused by the oscillation current of the conductor can be plotted under different meteorological parameters, as shown in Table 2, for further safety analysis of the transmission line during power system oscillation.

The oscillation angle varies between 0° and 360° during power system oscillation, as discussed in Subsection 2.1, which is called serious oscillation. Figure 4 plots the temperature increase response generated by the oscillation current. Considering different oscillation situations, the oscillation angle is assumed to change indistinctly, and the angle changes from 0° to 60°. Figure 5 plots the temperature increase response curve of weak oscillation.

As shown in Figures 4 and 5, the temperature increases rapidly when the power system oscillates at a constant period, regardless of the weather, and the conductor temperature exceeds the short-term maximum allowable temperature (100°C). The oscillation duration may be longer than the 20 s, and the conductor temperature reaches a high value in a real scenario. Thus, the temperature will exceed the safe temperature quickly during long-term oscillation, which is detrimental to the safe operation of transmission lines. Thus, the safety of transmission lines during oscillation can be analyzed on the basis of the flow chart in Figure 2. For example, Figure 4 in case 2 shows that the initial temperature of the transmission line is 62.9 °C, which is a safe condition. After 3 s of oscillation, the temperature is greater than T_N and less than T_E, requiring the release of a safety alarm signal for the dispatcher to take relevant measurements to recover power system stability as soon as possible to ensure the safe operation of the transmission line. After 14 s of oscillation, the temperature exceeds T_E, which is in an unsafe condition, and thus requires rapid intervention to avoid conductor breakage from overheating.

The oscillation period is assumed as a constant from Equation (5). However, the actual situation is different. On the basis of the analysis in Subsection 2.2, the oscillation period varies, and the first and the last oscillation periods are longer than the others. Accordingly, the first and last oscillation periods are assumed to be 0.4 s and the others are 0.2 s for the case study of an actual situation. Results show that the temperature increase response is identical with those plotted in Figures 4 and 5. Thus, the oscillation period has minimal influence on the temperature increase response of transmission lines. Consequently, we do not plot and analyze the transmission line condition in an actual oscillation situation.

5 CONCLUSION

The conductor temperature exceeds the upper limit of safety during long-term oscillation, which leads to conductor damage or faults and jeopardizes the safe operation of transmission lines. On the basis of the oscillation current during power system oscillation, the temperature increase response of the conductor flowing through the oscillation current is calculated. Then, the safety of transmission lines during power system oscillation is analyzed. Consequently, the following conclusions are drawn:

- Different parameters characterizing the safe operation of transmission lines are closely related to the conductor temperature. Therefore, the safety of transmission lines can be characterized by the conductor temperature for safety analysis during power system oscillation.
- The temperature increase response cannot be regarded as an adiabatic process because the oscillation duration is considerably longer than the thermal stability calibration time in a short circuit. The conductor temperature increase can be calculated using the transient thermal balance equation in an iterative way.
- The oscillation period has minimal influence on the conductor temperature increase response, which is mainly influenced by the amplitude of the oscillation current, the oscillation duration, and the meteorological environment of the transmission line.
- Comparison of the different safety analyses of the conductor under different ambient weather conditions during power system oscillation reveals that the conductor temperature exceeds the short-term maximum allowable temperature in a relatively short time (15–20 s). Therefore, a reasonable block time of distance relay or protection against overheating is necessary to ensure the safe operation of transmission lines and avoid conductor breakdown or cascading faults during power system oscillation.

REFERENCES

Arango, O.J., Sanchez, H.M., & Wilson, D.H. 2010. Low frequency oscillations in the Colombian power system–identification and remedial actions. CIGRE Session 105.

Chen, L., Sun, H., Zhu, X., & Cheng, S. 2014. Planning and operation schemes for reducing SSR risk of outward power transmission of large-scale thermal generation bases via transmission line with series compensation. *Power System Technology*, 38(1): 113–119.

CIGRE Working Group B2. 43. 2014. *Guide for Thermal Rating Calculation of Over-head Lines*. Paris: CIGRE.

Cui, J., Meng, Q., & Chen, Y. 1993. *Setting Calculation of Power System Relay Protection and Automatic Safety Device*. Beijing: China Electric Power Press.

Dong, M. Yang, D., & Huang, Y. 2008. Analysis on low frequency oscillation incidents measured by WAMS of Central China Power Grid during the 2008 ice hazard. *Central China Electric Power* 21(5): 22–25.

Hu, Y., Deng, C., Xie, X., Li, Y., & Liu, H. 2016. Additional damping control of DFIG series compensated transmission system under sub-synchronous resonance. *Power System Technology* 40(4): 1169–1173.

Huang, J., Chen, W., Dong, D., & Tan, L. 2015. EMTP-based smart optimization of TCSC parameters oriented for mitigating wind farm SSO. *Power System Technology* 39(9): 2411–2417.

Huang, Y., Wang, X., & Chen, K. 2016. SSI mechanism simulation validation and practical mitigation strategy of DFIG-based wind farms. *Power System Technology* 40(8): 2364–2369.

Kosterev, D.N., Taylor, C.W., & Mittelstadt, W.A. 1999. Model validation for the August 10, 1996 WSCC system outage. *IEEE Transactions on Power Systems* 14(3): 975–976.

Li, Y. 2016. Damping technology of low frequency oscillation from turbine side. Huazhong University of Science and Technology, Wuhan, Hubei, China.

Lu, X., Chen, L., Chen, Y., Min, Y., Hou, J., & Liu, Y. 2017. Ultra-low-frequency oscillation analysis of power system primary frequency regulation. *Automation of Electric Power Systems* 41(16): 64–70.

Miao, S. & Zhu, Y. 2015. *Electrical Systems of Power Plants*, 5thed. Beijing: China Electric Power Press.

Wang, M. 2007. *Operation and Design of Relay Protection for High Voltage Power Grid*. Beijing: China Electric Power Press.

Xu, D. 2004. Analysis and control of low frequency oscillations in interconnected power systems. North China Electric Power University, Beijing, China.

Xu, Z. 2002. *New Distance Relay Protection for Transmission Lines*. Beijing: China Water & Power Press.

Yang, D., Wang, W., & Gao, J. 2017. Online electromechanical oscillation analysis and damping modulation for power system using ambient data: I modal parameters identification based on ORSSI. *Proceedings of the CSEE* 38(4): 911–928.

Zhang, P. & Bi, T. 2016. Mechanism analysis of transient disturbance risk of sub synchronous oscillation caused by HVDC. *Proceedings of the CSEE* 36(4): 961–968.

Emerging Developments in the Power and Energy Industry – Dufo-López, Krzywanski & Singh (eds)
© *2020 Taylor & Francis Group, London, ISBN 978-0-367-27169-5*

Neutral point potential balance strategy for NPC three-level inverter based on FOC-ADRC

Kangning Jiang & Lingzhi Yi
Hunan Province Engineering Research Center for Multi-Energy Collaborative Control Technology, Xiangtan University, Hunan Province, China

Yanshen Hu
Shenzhen MOSO Electric Co., Ltd., Guangdong Province, China

ABSTRACT: Aiming at the neutral-point potential fluctuation inherent in the neutral-point clamped three-level inverter, a first-order model is established for the neutral-point voltage, and a neutral voltage balance strategy based on the fractional-order active disturbance rejection control is proposed. The balance strategy can estimate the compensation neutral-point potential disturbance in time on the basis of simplified parameter tuning. The simulation and comparison experiment results validate that the anti-interference ability and compensation speed of the fractional active disturbance rejection control are better than the traditional active disturbance rejection controller and the classic PI controller.

Keywords: clamped three-level inverter, neutral point potential, active disturbance rejection control, fractional-order controller

1 INTRODUCTION

Three-level inverters are widely used in medium and high power photovoltaic power generation because of their high power density, low cost and low interference (Lingzhi YI & Qingzhong GUI 2017) (FENG Teng & KANG Long-yun 2018). However, there are inherent problems in the neutral-point potential, which may lead to safety problems such as output waveform distortion and switching overvoltage. For the DC side neutral-point potential fluctuation, there are usually two solutions including improving the hardware circuit voltage equalization or changing modulation strategy (LI Minyu & MA Xiaojun 2018). But the hardware circuit voltage equalization will increase the production cost, and the modulation strategy equalization does not realize the direct compensation of the potential.

The active disturbance rejection control (ADRC) theory proposed by the Chinese Academy of Sciences Han Jingqing scholars that overcomes the inability to solve overshoot, uncertainty and systemic coupling which is the inherent shortcomings of classical PID control (Han Jingqing 2008). In the field of motor systems and equipment manufacturing caused widespread concern and research (LIU Chunqiang & LUO Guangzhao 2017) (BI Yongtao & WANG Yuhang 2015). In literature (CAI Binjun & ZHU Jianlin 2012), the active disturbance rejection control technology is applied to the PV inverter for current tracking control to suppress various disturbances, but it relies too much on the accurate inverter model, and the NLSEF parameters need to be determined by experience.

In this paper, a control method for directly compensating the neutral-point voltage is proposed. By establishing the DC bus voltage state space model, the neutral-point potential fluctuation is regarded as the system disturbance that can be suppressed by optimizing the voltage loop control.

Establishing a midpoint balance strategy for fractional-order active disturbance rejection control (FOC-ADRC) based on voltage loop. A simulation experiment and a 15kW principle prototype was designed and developed. Experimental results show that this method can suppress and compensate for such disturbances in neutral-point fluctuations without complex parameter tuning, and is superior to traditional PI controller and classical ADRC controller.

2 MATHEMATICAL MODEL ANALYSIS

The fluctuation of the neutral-point potential U_N, that is, the difference ΔU of the divided capacitor voltages depends on the neutral-point current i_o of the inflow and outflow neutral-point potentials. The DC side voltage dividing capacitor $C_1 = C_2 = C$, the expression of the neutral-point potential U_N can be obtained from Figure 1:

$$U_N(t) = \Delta U_o + \frac{1}{C}\int_0^t i_o(t)dt \tag{1}$$

ΔU_o is the difference between the divided capacitor voltages at the initial time, and usually ΔU_o can be ignored. The DC side equivalent model of the T-type three-level inverter is shown in Figure 2. Among them, U_{dc} and i_c are bus voltage and current respectively; i_i is the input current of NPC inverter; R_{ac} is the sum of C_1, C_2, equivalent resistance of inverter loss and line loss.
According to Figure 2 can be listed:

$$\frac{du_{dc}}{dt} = \frac{2}{C}i_s - \frac{2}{R_{ac}C}u_{dc} - \frac{2}{C}i_{in} + \frac{2}{C}i_o \tag{2}$$

By the relationship of variables in formula (2), let

$$\begin{cases} x = u_{dc}, \ b = 2/C, \ \omega = i_o/C, \ u = -i_i \\ f_m = \dfrac{2}{C}i_s - \dfrac{2}{R_{ac}C}u_{dc} \end{cases} \tag{3}$$

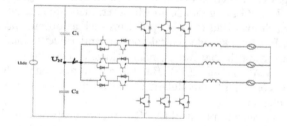

Figure 1. Three-level inverter main circuit topology.

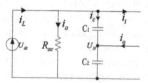

Figure 2. Equivalent circuit of DC side.

The equation of state for the ADRC control system can be listed:

$$\begin{cases} \dot{x} = f_m + bu + w \\ y = x \end{cases} \tag{4}$$

Where f_m is the dynamic model of u_{dc}; x is the state variable; w is the neutral-point potential fluctuation, which is regarded as special interference in FOC-ADRC; u is the control quantity.

3 STRUCTURE OF FOC-ADRC

The main parts of the traditional ADRC include the tracking differentiator (TD), the extended state observer (ESO), and the nonlinear state error feedback device (NLSEF), in which the transition process TD gives a reasonable control signal to solve the response speed in the traditional PID control. The contradiction between overshoot. The ESO is the core control part of the ADRC and is capable of estimating and compensating for system disturbances in real time. NLSEF can reduce the error of the controlled object in the form of an exponent. Among them, NLSEF adopts the speed control integrated function (fhan), which needs to set a large number of parameters. Three of them can only rely on experience to try and make up, which affects its practical application. Therefore, this paper chooses a fractional-order controller (FOC) instead of NLSEF, and together with TD and ESO constitute the fractional-order active disturbance rejection control (FOC-ADRC).

3.1 Fractional-order controller

A number of scholars have introduced fractional calculus into the field of automatic control in order to improve the performance of control system (M.K. Hole & M.D. Padil 2012). Among them, the FOC needs to be set with fewer parameters than other fractional-order controllers, and is superior to linear PID controllers in improving the dynamic performance and robustness of the system. The FOC needs to obtain its fractional differential equation. In this paper, the fractional operator s^γ is discretized by recursion method to obtain the approximate solution. The FOC is derived by the Bird's ideal transfer function and the voltage outer loop transfer function.

The Bird's ideal transfer function can be expressed as:

$$L(s) = \left(\frac{\omega_{gc}}{s} \right)^\gamma, \gamma \in R \tag{5}$$

Where ω_{gc} is the cutoff frequency satisfying $| G(j\omega_{gc}) | = 1$, and γ is the slope of the curve. The system characteristics are as follows:

$$\begin{cases} \varphi_m = \pi - \pi\gamma/2 \\ G_m = \infty \end{cases} \tag{6}$$

Where φ_m and G_m are the maximum values of phase margin and gain margin, respectively; Figure 3 corresponds to the characteristic relationship between system amplitude, phase and frequency. In the dq rotating coordinate system, the simplified voltage control diagram of the NPC three-level inverter is shown in Figure 4.

Figure 4 shows the expression of the the outer voltage loop (controlled object):

$$P(s) = \frac{1}{CT_{sv}s^2 + Cs} \tag{7}$$

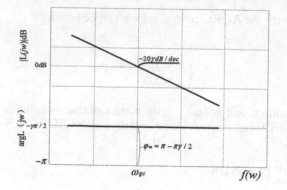

Figure 3.   Relationship of amplitude, phase and frequency.

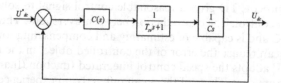

Figure 4.   Voltage outer loop control.

The Bird's ideal transfer function $L(s)$ can be determined by (5) and (6), and the expression of FOC can be derived by combining (7):

$$C(s) = \frac{L(s)}{P(s)} = \frac{\omega_{gc}^{\gamma}(CT_{svs}s^2 + Cs)}{s^{\gamma}}, \gamma \in R \tag{8}$$

The recursive approximation method is used to discretize the fractional operator s^{γ} (J.Q. Han 2007).

$$s^{\gamma} \approx \left(\frac{2}{T}\right)^{\gamma}\left(\frac{1-z^{-1}}{1+z^{-1}}\right)^{\gamma} = \left(\frac{2}{T}\right)^{\gamma}\lim_{n\to\infty}\frac{A_n(z^{-1}, \gamma)}{A_n(z^{-1}, -\gamma)} \tag{9}$$

Considering the accuracy and complexity, in equation (9), n takes 3, and the corresponding expression for A_3 of variables z^{-1} and γ is:

$$\begin{cases} A_3(z^{-1}, \gamma) = -\frac{1}{3}\gamma z^{-3}(1-z) - \gamma z^{-1} + 1 \\ A_3(z^{-1}, -\gamma) = \frac{1}{3}\gamma z^{-3}(1-z) + \gamma z^{-1} + 1 \end{cases} \tag{10}$$

3.2   Tracking differentiator

The tracking differentiator (TD) can extract more accurate differential signals and arrange the signal softening transition process, which is the key to the ADRC's superiority of traditional PID control in balancing fastness and overshoot. The TD transition process uses the speed control integrated function, which is denoted as fhan (x_1-x, x_2, r, h), and its algorithm formula is:

$$\begin{cases} d = rh^2, a_0 = hx_2, y = x_1 + a_0 \\ a_1 = (d^2 + 8d|y|)^{\frac{1}{2}} \\ a_2 = a_0 + sign(y)(a_1 - d)/2 \\ s_y = (sign(y + d) - sign(y - d))/2 \\ a = (a_0 + y - a_2)s_y + a_2 \\ s_a = (sign(a + d) - sign(a - d))/2 \\ fhan = -r(a/d - sign(a))s_a - rsign(a) \end{cases} \qquad (11)$$

Where h is the integral sampling step size; r is the speed factor; Tracking speed and noise filtering ability, take $h = 0.00005$. In addition, $u(k) \in [-r, r]$, the control quantity $u(k)$ is equal to fhan $(x_1 - x_2, r, h)$, which makes x_1 close to u_{dc}. If $u(k)$ is equal to fhan (x_1, x_2, r, h), then x_1 will converge to zero.

3.3 Extended state observer

The extended state observer (ESO) can expand all the disturbances that have an influence on the controlled system into a new state variable. And construct a higher-order observer than the controlled system to observe the expansion amount through the feedback mechanism. It can be seen from Figure 2 that the mathematical model of the neutral-point potential is the first order, the neutral-point potential fluctuation is regarded as the system disturbance, the bus voltage u_{dc} is taken as the input. The second-order ESO is designed to estimate the neutral-point potential fluctuation in time, as follows:

$$\begin{cases} e(k) = z_1(k) - u_{dc} \\ z_1(k+1) = z_1(k) + h(z_2(k) + bu - \beta_1 e) \\ z_2(k+1) = z_2(k) + h(-\beta_2 fal) \end{cases} \qquad (12)$$

$fal(e, \gamma, h)$ is a nonlinear power function whose expression is as follows:

$$fal(e, \gamma, h) = \begin{cases} |e|^\gamma sign(e), & |e| > h \\ \frac{e}{h^{\gamma-1}}, & |e| \le h \end{cases} \qquad (13)$$

The parameters β_1 and β_2 can be determined according to the relationship between the Fibonacci sequence and the ESO parameters (HAN Jing-qing 2008):

$$\begin{cases} f_0 = 1, f_1 = 1, f_{n+1} = f_n + f_{n-1}, n = 1, 2, 3, \cdots \\ \frac{1}{1^0 h}, \frac{1}{(3h)^1 h}, \frac{1}{(8h)^2 h}, \cdots \\ \beta_1 = \frac{1}{h}, \beta_2 = \frac{1}{3h^2} \end{cases} \qquad (14)$$

3.4 Neutral-point potential balance control strategy

This paper proposes a neutral-point potential balancing strategy based on fractional-order active disturbance rejection control (FOC-ADRC), which only acts on the voltage outer loop of the NPC three-level inverter. The current inner loop still uses the grid-based feedforward decoupling PI control. The control block diagram is shown in Figure 5, where e_d is the grid voltage, u_{dc} is the bus voltage, i_d and i_q are the current components in the dq coordinate system, and the neutral-

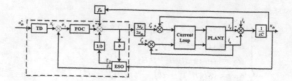

Figure 5.   Control block diagram of FOC-ADRC.

point potential is disturbed by $w = i_o/C$. The f_m is the dynamic compensation of the system and can be shared the pressure of ESO.

4   SIMULATION AND EXPERIMENTAL VERIFICATION

In Matlab/Simulink, the actual parameters and model simulations of the NPC three-level inverter are used. The simulation results are shown in Figure 6 including a comparison of the traditional PI controller, ADRC controller and FOC-ADRC controller for the neutral-point voltage balance control. The neutral-point voltage is represented by the sampled voltage of C_1. In the startup phase, the neutral-point potential of the inverter system based on FOC-ADRC controller is first stabilized at $u_{dc}/2$, and there is no large startup fluctuation. But the stability of the inverter system based on the traditional PI controller is the worst. The voltage disturbance is added at 0.3s. The neutral-point voltage of the inverter system based on the traditional PI controller fluctuates from 300V to 279V instantaneously, and it takes 0.23s to restore the steady state. The neutral-point voltage of inverter system based on the ADRC fluctuates to 288V, recovery time is 0.16s. But the inverter system based on FOC-ADRC controller only fluctuates to 292V, and only consumes 0.019s to recover. It can be seen that the voltage fluctuation of the point in the traditional PI control scheme is more obvious, and it is difficult to realize the rapid estimation of the disturbance size and timely recovery. However, the two ADRC schemes are less affected by the voltage disturbance. The FOC-ADRC can be more accurate estimation of disturbance and timely compensation, the neutral-point voltage recovery is more rapid, with better anti-interference ability and steady-state performance.

In order to further verify the effectiveness of the proposed method for the neutral-point balance control, a 15KW three-phase NPC three-level inverter prototype based on the main control chip DSP28335 was built in the laboratory. The specific parameters are: input voltage 600V; output 220V/50Hz; switching frequency 18kHz; voltage dividing capacitor $C_1 = C_2 = C = 1.2mF$; control parameter $h = 5 \times 10^{-5}s$; cutoff frequency ω_{gc} is 1.5×10^5 rad/s; order the number is $\gamma = 0.5$ and the phase margin is $3\pi/4$. It is used to verify the anti-interference ability of the inverter system based on FOC-ADRC controller and the inverter system based on traditional PI controller. The source of the neutral-point potential disturbance in the experiment is the specific voltage disturbance and the normal working process voltage of the inverter fluctuation.

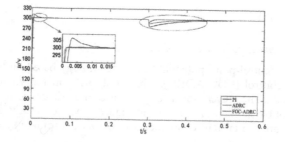

Figure 6.   Neutral-point voltage simulation.

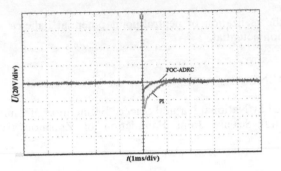

Figure 7. Output waveform for specific disturbance.

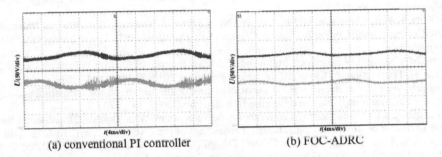

(a) conventional PI controller   (b) FOC-ADRC

Figure 8. Divided capacitor waveform of two methods under normal working.

Figure 7 shows a comparison of experimental waveforms for a given neutral-point potential disturbance based on a conventional PI controller and FOC-ADRC controller. For a given transient voltage disturbance, the FOC-ADRC can ensure that the voltage fluctuation of the voltage dividing capacitor C_1 is smaller, the steady-state recovery time is shorter, and the steady-state performance is better.

Figure 8 shows the partial voltage waveforms of the two methods for the neutral-point potential fluctuation inherent in normal operation. The inverter system based on the traditional PI controller has poor balance ability for the neutral-point potential, and the neutral-point potential has significant fluctuations. In this paper, the inverter system based on FOC-ADRC controller can realize the basic equalization of two voltage dividing capacitors, and the neutral-point potential fluctuation is almost zero, which can ensure the smooth operation of the bus voltage.

5 CONCLUSION

Aiming at the imbalance of the neutral-point potential inherent in the three-level inverter, this paper considers the neutral-point potential fluctuation as the system disturbance, and proposes a neutral-point balance strategy based on FOC-ADRC. Fractional-order controller is used to optimize the NLSEF of active disturbance rejection control and simplifies the parameter tuning steps. Simulation and experimental results show that the FOC-ADRC controller has stronger anti-interference ability and dynamic performance than the conventional ADRC controller and the traditional PI controller. It can estimate and compensate the system disturbance caused by the neutral-point potential fluctuation in time, and improve system stability based on the realization of neutral-point potential balance. FOC-ADRC has the engineering application potential.

ACKNOWLEDGMENT

The project Supported by National Natural Science Foundation of China (NSFC) (61572416) and National Natural Science Joint Foundation of Zhuzhou City, Hunan Province (2016JJ5033).

REFERENCES

Lingzhi YI, Qingzhong GUI, Zhen WANG. A Novel QuasiZ Source Cascaded Multilevel Photovoltaic Power System Applied to Smart Microgrid[A]. The 10th AsiaPacific Power and Energy Engineering Conference, 2017: 35–42

FENG Teng, KANG Long-yun, HU Bi-hua, WANG Ze-feng, A Neutral-point Potential Balancing Predictive Control Method Based on Medium Vectors Being Synthesized[J]. Power Electronics, 2018, 52(04): 8–10.

LI Minyu, MA Xiaojun. Research on Full Range Accurate Balance of Neutral Point Voltage for T-Type Inverter[J]. Transactions of China Electrotechnical Society, 2018, 33(08): 1814–1826.

Han Jingqing. Active disturbance rejection control technique-the technique for estimating and compensating the uncertainties[J]. National Defense Industry Press, 2008: 197–270.

LIU Chunqiang, LUO Guangzhao, TU Wencong. Servo Systems With Double Closed-loops Based on Active Disturbance Rejection Controllers[J]. Proceedings of the CSEE, 2017, 37(23): 7032–7039.

BI Yongtao, WANG Yuhang. Attitude Control Design of Missiles with Dual Control Based on Model Predictive Control and Active Disturbance Rejection Control[J]. Journal of Astronautics, 2015, 36(12): 1373–1383.

CAI Binjun, ZHU Jianlin. ADRC current track control of PV grid-connected inverter[J]. Electric Power Automation Equipment, 2012, 32(03): 104–108.

M.K. Hole, M.D. Padil, V.A. Vyawahare, Stability analysis of fractional-order systems[J]. International Journal of Computer Applications, 2012, 11: 12–18.

J.Q. Han, Active disturbances rejection control technique[J]. Frontier Science, 2007(01): 24–31.

HAN Jing-qing, Parameters of the Extended State Observer and Fibonacci Sequence[J]. Control Engineering of China, 2008(S2): 1–3.

Emerging Developments in the Power and Energy Industry – Dufo-López, Krzywanski & Singh (eds)
© 2020 Taylor & Francis Group, London, ISBN 978-0-367-27169-5

Probabilistic power flow calculation of power system considering DGs based on improved LHS

Zhi Li
College of Information Engineering, Xiangtan University, Xiangtan, Hunan Province, China

ABSTRACT: Probabilistic load flow (PLF) calculation is an important tool for power system steady state performance analysis. Most of the existing PLF methods require to know the probability distribution functions of input random variables, which are very difficult to be modeled accurately. In the paper, a PLF method based on gravitational search algorithm-(GSA) and Latin hypercube sampling (LHS) was proposed. It has many advantages, such as high accuracy and fast computation. It is unconstraint by the type of the input random variables probability distributions. Sequential quadratic programming (SQP) algorithm is incorporated in GSA as a local search mechanism to improve the performance of the conventional GSA algorithm, The validity of proposed method was tested on IEEE 14 bus system and IEEE 118 bus system. The simulation results show that the proposed method is effective, which not only has the advantages of fast computation, high accuracy and good robustness, but also can deal with the correlation between input random variables flexibly. The proposed method is of some practical engineering value.

Keywords: probabilistic load flow, Latin hypercube sampling, sequential quadratic programming, gravitational search algorithm

1 INTRODUCTION

By introducing the uncertainty of load, generator and network structure in the calculation of power flow, probabilistic power flow analysis obtains the probability distribution results of nodal voltages, branch load flow voltage limit, line overload and so on. Probabilistic load flow and its calculation method were first put forward in document (B. Borkowska 1974) in 1974. The basic algorithms are divided into analytical and simulation methods (Hu Zechun &Wang Xifan 2009, Ding Ming et al.2001): the simulation method needs a large number of simulation calculations. The algorithm is more time-consuming than the analytic method. However, it is widely accepted and used as the standard to verify the other probabilistic power flow methods (P. W. Sancer& G. T. Heydt 1978, G. T. Heydt 1975). Latin hypercube sampling (LHS) (Zhang Yibin & Wang Weisheng 2004) can improve the computational efficiency without changing the distribution of the output random variables, so it is widely used in the probability analysis of power systems, especially in the probabilistic power flow calculation.

Gravitational search algorithm (GSA) is a swarm intelligence optimization algorithm proposed by (Echat Reshedi Esmat R, Hossein N, Saeid S 2009) in 2009. GSA mimics the universal gravitational phenomena in the universe, and use the universal gravitation among the particles in the swarm to guide the search intelligently. The algorithm has the advantages of simplicity and versatility. While GSA has many shortcomings of poor local search ability and premature convergence,. In order to overcome its shortcomings, the sequential quadratic programming (SQP) (Birla D Maheshwari& R.P. Gupta 2006) algorithm with strong local optimization ability is introduced in the basic GSA, and an improved gravitational search algorithm (IGSA) is proposed. The classical algorithm SQP as a single operator improves the local search capability of the basic GSA effectively, and also overcomes its precocity and convergence phenomenon. Numerical simulation results show that the IGSA algorithm is superior to the basic GSA in terms of convergence speed and optimization accuracy, and achieves satisfactory optimization results.

This paper proposes an improved theory of universal gravitational search and Latin hypercube, which controls the input variables from the LHS sampling. The simulation results of IEEE-14 and IEEE-118 node test system are analyzed to verify the accuracy, rapidity and practicability of the proposed method.

2 IMPROVED LATIN HYPERCUBE SAMPLING

2.1 Sample generation

Latin hypercube sampling allows uniform distribution of samples and covers the entire sampling space. The input variable X has a sampling scale of N, and its corresponding cumulative probability distribution function is:

$$Y = F_{(X)} \ Y \in [0, 1] \tag{1}$$

Therefore, the sampling process is to divide Y into equal parts of N, taking the middle point of each interval as the sampling value of Y, and the sampling value of the random variable X can be obtained through the inverse cumulative distribution function:

$$X = F^{-1}(\frac{l - 0.5}{N}) \ l = 1, 2, ..., N \tag{2}$$

Yu Han combined Gram-Schmidt sequence orthogonalization and LHS algorithm together to sequence (Gram-Schmidt Latin hypercube sampling, GSLHS). This method realizes the correlation of different precision according to the requirement of processing problem, but the amount of calculation is large. Therefore, a new method is proposed in this paper, and the universal gravitational search algorithm is put forward. It is applied to Latin hypercube method, and the sampling of LHS is improved.

2.2 Ordering in improved latin hypercube sampling

2.2.1 Gravitational search algorithm

The improved gravitational search algorithm proposed is used to sample and study the random variables in probabilistic power flow. This method is similar to the approach adopted in the literature (Yu Han et al.2009), which is aiming at reducing the correlation between the sampled values in the random variable, the objective function is presented as follows:

$$\rho_{rms} = \frac{\sqrt{\sum_{i=1}^{K} \sum_{j=1}^{K} \rho_{ij}^2 - K}}{\sqrt{K(K - 1)}} \tag{3}$$

where ρ_{ij} is the element of setting the correlation coefficient matrix, K is the quantity of input random variables obtaining the features of correlation.

The gravitational force between particle i and particle j can be expressed as:

$$F_{ij}^d(t) = G(t)\frac{M_i(t) \times M_j(t)}{R_{ij}(t) + \varepsilon}(x_j^d(t) - x_i^d(t)) \tag{4}$$

Where $G(t)$ is universal gravitational constant in the period of t, $M_i(t)$ and $M_j(t)$ indicate the inertial mass of the particle i and j respectively, $R_{ij}(t)$ is the Euclidean distance between particle i and particle j, ε represents a very small constant, $x_i^d(t)$ and $x_j^d(t)$ are the positions of the particles i and the particles j in the d-dimensional space respectively.

In the period of t, the gravitation and acceleration of the particle i in the d-dimensional space are as follows:

$$F_i^d(t) = \sum_{j=1, j \neq i}^{N} rand_j \times F_{ij}^d(t) \tag{5}$$

$$a_i^d = \frac{F_i^d(t)}{M_i(t)} \tag{6}$$

Where $rand_j$ is a random number.

Particles will update the speed and position in each iteration according to the acceleration, that is:

$$v_i^d(t+1) = rand_j \times v_i^d(t) + a_i^d(t) \tag{7}$$

$$x_i^d(t+1) = x_i^d(t) + v_i^d(t+1) \tag{8}$$

2.2.2 Overview of SQP

The SQP method seems to be the best nonlinear programming method for constrained optimization problems. SQP is a direct way to handle constraints in an explicit manner.

It outperforms all other non-linear programming methods and perfume better than a large number of test problems in terms of efficiency, accuracy, and percentage of successful solutions. This method is very similar to Newton's constrained optimization method, just like unconstrained optimization. In each iteration, the Hessian of the Lagrangian function is approximated using the Broden-Fletcher-Goldfab-Shanno (BFGS) quasi-Newton correction method. A quadratic programming (QP) subproblem is then generated using the approximated results, the solution of which is used to form the search direction of the line search process. Since the minimization is non-convex, SQP ensures a local minimum of the initial solution.

Each iteration in the SQP algorithm involves three major steps: (i) solve the QP subproblem, and obtain the search direction; (ii) line search and merit function calculation; (iii) updating of the Hessian matrix of the Lagrangian function, using BFGS formula.

2.2.3 Improved GSA Algorithm

As a probability-based multipoint search technique, GSA has a disadvantage, sometimes converging to values that may not be optimal. On the other hand, the GSA searches for a large solution space with good convergence, making it close to the global optimum. The NLP method, such as SQP, is a single-point search method. If the initial selection is closer to the local optimum, there is a disadvantage of being trapped in the local optimum. The NLP method gives the global optimal solution if an appropriate initial selection is made.

Taking advantage of both GSA and SQP methods while overcoming the shortcomings of these methods, an SQP hybrid GSA is proposed. The SQP program is incorporated into the GSA as a local search mechanism to improve convergence. First, the GSA algorithm is executed normally, and the optimal fitness for each generation is selected in each iteration. From this optimal fitness, the corresponding agent is set to the initial value of the variable of the SQP method. The SQP routine is then executed depending on the adopted probability of local search α_{LS}, providing an improvement in the optimal fitness obtained from the GSA in the current iteration. In this way, the IGSA algorithm gives the global optimal solution.

2.2.4 *The improved GSA algorithm flow*

(1) Enter the power grid data, calculate the fitness value of each particle in the population and select the current optimal individual position;

(2) Generate initial population of N agents, set the iteration counter $t = 1$

(3) Evaluate the fitness for each agent, update the best(t) and worst(t), determinate corresponding best agent, X *best*.

(4) Generate the best population $\hat{g}_k$, and best fitness of population $f(\hat{g}_k)$.

(5) If r *and* $> \alpha_{LS}$, execute next step, otherwise execute step(7).

(6) Local search: SQP execution with $Xbest$ as point; New best output $f(\hat{g}_{k.sqp})$ and corresponding variable $\hat{g}_{k.sqp}$ obtain.

(7) Calculate G(t), M(t), F(t) and a(t).

(8) Update velocity and position of each agent.

(9) The result is output if $f(\hat{g}_{k.sqp}) < f(\hat{g}_k)$, otherwise return to step (2).

3 PROBABILISTIC MODEL DESCRIPTION

Distributed power supply used in this paper includes wind turbines and photovoltaic generators. The output of wind turbines varies with wind speed. The probability model of wind speed is based on *Weibull* distribution. It is simple in form and match the actual wind speed distribution, its probability density function is:

$$ f(v) = \frac{k}{c} \left(\frac{v}{c}\right)^{k-1} \exp\left[-\left(\frac{v}{c}\right)^k\right] \tag{9} $$

where, v is wind speed; k and c are 2 parameters of *Weibull* distribution and $k = 9.6372$, $c = 2.4531$; in this case, the *Weibull* distribution is adopted.

The output power of photovoltaic system is closely related to the intensity of illumination. Because of the randomness of light intensity, the output power of photovoltaic power generation is also random. According to statistics, in a certain period of time (1 hours or hours), the intensity of sunlight can be approximately regarded as the *Beta* distribution, and its probability density function is:

$$ f(r) = \frac{\tau(\alpha + \beta)}{\tau(\alpha) + \tau(\beta)} \left(\frac{r}{r_m}\right)^{\alpha-1} \left(1 - \frac{r}{r_m}\right)^{\beta-1} \tag{10} $$

where, r, r_{max} represent the actual and maximum intensity of illumination, α, β are the shape parameters of the *Beta* distribution of illumination; and τ is *Gamma* function.

The power load is sometimes time variant, and it is proved by long-term practice that the random model of the active power and reactive power in random flow calculation is accord with the normal distribution (Justus C G et al. 1976), and its probability density function is as follows:

$$ f(P) = \frac{1}{\sqrt{2\pi}\sigma_P} \exp\left[-\frac{(P - \mu_P)^2}{2\sigma_P^2}\right] \tag{11} $$

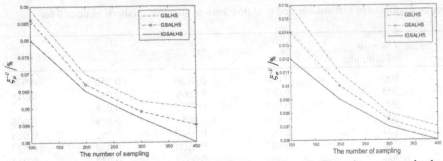

a The voltage expected value of the average relative error   Figure. b. The voltage expected value of the average
relative error

Figure 1.   The voltage of the average relative error map.

$$f(Q) = \frac{1}{\sqrt{2\pi}\sigma_Q} \exp\left[-\frac{(Q - \mu_Q)^2}{2\sigma_Q^2}\right] \tag{12}$$

where, μ_P and μ_Q are the average values of active and reactive power respectively, σ_P and σ_Q denote standard deviations of active and reactive power respectively.

The node injection power equation and branch load flow equation of the probabilistic load flow model are presented as follows:

$$\begin{cases} X = f(W) \\ Z = g(x) = g[f(W)] \end{cases} \tag{13}$$

In the formula, W is injected power of node, X is injected power variable of node, Z is the flow variable of system branch, F and G are node injection power equation and branch flow equation respectively. The concrete flow chart of flow calculation is shown in Figure 1.

The accurate values and standard deviations of the output random variables are represented by μ_{arc} and σ_{arc} respectively. The expected values and standard deviations of the output variables obtained by the probability flow under the small scale sampling are μ_{sim} and σ_{sim} respectively. The accuracy of the calculation results is calculated by using the relative error ξ_μ and ξ_σ of the expected value of output random variable and standard deviation. The formula is as follows:

$$\xi_\mu = \left|\frac{\mu_{sim} - \mu_{arc}}{\mu_{arc}}\right| \times 100\% \tag{14}$$

$$\xi_\sigma = \left|\frac{\sigma_{sim} - \sigma_{arc}}{\sigma_{arc}}\right| \times 100\% \tag{15}$$

4   EXAMPLE ANALYSIS

4.1   *IEEE-14 bus test system*

In order to verify the universality of this method, the paper takes the IEEE-14 node and the IEEE-14 node system as an example to compare the convergence effect of CSMCS, GSLHS, GSALHS and the improved method of this paper respectively under local correlation test. The

Table 1. Error comparisons of bus voltage and load flow between the two methods at node 9 for IEEE14-bus system.

The method of sampling	The scale of sampling	$\xi_\mu^{-U}/\%$	$\xi_\sigma^{-U}/\%$	$\xi_\mu^{-P}/\%$	$\xi_\sigma^{-P}/\%$
IGSALHS	100	0.080	0.0120	0.050	0.026
	200	0.065	0.0090	0.032	0.018
	300	0.057	0.0070	0.028	0.010
	400	0.050	0.0060	0.019	0.006
GSALHS	100	0.086	0.0140	0.052	0.028
	200	0.067	0.0100	0.039	0.022
	300	0.059	0.0075	0.030	0.013
	400	0.055	0.0064	0.023	0.012
	100	0.089	0.0160	0.073	0.055
	200	0.070	0.0110	0.072	0.046
GSLHS	300	0.062	0.0080	0.060	0.037
	400	0.060	0.0070	0.046	0.033

IEEE-14 node 9 is connected to the wind farm (Allan R N et al 1976), and the bus 1 is the balance bus. The output characteristics of the wind farm can be expressed as:

$$P_W = \begin{cases} 0, & v < v_{ci} \text{ or } v > v_{co} \\ P_{WR}\frac{v - v_{ci}}{v_t - v_{ci}}, & v_{ci} \leq v < v_t \\ P_{WR}, & v_t \leq v \leq v_{co} \end{cases} \tag{16}$$

where, P_{WR} is the rated capacity of the wind farm, the value is 100 KW; v_{ci}, v_t and v_{co} are the cut-in wind speed, the rated wind speed and the cut-out wind speed of the wind turbine, respectively, and their corresponding values are $v_{ci} = 2.5$ m/s, $v_t = 13$ m/s and $v_{co} = 25$ m/s.

The IEEE-14 node system is shown as below. Assuming that the system has 21 random variables, and they belong to the active power, reactive power and wind speed of the load, Load output obeys normal distribution, while the voltage qualified range is (0.95,1.05)p.u.

The simulation results can be obtained from simulation below:

1) The expectation error and standard deviation of the three methods are shown as Figure 1. The expected value error and the standard value error of the output variables of the three methods decrease with the increase of the sampling scale, and the fluctuation of the calculated results decreases with the increase of the sampling scale, however under the same sampling scale, the error of IGSA can be reduced greatly and the accuracy can be much improved.

It can be seen from the Table 1 that under the same sampling scale, the accuracy of IGSALHS is higher than GSLHS and GSALHS when calculating the 9 node voltage and 9-10 branches.

2) Figure 2 and Figure 3 give the probability cumulative distribution (CDF) curves of the node 9 voltage amplitude corresponding to the different output variables calculated by the IGSALHS, GSALHS, GSLHS, and MCS algorithms for the sampling scale, respectively, and the probability accumulative distribution (CDF) of the power of the line 9-10. It can be seen from the Figure 6 and 7 that the calculation results of IGSALHS is closer to the reference curve, and better than the approximate simulation of GSLHS and GSALHS.

Table 2 is a time comparison analysis, where the computation time of the IGSALHS method is the average of the 100 calculation results. It can be seen that the IGSALHS algorithm is slow because of its randomness, but the accuracy is much improved.

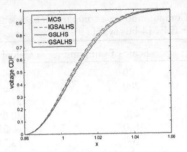

Figure 2.   Comparisons of CDF voltage at node 9.

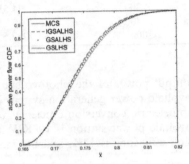

Figure 3.   Comparisons of CDF active power flow of line.

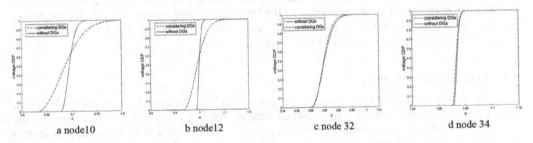

| a node10 | b node12 | c node 32 | d node 34 |

Figure 4.   CDF comparison of the voltage at some nodes.

4.2   IEEE-118 bus test system

Under the IEEE-118 node system, assuming that the probabilistic load flow calculated by the 10000 times MCS algorithm is accurate. The wind farm is injected into node 9, the bus 1 is the balance bus and the node 11-13 is connected to the photovoltaic power station. The parameters of the wind farm are in accordance with the IEEE-14 node system. The random output of the photovoltaic cells is obeyed *Beta* distribution, and the probability density function is as follows:

$$f(P_{PV}) = \frac{\tau(\alpha + \beta)}{\tau(\alpha) + \tau(\beta)} \left(\frac{P_{PV}}{P_{\mathrm{m}}}\right)^{\alpha-1} \left(1 - \frac{P_{PV}}{P_{\mathrm{m}}}\right)^{\beta-1} \tag{17}$$

Table 2. Computation time comparison between the two methods for IEEE14-bus system 9-10.

Method	Scale of sampling	Calculation time/s
IGSALHS	100	8.96
	200	14.37
	300	21.32
	400	30.04
GSALHS	100	8.32
	200	12.37
	300	20.32
	400	28.74
GSLHS	100	10.32
	200	15.66
	300	24.63
	400	30.52
MCS	10000	1287.65

where, P_m is the maximum output power of the photovoltaic generator, $P_{PV} = rA\eta$ is the actual output power of photovoltaic power generation system when time is t, A is the total light area and the η is the photoelectric conversion efficiency. $\alpha = 2.80$, $\beta = 6.32$. The maximum output power of photovoltaic power station is 100KW, the illumination area is 1000 m^2, the photoelectric conversion efficiency is 14%, and the maximum radiation intensity is 28 $W \cdot m^{-2}$.

4.3 Change of node voltage

The results of the voltage cumulative distribution curve (CDF) of the nodes are shown in Figure 5. It can be seen from the diagram that the voltage fluctuation of the node increases after the DGs injected into the system, and the farther the node is from the distributed power supply, the smaller the range of voltage fluctuation will be; the closer the node nears to the DGs, the more obvious the voltage out-of-limit.

4.4 Change of branch flow

The cumulative probability distribution of change of branch flow is respectively shown in Figure 6. From figure(a), there is more out-of-limit probability occur when considering the correlation between the branches for the active power flow of line 9-10 compared with no-considering, the former arrives at 32.16%, and the latter is only 3.22%. Another example is the line 19–20 of its out-of-limit probabilities being 36.52% and 1.76% under the two conditions. Through the above comparison, it can be seen that after the access to DGs, the active power

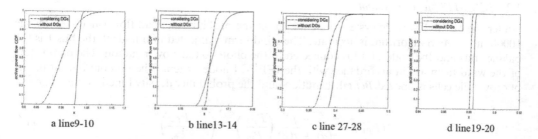

| a line9-10 | b line13-14 | c line 27-28 | d line19-20 |

Figure 5. CDF comparison of active power flow of some lines.

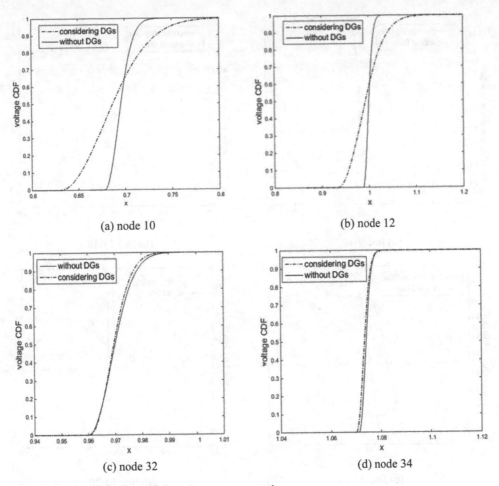

(a) node 10

(b) node 12

(c) node 32

(d) node 34

Figure 6.   6CDF comparison of the voltage at some nodes.

flow of the different branches of the system is different. The closer to the DGs access point, the heavier flow of the original power system will be, the more serious the fluctuation of the active power flow, namely suffering the greater the impact of the power of the DGs; the farther from the DGs access point, the lighter flow of the original power system will be, and smaller the influence of active power flow of its branch wind power will suffer from the access to DGs.

5   CONCLUSION

(1) In this paper, a hybrid GSA-SQP optimization algorithm has been proposed and success-fully applied to Latin hypercube sampling probabilistic power flow calculation. Simulation results show that the algorithm provides effective and robust high-quality solution. Moreover, the results obtained using IGSA is either better or comparable with those obtained using other techniques compared in the literature. The proposed method is suitable to find the global optimal solution for the probabilistic power flow calculation, for it benefits from the advantages of both GSA and SQP methods.

(2) The randomness of distributed generation increases the fluctuation of the node voltage, and the probability of the voltage limit of the node becomes greater. The closer to the access point of the distributed generation, the more obvious the influence is.

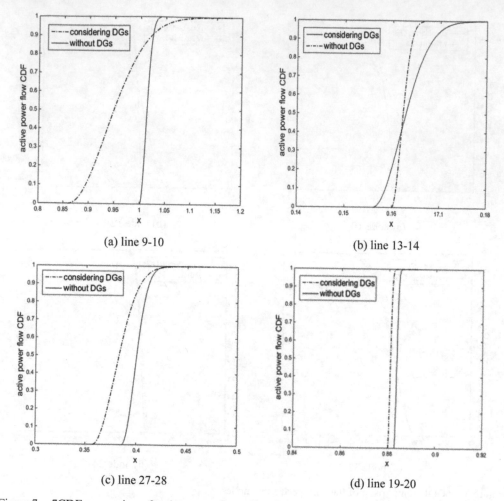

(a) line 9-10 (b) line 13-14

(c) line 27-28 (d) line 19-20

Figure 7. 7CDF comparisonof active power flow of some lines.

(3) The access of distributed generation has great influence on the power flow of branches, which increases the overload probability of branch currents. The extent of their impact on distributed generation is related to their distance from the distributed generation injected points and the original flow.

REFERENCES

B. Borkowska 1974. Probabilistic load flow. IEEE Transactions on Power Apparatus and Systems, vol. 3 no.3, pp.751–755.

Hu Zechun & Wang Xifan 2009. Error analysis of the probabilistic load flow based on cumulant method Power System Technology, 339(18): 32–37.

Ding Ming et al. 2001. Probabilistic load flow analysis based on Monte Carlo Simulation. Power System Technology, 25 (10: 10–14,22.

P.W. Sancer & G.T. Heydt 1978. A generalized stochastic power flow algorithms, IEEE Summer Meeting,16–21; Los Angeles, California, USA.

G.T. Heydt 1975. Stochastic Power Flow Calculations; IEEE Summer power Meeting, San Francisco, California, USA.

Zhang Yibin & Wang Weisheng 2004. Probability distribution of power output for wind power field and its application. Electrical Equipmen, 5(8): 38–40.

Esmat R, Hossein N, Saeid S 2009. GSA: A gravitational search algorithm. Inform Sci, 179(13): 2232–2248.

Birla D Maheshwari & R.P. Gupta 2006. A new nonlinear directional over current relay coordination technique, and banes and boons of near-end faults based approach, IEEE Trans. Power Delivery. 21, (3), pp. 1176–1182.

Yu Han et al. 2009. A probabilistic load flow calculation method with Latin hypercube sampling. Automation of Electric Power Systems, 33 (21): 32–35,81.

Justus C.G. et al. 1976. A Nationwide assessment of potential output from wind-powered generators. Journal of Applied Meteorology,15(7): 673–678.

Allan R.N. et al 1976. Numerical techniques in probabilistic load flow problems. International Journal for Numerical Methods in Engineering,10: 853–860.

Emerging Developments in the Power and Energy Industry – Dufo-López, Krzywanski & Singh (eds)
© 2020 Taylor & Francis Group, London, ISBN 978-0-367-27169-5

A new sensorless vector control method for PMSM based on adaptive fuzzy sliding mode observer and Kalman filter

Siqi Peng & Dongxiang Li
College of Information Engineering, Xiangtan University, Xiangtan, Hunan Province, China

ABSTRACT: A novel sensorless vector control method is proposed to PMSM based on sliding mode observer (SMO),which uses fuzzy control and a Kalman filter, overcoming the problem of chattering and phase delay in conventional SMO. In this paper, the position and speed of the rotor is obtained by an adaptive fuzzy SMO and a Kalman filter, the Lyapunov function is used to prove the reliability of the adaptive fuzzy SMO. A fuzzy control system is designed to adjust the gain of the sliding mode dynamically, which is able to reduce the chattering problem and improve the dynamic performance of the system. Meanwhile, a Kalman filter is designed to eliminate high frequency components and the system measurement noises in the back EMF, which is able to avoid the problem of phase delay and improve the observation precision of the system. The simulation results demonstrate the feasibility of the proposed method.

1 INTRODUCTION

Permanent magnet synchronous motor (PMSM) is widely used for speed-control in high precision and high-performance systems due to its advantages of small size, high power density and high efficiency (Ying. T et al. 2018; Q. Wang et al. 2014). In the conventional vector control system of PMSM, the position and speed of rotor is obtained through the mechanical sensors such as photoelectric encoder, absolute position encoder, which is not only increases the volume of the system and overall cost, but also reduces the reliability of the system operation. Therefore, it is very worthwhile to study sensorless control methods in order to work out these problems (J. Lee et al. 2010; G.O. Suvire & P.E. Mercado, 2012; Z. Shang et al. 2007).

At present, there is no method which is able to realize the stable operation of PMSM in full speed range. The sensorless control algorithms of PMSM mainly include high-frequency (HF) signal injection method, stator flux estimation method, model reference adaptive system (MARS), sliding mode observer (SMO) method, Kalman filter method, artificial intelligence estimation method, etc. (S.C. Agarlita et al. 2012; J.H. Feng & J.F. Xu, 2006; V. Smidl & Z. Peroutka, 2012; L. Yuan et al. 2013; X.B. Kong & X.J. Liu, 2014; A. Piippo et al. 2008). In recent years, the SMO method has attracted much attention due to its advantages such as simple calculation, good robustness and easy implementation, but the chattering problem limits the development of this method. In (O. Barambones & P. Alkorta, 2014; Z. Qiao et al. 2013), the saturation function and Sigmoid function were respectively adopted as switching functions, which reduced the chattering problem and improved the estimation accuracy of rotor position and speed, but the stability of the system is not reliable. A novel sliding mode observer method with estimated back-EMF feeding back to the calculation of stator current observation was designed in reference (W.Q. Lu et al. 2010), which has the features of a wide speed range and robustness. A SMO (G. Wang et al. 2013) based on the extended EMF of IPMSM was designed for sensorless operation, which is robust to load disturbance, but the estimation accuracy of rotor position is affected by low pass filter. A fuzzy sliding mode observer was designed in reference (H.S. Zhang et al. 2014), continuous sliding mode gain is obtained through fuzzy control and restrains the chattering problem of the system, but the continuous sliding mode gain reduces the system robustness. A SMO based on two-stage filter (W. Ding et al. 2012) was designed to obtain the rotor position and speed of PMSM, which improves the correctness of SMO method. To achieve the information about rotor position and speed more effectively, a new cascade sliding mode observer (M. Comanescu, 2016) was constructed, which solves the problem of phase delay in the conventional SMO.

In this study, a new sensorless vector control method of PMSM based on adaptive fuzzy SMO and Kalman filter is proposed. Firstly, the sliding mode gain is adjusted in real time by the fuzzy control system which is designed according to the estimated error value and the change rate of error. Secondly, the back EMF is obtained by a designed Kalman filter, which eliminates the high frequency components and the system measurement noises. The proposed method not only reduces the chattering problem and avoids the phase delay problem, but also enhances the stability of the system and has a larger speed range. Lastly, the simulation results show the feasibility of the proposed method.

2 ADAPTIVE FUZZY SMO DESIGN

2.1 Mathematical model of SPMSM

The mathematical model for SPMSM in the $\alpha\beta$ frame is given by

$$
\begin{cases}
\frac{di_\alpha}{dt} = -\frac{R}{L_s} i_\alpha + \frac{1}{L_s}(u_\alpha - e_\alpha) \\[2mm]
\frac{di_\beta}{dt} = -\frac{R}{L_s} i_\beta + \frac{1}{L_s}(u_\beta - e_\beta) \\[2mm]
e_\alpha = -\psi_f \omega_r \sin\theta \\[2mm]
e_\beta = \psi_f \omega_r \sin\theta
\end{cases}
\tag{1}
$$

Where i_α, i_β are the $\alpha\beta$ stator current components, e_α, e_β are the $\alpha\beta$ back EMF, R is stator resistance, L_s is the stator inductance, Ψ_f is the flux leakage, ω_r is angular velocity in electrical, θ is the rotor angular position.

2.2 Mathematical model of SPMSM

According to the mathematical model and the theory of synovial variable structure, the sliding mode surface is defined as follows

$$
s = \begin{bmatrix} s_\alpha & s_\beta \end{bmatrix} = \begin{bmatrix} \hat{i}_\alpha - i_\alpha & \hat{i}_\beta - i_\beta \end{bmatrix} = 0
\tag{2}
$$

Where S_α, S_β are the $\alpha\beta$ stator current error, $\hat{i}_\alpha$, $\hat{i}_\beta$ is the estimation of stator current, i_α, i_β are the actual stator current. In order to restrain the chattering problem, the sigmoid function is used to replace the sign or saturation functions in the conventional SMO. The sigmoid function can be expressed as follows

$$
F(x) = \left(\frac{2}{1 + e^{-ax}} \right) - 1
\tag{3}
$$

The mathematical model of adaptive fuzzy SMO from (1) is

$$\begin{cases} \frac{di_a}{dt} = -\frac{R}{L_s}\hat{i}_a + \frac{1}{L_s}[u_a - K_s F(s_a)] \\ \frac{di_\beta}{dt} = -\frac{R}{L_s}\hat{i}_\beta + \frac{1}{L_s}[u_\beta - K_s F(s_\beta)] \end{cases} \tag{4}$$

Where K_s is the gain of the sliding mode, which is adjusted by a fuzzy control system. Equation (4) minus (1), the estimation error of stator current is expressed as follows

$$\begin{cases} \frac{ds_a}{dt} = -\frac{R}{L_s}s_a + \frac{1}{L_s}e_a - \frac{1}{L_s}K_s F(s_a) \\ \frac{ds_\beta}{dt} = -\frac{R}{L_s}s_\beta + \frac{1}{L_s}e_\beta - \frac{1}{L_s}K_s F(s_\beta) \end{cases} \tag{5}$$

The reliability of the proposed observer is proved by a Lyapunov function. The Lyapunov function can be expressed as follows

$$V = s^T s/2 \tag{6}$$

The reliability condition of the adaptive fuzzy SMO is considered

$$\dot{V} = s^T \dot{s} \leq 0 \tag{7}$$

The $\dot{V}$ can be expressed as follows

$$\dot{V} = s^T \dot{s} = s_a \dot{s}_a + s_\beta \dot{s}_\beta$$
$$= \frac{1}{L_s}(\hat{i}_a - i_a)[e_a - K_s F(\hat{i}_a - i_a)] + \frac{1}{L_s}(\hat{i}_\beta - i_\beta)[e_\beta - K_s F(\hat{i}_\beta - i_\beta)] - \frac{R}{L_s}\left[(\hat{i}_a - i_a)^2 + (\hat{i}_\beta - i_\beta)^2\right] \tag{8}$$

Where

$$-\frac{R}{L_s}\left[(\hat{i}_a - i_a)^2 + (\hat{i}_\beta - i_\beta)^2\right] < 0$$

Therefore, K_s should satisfy the following equation

$$K_s > \max(|e_a|, |e_\beta|) \tag{9}$$

When the gain of the fuzzy sliding mode is greater than the maximum of the back EMF, the reliability of the adaptive fuzzy SMO can be guaranteed.

Based on equivalent control theory, the back EMF can be expressed as follows

$$e_a = K_s F(\hat{i}_a - i_a)$$
$$e_\beta = K_s F(\hat{i}_\beta - i_\beta) \tag{10}$$

2.3 Fuzzy sliding mode gain design

The gain value of sliding mode in conventional SMO is always given by experience, and it is difficult to change in real time. The gain of fuzzy sliding mode is dynamically adjusted by a fuzzy control system, which can reduce the chattering problem and offer the system has a better dynamic performance. The block diagram of the adaptive fuzzy SMO is shown in Figure 1.

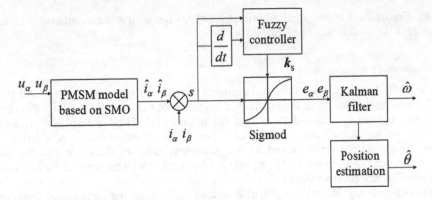

Figure 1.   Block diagram of the proposed adaptive fuzzy SMO.

Table 1.   Fuzzy reasoning rule.

$s/\dot{s}$	NB	Z	PB
NB	NH	NB	NM
Z	NS	Z	PS
PB	PM	PB	PB

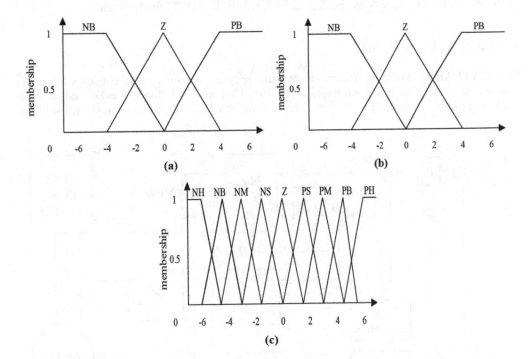

Figure 2.   Membership function of the input and output.

   This paper uses MATLAB fuzzy control toolbox to design the gain of fuzzy sliding mode. The inputs of the fuzzy control system are s and $\dot{s}$, $\dot{s} = \mathrm{d}s/\mathrm{d}t$, the output of this system is K_s. If $s\dot{s} > 0$, K_s should be increased; if $s\dot{s} < 0$, K_s should be reduced, Table 1 lists fuzzy reasoning rule of K_s. Figure 2a is the membership function of s, Figure 2b is the membership function of $\dot{s}$, Figure 2c is the membership function of K_s.

2.4 *Kalman filter observer design*

The back EMF from equation (10) still contains high frequency components and the noise caused by measuring can affect the estimation accuracy of the back EMF. To solve this problem, a Kalman filter is designed to extract the back EMF. The Kalman filter is a random observer of the nonlinear system, which can estimate the state of the system online. It can eliminate the estimation error caused by the parameter error of the rotor and the high frequency component of the back EMF, by which phase delay which is caused by low-pass filter can be solved by the Kalman filter.

Since the electromagnetic time constant is much smaller than the mechanical time constant, it can be assumed that the equationd $\omega_r/dt = 0$. The state equation of the designed Kalman filter observer is as follows

$$
\begin{cases}
\frac{d}{dt}\hat{e}_\alpha = -\hat{\omega}_r\hat{e}_\beta - k_l(\hat{e}_\alpha - e_\alpha) \\[2mm]
\frac{d}{dt}\hat{e}_\beta = \hat{\omega}_r\hat{e}_\alpha - k_l(\hat{e}_\beta - e_\beta) \\[2mm]
\frac{d}{dt}\hat{\omega}_r = (\hat{e}_\alpha - e_\alpha)\hat{e}_\beta - (\hat{e}_\beta - e_\beta)\hat{e}_\alpha
\end{cases}
\tag{11}
$$

Where $\hat{e}_\alpha$ and $\hat{e}_\beta$ are the back EMF estimated by the Kalman filter, $\hat{\omega}_r$ is the estimated value of the rotor angular velocity by the Kalman filter, k_l is the gain of the Kalman filter.

3 SIMULATION RESULTS

The MATLAB simulation software is used to verify the feasibility and effectiveness of the above design system. The parameters of the SPMSM are as follows: the DC side voltage 311V, the phase resistance 2.875Ω, the direct axis inductor $L_d = L_q = 8.35$mH, the rotor inertia

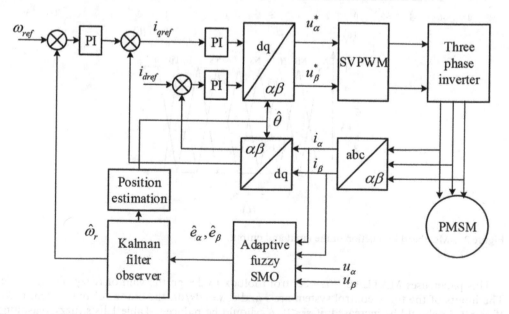

Figure 3.   Block diagram of sensorless control system for SPMSM.

35

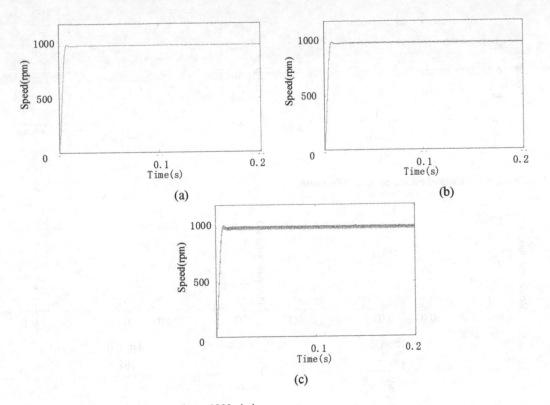

Figure 4. Speed estimation results at 1000 r/min.

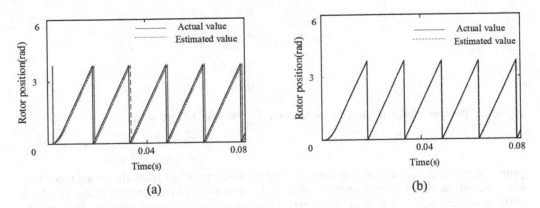

Figure 5. Comparison diagram of rotor actual position and rotor estimation position.

J = 0.001kg·m2, the rotor flux $\Psi_r = 0.175$Wb, the pole pairs 4. Figure 3 is the block diagram of sensorless control system for SPMSM.

Figure 4 shows the results of speed response waveform at 1000 r/min. Figure 4a is the actual speed, Figure 4b is the estimated speed of conventional SMO method, Figure 4c shows the results of proposed method. We can see that the proposed method can effectively estimate the speed of the rotor. Compared with the conventional SMO method, the chattering phenomenon of the sliding mode variable structure system can be obviously reduced.

Figure 5 shows the comparison of rotor actual position and rotor estimation position when the given speed at 1000 r/min. Figure5a shows the comparison of rotor actual position and rotor estimation position by conventional SMO method, Figure 5b shows the results of

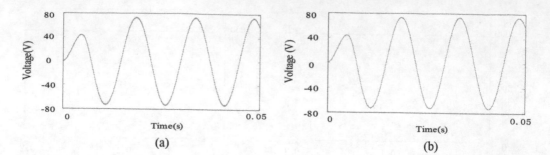

Figure 6.   Values of estimated $\hat{e}_\alpha$ at 1000 r/min.

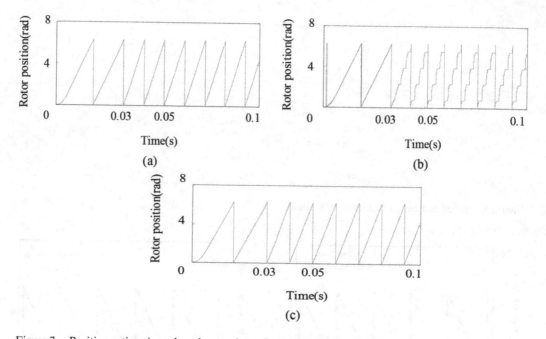

Figure 7.   Position estimation when the speed step from 1000 r/min to 1500 r/min.

proposed method. We can see that the proposed method can better track the actual rotor position, the phase delay is very small because of the omission of the lowpass filter. The back EMF which extracted by Kalman filter makes the rotor position waveform smoother and the rotor position estimation accuracy higher.

In order to better reflect the advantages of the Kalman filter observer in extracting the back EMF. Figure 6 is the results of the estimated $\hat{e}_\alpha$ when the given speed at 1000 r/min. It can be seen from Figure 6, the back-EMF waveform observed by Kalman filter is smoother, and it can effectively eliminate the measuring noise and high frequency components.

Figure 7 shows the results of position estimated when the speed increases from 1000 r/min to 1500 r/min in 0.03 seconds. It can be seen that when the rotor speed goes up, the conventional SMO method has a state of instability due to the previous sliding mode gain could not adapt to the new work condition. The proposed method uses fuzzy control to dynamically adjust the gain of the sliding mode, which has larger speed range and better dynamic performance.

4 CONCLUSIONS

A novel sensorless vector control method for PMSM based on adaptive fuzzy SMO and Kalman filter is proposed in this paper. The position and speed of the rotor can be effectively estimated through the proposed method. The gain of sliding mode achieved by the fuzzy control can effectively reduce the chattering problem and improve the speed range of the system. The designed Kalman filter observer can effectively estimate the value of the back EMF and estimate the high frequency components and noise interference in the back EMF. The dynamic performance of system can be improved due to omitting the use of lowpass filter. The simulation results verify the feasibility of the proposed sensorless control system.

REFERENCES

Y. Tang, Y.J. He, F. Wang, et al, 2018.: 'Back-EMF-based sensorless control system of hybrid SRM for high-speed operation', *IET Electr. Power Appl.*, 12, (6), pp. 867–873.

Q.L. Wang, X. Zhang, C.W. Zhang, 2014.: 'Double sliding- mode model reference adaptive system speed identification for vector control of permanent magnet synchronous motors', *Proc of the CSEE.*, 34, (6), pp. 897–902.

J. Lee, J. Hong, K. Nam, et al, 2010.: 'Sensorless control of surface-mount permanent-magnet synchronous motors based on a nonlinear observer', *IEEE Transactions on Power Electronics.*, 25, (2), pp. 290–297.

G.O. Suvire & P.E. Mercado, 2012.: 'Active power control of a flywheel energy storage system for wind energy applications', *Renewable Power Generation Iet.*, 6, (1), pp. 9–16.

Z. Shang, R.X. Zhao and R.Z. Dou, 2007.: 'Research on sensorless control method of PMSM based on an adaptive sliding mode observer', *Proc of the CSEE.*, 27, (3), pp. 23–28.

S.C Agarlita., I. Boldea. & F. Blaabjerg, 2012.: 'High-frequency-injection-assisted "active-flux"-based sensorless vector control of reluctance synchronous motors, with experiments from zero speed', *IEEE Trans on Industry Applications.*, 48, (6), pp. 1931–1939.

J.H Feng & J.F. Xu, 2006.: 'Permanent magnet synchronous machines direct torque control system based on adaptive stator flux observer', *Proc of the CSEE.*, 26, (13), pp. 151–157.

V. Smidl & Peroutka. Z, 2012.: 'Advantages of square-root extended Kalman filter for sensorless control of AC drives', *IEEE Transactions on Industrial Electronics.*, 59, (11), pp. 4189–4196.

L. Yuan, F. Xiao & J.Q. Shen, et al, 2013.: 'Sensorless control of high-power interior permanentmagnet syn chronous motor drives at very low speed', *IET Electr. Power Appl.*, 7, (3), pp. 199–206.

X.B. Kong & X.J. Liu, 2014.: 'Efficient nonlinear model predictive control for permanent magnet syn chronous motor', *Acta Automatica Sinica.*, (09), pp. 1958–1966.

A. Piippo, M. Hinkkanen & J. Luomi, 2008.: 'Analysis of an adaptive observer for sensorless control of interior permanent magnet synchronous motors', *IEEE Trans on Industrial Electronics*, 55, (2), pp. 570–576.

O. Barambones & P. Alkorta, 2014.: 'Position control of the induction motor using an adaptive sliding-mode con troller and observers', *IEEE Transactions on Industrial Electronics.*, 61, (12), pp. 6556–6565.

Z. Qiao, T. Shi & Y. Wang, et al, 2013.: 'New sliding-mode observer for position sensorless control of perma nent- magnet synchronous motor', *IEEE Transactions on Industrial Electronics.*, 60, (2), pp. 710–719.

W.Q. Lu, Y.W. Hu & X.Y. Du, et al, 2010.: 'Sensorless vector control using a novel sliding mode observer for PMSM speed control system', *Proc of the CSEE.*, 30, (33), pp. 78–83.

G. Wang, R. Yang & D. Xu, 2013.: 'DSP-based control of sensorless IPMSM drives for wide-speed-range opera tion', *IEEE Transactions on Industrial Electronics*, 60, (2), pp. 720–727.

H.S. Zhang, P. Wang & B.C. Han, 2014.: 'Rotor position measuring method for magnetic levitation high speed PMSM based on fuzzy sliding mode observer', *Proceedings of the CSEE.*, 29, (7), pp. 147–153.

W. Ding, D.L. Liang & Z.Q. Luo, 2012.: 'Position sensorless control of PMSM using sliding mode observer with two-stage filter', *Electric Machines and Control*, (11), pp. 1–10.

M. Comanescu, 2016.: 'Speed, rotor position and load torque estimation of the PMSM using an extended dynamic model and cascaded sliding mode observers', *International Symposium on Power Electronics, Electrical Drives, Automation and Motion. IEEE*, pp. 98–103.

Emerging Developments in the Power and Energy Industry – Dufo-López, Krzywanski & Singh (eds)
© 2020 Taylor & Francis Group, London, ISBN 978-0-367-27169-5

A novel soft-switching bidirectional DC-DC converter based on LLC resonance

Lingzhi Yi & Qingping Li
Hunan Province Engineering Research Center for Multi-Energy Collaborative Control Technology,
Xiangtan University, Hunan Province, China

Yanshen Hu
Shenzhen MOSO Electric Co., Ltd., Guangdong Province, China

Xin Long
XEMC Wind Power Co.,Ltd., Hunan Province, China

ABSTRACT: In order to further improve the power density and efficiency of the battery side bidirectional DC-DC converter in the energy storage system, a novel soft switching bidirectional DC-DC converter based on LLC resonance is proposed. The converter effectively reduces the transformer turns ratio and improves conversion efficiency. The converter can implement bidirectional LLC characteristics in an asymmetric half-bridge topology, and all switches in the converter can be soft-switched. At the same time, the converter has a simple structure and can adopt synchronous rectification technology, and has the advantages of high efficiency and low cost. The realization process of the converter soft switch is introduced, and the resonance characteristics and related parameters of the soft switch are analyzed. Finally, a simulation model was built with a high voltage side of 350 ~ 400 V and a low voltage side of 45 ~ 50 V, which verified the effectiveness and practicability of the converter.

1 INTRODUCTION

With the increasing demand for electricity in energy storage systems, electric vehicles, and renewable energy systems, resonant converters have developed rapidly. Multi-element resonant converters have been proposed since the 1990s (Severns R P 2002). Multi-element resonant soft-switching technology has been widely used, especially the three-element resonant circuit that can realized full soft-switching of power supply has been deeply studied. Various half-bridge, full-bridge, multi-level LLC series resonant converters (LLC Series Resonant Convert, LLC-SRC) have been proposed and studied (Yang B et al 2002, Jeong Y et al 2017), and the parameter optimization design method of LLC resonant converters have been proposed from different angles (Li Z & Wang H 2017, Canales F et al 2002). At the same time, the conversion efficiency is further improved by combining synchronous rectification technology.

Currently, soft-switching technology is widely used in bidirectional DC-DC. In (Lu Y et al 2018), the asymmetric bidirectional DC-DC constructed by push-pull and full-bridge is analyzed, and the soft-switching is realized by adding a MOSFET on the push-pull side. The bidirectional soft-switching mode of the symmetric half-bridge was analyzed in paper (Wu X et al 2011, Chen W et al 2010). The application of LLC resonant circuit in full-bridge bidirectional DC-DC is analyzed in paper (Kim E S et al 2015). However, it is required that in the household energy storage system, since the bidirectional DC-DC is mainly used for the coupling between the energy storage unit and the DC bus of the system. Low-voltage high-current topology to deal with the rapid charge and discharge of the battery.

Therefore, in view of the actual starting point of the household energy storage system, this paper combined LLC with push-pull half-bridge and double-voltage half-bridge to study a novel soft-switching bidirectional DC-DC based on LLC resonance that can implement the full soft-switching of power MOSFET. Compared with the traditional power converter, the topology has the advantages of simple structure, simple control, low cost and high work efficiency. The bidirectional working mode of the circuit is analyzed. The correlation calculation and formula derivation of the resonance parameters are given. The realization conditions of the soft-switching are also analyzed. Finally, a simulation model was built to verify the feasibility and effectiveness of the circuit.

2 CIRCUIT AND PRINCIPLE

2.1 *Main circuit topology*

The main circuit structure of a novel soft-switching bidirectional DC-DC converter based on LLC resonance is shown in Figure 1. In the figure, S_1 & S_2 is the high-voltage side main switches, and S_3 & S_4 is the low-voltage side main switches; different with the conventional LLC series resonant converter, the high-voltage side adopt a double-voltage half-bridge structure, and the low-voltage side adopt a push-pull half-bridge structure, except for C_r & L_r in the resonant network, the other two resonant element with transformer magnetizing inductance L_{m1} and an additional inductance L_{m2} form a bidirectional LLC resonant network; in the circuit, D_1, D_2, D_3, D_4 and C_{o1}, C_{o2}, C_{o3}, C_{o4} are the body diode and parasitic capacitance of the switches S_1, S_2, S_3, S_4 respectively.

The converter can be switched between the forward power flow and the reverse power flow through the drive control of the switch to realize the bidirectional transmission of energy; the buck operation is perform when working in the forward power flow mode, and the high-voltage side is square wave generation, the low-voltage side is the rectification network. When working in the reverse power flow mode, it is boost operation. At this time, the low-voltage side is a square wave generation network, and the high-voltage side is a rectification network. Transformer T_R is a push-pull transformer with two center-tap windings of the same number of turns and a concentrate winding. Since the leakage inductance value of the push-pull transformer is relatively small, in order to simplify the analysis, subsequent calculations will ignore the influence of the leakage inductance on the circuit.

2.2 *Forward buck and reverse boost principle*

When the converter is buck operation, it is a forward power flow, as shown in Figure 1. In the figure, S_1 & S_2 worked in the switch state, and S_3 & S_4 worked in the rectification state. The additional magnetizing inductance L_{m2} is input to the square wave clamp and does not participate in the resonance, at this time the LLC resonant network composed of L_{m1}, L_r and C_r.

The main switches drive signal and waveform of the converter are shown in Figure 2. In this figure, V_{gs} is the drive signal of the switches, V_{ds} is the drain-source voltage waveform of the switches, i_{SX} is the current waveform through the switches, i_{LX} is the current through the resonant

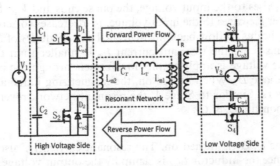

Figure 1.   Novel soft-switching bidirectional DC-DC converter based on LLC resonance.

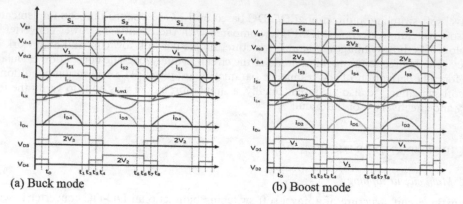

(a) Buck mode   (b) Boost mode

Figure 2.   Operation principles of the proposed LLC resonant circuit.

network, i_{DX} is the current through the switch body diode and V_{DX} is the voltage across the body diode. The converter adopt a control strategy similar to that of the traditional LLC series resonant converter, which is variable frequency modulation. In the buck operation S_1 & S_2 are alternately turned on in one switching cycle, and each switch transistor has a duty cycle of about 50% (with dead time). In the boost operation S_3 & S_4 works in the switching state. In order to simplify the analysis, it is assumed that the circuit has entered a steady state. One switching cycle of the converter can be divide into eight modes, and the last four modes are similar to the first four modes. The equivalent circuits of the first four modes of working stages are shown in Figure 3, and the working principles of the working modes are described as follows: stage 1-4 is the first four modes of buck operation, stage 5-8 is the first four modes of boost operation.

Buck operation

Stage 1 $[t_0, t_1]$: At time t_0, S_1 is turned on. The magnetizing inductance L_{m1} is clamp by the output voltage, the inductor current i_{Lm1} increases linearly, the resonant current i_{Lr} rises in a sinusoidal form, and the current flowing through the rectifying side body diode depends on the difference between i_{Lr} and i_{Lm1}. At this time, S_2 is turned off and no current flows through the half bridge branch. The voltage is the input voltage value, and the voltage value of the capacitor C_2 is $1/2V_1$.

Stage 2 $[t_1, t_2]$: The resonant current i_{Lr} passes through the peak and falls to equal i_{Lm1}, and the rectified side body diode current is reduced to zero and D_3 & D_4 turned off under ZCS condition. At this time, L_{m1} is no longer clamp by the output voltage, thereby participating in the resonance process.

Stage 3 $[t_2, t_3]$: At time t_2, S_1 is turned off. Current i_{S1} begins to charge C_{o1} of S_1 and gradually decreases, and resonant current i_{Lr} remains unchanged, so that C_{o2} of S_2 in the other half of the bridge branch needs to be continuously discharged to satisfy the current balance relationship, and V_{ds2} begins to gradually decrease. At this time, both C_{o1} and C_{o2} participate in resonance, and the currents i_{S1} and i_{S2} through S_1 and S_2 are the same size and opposite directions. When V_{ds1} rises to the input voltage, the phase ends and V_{ds2} falls to zero.

Stage 4 $[t_3, t_4]$: When V_{ds1} rises to the input voltage, i_{S1} decreases to zero. In the other half of the bridge branch, i_{S2} begins to flow back through the body diode D_2 of S_2 to feed back to the input terminal. At this time, the resonant current i_{Lr} is smaller than the excitation current i_{Lm1}, so that a reverse voltage is apply to the transformer through the resonant network, the rectifying side body diode D_3 When turned on, the magnetizing inductance L_{m1} is clamp out of the resonant network. Similar to Stage 1, the resonant network reverts back to the resonance of the two components L_r & C_r.

Boost operation

Stage 5 $[t_0, t_1]$: At time t_0, S_3 is turned on. The resonant current i_{Lr} also rose in a sinusoidal form. At this time, since the inductor L_{m2} is clamp by the output voltage, the inductor current i_{Lm2} linearly increase. The current flowing through the rectifying side body diode depends on

41

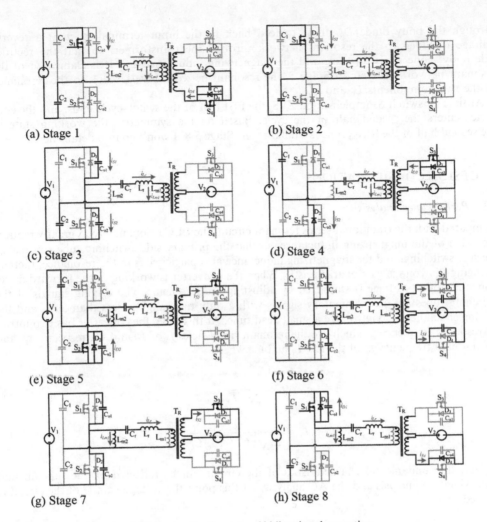

(a) Stage 1	(b) Stage 2
(c) Stage 3	(d) Stage 4
(e) Stage 5	(f) Stage 6
(g) Stage 7	(h) Stage 8

Figure 3.   Equivalent circuit diagram of each mode of bidirectional operation.

the difference between i_{Lr} and i_{Lm2}, and the current will reach a peak at this stage, and the current rise speed is greater than the current fall speed due to the influence of the rectifying side junction capacitance. At this time, the terminal voltage of the capacitor C_1 is also $1/2V_1$, and S_4 remains off and no current flows through the push-pull branch, and the terminal voltage is 2 times the input voltage V_2.

Stage 6 [t_1, t_2]: At time t_1, when the resonant current i_{Lr} falls through the peak and becomes equal to i_{Lm2}, the rectified body diode current drops to zero, and D_1 & D_2 achieve zero current turn-off. At the same time, due to the energy transfer between the junction capacitors, the resonance current i_{Lr} will superimpose the high-frequency resonance with a small amplitude on the basis of the original current.

Stage 7 [t_2, t_3]: At time t_2, S_3 is turned off. Current i_{S3} begins to charge C_{o3} of S_3. Due to the coupling effect of transformer T_R, C_{o4} of S_4 in the other half of the bridge branch needs to be continuously discharged to satisfy the voltage balance relationship, and V_{ds4} start to fall. At this time, C_{o3} and C_{o4} participate in the resonance at this stage, and the currents i_{S3} and i_{S4} through S_3 and S_4 are the same size and opposite directions. When V_{ds3} rise to 2 times the input voltage, ie $2V_2$, the phase ends and V_{ds4} drops to zero, providing zero voltage conditions for the turn-on of the switch.

Stage 8 [t_3, t_4]: At time t_3, when V_{ds3} rose to 2 times the input voltage, i_{S3} also decreased to zero. At this time, the other half of the bridge branch i_{S4} begins to flow through the body diode D_4 of S_4 to feed back to the input terminal, so that a reverse voltage is added to the resonant network through the transformer, so that the rectifier side body diode D_1 is turn on, and the magnetizing inductance L_{m2} It is clamp out of the resonant network. Similar to Stage 5, the resonant network reverts back to the resonance of the two components L_r and C_r.

At time t_4, switch S_4 implemented ZVS, the first half of the switch cycle ends, and the converter enters the second half of the cycle. Based on the symmetry, the working state in the second half of the $[t_4, t_8]$ time is similar to the Stage 5-8. I won't go into details here.

3 DESIGN CONSIDERATIONS

3.1 *Performance analysis*

Compared with the traditional series resonant circuit, the LLC resonant circuit greatly reduces the value of the magnetizing inductance, so that the primary side switching tube can realize the soft-switching, and the magnetizing inductance also participates in the resonance, thereby changing the voltage gain characteristic. When the converter is working in the forward direction, the output voltage is stabilized by adjusting the frequency. The driving signals of the switching tubes are complementary signals with a duty ratio slightly less than 50%, and the driving signal needs to introduce a small dead time. At this time, the converter has two intrinsic resonant frequencies, which are the resonant frequencies f_r of L_r and C_r, respectively, and the resonant frequency f_m of the L_m, L_r and C_r, where

$$f_r = \frac{1}{2\pi\sqrt{L_r C_r}} \tag{1}$$

$$f_m = \frac{1}{2\pi\sqrt{(L_r + L_m)C_r}} \tag{2}$$

In fact, the fundamental characteristics of the circuit can be reflected by the fundamental approximation analysis and the AC analysis. At this point the voltage gain G of the circuit is defined as:

$$G = \frac{1}{\sqrt{\left(1 + \frac{1}{m} - \frac{1}{m\omega^2}\right)^2 + Q^2\left(\omega - \frac{1}{\omega}\right)^2}} \tag{3}$$

Where

$$m = \frac{L_m}{L_r}, \omega = \frac{f_s}{f_r}, Q = \sqrt{\frac{L_r}{C_r}}/R_V \tag{4}$$

4 SIMULATION RESULTS

A simulation mode prototype of 500 VA, suitable for household energy storage system, using the novel soft-switching bidirectional DC-DC converter topology based on LLC resonance proposed in this paper, the simulation waveforms verified the working principle and soft-switching characteristics of the topology. Actual parameters can be optimized accordingly. The control mode adopts Pulse Frequency Modulation (PFM), adjust the gain in this way.

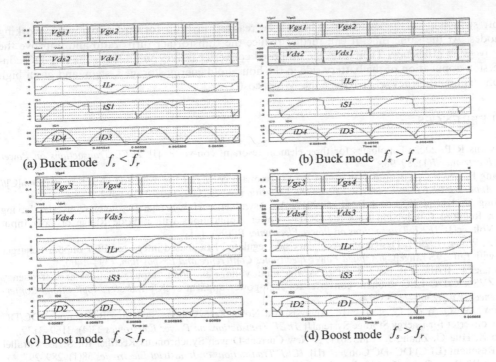

(a) Buck mode $f_s < f_r$ (b) Buck mode $f_s > f_r$

(c) Boost mode $f_s < f_r$ (d) Boost mode $f_s > f_r$

Figure 4.   Simulation waveforms.

Simulation Components Parameter as follows: S_1, S_2, S_3, S_4: MOSFET; T_R:4:4:16; C_1,C_2:2 uF; C_r:68 nF; L_r:35 uH; L_{m1}:220 uH; L_{m2}:220 uH.

The simulation waveforms are shows that the converter achieve ZVS & ZCS in Figure 4.

Figure 4-a shows the converter worked in buck mode and the switches frequency is less than resonant frequency ($f_s < f_r$). In this steady stage, before the switch S_1 and S_2 turned on, the voltage V_{ds1} and V_{ds2} across the switch has dropped to zero, in addition, before the body diode D_3 and D_4 shut down, the current i_{D3} and i_{D4} has dropped to zero.

Figure 4-b shows the converter worked in buck mode and the switches frequency is greater than resonant frequency ($f_s > f_r$). In this steady stage, before the switch S_1 and S_2 turned on, the voltage V_{ds1} and V_{ds2} across the switch has dropped to zero, in addition, before the body diode D_3 and D_4 shut down, the current i_{D3} and i_{D4} has dropped to zero.

Figure 4-c shows the converter worked in boost mode and the switches frequency is less than resonant frequency ($f_s < f_r$). In this steady state, before the switch S_3 and S_4 turned on, the voltage V_{ds3} and V_{ds4} across the switch has dropped to zero, in addition, before the body diode D_1 and D_2 shut down, the current i_{D1} and i_{D2} has dropped to zero.

Figure 4-d shows the converter worked in buck mode and the switches frequency is less than resonant frequency ($f_s > f_r$). In this steady state, before the switch S_3 and S_4 turned on, the voltage V_{ds3} and V_{ds4} across the switch has dropped to zero, in addition, before the body diode D_3 and D_4 shut down, the current i_{D1} and i_{D2} has dropped to zero.

5 CONCLUSION

In this paper, a new bidirectional DC-DC converter topology based on LLC is proposed for the actual needs of household energy storage systems. The forward and reverse working modes of the converter are analyzed, and the equivalent circuit of the model is given. At the same time, the relevant simulation model is built. The simulation experiments show that the converter can combine the advantages of the traditional LLC series resonant converter and the traditional PWM push-pull circuit and the double-voltage half-bridge circuit. The natural soft switching of the switching tube can be realized in both forward and reverse working modes. At the same time, it can further reduce the transformer turns ratio and improve the conversion efficiency. Therefore, the proposed converter is very suitable for the DC-DC conversion application of the battery side in the household energy storage system, and has a high conversion efficiency while reduce the system cost.

REFERENCES

Severns R P. 2002. Topologies for three-element resonant con-verters[J]. *IEEE Transactions on Power Electronics* 7(1): 89–98.

Yang B, Lee F C, Zhang A J, et al. 2002. LLC resonant converter for front end DC/DC *conversion*[C]// *Applied Power Electronics Conference and Exposition*. Apec 2002: 1108–1112 vol.2.

Jeong Y, Kim J K, Lee J B, et al. 2017. An Asymmetric Half-Bridge Resonant Converter having a Reduced Conduction Loss for DC/DC Power Applications with a Wide Range of Low Input Voltage[J]. *IEEE Transactions on Power Electronics* PP(99): 1–1.

Li Z, Wang H. 2017. Design of a secondary side regulated LLC based integrated PEV onboard charger with full ZVS range[C]//*Applied Power Electronics Conference and Ex-position*. 1394–1400.

Canales F, Barbosa P, Lee F C. 2002. A wide input voltage and load output variations fixed-frequency ZVS DC/DC LLC resonant converter for high-power applications[C]//*Industry Applications Conference*, IEEE,2306–2313 vol.4.

Lu Y, Wu Q, Wang Q, et al. 2018. Analysis of a Novel Zero-Voltage-Switching Bidirectional DC/DC Converter for Energy Storage System[J]. *IEEE Transactions on Power Electronics* 33(4): 3169–3179.

Wu X, Hua G, Zhang J, et al. 2011. A New Current-Driven Synchronous Rectifier for Series–Parallel Resonant (LLC) DC–DC Converter[J]. *IEEE Transactions on Industrial Electronics* 58(1): 289–297.

Chen W, Rong P, Lu Z. 2010. Snubberless Bidirectional DC–DC Converter With New CLLC *Resonant* Tank Featuring Mini-mized Switching Loss[J]. *IEEE Transactions on Industrial Electronics* 57(9): 3075–3086.

Kim E S, Park J H, Joo J S, et al. 2015. Bidirectional DC-DC con-verter using secondary LLC *resonant* tank[C]//*Applied Power Electronics Conference and Exposition*. IEEE, 2104–2108.

Emerging Developments in the Power and Energy Industry – Dufo-López, Krzywanski & Singh (eds)
© 2020 Taylor & Francis Group, London, ISBN 978-0-367-27169-5

The deadbeat current control of MMC rectifier with wide range output voltage

Lingzhi Yi, Xiaohui Huang & Luo Liu
Hunan Province Engineering Research Center for Multi-Energy Collaborative Control Technology, Xiangtan University, Hunan Province, China

Jie Li
State Grid Xiangtan Power Supply Company, Hunan Province, China

ABSTRACT: Modular Multilevel Converter (MMC) has become the core topology in high voltage and high power occasion. Expanding the voltage output range of the MMC rectifier will further to broaden the application area of the MMC. Existing MMC rectifier control methods have many disadvantages, such as complex control structures and the need for a large number of PI regulators. To overcome those shortcomings, this paper proposed a new control strategy for MMC Rectifier with Wide Range Voltage Output. The proposed control method avoids the use of a large number of PI regulators, greatly simplifying the control structure of the system. By building up an MMC rectifier model in Matlab/Simulink, the correctness and effectiveness of the proposed control strategy are verified.

1 INTRODUCTION

Modular multilevel converter (MMC) is a new type of voltage source converter (VSC) topology that has emerged in recent years and is a major breakthrough in the topology of multilevel converters. Compared with traditional two-level or three-level voltage source converters, MMC has advantages such as high output voltage level, low voltage stress of power devices, and fault-tolerant capacity. It has been widely used in high voltage applications (Akagi H, et al. 2007, Hagiwara M, et al. 2010, Nami A, et al. 2015). Wide-range output rectifiers for high-power applications have been difficult to achieve in the industry. The emergence of MMC brings a good solution for this. Motor drive system based on back to back MMC has been studied in the literature (Guan M, et al. 2017, Y. S. Kumar & G. Poddar 2018). Through by providing a varying DC voltage for the motor side MMC, the problem of violent capacitor voltage fluctuation is solved when the motor side MMC operates at low frequency. The grid sides MMC as a rectifier requires a wide voltage range output. MMC based DC Ice-Melting Application has been studied in the literature (Li B, et al. 2016, Guo Y, et al. 2017). DC Ice-Melting Application also requires a wide range of output voltages to supply the melting load. Expanding the voltage output range of the MMC rectifier will further to broaden the application area of the MMC. But the control methods used in the above articles are complicated.

Deadbeat control has many advantages, such as simple control algorithm, a small amount of calculation, fast calculation speed, high tracking accuracy, and fast dynamic response (Fu X, et al. 2009). The use of deadbeat control for MMC controller design can greatly reduce coordinate transformation and PI regulators while reducing the computational complexity of the control system (Deng X, et al. 2014, Rong F, et al. 2017). Therefore, this paper proposed a Deadbeat current control method for wide voltage range output MMC rectifier to optimizer the control system.

2 CIRCUIT TOPOLOGY AND OPERATING PRINCIPLE

Figure 1 shows the topology of the MMC rectifier, which is composed of an MMC based on full-bridge sub-modules(SMs). The rectifier is composed of six arms, each consisting of N series-connected full-bridge SMs and an arm inductor (L).

The structure of a full-bridge SM is also shown in Figure 1. It is composed of four IGBTs and four anti-parallel diodes. According to the conduction conditions of IGBT and diode, we can get the working states of FBSM in Table 1. The radically difference between FBSM and half bridge sub-module (HBSM) is that the output voltage of FBSM can be negative in any current direction. This could make FBSM a lot more flexibility. The voltage of FBSM can alternately change between U_c, 0 and $-U_c$. Therefore, the full-bridge MMC has the advantage of a very wide DC output voltage range.

Figure 2 shows the single-phase equivalent circuit of MMC rectifier. u_j ($j = a, b, c$) denote the three-phase grid voltages; i_j denote the three-phase grid currents; u_{pj} and u_{nj} denote the upper-arm and lower-arm voltages, respectively; i_{pj} and i_{nj} denote the upper-arm and lower-arm currents, respectively; L denote the inductor of each arm; i_{cirj} and denote the circulating current between the bridge arm and DC-side.

The state equations of MMC rectifier can be expressed as

$$\begin{cases} \frac{1}{2}U_{dc} - u_{pj} - u_{sj} = L\frac{di_{pj}}{dt} \\ \frac{1}{2}U_{dc} - u_{nj} + u_{sj} = L\frac{di_{nj}}{dt} \end{cases} \tag{1}$$

According to Kirchhoff's voltage law and the strict symmetry of upper and lower arms, i_j and i_{cirj} can be expressed as

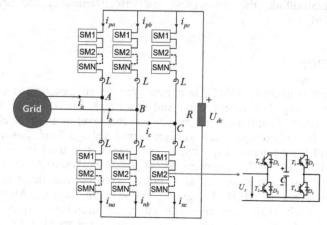

Figure 1. Circuit topology of rectifier based on MMC.

Table 1. Working states of the individual FBSM.

	T1	T2	T3	T4	I_{SM}	U_{SM}	FBSM states
Normal operation	1	0	0	1	-	U_C	Positively inserted
	0	1	1	0	-	$-U_C$	Negatively inserted
	1	0	1	0	-	0	Bypass
	0	1	0	1	-	0	Bypass
Dc fault	0	0	0	0	>0	U_C	Blocked
	0	0	0	0	<0	$-U_C$	Blocked

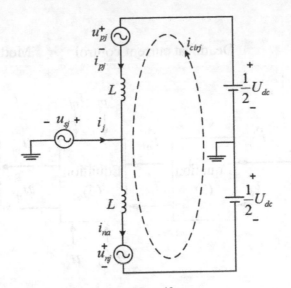

Figure 2.   Single-phase equivalent circuit of MMC rectifier.

$$\begin{cases} i_j = i_{nj} - i_{pj} \\ i_{cirj} = \frac{i_{nj}+i_{pj}}{2} \end{cases} \tag{2}$$

(2) can be rewritten as

$$\begin{cases} i_{pj} = \frac{2i_{cirj}-i_j}{2} \\ i_{nj} = \frac{2i_{cirj}+i_j}{2} \end{cases} \tag{3}$$

3   DEADBEAT CONTROL STRATEGY OF MMC

The basic idea of deadbeat control can be described as: Calculate the control value of the next switching cycle based on the system's state equation, the output feedback signal and the required next-time reference of output values. The overall control block diagram of the dead-beat control is shown in Figure 3. The entire control structure is divided into three parts: circulating current control, deadbeat current control, and modulation strategy.

The role of the Circulating current is to charge the capacitor voltage of sub-modules and compensate the lost energy. The main component of the circulating current is DC current. In the circulating current control part, $\bar{u}_{cj}$ denote the average measured values of each phase capacitor voltage. u_{cj}^* denote average value references of each phase capacitor voltage. i_{cirj}^* denote the references of circulating current. After u_{cj}^* and $\bar{u}_{cj}$ are compared, the value of i_{cirj}^* can be obtained through a PI regulator. i_{cirj}^* can be expressed as

$$i_{cirj}^* = (u_{cj}^* - \bar{u}_{cj})\left(K_p + \frac{K_i}{s}\right) \tag{4}$$

According to equation (3), the upper-arm current reference value i_{pj}^* and lower-arm currents reference value i_{nj}^* can be obtained. Therefore, tracking of the grid current reference value translates to tracking the arm current reference value.

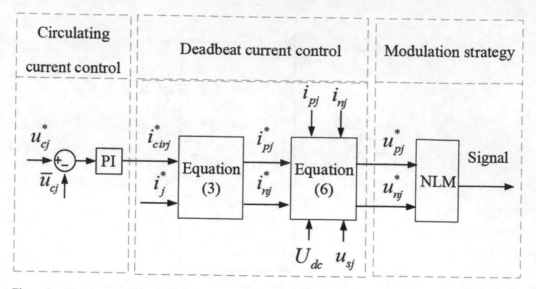

Figure 3. The overall control block of MMC.

At the K-th sampling time, by discretizing equation (1), the following equations can be obtained.

$$\begin{cases} u_{pj}(K) = \frac{1}{2}U_{dc}(K) - u_{sj}(K) - L\frac{i_{pj}(K+1)-i_{pj}(K)}{T} \\ u_{nj}(K) = \frac{1}{2}U_{dc}(K) + u_{sj}(K) - L\frac{i_{nj}(K+1)-i_{nj}(K)}{T} \end{cases} \qquad (5)$$

Where T is the sampling period; $u_{pj}(K)$ is the sum of all SM module voltages that the upper-arm should conduct at the K-th sampling time; $u_{nj}(K)$ is the sum of all SM module voltages that the lower-arm should conduct at the K-th sampling time; $U_{dc}(K)$ is the DC-side voltage at the K-th sampling time; $u_{sj}(K)$ is the grid voltage at the K-th sampling time; L is the arm inductor; $i_{pj}(K)$ is the upper-arm current at the K-th sampling time; $i_{nj}(K)$ is the upper-arm current at the K-th sampling time; $i_{pj}(K+1)$ is the desired output current of upper-arm at the beginning of the K+1-th control cycle; $i_{nj}(K+1)$ is the desired output current of lower-arm at the beginning of the K+1-th control cycle.

As can be seen from equation (5), at the beginning of each control cycle, based on the sampled output current value and the desired output current value, the sum of all SM module voltages in the upper and lower arms can be calculated separately. Therefore, the output current can be quickly tracked. The currents $i_{pj}(K+1)$ and $i_{nj}(K+1)$ in equation (5) are replaced by i_{pj}^* and i_{nj}^* respectively. The voltage $u_{pj}(K)$ and $u_{nj}(K)$ are replaced by u_{pj}^* and u_{nj}^* respectively. In this way, the actual current can track the reference current.

$$\begin{cases} u_{pj}^* = \frac{1}{2}U_{dc}(K) - u_{sj}(K) - L\frac{i_{pj}^*-i_{pj}(K)}{T} \\ u_{nj}^* = \frac{1}{2}U_{dc}(K) + u_{sj}(K) - L\frac{i_{nj}^*-i_{nj}(K)}{T} \end{cases} \qquad (6)$$

Where, u_{pj}^* and u_{nj}^* are the output voltage values of upper and lower arm in the control period T, which can be obtained by the nearest level modulation method. This situation different from traditional MMC rectifiers. $U_{dc}(K)$ in equation(6) is no longer a fixed value, but a value that needs to be changed by the need. The value of U_{dc} can vary from 0 to the nominal value by adjusting the DC bias value U_{dc_ref}.

4 NLM FOR MMC RECTIFIER WITH WIDE OUTPUT VOLTAGE RANGE

Unlike the half-bridge sub-module with no negative-level output, the full-bridge sub-module can use its negative-level output characteristics to achieve a wide range of output voltage. Such an operation needs to be implemented by adjusting the DC bias reference value. Figure 4 shows the capacitor voltage control flow chart (Lin W, et al. 2016). Where, i_{arm} is the arm current, and u_{arm_ref} is the arm voltage reference value.

According the value of u_{arm_ref}, we can calculate the number of SM should be inserted at any sampling time.

$$N_{ref} = round\left(abs\left(\frac{u_{arm\_ref}}{U_c} \right) \right) \tag{7}$$

Where, U_c is the rated value of capacitor voltage; Round(x) is the rounding function; abx(x) is the absolute value of x.

When the cell(FBSM) is conducted, it outputs a positive level. When the cell is turned on, it outputs a negative level. The conduction direction of the insert cells is determined according to u_{arm_ref}. When u_{arm_ref} is positive, N_{ref} cells with positive level will be inserted. When u_{arm_ref} is negative, N_{ref} cells with negative level will be inserted. After determining the number of cells should be inserted and knowing whether the output is positive or negative, it is necessary to select N_{ref} cells among all N cells according to the direction of arm current. When u_{arm_ref} is positive and i_{arm} is positive, positively insert N_{ref} cells from all the N cells with the lowest capacitor voltages. When u_{arm_ref} is positive and i_{arm} is negative, positively insert N_{ref} cells from all the N cells with the highest capacitor voltages. In the same time, the rest of N-N_{ref} cells is bypassed. When u_{arm_ref} is negative, the situation is similar with u_{arm_ref} is positive.

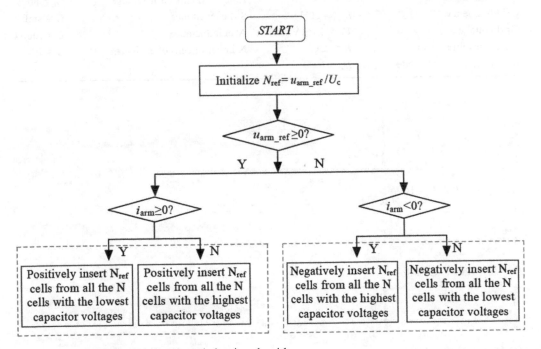

Figure 4.   MMC sub-module voltage balancing algorithm.

5 SIMULATION RESULTS

In order to verify the correctness and effectiveness of the proposed control strategy, a simulation model was built in Matlab/Simulink. The simulation parameters are shown in Table 2.

In order to verify the wide range voltage output capability of the MMC rectifier, the DC current reference values are set to 8000V, 800V and 80V, respectively. The DC voltage, DC current, grid current, sub-module capacitor voltage and the total voltage of all upper arm sub-module are measured.

Corresponding to three different voltages level, the output voltage, the output current, the grid voltage, and the sub-module capacitor voltage waveform are given in Figure 5 respectively. In Figure 5(a), the output voltage rises smoothly from 0V to 8000V in 0.5 seconds, and stabilizes at 8000V. In Figure 5(b), the output voltage rises smoothly from 0V to 800V in 0.5 seconds, and stabilizes at 800V. In Figure 5(c), the output voltage rises smoothly from 0V to 80V in 0.5 seconds, and stabilizes at 80V. It can be seen that the rectifier achieves a wide voltage range output

After entering the steady state after 0.5 seconds, all of the DC currents in Figure 5(a), Figure 5(b), Figure 5(c) are maintained at 160A. It is indicates that the rectifier has large current output capability under different voltage output conditions. When the grid voltage remains the same, the reduced energy that consumed by the rectifier side will cause the grid current to decrease. When the output voltage is 8000V, the grid current amplitude is maintained at 250A. When the output voltage is 800V, the grid current amplitude is

Table 2. Simulation parameters.

Quantity	Value	Quantity	Value
Output DC voltage	0-8000V	Rated SM capacitor voltage	$U_c = 800V$
Grid frequency	$f_{grid} = 50Hz$	SM capacitance	$C = 4mF$
Grid voltage	$V_g = 4.16kV$	Arm inductance	$L = 10mH$
Rated current of grid	$I_g = 250A$	Number of cells of each arm	$N = 10$
Rated current of DC-side	$I_{dc} = 160A$		

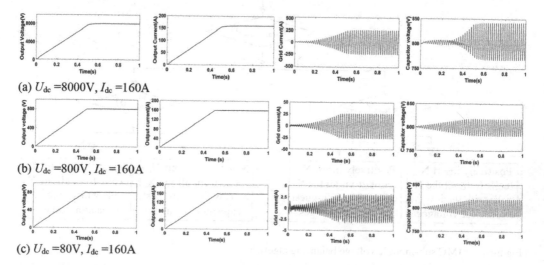

(a) U_{dc} =8000V, I_{dc} =160A

(b) U_{dc} =800V, I_{dc} =160A

(c) U_{dc} =80V, I_{dc} =160A

Figure 5. Simulation results for MMC rectifier.

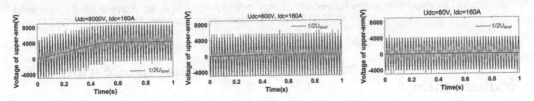

Figure 6.   Upper-arm voltage of MMC rectifier.

maintained at 25A. When the output voltage is 80V, the grid current amplitude is maintained at 2.5A. In Figure 5, the capacitor voltage fluctuations did not exceed $\pm40V$ (5%). Capacitance voltage fluctuations decrease when output power decreases.

Figure 6 shows the upper-arm voltage of the MMC rectifier. It can be seen from the Figure 6 that by adjusting the DC bias command, the bridge arm voltage can be adjusted. This is achieved the purpose of regulating the output DC voltage.

6 CONCLUSION

This paper presents a deadbeat control strategy for an MMC rectifier with a wide range of output voltages. The proposed control method achieves a wide range of output voltages while capacitor voltage fluctuations are controlled to a desired range. The proposed method does not need to adopt the double loop control structure. Compared with the traditional control method, the proposed method greatly simplifying the control structure. At the same time, the number of PI regulators is reduced, and the dynamic response of the system is improved.

The project supported by National Natural Science Foundation of China (61572417) , Scietific Research Fund of Hunan Provincial Education Department(15C327) and Key Research and Development Project of Industry-University-Research of Xiangtan University (16PYZ022).

REFERENCES

Akagi H, et al. 2007. Control and Performance of a Transformerless Cascade PWM STATCOM With Star Configuration. IEEE Transactions on Industry Applications 43(4):1041–1049.

Hagiwara M, et al. 2010. A Medium-Voltage Motor Drive With a Modular Multilevel PWM Inverter. IEEE Transactions on Power Electronics 25(7):1786–1799.

Nami A, et al. 2015. Modular Multilevel Converters for HVDC Applications: Review on Converter Cells and Functionalities. IEEE Transactions on Power Electronics 30(1):18–36.

Guan M, et al. 2017. Back-to-back hybrid modular multilevel converters for ac motor drive. IECON 2017-43rd Annual Conference of the IEEE Industrial Electronics Society. China: Beijing.

Y.S. Kumar, G. Poddar. 2018. Medium-Voltage Vector Control Induction Motor Drive at Zero Frequency Using Modular Multilevel Converter. IEEE Transactions on Industrial Electronics 65(1):125–132.

Li B, et al. 2016. Control and Analysis of the Modular Multilevel DC De-Icer With STATCOM Functionality. IEEE Transactions on Industrial Electronics 63(9):5465–5476.

Guo Y, et al. 2017. Control Strategies of DC Ice-melting Equipments for Full-bridge Modular Multilevel Converters. Automation of Electric Power Systems 41(5):106–113.

Fu X, et al. 2009. Deadbeat Decoupling Control of VSC-HVDC Systems for Grid Connection of Wind Farms. Transactions of China Electrotechnical Society 24(11):157–164.

Deng X, et al. 2014. Study of control strategy for MMC-HVDC system based on deadbeat current control[J]. Power System Protection and Control 2014(8):34–39.

Rong F, et al. 2017. The Deadbeat Control Strategy of Modular Multilevel Converter[J]. Proceedings of the CSEE 37(6):1753–1763.

Lin W, et al. 2016. Full-Bridge MMC Converter Optimal Design to HVDC Operational Requirements. IEEE Transactions on Power Delivery 31(3):1342–1350.

Emerging Developments in the Power and Energy Industry – Dufo-López, Krzywanski & Singh (eds)
© 2020 Taylor & Francis Group, London, ISBN 978-0-367-27169-5

Control of active magnetic bearing system in accordance with state feedback theory

Chong Feng & Yibing Zhang
School of Information Engineering, Xiangtan University, Xiangtan, Hunan Province, China

ABSTRACT: In this study, State Feedback Control (SFC) method is proposed to control an active magnetic bearing system (AMB). The AMBs employed in this study have two four-pole bearings. The distances between the two can be controlled using the electrical signal from the controller. The magnetic forces are generated under the feedback loop control. The designation of the controller can optimize the control of traditional PID, and the parameters of each state can be detected in the control process. The simulation and experimental results show the high quality of the control strategy, suggesting that the control algorithm proposed in this study is feasible and has strong robustness.

Keywords: state feedback control, AMB system, PID, strong robustness

1 INTRODUCTION

AMBs are the typical electromagnetic products, which provide stable suspension for rotors. AMBs are used in a wide range of practical applications such as energy storage, high-speed-turbine, CNC machines, artificial blood pumps, mixers, etc. Since AMBs are capable of levitating rotor without mechanical contact, friction or lubrication, there exists no wear on the bearing surfaces and no contamination of the environment caused by bearing particles (E.H. Maslen & G.Schweitzer & H.Bleuler 2009). It has important research implication in energy saving and environmental protection (G.Schweitzer & H.Bleuler & A.Traxler. 1994).

The design of active magnetic bearing control system has been broadly studied. About a decade ago, more than 90% of active magnetic bearing system controller designs were based on a decentralized Proportional-Integral-Derivative (PID) position control. In the decentralized design, the AMB system is decoupled into single-input and single-output (SISO) loops (A.Chiba & T.Fukao & O.Ichikawa 1994). However, with PID controllers, a velocity feedback signal is used, and the velocity signal is often difficult to measure in an AMB system. Due to the system and sensor noises and the small size of the air gap between the rotor and the stator, the numerical differentiation of the position signal is often poor (B.Polajžer, & J.Ritonja & G. Štumberger. 2006).

To improve the stability and performance of the AMB, modern robust control that overcomes the inherent shortcomings of classical PID control approaches (i.e., the inability to solve overshoot, uncertainty and systemic coupling) were applied to the AMB system with attractive results. The AMB system can be controlled in two modes, i.e., the current controlled mode and the voltage-controlled mode. In general, the voltage-controlled mode is prioritized compared with the current controlled mode. On the one hand, the control input for the voltage-controlled mode is the coil winding voltage, which is truer compared to the coil current for the current-controlled mode. On the other hand, the power amplifier for the current-controlled mode is generally more complicated and expensive than that for the voltage-controlled mode. Therefore, the voltage-controlled AMB system is used in this study to reject static load disturbance in accordance with State Feedback theory.

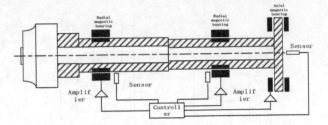

Figure 1. The structure of magnetic bearing system.

2 MODELING AND ANALYSIS OF AMB SYSTEM

2.1 *Classical control mathematical model of AMB*

To establish the mathematical model of magnetic bearing, the force analysis is needed. The first step is to deduce the expression of the electromagnetic force in the magnetic bearing. Subsequently, the expression of the external force acting on the rotor at any position is yielded, and the kinematics equation is obtained according to Newton's law of motion. Finally, the system transfer function of the model through Laplace change can be yielded. Without considering the coupling of magnetic degrees of freedom, the single degree of freedom can be analyzed.

Regardless of the gravity and magnetic leakage phenomenon under vacuum conditions, in accordance with the theory of electromagnetic field, B is magnetic density vector, and it yields:

$$F = \frac{B^2 S}{\mu_0} \tag{1}$$

$$F_{sum} = F_1 - F_2 = K\left(\frac{(i+\Delta i)^2}{(x+\Delta x)^2} - \frac{(i+\Delta i)^2}{(x+\Delta x)^2}\right), K = \mu_0 N^2 S \tag{2}$$

Using Taylor expansion of (2), it yields:

$$F_{sum} = k_i i_b + k_x x = m\frac{d^2 x}{dt^2} \tag{3}$$

where, k_i, k_x are the force-current factor and force-displacement factor, respectively.

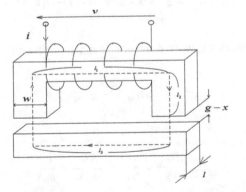

Figure 2. Simplified cores with a winding.

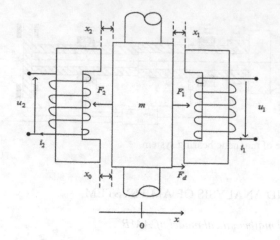

Figure 3.   Schematic of a one-DOF magnetic bearing.

Laplace transform is used for (3), so the transfer function of the system:

$$G(s) = \frac{X(s)}{I(s)} = \frac{k_i}{ms^2 - k_x} \tag{4}$$

2.2   State control mathematical model of AMB

For the further study of the control strategy, a simplified magnetic bearing system, i.e., the two-pole, single degree-of-freedom (DOF) magnetic bearing shown in Figure 3 is to be considered.

Assume that the levitated shaft or rotor only moves in the x-direction, no bending of the shaft occurs, and no gravity acts on the shaft. The shaft is at an unstable equilibrium point, and active feedback control is needed to make this point stable.

Figure 3 suggests that the magnitude of magnetic flux in iron core:

$$\psi = \frac{Ni}{2R_g} = \frac{Ni\mu_0 wl}{2\ g} = \frac{K}{N}\frac{i}{2g}, \quad K = \mu_0 N^2 S \tag{5}$$

The energy stored in a magnetic field in a given volume V is defined as:

$$W_\psi = \frac{1}{2}\int_V H \bullet B dV = \frac{1}{2}\int_V \frac{B^2}{\mu_0} dV \tag{6}$$

Besides, the relation between the work done in moving an object and the force applied to that object is:

$$dW_\psi = F dg \tag{7}$$

Thus:

$$F = \frac{dW_\psi}{dg} = \frac{SB^2}{\mu_0} = \frac{(SB)^2}{\mu_0 S} = \frac{\psi^2}{\mu_0 S} \tag{8}$$

Substituting the flux from (5) into the previous equation, it yields:

$$F = \frac{\psi^2}{\mu_0 S} = \frac{1}{\mu_0 S}\left(\mu_0 N S \frac{i}{2g}\right)^2 = \frac{\mu_0 N^2 S}{4}\left(\frac{i}{g}\right)^2 = \frac{K}{4}\left(\frac{i}{g}\right)^2 \tag{9}$$

The Newtonian motion equation of the system is written as:

$$m\ddot{x} = F_1 + F_2 - F_d = \frac{K}{4}\left(\frac{i_1}{x_1}\right)^2 - \frac{K}{4}\left(\frac{i_2}{x_2}\right)^2 + F_d \tag{10}$$

Or:

$$\ddot{x} = \frac{K}{4m}\left(\frac{i_1}{x_1}\right)^2 - \frac{K}{4m}\left(\frac{i_2}{x_2}\right)^2 + \frac{1}{m}F_d \tag{11}$$

where the total force on the rotor is a nonlinear function of the currents in the coils and the air gaps between the coils and the rotor surfaces.

According to Faraday's Law, the back-EMF is induced in the coil because of a change in air gap flux. Kirchhoff's Voltage Law for the coil circuits is expressed as:

$$u = iR + L_s\frac{di}{dt} + N\frac{d\psi}{dt} \tag{12}$$

Thus, the voltages in Figure 3 are:

$$u_1 = i_1 R + L_s\frac{di_1}{dt} + \frac{K}{2}\frac{d}{dt}\left(\frac{i_1}{x}\right); \quad u_2 = i_2 R + L_s\frac{di_2}{dt} + \frac{K}{2}\frac{d}{dt}\left(\frac{i_2}{x}\right) \tag{13}$$

It has two inputs to the system, u_1 and u_2 controlling the forces on the rotor. The system has four outputs: i_1, i_2, x_1 and x_2. The states of the nonlinear system will be represented by the column vector z as:

$$\dot{z} = \begin{bmatrix} \dot{z}_1 \\ \dot{z}_2 \\ \dot{z}_3 \\ \dot{z}_4 \end{bmatrix} = \begin{bmatrix} \dot{x} \\ \ddot{x} \\ \dot{i}_1 \\ \dot{i}_2 \end{bmatrix} = \begin{bmatrix} z_2 \\ \frac{K}{4m}\left(\frac{z_3}{x_0-z_1}\right)^2 - \frac{K}{4m}\left(\frac{z_4}{x_0-z_1}\right)^2 + \frac{1}{m}F_d \\ \frac{2(x_0-z_1)}{2L_s(x_0-z_1)+K}\left[-Rz_3 - \frac{K}{2(x_0-z_1)^2}z_2z_3 + u_1\right] \\ \frac{2(x_0-z_1)}{2L_s(x_0-z_1)+K}\left[-Rz_4 - \frac{K}{2(x_0-z_1)^2}z_2z_4 + u_2\right] \end{bmatrix} \tag{14}$$

2.3 The linearized MIMO model

To obtain the linearized model, (14) can be transformed into a four-order multiple input multiple output (MIMO) system using Jacobian transformation.

$$\dot{z} = \begin{bmatrix} \dot{z}_1 \\ \dot{z}_2 \\ \dot{z}_3 \\ \dot{z}_4 \end{bmatrix} = \begin{bmatrix} \dot{x} \\ \ddot{x} \\ \frac{di_1}{dt} \\ \frac{di_2}{dt} \end{bmatrix} = \begin{bmatrix} 0 & 1 & 0 & 0 \\ \frac{2k_s}{m} & 0 & \frac{k_i}{m} & -\frac{k_i}{m} \\ 0 & -\frac{k_i}{L} & -\frac{R}{L} & 0 \\ 0 & \frac{k_i}{L} & 0 & -\frac{R}{L} \end{bmatrix}\begin{bmatrix} x \\ \dot{x} \\ i_1 \\ i_2 \end{bmatrix} + \begin{bmatrix} 0 & 0 \\ 0 & 0 \\ \frac{1}{L} & 0 \\ 0 & \frac{1}{L} \end{bmatrix}\begin{bmatrix} u_1 \\ u_2 \end{bmatrix} + \begin{bmatrix} 0 \\ 1/m \\ 0 \\ 0 \end{bmatrix}F_d \tag{15}$$

Where:

$$k_s = \frac{Ki_0^2}{2x_0^3}; \quad k_i = \frac{Ki_0}{2x_0^2}; \quad L = \frac{K + 2x_0 L_s}{2x_0} \tag{16}$$

A block diagram representation of the linearized MIMO is shown in Figure 4

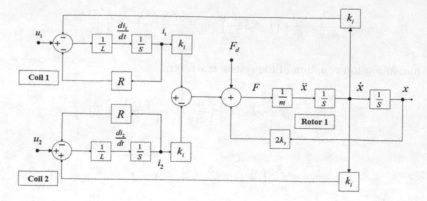

Figure 4. Linearized MIMO system.

3 CONTROLLER DESIGN

In this study, we will design two types of controllers of AMB system, i.e., PID controller and state space controller. The following physical and derived parameters for the AMB used in the simulations are listed below. These are the same as we used in our material experiment.

Figure 5 & Figure 6 show the expression of the PID control and State Space voltage control respectively. By setting $F_d = 0$, we can derive the open-loop transfer function from u to x:

$$G(S) = \frac{X(s)}{U(s)} = C(sI - A)^{-1}B_1 = \frac{2k_i}{mLs^3 + mRs + 2(k_i^2 - Lk_s)s - 2k_sR} \tag{17}$$

The system is naturally unstable. By substituting the parameters, it yields:

$$G(s) = \frac{1}{D(s)} \begin{bmatrix} G_{11}(s) & G_{12}(s) \\ G_{21}(s) & G_{22}(s) \\ G_{31}(s) & G_{32}(s) \\ G_{41}(s) & G_{42}(s) \end{bmatrix}; \text{where: } D(s) = (s - 205.26)(s + 197.93)(s + 21.79)(s + 14.47)$$

$$\tag{18}$$

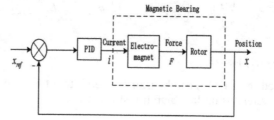

Figure 5. PID loop plant.

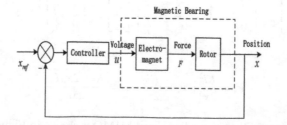

Figure 6. Voltage-controlled loop plant.

Table 1. Physical parameters for the AMB.

Parameter Name	Parameter Symbol	Value	Unit
Nominal Air Gap	x_0	0.0004	m
Bias Current	i_0	2.5	A
Coil Resistance	R	1.7	Ω
Rotor Mass	m	50	kg
Coil Self-Inductance	L_S	0.0783	H
Magnetic Constant	K	0.0000313	H • m
Magnetic pole angle	θ	0.2138	rad
Static Disturbance Force	F_d	50	N

Table 2. Derived parameters for the AMB.

Parameter Name	Parameter Symbol	Value	Unit
Force-Displacement Constant	k_s	1529600	N/m
Force-Current Constant	k_i	244.7	N/A
Total Inductance	L	0.1175	mH

$$G_{11}(s) = 41.6720(s + 14.47); \quad G_{12}(s) = -41.6720(s + 14.47)$$
$$G_{21}(s) = 41.6720s(s + 14.47); \quad G_{22}(s) = -41.6720s(s + 14.47)$$
$$G_{31}(s) = 8.5137(s - 227.15)(s + 224.24)(s + 17.39); \quad G_{32}(s) = 86827.7281s$$
$$G_{41}(s) = 86827.7281s; \quad G_{42}(s) = 8.5137(s - 227.15)(s + 224.24)(s + 17.39)$$

(19)

4 SIMULATION AND EXPERIMENTAL VERIFICATION

In MATLAB/Simulink, the actual parameters and model simulations of the mentioned AMB system are used. To stabilize the AMB system, the controller is designed in state feedback with pole-placement using MATLAB's place () function. While the design of the traditional PID controller using MATLAB's SISOTOOL toolbox. The simulation results are shown in Figure 7 & Figure 8, which include the stability analysis of two control algorithms.

Figure 7 shows the close-loop step response with a random number disturbance of the AMB system with PID controller. It is suggested that in the presence of disturbances, the output can track the input well, the adjustment time is short. According to the Bode diagram, the system has a good stability margin, the steady-state characteristics of the system are ideal. However, other state signals of the system (e.g., velocity, control voltage, control current etc.) are hard to measu re, which usually limits the further improvement of the control effect.

Figure 8 shows the disturbance rejection process of the rotor position signal and rotor velocity signal, respectively. When a static force of 50 N and a high frequency sinusoidal disturbance of 1000 Hz were applied at 0.01 second, the ripple of the latter is nearly 2×10^{-7}m. For removing the steady state error, we added integral feedback in the position terms to the state feedback gain matrix, after a small amount of tuning, the integral feedback can remove the steady-state error and does so with very small currents and control voltages. The error signal between the output signal and the reference signal is shown. It approaches to zero after about 0.06 s. the rotor is self–stable whenever having outside disturbance. As we can see, it shows that the anti-interference ability of the system is strong. Furthermore, we can get signals of

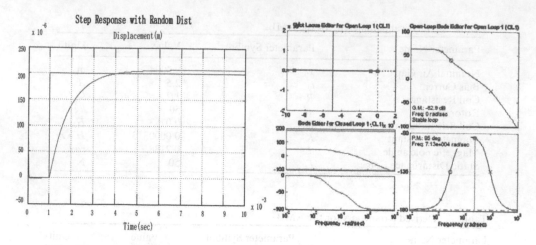

Figure 7.   simulation results of PID control modes.

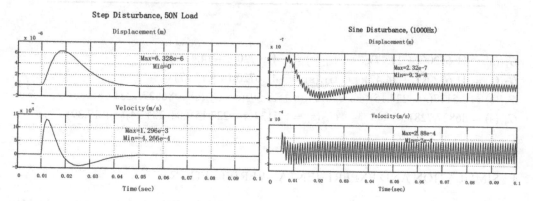

Figure 8.   simulation results of state space control modes.

various states like rotor velocity, voltage and current accurately, which contributes to the further design of controller and to the sensorless control.

Since rotating machinery is never perfectly balanced, we use a sinusoidal load disturbance to model the unbalance condition in rigid rotors. This is done in the presence of the state feedback plus integral control without altering the gains.

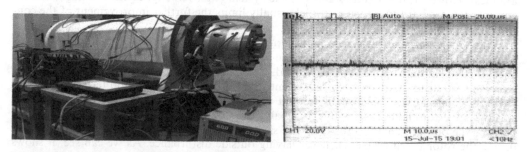

Figure 9.   Experimental prototype and experimental results.

59

5 CONCLUSION

Given that the state signal is hard to measure and the poor capacity of resisting disturbance in the design of the active magnetic bearing controller, this study has proposed a method based on state space theory to optimize the stability characteristics of the system. Compared with conventional PID control, simulation and experimental results demonstrate that the State Feedback Control (SFC) control method is superior in trajectory tracking, with the improved disturbance rejection and the higher dynamic performance. It can make up for the system disturbance caused by the out force and unbalance condition in rigid rotors in time, improve system stability and make it strongly robust. It can also precisely measure the state signal of system in the control process, which shows the potential in engineering application. The future research considers the optimal design of sensorless control for best performance while reducing the control effort.

REFERENCES

E.H. Maslen & G. Schweitzer & H. Bleuler, et al. 2009. Magnetic Bearings—Theory, Design and Application to Rotating Machinery[M].

G. Schweitzer & H. Bleuler & A. Traxler. 1994. Active magnetic bearings: basics, properties and applications of active magnetic bearings[M].

A. Chiba,T. Fukao & O. Ichikawa, et al. 1989. Magnetic Bearings and Bearingless Drives[J]. Newnes, 14(3): 158–162.

B. Polajžer & J. Ritonja & G. Štumberger, et al. 2006. Decentralized PI/PD position control for active magnetic bearings[J]. Electrical Engineering, 89(1): 53–59.

R.A. Rarick. 2007. Control Of An Active Magnetic Bearing with and without Position Sensing[J].

T. Dever & G. Brown & K Duffy,et al. 2013. Modeling and Development of Magnetic Bearing Controller for High Speed Flywheel System[C]//International Energy Conversion Engineering Conference.

T. Ishikawa & K.I. Matsuda & R. Kondo,et al. 2009. 5-DOF Controlled Self-Bearing Motor[J]. Jsdd. 3(4): 483–493.

Emerging Developments in the Power and Energy Industry – Dufo-López, Krzywanski & Singh (eds)
© 2020 Taylor & Francis Group, London, ISBN 978-0-367-27169-5

Impact of large-scale electric vehicle charging on a power grid and its optimal scheduling strategy

Caixue Chen & Gongkai Zhang*
School of Information Engineering, Xiangtan University, Xiangtan, Hunan Province, China

ABSTRACT: An electric vehicle is a moving load, the charging process of which is random with respect to spatial and temporal distribution. Disordered charging of large-scale electric vehicles will affect the stability of the grid. In this article the Monte Carlo method is used to simulate the regular mode of disordered charging of large-scale electric vehicles, and then the load curves of electric vehicles charging under different permeability conditions are analyzed. Taking the IEEE33 node system as an example, in this article the forward and backward power flow algorithm is used to calculate the impact of random access of electric vehicles to the distribution system under different permeabilities. For the benefit of the users and the security of the power grid, to realize the "peak clipping and valley filling" of an electric vehicle charging load and the optimization goal of the lowest user charging fee, it is very important that we establish a multiobjective optimization model. A genetic algorithm is used to solve the optimization of the multiobjective model. The simulation results show that the model is practical and the algorithm is effective.

Keywords: electric vehicle, forward and backward power flow algorithm, genetic algorithm, Monte Carlo, peak clipping and valley filling

1   INSTRUCTIONS

The profusion of fuel-powered vehicles has imposed tremendous pressure on global resources and the environment (Li et al. 2015; Huang & Yu 2018). Human society is gradually facing a severe energy shortage and environmental pollution (Sun et al. 2013). Electric vehicles using electricity as the main energy source have attracted more and more attention because of their huge advantages in energy saving and emission reduction (Li et al. 2012). According to the Ministry of Industry and Information Technology's electric vehicle development strategy research report, China's electric vehicle ownership will reach 60 million in 2030 (Hu et al. 2012). But the disordered charging of large-scale electric vehicles will cause safety problems such as voltage offset, network loss, and supply and demand balance to the power grid, as well as a certain pressure on users in terms of economic benefits (Zhang et al. 2016; Chen et al. 2018). Therefore, quantitative assessment of the impact of large-scale electric vehicles on the grid after access to the grid and research on charging control strategies aimed at reducing the negative impacts have become increasingly hot pursuits (Tian et al. 2012; Yu & Liu 2015).

This article focuses mainly on private electric vehicles. Based on the Monte Carlo method, which is used to analyze the average vehicle-miles of travel of electric vehicles and to establish the corresponding mathematical model simulation to describe the load conditions generated during the disordered charging of electric vehicles, the mathematical model of network loss is established. The peak-to-valley electricity price period optimization model with the optimal effect of "peak clipping and valley filling" is established. The genetic algorithm is used to optimize the time-zone formulation scheme, and finally a case analysis is carried out.

2 FORECAST OF ELECTRIC VEHICLES AVERAGE VEHICLE-MILES

2.1 *Forecast of electric vehicle driven distance*

The average vehicle-miles of travel of private electric vehicles are obtained by fitting the US traffic data (NHTS2009) (Zhan et al. 2012). The daily mileage of electric car users (Figure 1) is approximately equal to the lognormal distribution. The probability density function is as follows:

$$f_D(x) = \frac{1}{x\sigma_D\sqrt{2\pi}}\exp[-\frac{(\ln x - \mu_D)^2}{2\sigma_D^2}] \tag{1}$$

where μ_D is the average of the logarithms of the probability density function variables x, with $\mu_D = 3.20$; σ_D is the standard deviation of the logarithm of the probability density function variable, with $\sigma_D = 0.88$.

2.2 *Forecast of electric vehicle trip end time*

Considering the time characteristics of electric vehicle charging, the last travel time of electric vehicle users is taken as the starting time of electric vehicle charging. The probability density function (Figure 2) is as follows:

$$f_t(x) = \begin{cases} \frac{1}{\sigma_t\sqrt{2\pi}}\exp[\frac{(x-\mu_t)}{2\sigma_t^2}], (\mu_t - 12) < x < 24 \\ \frac{1}{\sigma_t\sqrt{2\pi}}\exp[\frac{(x+24-\mu_t)}{2\sigma_t^2}], 0 < x < (\mu_t - 12) \end{cases} \tag{2}$$

where μ_t is the expected value of the probability density function variable, with $\mu_t = 17.6$; σ_t is the standard deviation of the above probability density function variable, with $\sigma_t = 3.4$.

3 LOAD CALCULATION FOR DISORDERED CHARGING

The electric vehicle is charged by means of the conventional constant power charging mode, and the power is fixed at 15 kW·h per 100 km. Every day, the charging power is the same as the driving power consumption of the day, the maximum mileage of each vehicle is 200 km, the mileage is generated according to the lognormal distribution, and the capacity of the battery is calculated based on 200 km. The charging power is 3 kW; the battery capacity is 30 kW·h; and the number

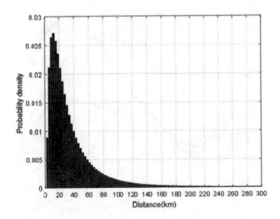

Figure 1.   Forecast of the daily range of electric vehicles.

62

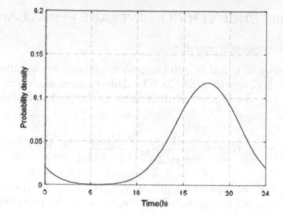

Figure 2.   The probability density function at the end of the last trip.

of electric vehicles is 1000, 2000, and 5000. The corresponding permeability rates are 11.21%, 23.89%, and 56.90%, respectively. The simulation curves are shown in Figure 3 and Figure 4.

Figure 3 shows the total power distribution curve of the electric vehicle under different permeability environments, and Figure 4 shows the load curve of the power distribution system

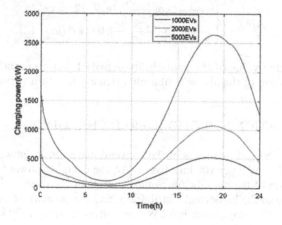

Figure 3.   Charging load of electric vehicles.

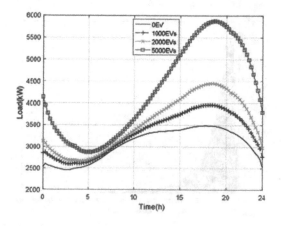

Figure 4.   Total load of grid.

bus when a large-scale electric vehicle is charged. It can be seen from Figure 3 that starting from 10:00 hours, the charging load of electric vehicles gradually increases. From 16:00 hours to 20:00 hours, the charging load of electric vehicles is at a peak, and at the same time, the peak of the grid visibly appears, which will make a great impact on the grid.

4 IMPACT OF DISORDERED CHARGING ON THE POWER GRID

4.1 Power distribution system and calculation

In this article, the IEEE33 node system is used as an example to study the impact of electric vehicle access on the power distribution system. The power distribution system has 32 branches and one power supply, the network voltage reference value is 12.66kV, the three-phase power reference value is 10 MVA, and the total network load is 3715 kW + j2300kVar . The 0 node is the power point, whose standard voltage is 1.05.

Based on Figure 5, the power mathematical formula of the m–n branch is calculated as follows:

$$\overset{*}{S}_{mn} = \overset{*}{S}_n + \sum_{k=1}^{y} \overset{*}{S}_{mk} + I_{mn}^2 \overset{*}{Z}_{mn} \tag{3}$$

$$\begin{cases} \overset{*}{S}_{mn} = P_{mn} + jQ_{mn} \\ \overset{*}{Z}_{mn} = R_{mn} + jX_{mn} \end{cases} \tag{4}$$

where $\overset{*}{S}_{mn}$ is the current on the m–n branch, y is the lower level nodes connected to node n, $\overset{*}{S}_n$ is the input power of node n, I_{mn} is the injection power of node n, and $\overset{*}{Z}_{mn}$ is the impedance on the branch road.

The formula for the forward trend of the branch calculates the real part and the imaginary part separately, as shown in Eq. (5):

$$\begin{cases} P_{mn} = P'_n + \frac{P_n'^2 + Q_n'^2}{U_j^2} R_{mn} \\ Q_{mn} = Q'_n + \frac{P_n'^2 + Q_n'^2}{U_j^2} X_{mn} \end{cases} \tag{5}$$

where U_j is the voltage amplitude of node j.

And $P'_n = P_n + \sum_{k=1}^{y} P_{nk}, \quad Q'_n = Q_n + \sum_{k=1}^{y} Q_{nk}.$

$$\begin{cases} u_n = u_m - \frac{(P_{mn}R_{mn} + Q_{mn}X_{mn})u_m - (P_{mn}X_{mn} + Q_{mn}R_{mn})f_m}{u_m^2 + f_m^2} \\ f_n = f_m - \frac{(P_{mn}R_{mn} + Q_{mn}X_{mn})f_m + (P_{mn}X_{mn} - Q_{mn}R_{mn})u_m}{u_m^2 + f_m^2} \end{cases} \tag{6}$$

The return formula of the node voltage according to Ohm's law is shown in Eq. (6), where $U_m^* = u_m + nf_m$ is the voltage of node m.

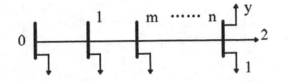

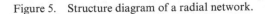

Figure 5. Structure diagram of a radial network.

4.2 *Results and analysis of simulation*

The voltage level curve of each node of the distribution system under different permeabilities is calculated via the forward and backward power flow algorithm.

According to Chinese policy, the voltage deviation of a 10-kV distribution network is under ±7%. As can be seen from Figure 6, we set the voltage limit. It can be seen from the above that when electric vehicles are connected to the power grid at high permeability, the fluctuation range of the distribution system node can easily exceed ±7%. Excessive voltage offset will affect the safety and stability of the power distribution system.

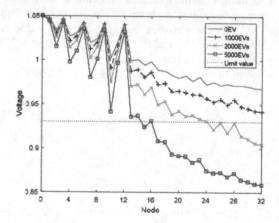

Figure 6. Voltage offset of each node.

Table 1. Voltage excursion and off-limit scaling parameters.

Number of electric vehicles	Maximum downward offset voltage	The ratio exceed the vehicles
0	0.97	0
1000	0.94	0
2000	0.90	24.12%
5000	0.86	51.52%

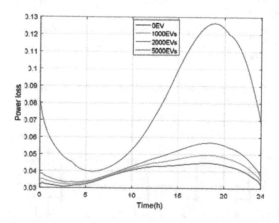

Figure 7. Power distribution system network loss rate.

It can be seen from Figure 7 that the peak-to-peak phenomenon caused by the disordered charging of large-scale electric vehicles on the distribution system from 16:00 to 20:00 hours causes a higher network loss rate. To improve the safety and reliability of the power distribution system, some methods must be used to optimize the scheduling of electric vehicles.

5   OPTIMIZED SCHEDULING OF ELECTRIC VEHICLES

5.1   *Establishment of multiobjective functions*

To realize the optimal scheduling of electric vehicles, in this article we establish the optimization goal of considering the user layer and the power grid layer. The objective function $f1$ is established to realize the minimum charging cost for the users. To alleviate the problem of "peaking on the peak" of the power distribution system, the objective function $f2$ of the total load of the balance is established, thus forming a multiobjective function. Finally, multiobjective optimization is achieved using the genetic algorithm.

$$\begin{cases} f_1 = \min \sum_{\tau=1}^{n} \sum_{i=1}^{N} C_{i\tau} M_{\tau} p_i \Delta t \\ f_2 = \min \sum_{\tau=1}^{n} \left[(P_0 + \sum_{i=1}^{N} C_{i\tau} p_i) - P_{av} \right]^2 \end{cases} \tag{7}$$

In Eq. (7), τ is the random period; there are a total of 96 time periods, denoted by n; and i indicates the random vehicle. The total number of vehicles is N; $C_{i\tau}$ represents whether car i is charged in the first period, with 1 representing charging and 0 not charging; M_{τ} is the electricity price of the period; p_i is the charging power of the random vehicle; Δt represents the possible charging time of the vehicle; P_0 indicates the basic load of each period of the grid; and P_{av} indicates the average daily load.

When the electric vehicle arrives at the charging station, the remaining state of charge (SOC) is used as the initial charge quantity SOC_i, and each electric vehicle must be fully charged when it leaves; that is, the amount of charge at the charging station is $1 - SOC_i$. To this end, the power constraint is established as follows:

$$p_i = \frac{(1 - SOC_i) \times D_i}{T_i} \tag{8}$$

where T_i is the charging time of the electric vehicle, D_i is the storage capacity of the battery, and p_i is the charging power of the electric vehicle.

Considering the power constraint of the substation, the total load power of the substation cannot exceed the upper limit of the power of the substation. The power constraints of the substation are established as follows:

$$P_{base} + \sum_{i=1}^{N} C_{i\tau} p_i < P_{sta} \tag{9}$$

In Eq. (9), P_{base} is the original residential electricity load of large-scale electric vehicles when they are not connected to the distribution network; P_{sta} is the maximum load on the distribution system in the area.

5.2 Results and analysis of simulation

Taking 500 electric vehicles as an example, the electric vehicle charging power is 3 kW, the battery capacity is 30kW·h, and the genetic algorithm terminates the evolutionary algebra to 800. When considering the lowest user charging cost, the simulation effect is shown in Figure 8. When considering only the minimum peak-to-valley difference in a charge load, the simulation effect is shown in Figure 9.

When considering only the charging cost, the disordered charging cost is about 2014.9 yuan, and the ordered charging cost is only 1205.6 yuan; the difference is 809.3 yuan. When the peak-to-valley difference is the smallest, the peak-to-valley difference is 25.45%.

When considering the user layer charging cost minimum and "sharp peak filling" of the distribution system load, the simulation graphs of the example are as above.

It can be seen from Figure 10 and Table 2 that the valley value of the disordered charging is 3 kW, and when it reaches the peak value, it is about 325 kW. After the optimization of the ordered charging, the effect of "cutting the peak and filling the valley" can be seen from Figure 11. The peak-to-valley difference was directly reduced by 14.23%. The user layer was optimized with the control strategy of time-of-use electricity price, saving about 759.2 yuan. It can be seen that after optimized scheduling, large-scale electric vehicle charging has significantly improved the security of the power grid and the users' cost.

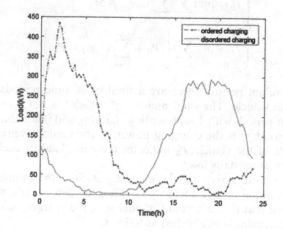

Figure 8.   Charging load of the lowest. user charging cost.

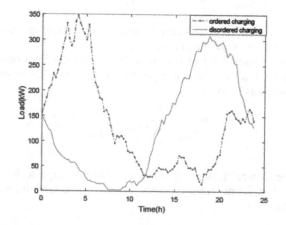

Figure 9.   Total load of the minimum peak-to-valley difference.

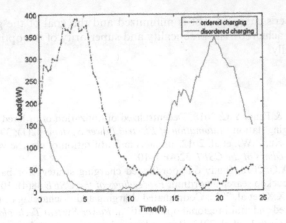

Figure 10.   Charging load of electric vehicles.

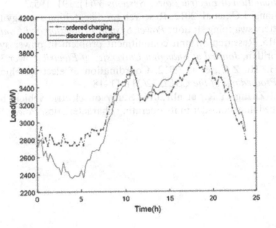

Figure 11.   Total load of distribution system.

Table 2.   Comparison of ordered charging and disordered charging parameters.

Forms of Charging	Value of charging valley (kW)	Value of the charging peak (kW)	Rate of charge peak-to-valley difference	Charging fee (yuan)
Disordered	2364.6	3986.2	40.68%	2107.9
Ordered	2736.8	3721	26.45%	1348.7

6 CONCLUSION

In this article, the spatial and temporal distributions of electric vehicles are analyzed, the load curve of large-scale electric vehicles connected to the power grid under different penetration rates is obtained, and the peak-to-valley phenomenon caused by electric vehicle charging is seen. Based on the IEEE33-node power distribution system, the voltage-offset and network loss caused by large-scale electric vehicles connected to the grid are obtained by using the for-ward-pushing power flow algorithm. This result is an important basis for electric vehicles to access the power grid. The establishment of multiobjective optimization functions is more practical and comprehensive. Finally, using the genetic algorithm to achieve the goal optimization, the user's charging cost is minimized and the goal of the grid's "peak clipping and valley filling" is achieved. The practicality and superiority of the optimization method are verified experimentally.

REFERENCES

Chen, S., Wang, X.N., & Feng, Y.C. 2018. Decentralized optimization of ordered charging scheduling in electric vehicle charging station. *Automation of Electric Power Systems* 42(1): 39–46.

Hu, Z.C., Song, Y.H., Xu, Z.W., et al. 2012. Impacts and utilization of electric vehicles integration into power system. *Proceedings of the CSEE* 32(4): 1–10.

Huang, M.L. & Yu, A.Q. 2018. Study on coordinated charging strategy for battery swapping station based on improved cuckoo search algorithm. *Proceedings of the CSEE* 38(4): 1075–1083.

Li, M., Su, X.L., Yan, X.X., et al. 2015. Coordinated charging and discharging of plug-in electric vehicles based on multi-layered and multi-regional optimization. *Power System Technology* 39(12): 3556–3562.

Li, Q.S., Xiao X.N., & Guo, J., et al. 2012. Research on scheme for ordered charging of electric vehicles. *Power System Technology* 36(12): 32–38.

Sun, X.M., Wang, W., Su, S., et al. 2013. Coordinated charging strategy for electric vehicles based on time-of-use price. *Automation of Electric Power Systems* 37(1): 191–195.

Tian, W.Q., He, J.H., Jiang, J.C., et al. 2012. Research on dispatching strategy for coordinated charging of electric vehicle battery swapping station. *Power System Protection and Control* 40(21): 114–119.

Yu, A.Q. & Liu, T. 2015. Research on unit commitment problem in power system based on adaptive quantum genetic algorithm. *Journal of Shanghai University of Electric Power* 31(1): 24–28.

Zhan, K.Q., Song, Y.H., Hu Z.C., et al. 2012. Coordination of electric vehicle charging to minimize active power losses. *Proceedings of the CSEE* 32(31): 11–18.

Zhang, Y., Rong, Z.P., Zhang, Y.X., et al. 2016. Study on electric vehicle battery swapping station space-time double scheduling consider in its operating characteristics. *Power System Technology* 40(9): 2616–2624.

Emerging Developments in the Power and Energy Industry – Dufo-López, Krzywanski & Singh (eds)
© 2020 Taylor & Francis Group, London, ISBN 978-0-367-27169-5

Research on active response control strategy of smart building micro-grid with energy storage system

S. Li* & Y. He
College of information Engineering, Xiangtan University, Xiangtan, Hunan Province, China

C. Liu
Northeast Electric Power University, Jilin, Jilin Province, China

N. Li
Xi'an University of Technology, Xi'an, Shanxi Province, China

Y. Li
Trinity International Consulting Company, Beijing, China

ABSTRACT: The controllable load and energy storage devices are controlled to supplement the power while renewable energy generation is insufficient in smart building micro-grid. To improve the efficiency of the control system, this paper proposes a new active response control strategy for considering energy storage system and controllable load of smart building. The energy storage system of the smart building is constructed by two parts, including the small battery of home power supply system and a centralized energy storage system. The schedulable value of the building controllable load is determined by the particle swarm optimization algorithm based on the prediction of the renewable power generation and building load in a time interval. Then, active response control strategy is proposed to coordinate control building controllable loads and building energy storage systems, which effectively improve the power quality of the distribution network. The effectiveness of the proposed method is verified in the experimental case.

Keywords: Smart building micro-grid, Controllable load, Energy storage system structure

1 INTRODUCTION

Nowadays, building load accounts for 40% of the total electricity consumption of the city in some large cities, and most of the building load are controllable in the smart building micro-grid (Jia et al. 2018). It has great potential for the micro-grid optimal dispatching to actively increase the proportion of the distributed renewable power generation.

Demand response technologies and energy storage equipment have been proposed to balance the fluctuation rate of renewable power generation (Zhai et al. 2014). Especially, the scheduling potential of the controllable load is deeply explored in the current years, which make the user terminal be actively involved in the grid response through the time-sharing price strategy and the incentive policy (Chen et al. 2013). Then, with the development of extensive data and prediction technologies in the field of electric power research, power and load coordinated scheduling technology has been explored. Strategies for controlling smart building load mainly include time-sharing price strategy and incentive-type policy (Li et al. 2018). However, price based on demand response strategy is not applicable to emergency situations.

---

*Corresponding author: lishuaihu2010@126.com

The controllable load has the limited adjustment range which cannot meet the demand for balancing the fluctuation of the micro-grid with the large-scale renewable power generations (Tan et al. 2014). The incentive demand response strategy has significantly effects on the residents' electricity comfort. Moreover, these two methods are susceptible to subjective consciousness, which is not completely excavated the scheduling potential of building loads.

To address the problem, energy storage was applied in the smart building micro-grid (Ai et al. 2014). The application of the household distributed energy storage can effectively promote the implementation of incentive policies and improve the coordination of power source and load. The building centralized energy storage system has a flexible power configuration function, which is integrated with multiple battery modules. Building energy storage dispatches control system will control the internal intelligent switch action to configure the optimal capacity of the electric energy for the dispatch. On the basis of that, this paper proposes a building energy storage system suitable for flexible building load scheduling. The energy storage system of the smart building is constructed in two parts to make the system have flexible scheduling capability, including the small battery of home power supply system and a centralized energy storage system. And then, based on the reliable prediction of the renewable power generation and building electricity load in a time interval, the particle swarm optimization algorithm is applied to determine the scheduler margin of building controllable load. Finally, the proposed control strategy is adopted to coordinate control the building controllable load and energy storage system to effectively improve grid power quality and the stability of the smart building micro-grid.

The rest of this paper is organized as follows. Section II proposes an internal structural model of a smart building micro-grid energy storage system with flexible scheduling capability. The maximum schedulable indicator of building controllable loads in the forecasting interval is established in Section III. The active response control strategy to control building controllable loads and building energy storage systems in response to micro-grid scheduling is proposed in Section IV. This proposed method is tested on the MATPOWER 14-node test system in Section V.

2 ENERGY STORAGE SYSTEM MODEL OF THE SMART BUILDING

As above mentioned, the energy storage system is constructed by two parts, including the small battery of home power supply system and a centralized energy storage system in the smart build. Their models are as follows.

2.1 *Battery model of home distributed power supply system*

In general, the family distributed power supply system of the smart building consists of a battery, a bidirectional AC/DC inverter, and a signal switching circuit. When the power supply is insufficiently, the smart meter controls will send a discharge signal to the battery of the home distributed power supply system. When the power generation of the renewable generation is too high, the battery will change to the charging mode. In order to improve the battery life and limit the battery charge and discharge depth, the specific constraints are as follows (Tan et al. 2014):

$$
\begin{cases}
P_{Ai}^{char}(\min) \leq P_{Ai}^{char}(t) \leq P_{Ai}^{char}(\max) \\
P_{Ai}^{dischar}(\min) \leq P_{Ai}^{dischar}(t) \leq P_{Ai}^{dischar}(\max) \\
k_1 \leq Ai_{stop}^{char} \leq 1; \quad 0 \leq Ai_{stop}^{dichar} \leq k_2 \\
\varphi = \lambda \cdot g(t) \\
g(t) = \begin{cases} 1 & (SOC_{Ai}(t) \leq Ai_{stop}^{char}) \\ 0 & (SOC_{Ai}(t) \geq Ai_{stop}^{dischar}) \end{cases} \\
SOC_{Ai}(t) = SOC_{Ai}(t_0) + \dfrac{\varphi \eta_c \int_{t_0}^{t} P_{Ai}^{char}(t)dt}{E_{ess}(Ai)} + \dfrac{(1-\varphi)\int_{t_0}^{t} P_{Ai}^{dischar}(t)dt}{\eta_d E_{ess}(Ai)}
\end{cases} \quad (1)
$$

Where $SOC_{Ai}(t)$ is the state of charge of the battery at the ith node at time t, Ai_{stop}^{char} is the battery stop charging condition, and $Ai_{stop}^{dischar}$ is the battery termination discharge condition. $P_{Ai}^{char}(t)$ is the charging power of the battery i at the time t, and $P_{Ai}^{dischar}(t)$ is the discharging power of the battery i at the time t. η_c and η_d are the charging and discharging efficiency of the battery respectively. When $\lambda = 1$, the battery i participates in scheduling, and when $\lambda = 0$, it does not participate in scheduling. When $\varphi = 1$, the battery is charged, and when $\varphi = 0$, the battery is discharged;

2.2 Model of the centralized energy storage system

The battery compartments are connected by intelligent controllable switches. The specific internal structure is as Figure 1 shown.

The centralized energy storage system contains B model batteries and C model batteries. B and C are the same type of battery, but the capacity of C is several times that of B. Two different capacity levels are set to enable the battery to be deeply charged and discharged while participating in the micro-grid scheduling, improve the flexible scheduling performance of the system, and improve the service life of the battery. Through the building dispatch terminal signal, the on/off state of the intelligent switch connected to the building energy storage device is controlled, thereby, controlling different types of power supply combination access. In the Table 1, 1 indicates that the switch is closed, and 0 indicates that the switch is open.

It is assumed that the new energy generation and smart building electricity reliable prediction confidence interval T can be accurately predicted. According to the actual situation, the intelligent building dispatching center will comprehensively evaluate the dispatching capacity within the controllable load interval of the building, and formulate the electric energy that needs to be added to the centralized energy storage system in the interval. The intelligent

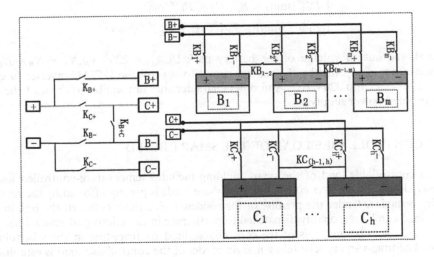

Figure 1.   Internal switch structure between battery packs.

Table 1.   Power type under different switch states.

Switch status			
KB+B-	KC+C-	KB+C	Power type
1	0	0	B
0	1	0	C
1	1	1	B+C

information platform can effectively control the intelligent energy in the centralized energy storage system. Control the action of the switch.

In the confidence interval T, it is predicted that the new energy generation fluctuation will lead to the additional energy $W_{DS}(t, t + T)$ in this interval. At the same time, the controllable load of the building and the energy $W_{AC}(t, t + T)$ that can be dispatched by the family distributed generation in this interval, the centralized building energy storage system needs to participate in the dispatching of the energy $W_D(t, t + T)$(the difference between the two mentioned above).

$$
\begin{cases}
W_D(t, T + t) = \phi(N_B \cdot S_A + N_C \cdot S_C) - S_0 \\
N_C = \dfrac{W_D(t, T + t)}{\phi \cdot S_C} + S_1 \\
N_B = \dfrac{S_1 + S_0}{\phi \cdot S_C} \\
N_N = N_B + N_C
\end{cases}
\tag{2}
$$

In the formula (2), N_B and N_C are the number of batteries of type B and C in the energy storage system, respectively; S_A and S_B are the maximum energy of storage of batteries A and B, respectively; S_0 is the electric energy that is added to make the B type battery participate in the response in an integer number of times; ϕ is the battery discharge depth (the general storage discharge depth is 80% (Ai et al. 2014)).

The charge and discharge power constraints of the centralized energy storage system are as follows:

$$
\begin{cases}
P_{LY}^{char}(\min) \leq P_{LY}^{char}(t) \leq P_{LY}^{char}(\max) \\
P_{LY}^{dischar}(\min) \leq P_{LY}^{dischar}(t) \leq P_{LY}^{dischar}(\max)
\end{cases}
\tag{3}
$$

In general, the output voltage of the battery pack:$U_{battery} = 220 \cdot N_N, N_N = N_B + N_C$; The PWM waveform is adjusted by the control circuit to control the DC/AC inverter to convert N_N time 220V DC into 220V AC. And other characteristics are similar to those of the household distributed power supply.

3 THE CONTROLLABLE LOAD OF THE SMART BUILD

The main controllable load of the smart building includes temperature-controlled load and time-delayed load. The power consumption of these loads is greatly affected by the residential electricity behavior. Under the premise of the residents' electricity comfort, the building controllable load can be reasonably dispatched to participate in the micro-grid scheduling. In different periods, feasible margins of the controllable load participating in the scheduling are different. The time-varying scheduling margin model of the controllable load is established as follows:

$$
\begin{cases}
H_{TCL}^p(t) = \left[h_1^p(t), h_2^p(t), \cdots, h_y^p(t) \right] \\
K_{DL}^p(t) = \left[k_1^p(t), k_2^p(t), \cdots, k_x^p(t) \right] \\
H_{TCL}^r(t) = \left[h_1^r(t), h_2^r(t), \cdots, h_y^r(t) \right] \\
K_{DL}^r(t) = \left[k_1^r(t), k_2^r(t), \cdots, k_x^r(t) \right]
\end{cases}
\tag{4}
$$

K_{DL}^p is the building delay load operation state matrix, which consists of y elements, and $h_i^p(t)$ is the predicted operating state of the ith delay load in the building.H_{TCL}^p is the building temperature control load operating state matrix, which consists of x elements, and $h_j^p(t)$ is the predicted operating state of the jth temperature-controlled load in the building. H_{TCL}^r and H_{DL}^r are the state matrices of the temperature-control and the time-delayed load after adopting the dispatching strategy.

$$W_{TCL}(t, t+T) =_t^{t+T} \sum_{j=1}^{h_y} (h_j^p(t) - h_j^r(t)).P_{TCL}^j(t)dt$$

$$W_{DL}(t, t+T) =_t^{t+T} \sum_{i=1}^{k_x} (k_i^p(t) - k_i^r(t)).P_{DL}^i(t)dt \tag{5}$$

$$W_{ACL}(t, t+T) = W_{TCL}(t, t+t) + W_{DL}(t, t+T)$$

The temperature-controlled load in the building is mainly the air conditioning load and the refrigerator load. This type of load participation scheduling is mainly affected by the external temperature, in order to prevent the load rebound phenomenon. Based on the thermal principle in the literature (Wang et al. 2012), the optimal turn on and turn off time is established. Time-delayed load participation scheduling is mainly restricted by the characteristics of residential electricity use. The potential for participation in scheduling is different at the different periods, and the response time is postponed or advanced without affecting the residents' electrical comfort. Considering the economic feasibility index, the time delay load cannot be interrupted during the completion of one work cycle.

Based on the thermodynamic equation (Lu et al. 2012), the optimal on-off time of building temperature-controlled load participating in Smart building micro-grid dispatching is established: T_{on}^i, T_{off}^i. Based on the electrical characteristics of buildings, the optimal delay time and early response time for building delay load are: $T_{s\_shif}^j, T_{t\_shif}^j$.

$$\max \alpha_t^T = (W_{ACL}(t, t+T))$$

$$s.t \ T_{TCL}^i(+) \leq T_{off}^i$$

$$T_{TCL}^i(-) \leq T_{on}^i \tag{6}$$

$$0 \leq T_{DL}^j(+) \leq T_{s\_shif}^j$$

$$0 \leq T_{DL}^j(-) \leq T_{t\_shif}^j$$

In the formula (6), α_t^T is the maximum dispatchable margin of controllable loads for buildings in T interval, $T_{TCL}^i(+)$ ($T_{TCL}^i(-)$) is the turn off (turn on) time when the ith temperature control load participates in the response, and $T_{DL}^j(+)$ ($T_{DL}^j(-)$) is the time when the jth delay load participates in the response delay response (early response).

4 ACTIVE RESPONSE CONTROL STRATEGY

Smart buildings adopt the hierarchical control strategy to actively respond to the optimal dispatching of the micro-grid (Ai et al. 2014). Considering the cost of smart buildings participating in micro-grid dispatching, the strategy of smart building response scheduling is divided into three levels. Priority is given to the controllable load in the building to participate in the response, and the household distributed power supply participates in the response when the controllable load response margin of the building is insufficient. When the above schemes are insufficient to meet the scheduling requirements, the centralized energy storage system of the building is controlled to participate in the micro-grid scheduling.

The battery of the family distributed power supply system is the second layer in the scheduling structure of the building dispatching system. Its service life is closely related to its depth of discharge and the number of charge and discharge cycles. In order to maximize the benefits of home distributed power usage. Constraining it and scheduling time must complete a complete

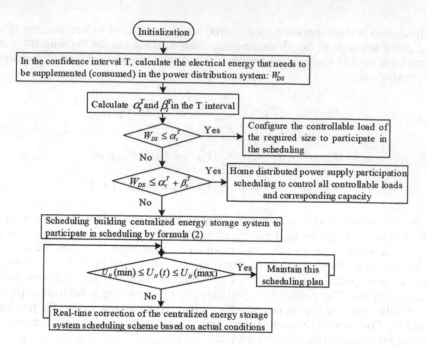

Figure 2. The flowchart of the proposed control strategy.

charge, discharge or electrical cycle. That is, the household distributed power supply participates in the scheduling in the confidence interval of $[t\text{-}T,\ t]$. If a complete charge and discharge is not completed, the state continues to the $[t,\ t+T]$ confidence interval until completion.

From the formula (1), the household distributed power to dispatch margin can be derived as:

$$W_{LS}(t,t+T) = \sum_{i=1}^{M}(SOC_{Ai}(t) - SOC_{Ai}(t+T)) \cdot E_{ess}(Ai) \qquad (7)$$

Let β_t^T be the maximum controllable margin of the household distributed power supply within the interval. Figure 2 shows the flowchart of the proposed method.

Step1: Based on the reliable prediction of new energy generation and building electricity load data in the T interval, the energy that needs to be supplemented (consumed) in this interval due to fluctuations in new energy generation is: $W_{DS}\ (t,t+T)$.

Step2: The maximum dispatching margin α_t^T of building controllable load in confidence interval T is solved by using particle swarm optimization (PSO) through formula 6. Based on the actual situation, the state matrix information of controllable load is updated after response to dispatch strategy.

Step3: According to home distributed power supply system constraints, the maximum adjustable margin β_t^T in the interval is solved.

Step4: Adopt layered control strategy to control the controllable load, family distributed power supply and building centralized energy storage system to participate in dispatching.

Step5: The node voltage of the distribution network is detected in real time. If it deviates from the normal safety setting value, the scheduling scheme of the centralized energy storage system is re-allocated according to the actual situation.

5 CASE STUDY

This paper uses the actual household wind power data of 100 residential buildings in a certain area of Guangdong Province and the actual wind speed data of each wind

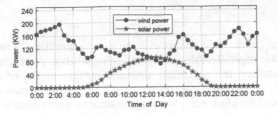

Figure 3.　Renewable energy daily power generation curve.

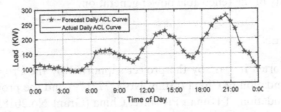

Figure 4.　The actual and predicted capacity of ACL.

farm every 15 minutes. The MATPOWER 14-node test system of MATLAB2017b simulation platform is used to test the feasibility of the proposed scheme.

The segmentation function in (Wu et al. 2013) is used to establish the output characteristics of the wind turbine. The cut-in wind speed is $v_0 = 3$ m/s, the rated wind speed is $v_1 = 8$ m/s, the cut-out wind speed is $v_2 = 14$ m/s, and the wind power generation system capacity is 200 KW. Applying the measured light data in this area, the photoelectric conversion efficiency is 14%, the battery area is 2 m^2, and the photovoltaic array capacity is 100 kW. The test system photovoltaic power station model is constructed.

Wind power generation in this area has obvious anti-peak characteristics. Wind power generation is sufficient during the period from 0:00 to 5:00. Based on reliable predictions in the area, the centralized energy storage system of the building is charged to store this part of the energy. The peak period of power consumption will occur during the three time periods [7:00, 8:30], [12:30, 2:30], [20:00, 22:00] based on the building load power characteristics. Taking the confidence interval T=30 min, 50 households in the 100 building systems are equipped with household distributed power, and the capacity of the distributed power supply is 3 kw.h;

In [18:00, 22:00], the photovoltaic system has stopped generating electricity, the wind power output is insufficient, and it is at the peak of the daily peak load of the building load. The particle swarm optimization (PSO) algorithm is used to solve the optimal schedulable margin of the controllable load and the distributed power supply of the home. Based on the hierarchical control strategy, the building controllable load and the building energy storage system are allocated to participate in the scheduling. The specific scheduling data are shown in Table 2. The node voltage in the test system connected to the building load is always maintained within the safe value, which verifies the reliability of the scheduling scheme.

Table 2.　Peak electricity dispatch schedule at night.

Time	W_{ACL}(kw.h)	W_{LS}(kw.h)	W_{DS}(kw.h)	$U_N(min)$	$U_N(max)$
18:00-18:30	38.7	31.5	82.5	0.952	1.032
18:30-19:00	39.2	27.6	90.5	0.956	1.041
19:00-19:30	42.7	32.6	107	0.961	1.053
19:30-20:00	51.3	21.3	134	0.943	1.028
20:00-20:30	54.6	0	145	0.957	1.021
20:30-21:00	46.2	0	127.5	0.943	1.017
21:00-21:30	37.5	0	88	0.957	1.046
21:30-22:00	30.6	0	30.5	0.975	1.053

6 CONCLUSION

The new smart building micro-grid energy storage system structure is proposed in this paper to make the system have more flexible configuration capability. According to the characteristics of residential electricity, the maximum dispatch margin model of building controllable load is established. The hierarchical control strategy is used to reasonably configure the building controllable load and building energy storage system to participate in the dispatch. Within the reliable prediction range of new energy power generation, it is possible to flexibly configure the building energy storage system and the controllable load to participate in the response to absorb the volatility of the renewable power generation.

ACKNOWLEDGMENT

This work was supported in part by the project supported by the Program of the National Natural Science Foundation of China (Grant No.51777179) and the project supported by the Natural Science Foundation of Hunan Province, China (Grant No.2018JJ2402), and the project supported by The China Scholarship Council (CSC) State Scholarship Fund International Clean Energy Talent Project (Grant No. [2018]5046).

REFERENCES

Jia, Y., Yi, L., Li, S., 2018. Optimization of building load power dispatching based on multi-objective molecular dynamics theory. *Power Grid Technology* 42(05): 1549–1555.

Zhai, S., Wu, Y., Cui, Y., Yi, L., Wang, J., 2014. Battery energy storage stabilizes short-term wind power fluctuations. *Automation of Electric Power Systems* 38(02):17–22+58.

Chen, W., Ai, X., Wu, T., Liu, H., 2013. Summary of the influence of photovoltaic grid-connected power generation system on power grid. *Electric Power Automation Equipment* 33(02):26–32+39.

Li, S., He, Y., Li, H., Peng, H., 2018. Preventive control considering aggregated controllable loads for voltage stability in electrical power systems. *U.P.B. Sci. Bull., Series C. Vol.* 80, ISSN 2286-3540.

Tan, X., Wang, H., Zhang, L., Zou, L., 2014. Multi-objective optimization configuration method and evaluation index of micro-grid composite energy storage. *Automation of Electric Power Systems* 38(08):7–14.

Ai, X., Zhao, Y., Zhou, S., 2014. Model and simulation of direct load control in distribution network adapted to clean energy consumption. *Proceedings of the CSEE* 34(25):4234–4243.

Wang, C., Liu, M., Lu, N., 2012. A power fluctuation smoothing method for micro-grid tie line using residential temperature control load control. *Proceedings of the CSEE* 32(25):36–43+8.

Lu, N., 2012. An evaluation of the HVAC load potential for providing load balancing service. *IEEE Trans on Smart Grid* DOI:10.1109/TSG.2012.2183649.

Wu, X., Wang, X., 2013. Confidence interval estimation of new energy portfolio output. *Automation of Electric Power Systems* 37(16):7–12.

Emerging Developments in the Power and Energy Industry – Dufo-López, Krzywanski & Singh (eds)
© 2020 Taylor & Francis Group, London, ISBN 978-0-367-27169-5

Innovative floating support technologies for submarine cable landing

Zhigang Zhang, Yijun Dong & Zhoulong Yuan
Zhoushan Power Supply Company of State Grid Zhejiang Electric Power Company, Zhoushan Zhejiang, China

Zhen Liu*
Shandong Provincial Key Laboratory of Ocean Engineering, Ocean University of China, Qingdao, China
Department of Ocean Engineering, College of Engineering, Ocean University of China, Qingdao, China
Qingdao National Engineering Laboratory for Subsea Equipment Test and Detection Technology, Qingdao, China

ABSTRACT: Traditional floating support technology based on the pneumatic tire for submarine cable landing is outdated compared to the rapid development of submarine cables with higher voltage levels. It is necessary to update the floating support technology for the submarine cable landing to ensure construction efficiency and quality. An innovative technology using the floating pillow for the support of the cable during its landing deployment is proposed in this article. A matched rapid gas inflating and deflating technology is also presented. Applications of these technologies are expected to guarantee the reliability and economy of the submarine cable landing constructions.

Keywords: cable landing, floating pillow, floating support technology, rapid inflating and deflating

1 INTRODUCTION

The rapid development of offshore islands in China (Shen 2016) had led to an increasing demand for higher reliability and efficiency of the electricity supply to the island micro grid and its connection to the main grid (Wu et al. 2015). Compared to the former low-voltage level submarine cables in traditional applications, the present cables have a much higher voltage level and cross-sectional area (Qiu et al. 2018), which also achieve a higher level of reliable electrical properties and construction technologies.

Currently, the cable laying technologies based on the advanced professional vessels are relatively mature. On the other hand, the cable landing technologies are still immature, especially in the use of tires for cable floating support (Zhang et al. 2014; Zhao & Chen 2011). This technology needs to tie the tires to the cable for connection and support (Chen 2009), which can readily cause cable twisting. It is also difficult to remove the tires from the cable. This kind of cable landing technology cannot satisfy the present demand for continuous and abundant laying engineering quantities of the submarine cables with larger cross-sectional areas (Jiang 2008), which could seriously affect the general construction.

---

*Corresponding author: liuzhen@ouc.edu.cn

The tire floating technology has already been eliminated in foreign countries. A more professional floating support facility that could be inflated and deflated rapidly has been employed for most cable landing constructions. In some cases, the GPS modules are also installed in some devices for precise location. China is the largest electrical cable country in terms of production and usage. It is also necessary to develop a professional floating cable supporter during its landing construction with corresponding engineering technologies to improve the engineering efficiency and quality of the cable landing construction (Chen & Zhu 2005).

It can be found that the most popular floating supporters use gasbags with various shapes and materials. According to its shape, we can define its type as the floating pillow, the floating ball, and the gas-inflating tube, respectively. All these floating supporters could be inflated or deflated rapidly to realize the functions of cable floating assistance and fast separation from the cable. It also has been proved that the gasbag has the following advantages: low cost, high efficiency, short production period (Xin et al. 2014), safety, flexibility, and reusability. The most important key factor is that it is more suitable for cables with larger cross-sectional area.

2 CONSTRUCTION TECHNOLOGIES FOR CABLE LANDING

2.1 Landing technologies for different terminals

The cable landing includes mainly two parts: initial and end terminal landing. For the initial terminal, the vessel is anchored at the nearest location on the cable routing to the landing spot to reduce the cable landing distance by using the GPS locating device. The anchors should avoid touching other original submarine cables nearby. The cable terminal is then pulled out from the back-twist frame and dives into the seawater from the vessel bow, as shown in Figure 1. The floating supporter supports the cables on the sea surface with additional buoyancy. Obviously, a shorter landing length is better. As the cable is pulled longer from the back-twist frame, more supporters will be installed on the cable. A winch is installed on the beach in advance to drag the cable to the landing spot. As the cable terminal arrives, the supporters will be removed from the cable and make the cable sink at the predefined routing locations.

For the ending terminal landing, two-head landing methods are always employed. After the vessel is anchored, the cable is similarly pulled out to enter the sea water, which will generate an "Ω" shaped curve. An adjustable head is installed on the ending terminal, which is also connected to the dragging wire from the winch. After the winch begins to operate, as shown in Figure 2, the vessel will also move to get out of the routing. With the assistance of the location device, the divers should make sure that the routing channel is cleaned up and remove the supporters from the cable to submerge it. It can be seen that the floating supporters aim to provide the additional buoyancy to make the cable float on the sea surface instead of laying it on the seabed to reduce friction and resistance significantly (Tolan 1977; Lontz 1946).

Figure 1.   Cables and tied gasbag supporter.

Figure 2.   Cables landing with floating supporters.

2.2 *Challenges for cable landing*

During the cable landing, the floating cable is under the joint action of the floating force, gravity, buoyancy, wind and current forces, and wave forces, which could lead the cable away from the designed routing. Therefore, analysis of the floating state is necessary and important for submarine cable laying construction. In order to increase the construction efficiency and reduce the costs of the cable landing, application of the floating supporters would be a feasible solution for rapid and convenient cable landing. On the other hand, employment of professional supporters is usually restricted by the actual sea conditions. Optimizing the vessel type and equipping assorted facilities such as gas inflating and deflating devices can improve the supporter performance and increase the adaptation level of the sea conditions (Vattenfall 2016).

In the preliminary study, the engineers could employ the GIS software to calculate and fix the landing point for the evaluation of the expected length and the cable's bending radius (Li Zhou et al. 2009). It is difficult to analyze theoretically the further bending forces and torques of the floating cable (Zajac 1957). Physical modeling experimental tests in the wave flume or tank could provide more reliable and valuable information to designers and engineers. In addition, with the rapid development of computers, the computational fluid dynamics software could be utilized for further complicated hydrodynamics analysis of the cable landing. The numerical simulations can help to not only optimize the shape floating supporters but also predict the supporter number and their reasonable distances.

3   CURRENT STATUS OF CABLE FLOATING TECHNOLOGIES

In most domestic cable landing construction, as presented in Figure 3, the inflated tires are applied as the temporary floating supporter to provide buoyancy during cable landing. In the construction, the cables will be shipped to the deploying point and tied with the tires while being pulled out. After the cable is dragged to the beach by the winch, the tires will be separated from the cable manually by engineers on the transportation boats. Meanwhile, the cable laying vessel can move slowly and adjust the cable submerging speed and direction to the desired routing.

In current oversea cable laying constructions, during the initial and ending landing of the terminals landing, there are many methodologies to assist the cable to float on the sea surface (Worzyk 2009). The cable will be submerged to the seabed as the terminal has been successfully connected to the onshore point and the cable curve fits the preset routing well. As mentioned in the previous section, there are many types of floating gasbag supporters that could be utilized in the engineering application, as illustrated in Figure 4.

Figure 3.   Domestic cable landing method.

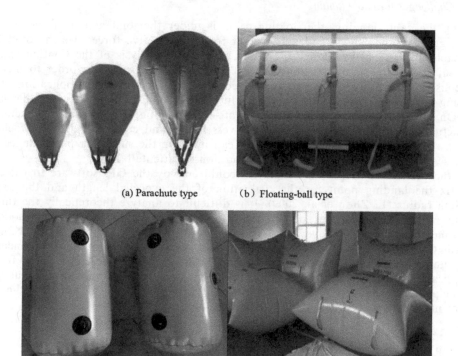

(a) Parachute type   (b) Floating-ball type

(c) Floating-pillow type   (d) Pillow type

(e) Floating-bridge type

Figure 4.   Oversea floating gasbag supporters (Doowin Rubber & Plastic 2015).

Although the inflated tires have been employed in the engineering application of cable landing for many years, its disadvantages are obvious compared to the foreign floating gasbag-type supporters manufactured using fabrics with a high-strength PVC coating, as summarized in Table 1. It can be seen that the tires have negative effects on both construction progress and quality. Therefore, it is essential to improve the domestic outdated floating supporter with the corresponding updated methodologies.

4 FLOATING-PILLOW SUPPORTER FOR CABLE LANDING

The floating-pillow supporter could provide consistent buoyancy to assist transport and installation of the cables or steel tubes in the shallow water. The floating pillow is the most widely used buoyancy assistance device in foreign cable landing applications. There are mainly three types for this kind of supporter, as shown in Figure 5: twin-chamber type, twin-boom type, and vertical type. Most of these supporters are inflatable bags made of fabrics with high-strength PVC coating, which have high resistance to friction and ultraviolet rays and have passed certification tests. The decompression valves and inflation/deflation valves are always installed on the air bags.

The twin-chamber type supporter has two individual chambers that are made of inflatable bags. The connecting piece between two chambers can be used to support the cable or other tubes so that two chambers can clamp these items and provide buoyancy. In most cases, ropes are tied on the top of the cables or tubes for further fixation.

For the twin-boom type supporter, two inflatable bags are connected directly by the ropes instead of the connecting pieces. This change makes the room between two bags more flexible. Compared to the original pieces, the lengths of the ropes can be easily adjusted to adapt the diameter variation of the cables or tubes. Similarly, the top ropes can fix the items after they are put between two bags.

The vertical type supporter has only one cylinder-shape inflatable bag. It is more like a type of enclosed buoyancy unit. The bag can stand vertically in the water because of its shape. It has only one lifting point on the bag to withstand vibrations with large angles, like the parachute type air-lift bags.

5 RAPID INFLATION/DEFLATION SYSTEM

5.1 *System structure principle*

The rapid inflation/deflation system is a new design of the research team, which owns all the copyright and intellectual property rights. The system, as illustrated in Figure 6, consists of the gas source, the control panel and valve, the blocking valve, and the gas input and recycling pipelines.

Table 1. Comparison of floating supporters.

| Floating Supporter | Material | | Construction | | Summary |
	Type	Performance	Routing	Inflation and Deflation	
Inflated tires	Rubber	Low resistance to corrosion, friction, puncture, and relatively heavy	Controlled and rechecked by professional divers	Difficult to operate	Low efficiency and high costs
Professional gasbag	Fabrics with high-strength PVC coating	High resistance to corrosion, friction, ultraviolet rays, and high-intensity light	Convenient location capability with built-in GPS device	Easy to operate	High efficiency and automation (Maximilian 1963)

(a) Twin-chamber type   (b) Twin-boom type   (c) Vertical type

Figure 5. Floating pillow supporters (Doowin Rubber & Plastic 2015).

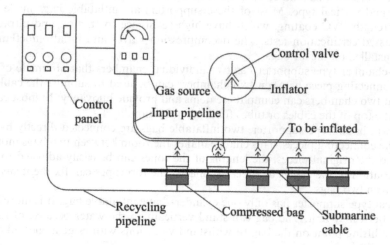

Figure 6. Schematic of rapid inflation/deflation system.

The function of the gas source is to provide dry and clean compressed air. The control valve can be opened or closed under an instruction of the control panel to begin or shut down the compressed air transportation automatically. The blocking valve is also an automatic switch to shut down the air input as the volume/pressure inside the bag reaches the preset values, which keeps the air from leaking under the desired pressure value. The control panel has two modes of operation: automatic and manual modes. When the electrical controlling devices experience failure, the manual mode can be switched on to keep the system working, which is expected to improve the system stability (Sun et al. 2013).

5.2 *Principle components and functions*

Rubber-bag with cord threads The air chambers of the floating supporter are the rubber-bag with cord threads, as shown in Figure 7. The bag material is constructed from high tensile strength Trevira® polyester 3/3 base cloth with heavy-duty UV stabilized PVC coating and natural rubber mixture, as illustrated in Figure 7a (Unique Seaflex 2018). The thickness of the material is approximately 4–5 mm, and the weight of an individual floating supporter is approximately 15 kg. The air bag is installed with a stainless-steel inflation valve. The proto-type of this floating supporter is presented in Figure 7b. It has the advantages of higher resistance to friction, puncture, and corrosion. The bag has a better stress and can withstand higher air pressure.

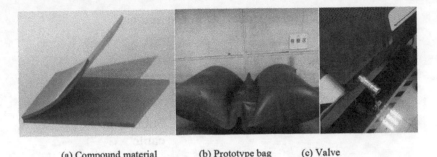

(a) Compound material   (b) Prototype bag   (c) Valve

Figure 7.　Innovative floating supporter (twin-chamber type).

Inflation/deflation valve The inflation/deflation valve is another key component in the floating supporter. The gas tightness and inflation/deflation speed are determined mainly by the structure and function of this valve. The specially designed valve with a stainless-steel quick-plug joint is applied as shown in Figure 8. The joint has excellent tightness and is easy to operate, which can increase the service life of the valve significantly. Furthermore, the valve can shut down the connection with the gas pipeline immediately after failure occurs to avoid the air leaking. The gas pipeline is a flexible tube which is generally inflationary under variation of air pressure. The pipeline has good capability of elongation and coiling, which can avoid air blocking during the tube winding and is good for recovery and storage.

5.3　*System operating principle*

As the system is out of operation, transportation of gas from the source to the pipeline is shut down. Only the floating supporters on the recovery side are inflated and well prepared. When the cable landing operation begins, the control panel could calculate the air volume for an individual air chamber according to the cable length in advance. After the gas source is switched on, the high-pressure air will be transported to the pipelines, which will be in the ready-for-inflation state.

During inflation, the cable terminal is pulled out from the cable tray through the back-twist frame and dives into the sea through the entry channel. As the cable continues to be pulled out, the engineers can connect the cable with the floating supporters by filling it into the connecting piece between two air chambers and tying them. Meanwhile, the engineers plug the

Figure 8.　Stainless-steel joint and flexible pipeline for the inflation/deflation valve.

84

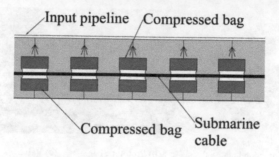

Figure 9.   Schematic of the inflation process.

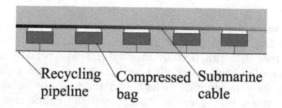

Figure 10.   Schematic of the deflation process.

gas transportation pipeline to the supporters to be inflated and switch on the control valve. The inflation process could be finished in several seconds for an individual bag. The valve will be switched off to shut down the gas input automatically, and the pipeline can be disconnected with the gasbag subsequently. All these processes could be finished as the cable is being pulled out, and the cable will keep floating on the surface as the length of the cable extends. A simplified schematic of the inflation process is shown in Figure 9.

During deflation, all the processes are carried out after the cable lands successfully and the cable curve has been adjusted following the designed routing. The gas source is powered off and the pipeline is in the low-pressure state. By remoting the valve on one air bag of the floating supporter to be switched on, air will be released and this bag will be compressed under environmental air pressure. Consequently, the cable can slide into the sea and fall to the seabed (Wilson 1971). After the cable lies in position, the pipelines and the floating supporters can be rolled back to the cable laying vessel and the left air-charged bag can be deflated on the vessel. A simplified schematic of the deflation process is shown in Figure 10.

6   CONCLUSIONS

Submarine cable laying creates serious demands on the facilities and methodologies of cable landing operations. Domestic traditional cable landing methods and devices are outdated and cannot satisfy the increased standard of higher-level cables with higher voltages.

The cable laying methodologies and facilities are generally reviewed and the key challenges have been described. It also has been pointed out that the premanufactured standard floating-pillow type cable supporters can meet the demand for rapid and safe cable landing operations with lower costs. Three types of the floating-pillow supporters are also reviewed.

An innovative airbag inflation/deflation system for the twin-chamber type floating-pillow supporters is proposed in this article. The basic structure principle, the key components, and the operation principle of this system are introduced.

It can be imagined that the standard production and operation of the airbag type floating supporters for cable landing could improve cable laying construction by increasing the efficiency and reliability and decreasing construction costs. The inflation/deflation system for the cable landing supporters also has a brilliant future in engineering applications.

ACKNOWLEDGEMENT

The research is financially supported by the Science and Technology Project of State Grid Zhejiang Electric Power Company:Research on the Key Technologies of the New Floating Devices for Submarine Cable Landing operation.

REFERENCES

AB. Vattenfall. Available at: https://corporate.vattenfall.se/om-oss/var-verksamhet/var-elproduktion/vindkraft/lillgrund-vindkraftpark/, June 9, 2016/October 23, 2018.

Chen, K. 2009. Laying and protection of submarine cables of Hainan Interconnection Project. *Southern Power System Technology* 5: 25–26.

Chen, L., & Zhu, X. 2005. 500 kV AC submarine cable link from mainland Canada to Vancouver Island. *International Electric Power for China* 1: 48–50.

Doowin Rubber & Plastic. Available at: www.doowin-dooflex.com, 2015/October 23, 2018.

Jiang, P. 2008. Research and application of extra long large cable mechanical laying technology. *Electric Power Construction* 5: 22–25.

Li, C., Zhou, X., & Yin, X. 2009. Application of GIS technology in the submarine cable route planning. *Ship Electronic Engineering*: 188–191, 202.

Lontz, D. M. Float U.S. Patent 2,395,892[P], March 5, 1946.

Maximilian, S. 1963. Cable float: U.S. Patent 3,074,083, January 22, 1963.

Qiu, W., Bao, J., Yu, L., & Han, G. 2018. Submarine cable and its technical difficulties. *Journal of Shenyang Institute of Engineering (Natural Science)* 1: 41–44.

Shen, G. 2016. Research on key equipment for submarine cable construction by burying-after-laying Process. *Building Construction* 38(7): 963–965.

Sun, X., Li, H., & Qi, Y. 2013. Improvement and application of central tire inflation system. *Automobile Technology* 2013: 28–30.

Tolan, P. 1977. Hawser float assembly. U.S. Patent 4,063,322, December 12, 1977.

Unique Seaflex. 2018. Available at: https://www.uniquegroup.com/, April 2018/October 23, 2018.

Wilson, W. 1971. Float device for pipelines. U.S. Patent 3,594,835, July 27, 1971.

Worzyk, T. 2009. *Submarine Power Cables: Design, Installation, Repair, Environmental Aspects*. Berlin: Springer Science+Business Media.

Wu, A., Yuan, Z., & Gong, Y. 2015. Overview of construction technology for deep-buried laying of submarine cables in China. *Zhejiang Electric Power* 3: 57–62.

Xin, X., Zhang, X., Liu, Y., Huang, H., & Jing, X. 2014. Feasibility study of large jacket launch aided with gasbags. *Petroleum Engineering Construction* 5: 26–30.

Zajac, E. 1957. Dynamics and kinematics of the laying and recovery of submarine cable. *Bell System Technical Journal* 36(5): 1129–1207.

Zhang, Z., He, X., & Ding, Z. 2014. The study of 110 kV submarine cable laying and protection in Shengsi Interconnection Project. *China Electric Power (Technology Edition)*, 5: 25–28.

Zhao, J., & Chen, Z. 2011. Research on submarine cable projects at home and abroad. *East China Electric Power* 9:1477–1481.

Emerging Developments in the Power and Energy Industry – Dufo-López, Krzywanski & Singh (eds)
© 2020 Taylor & Francis Group, London, ISBN 978-0-367-27169-5

Fault detection and diagnosis of analog circuits based on LOGFA algorithm optimized ELM

L.J. Wang & Y.M. Liang

College of Information Engineering, Xiangtan University, Xiangtan, Hunan, China
Key Laboratory of Intelligent Computing & Information Processing (Xiangtan University), Ministry of Education, Xiangtan, Hunan, China

ABSTRACT: The development of fault detection and diagnosis is of great significance to ensure the normal operation of the system and improve the reliability of the equipment. Extreme learning ma-chine (ELM) is widely used in the field of fault diagnosis because of its fast learning speed and high test accuracy, but its inherent randomness has a great influence on its generalization ability and diagnostic accuracy. To solve this problem, we propose a novel analog circuit fault diagnosis method based on the LOGFA algorithm and extreme learning machine in this paper. The internal parameters of the extreme learning machine are optimized by using the optimization capabilities of the LOGFA algorithm. In addition, the Sallen-Key low-pass filter is chosen as the test circuit. The simulation results show that the proposed method effectively improves the accuracy of analog circuit fault diagnosis.

1 INTRODUCTION

With the rapid development of modern electronic technology, the overall scale and structural functions of the circuit are gradually modularized and integrated. Due to the complex characteristics of the analog circuits, such as the continuity of the input and output signals, the finiteness of the testable node and the nonlinearity of the circuit, it is complicated to adopt mathematical methods for fault checking. However, fault diagnosis of the analog circuit is an indispensable part to ensure the normal operation of the system. Therefore, researchers and experts have devoted themselves to the study of the fault diagnosis of the analog circuit.

Recently, many various methods of fault diagnosis have been developed, such as wavelet transform (Aminian, F. & Aminian, M. 2001), fault dictionary method (Marin, C.V. et al. 2011), support vector machines (Cui, J. & Wang, Y. 2011, Liu, Y. et al. 2009), neural networks (Alessandra, F. et al.1999, Kavithamani, A. et al. 2013), and so on (El-Yazeed, M.F.A & Mohsen, A.A.K. 2003, Luo, H. et al. 2011, Song, L.et al. 2012). However, when the wavelet transform is used for fault extraction, since the optimal wavelet base cannot be accurately selected, it cannot guarantee that the optimal fault characteristics can be obtained after wavelet transform. The fault dictionary method requires a large amount of independent data to build a fault dictionary, and can only identify and diagnose faults that have been saved in the dictionary. The neural networks have strong self-learning ability and excellent classification ability, which is the focus of scholars. But building neural networks usually requires a lot of sample data. Similarly, other diagnostic methods also have different limitations.

Based on the LOGFA algorithm and extreme learning machine, a new analog circuit fault diagnosis method is developed in this paper. The experimental results of the low-pass filter circuit exhibit that the proposed algorithm has excellent fault diagnosis and classification capabilities.

This paper is organized as follows. In Section 2, the improved firefly algorithm (LOGFA algorithm) is briefly described. Section 3 gives the notion of extreme learning machine algorithm. In Section 4, an algorithm for the analog circuit fault diagnosis is analyzed in detail. The simulation results are given in Section 5. Finally, the conclusions are presented in Section 6.

2 LOGFA ALGORITHM

The firefly algorithm is a kind of stochastic optimization algorithm constructed by simulating the group behavior of fireflies, which is proposed by Cambridge University scholar Yang in 2009. It has good convergence speed and convergence precision and is easy to implement. However, due to the short development time of the algorithm, there are defects that the accuracy of the solution is too low and it is easy to fall into local convergence. In Ref (Yi, L.Z. et al. 2017), Yi proposes an improved firefly algorithm based on the law of universal gravitation, which improves the problem that it is easy to fall into the local optimal solution. Therefore, in this paper, we use the improved firefly algorithm as the LOGFA algorithm.

2.1 The evolutionary model of LOGFA

Definition 1: The gravity between firefly i and j can be calculated by

$$F_{ij} = G \frac{m_i(t)m_j(t)}{r_{ij}^2} \tag{1}$$

where $G = G_0 e^{-\gamma r_{ij}^2}$ is the gravitational constant of the time t, and G_0 is the original gravitational constant. γ is the light intensity absorption factor. r_{ij} represents the spatial distance between firefly i and j, and m_{ij} indicates the quality of the firefly i.
  Definition 2: The quality m_i is defined by

$$m_i = \frac{f(x_i) - f_{worst}(x_i)}{f_{best}(x_i) - f_{worst}(x_i)} \tag{2}$$

where $f(x_i)$ indicates the target fitness value of the firefly i at x_i. If it is to solve the minimum problem, where $f(x_i)$ represents the target fitness value corresponding to the location of the firefly i. If it is to solve the minimum problem, then $f_{best}(x_i) = \min_{i \in \{1,2,..,n\}} f(x_i)$, $f_{worst}(x_i) = \max_{i \in \{1,2,..,n\}} f(x_i)$. If it is to solve the maximum problem, then $f_{best}(x_i) = \max_{i \in \{1,2,..,n\}} f(x_i)$, $f_{worst}(x_i) = \min_{i \in \{1,2,..,n\}} f(x_i)$.

  Location updates can be expressed as

$$\begin{cases} a_i = \frac{F_{ij}(t)}{m_i(t)} \\ x_i(t+1) = x_i(t) + \frac{1}{2}a_i + \alpha(R - \frac{1}{2}) \\ \alpha = \alpha \cdot \Delta\alpha \end{cases} \tag{3}$$

where, R is a random factor and is evenly distributed over [0, 1]. The parameter $\alpha \in$ [0, 1] is the step factor, and $\Delta\alpha$ is the step attenuation coefficient, which is in the range of [0.95, 1].

2.2 The strategy of Gaussian variation

If the firefly population does not evolve over successive iterations, it can be judged that it has fallen into the local optimal region. Then, the Gaussian variation factor is used to disrupt the firefly population and restore its ability to evolve. The operation is as follows:

$$x_i = x_i + x_i \cdot N(0,1) \tag{4}$$

where, N is a random vector obeying a Gaussian distribution with the mathematical expectation of 0 and variance of 1.

3   EXTREME LEARNING MACHINE

Extreme learning machine (ELM) is a single hidden layer learning algorithm originating from neural networks. It is a new machine learning algorithm proposed by Professor Huang of Nanyang Technological University in 2004 (Huang, G.B. et al. 2004). For a single hidden layer neural network, assuming that $x_i = [x_{i1}, x_{i2}, \dots, x_{in}]^T \in R^n$ is the input data of the neural network. $t_i = [t_{i1}, t_{i2}, \dots, t_{im}]^T \in R^m$ is the output data of the neural network. The mathematical model of single hidden layer feedforward neural network (SLFN) is

$$\sum_{i=1}^{L} B_i g(a_i \cdot x_j + b_i) = t_j , \quad j = 1, 2, \dots N \tag{5}$$

where, $g(x)$ is the excitation function. L represents the number of hidden layer nodes. $a_i = [a_{i1}, a_{i2}, \dots, a_{in}]^{Tn}$ is the input weight factor connecting input neurons with ith hidden layer node. b_i is the threshold of the ith hidden layer node. B_i is the output weight factor connecting output neurons with ith hidden layer node.
The equation (5) can be simplified as follows

$$HB = T \tag{6}$$

where

$$H = \begin{bmatrix} g(a_1 \cdot x_1 + b_1) & g(a_2 \cdot x_1 + b_2) & \dots & g(a_L \cdot x_1 + b_L) \\ g(a_1 \cdot x_2 + b_1) & g(a_2 \cdot x_2 + b_2) & \dots & g(a_L \cdot x_2 + b_L) \\ & & \vdots & \\ g(a_1 \cdot x_N + b_1) & g(a_2 \cdot x_2 + b_2) & \dots & g(a_L \cdot x_N + b_L) \end{bmatrix}_{N \times L} \tag{7}$$

$$B = \begin{bmatrix} B_1^T \\ \vdots \\ B_L^T \end{bmatrix}_{L \times m} , \quad T = \begin{bmatrix} t_1^T \\ \vdots \\ t_N^T \end{bmatrix}_{N \times m} \tag{8}$$

When the excitation function $g(x)$ is infinitely differentiable in any interval, the output weight factor can be obtained according to equation (9). And the solution is shown in equation (10).

$$\min \|HB - T\| \tag{9}$$

$$\hat{B} = H^+ \cdot T \tag{10}$$

where, H^+ is the Moore-Penrose (MP)generalized inverse matrix of matrix H.
The steps of the ELM algorithm can be summarized as follows
Step1: Given a training set $N = \{(x_j, t_j)| x_j \in R^n, t_j \in R^m, j = 1, 2, \dots, N\}$, the excitation function $g(x)$ and the number of hidden layer nodes L. Then the input weight factor a_i and the threshold b_i are randomly generated. And the subscript $i = 1, 2, \dots, N$.
Step2: Calculating the hidden layer input matrix H.
Step3: Calculating the output weight factor $B = H^+ T$.

4 ELM ALGORITHM BASED ON LOGFA ALGORITHM OPTIMIZATION

The extreme learning machine has the advantages of higher training speed and fewer parameters to be set, but the input weight and the bias parameter of the hidden layer are generated by the random function, which reduces the data classification accuracy of the extreme learning machine to some extent. Therefore, the improved firefly algorithm is used to optimize the parameters (a_i, b_j and β) of ELM to obtain the optimal network model. The proposed diagnostic algorithm flow proposed mainly includes two parts, which are data acquisition processing part and the optimization training classification model part. First, the faults of the analog circuit are established and the data is sampled. Then, the obtained data is extracted by principal component analysis (PCA), which are the training data and the test data. And then using the training data to train the extreme learning machine, and finally the optimal fault classification model is obtained.

The detailed steps of the algorithm are as follows:

Step 1: Initialize the parameters including the number of fireflies n, the number of subpopulations m, the initial attractiveness β_0, the light absorption parameter γ, and the random factor a. The initial position of the firefly individual x_{ij} is randomly generated ($i = 1, 2, \ldots, n$; $j = 1, 2, \ldots, m$).

Step 2: Build the network architecture of extreme learning machine. Initialize the number of neurons in the hidden layer and set the search interval in the hidden layer neurons.

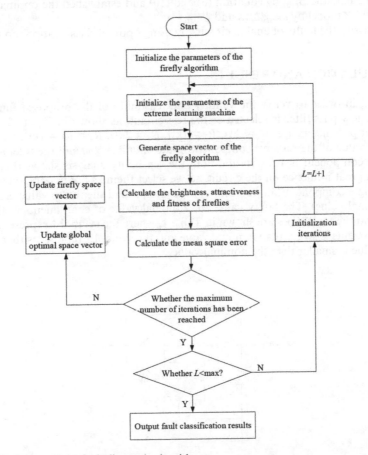

Figure 1.   Flow chart of the fault diagnosis algorithm.

Step 3: In order to obtain the optimal weight W between the input layer and the hidden layer, calculating the size of the firefly $SizePop$.

$$SizePop = L \times n_i + L \tag{11}$$

where n_i is the number of the input layer neurons, and L is the number of the hidden layer neurons;

Step 4: Update the spatial location of all firefly individuals according to the equation (3).

Step 5: Calculate the output weight matrix B according to equation (10).

Step 6: Calculate the output matrix $\hat{T}$ of the training data.

Step 7: Calculate the mean squared error between the matrixes $\hat{T}$ and T.

$$RESM = \sqrt{\frac{1}{n} \sum_{i=1}^{n} (\hat{T} - T)^2} \ , \ i = (1, 2, ..., n) \tag{12}$$

Step 8: Update the global optimal space vector. Update the number of iterations and determine whether the conditions for jumping out of the iteration are met. If not, skip back to step 3 to continue the iteration loop.

Step 9: To determine if the value of L is equal to the value of maximum search range of the hidden layer. If not, initialize the number of iterations. Let $L = L + 1$ and jump back to step 3. In contrary, exit Firefly and output the a_{best} and b_{best}.

Step 10: Calculate the B_{best} by equation (6) and (10) and established the optimal ELM classification network model by a_{best}, b_{best} and B_{best}.

Step 11: Classify the faults of analog circuits by using optimal classification models.

5   THE SIMULATION AND RESULTS

In this section, in order to verify the validity and feasibility of the proposed fault diagnosis algorithm, the low-pass filter is selected as the test circuit, as shown in Figure 2. The corresponding circuit parameters are $R_1 = 1\text{k}\Omega$, $R_2 = 1\text{k}\Omega$, $R_3 = 2\text{k}\Omega$, $R_4 = R_5 = 1\text{k}\Omega$, $C_1 = C_2 = 5\text{nF}$. The tolerance value of the resistive component in the circuit is 5%, and the tolerance value of the capacitive component is 10%. In addition, the sensitivity analysis shows that R_2, R_3, C_1 and C_2 have a great influence on the circuit, so we select them as faulty components. Assume that only a single component failure occurs in the circuit, and the faulty value is $X \pm 50\% \ X$. (X is the nominal value). The faulty value and nominal value of components in the low-pass filter circuit and the fault types are shown in Table 1, where the symbol ↑ represents the faulty value of components is larger than nominal value and the symbol ↓ represents that the components' fault value is smaller than the nominal value.

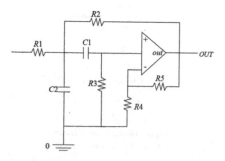

Figure 2.   The circuit of Sallen-Key low-pass filter.

91

Table 1. Faults type and its code.

Fault code	Fault class	Nominal value	Faulty value
1	$R_2\uparrow$	3kΩ	4.5kΩ
2	$R_2\downarrow$	3kΩ	1.5kΩ
3	$R_3\uparrow$	2kΩ	3kΩ
4	$R_3\downarrow$	2kΩ	1kΩ
5	$C_1\uparrow$	5nF	7.5nF
6	$C_1\downarrow$	5nF	2.5nF
7	$C_2\uparrow$	5nF	7.5nF
8	$C_2\downarrow$	5nF	2.5nF
9	NF		

The specific information of each parameter in the diagnostic classification algorithm is set as follows. The number of hidden layer nodes in the extreme learning machine is six, the number of firefly population is 100, the maximum number of iterations is 200, the step factor $\alpha = 0.5$, and the light intensity absorption factor $\gamma = 1$. And the excitation signal of the circuit is 2V sinusoidal AC voltage signal, and the operational amplifier adopts 3554AM operational amplifier chip. The sampling frequency is 20. Then, the output data is collected by using the Monte Carlo analysis and the collected data is processed using principal component analysis to obtain the data to be tested. Finally, the fault diagnosis results are shown in Figure 3 and Table 2. As shown in Table 2, each row and column represent the classification of the fault type. If the value on the diagonal of the table is 20, it means that the sampling data of the corresponding fault type is all accurately identified. Otherwise, it means that an error has occurred. For instance, the number in the first column of the first row is 20, which means that all 20 samples of fault type $R_2\uparrow$ are accurately classified. However, the number in the eighth row and the eighth column is 19, indicating that 19 test data are correctly diagnosed for 20 samples of $C_2\downarrow$, and another one test data is classified into $R_3\uparrow$ by the wrong identification. After calculation, the classification accuracy of the proposed algorithm is 98.33%.

In recent years, researchers have proposed a lot of different methods for fault diagnosis of analog circuits. In Ref. (Liu, H. et al. 2009), the authors proposed an analog circuit fault diagnosis algorithm of a radial basis function neural network composed of Bagging algorithm. And in Ref. (Song, G. et al. 2007), the authors combine genetic algorithm with the neural network, and use the genetic algorithm to optimize the parameters of the neural network in

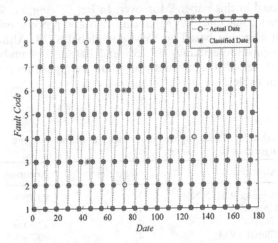

Figure 3. Caption of a typical figure.

92

Table 2. Faults type and its code.

Fault class	$R_2\uparrow$	$R_2\downarrow$	$R_3\uparrow$	$R_3\downarrow$	$C_1\uparrow$	$C_1\downarrow$	$C_2\uparrow$	$C_2\downarrow$	NF
$R_2\uparrow$	20								
$R_2\downarrow$		19							
$R_3\uparrow$			20						
$R_3\downarrow$				19					
$C_1\uparrow$					20				
$C_1\downarrow$						20			
$C_2\uparrow$							20		
$C_2\downarrow$								19	
NF									20

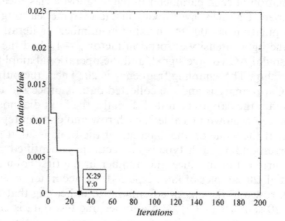

Figure 4. Curve of evolution.

training stage, and proposes an analog circuit fault diagnosis algorithm based on genetic algorithm and wavelet transform. Although the method proposed in the mentioned article does improve the accuracy of fault diagnosis in analog circuits, it is still lower than the fault diagnosis algorithm proposed in this paper. Moreover, In Ref (Zhang, A. & Chen, C. 2014), the authors use support vector machines (SVM) as classifiers and propose two kinds of algorithms of analog circuit fault diagnosis based on SVM and global SVM. Although these two algorithms have good classification accuracy (97.85% and 100%), the number of fault types that can be identified and classified is significantly less than the algorithm proposed in this paper. All results are shown in Table 3.

Table 3. The comparison of other algorithms with the algorithm of this paper.

Algorithm	categories	Accuracy
Bagging RBF NN	9	93.75%
Genetic-WN	9	95%
SVM	4	97.85%
Global SVM	4	100%
LOGFA-ELM	9	98.33%

6 CONCLUSION

In this paper, an analog circuit fault diagnosis algorithm based on improved firefly algorithm and extreme learning machine is proposed. The optimization characteristics of the improved firefly algorithm are used to optimize the internal parameters of the EML, and further improve its capabilities of fault diagnosis and classification. Furthermore, the experimental results of low-pass filter circuit indicate that the proposed algorithm effectively improves the accuracy of analog circuits fault diagnosis, which verifies the effectiveness and practicability of the proposed algorithm.

ACKNOWLEGEMENTS

This work was supported by the National Natural Science Foundation of China under Grant nos. 61402391.

REFERENCES

Aminian, F. & Aminian, M. 2001. Fault Diagnosis of Analog Circuits Using Bayesian Neural Networks with Wavelet Transform as Preprocessor. *Journal of Electronic Testing* 17(1): 29–36.

Marin, C.V. et al. 2011. A dictionary approach to fault diagnosis of analog circuits. *In: Proc. AFRICON, Livingstone*: 1–5.

Cui, J. & Wang, Y. 2011. A novel approach of analog circuit fault diagnosis using support vector machines classifier. *Measurement* 44(1): 281–289.

Liu, Y. et al. 2009. Fault diagnosis of analog circuit based on support vector machines. *IEEE International Conference on Communications Technology and Applications*: 40-43.

Alessandra, F. et al.1999. A Neural Network Diagnosis Approach for Analog Circuits. *Applied Intelligence* 11(2): 169–186.

Kavithamani, A. et al. 2013. Soft Fault Classification of Analog Circuits Using Network Parameters and Neural Networks. *Journal of Electronic Testing* 29(2): 237–240.

El-Yazeed, M.F.A & Mohsen, A.A.K. 2003. A Preprocessor for Analog Circuit Fault Diagnosis Based on Prony's Method. *AEUE - International Journal of Electronics and Communications* 57(1): 16–22.

Luo, H. et al. 2011. A SVDD approach of fuzzy classification for analog circuit fault diagnosis with FWT as preprocessor. *Expert Systems with Applications* 38(8): 10554–10561.

Song, L. et al. 2012. Analog circuit diagnosis based on particle swarm optimization radial basis function network. *Application Research of Computers* 29(1): 72–71.

Yi, L.Z. et al. 2017. An improved firefly algorithm based on newton's law of universal gravitation. *Journal of Advances in Information Technology* 8(4): 219–224.

Huang, G.B. et al. 2004. Extreme learning machine: a new learning scheme of feedforward neural networks. *Proceedings of International Joint Conference on Neural Networks, Budapest, Hungary*: 985–990.

Liu, H. et al. 2009. Analog circuit fault diagnosis using bagging ensemble method with cross-validation. *International Conference on Mechatronics and Automation*: 4430–4434.

Song, G. et al. 2007. Fault Diagnosis Approach of Analog Circuits Based on Genetic Wavelet Neural Network. *International Conference on Electronic Measurement and Instruments* 3: 675–679.

Zhang, A. & Chen, C. 2014. Fault diagnosis based semi-supervised global LSSVM for analog circuit. *International Conference on Mechatronics and Control*: 744–748.

Emerging Developments in the Power and Energy Industry – Dufo-López, Krzywanski & Singh (eds)
© 2020 Taylor & Francis Group, London, ISBN 978-0-367-27169-5

Research on compensation strategy of UPQC based on super capacitor

Qian Ma & Wenli Wang
School of information engineering, Xiangtan University, Xiangtan, Hunan, China

ABSTRACT: This paper first analyzes the voltage compensation issue of super capacitor-based unified power quality controller (SC-UPQC). On one hand, taking rated voltage of load as the reference, the constraint condition of the compensation strategy of SC-UPQC is demonstrated. On the other hand, the concept of the *change rate of the amplitude of series voltage compensation* is introduced. Taking the rated voltage of load as the reference, the range of compensation voltage and current in the series and parallel units of UPQC is further analyzed. Then, as the SC-UPQC is able to solve the problem of insufficient energy storage of active power, a new compensation strategy based on a parallel unit with constant apparent power after fault occurrence is proposed. By choosing appropriate parallel compensation current, the proposed compensation strategy can achieve the reasonable distribution of active and reactive power between power grid and series-parallel unit. Therefore, the compensation capacity of the series-parallel unit is optimized, and can compensate the voltage in the faults like voltage swells and sags. A simulation model is established in Matlab/Simulink. The simulation results verify the effectiveness of the proposed compensation strategy.

Keywords: unified power quality controller, Super-capacitor, constant apparent power, compensation capacity

1 INTRODUCTION

The harmonic current generated by non-linear loads such as arc furnace and power electronics device greatly pollutes power grid and reduces power factor. Super capacitor-based unified power quality controller (SC-UPQC) consists of a series active filter, a parallel active filter and a super capacitor on DC side (Y. Shi, Z. Wu & X. Dou, et al, 2015; Z. Li, L. Xia & Z. Wu, et al, 2011; A. E. Leon, S. J. Amodeo & J. A. Solsona, et al, 2011).

Currently, little research explores the constraint conditions of compensation strategy, thereby leading to the ignorance of some compensation conditions in analyzing the compensation of SC-UPQC. For example, literature (V. Khadkikar, 2012; F. Wu, J. Zheng & J. Mei, 2012; H. Ma, G. Xu & S. Song, et al, 2014) only focused on compensating for voltage amplitude, while ignoring the voltage phase compensation. In this paper, taking the rated voltage of load taken as the reference, the constraints of the compensation strategy for SC-UPQC are analyzed, so that the physical meaning of each compensation quantity is clarified before and after the compensation.

The selection of compensation voltage and current is critical for a compensation strategy for SC-UPQC. The amplitude range of series-parallel compensation voltage and current was not clarified in literature (X. Li, M. Xia & Y. Mao, et al, 2013; X. Zhang & X. Tong, 2016; S. Devassy & B. Singh, 2015), leading to a large compensation capacity of UPQC. Thus, this paper takes the rated voltage of load as the reference, introduces the change rate of the amplitude of series voltage compensation, and analyzes the range of compensation voltage and current. The compensation strategy of SC-UPQC is investigated in that range.

In terms of the compensation strategy of SC-UPQC, literature (B. Ambati & V. Khadkikar, 2014; S. Devassy & B. Singh, 2017) still adopted the compensation strategy of traditional topology, which weakened the compensation performance of SC-UPQC. Specifically, the compensation strategy of UPQC-Q in literature (S. Devassy & B. Singh, 2017) eliminated the active power circulation and ensured no capacity overshoot of series-parallel unit in the process of power quality compensation. However, the series compensation voltage was always perpendicular to the power supply voltage under the premise that the load voltage amplitude remained constant before and after the compensation. Therefore, it is possible to compensate only voltage sag but the voltage swell. Therefore, it is urgent to design an appropriate compensation strategy for SC-UPQC.

A novel compensation strategy based on a parallel unit with constant apparent power after fault occurrence is proposed in this paper. The proposed strategy takes the rated voltage of load as the reference and incorporates the change rate of the amplitude of series compensation voltage. Then the constraint conditions of the proposed compensation strategy are clarified. The proposed strategy could achieve the reasonable distribution of active and reactive power between the power grid and the series-parallel unit, which not only optimizes the compensation capacity of SC-UPQC, but also could deal with various voltage faults.

2   CONSTRAINT ANALYSIS OF COMPENSATION STRATEGY OF SC-UPQC

The rated voltage of load is taken as the reference. The constraint conditions of the compensation strategy of SC-UPQC are analyzed in this paper, and the phasor diagram is shown in Figure 1.

As shown in Figure 1, the rated voltage of load U_L is taken as the reference; the grid voltage U_S and the current I_S are equal to the load voltage U_L and the current I_L, respectively; U'_S indicates the grid voltage after a fault. It can be seen from Figure 1 that the amplitude and phase of U'_S have jumped, which accords with the actual situation and has a definite physical meaning. As shown in Figure 1, the grid voltage after a fault is compensated by the series compensation unit, so that the load voltage can remain unchanged. Therefore, whether the load voltage after compensation is consistent with the load reference voltage can be clearly seen from Figure 1.

Figure 1 shows that the compensated U'_S and I'_S are in the same phase, i.e. the power factor is 1 and the power factor angle is 0° on the grid side. The compensated load voltage is consistent with the reference, that is, both the phase difference δ' and the amplitude difference ΔU_L of the load voltage before and after the compensation are both zero. Since the SC-UPQC can solve the problem of insufficient active energy storage, the active power on the grid side, the energy storage unit side and the load side needs to be kept constant after a voltage fault, The equation can be expressed as

$$P_{sc} = P'_S - P_L \qquad (1)$$

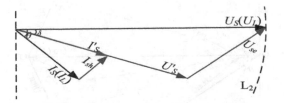

Figure 1.   Phasor diagram based on rated load voltage.

where P'_S is the active power on the grid side after a voltage fault, P_L is the active power on the load side, and P_{sc} is the active power of the energy storage unit. In summary, the constraint conditions of the compensation strategy of SC-UPQC can be expressed as

$$\theta' = 0; \ \delta' = 0; \ \Delta U_L = 0; \ P_{sc} = P'_S - P_L \tag{2}$$

3 NOVEL COMPENSATION STRATEGY OF SC-UPQC

3.1 *Phasor analysis of the proposed compensation strategy*

A novel compensation strategy based on a parallel unit with constant apparent power after fault occurrence is proposed. The phasor diagram is shown in Figure 2.

It can be known from the previous analysis that P_{sc} is not equal to 0 in the SC-UPQC topology. As a result, to make the apparent power of parallel unit stay constant during compensation, the selection of the value of the apparent power becomes a key to the optimization of UPQC capacity. As shown in Figure 2, the series compensation current I'_{sh} is made perpendicular to the grid voltage U'_S after a fault, so that the power factor on the grid side remains to be 1, and the value of the I'_{sh} is minimized under a voltage fault. Since the amplitude of the load voltage is constant before and after the compensation, the apparent power of the parallel unit is also minimized under the voltage fault. Since I'_{sh} must be perpendicular to U'_S, the maximum value of I'_{sh} ($I'_{sh\text{-}max}$) can be easily obtained within the determined range (as shown in Figure 2). $I'_{sh\text{-}max}$ represents the maximum of the compensation current of the parallel unit, and the resulting apparent power is the minimum capacity required to design the parallel unit. To ensure that the apparent power is constant when the parallel unit is compensated, a circle centered at point P and with a radius of $I'_{sh\text{-}max}$ is drawn, to have intersections with OA' and OB' at points A and point B, respectively.

On the arc AB, the grid current gradually increases from OB to OA. As shown in Figure 2, OA is the maximum of the grid current on the arc AB ($I''_{S\text{-}max}$), and at this time, the fault voltage U'_S coincides with OA', and the compensation voltage of the series unit is the maximum compensation voltage U_{max}, The resulting apparent power is the minimum capacity required to design the series unit.

As shown in Figure 1, according to the cosine theorem, the compensation voltage amplitude of the series unit is given by

$$U_{se} = \sqrt{U^2 s + U'_S{}^2 - 2U_S U'_S \cos \delta} = U\sqrt{1 + K^2 - 2K \cos \delta} \tag{3}$$

where $U_S = U_L = U$, $U'_S = KU$, and K denotes the change rate of the amplitude of grid voltage sag/swell ($K \geq 0$).

Next, x is defined as the change rate of the amplitude of series compensation voltage, and then the series compensation voltage can be expressed as

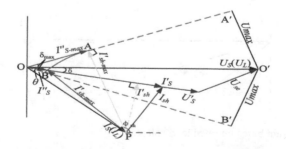

Figure 2. Phasor diagram of constant apparent power based on parallel unit after fault.

97

$$U_{se} = xU \Rightarrow \sqrt{1 + K^2 - 2K\cos\delta} = x \qquad (4)$$

where $0 \le x \le X_{max}$, and U_{se} takes the maximum value when $x = X_{max}$, that is $U_{max} = X_{max}U$. According to the geometric relations in Figure 2, we have

$$\sin|\delta| = \frac{xU}{U} = x \qquad (5)$$

According to the relationship between the parallel compensation current $I'_{sh\text{-}max}$, the load current I_L and the compensated grid current I''_S as shown in Figure 2, the current I''_S flowing through the series unit can be calculated by:

$$I''_S = I_L \cos(\theta - \delta) - I_L\sqrt{\sin^2(\theta + \delta_{max}) - \sin^2(\theta - \delta)} \qquad (6)$$

Using Eq. (3) and Eq. (6), the apparent power of series units can be given by:

$$S'_{se} = U_{se}I''_S = UI_L\sqrt{1 + K^2 - 2K\cos\delta}(\cos(\theta - \delta) - \sqrt{\sin^2(\theta + \delta_{max}) - \sin^2(\theta - \delta)}) \qquad (7)$$

By combining Eq. (4) with Eq. (7), we have

$$S'_{se} = UI_L x(\cos(\theta - \delta) - \sqrt{\sin^2(\theta + \delta_{max}) - \sin^2(\theta - \delta)}) \qquad (8)$$

Eq. (8) is an increasing function about x. In summary, the apparent power of UPQC is calculated by:

$$S'_{UPQC} = S'_{sh} + S'_{se} \qquad (9)$$

3.2 Comparative analysis of compensation capacity

For the SC-UPQC topology, the apparent power on the parallel unit of the SC-UPQC under a traditional compensation strategy can be expressed as:

$$S_{sh} = U_L I_{sh} = UI_L\sqrt{1 + \frac{\cos^2\theta}{K^2} - 2\frac{\cos\theta\cos(\theta - \delta)}{K}} \qquad (10)$$

The apparent power on the series unit of the SC-UPQC can be expressed as:

$$S_{se} = U_{se}I'_S = \frac{UI_L\cos\theta\sqrt{1 + K^2 - 2K\cos\delta}}{K} \qquad (11)$$

According to Eq. (4), K is given by

$$K = \cos\delta \pm \sqrt{x^2 - \sin^2\delta} \qquad (12)$$

By combining Eq. (10), Eq. (11) and Eq. (12), we have

$$S_{sh} = \begin{cases} UI_L\sqrt{1 + \dfrac{\cos^2\theta}{2\cos^2\delta + 2\cos\delta\sqrt{\cos^2\delta + x^2 - 1} + x^2 - 1} - 2\dfrac{\cos\theta\cos(\theta-\delta)}{\cos\delta + \sqrt{\cos^2\delta + x^2 - 1}}} & K = \cos\delta + \sqrt{x^2 - \sin^2\delta} \\[4mm] UI_L\sqrt{1 + \dfrac{\cos^2\theta}{2\cos^2\delta - 2\cos\delta\sqrt{\cos^2\delta + x^2 - 1} + x^2 - 1} - 2\dfrac{\cos\theta\cos(\theta-\delta)}{\cos\delta - \sqrt{\cos^2\delta + x^2 - 1}}} & K = \cos\delta - \sqrt{x^2 - \sin^2\delta} \end{cases}$$

$$\tag{13}$$

$$S_{se} = \begin{cases} \dfrac{UI_L x\cos\theta}{\cos\delta + \sqrt{\cos^2\delta + x^2 - 1}} & K = \cos\delta + \sqrt{x^2 - \sin^2\delta} \\[4mm] \dfrac{UI_L x\cos\theta}{\cos\delta - \sqrt{\cos^2\delta + x^2 - 1}} & K = \cos\delta - \sqrt{x^2 - \sin^2\delta} \end{cases}$$

$$\tag{14}$$

Therefore, the total apparent power of UPQC is obtained expressed as:

$$S_{UPQC} = S_{sh} + S_{se} \tag{15}$$

Assuming that $x = X_{max} = 0.5$, and $\theta = 45°$, we can obtain $\delta_{max} = 30°$ from Eq. (5). Therefore, S'_{sh} is a constant, and S'_{se}, S_{sh}, and S_{se} are the functions of one variable, i.e. angle δ, .

According to the proposed compensation strategy, it is required to find the minimum apparent power of series-parallel units under various voltage faults. So it is necessary to find the maximum apparent power on series-parallel units. Figure 3 illustrates the waveform of series units, parallel units and total apparent power with different compensation strategies.

In Figure 3a, the maximum apparent power of the parallel units obtained by traditional compensation strategy and the proposed strategy are 1.22 pu and 0.966 pu, respectively. In Figure 3b, the maximum apparent power of the series units obtained by traditional compensation strategy and the proposed strategy are 0.707 pu and 0.13 pu, respectively. In Figure 3c, the maximum total apparent power obtained by traditional strategy and the proposed strategy are 1.85 pu and 1.1 pu, respectively. In summary, the maximum apparent power required on the series-parallel unit of the proposed compensation strategy is smaller than that of the traditional compensation strategy.

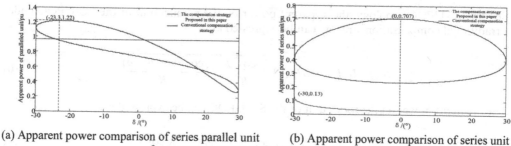

(a) Apparent power comparison of series parallel unit   (b) Apparent power comparison of series unit

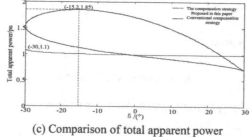

(c) Comparison of total apparent power

Figure 3.   Comparison of apparent powers under different compensation methods.

4 SIMULATION RESULTS AND DISCUSSION

To verify the proposed compensation strategy, simulation was carried out on Matlab/Simu-link platform. The simulation parameters are shown in Table 1.

The three constraints in Eq. (2) were simulated and verified, i.e. the amplitude and phase of the load voltage before and after the compensation remained unchanged, and the power factor of the grid side after the compensation was equal to1. The simulation results are shown in Figure 4.

For the convenience of comparison and analysis, the simulation settings were as follows: the grid voltage worked in the period of 0-0.1 s, the grid voltage fault occurred in the period of 0.1 ~ 0.3 s, and the condition was: $k = 0.538$, $\delta = 15$. (As shown in Figure 2, $U'_S < U_S$, δ is positive). The SC-UPQC came into operation in the period of 0.2 ~ 0.3 s and 0.4 ~ 0.5 s. The waveform of the grid voltage U_S is shown in Figure 4b.

Figure 4a shows the load voltage waveform. As indicated by Figure 4a, before the SC-UPQC came into operation, the load voltage changed with the grid voltage. After the SC-UPQC came into operation, the compensated voltage was consistent with the voltage before the fault. It can be seen that the proposed compensation strategy was applicable to different grid faults of amplitude transients and phase jump.

For the constraints in Eq. (2), the power on the grid side and on the series-parallel unit were both simulated and analyzed. The simulation results are shown in Figure 5.

It can be seen from Figure 5 that before the SC-UPQC was connected to the grid, no matter whether a voltage fault occurred or not, the parallel unit always provides reactive compensation and the grid provided all the active power of the load. In the period of 0.2-0.3 s, SC-UPQC started to work when the voltage sagged, and the active power P_S and reactive power of the grid were 0.21 kW and 0, respectively. The active power P_{sh} and reactive power Q_{sh} of the parallel unit were 6.69 kW and 6.97 kVar, respectively. The active power P_{se} and reactive power Q_{se} of the series unit were 0.17 kW and 0.1 KVar, respectively. The active power generated by the series-parallel unit was provided by the super-capacitor.

Table 1. The simulation parameters of UPQC.

Parameter	Value
Grid voltage	380V/50Hz
DC capacitor	5600uF
Series inductance and equivalent resistance	2mH $R_C=10\%wL_C$
Parallel inductance and equivalent resistance	1mH $R_C=10\%wL_C$
Power factor angel	45°
Rated capacity of load	10K V·A
Rated current of SC	100 A

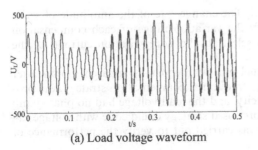

(a) Load voltage waveform

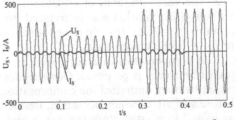

(b) Power supply voltage and current waveform

Figure 4. Voltage and current waveforms obtained by simulation.

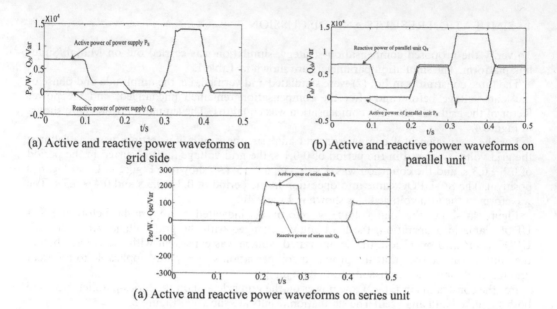

(a) Active and reactive power waveforms on grid side

(b) Active and reactive power waveforms on parallel unit

(a) Active and reactive power waveforms on series unit

Figure 5. Active and reactive power waveforms obtained by simulation.

In the period of 0.4-0.5 s, the SC-UPQC started to work when the voltage swelled. The active power P_S and reactive power of the grid were 0.55 kW and 0, respectively. The active power P_{sh} and reactive power Q_{sh} of the parallel unit were 6.69 kW and 6.97 kVar, respectively. The active power P_{se} and reactive power Q_{se} of the series unit were 0.17 kW and 0.1 kVar, respectively. The active power generated by the parallel unit was provided by the super capacitor while the active power absorbed by the series unit was absorbed by the super capacitor.

5 CONCLUSION

The topology of SC-UPQC is proposed and analyzed in this paper. The following conclusions are obtained.

1. Taking the rated voltage of load as the reference, four constraint conditions of the compensation strategy of SC-UPQC were analyzed. The clarification of constraint conditions makes it is not easy to ignore a certain parameter of compensation while analyzing the strategy of SC-UPQC.

2. The variation range of UPQC was analyzed. The change rate of the amplitude of series voltage compensation was introduced, to clarify the variation range of each compensation amount in UPQC. It is therefore convenient to select parallel compensation current for the proposed compensation strategy.

3. A compensation strategy based on a parallel unit with constant apparent power after fault occurrence is proposed. Compared with the traditional compensation strategy, the proposed strategy optimized the compensation capacity, and the load voltage had no phase jump before and after the compensation. Besides, the proposed strategy could cope with voltage sag and swell in a satisfactory manner. Simulation was carried out to verify the performance of the proposed strategy.

REFERENCES

Y. Shi, Z. Wu and X. Dou, et al, 2015. Study on compound control technology of single-phase dynamic voltage restorer [J]. Transactions of China Electro technical Society, 30(17): 85–95.

Z. Li, L. Xia and Z. Wu, et al, 2011. Optimum design of output filter for UPQC [J]. Electric Power Automation Equipment, 31(7): 56–61.

A. E. Leon, S. J. Amodeo and J. A. Solsona, et al, 2011. Non-linear optimal controller for unified power quality conditioners[J]. IET Power Electronics, 4(4): 435–446.

V. Khadkikar, 2012. Enhancing electric power quality using UPQC: a comprehensive overview [J]. IEEE Transactions on Power Electronics, 27(5): 2284–2297.

F. Wu, J. Zheng and J. Mei, 2012. Multi-objective control strategy of multi-converter unified power quality controller [J]. Electric Power Automation Equipment, 32(10): 88–93.

H. Ma, G. Xu and S. Song, et al, 2014. Quantitative analysis of harmonic current responsibility in distribution network[J]. Electric Power Automation Equipment, 34(6): 44–49.

X. Li, M. Xia and Y. Mao, et al, 2013. A micro grid control strategy based on improved structure of UPQC [J]. Automation of Electric Power System, 37(22): 22–27.

X. Zhang and X. Tong, 2016. Capacity optimization design and control of unified power quality conditioner[J]. Power System Protection and Control, 44(22): 116–121.

S. Devassy and B. Singh, 2015. Dynamic performance of solar PV integrated UPQC-P for critical loads [C]. Annual IEEE India Conference (INDICON), New Delhi, pp. 1–6.

B. Ambati and V. Khadkikar, 2014. Optimal Sizing of UPQC Considering VA Loading and Maximum Utilization of Power-Electronic Converters[J]. IEEE Transactions on Power Delivery, 29(3): 1490–1498

S. Devassy and B. Singh, 2017. Modified pq-Theory-Based Control of Solar-PV-Integrated UPQC-S[J]. EEE Transactions on Industry Applications, 53(5): 5031–5040.

Emerging Developments in the Power and Energy Industry – Dufo-López, Krzywanski & Singh (eds)
© 2020 Taylor & Francis Group, London, ISBN 978-0-367-27169-5

Stability performance of an innovative submarine cable laying vessel during cable landing

Shanjun Bao, Jiancheng Wang, Zhaogang Ding & Zhigang Zhang
Zhoushan Power Supply Company of State Grid Zhejiang Electric Power Company, Zhoushan, China

Zhen Liu*
Shandong Provincial Key Laboratory of Ocean Engineering, Ocean University of China, Qingdao
Department of Ocean Engineering, College of Engineering, Ocean University of China, Qingdao, China
Qingdao National Engineering Laboratory for Subsea Equipment Test and Detection Technology, Qingdao, China

ABSTRACT: A submarine is used to connect the main grid and the island grid for electricity supply. The stability performance of the cable laying vessel under complex ocean environmental conditions is important for the construction operation. In this article the effects of the cable landing operation on the stability of the submarine cable laying vessel are studied numerically. A numerical model based on ANSYS-AQWA is established for the vessel, the mooring lines, and the landing cable, which has been validated by corresponding experimental data. It is found that a mooring angle of 45° for four anchoring lines and a reasonable length extension of the cable during landing will ensure the safety of the cable and maintain the stability of the vessel.

Keywords: Cable landing, Cable laying vessel, Mooring system, Numerical study, Stability performance

1   INTRODUCTION

China is a large marine country with a vast sea area and a long coastline of more than 180,000 km (Fan 2008). With the rapid and deep developments of the marine economy and maritime power strategy, the requirement for island electricity has increased dramatically. Currently, the electricity supply is provided by means of the submarine cables connected to the grid on the mainland. The cables may also be utilized for the interconnections among different islands.

Zhoushan is the largest archipelago in China, which occupies more than one-third of China's total islands. There are 84 islands with permanent resident populations of more than 1000 in the Zhoushan archipelago, which comprises the most rapid developing islands in China (Li 1997). A huge demand for electricity has led to the construction of interconnecting submarine cable networks to ensure an electricity supply for development of the economy. Therefore, construction safety, efficiency, and reliability during the cable deployment have become key issues, which will also guarantee that the cable laying will fit the designed routing.

As is well known, a safe and reliable cable laying vessel is important to the extension of the offshore construction period, guarantee of construction quality, and enhancement of economic returns (Wang 2003). The anchoring stability and mooring system of the construction vessels have been studied to gain knowledge that can improve the performance of the vessels. Effects of superposed wave frequency line motions on the low-frequency surge damping of the

---

*Corresponding author: liuzhen@ouc.edu.cn

mooring system have been investigated numerically (Huse & Matsumoto 1989) and were validated by a series of experimental studies. A dynamic stiffness approach was utilized for dealing with the dynamics of the mooring cables (Sarkar & Taylor 2000). Effects of the motions of a vessel operating on a very low keel clearance in extremely shallow water conditions on the tensions of mooring lines were studied (Cavefors & Oscarsson 2016). A proper mooring plan including the form of anchors, clump weights, chains, and cables is very important to achieve stability of the vessels. An analysis tool was proposed to determine the stiffness characteristics of a multicomponent cable, which could also demonstrate how all the components could be included in the dynamic analysis of an offshore construction vessel moored by a multileg anchoring system (Ansari 1980). Damping effects of the low-frequency motions on the dynamic performance of the mooring lines have been investigated (Liu 1997). A dynamic analysis was conducted on the response of the offshore maritime terminals (Ansari & Khan 1986). A statistical estimation of total second-order responses of moored floating structures subjected to stationary Gaussian random waves was presented which included slow drift oscillations, where the interference effects of first- and second-order responses were also analyzed (Kato et al. 1990). Furthermore, the nonlinear fluid-drag force was linearized using the statistical linearization technique in the dynamic analysis of the mooring lines in random wave climates (Sarkar & Taylor 2002).

Currently, it has been very popular to conduct a dynamic analysis of the offshore structures including the platforms and vessels using computational fluid dynamics (CFD) programs and commercial software packages. The commercial CFD software ANSYS-AQWA plays a dominant role especially in the application of studies of the dynamic response of the vessels and other floating structures. The motion responses of a liquefied natural gas (LNG) vessel under the joint effects of the wind, waves, and currents under a mooring status (Hu et al. 2012). A preliminary estimation of intensity response of a deep-see vessel under still water and incident regular wave conditions was conducted using ANSYS-AQWA (Liu & Li 2015). An integration numerical analysis for the dynamic response of a classic beam structure under irregular wave conditions within the JONSWAP spectrum (Roy et al. 2017). In a time-domain study, ANSYS-AQWA was employed to conduct a nonlinear dynamic analysis of a completely integrated model (Jameel et al. 2013).

In this study, an innovative cable laying vessel is introduced, which is specially designed for cable connecting construction in Zhoushan. A numerical model for the dynamic analysis of the vessel connecting the mooring lines and the cables to be landed is established based on ANSYS-AQWA. Effects of the operating condition for the cable landing on the stability performance of the vessel under various ocean environmental dynamic conditions are studied numerically, which will suggest the direction for further optimization of the vessel design under cable-landing construction conditions.

2 CABLE LAYING VESSEL

2.1 Innovative cable laying vessel

The bathymetry around Zhoushan archipelago is generally flat with a very small bottom slope. The water depth is usually 5–13 m, and the maximum local depth is approximately 50 m. According to the local ocean environmental hydrodynamic conditions, the design criteria include the wind speed of 13.8 m/s, 4 knots for the current, and a significant wave height of 1.5 m.

An innovatively designed cable laying vessel, as shown in Figure 1, is utilized to carry at most 5000-ton submarine cables for long-distance transportation and laying following the desired routing. The length, width, and height of the vessel are 107 m, 32 m, and 6.5 m, respectively. The maximum draft of the vessel is 4.8 m. There are eight winches to withdraw and release the mooring lines for positioning and anchoring flexibly. The cable is released horizontally on the deck and its end is connected at the stern. All the cable laying, connecting, and releasing operations could be conducted automatically on this vessel.

104

Figure 1. Innovatively designed cable laying vessel (Chen et al. 2018).

Based on the designed purpose of the vessel, the maximum operating conditions include operating water depth of 100 m, current speed of 4.0 knots, and significant wave height of 1.25 m. The test cases in the following numerical simulations are also designed based on these conditions.

2.2 *Principles of mooring-line design*

The main functions of the mooring system include (1) maintaining stability of the vessel and guaranteeing cable laying under the operating conditions and (2) ensuring the hull of the vessel will not be damaged under extreme sea conditions (e.g., typhoon impact, storm surge, and extremely large wave height) (Xu & Dong 1994).

The mooring system is one of the most important parts of the cable laying vessel. The design of the mooring system should follow these principles:

1. Stability: Under the joint effects of the wind, current, and waves, the vessel can adapt to different sea levels and wave heights to maintain its dynamically stable floating status. The system can restrict motion responses such as the surge and pitch in acceptable ranges.
2. Survivability: The mooring tension forces should not exceed the designed breaking force of the mooring lines, and the lines should not be broken at all.
3. Economy: The mooring radius should be as small as possible to reduce the occupied sea area and material consumption so as to reduce costs.

Two typical mooring systems are usually employed for offshore engineering: catenary type and tension type (You et al. 2010; Zhao et al. 2013). The tension type mooring system fixes the floating platform by using tensioned lines, whose restoring forces are provided by the axial stiffness of the lines. There are no heavy laying sections for the tension lines, which reduces the costs significantly. Furthermore, the tension forces of the lines are much higher than those of the catenary-type lines, which leads to a higher requirement for material strength and safety. For the catenary-type system, the lines can realize the positioning of the vessel. It is more suitable for relatively shallow water zones (Li et al. 2010). In considering the operating characteristics of two types of mooring system, the catenary type has been chosen as the mooring system for further numerical studies.

3 NUMERICAL MODEL

3.1 *Governing equations*

In offshore engineering, the ideal fluid is assumed to be incompressible, inviscid, and irrotational for large-scale structures (the characteristic length is larger than 0.2 times the incident wave length). The 3D potential theory can be utilized, which ignores high-order velocity potentials. Based on this theory, the velocity potential ϕ satisfies the Laplace equation in the calculation regions:

$$\nabla^2 \phi = 0 \tag{1}$$

If the complex form of the velocity potential is induced, its definition can be written as

$$\varphi = \mathrm{Re}(\phi e^{-i\omega t}) \tag{2}$$

where ω is the incident wave frequency and t is the time.

Furthermore, the velocity potential can be divided into three parts: the incident potential φ_I, the diffraction potential φ_D, and the radiation potential φ_R. φ_I, φ_D, and φ_R represent the disturbances from the incident waves, the objects to the flow field, and the motion of the objects to the flow field, respectively. Each component of the velocity potential should satisfy the Laplace equation.

The incident potential φ_I can be written as

$$\phi_I = \frac{Ag}{\omega} \frac{\cosh k(z+d)}{\cosh kd} e^{[ik(x\cos\beta + y\sin\beta)]} \tag{3}$$

where A is the wave amplitude and k is the wave number. h, g, and β represent the water depth, gravitational acceleration, and the angle between the incident wave direction and the x-axis.

The governing equation of the diffraction potential can be written as

$$\nabla^2 \phi_D = 0 \tag{4}$$

For the free surface condition,

$$\frac{\partial \varphi_D}{\partial z} - \frac{\omega^2}{g} \varphi_D = 0 (z = 0) \tag{5}$$

For the bottom condition,

$$\frac{\partial \varphi_D}{\partial n} = -\frac{\partial \varphi_I}{\partial n} \tag{6}$$

The diffraction condition at the infinite far field can be written as

$$\lim_{R \to \infty} \sqrt{R} \left(\frac{\partial \phi_D}{\partial R} - iK\phi_D \right) = 0 \tag{7}$$

where $R = \sqrt{x^2 + y^2}$.

Considering six degrees of freedom (DOF), the radiation potential φ_R can be written as

$$\phi_R = i\omega \sum_{j=1}^{6} \varepsilon_j \phi_j \tag{8}$$

where ε_j represents the motion of the jth DOF, and φ_j represents the radiation potential according to the unit motion of the jth DOF.

The governing equation of the radiation potential can be written as

$$\nabla^2 \varphi_R = 0 \tag{9}$$

For the free surface condition,

$$\frac{\partial \varphi_D}{\partial z} = \frac{\omega^2}{g} \varphi_R(z = 0) \tag{10}$$

For the bottom condition,

$$\frac{\partial \varphi_R}{\partial z} = 0 (z = -d) \tag{11}$$

For the surface of the object,

$$\frac{\partial \varphi_j}{\partial n} = n_j \ (j = 1, \ 2, \ \dots, \ 6) \tag{12}$$

where n is the unit vector, and its direction is perpendicular to the surface of the floating body.

The diffraction condition at the infinite far field can be written as

$$\lim_{R \to \infty} \sqrt{R} \left(\frac{\partial \phi_D}{\partial R} - ik\phi_D \right) = 0 \tag{13}$$

As the incident potential of the waves is known, the total velocity potential and the pressure on the floating body can be derived from the diffraction and radiation potentials under the definite condition. Subsequently, the wave force on the large-scale body can be calculated and derived. Therefore, the wave forces on the floating body can be written as

$$F_{\omega j} = F_{wj}{}^k + F_{wj}{}^d = \rho i \omega \iint_{So} \phi_I n_j dS + \rho i \omega \iint_{So} \phi_D n_j dS = \rho i \omega \iint_{So} \phi_I \frac{\partial \phi_j}{\partial n} - \phi_j \frac{\partial \phi_i}{\partial n} dS \tag{14}$$

where $F_{\omega j}{}^k$ is defined as the Froude–Krylov force and $F_{\omega j}{}^d$ is the radiation force. The sum of the above two components is the total wave force.

According to the API code, the wind force F_w and the water flow force F_{cs} can be estimated as follows:

$$F_w = C_w \sum (C_s C_h A_1) V_w^2 \tag{15}$$

$$F_{cs} = C_{ss} C_d A_2 V_c^2 \tag{16}$$

where C_w is the wind force coefficient, which is usually can be defined as 0.615 Ns^2/m^4. C_s is the hull shape coefficient, and C_h is the vessel height coefficient. A_1 and V_w represent the vertical projected area on the direction of the incident wind and the designed wind speed, respectively. C_{ss} is the flow force coefficient, which is usually defined as 515.62 Ns^2/m^4. C_d is the resistance coefficient. For the cylinder, $C_d = 0.5$; and for the plate, $C_d = 1.5$. A_2 and V_c represent the vertical projected area on the direction of the incident water flow and the designed current speed, respectively.

3.2 Set-up in ANSYS-AQWA

The vessel model is established in ANSYS-APDL, as presented in Figure 2. The characteristics of each component can be given based on the reasonable model establishment, and therefore the quality and the center of mass can be determined. During the mesh generation, in order to

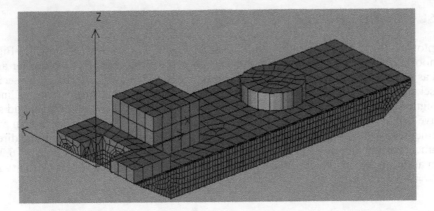

Figure 2. 3D numerical model of the cable laying vessel.

ensure the calculation speed and save computational time cost, the large-scale grids for the upper parts are coupled with the small-scale grids for the lower parts. The numerical model with proper meshes can be used to generate an executive aqwa file using the command anstoaqwa for the frequency-domain and time-domain calculations. The largest length of the quadrangle mesh side is 1.0 m. There are a total of 5384 meshes and 216,000 recording points. The calculation time step is 0.05 s, and total calculation period is 10,800 s.

3.3 *Validation of the numerical model*

The numerical model has been validated for its prediction capability of the mooring system. An experimental test of the vessel model in the wave basin was employed for comparison of the vessel motion and the tension on the line (Chen et al. 2018). Up to eight lines were used for the experimental tests. The six DOF optical sensor and the tension-force load-cell were utilized for the measurement of the vessel motions and the tensions on the mooring lines. The irregular wave condition was applied with a significant wave height of $H = 0.175$ m and averaged period of $T = 1.75$ s. The JONSWAP spectrum is utilized with the peak factor of $\gamma = 3.3$.

Comparisons of the numerical and experimental results are shown in Figure 3. As illustrated in Figure 3a, the numerical predictions of the amplitudes of the vessel heaving motion are slightly smaller than those derived from the experimental tests, and the averaged error is approximately 6.5%. For the tension on one of the mooring lines, as shown in Figure 3b, the numerical results of the variation amplitudes are also smaller than experimental data, where the averaged error is approximately 5.3%. Generally, the numerical model established in the present study has shown good capability for the prediction of both the vessel motions and the tension forces on the mooring lines.

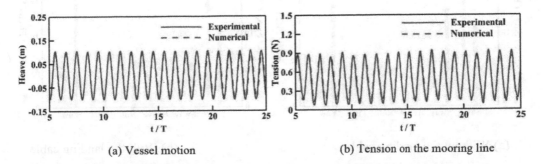

| (a) Vessel motion | (b) Tension on the mooring line |

Figure 3. Comparison of numerical and experimental results of a vessel with a mooring line.

4 RESULTS AND DISCUSSIONS

The deployment of the mooring lines and the landing cable of the vessel is shown in Figure 4. Four mooring lines are extended from the four corners of the vessel. The mooring angle is defined as α, as given in Figure 4. The submarine cable is pulled out from the center section of the vessel side, and its length from the vessel to the anchoring point on the coastal line is L. The designed conditions include wind speed of 20.7 m/s, current speed of 2.6 m/s, and significant wave height of $H_s = 2.6$ m with a significant period of $T_s = 7.0$ s. Considering an extreme environmental assumption, the wind, current, and waves are incident from the same direction. The diameter of the mooring line is $D_M = 0.05$ m, and its elasticity modulus is $E_M = 6.9 \times 10^3$ MPa. In addition, the diameter of the submarine cable is $D_C = 0.18$ m, and its elasticity modulus is $E_C = 4.5 \times 10^4$ MPa.

4.1 *Effects of mooring angle on stability performance*

Two mooring angles are considered in this section: $\alpha = 30°$ and $\alpha = 45°$. The significant period is $T_s = 7.0$ s. The cable length L is 210 m, where the distance between the coastal line and the vessel is 200m. Maximum amplitudes of the motions in six DOF of the vessel without or with the landing cable are compared in Figure 5.

As shown in Figure 5a, if the vessel is not under the cable landing status, the maximum surge amplitude for $\alpha = 45°$ is 30% larger than that for $\alpha = 30°$. On the other hand, the

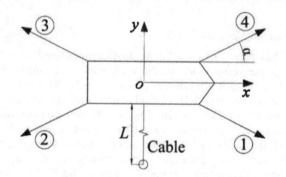

Figure 4.   Deployment of the mooring lines and the submarine cable.

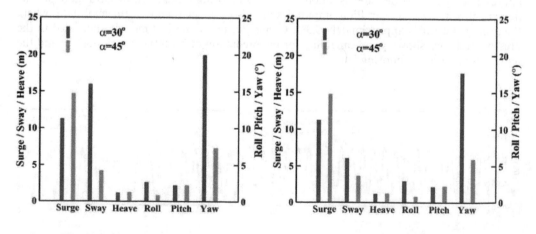

(a) Without the landing cable   (b) With the landing cable

Figure 5.   Maximum amplitudes of motions of the vessel.

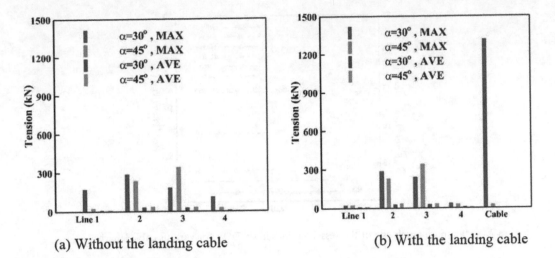

(a) Without the landing cable   (b) With the landing cable

Figure 6.   Tension forces on the mooring lines.

maximum sway amplitude for $\alpha = 30°$ is 2.3 times that for $\alpha = 45°$. In addition, the maximum yaw amplitude for $\alpha = 30°$ is approximately 2.5 times that for $\alpha = 45°$. If the vessel is landing the cable, as shown in Figure 5b, the cable acts as a mooring line to provide more stability to the vessel, which significantly reduces the translational motions. On the contrary, the landing cable has little effect on the yaw motion, which may be caused by the laying direction of the cable from the vessel. From the aspect of the vessel itself, the mooring angle of $\alpha = 45°$ demonstrates a better performance for vessel stabilization.

The tensions on the mooring lines without or with the landing cable are shown in Figure 6. As can be seen in Figure 6a, without the landing cable, the averaged tensions for each line are quite small. In most cases, the maximum tensions for $\alpha = 30°$ are larger than those for $\alpha = 45°$ except line 3. All these values are much less than the safety threshold of the mooring line (946.1 kN). Under the condition with the landing cable in Figure 6b, the averaged tensions on the mooring lines are similar to those in Figure 6 (a). The maximum tension on line 1 has been reduced significantly. For the landing cable itself, the maximum tension for $\alpha = 45°$ is small and safe enough. On the other hand, the maximum tension for $\alpha = 30°$ exceeds 1300 kN, which is not acceptable for the submarine cable. Obviously, the mooring angle of $\alpha = 45°$ shows a better performance on both the vessel stability and the tension on the submarine cable.

4.2   *Effects of landing-cable length*

Considering the typical engineering operation and environmental conditions, four lengths of landing cables are investigated in this study: 200 m, 210 m, 220 m, and 230 m. The effects of the landing-cable length on the maximum amplitude of the vessel motions are shown in Figure 7. It should be pointed out that the vessel is generally stable and the significant motions can be observed only for several moments. The maximum amplitude of surge is close to 15.0 m, and the value for yawing is 12°. The variation trend of translational motions is generally mild. As the landing-cable length decreases, the motion amplitudes decreases because of the mooring effects from the cable which acts like a mooring line. The rotational motions are affected by the landing cable for the shorter length. For the longer length, the stability of the vessel is maintained mainly by the mooring lines, similar to the case without any landing cables.

Figure 8 presents the tension forces on the mooring lines for various landing-cable lengths. Lines 2 and 3 face the incident waves and are responsible for maintaining the stability of the vessel and withstanding most tensions. As shown in Figure 8a, the values of averaged tensions for four lines are similar for various landing-cable lengths. The values of two pairs of mooring lines at both sides of the vessel are approximately 6.5 kN and 33.0 kN, respectively. For the maximum tension in Figure 8b, the greatest value at the facing-wave side for line 3 is achieved

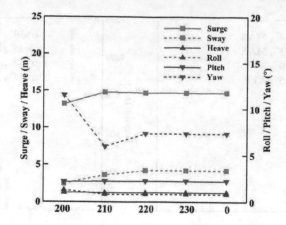

Figure 7.   Maximum amplitude of the vessel motions in six DOF for various cable lengths.

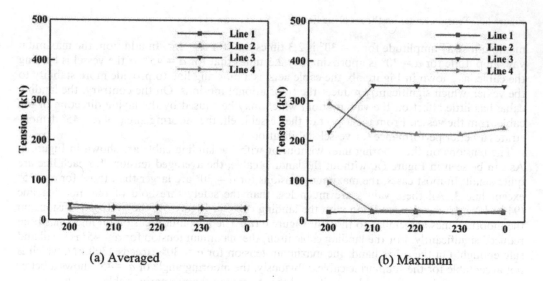

(a) Averaged   (b) Maximum

Figure 8.   Tension forces on the mooring lines for various cable lengths.

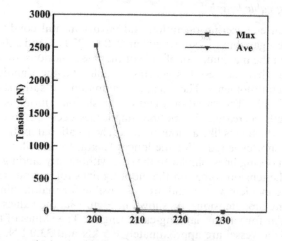

Figure 9.   Tension forces on the land cable for various cable lengths.

at 345 kN under a cable length of $L = 210$ m. As the cable length decreases to 200 m, the variation characteristics are different from that of the other lengths.

Tensions on the landing cable are shown in Figure 9. It should be noted that the averaged and maximum tension values for all the cable lengths are very small. The averaged values are approximately 10 N. On the other hand, the maximum tensions is extremely high for $L = 200$ m, which is 2350 kN. This phenomenon demonstrate that the maximum tensions are caused by a short-term shock because of the freak waves. It indicates that the value of $L = 210$ m is an important threshold for the landing cable.

4 CONCLUSIONS

The submarine cable is used for the island electricity supply, which has been in huge demand because of the rapid development of the Chinese marine economy. The cable laying vessel can deploy the submarine cables automatically and rapidly following the designed routing. Therefore, the stability of the vessel has a significant effect on the cable laying operation, especially under complex ocean environmental conditions.

A numerical model based on the commercial computational fluid dynamics software ANSYS-AQWA was established in this study for the investigation of the floating system of the vessel, the mooring lines, and the landing cables. The model has been validated by corresponding experimental data. Effects of the cable landing on the stability of the vessel were first investigated numerically.

For the four-mooring-line system, the mooring angle of 30° causes a larger maximum amplitude of the motions in six DOF and extremely high tension forces on the landing cable. On the other hand, the mooring angle of 45° shows a better performance especially on reducing the tensions on the landing submarine cable. Although a shorter length of the landing cable can reduce the motions of the vessel slightly, thus acting more like an additional mooring line, the tension forces will be significantly increased. Therefore, a proper longer landing cable is strongly suggested for cable laying operations.

ACKNOWLEDGEMENT

The research is financially supported by the Science and Technology Project of State Grid Zhejiang Electric Power Company: Research on the Key Technologies of the New Floating Devices for Submarine Cable Landing operation.

REFERENCES

Ansari, K.A. 1980. Mooring with multicomponent cable systems. *Journal of Energy Resources Technology* 102(2): 62–69.

Ansari, K.A., & Khan, N.U. 1986. The effect of cable dynamics on the station-keeping response of a moored offshore vessel. *Journal of Energy Resources Technology* 108(1): 52–58.

Cavefors, J. & Oscarsson, A. 2016. Vessel motions and mooring line tensions in very shallow water. Göteborg: Chalmers University of Technology.

Chen, Z., Wang, Y., Zheng, H., et al. 2018. Numerical study of operating stability of an innovative cable laying vessel under extreme conditions. *2018 Oceans IEEE Paper*.

Fan, X. 2008. Strategy choice of the preservation and management of Chinese coastline. *Geological Survey and Research* 31(1): 28–32.

Hu, Y., Hu, Z., Liu, Y., & Liu, J. 2012. Analysis of the large LNG ships moored against a quay based on AQWA. *Ship Science and Technology* 34(2): 70–73.

Huse, E. & Matsumoto, K. 1989. Mooring line damping due to first-and second-order vessel motion. *Offshore Technology Conference*.

Jameel, M., Ahmad, S., Islam, A.B.M.S., et al. 2013. Non-linear dynamic analysis of coupled spar platform. *Journal of Civil Engineering and Management* 19(4): 476–491.

Kato, S., Kinoshita, T., & Takase, S. 1990. Statistical theory of total second order responses of moored vessels in random seas. *Applied Ocean Research* 12(1): 2–13.

Li, Z. 1997. The characteristics and exploitation of the island resources of Zhejiang province————taking the Zhoushan archipelago as an example. *Journal of Natural Resources* 12(2): 139–145.

Li, Z., Xu, X., & Wang, H. 2010. Development of offshore platform mooring systems. *Oil Field Equipment* 39(5): 75–78.

Liu, X., & Li, J. 2015. The analysis of some deep-sea functional ships' structural strength based on AQWA. *Shipbuilding Vocational Education* 3(4): 48–51.

Liu, Y. 1997. Dynamic analysis of mooring cables and their damping effect on the low-frequency motion of floating platforms. Göteborg: Chalmers University of Technology.

Roy, S., Ghosh, V., Dey, S., et al. 2016. A coupled analysis of motion and structural responses for an offshore spar platform in irregular waves. *Ships and Offshore Structures* 12(Sup1.): S296–S304.

Sarkar, A. & Taylor, R.E. 2000. Effects of mooring line drag damping on response statistics of vessels excited by first-and second-order wave forces. *Ocean Engineering* 27(6): 667–686.

Sarkar, A. & Taylor, R.E. 2002. Dynamics of mooring cables in random seas. *Journal of Fluids and Structures* 16(2): 193–212.

Wang, Y. 2003. General design of shallow water pipe and cable laying vessel. *Coastal Engineering* 22(3): 38–45.

Xu, X. & Dong, Y. 1994. Dynamic response analysis of single point mooring system. *Journal of China University of Petroleum (Edition of Natural Science)* 5: 74–78.

You, Y., Li, W., Liu, W., et al. 2010. Development status and perspective of marine energy conversion systems. *Journal of Modern Power Systems and Clean Energy* 34(14): 1–12.

Zhao, Z., Huang, M., Bi, Y., et al. 2013. Research and design of pendulum wave energy conversion device. *Journal of Ocean Technology* 32(3): 101–105.

Emerging Developments in the Power and Energy Industry – Dufo-López, Krzywanski & Singh (eds)
© 2020 Taylor & Francis Group, London, ISBN 978-0-367-27169-5

Evaluation on performance of HESS-based grid-connected photovoltaic power system

Caixue Chen & Xu Liu
School of Information Engineering, Xiangtan University, Xiangtan, China

ABSTRACT: This paper expounds the modeling, simulation and performance evaluation of a grid-connected photovoltaic power system based on hybrid energy storage system (HESS). After adopting an active connection method to connect the HESS with DC bus, an energy compatibility control measure based on the DC bus voltage adjustment and the state of charge is put forward, to ensure the system efficiency and safety. Moreover, by analyzing simulated scenarios with the step change in solar irradiance and grid fault, the dynamic performance of the system in the critical working condition is evaluated. Finally, the evaluation results on simulations based on PSCAD/EMTDC are given.

Keywords: grid-connected photovoltaic, hybrid energy storage system, Energy coordination, droop control, DC bus voltage adjustment

1 INTRODUCTION

Energy crisis and environmental crisis has dramatically promoted the development of distributed generation (DG) with renewable energy sources like wind power and solar energy. At the same time, there are significant challenges to DG, especially intermittent impact and technical problems created by grid connection such as dramatic change in solar irradiance, sudden change of load conditions and faults of the point of common coupling (PCC) in some cases. Thus, with the development of energy network and electric car charging techniques, requirements have been raised on the management of the power converter DC bus voltage (L. Xu & D. Chen 2011).

When interference occurs to the DG system, such as environmental change, load change and grid faults, the DC bus voltage will fluctuate, resulting in overshoots or undershoots and voltage drop. Meanwhile, misadjustment to the DC bus voltage may make the DG system low in efficiency (M. A. Tankari et al. 2011). Therefore, in some cases, it's essential to manage the power fluctuation of the DG system using rapid dynamic compensation systems, such as battery energy storage system (BESS) and HESS.

Among traditional distributed power generation tools, battery has a high power density and high charging/discharging switching speed. But if it is used for power smoothing, a large capacity is needed on one hand, and on the other hand, frequent charging and discharging is necessary, which will reduce the battery lifetime (STROE D et al. 2017). Supercapacitors are a type of power energy-storage component, able to support rapid high-power charging/discharging. Using battery with supercapacitors to constitute an HESS to smooth the fluctuation of PV power can help to reduce the frequency of charging and discharging, thereby extending the battery lifetime (Dai P et al. 2017). Thus, in this paper, HESS is applied to the grid-connected photovoltaic power system, and the system performance is evaluated in the case where the DC bus voltage by control the buck-boost converter and grid-side voltage-source converter (VSC).

2 SYSTEM STRUCTURE AND MODELING

The structure of the HESS-based grid-connected photovoltaic power system is shown in Figure 1. This system consists of the photovoltaic model, DC-DC boost converter with Maximum Power Point Tracking (MPPT), VSC used for grid interference, and HESS model connected to the DC bus via the DC-DC buck/boost converter. The grid is expressed by an ideal three-phase voltage source and is assumed to be an infinite rigid AC system.

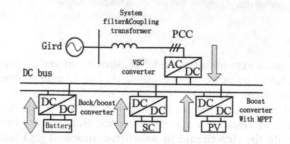

Figure 1.   Structure of HESS-based grid-connected photovoltaic power system.

2.1   *Photovoltaic array model*

The photovoltaic model considered in PSCAD is based on the description in (LIU Dongran et al. 2011), the output current of the photovoltaic system is

$$I = I_{ph} - I_d \left[\left(\exp \left(\frac{q}{AkT} (U + R_s I) \right) - 1 \right) \right] - \frac{U + R_s I}{R_{sh}} \tag{1}$$

Where: R_s and R_{sh} are equivalent series-parallel impedances; I_{ph} and I_d are the photo-generated current and the reverse saturated leakage current of the diode; U and I are the terminal voltage and output current of the photovoltaic system, respectively; T is photovoltaic Temperature; q is the amount of electron charge; *A* is a constant; k is the Boltzmann constant.

2.2   *Lead-acid battery model*

The lead-acid battery model considered is based on the dynamic model described in (TREMBLAY et al. 2007). The two important parameters describing the behavior of the battery are the terminal voltage V_{bat} and the state of charge SOC. The battery current can be obtained according to the calculation formula. I_{bat} is as follows Shown.

$$\begin{cases} V_{bat} = V_0 + R_b I_b - KQ \frac{1}{\int I_b dt + Q} + Ce^{B \int I_b dt} \\ SOC = 100 \left(1 + \frac{\int I_b dt}{Q} \right) \end{cases} \tag{2}$$

Where: R_b is the internal resistance of the battery; V_b is the battery terminal voltage; V_o is the open circuit voltage of the battery; i_b is the battery charging current; K is the battery polarization voltage; Q is the battery capacity; B and C are the fitting factors.

115

2.3 Supercapacitor model

The supercapacitor model considered is based on the model described in (R.L. Spyker & R.M. Nelms 2000), where the two important parameters describing the behavior of the battery are the energy storage E and the state of charge SOC, as shown below.

$$\begin{cases} E = \dfrac{CU_{max}^2}{2} \\ SOC = 100\left(\dfrac{Q_{init} - \int_0^t I(t)dt}{Q_T}\right) \end{cases} \tag{3}$$

Where Q_{init} is the initial capacity value, Q_T is the rated capacity, I is the discharge current, t is the discharge time, C is the supercapacitor value, and U_{max} is the maximum voltage that the supercapacitor can withstand.

3 COORDINATED CONTROLLER DESIGN

3.1 Control scheme of DC-DC boost converterl

The DC-DC boost converter can ensure the maximum power extracted from the photovoltaic system. The MPPT algorithm is adopted based on the perturbation and observation method, in which PI controller is used to adjust the load to ensure that the PV power is consistent with the reference value generated by MPPT.

3.2 Control scheme of grid-side VSC

The inverter adopts the PQ control strategy to transmit all power of DC bus to the grid, thus ensuring the constant voltage of DC bus and injecting the current at unit power factor to the grid. All the controlling methods are set based on the current-mode controlling scheme. According to this scheme, phase-locked loop (PLL) at PCC is used to extract the voltage, so $V_{ud}=|V_{grid}|$ and $V_{uq}=0$. In this way, the current applied to the VSC can directly be adjusted to avoid the fragility of VSC when a large and sudden current is applied, such as network fault and sudden load change. Synchronizing the scheme ensures that the d side of d and q axles aligns with the vector of the grid voltage ($v_q=0$), enabling P and Q to be proportional and to be controlled by i_d and i_q.

$$\begin{cases} P = \frac{3}{2}u_d i_d \\ Q = -\frac{3}{2}u_d i_q \end{cases} \tag{4}$$

In this example, the active power is represented by the DC bus voltage (V_{dc}), where the V_{dc} and Q errors are adjusted by the PI regulator to generate the reference current id, ref and $i_{q, ref}$, respectively, used by the closed-loop current controller.

$$\begin{cases} i_{d, ref} = (k_{pd2} + \frac{k_{id2}}{s})(V_{dc,ref} - V_{dc}) \\ i_{q, ref} = -(k_{pq3} + \frac{k_{iq3}}{s})(Q_{ref} - Q) \end{cases} \tag{5}$$

Where v_{ud}, ref and v_{uq}, ref are the out put voltage references of the current inner loop. Is the output of the d-axis current inner loop PI regulator, processing the error signal of $e_d = i_d - i_{d-ref}$, and the output of the q-axis current inner loop PI regulator, processing the error signal of $e_q = i_q - I_{q, ref}$. Figure 2 shows the overall VSC controlling scheme.

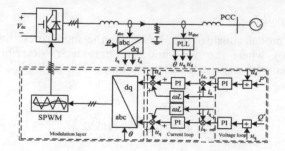

Figure 2.   Grid-side VSC controlling scheme.

$$\begin{cases} v_{ud,\,ref} = v_{ud} + u_d - \omega Lfiq \\ v_{uq,\,ref} = v_{uq} + u_q + \omega Lfiq \end{cases} \tag{6}$$

3.3   *Control scheme of grid-side VSC*

In this paper, an active connection method of HESS is used (Ma T et al. 2015). Both the supercapacitor and the battery are equipped with a DC-DC converter connected to the DC bus, respectively. In this system, the battery provides a long-time main energy buffer, while the super-capacitor is used for peak power smoothing.

3.3.1 *BESS charge and discharge control strategy*

Figure 3 shows the controlling algorithm of SOC and bus voltage designed to ensure the BESS safety and operation optimization. As the terminal voltage of the battery is associated with its state of charge (SOC) of the residual capacity, when the SOC is excessively low, the terminal voltage will markedly be reduced, which will affect the batter; when the SOC reaches a certain level, battery charging will be halted. Thus, it is required that SOC should work within a proper range to avoid BESS over-charging or discharging. In this case, SOC_H is set as 95% but for deep discharging, only 50% (SOC_L) capacity of the BESS is available. As the threshold values of the DC bus voltage can easily result in frequent charging and discharging of the BESS, in this paper, the hysteresis control method is used for boost, preventing the battery from working continuously at the charging/discharging mode, as shown in Figure 4. By taking the input of SOC and DC bus voltage for reference, this algorithm can provide an input for controlling the logic circuit, so as to confirm the working mode.

BESS charging and discharging are respectively controlled by the buck and boos circuits, as shown in Figure 5. Firstly, according to the reference voltage Vdc-ref provided based on the droop curve control strategy, the outer voltage loop and inner current loop are used to control

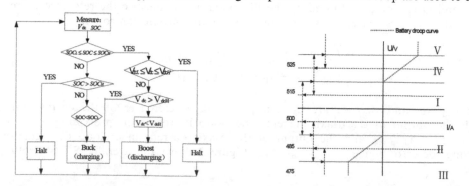

Figure 3.   Battery control algorithm flow chart.   Figure 4.   Battery control strategy.

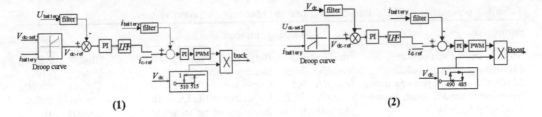

(1)   **(2)**

Figure 5.   Battery DC/DC controlling: (1) battery charging (2) battery discharging.

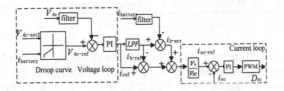

Figure 6.   Charging/discharging compatibility control of the supercapacitor.

the DC bus voltage, making the bus voltage in line with the supposed voltage and the output signal of the inner current loop's PI transmit to the PWM to complete the circuit. Then, based on the hysteresis and PWM techniques, the charging/discharging mode or halt of the operation is decided to prevent frequent switch between regions and reduce the fluctuation of DC bus voltage.

3.3.2 *Supercapacitor charge and discharge control strategy*

Based on the BESS charging/discharging control strategy, a new hybrid energy-storage compatibility controlling method is applied to the supercapacitor. This method fully exerts the fast response advantage of the supercapacitor, which can not only smooth the high-frequency part of the net power, but also smooth the unresponsive part of the battery power, that is, the current reference for the supercapacitor. Thus, this method helps to relieve battery stress and extend battery lifetime. The Dotted box in Figure 6 shows the diagram of the supercapacitor charging/discharging control strategy.

As shown in Figure 6, V_{dc-ref} and Iref are the voltage reference value and current reference value respectively; i_{b-ref} is the current reference value of the battery; i_{b-eer} is the deviation signal between the actual value of the battery and the reference value; V_b and V_{sc} respectively Indicates the terminal voltage of the battery and the supercapacitor; i_{sc} is the feedback current of the supercapacitor; D_{sc} is the duty cycle of the supercapacitor bidirectional DC-DC control circuit.

4   SYSTEM SIMULATION VERIFICATION

Through the simulations on PSCAD/EMTDC, the hybrid energy-storage compatibility control method is used to evaluate the overall performance of the system. Related parameters of the photovoltaic model, battery model and circuit components are shown in Table 1.

Evaluations are performed in the conditions of solar irradiance change (solar irradiance reduced by 60%) and grid fault (voltage decreases by 50%) through simulations. When t = 3s, a sudden change of the solar irradiance will occur and last for 0.5s. When t = 5s, the voltage will fall and last for 0.25s.

To evaluate the dynamic performance of the system under different circumstances, the system control parameters are always fixed in any case. Figure 7 shows the simulation results

Table 1. parameters of HESS-based grid-connected photovoltaic power system.

parameter		value	parameter		value
Module short circuit current	Isc,r/	8.2	Battery rated voltage	Ubat/V	240
Module open circuit voltage	Voc,r/V	36.3	Battery capacity	Q/Ah	210
Module maximum power current	Imp,r/A	7.5	Capacitor at PV terminal	Cpv/μF	10000
Module maximum power voltage	Vmp,r/V	28.7	Capacitor at the DC-link	Cpv/μF	50000
VSC total filter inductor	Lf/mH	0.264	Boost converter inductor	L1/mH	10
Buck/boost converter inductor	L2/mH	0.12	DC bus rated voltage	Udc/V	500
Grid rms voltage	Urms,grid/kV	10	Grid frequency	ω/Hz	50
Converters switching	frequency/kHz	2.5	Transformer voltage	ratio/kV	0.22/10
Transformer	MVA/MVA	0.1	Supercapacitor	Csc/μF	1000

of the HESS-based grid-connected photovoltaic power system, BESS-based grid-connected photovoltaic power system and grid-connected photovoltaic power system. According to the results, during the normal operation of the HESS-based grid-connected photovoltaic power system, the total power transmitted to the grid has declined by several percents, mainly caused by the HESS operation loss.

Without compensation from the super-capacitor, reduction of the solar irradiance leads to over 5% decrease of the DC bus voltage, as shown in Figure 7 (1). Similarly, when a step increase happens to the solar irradiance, the DC bus voltage will boost with the similar amplitude. Under such disturbance conditions, overshoot and undershoot will also occur to the grid-side active power, as shown in Figure 7 (2). When the system is affected by grid faults, the DC bus voltage will remain again at about 1 p.u by the system controller, but marked by significant overshoot and undershoot with the peak value over 5%.

With compensation from the supercapacitor, an improvement for dynamic performance is observed. As the DC bus voltage and power shown in Figure 7 (1) and (2), compared with the situation without compensation from the supercapacitor, the steady state time lasts shorter. Meanwhile, the super-capacitor can reduce the overshoot and undershoot of bus voltage to some extent, with the peak value of the DC bus steady state smaller than 1%. Under such disturbance conditions, the DC bus voltage fluctuation remains within a range of 5%.

Figure 8 shows the simulation results of batteries in HESS and BESS. As shown in Figure 8, the positive power indicates that the battery is discharged, while the negative power indicates that the battery is charged by absorbing power from the DC bus. It can be seen from Figure 7 (1) and Figure 8 that when the DC bus has a voltage overshoot which exceeds the preset range, the Buck-Boost circuit in the battery will be triggered to execute buck or boost operation. Since the DC bus voltage in HESS is kept within a fluctuation range of 5%, unnecessary charging and discharging of the battery is avoided, which benefits the optimal and safe operation of the battery.

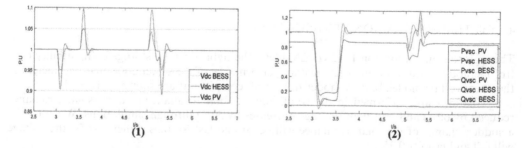
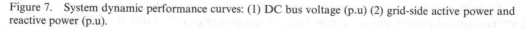

Figure 7. System dynamic performance curves: (1) DC bus voltage (p.u) (2) grid-side active power and reactive power (p.u).

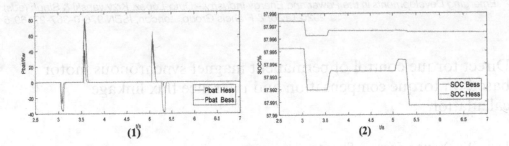

Figure 8.   Battery dynamic performance curves: (1) power (kW), (2) battery SOC (%).

5   CONCLUSION

This paper expounds the modeling and controlling of the HESS-based grid-connected photovoltaic power system. Targeting at the performance evaluation of the HESS-based grid-connected photovoltaic power system and based on various clinical disturbance factors of the system, an active HESS-based connection method is applied. This method aims at controlling the fluctuation of the DC bus voltage to ensure the efficiency of the grid-side VSC. Meanwhile, an HESS-based compatibility controlling strategy is put forward, which can further reduce the overshoot and undershoot as well as the dynamic steady time of the DC bus voltage to an acceptable range.

REFERENCES

L. Xu and D. Chen. 2011. "Control and operation of a DC microgrid with variable generation and energy storage," IEEE Transactions on Power Delivery, vol. 26, pp. 2513–2522.

M. A. Tankari, M.B.Camara, B. Dakyo, and C. Nichita. 2011. "Ultracapacitors and batteries integration for power fluctuations miitigation in wind-PV-diesel hybrid system," International Journal of Renewable Energy Research, vol. 1, pp. 86–95.

STROE D,KNAP V,SWIERCZYNSKI M, et al. 2017. Operation of a grid-connected lithium-ion battery energy storage system for primary frequency regulation: a battery lifetime perspective [J]. IEEE Transactions on Industry Applications, 53(1): 430–438.

Dai P, Cauet S, Coirault P. 2016. Disturbance rejection of battery/ultracapacitor hybrid energy sources [J]. Control Engineering Practice, 54: 166–175.

LIU Dongran, CHEN Shuyong, MA Min, et al. 2011. A review on models for photovoltaic generation system [J]. Power System Technology, 35(8): 47–52.

TREMBLAY, DESSAINT L A, DEKKICHE A I. 2007. A generic battery model for the dynamic simulation of hybrid electric vehicles [C] // IEEE Vehicle Power and Propulsion Conference (VPPC): 284–289.

R.L.Spyker, R.M. Nelms. 2000. Classical equivalent circuit parameters for a double-layer capacitor [J]. IEEE Trans. On Aerospace and Electronic Systems, 36(3): 829–836.

Ma T, Yang H, Lu L. 2015. Development of hybrid battery-supercapacitor energy storage for remote area renewable energy systems [J]. Applied Energy, 15356–62.

Emerging Developments in the Power and Energy Industry – Dufo-López, Krzywanski & Singh (eds)
© 2020 Taylor & Francis Group, London, ISBN 978-0-367-27169-5

Direct torque control of permanent magnet synchronous motor based on torque compensation and reference flux linkage calculation

Jiang Wu, Yibing Zhang & Shuang Chen
The College of Information Engineering, Xiangtan University, Xiangtan, Hunan Province, China

ABSTRACT: For traditional direct torque control system has the flux and torque ripple and other issues, a modified direct torque control (DTC) scheme for permanent magnet synchronous motor (PMSM) is investigated in this paper, which features in relatively lower flux and torque ripple and almost fixed switching frequency. The proposed SVM-DTC based on the compensation of the error torque by Feedforward compensation to minimize the torque and stator flux ripples. Modeling and experimental results show that the flux and torque ripples are greatly reduced when compared with those of the classic DTC. With the new scheme, very short sampling time is not essential. All the advantages of the basic DTC are still retained. In addition, fixed switching frequency at different operating conditions becomes possible.

Keywords: direct torque control, SVPWM, permanent magnet synchronous motor, Feedforward compensation

1 INTRODUCTION

With the development of PM material, permanent magnet synchronous motor (PMSM) have been applied in many fields where high performance and rapid torque response are required. Consequently many interests are focused on the study of the control strategies of PMSM (H.Zhu, X .Xiao & Y.Li 2011; X.Qiu, W X.Huang & F F.Bu; B.Lin, D.Sun & Y K.He 2014). It is well known that the most popular two control strategies are field oriented control (FOC) and the direct torque control (DTC), which have been invented in 1970's and 1980's respectively. Although both of them were firstly used with induction machines, they can also be well applied on PMSMs (M.Asad, A I.Bhatti & S.Iqbal 2012). Former studies have shown that DTC has many advantages over FOC in the aspects of faster dynamic response, lower parameter dependence, and simpler configuration due to the elimination of coordinate transformation and position sensors. But basic DTC strategies still have some drawbacks such as high sampling frequency, torque ripple and current distortion because of the variable switching frequency (Y.Zhang & J.Zhu. 2011; Z.Li, Y. Zhang & Li.Z, et al. 2015; H Q.Zhang, X S.Wang & P F.Wei, et al. 2012). Many researchers have devoted to the improvement of the steady state performance of the basic DTC, especially for induction machine (IM) since the first industrial product has been produced in 1996 by ABB (H.Zhu X.Xiao & Y.Li 2011; F.Niu, B.Wang & A S.Babel, et al. 2016). In reference (E.Monmasson & J P.Louis. 2003), the technology of space vector pulse-width-modulation (SVM) used in IM DTC system with low torque and flux ripples was reported. Nowadays some of the approaches have been introduced into the DTC improvement for PMSM, which multiple voltage vectors are used and fixed switching frequency operation are allowed so as to improve the torque ripple and current distortion, but generally they require more complicated modified control systems.

In this paper, the feedforward compensation and SVPWM technology are simultaneously introduced into the PMSM DTC control system. Based on the space voltage vector modulation of the SVM technology, a torque feedforward compensation controller is introduced to compensate the system torque error online to reduce the error caused by various factors on the disturbance of the control system. By introducing feedforward compensation, the dynamic performance of PMSM and the robustness of load variation are improved, which greatly reduces the torque ripple of PMSM and improves the stability of the system.

2   BASIC PRINCIPLES OF PMSM SVM-DTC CONTROL

2.1   *Mathematical model of PMSM*

Before establishing mathematical model of PMSM, assumptions are often made as follows: permanent saturation of the motor's iron core is neglected, eddy current and hysteresis loss are excluded, and motor's current is symmetrical three phase sine current. the voltage and flux linkage equations of PMSM are obtained in the (α–β) coordinate.

In this paper, the SPMSM voltage equation in the coordinate system is as follows:

$$\begin{cases} u_\alpha = R_s i_\alpha + L_s \frac{di_\alpha}{dt} - \omega_r \phi_r \sin \theta_r \\ u_\beta = R_s i_\beta + L_s \frac{di_\beta}{dt} + \omega_r \phi_r \cos \theta_r \end{cases} \tag{1}$$

Where, ϕ_α, ϕ_β are separately the components of the stator flux vector in the α and β axes, u_α, u_β and i_α, i_β are separately the stator voltage and current components of the stator voltage in the stator two-phase stationary coordinate system. R_s is stator resistance, L_s is the stator inductance, ω_r is the rotor speed, ϕ_r is the rotor permanent magnet flux linkage, θ_r is the rotor position angle.

$$\begin{cases} u_\alpha = R_s i_\alpha + \frac{d\phi_\alpha}{dt} = R_s i_\alpha + \frac{d(L_s i_\alpha + \phi_r \cos \theta_r)}{dt} \\ u_\beta = R_s i_\beta + \frac{d\phi_\beta}{dt} = R_s i_\beta + \frac{d(L_s i_\beta + \phi_r \sin \theta_r)}{dt} \end{cases} \tag{2}$$

From (2), making the following definition

$$\begin{cases} \phi_\alpha = L_s i_\alpha + \phi_r \cos \theta_r \\ \phi_\beta = L_s i_\beta + \phi_r \sin \theta_r \\ \phi_s = \sqrt{\phi_\alpha^2 + \phi_\beta^2} \end{cases} \tag{3}$$

The equation of motion for SPMSM is:

$$\frac{d\omega_r}{dt} = \frac{1}{J}(T_e - T_L - D\omega_r) \tag{4}$$

Where, T_e and T_L are separately Electromagnetic torque and Load torque of PMSM, J is moment of inertia, D is coefficient of friction. The electromagnetic torque T_e is given by:

$$T_e = \frac{3}{2} \cdot n_p (\phi_\alpha i_\beta - \phi_\beta i_\alpha) = \frac{3}{2} \cdot \frac{n_p}{L_s} \phi_s \phi_r \sin \delta \tag{5}$$

2.2 The Direct Torque Control Based On Space Vector Modulation (SVM-DTC)

In the conventional DTC system, the flux amplitude $\phi_{s,k}$ and phase angle θ_s can be calculated in the α-β coordinate system through CLARK transformation and some math operation after stator voltage and current are sampled. After a control period, the flux amplitude becomes $\phi_{s,k+1}$ and the phase angle becomes $\theta_s{}'$, with the included angle between θ_s and $\theta_s{}'$ is $\Delta\theta_s$. As is shown in Figure 1, defining $\phi_{s,k+1} = \phi_s^*$, $\phi_{s,k} = \phi_s$, then

$$\begin{cases} \phi_{sa,k+1} = |\phi_s^*|\cos(\theta_s + \Delta\theta_s) \\ \phi_{s\beta,k+1} = |\phi_s^*|\sin(\theta_s + \Delta\theta_s) \end{cases}, \qquad \begin{cases} \phi_{sa,k} = |\phi_s|\cos(\theta_s) \\ \phi_{s\beta,k} = |\phi_s|\sin(\theta_s) \end{cases} \tag{6}$$

Define the flux difference between $\phi_{s,k+1}$ and $\phi_{s,k}$ as ϕ_{sref}, that is to say, $\phi_{sref} = \phi_{s,k+1} - \phi_{s,k}$, then

$$\begin{cases} \phi_{saref} = \phi_{sa,k+1} - \phi_{sa,k} = |\phi_s^*|\cos(\theta_s + \Delta\theta_s) - |\phi_s|\cos(\theta_s) \\ \phi_{s\beta ref} = \phi_{s\beta,k+1} - \phi_{s\beta,k} = |\phi_s^*|\sin(\theta_s + \Delta\theta_s) - |\phi_s|\sin(\theta_s) \end{cases} \tag{7}$$

ϕ_{sref} can be calculated in equation (8).

In order to make up for error vector ϕ_{sref}, an equivalent reference voltage vector u_{ref} is needed. Through the discretization of equation:

$$\phi_s = \int u_s(t) - i_s(t)R_s dt, \phi_{sref} = \phi_{s,k+1} - \phi_{s,k} = u_{s,k}T_s - R_s i_{s,k} T_s \tag{8}$$

Then, the reference voltage equation is as follows:

$$\begin{cases} u_{aref} = \left(|\phi_s^*|\cos(\theta_s + \Delta\theta_s) - |\phi_s|\cos(\theta_s)\right)/T_s + R_s i_{sa,k} \\ u_{\beta ref} = \left(|\phi_s^*|\sin(\theta_s + \Delta\theta_s) - |\phi_s|\sin(\theta_s)\right)/T_s + R_s i_{s\beta,k} \end{cases} \tag{9}$$

Figure 1 is the block diagram of the PMSM SVM-DTC control system. The torque output by the speed regulator is used as the given torque T_e^*, the difference from the observed torque T_e is adjusted by PI to obtain the phase compensation angle $\Delta\theta_s$. According to equation (7), the stator flux linkage reference vector ϕ_s^* and the stator flux linkage observation value ϕ_s can be obtained, and then according to the formula (9), The stator voltage components u_{aref} and $u_{\beta ref}$ can be obtained, and then the control signal of the inverter switching state can be obtained by SVM modulation technology, thereby continuously adjusting the voltage vector, thereby obtaining a better stator flux linkage circle.

Since the parameters of the motor are basically constant, when the stator flux linkage is kept constant, the torque is only related to the torque angle between the stator flux linkage and the rotor flux linkage. The relationship between torque variation and load angle can be derived from equation (5).

$$\Delta T_e = \frac{3}{2} \cdot \frac{n_p}{L_s} \phi_s \phi_r \cos\delta \cdot \Delta\delta = K_T \cos\delta \cdot \Delta\delta \tag{10}$$

Since the control period is small, the value of s does not change much, so the change in torque is consistent with the change in load angle. The schematic diagram of the variation of the stator flux vector and the rotor permanent magnet flux vector in a control cycle is shown in Figure 1, where the magnitude of the flux vector remains unchanged.

According to the above analysis, the required change of electromagnetic torque can be reflected as the change of the vector flux linkage angle of the stator, and on the other hand, the change of the stator flux linkage reflected by the change of the stator flux vector angle, finally, it can be realized by the control compensation of the voltage vector.

Based on this idea, the PI controller can be used to control the torque loop, and the increment of the stator flux vector angle can be obtained in real time according to the torque error, and then the stator voltage vector required for the motor operation can be obtained. Among them, the input and output relationship of the torque loop PI controller is

$$\Delta\theta_s = \left(k_p + \frac{k_i}{s}\right)(T_e^* - T_e) \tag{11}$$

Where: k_p is the proportional coefficient; k_i is the integral coefficient. Liking the traditional control method, this method is also implemented on the stationary coordinate system. The control structure block diagram is shown in Figure 2. From the control structure, it is mainly composed of four parts, which are reference flux vector calculation module, reference voltage vector calculation module, torque and flux linkage observation module, and SVPWM modulation module. Wherein, the torque is given as the output of the speed controller, and the difference between the observed feedback torque and the torque loop PI controller adjusts the output to obtain the increment of the stator flux vector angle, and then according to the stator flux vector Given the amplitude and the angle of the current stator flux vector, the amount of change in the stator flux vector is calculated, and the reference stator voltage vector is calculated from the voltage equation of the flux linkage. Finally, the drive of the

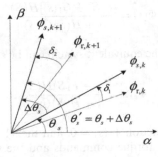

Figure 1.   Schematic diagram of flux vector variation circuit.

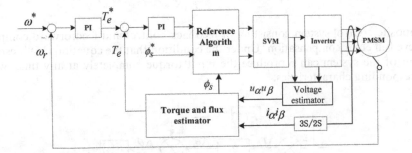

Figure 2.   SVM-DTC system structure for PMSM.

inverter is combined with the space voltage vector modulation output. The signal controls the motor.

2.3 *Scheme of torque compensation*

In order to reduce the torque ripple in the motor operation, the feedforward method is used to compensate the given torque in the literature. The block diagram of torque compensation is shown in Figure 3.

Where T_e^* is the motor given value, the torque compensation scheme is given in the literature, compares the given torque T_e^* with the actual torque T_e, and performs the PI operation on the deviation ΔT_e to obtain the compensation amount $\Delta\theta_s'$. The compensation amount is added to the feedforward compensation amount to obtain the corrected actual $\Delta\theta_s$, and the deviation of torque reference T_e^* and feedback values T_e is reduced by feedforward compensation. The compensation value does not directly use the torque deviation value, but introduces the PI controller to deal with the deviation signal, which makes the system adjustment process more moderate, eliminates the influence of single calculation error and measurement interference, and enhances the system robustness. $F(s)$ is the feedforward transfer function. $H(s)$ is the transfer function of PMSM. The paper assume that the transfer function of the PI controller is as follows:

$$G_1(s) = \frac{k_p s + k_i}{s} \tag{12}$$

Where, k_p is the proportional coefficient, k_i is the integral coefficient. The closed-loop transfer function of the block diagram of torque compensation is then:

$$\Phi(s) = \frac{[F(S) + G_1(s)]H(S)}{1 + H(S)G_1(s)} \tag{13}$$

The error transfer function of the equivalent system can be seen as follows:

$$\Phi_e(s) = \frac{1 - F(S)H(S)}{1 + H(S)G_1(s)} \tag{14}$$

In theory, the error transfer function $\Phi_e(s) = 0$, and then $1 - F(S)H(S) = 0$, the output of the system can entirely track the torque commands and the system has ideal response. The feedforward transfer function is then:

$$F(S) = \frac{1}{H(S)} \tag{15}$$

A composite control system consisting of feedback control and feedforward compensation will achieve full error compensation. Under the condition that the equation (15) is established, the output of the system can reproduce the input torque completely at any time, with ideal time corresponding characteristics.

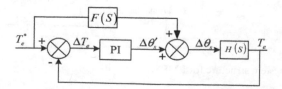

Figure 3.   Block diagram of torque compensation.

The existence of feedforward compensation F(s) is equivalent to adding an input signal $T_e^* \cdot F(S)$ to the system, and the error signal generated by feedforward compensation is compared with the error signal generated by the original input signal T_e^*, two error signal are equal in magnitude and opposite in direction. Equation (15) is called the full compensation condition for the error of the input signal. Since $H(S)$ generally has a more complicated form, the physical implementation of Equation (15) is difficult. In this paper, the partial compensation conditions that meet the tracking progress requirements are adopted to make the form of $F(S)$ simple and easy to implement. From the compensation principle, feedforward compensation does not change the characteristics of the feedback control system because the feedforward compensation actually uses the open loop control method to compensate the measurable disturbance signal. From the perspective of suppressing the disturbance, the feedforward control can reduce the burden of the feedback control, so the gain of the feedback control system can be made smaller to facilitate the stability of the system.

3 ANALYSIS OF SIMULATION AND EXPERIMENT

In order to verify the efficiency of the proposed controller, the closed-loop control system of PMSM is built on the platform of MATLAB/Simulink. Parameters of PMSM used in the simulation are given as follows:

Table 1. Main parameters of PMSM.

Parameter	Value	Parameter	Value
Armature resistance	1.2Ω	Rotor flux linkage	0.175Wb
Armature inductance of d axis	8.5mH	Rotor inertia	$0.0008kg \cdot m^2$
Armature inductance of q axis	8.5mH	Number of poles	4

In the simulation, the given stator flux linkage amplitude is 0.175 Wb, the given speed is 1000 r/min, and the initial value of the given torque is 0 Nm. When t = 0.2 s, the given torque is abruptly changed to 5 Nm, and the inverter switching frequency is set to 10 kHz. The simulation and experimental results are shown in Figures 4 to 11, comparing Figure 4 and Figure 5, due to the SVM and torque feedforward compensation technology, the voltage vector that fully compensates the stator flux linkage error can be generated. The stator flux linkage and torque ripple of the motor are significantly smaller, and the flux trajectory is smoother. Figure 8 and Figure 9 are the

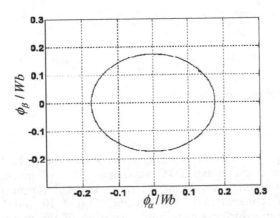

Figure 4.   Stator flux linkage with compensation.

126

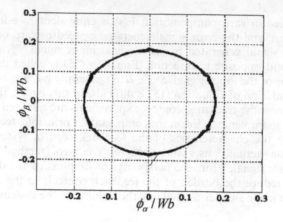

Figure 5.   Stator flux linkage without compensation.

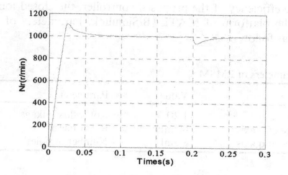

Figure 6.   Speed response with compensation.

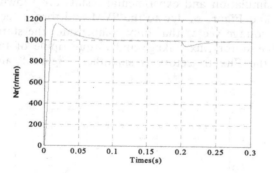

Figure 7.   Speed response without compensation.

electromagnetic torque waveforms at steady state. When the given torque is 5 Nm, the torque ripple of the conventional DTC scheme is 10% of given value, The torque ripple of modified solution is 4% of the set value, and the torque ripple is reduced by 6% compared to the conventional DTC scheme. Figure 10 and Figure 11 show the stator current waveform during steady-state operation of the motor. When t = 0.02 s,

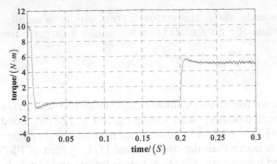

Figure 8.   Torque response with compensation.

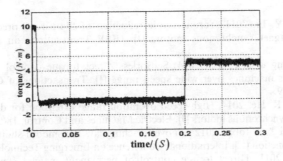

Figure 9.   Torque response without compensation.

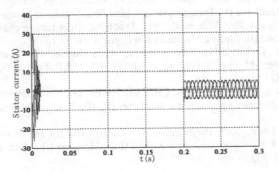

Figure 10.   Stator current with compensation.

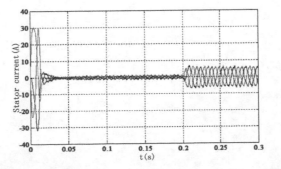

Figure 11.   Stator current without compensation.

128

the motor enters a steady state, the stator current ripple is small, and the waveform is relatively smooth, while the current DTC scheme has a larger current ripple.

4 CONCLUSION

In order to reduce the torque and flux linkage ripple efficiently and make the inverter switching frequency constant simultaneously, direct torque control of PMSM based on torque compensation is studied. By using SVPWM and torque compensation, the dynamic and static performance of the control system is better than the conventional DTC system of PMSM. Simulation results show that the proposed scheme can effectively reduce torque and flux linkage fluctuations.

REFERENCES

H. Zhu & X.Xiao & Y. Li. 2011, Torque ripple reduction of the torque predictive control scheme for perma nent magnet synchronous motors [J]. IEEE Transactions on Industrial Electronics, 59(2): 871–877.

X. Qiu & W X. Huang & F F. Bu. 2014, Sensorless direct torque control of interior permanent magnet synchronous machines over wide speed range [J]. Transactions of China Electrotechnical Society, 29(9): 92–99.

B. Lin & D. Sun & Y K. He. 2014, High-speed operation range extension for direct torque controlled permanent magnet synchronous motors [J]. Electric Machines and Control. 18(9): 9–16.

M. Asad & A I. Bhatti & S. Iqbal. 2012 A novel reaching law for smooth sliding mode control using inverse hyperbolic function [C]. International Conference on Emerging Technologies. IEEE.

Y. Zhang & J. Zhu. 2011, Direct torque control of permanent magnet synchronous motor with reduced torque ripple and commutation frequency [J]. IEEE Transactions on Power Electronics, 26(1): 235–248.

Z. Li & Y. Zhang & Li Z., et al. 2015, A Direct Torque Control Strategy of Induction Motor Drives with Simple Duty Ratio Regulator [J]. Transactions of China Electrotechnical Society, 30(1): 72–80.

H Q. Zhang & X S. Wang & P F. Wei, et al. 2012, Study on direct torque control algorithm based on space vector modulation [J]. Electric Machines and Control, 16(6): 13–18.

H. Zhu & X. Xiao & Y. Li. 2011, Torque Ripple Reduction of the Torque Predictive Control Scheme for Per manent-Magnet Synchronous Motors [J]. IEEE Transactions on Industrial Electronics, 59(2): 871–877.

F. Niu & B. Wang & A S. Babel, et al. 2016, Comparative Evaluation of Direct Torque Control Strategies for Permanent Magnet Synchronous Machines [J]. IEEE Transactions on Power Electronics, 31(2): 1408–1424.

E. Monmasson & J P. Louis. 2003, Presentation of a control law for IM drive based on the dynamic reconfiguration of a DTC algorithm and a SVM-DTC algorithm [J]. Mathematics & Computers in Simulation, 63(3): 321–333.

Emerging Developments in the Power and Energy Industry – Dufo-López, Krzywanski & Singh (eds)
© 2020 Taylor & Francis Group, London, ISBN 978-0-367-27169-5

A novel railway power conditioner based on modular multilevel converter with integrated super capacitor energy storage system

Qian Ma & Min Zhang
The College of Information Engineering, Xiangtan University, Xiangtan, China

ABSTRACT: To comprehensively manage power quality problems in traction power supply system, railway static power conditioner based on modular multilevel converter (MMC-RPC) is applied. Although the new MMC-RPC is of various advantages, the problem of a large voltage fluctuation still exists in MMC, especially during low-frequency running of ac locomotives. To solve this problem and to absorb regenerative braking energy during regenerative braking, a novel railway static power conditioner based on modular multilevel converter with integrated super capacitor energy storage system is proposed in this paper. Through analyzing energy flow relationship and the generation principle of voltage fluctuation components in the system, a combined control strategy capable of controlling charging and discharging of SC and simultaneously suppressing the voltage fluctuation is presented. Then simulation results show that the SC-MMC-RPC, under the control of ESS, can not only save energy, but also suppress voltage fluctuation. This reduces the use of large-capacity capacitors and then improves economy comprehensively.

Keywords: Railway static power conditioner, modular multilevel converter, voltage fluctuation, regenerative braking energy, super capacitor, energy storage system, combined control strategy

1 INTRODUCTION

At present, China has the fastest and longest high-speed railway in the world, and its constructers focus on improving load capacity and operational speed (Xia Yankun et al, 2014; Li Qunzhan et al, 2006). MMC-RPC is proposed to solve the negative sequence current problem of railway traction system (Zhao Yan et al, 2014; Jing Long et al, 2015; Song Pinggang et al, 2015).

In the braking process, a great regenerative braking energy was generated by high-speed ac locomotives (Luo A & Wu C, 2011). When two power supply arms are operating in the regenerative braking mode and in the traction mode, separately, the MMC-RPC transfers regenerative braking energy from the braking side to the traction side. Through this, the utilization of regenerative braking energy is realized. When both power supply arms are operating in regenerative braking mode, the MMC-RPC can only send regenerative braking energy back to the grid. In this situation, the utilization rate of regenerative braking energy is zero and regenerative braking energy has a huge impact on the grid. Therefore, the problem about regenerative braking energy can't be completely solved through the MMC-RPC.

Furthermore, an inner structure problem of MMC limiting MMC's application exists (Li Z et al, 2013; Zhou Yuebin et al, 2012). The problem is that there will always be voltage fluctuations in the sub-module (SM) capacitor. This indicates that the capacity of SM capacitor must be large enough to ensure that voltage fluctuations of SM capacitor remains within the limit range. Otherwise, system failure may be caused. In the practical application, the excessive use of large-capacity capacitors not only occupies a large space of the system, but also reduces system reliability and increases the overall cost greatly.

In response to the deficiencies above, a novel railway static power conditioner based on modular multilevel converter with integrated super capacitor energy storage system is proposed in this paper (SC-MMC-RPC). The MMC-RPC in this topology is used to control negative sequence current, while the SC is used to store and release regenerative braking energy (M. Chymera et al, May 2011; F. Ciccarelli et al, 2011), and eliminate voltage fluctuations in MMC. In this paper, through analyzing the working principle of regenerative braking and the voltage fluctuations in MMC, to save energy and eliminate voltage fluctuations, the corresponding ESS control strategy is proposed.

2 TOPOLOGY OF RPC BASED ON MMC WITH INTEGRATED SUPER CAPACITOR ENERGY STORAGE SYSTEM

A new MMC-RPC compensation system based on super capacitor energy storage system (SC-MMC-RPC), as shown in Figure 1.a, is proposed in this paper. The RPC is used to govern negative sequence current and simultaneously compensate reactive power dynamically. Moreover, ESS energy storage system is used to store and release regenerative braking energy, with each phase of MMC consisting of 2N same sub-modules. The SC-MMC-RPC system, compared with the MMC-RPC system, has a parallel super capacitor in each sub-module for energy storage, as shown in Figure 1.b. In the sub-module structure, the switch S1 and S2 on the left control the charging and discharging of the sub-module capacitor (C_c), with the switch S3 and S4 on the right controlling the charging and discharging of super capacitor.

3 SC-MMC-RPC OPERATING PRINCIPLE ANALYSIS

3.1 *Regenerative braking energy control mode*

As ac/dc/ac electric locomotives are generally adopted in high-speed railway, with their power factor being close to 1. In order to facilitate analysis, the traction load is regarded as a resistive load. It is assumed that the load power of power supply arms α and β is $P_{\alpha L}$ and $P_{\beta L}$, respectively, while the port power of both traction transformers is $P_{\alpha s}$ and $P_{\beta s}$, respectively. The compensation power of two RPC ports are $P_{\alpha c}$ and $P_{\beta c}$. The real-time power of SC energy storage system port is P_{sc}, and the maximum power is P_{scm}.

When operating in the mode, regenerative braking power of α arm $P_{\alpha L}$ is greater than the load power of β arm $P_{\beta L}$ and $P_{\alpha L} \leq P_{\beta L} + P_{scm}$, as shown in Figure 2.a. If there is no SC energy storage system, α arm will transfer part of regenerative braking energy ($P_{ac} = P_{\beta L}$) to β arm through the RPC, and the remaining regenerative braking energy will be sent back to the

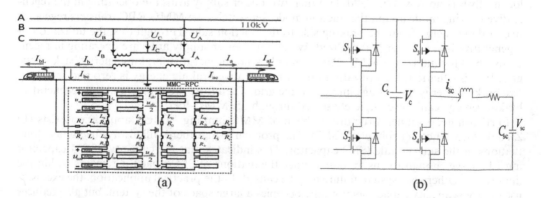

Figure 1. a) Topology of MMC-RPC with integrated super capacitor ESS b) SM with integrated super capacitor.

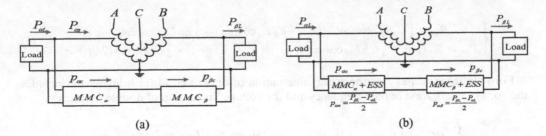

Figure 2.   Schematic diagram of energy flow during regenerative braking. a) MMC-RPC b) MMC-RPC with Integrated Super Capacitor ESS

grid, causing a big impact on the grid. In this situation, the regenerative braking energy that can be utilized is $P_{\beta L}$. Then, Figure 2.b shows that after the addition of SC energy storage system, α arm would transfer part of regenerative braking energy ($P_{ac}=P_{\beta L}$) to β arm through RPC. Also, the remaining regenerative braking energy ($P_{\alpha L}-P_{\beta L}$) would be stored by super capacitor, along with the regenerative braking energy can be used to increase $P_{\alpha L}$ in this situation.

3.2 Voltage fluctuation analysis of MMC

The average value of the capacitor voltage can be controlled in a balance condition through voltage regulation method, with the voltage value fluctuating around the reference value. Combined with the characteristics of compensation current of MMC-RPC, taking A-phase as example, it is assumed that the ac side output voltage and current of SPH-MMC are:

$$\begin{cases} e = e_a - e_b = U_{Am} \cos(\omega t) \\ i_c = I_{cm} \cos(\omega t + \varphi) \end{cases} \tag{1}$$

The ignorance of loss of equivalent resistance-inductive loads on the bridge arms has simplified the analysis. Meanwhile, considering simplicity, the voltage and current of the upper and lower bridge arms of MMC can uniformly be described as:

$$\begin{cases} u_{p,n} = \frac{1}{2} U_{dc} \pm e = \frac{1}{2} U_{dc} \pm E_m \cos(\omega t) \\ i_{p,n} = \frac{1}{2} I_{dc} \pm \frac{i_c}{2} = \frac{1}{2} I_{dc} \pm \frac{1}{2} I_{cm} \cos(\omega t + \varphi) \end{cases} \tag{2}$$

The instantaneous power of upper and lower bridge arms can be obtained from formula (2):

$$\begin{cases} P_{ap} = u_{ap} i_{ap} = \frac{1}{4} U_{dc} I_{dc} - \frac{1}{2} I_{dc} e + \frac{1}{4} U_{dc} i_c - \frac{1}{2} i_c e \\ P_{an} = u_{an} i_{an} = \frac{1}{4} U_{dc} I_{dc} + \frac{1}{2} I_{dc} e - \frac{1}{4} U_{dc} i_c - \frac{1}{2} i_c e \end{cases} \tag{3}$$

According to the law of energy conservation, if the converters have no power loss, the power balance will be achieved on the ac side and dc side of the MMC:

$$U_{dc} I_{dc} = \frac{1}{2} E_m I_{cm} \cos(\varphi) \tag{4}$$

With the substitution of formula (1) and (4) into formula (3), the instantaneous power of the upper and lower bridge arms can be rewritten into formula (5), and formula (6) is the corresponding energy flow.

$$\begin{cases} P_{ap} = \frac{1}{4} U_{dc} I_{cm} \cos(\omega t + \varphi) - \frac{1}{2} E_m I_{dc} \cos(\omega t) - \frac{1}{4} E_m I_{cm} \cos(2\omega t + \varphi) \\ P_{an} = u_{an} i_{an} = -\frac{1}{4} U_{dc} I_{cm} \cos(\omega t + \varphi) + \frac{1}{2} E_m I_{dc} \cos(\omega t) - \frac{1}{4} E_m I_{cm} \cos(2\omega t + \varphi) \end{cases} \tag{5}$$

Formula (6) indicates that the energy fluctuation of upper and lower bridge arms includes the voltage fluctuation of base frequency and the voltage fluctuation of double frequency.

$$\begin{cases} W_{ap} = \int_0^t P_{ap} dt = \frac{U_{dc} I_{cm} \sin(\omega t + \varphi)}{4\omega} - \frac{E_m I_{dc} \cos(\omega t)}{2\omega} - \frac{E_m I_{cm} \cos(2\omega t + \varphi)}{8\omega} + W_0 \\ W_{an} = \int_0^t P_{an} dt = -\frac{U_{dc} I_{cm} \sin(\omega t + \varphi)}{4\omega} + \frac{E_m I_{dc} \cos(\omega t)}{2\omega} - \frac{E_m I_{cm} \cos(2\omega t + \varphi)}{8\omega} + W_0 \end{cases} \tag{6}$$

4 COMBINED CONTROL STRATEGY RESEARCH OF SC-MMC-RPC

The combined control strategy proposed in this paper consists of two parts, namely charging and discharging control of SC as well as voltage fluctuation control. However, a lot of methods are provided by a lot of literatures to regulate active power and compensate reactive power of RPC. This makes the related parts not be described in this paper.

4.1 Control strategy of SC

SC controls the power flow between the dc side of the sub-module capacitor and the super capacitor. Under regenerative braking, SC absorbs regenerative braking energy and releases it under the heavy load operating condition. It is assumed that the total power required to be supplied to the traction load from the super capacitor is P_{SC}. Each super capacitor, considering the balanced operation of the system, needs to provide the same power. So, there are:

$$\begin{cases} p_{SC} = 8n \cdot \bar{v}_{SC} \cdot i_{SC,k,j} \\ i_{SC,k,j} = p_{SC}/(8n \cdot \bar{v}_{SC}) \end{cases} \tag{7}$$

$\bar{v}_{SC}$ is average voltage of super capacitor, $i_{sc,\,k,\,j}$ and is the feedforward reference current of the j-th phase k-th energy storage system of bidirectional DC/DC converter.

4.2 Voltage fluctuation elimination control strategy

To restrain the voltage fluctuations of the SM capacitors, the super capacitor energy storage system is used to absorb the fluctuation power. Based on an operating principle mentioned above, voltage fluctuation of MMC has been analyzed. Containing the voltage fluctuation components of base frequency and double frequency, the capacitor voltage of the SM will affect the voltage fluctuation of the capacitor. Super capacitors are used to absorb these two parts of fluctuation components.

$$\begin{cases} P_{ff} = \frac{P_{ap}}{n} = \frac{u_{ap} i_{ap}}{n} \\ i_{SC,fluc} = \frac{P_{ff}}{V_{sc}} \end{cases} \tag{8}$$

Using the direct current control method to control the bi-directional DC/DC converter, and the reference current is the sum of $i_{sc,k,j}$ and $i_{sc,\,fluc}$, that is $i^*{}_{sc} = i_{sc,k,j} + i_{sc,fluc}$.

5   SIMULATION RESULTS

To verify the proposed SC-MMC-RPC system, an MMC-RPC with integrated super capacitor ESS is built in Matlab/Simulink.

5.1   *Simulation of regenerative braking energy control mode*

Pretend that P_{aL} = -7MW, $P_{\beta L}$ = 2MW, P_{aL} > $P_{\beta L}$ + P_{scm}, as shown in Figure 4.a. P_{aL} = -7MVA illustrates that the α arm is operating in regenerative braking state. $P_{\beta L}$=2MW means that the β arm is operating in traction state. As shown in Figure 4.a, SC-RPC is put into operation after 0.36s. Moreover, part of regenerative braking energy P_{aL} is transmitted to support the load of β arm (that is, $P_{\beta L}$), while the other part of regenerative braking energy is stored by super capacitor as P_{scm}. At this moment, this makes P_{as} = -1MW send back to the grid. $P_{\beta s}$ = 0 illustrates that β arm needs not supply the load of β arm, and all the required energy is supplied by α arm with its regenerative braking energy. According to the calculation, the charging power of super capacitor is P_{sc} = P_{scm} = 4MW, with Figure 4.b showing that the super capacitor is operating in the charging state.

5.2   *Simulation of voltage fluctuation suppression*

Based on the voltage fluctuations of SM of one power supply arm depicted in Figure 5.a, the SM is operating without the voltage fluctuation elimination control strategy. Large low frequency fluctuations, due to the low-frequency operation of MMC, appeared. However, the average value of capacitor voltage stayed constant due to the voltage regulation method. As shown in Figure 3, the fluctuations are fluctuating at a range of 190V-210V, occupying 10% of the reference voltage value of capacitor .

Figure 5.b illustrates the MMC ESS system operation performance when applying the voltage fluctuations elimination control. It is clearly shown in Figure 5.b that the large low-frequency fluctuations actively compensate with the help of the proposed voltage fluctuation elimination control. The super capacitors absorb the fluctuation power in real time, thus leaving a very small voltage fluctuation in submodules capacitor voltage.

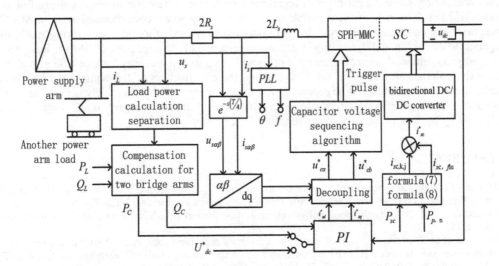

Figure 3.   Block diagram of the controller system of SC-MMC-RPC.

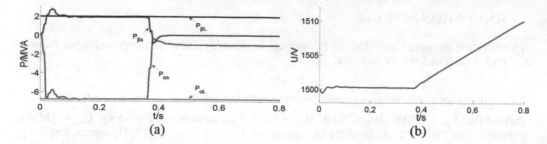

Figure 4. a) The low voltage side power of traction transformer b) The waveform of super capacitor voltage.

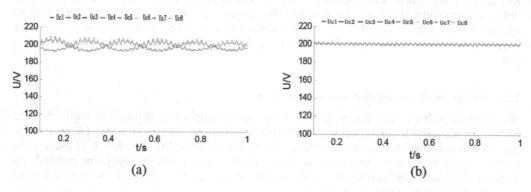
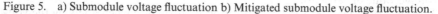

Figure 5. a) Submodule voltage fluctuation b) Mitigated submodule voltage fluctuation.

6 CONCLUSION

MMC-RPC, though capable of solving the negative sequence current problem effectively, can't completely solve the problem of regenerative braking energy and internal voltage fluctuations of MMC. In view of these problems, a new railway power conditioner based on MMC integrated super capacitor energy storage system is proposed. Firstly, the working mode of regenerative braking and the voltage fluctuation principle of the capacitor of SM are analyzed. Secondly, a combined control strategy is proposed to control the energy flow of ESS and eliminate the voltage fluctuation of the SM. Finally, the effectiveness of the control strategy was verified in matlab/simulink. The regenerative braking energy can be effectively utilized and the usage of large-capacity capacitors can be greatly reduced through controlling ESS effectively, followed by the comprehensive improvement of the economy .

REFERENCES

Xia Yankun et al. 2014, Negative sequence compensation of high-speed and heavy-haul electric railroad with V connection transformer [J]. Electric Power Automation Equipment, 34(2): 73–78.
Li Qunzhan et al. 2006, High-speed railway electrification project [M]. Southwest Jiao Tong university press: 253.
Zhao Yan et al. 2014, Application of three phase modular multilevel converter (MMC) in cophase traction power supply system [C]//IEEE Conference and Expo, Transportation Electrification Asia-Pacific (ITEC Asia-Pacific), Beijing, China, DOI: 10.1109/ITEC-AP. 6941064.
Jing Long et al. 2015, A power quality compensating method for traction power supply system based on MMC[J]. Automation of Electric Power Systems, 39(11): 173–179.
Song Pinggang et al. 2015, Direct power control strategy of railway static power conditioner based on modular multilevel converter [J].Power System Technology, 39(9): 2511–2518.

Luo A&Wu C, 2011,Railway static power conditioners for high-speed train traction power supply systems using three-phase VV transformers [J]. IEEE Transactions on Power Electronics, 26(10): 2844–2856.

Li Z et al. 2013, An inner current suppressing method for modular multilevel converter [J] .IEEE Transactions on Power Electronic, 28(11), 4873–4879.

Zhou Yuebin et al. 2012, Analysis of sub-module capacitor voltage ripples and circulating currents in modular multilevel converters [J].Proceedinds of the CSEE, 32(24): 8–14.

M. Chymera et al. May 2011, "Simplified Power Converter for Integrated Traction Energy Storage," IEEE Trans. Vehicular Technology, vol. 60, no. 4, pp. 1374–1383.

F. Ciccarelli et al.2011, Supercapacitors-based energy storage for urban mass transit systems. Power Electronics and Applications, Proceedings of the 2011–14th, pp. 1–10.

Emerging Developments in the Power and Energy Industry – Dufo-López, Krzywanski & Singh (eds)
© 2020 Taylor & Francis Group, London, ISBN 978-0-367-27169-5

Investment and financing problems of the energy internet and suggestions

Xue-feng Zheng
School of Government, Peking University, Beijing, China

Hong-yu Jia*
State Grid Information and Communication Industry Group, Beijing, China

ABSTRACT: At present, energy Internet construction has entered a new stage of practical operation. The energy Internet has broadened the field of investment and financing and extended the industrial chain of investment and financing. The government's support policies for energy Internet investment and financing have continuously improved, and green finance has gradually formed scale. At the same time, domestic and international energy Internet investment and financing has the problems of large investment demand, less government investment, and weak investment incentive mechanisms; the form of financing is mainly green credit; the funds are mainly from commercial banks; there is a single financing channel; and the financing cost is high. The cooperative financing mechanism is not perfect and there is no practical profit model. In the future, we will continue to promote energy Internet investment and financing. We need to actively build a green financial system, explore multilevel financing, expand investment and financing channels, innovate business models, and actively build a "destiny community" to achieve intergovernmental collaboration.

Keywords: energy Internet, financing model, investment and financing, investment risk

1 INTRODUCTION

In 2004, the term "energy Internet" first appeared in *The Economist*. In 2011, Rimi Rifkin elaborated on the concept of "energy Internet" in "The Third Industrial Revolution" (Rifkin 2011). In September 2015, at the UN Development Summit, President Xi Jinping proposed that China build a "global energy Internet" to promote global electricity demand in a clean and green way. The Paris Agreement reached at the end of 2015 at the United Nations Climate Change Conference brought significant historical opportunities for the development of the global energy Internet. In 2017, the new UN secretary-general Guterres said that he will implement the global energy Internet in the UN 2030 Agenda for Sustainable Development, and guide member states to participate in and promote global networking. In 2017, 55 of China's first group of energy Internet demonstration projects have been started, marking the transition of the energy Internet into practical operation. The energy Internet investment space is huge, and the expected return is stable and considerable, but it still faces institutional obstacles and financing difficulties. It is necessary for countries to jointly explore investment and financing platforms and methods.

*Corresponding author. Email:1565648716@qq.com

2 ENERGY INTERNET INVESTMENT AND FINANCING FEATURES

2.1 *Energy Internet broadens the field of investment and financing*

During the "Thirteenth Five-Year Plan" period, grid investment is expected to reach 2.38 trillion yuan, which will drive GDP growth by about 0.6 percentage points per year. At the same time, it will drive the innovation and development of new emerging industries such as new energy, new materials, smart manufacturing, and electric vehicles, nurturing huge business opportunities. Some international organizations have proposed an energy vision. By 2050, fossil energy will account for only 12% to 20% of the entire energy structure, while renewable energy will account for 65% to 79% of the overall energy structure. According to forecasts to 2030, by that year, one quarter of the world's energy supply will consist of low-carbon energy.

According to the data released by the National Energy Administration on July 27, 2015, the proportion of clean energy in China increased further in the first half of the year. Non–fossil fuel energy power generation increased by 16% year-on-year, accounting for 22.9% of the national power generation, up 3 percentage points year-on-year. The National Energy Administration previously estimated that the pilot project of the energy Internet will drive more than 40 billion yuan of investment this year. The energy Internet market has great potential and will greatly expand the investment and financing sector and develop more new business opportunities. Under the current global background of actively coping with climate change and environmental pollution, clean development is the trend of the future, and it is also a key investment area of the world. According to the data, nearly half of the global social capital investment in infrastructure is currently invested in clean energy development projects. The energy Internet directs capital to the clean energy sector.

2.2 *Energy Internet extends the investment and financing industry chain*

As the global energy Internet covers the global energy infrastructure, it involves power, power grid, trading, equipment, scientific research, information, consumption. and other fields. It has large investment demand, a long industrial chain, and strong driving force. It has a strong driving effect on economic growth. At the same time, it will also vigorously promote the development of strategic emerging industries such as high-end equipment manufacturing, new energy, electric vehicles, energy conservation, and environmental protection; greatly extend the investment and financing chain; and have a significant role in promoting economic development, structural adjustment, and industrial upgrading.

The global energy Internet may spawn a trillion-dollar industry sector with numerous business opportunities. The first is the power equipment sector, including advanced, efficient, and safe clean energy power generation equipment; UHV power transmission and transformation equipment; energy storage equipment; and large power grid control and modern communication equipment. The second is the grid sector, including UHV AC synchronous grids, UHV DC transmission engineering, smart grids, and so on. The third is the power trading sector. Focusing on the global energy Internet, we will explore the establishment of a global power trading market and advocate the wider and more balanced and orderly flow of energy through the hierarchical cooperation of the existing power trading markets. The fourth is the intelligent power board to establish a new power supply and consumption relationship between power grid and customer energy flow, information flow and business flow, and realize emerging energy consumption such as distributed energy, electric vehicle charging and discharging, and the smart home.

2.3 *Government support policies continue to improve*

In 2015, the Chinese government promulgated the "Guiding Opinions of the State Council on Actively Promoting the 'Internet +' Action" (Guo Fa [2015] No. 40), which requires the use of Internet technology to promote energy production intelligence, build a distributed energy network, explore new modes of energy consumption, and develop grid-based communication facilities and new services. The opinions require protection and support in terms of network, completeness, standards, environment, credit, talents, finance and taxation, and financing.

In 2016, the Chinese government promulgated the "Guiding Opinions on Promoting the Development of 'Internet +' Smart Energy" (Development and Reform Energy [2016] No. 392) requires flexible energy production and consumption infrastructure, multienergy collaborative integrated energy network construction, flexible green energy, in-depth research and exploration of transaction models, energy big data services, key technologies of the energy Internet, and energy Internet standards systems. The document will include the energy Internet in major engineering packages, increase the capital investment in the central and local budgets, and guide more social capital to enter. In March 2017, the "Guidance Opinions on Supporting the Development of Green Bonds" issued by the China Securities Regulatory Commission and the "Guidelines for Green Debt Financing Tools" issued by the Association of Dealers provided guidelines for nonfinancial companies to issue green bonds and other green debt financing instruments. It has clarified the green channel for issuing green bonds and debt financing instruments.

2.4 *Green finance beginning to scale*

According to the data, some international green finances have begun to take shape. For example, the US green fund has more than doubled from 2007 to 2010, providing a large amount of financing for green industries. There are also the Standard & Poor's Global Clean Energy Index, the Nasdaq US Clean Index, and the FTSE Japan Green 35 Index.

The scale of green bond issuance also rose from almost blank in 2015 to 240 billion yuan in 2016, accounting for nearly 40% of the global total. In the first half of 2018, the number of green bonds issued was 48, and the issuance scale was 52.372 billion yuan. The number of issuances increased by 26% compared with the same period of last year. The domestic entity issued 4 green bonds overseas, with a scale of approximately RMB 13.730 billion. The size of green bonds issued in China and abroad accounted for about 13.50% of the global green bond issuance.[1] In addition, green industry development funds, green trusts, and green insurance have also achieved significant growth.

Developing a green index system is an important means to expand the influence of green financial products and improve market liquidity. Fang Xinghai, vice chairman of the China Securities Regulatory Commission, said at the Boao Forum for Asia in March 2017 that the China Securities Regulatory Commission is studying the possibility of encouraging the issuance of green financial product indices, such as the green stock index and the green bond index (Fang 2017). In terms of stock index, there are already 19 in China as of 2016 (Zhong 2017); tracking 18 products of the Green Index, the scale is close to 10.7 billion yuan (China Securities Index: Green Finance and Green Stock Index Research 2017). By the end of 2016, China had a total of 265 energy-saving and environmental protection and green investment funds, including 159 equity investment funds, 33 venture capital funds, 28 securities investment funds, and 45 other types of funds (An 2017).

3 ENERGY INTERNET INVESTMENT AND FINANCING DIFFICULTIES

The construction of a global energy Internet not only needs a platform for mutual cooperation, but also needs to build a multiparty-recognized co-construction mechanism to share the achievements brought about by the construction of the global energy Internet.

3.1 *The investment gap is large and the investment incentive mechanism is weak*

The Global Infrastructure Hub (GIH) report "Global Infrastructure Investment Outlook 2040" states that global infrastructure (road, rail, airport, seaport, electricity, water,

---

1. According to CBI, the global green bond issuance in the first half of 2018 was US$73.7 billion.

telecommunications) investment demand will reach $94 trillion in 2016–2040. In order to meet this investment demand, there is a global need to increase the proportion of infrastructure investment in GDP to 3.5%, and the current investment trend is 3%. The infrastructure investment gap in the Americas and Africa is the largest.

The Financial Research Institute of the Development Research Center of the State Council issued a forecast that during the "Thirteenth Five-Year Plan" period, China's green investment demand will reach 20,000 to 4 trillion yuan per year. According to the prediction of Liu Zhenya, the chairman of the former State Grid Corporation of China, according to the "Ultra High Voltage + Smart Grid + Clean Energy," by the time the global energy Internet is initially built in 2050, the cumulative investment will exceed 50 trillion US dollars (Liu 2015). However, the current annual budget of the government can only be invested in about 30 billion yuan in green areas such as energy conservation, environmental protection, and clean energy. In all green investments, government funding is expected to account for only 10% to 15%, and the remaining 85% to 90% require social funding. Among them, the fiscal year can invest about 300 billion yuan in green industry, up to 15%; that is, more than 85% of green investment needs to come from social capital.

At present, the green investment return period is relatively long, with semipublic investment characteristics, and there are some market failures, which have led to a decline in the investment enthusiasm of social capital. How to stimulate the effective investment of social capital is a key issue in the construction of the energy Internet.

3.2 *Less financing mode, single channel, high cost*

Under the global energy Internet investment and financing structure, Chinese financial institutions are now more inclined to support sovereign guaranteed projects than project financing projects. Paul Frankel said that for a variety of reasons, most of the current corporate financing is still from commercial banks. Many financial institutions believe that renewable energy is risky, expensive, and technically unreliable, so they impose high interest rates. If it is operated in this way, it is impossible to achieve effective financing.

At present, green credit accounts for more than 95% of green finance, equity investment accounts for only 2.6%, and funds account for only 0.1%. In addition, financing difficulties, financing, slow financing, and fewer financing channels have become a constraint for most companies to "go global." Banks themselves also face risk factors such as high financing costs and frequent exchange rate fluctuations. In the face of the dilemma, it is urgent to consider financial innovation. By leveraging the advantages of the company's own resources, we will strengthen cooperation between the parties and explore an open and innovative investment and financing model.

3.3 *The international financing cooperation mechanism is not perfect and there is no practical profit model*

As a global energy Internet construction infrastructure, financing cooperation is based on bilateral communication between China and host countries and related countries. In addition to the relatively complete bilateral communication mechanism established by the China–Pakistan Economic Corridor, the other five economic corridors have not yet formed a normalized financing coordination mechanism, and the financing cooperation between China and other relevant capital surplus countries also needs to be deepened.

At present, the profitability of global infrastructure investment comes mainly from three aspects: government payment, consumer payment, and property appreciation brought about by facility development. However, at present, some countries, especially those along the "Belt and Road" or low-income countries, cannot afford to pay for the government or consumers; or infrastructure construction has long been replaced by government agencies, and the large debt scale has made the government unable to provide new services. Consumers do not recognize the concept of paying for the facilities; or countries where land is privately owned are unable to provide land for the construction operators to develop and obtain long-term benefits. The government has no ability to pay, consumers are unwilling to pay, and there is a lack of public resources such as land. Chinese investors can't find a viable profit model in infrastructure projects, and entry will be very difficult. In reality, the international debt ratio, especially the national infrastructure along the "Belt and Road," tends to be higher (Zheng & Liu 2018).

4 ENERGY INTERNET INVESTMENT AND FINANCING COUNTERMEASURES

4.1 *Building a green financial system*

Establishing a sound green financial system requires supporting policies such as finance and environmental protection, with related laws and regulations, and solving the externalities of the project environment by establishing appropriate incentive and restraint mechanisms. At the same time, financial institutions and financial markets are also required to intensify their efforts to solve problems such as maturity mismatches, information asymmetry, and lack of products and analytical tools faced by green investment and financing through the development of new financial instruments and service means.

Global energy Internet construction also needs to build a green channel focusing on investors' preferential green investment loan mechanisms, credit enhancement, approval, etc.; establish an environmental information disclosure mechanism; disclose corporate ecological environmental protection information; establish an ecological environmental information database; and provide more guidance. More funds are invested in enterprises with excellent environmental protection technology, high grades, and good credit. At the same time, it is necessary to establish a green financial risk prevention mechanism, introduce third-party environmental rating agencies, and develop a risk assessment mechanism. In addition, a green financing review and supervision mechanism needs to be built to supervise project filing, use of funds, and evaluation of effects. Green financial reform and innovation pilot zones need to be actively carried out to promote China's green transformation.

4.2 *Expand investment and financing channels and explore multilevel financing*

The global energy Internet is a giant capital-intensive system with huge capital needs. To this end, we must rely on the international capital market; give full play to the credit advantages of all parties; and comprehensively use various equity financing and debt financing, direct financing and indirect financing, and traditional financing and innovative financing tools to build an economic, reliable, diversified, and flexible financing security system to achieve diversification of equity financing entities, diversification of debt financing channels, and globalization of capital resource allocation. The areas that need to be broken into in green financial reform are green bonds, green funds, and green public–private partnership (PPP).

From the perspective of global infrastructure investment experience, PPP is also an ideal project investment model. First, actively build a comprehensive service platform and explore the establishment of different levels of energy interconnection PPP centers; second, cultivate investment entities with leading capabilities through various means to participate and share risks; third, according to the investment policy and power of the host country regulatory requirements, economic strength, technical level, project investment scale, etc., promote flexible use of BOO (build–own–operate), BOT (build–operate–transfer), BOOT (build–own– operate–transfer), BTO (build–transfer–operate), BLT (build–lease–transfer), POT, and other different investment methods.

In addition, consider exploring diversified equity and bond financing to explore the formation of a global energy Internet development fund.

4.3 *Actively innovating business models*

On the basis of providing financial services for UHV grids and distribution network construction, we will build intelligent facilities for energy production and consumption, integrated energy networks, and information and communication infrastructure and create an open and shared energy Internet ecological environment and develop energy storage. And the new models of electric vehicle application, the development of smart energy and new models of value-added services, fostering green energy and flexible trading market models, etc., actively strive for more and more comprehensive financial business qualifications in the fields of energy finance, third-party integrated energy services, etc. Achieve breakthroughs. Actively promote content and form innovation, industrial chain financial innovation, Internet financial innovation, and international business innovation and explore content innovation.

4.4 *Actively build a community of destiny and achieve synergy between governments*

Under the support of an open and innovative investment and financing model, countries will build a global energy Internet cooperation platform. As a major innovation in the energy field, the global energy Internet will greatly promote the development and consumption of renewable energy, and integrate "one pole together," large energy bases of various continents and various distributed power sources and fundamentally solve major problems affecting human survival and development, such as the energy crisis, environmental pollution, and rising sea levels.

Strengthen the policy coordination between Chinese and foreign governments in the field of infrastructure and public services, and consider investment and financing cooperation an important topic of bilateral cooperation mechanisms such as Sino–foreign strategic dialogue, and guide project implementation. Promote bilateral investment treaty negotiations with countries along the route, and integrate infrastructure investment into the protection of bilateral investment agreements. Strengthen the interaction between Chinese and foreign governments. At the same time, make full use of "two-track diplomacy" channels such as think tanks, chambers of commerce, industry associations, and nongovernmental organizations to participate in the pre-project design process and jointly promote project cooperation.

5 CONCLUSIONS

The global energy Internet is a global energy infrastructure. It has a large investment demand, a long industrial chain, and strong driving force. It has a strong driving force for economic growth and will continue to generate new technologies, new formats, and new industries. It is the future development of the real economy. The forces nouvelles will create new demands for the development of the financial industry, open up new areas, and provide new opportunities. At the same time, the development of the financial industry can provide long-term and stable financial guarantees for major energy Internet investment projects. It is necessary to actively explore the establishment of a new business models based on the global energy Internet, covering all aspects of energy production, transmission, and consumption, and promote the construction of a new investment and financing platform, relying on the international capital market to achieve diversification of financing entities, diversification of financing channels and global allocation of funds to provide competitive financial solutions.

REFERENCES

An, Guojun. 2017. China's green fund development: promising prospects. *The Banker*. Available at: http://funds.hexun.com/2017-08-14/190436840.html, August 14, 2017.

China Securities Index. Green Finance and Green Stock Index Research. Available at: http://heneng.juhangyc.com/201702/weixin_3685034.html.

Fang, Xinghai. 2017. Green bond applications require third-party certification. *China Securities Journal* Available at: http://greenfinance.xinhua08.com/a/20170410/1698288.shtml, April 20, 2017.

Liu, Zhenya. Building a global energy Internet to promote clean and green energy development. *People's Daily*, October 22, 2015.

Rifkin, J. 2011. *The Third Industrial Revolution: How Lateral Power Is Transforming Energy, the Economy, and the World.* New York: Palgrave Macmillan.

Zheng, Xuefeng & Liu, Lifeng. 2018. "One belt and one road" investment and financing problems and countermeasures. *China Economic and Trade Guide* 15: P34–336.

Zhong, Rongsa. 2017. Promoting the fund industry to practice responsible investment. *China Securities Journal.* Available at: http://greenfinance.xinhua08.com/a/20170410/1698454.shtml.

Emerging Developments in the Power and Energy Industry – Dufo-López, Krzywanski & Singh (eds)
© 2020 Taylor & Francis Group, London, ISBN 978-0-367-27169-5

Research on a flexible AC/DC microgrid based on a power-coordinated control strategy

Di Wu, Houtao Sun, Luyao Chen, Dan Hong, Zhuofei Yu & Jian Luo
NARI Group Corporation, Nanjing, China

ABSTRACT: A flexible multiterminal AC/DC microgrid is based on DC network and power optimization control technology. A system architecture that adopts a star structure in the local and remote annular structure to improve power supply reliability as well as the system operational mode that includes load transfer in normal situations and fault support was proposed. Then, a coordinate control strategy of power flow was introduced for optimize output power of the distribution transformer, and a corresponding $P–U$ droop control strategy was analyzed. Meanwhile, an optimization method of control parameters of a current inner-loop controller based on a high-order primary system model was proposed. Finally, an AC/DC microgrid case was modeled in PSCAD/EMTDC to optimize control parameters of a voltage source converter (VSC) and simulate the system transient operating conditions. Simulation results indicate that the system could realize AC and DC load transfer at the same time, and the proposed method is able to regulate power flow optimally.

Keywords: coordinating control, flexible AC/DC microgrid, high-order priority system model order, optimized droop

1 INTRODUCTION

The multiterminal AC/DC microgrid is a complex power system combining a hybrid power network and power flow optimization control strategy. It is also a typical application scenario of multiterminal flexible-HVDC (Tang 2013; Zhou, 2014). The flexible multiterminal AC/DC microgrid is based on an AC distribution network, using a voltage source converter to link up different AC distribution areas and achieve the goal of a "closed-loop design" and "closed-loop operation." It regulates and controls power flow according to the flow optimization target, which could solve a power supply shortage by tapping power supply potentials of each distribution transformer, instead of expanding the power supply capacity for existing distribution networks or providing other special power supply cables. The main purpose of the AC/DC microgrid is to improve the reliability of the distribution grid and increase the efficiency of the distribution transformer.

More and more researchers focus their studies on DC distribution, and the interest in AC/DC network is increasing as well. However, most research is directed to the power flow controller and powerflow calculation of an AC and DC hybrid network (Wang 2013; Xue 2013; Liu 2016; Peng 2016), few of them focus on self-adaption coordinated power flow control. Wu (2017) and Zhang (2015) presented a hierarchical control strategy of power flow for an AC and DC hybrid grid, and the $P–U$ droop control algorithm was analyzed as well; however, the power flow control method could not implement self-adaption regulation and control.

This article first presents a hybrid system architecture and operation mode that considers both flow transfer and AC load support. Second, a self-adaption control strategy for regulating the power flow at the DC side is proposed that could regulate the power flow at the DC side according to the available capacity of AC grid. Third, a bidirectional $P–U$ droop control algorithm and controller are put forward. In addition, a control parameter optimization method of inner-loop controller and outer-loop controller based on a high-order primary system model is presented. Finally, a study case is modeled in PSCAD/EMTDC to verify the proposed method.

2 SYSTEM ARCHITECTURE

The closed loop operation of the AC distribution network is usually applied to cope with electric vehicles (EVs) and other random heavy loads where the short circuit capacity increases and the stability margin decreases. By using a multiterminal flexible AC/DC microgrid, cable and transformer overload caused by a large-capacity load could be avoided. Furthermore, the reliability and security of a power supply could be improved. More importantly, the constraint of a heavy load accessing an existing AC distribution system could be relieved significantly, and power supply efficiency could be improved.

Based on the AC distribution network, AC/DC converters that supply DC power are installed at the AC feeder side to form an independent microgrid. Multiterminal AC/DC microgrids are connected through a DC network to form an AC/DC hybrid microgrid. A hybrid type system architecture is adopted for an AC/DC microgrid that could increase the reliability and flexibility of the power supply. For DC loads located in different geographical areas, an annular series loop is adopted remotely to reduce the transmission line length of the DC load supply when the distance among AC/DC microgrids is great, and also the design capacity of the DC line could be reduced because the power comes from two sides of the DC network. Also, the AC/DC microgrid could a adopt a local star topology when the distance between AC/DC microgrids is small to reduce the rated current of the DC line and increase the operating flexibility; it could also increase reliability when a single voltage source converter (VSC) is faulty. A typical AC/DC microgrid is shown in Figure 1 and the DC voltage is usually 600–900 V.

The main operation modes contain a power supply at a single end, load power transfer at multiple ends, urgent support, and fault isolation. If the grid working condition changes, the

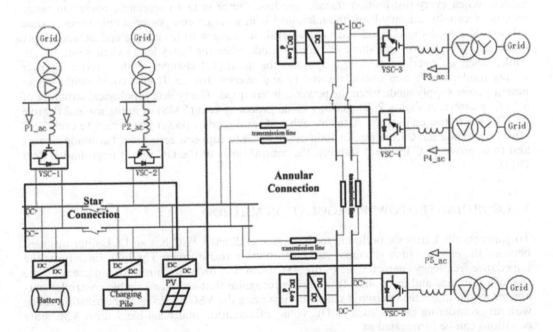

Figure 1.   System architecture.

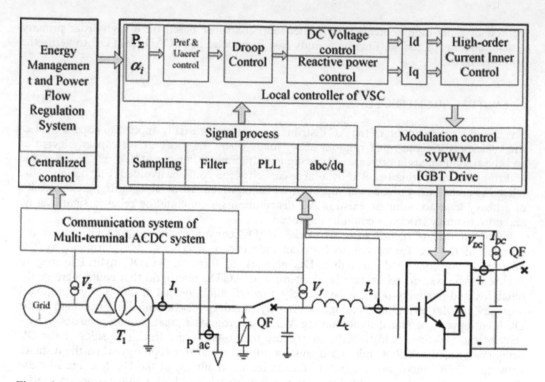

Figure 2.   Main circuit topology.

microgrid could switch modes automatically, fitting to the new condition, developing a switching logic corresponding to the relationship between working conditions and operation modes. When every distribution transformer load factor is in an economic operation range without access by a heavy load, the microgrid is in a single end power supply mode. While a heavy DC load or AC load accesses the grid, it changes to a multiple-end power supply mode, and then returns to a single-end supply mode when the heavy load switches out. Meanwhile, when an overload condition appears, the microgrid changes to the overload power supply mode, which uses overload capacity and energy storage. It also could switch to an urgent power supply mode when the power is interrupted. The typical topological structure of a VSC is shown in Figure 2. The energy management system (EMS) of multiterminal flexible AC/DC microgrid collects the status of each VSC and sends a power instruction to each VSC after flow allocation calculation. In addition, in order to suppress zero-order harmonic current and to improve VSC transient stability, the neutral point of the DC side is ungrounded (Wu 2017).

3   COORDINATED POWER REGULATION METHOD

To suppress the harmonic pollution to the power grid, each VSC has an LCL filter installed between the commutation reactance and commutation transformer. The LCL filter turns the impedance-frequency characteristic of the VSC from first order to third order, decreasing the risk of resonance and instability. It is vital to recognize that an unreasonable controller parameter would make the microgrid unstable. Assuming the VSC works in a linearization range without considering factors such as DC voltage fluctuation and dead-band time, VSC state equations can be formulated as

$$\begin{cases} L_1 \cdot \frac{dI_1}{dt} = U_f(t) - U_s(t) \\ C \cdot \frac{dU_f}{dt} = I_2(t) - I_1(t) \\ L_2 \cdot \frac{dI_2}{dt} = U_c(t) - U_f(t) \end{cases} \tag{1}$$

where U_f is the filter bus voltage, U_S is the system voltage, U_C is the converter bus voltage, I_1 is the system current, and I_2 is the converter valve current.

Both U_S and U_C have effects on the converter valve current I_2; therefore, grid side current I_1 and filter bus voltage U_f need to be eliminated to obtain an equation for I_2. According to Equation (1) and the linear superposition theorem, equations can be presented as:

$$\frac{U_s(s)}{I_2(s)} = -\left[L_1 L_c C s^3 + (L_1 + L_c)s \right]$$
$$\frac{U_c(s)}{I_2(s)} = \frac{L_1 L_c C s^3 + (L_1 + L_c)s}{L_1 C s^2 + 1} \tag{2}$$

The AC/DC microgrid could transfer redundant capacity of each distribution transformer to the heavy load and optimize power flow. The active power real-time value of each distribution transformer is sent to the EMS, then the sum of load P_{all} can be obtained. The regulation coefficient of each transformer is

$$\alpha_i = \frac{S_i - P_{i\_ac}}{S_\Sigma - \sum P_{i\_ac}} \tag{3}$$

The flow control strategy is as follows. While DC loads such as charging piles are located at different places, the real active power of the DC load cannot be measured directly. To regulate the active power of different microgrids, the controller adopts the active power of the VSC as the load power because the power is balanced when the DC voltage is stable. And based on the communication system, the sum of the real power of the DC load can be obtained as P_{all}. P_{all} multiplied by the regulation coefficient α_i is the power flow target value P_{ref} of each distribution converter, and $S_{T\text{-}i}$ is the total capacity of transformer i. The coordinated droop control method has no need of real-time monitoring of the DC load all over the DC network; only the DC power of each converter needs to be measured. When a heavy DC load accesses the system, the power vacancy could be regulated from other converters where the converter was in a light load condition. Likewise, when a heavy AC load occurs, the active power vacancy would be regulated from other transformers through a DC network.

4 VOLTAGE-POWER DROOP CONTROL STRATEGY

4.1 P–U outer loop controller

Each converter uses active power (P) to DC voltage (U_{DC}) droop control and reactive power (Q) to AC voltage (U_{AC}) droop control. A multiterminal AC/DC microgrid might be a reversal of the active power when in an AC and DC shift supply. The P–U_{DC} droop control method proposed in this article aims at the bidirectional active flow problem. In this droop control method, the independent variable is U_{DC} and the dependent variable is P. When the load changes, U_{DC} is the controlled variable that regulates U_{DC} to achieve a balance between U_{DC}, P, and load. In the AC shift supply mode the power flows from the DC side to the AC side in the first quadrant as shown in Figure 4, and the converter works as an inverter. When the power flowing to the AC side increases, the control system would boost the DC voltage according to the droop characteristic that achieves a new P–U_{DC} balance and slow down active power for the load on the AC side. In the DC shift supply mode the power flows from the AC side to the

DC side in the fourth quadrant, and the converter works as a rectifier. When the power flowing to the DC side increases, the control system would reduce the DC voltage according to the droop characteristic that achieves a new P–U_{DC} balance and promote active power for the load on the DC side.

If the reactive load increases the AC voltage might decrease when a multiterminal AC/DC microgrid is connecting with a weak grid. The U_{AC}–Q droop control strategy proposed in this article aims at reducing the voltage security risk. When the AC voltage decreases, the control system would increase the reactive power flowing to the AC side to boost the AC voltage to achieve a new Q–U_{AC} balance.

To realize the droop control strategy depicted in Figure 3 and Figure 4, this article adds a P–U_{DC} deviation proportion droop controller to the DC voltage outer-loop controller. The Q–U_{AC} control is similar to the P–U_{DC} droop controller. The proposed architecture of the P–U_{DC} and Q–U_{AC} droop controller is shown in Figure 5. For engineering applications, selection of the droop coefficient should take into account the fast step response and the oscillation stability. At the same time, to avoid the differential state caused by a proportional droop control, the integral element is needed to remove static errors.

The step response and the oscillation stability need to be considered when selecting an appropriate droop coefficient. As the droop control is a single proportional element, its target feedback control quantity has a static error. Taking the P–U_{DC} droop as an example, if the converter works in the first quadrant, as depicted in Figure 4, the U_{DC} decrease corresponds to the P increase. At this point, there is a certain deviation between U_{DC} and U_{ref} as depicted in the dashed part of Figure 4, which means there must be a deviation of the reference voltage after the deviation between P and P_{ref} is multiplied by the droop coefficient. That is to say, there is static error between the instantaneous active power and desired power.

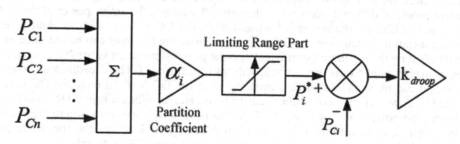

Figure 3.   Droop control strategy.

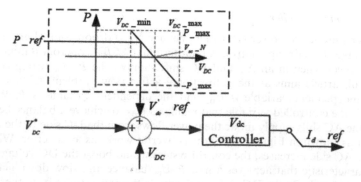

Figure 4. P–U_{DC} droop control.

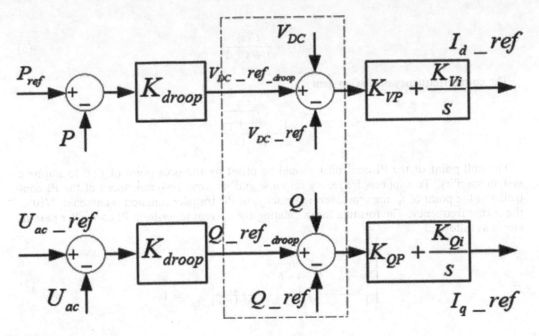

Figure 5. $P–U_{DC}$ and $Q–U_{AC}$ droop.

4.2 High-order primary system model for VSC

By using the PI controller to realize current inner-loop control, take I_2 as the feedback control component. The inner current control system is shown in Figure 6.

According to Equations (1) and (2) and Figure 6, the LCL high-order primary model based open-loop transfer function of the inner current loop controller is presented as

$$C_o(s) = \frac{K_{IP}s + K_{Ii}}{s} \cdot \frac{K_{PWM}\left(L_2 C_f s^2 + 1\right)}{\frac{L_1 L_2 C_f s^4}{f_{PWM}} + L_1 L_2 C_f s^3 + \frac{(L_1+L_2)s^2}{f_{PWM}} + (L_1 + L_2)s} \tag{4}$$

The closed-loop transfer function of the current controller is

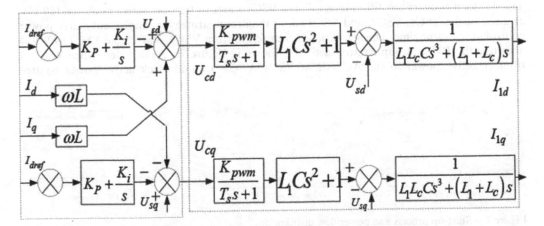

Figure 6. dq inner current control system.

$$C_i(s) = \frac{C(s)}{1 + C(s)} \tag{5}$$

The corner frequency of VSC system is

$$f_n = \frac{1}{2\pi} \cdot \sqrt{\frac{\frac{1}{L_1} + \frac{1}{L_2}}{C}} \tag{6}$$

The null point of the PI controller should be offset by the pole point of $C(s)$ to improve system stability. To suppress frequency response and improve response speed of the PI controller at the point of f_n, the crossover frequency f_c of the transfer function is chosen as 1/10 of the corner frequency. The formula for calculating the current inner-loop PI controller parameter is as follows:

$$\begin{cases} \frac{1}{2\pi} \cdot \frac{K_{Ii}}{K_{Ip}} = \frac{1}{2\pi} \cdot \sqrt{\frac{\frac{1}{L_1} + \frac{1}{L_2}}{C}} \\ \left| \frac{K_{Ii}s + K_{Ip}}{s} \cdot \frac{L_2 C s^2 + 1}{L_1 L_2 C s^3 + (L_1 + L_1)s} \cdot \frac{K_{PWM}}{1 + T_s s} \right|_{s = j \cdot f_{1c}} = 1 \end{cases} \tag{7}$$

5 CASE STUDY

5.1 *Simulation parameter*

This article takes a four-terminal multiterminal flexible AC/DC microgrid as the study case modeled in PSCAD/EMTDC, shown in Figure 8. The rated capacities of these four VSC converter stations are all 400 kVA, U_s = 400V and U_{DC} = 800V, and the pulse width modulation (PWM) frequency is 1.35 kHz. The commutation reactance is 0.01Ω and 0.5 mH, the DC-side capacitor is 1500 μF, and the DC line reactance is 0.02Ω. By using the high-order primary model parameter optimization method, the droop coefficient is chosen as 9e-4, the voltage outer loop PI controller parameter is chosen as 3/30, and the current inner-loop PI controller parameter is chosen as 0.3/5. Four distribution transformer capacities are 800 kVA, 630 kVA, 630 kVA, 800 kVA, and the power flow regulation coefficient is 1.24, 1, 1, and 1.24.

5.2 *AC load transfer*

The AC load transfer condition is as follows: VSC4 starts to transfer active power of 150 kW at 2 s, and the VSC1, VSC2, and VSC3 adopt the droop strategy to implement active power control at the same time. The active power transferred from VSC1, VSC2, and VSC3 to VSC4 adopting the P–U_{DC} droop control method is shown in Figure 7a. As depicted, when VSC1–3 transfer active power to the AC side of VSC4, VSC1–3 could track active power control

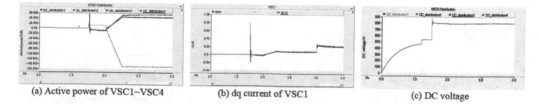

(a) Active power of VSC1~VSC4 (b) dq current of VSC1 (c) DC voltage

Figure 7. Start-up process and power flow distribution.

instructions on the basis of control of DC voltage independently. The active power of VSC1–VSC3 is 54 kW, 43.5 kW, and 43.5 kW, and the active power ratio of each microgrid is 1.24:1:1, which is close to the power flow regulation coefficient (1.27:1:1). The reason for the deviation between the real-time value and target value of the active power of VSC1–VSC3 is that the droop control is in control mode with steady error, and the DC line reactance has an effect on the DC voltage of each microgrid and thus the active power could not accurately reach the control results. In addition, VSC4 could realize accurate active power control based on power load distribution on the basis of high-order current control as seen in Figure 7b. Figure 7c is the DC voltage of VSC1–4. It is denoted that DC voltage rapidly reaches the rated voltage (800 V) after start-up, and transient voltage changes are less than 10%, which has no influence on system protection. When VSC1–3 transfer rated active power, the DC voltage quickly reaches a stable condition without an oscillatory instability problem after the DC voltage undergoes a short period of transient surge.

5.3 DC load switching in

The DC load is set near VSC3 and switched in DC line at 4 s. The system transient process is simulated when VSC1–VSC3 transfer power flow based on the steady control coefficient. The base direction of active power flow of VSC1, VSC2, and VSC3 is defined from its AC side to DC side, a DC load of 2.66Ω switches in the DC line close to VSC3 in 4 s, and the power flow control target of VSC4 remains unchanged. Active power and DC voltage of VSC1–VSC4 are shown in Figure 8a–d. As depicted, the output power of VSC1, VSC2, and VSC3 is 146.5 kW, 121 kW and 117 kW respectively, and the active power ratio is 1.21:1:0.97, which means the system flow regulation strategy realized the power reallocation according to the power instruction although little deviation existed. Also, the dynamic response ability of the DC voltage of a flexible AC/DC microgrid remained the same.

Accessing a heavy load has little effect on DC voltage, with only a small DC voltage value drop. When VSC1–VSC3 transfer active power from the AC side to the DC side, the DC voltage-power droop regulating process is as shown in Figure 8. For a DC load near VSC3, the power drop regulating process is located at the lower half of Figure 2; thus the DC voltage of VSC3 is

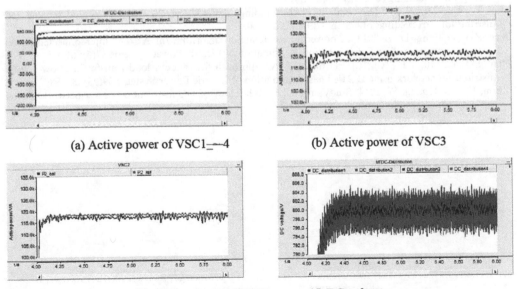

    (a) Active power of VSC1–4   (b) Active power of VSC3

    (c) Active power of VSC2   (d) DC voltage

Figure 8. Simulation results.

lower than the rated DC voltage, and the active power output is a little bigger than the power instruction, shown by the red line of Figure 8b. Because of the DC line reactance, the DC voltage of VSC1 and VSC2 is higher than the rated voltage, and the active power of VSC1 and VSC2 is smaller than the power instructions shown in Figure 8c, taking VSC2 as a typical example.

6 CONCLUSION

In this article, a microgrid power flow coordinating control strategy of self-adapted style and operation modes of a flexible AC/DC microgrid have been designed that could realize active power flow regulation and control. Based on independent control of DC bus voltage by the microgrid's VSC, it could carry out power flow control in different cases of AC load transfer or DC load switching in. Also, the active power of each VSC could be regulated according to each flow control coefficient, which was determined by each transformer's capacity and its AC load. A bidirectional P–U_{DC} droop controller and a current inner-loop controller based on a high-order primary system model have also been proposed. The control parameters optimization method has been presented as well based on high-order system modeling. A case has been modeled in PSCAD/EMTDC to verify the proposed method. Simulation results indicated that a flexible AC/DC microgrid could realize optimal power flow control. Each VSC could carry out control DC voltage independently under active power flow closed-loop control without an instability problem, and the control reliability has been validated as well.

REFERENCES

Liu, Y.H., Pei, W., Yang, Y.H., et al. 2016. Power flow calculations for AC/DC hybrid distrbution network with multi-terminal DC. Electric Power Construction 37(5): 83–90.

Peng, K., Xian, R.C., Zhang, X.H., et al. 2016. Hierarchical power flow control strategy and algorithm for multi-terminal interconnected AC/DC distribution network. Automation of Electric Power Systems 40(14): 72–77.

Tang, G.F., Luo, X., & Wei, X.G. 2013. Multi-terminal HVDC and DC-grid technology. Proceedings of the CSEE 33(10): 8–17.

Wang, C.S., Sun, C.B., Peng, K., et al. 2013. Study on AC-DC hybrid power flow algorithm for microgrid. Proceedings of CSEE 33(3): 8–15.

Wu, J., Wu, D., Zhu, J.D., et al. 2017. Grounding method design of multi-terminal flexible DC distribution. Proceedings of CSEE 37(9): 2551–2560.

Xue, Z.Y. & Fang, D. Z. 2013. A power flow calculation algorithm for AC/DC interconnected power systems based on bi-directional iteration. Automation of Electric Power Systems 37(5): 61–67.

Zhang, X.Y., Hu, T.L., Xu, R.L., et al. 2015. An approach for random load transfer in a low-voltage distribution network using DC tie lines. Automation of Electric Power Systems 39(23): 83–88.

Zhou, F.Q. & Huang, W. 2014. Study on the key technology of DC distribution power network. Power System Protection and Control 42(22): 62–67.

Emerging Developments in the Power and Energy Industry – Dufo-López, Krzywanski & Singh (eds)
© 2020 Taylor & Francis Group, London, ISBN 978-0-367-27169-5

Fluid-structure interaction and performance analysis of an ocean current turbine

Zhigao Dang*, Zhaoyong Mao & Wenlong Tian
School of Marine Science and Technology, Northwestern Polytechnical University, Xi'an, China

ABSTRACT: In this paper, a horizontal-axis ocean current turbine is taken as investigation object, to study the Fluid-Structure Interaction (FSI) effect at working conditions. With the help of finite element analysis software ANSYS, the pressure on the turbine is calculated in the flow field, which provide input loadings for static structural analysis before prestress modality analysis. The results show that the phenomenon of obvious stress concentration will occur at the root of the blade during wind turbine operation. The maximum stress will increase with the increase of flow velocity, which inevitably influence the modal characteristics of the turbine. The ocean current turbine generated power and thrust force are also investigated, and the results show that the optimal power is obtained when the Tip-Speed Ratio (TSR) is 5.

Keywords: Fluid-Structure Interaction, Ocean Current Turbine, Prestress Modality Analysis, Power Coefficient, Tip-Speed Ratio

1 INTRODUCTION

There are abundant renewable energy sources in the ocean, including solar energy (Avery et al. 1999), wave energy (Aderinto et al. 2018), current energy (Akimoto et al. 2013), thermal energy (Li et al. 2018), salinity energy (Arias et al. 2018) and so on. Marine current energy has only been investigated for about more than 20 years, however, it is one of the fastest growing renewable energy sources (Li et al. 2016). There are two kinds of current power generation devices, including impeller type and non-impeller type, according to the structures of current energy capture components (Day et al. 2015). Impeller type ocean current power generation device mainly relies on the impeller to convert ocean current energy into mechanical energy of impeller rotation, including vertical axis turbines and horizontal axis turbines. The research on vertical axis ocean current power generation device is earlier than that of horizontal axis devices all over the world. As for horizontal axis ocean current power generation device, there are several successful cases for commercial application of current power generation device (Fraenkel 2007). Figure 1 shows the first array of ocean current generators in the world in Norway built by Hammerfest Strøm Company.

However, the investigation for horizontal axis ocean current power generation device starts late in China, and there is no successful commercial application of current power generation device at present. Several universities and institutes are in the prototype test stage, therefore, the research on this aspect is profound and challenging. As a primary energy conversion device of ocean current energy, the impeller of ocean current power generation is the core component. The hydrodynamic performance of the impeller is the key to the effective development and utilization of ocean current energy, which would be solved in this paper. The layout of this paper is as follows: Section 1 gives a brief introduction. In Section 2, the model of

---

*Corresponding author: zhigao_dang@mail.nwpu.edu.cn

Figure 1.   HS1000 ocean current turbine arrays.

ocean current turbine is introduced and numerical methods for hydrodynamic analysis is given. Then, results and discussions for FSI analysis of the ocean current turbine and working performance are demonstrated in Section 3. Finally, conclusions are presented in Section 4.

2   MODELING AND METHODS

2.1   *Physical model*

The commercial CFD software of ANSYS 15.0 is used to solve the incompressible RANS equation, and the turbulence terms is modelled by the shear stress transport (SST) turbulence model. A rotating reference model is applied to simulate the rotation of the turbine for saving computation time, while the outer fluid domain remains stationary (Maizi et al. 2011).

As shown in Figure 2, the computational domains include three subdomains, turbine part (solid), rotational fluid domain (fluid) and stationary fluid domain (fluid). In the calculation of fluid field characteristics, the solid part is suppressed, then the pressure fluctuation of the fluid is applied on the solid turbine to simulate the FSI process. A uniform and steady velocity of 0.5m/s is applied at the inlet of the computation domain. A pressure outlet is applied at the outlet of the domain. Then symmetry boundary conditions are applied at the side wall of the stationary fluid domain to improve the stability of the numerical simulations. No-slip

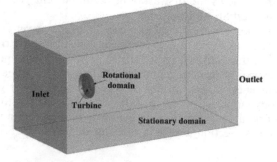

Figure 2.   Computational domains and boundary conditions.

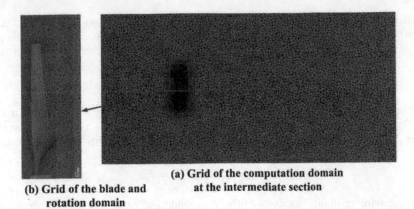

(b) Grid of the blade and
rotation domain

(a) Grid of the computation domain
at the intermediate section

Figure 3. Computing mesh at different positions: (a) intermediate section, and (b) blade and rotation
domain.

boundary conditions is applied at the surfaces of the turbines. The interface is established at
the overlap faces between the stationary fluid domain and the rotational fluid domain.

The grids for the model in Figure 2 are generated using the ANSYS Workbench grid generation utility, which is shown in Figure 3. The grid near the blade, where high velocity gradient
gradients are expected, is refined with a higher grid resolution. The convergence criterion for
the scaled residuals are 1×10^{-5}, and the maximum number of iterations is set as 1500, which
enables all the residuals to meet convergence in the simulation.

2.2   Governing equations

The computational model is simplified to a steady incompressible flow without considering
the heat transfer effect in the flow process. Therefore, the governing equations in the flow process mainly consist of continuous equations and momentum equations. The RANS equation
is applied to simulate the flow field.

Continuous equation:

$$\frac{\partial(\rho u_i)}{\partial x_i} = 0 \qquad (1)$$

Momentum equation:

$$\frac{\partial(\rho u_i u_j)}{\partial x_i} = \frac{\partial p}{\partial x_i} + \frac{\partial}{\partial x_i}\left(\mu \frac{\partial u_i}{\partial x_j} + \frac{\partial u_j}{\partial x_i} - \rho\overline{u_i'u_j'}\right) \qquad (2)$$

where u_i, $u_j(i, j = 1, 2, 3)$ is the mean velocity component, x_i, x_j is the coordinate component, p is the average pressure of fluid, u is the dynamic viscosity coefficient of fluid, ρ is the
fluid density.

The term $-\rho u_i'u_j'$ in Equation (2) is called Reynolds stress, which represents the influence of
turbulence. For viscous flow, the term is unknown, so in order to solve the momentum equation, a turbulence model must be established to simulate the Reynolds stress term to make the
equation closed. Considering the wide range of turbine rotation speed, the $k - \omega$ SST turbulence model is adopted, which has better accuracy and algorithm stability in the near wall
region (Yang et al. 2009).

With non-negligible stress loads, the natural frequencies of structures will be affected by
structural stresses. For ocean current turbines, the blade length and mass are larger, the rotating prestressing force is stronger because of the rotating action during operation. In addition,

the FSI effect leads to the hydrodynamic non-uniformity of the wind turbine, and its modal characteristics will inevitably change. The static structural analysis is needed for the prestress modality analysis, and the equation for calculating the load F is

$$Kx = F \qquad (3)$$

where K is stiffness matrix, x is nodal displacement vector. The prestress effect matrix S is obtained through the static analysis of the nodal displacement vector x. Considering the influence of the prestress effect matrix S, the original modal analysis equation can be modified to

$$(K + S - \omega_1^2 M) \cdot \varphi_i = F \qquad (4)$$

Comparing with modality analysis without considering prestress, the prestress modality analysis basically has the same procedures. However, it should be noted that the initial stress state of the structure must be determined by means of loading.

3 RESULTS AND DISCUSSIONS

3.1 *Prestress modality analysis*

The pressure distribution loaded on the blade of the turbine is needed to be calculated to conduct structural static analysis. And the results of the structural static analysis provide a prerequisite for the further prestress modality analysis. The pressure loaded on the turbine blade at different flow velocities is shown in Figure 4. It is seen that the pressure distribution on the surface of the turbine blade is uneven, the pressure at the tip of the blade is largest, and the pressure at the center of the blade is smaller. What's more, the pressure at the leading edge is greater than that at the trailing edge, which fully reflects the space-time distribution characteristics of the pressure caused by the rotation of the turbine. Comparing the pressure distribution at different velocities, it is seen that larger pressure is observed in the fluid field with higher velocity of flow.

The equivalent stress at the blades after loading the pressure onto the blades is shown in Figure 5. It is noted that the stress concentration will occur at the root of the blade during wind turbine operation. As shown in Table 1, the maximum stress will increase with the current velocity, which inevitably influence the modal characteristics of the turbine. When the flow velocity is 0.5m/s, the maximum stress is 39.688MPa; when the flow velocity is increased

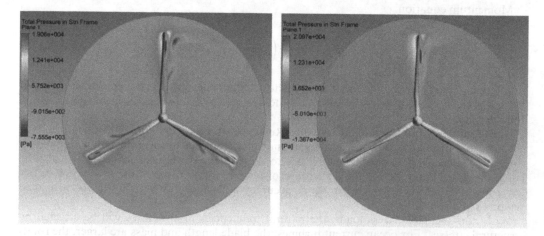

Figure 4. Pressure loaded on the turbine blade at different flow velocities: (a) 0.5m/s, and (b) 1m/s.

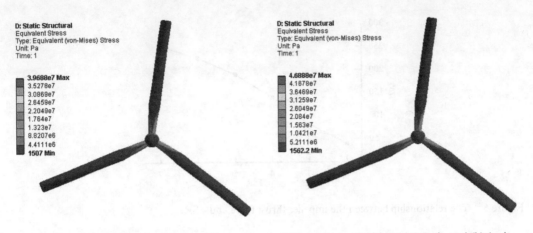

D: Static Structural
Equivalent Stress
Type: Equivalent (von-Mises) Stress
Unit: Pa
Time: 1

3.9688e7 Max
3.5278e7
3.0869e7
2.6459e7
2.2049e7
1.764e7
1.323e7
8.8207e6
4.4111e6
1507 Min

D: Static Structural
Equivalent Stress
Type: Equivalent (von-Mises) Stress
Unit: Pa
Time: 1

4.6888e7 Max
4.1878e7
3.6469e7
3.1259e7
2.6049e7
2.084e7
1.563e7
1.0421e7
5.2111e6
1562.2 Min

Figure 5.   Surface stress distribution of the turbine at different flow velocities: (a) 0.5m/s, and (b) 1m/s.

Table 1.   Maximum stress and the first six modes at different working conditions.

Flow velocity	Maximum stress	Natural frequency at the first 6 modes					
		1st	2nd	3rd	4th	5th	6th
0 m/s	-	75.861	79.936	80.081	83.732	84.007	84.064
0.5m/s	39.688MPa	75.863	79.939	80.084	83.734	84.009	84.067
1m/s	46.888MPa	75.864	79.939	80.085	83.735	84.01	84.068

to 1m/s, the maximum stress is 46.888MPa. Correspondingly, there is a trend of gradual increase of the natural frequencies of the turbines for the first six modes with the increase of flow velocities.

3.2   Turbine performance

The tip-speed ratio (TSR) for turbines is the ratio between the tangential speed of the tip of a blade and the actual speed of the wind, which is an important parameter representing the performance of turbines, shown as

$$\lambda = \frac{\omega r}{v} \tag{5}$$

where ω is the turbine rotational speed in radians/second, r is the turbine radius in meters, v is the wind speed specified in meters/second.

Correspondingly, the generated power defined as

$$P = 0.5 C_p \rho \pi r^2 v^3 \tag{6}$$

And thrust on the turbine in the axial direction defined as

$$T = 0.5 C_t \rho \pi r^2 v^2 \tag{7}$$

where C_p is the coefficient of power and C_t is the coefficient of thrust force.

In this section, the flow velocity is fixed at 1m/s, then the TSR is changed by changing the rotational speed ω, to investigate the characteristics of generated power P and the thrust force T. As the values of density of flow ρ, radius of the turbine r and flow

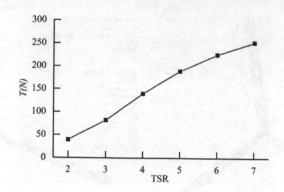

Figure 6.   The relationship between the impeller thrust force and TSR.

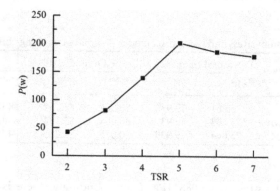

Figure 7.   The relationship between the impeller generated power and TSR.

velocity v are fixed, so the change regulations of C_p and C_t have the same characteristics with that of P and T.

Figure 6 shows the relationship between the impeller thrust and TSR. The impeller thrust is positive in the range of test. It is seen that the value of impeller thrust is increasing with the increase of TSR. Figure 7 shows the variation curve of impeller generated power with TSR. It can be found that the power of the impeller increases first and then decreases with the increase of TSR. The maximum generated power is obtained when the TSR is 5.

4   CONCLUSIONS

In this research, a horizontal axis ocean current turbine is designed to produce electricity by ocean current energy. Three-dimensional steady numerical simulations are carried out to investigate the influence of flow velocities. The fluid-structure interaction characteristics of the turbine is investigated, and then the working performance of the turbine is analyzed. The results demonstrate that:

(1) The maximum stress of the blade increases with the increase of wind speed. There is obvious stress concentration at the root of the blade, which is the dangerous location of the blade and is prone to fracture. Therefore, in practical engineering, the corresponding location can be strengthened when the blade is designed and manufactured.

(2) When modal analysis of the water turbine is carried out, the effect of rotation on the corresponding frequencies of wind turbine modes should be taken into account, so as to avoid the coincidence of the rotation frequency of turbine and the natural frequency of turbine, and to avoid resonance and damage to ocean current turbine.

(3) The ocean current turbine has the maximum generated power, obtained at TSR = 5.

In this paper, the turbine is supposed to rotate around a fixed axis and the influence of the structures that support the turbine is ignored. Actually, different installation modes have different influence to the working performance of the turbine, and the noise performance of the turbine should be also considered in the future research.

REFERENCES

Avery, W.H. & Berl, W.G. 1999. Solar energy from the tropical oceans. *Int. J. Hydrogen. Energ.* 24(4): 295–298.

Aderinto, T. & Li, H. 2018. Ocean wave energy converters: status and challenges. *Energies* 11(5): 1250.

Akimoto, H., Tanaka, K. & Uzawa, K. 2013. A conceptual study of floating axis water current turbine for low-cost energy capturing from river, tide and ocean currents. *Renew. Energ.* 57: 283–288.

Li, D., Yue, J., Zhang, L. & Duan, X. 2018. Numerical study on ocean thermal energy conversion system. *J. Renew. Sustain. Ener.* 10(4): 044501.

Arias, F.J. 2018. Ocean thermal energy conversion by deliberate seawater salinization. *Int. J. Energy Res.* 42(2): 499–507.

Li, W., Zhou, H., Liu, H., Lin, Y. & Xu, Q. 2016. Review on the blade design technologies of tidal current turbine. *Renew. Sust. Energ. Rev.* 63: 414–422.

Day, A.H, Babarit, A., Fontaine, A., He, Y.P, Kraskowski, M., Murai, M., Penesis, I., Salvatore, F. & Shin, H.K. 2015. Hydrodynamic modelling of marine renewable energy devices: A state of the art review. *Ocean Eng.* 108(1): 46–69.

Fraenkel, P. 2007. Marine current turbines: pioneering the development of marine kinetic energy converters. *P. I. Mech. Eng. A-J. Pow* 221(2): 159–169.

Maizi, M., Mohamed, M.H., Dizene, R. & Mihoubi, M.C. 2018. Noise reduction of a horizontal wind turbine using different blade shapes. *Renew. Energ.* 117: 242–256.

Yang, Y., Gu, M., Chen, S. & Jin, X. 2009. New inflow boundary conditions for modelling the neutral equilibrium atmospheric boundary layer in computational wind engineering. *J. Wind Eng. Ind. Aerod.* 97(2): 88–95.

Emerging Developments in the Power and Energy Industry – Dufo-López, Krzywanski & Singh (eds)
© 2020 Taylor & Francis Group, London, ISBN 978-0-367-27169-5

Research and analysis on the relationship between floating electrode discharge signal and ground current in electrical equipment

Feiyue Ma, Bo Wang & Pei Ding
State Grid Ningxia Electric Power Research Institute, Yinchuan, China

ABSTRACT: Experiments and researches are carried out for the changes of partial discharge and ground current signals caused by floated deficiency in the high-voltage electrical equipment such as GIS and transformer. The floated discharge model and the test platforms are applied to simulate the PD inside the GIS and oil-immersed transformer. The variation of partial discharge signals and ground current signals are observed by increasing the voltage under the two test conditions. By making the analyses such as the correlation between the two signals, the trend of the partial discharge intensity and ground current, the spectrum analysis of the ground current, to verify the fact that the ground current can effectively reflect the occurrence of the floated discharge in the power equipment. It is of great significance by monitoring the variation of the ground current and combing with the judgment of the partial discharge detection.

1 INTRODUCTION

Partial Discharge(PD)is a process of generating discharges around high-voltage conductors or other particle position during operation of an electrical equipment. The defect and some kinds of contamination inside the power equipment will result in uneven distribution of electric field strength, and even worse, high field strength regions. The defect and contamination may be brought in during procedure of production, manufacture and installation processes. PD occurs when the electric field strength reaches the breakdown field strength of insulation medium [QIU Yuchang.1994]. As long-term operational experience shows, PD is one of the most important factors to cause breakdown of high-voltage electrical equipment such as Gas Insulated Switchgear(GIS), transformer.

PD detection is of great significance for the evaluation of high voltage equipment operating status and maintenance strategy formulation. There are many advanced PD detection methods [LI Junhao. HAN Xutao. LIU Zehui. 2015], such as pulse current detection, UHF detection, ultrasonic detection, HF detection. With the development of PD detection technology, the PD source location and pattern recognition has come to reality. Scholars such as, Weidong Liu researched how to realize the detection and location of PD signal inside GIS by UHF detection technology [LIU Weidong. HUANG Yulong. WANG Jianfeng. 1999]; Junhua Liu studied how to realize the detection and location of PD signal inside GIS by ultrasonic and UHF detection technology [LIU Junhua.YAO Ming. HUANG Chengjun. 2009]; Xiaoxiang Zhang analyzed the relationship between discharge magnitude and UHF signal inside GIS using typical PD deficiency [ZHANG Xiaoxing.TANG Junzhong. TANG Ju. 2012.]; Dengwei Ding obtained the relationship between discharge severity and UHF signal inside GIS using different PD deficiency [Ding Dengwei, Gao Wensheng, Liu Weidong. 2014]. More and more experts and scholars have invested in the research of PD in recent years. However, not enough research on the correlation between partial discharge inside electrical equipment and ground current has been done.

In this paper, floating electrode discharge defects in transformers and GIS were simulated and the relationship between PD signal and ground current signal is analyzed in depth by studying the variation principle of partial discharge signal and ground current signal when generating suspended discharge defects. Furthermore, it is hoped that this grounding current characteristic and data can be combined with charging detection in the field and applied to the comprehensive diagnostic of the equipment finally.

2 TEST SYSTEM

Two test system platforms are built in this research, test schematic is as shown in Figure1 (GIS test system filled with SF6 gas insulated medium) and Figure2 (transformer test system filled with oil insulated medium). In GIS test platform, the diameter of central conductor is about 86mm, the thickness of spacer is about 50mm, the distance between two spacers about 102mm. As shown in the GIS test system, floated discharge model was placed between central conductor and shell around observation window to make it easy to be installed. As shown in the transformer test system, floated discharge model is set in a Plexiglas cone cavity filled with transformer oil. To obtain apparent discharge magnitude with conventional pulse current PD equipment; ground current with oscilloscope directly. At the same time, apply oscilloscope and popular PD equipment to collect UHF、AE and HF signals. The type of oscilloscope we use is Agilent DSO6104A, with sampling frequency of 4GHz, frequency bandwidth of 1GHz.

In the test system platform, the frequency bandwidth parameters of four types of sensors are shown as follow:

- UHF sensor bandwidth: 300-1500MHz。
- AE sensor bandwidth: 20-200kHz。
- HF sensor bandwidth: 0.5-50MHz。
- Ground current sensor bandwidth: 40Hz-5KHz。

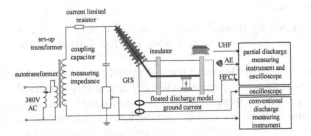

Figure 1.   Experiment platform for GIS filled with SF6.

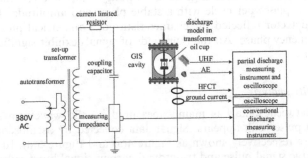

Figure 2.   Experiment platform for transformer.

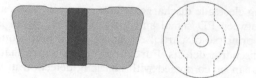

Figure 3.   Schematic of floated discharge model.

Specially a floated discharge model was designed for transformer electrical equipment, the model schematic is shown in Figure3. The model has a conical shape and chamfer edges made of epoxy resin. The upper diameter of the cone is 32mm, the lower diameter is 30mm, and the thickness is about 8mm. Copper column is embedded into the middle of epoxy resin, both upper and lower distance with epoxy resin surface is 1mm. As dotted line shown in Figure3, a concave channel is designed at the bottom of the model to ensure that the copper electrode is in the insulating medium. During test procure, the copper electrode is about 1.5mm in distance with high-voltage terminal.

3   TEST PROCEDURE

Increase HV source slowly, taking 5kV as a voltage interval to collect signal data, including ground current, apparent discharge, UHF, AE and HF signals. Reduce the increase rate with a voltage interval about 1kV, when the copper electrode generates significant PD signals.

Observe ground current data and various PD signals shown in oscilloscope, save signal data at time base of 10ms and 200ns in order to analysis correspondence with MATLAB.

4   RESULT AND ANALYSIS

4.1   Signal pattern analysis

The PD signal generated by floated discharge model is able to be detected significantly by four types of sensors as shown in Figure 4 when conduct GIS test system filled with SF6 gas insulated medium.

The PD signal generated by floated discharge model is able to be detected significantly by four types of sensors as shown in Figure 5 when the model was immersed in transformer oil.

Obvious discharge signals were detected by oscilloscope and PD detect instrument in Figure 4. Furthermore, there are two cluster periodic pulse signals with stable phase and amplitude. The pattern is recognized as floated discharge according to typical characteristics of spectrum library.

Under the two test environments, the discharge detector, oscilloscope and conventional PD detector have detected obvious discharge signal. Among them, the UHF and ultrasonic signals hold two clusters of pulses per cycle with a stable phase and amplitude. The pulse signal in conventional PD detector reflected the characteristics of symmetrical pulse and close amplitude in power frequency phase. All the four kinds of signals exhibit significant characteristics of floated discharge.

4.2   Signal relativity analysis

The correlation between signals is mainly measured by the waveform similarity and the moment when the pulse peak appears. Signal data collected by oscilloscope at time base of 10ms and 200ns are respectively shown as Figure 4(a) and Figure 5(a). In the figures above, the floating discharge signal pulse and the ground current signal burst pulse rising edge at the same time. In the position where each pulse appears in the UHF signal, a pulse appears in the

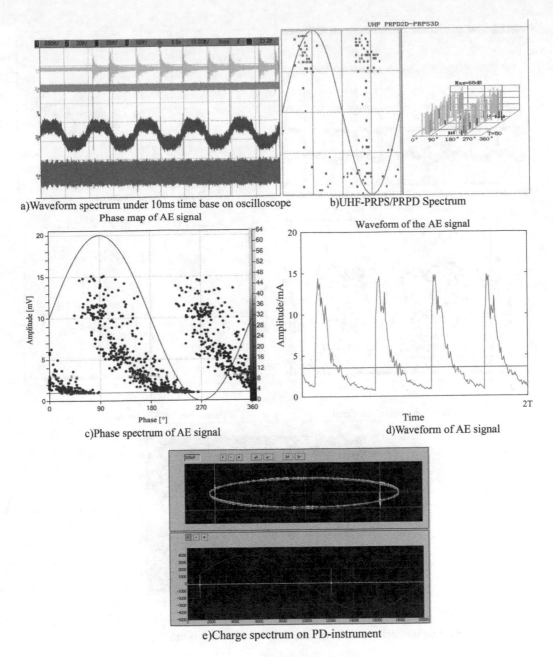

a)Waveform spectrum under 10ms time base on oscilloscope

b)UHF-PRPS/PRPD Spectrum

Phase map of AE signal

Waveform of the AE signal

c)Phase spectrum of AE signal

d)Waveform of AE signal

e)Charge spectrum on PD-instrument

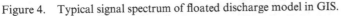

Figure 4.   Typical signal spectrum of floated discharge model in GIS.

ground current signal. Since the ultrasonic signal is attenuated during propagation, one single pulse rising edge is detected and the oscillation followed on the waveform, however, the position of the peak of the signal envelope spectrum coincides with the peak position of the ground current signal pulse in time .Which means the ground current sensor coupled the pulse signal inspired by the floated model and transmitted to the test loop. The ground current signal reflects the occurrence of partial discharge of the model with the discharge signal at the same time.

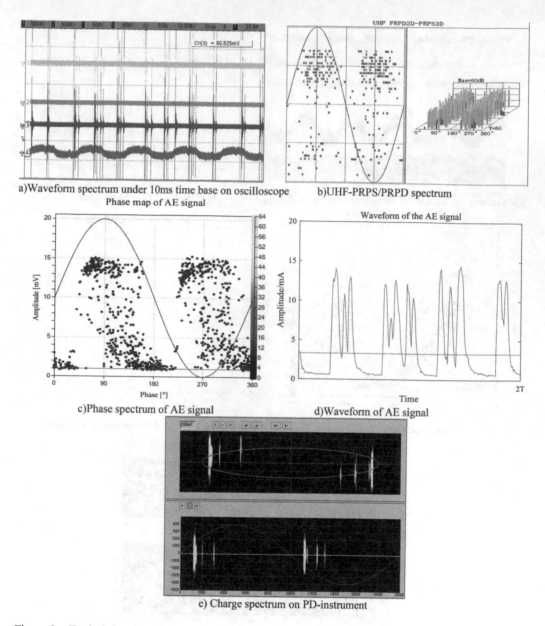

a)Waveform spectrum under 10ms time base on oscilloscope

b)UHF-PRPS/PRPD spectrum

c)Phase spectrum of AE signal

d)Waveform of AE signal

e) Charge spectrum on PD-instrument

Figure 5. Typical signal spectrum of floated discharge model in transformer.

4.3 *Ground current signal relativity analysis*

During the test, the trend of apparent discharge magnitude obtained by conventional pulse current PD equipment and grounding current changing with the increase of voltage level was recorded at same time. As shown in Figure 6 and Figure 7, floated discharge model is equivalent to the structure of voltage divider regardless the type of the insulating medium. The single charge and discharge is relatively stable, but the energy inspired by the floated discharge is significant with more than 600pC in apparent discharge. With the increase of high-voltage, floated discharge mode discharges more and more intense, the ground current signal shows an obvious increase respectively and then a steady trend finally.

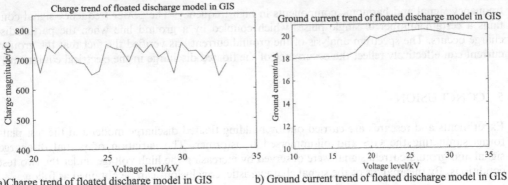

a)Charge trend of floated discharge model in GIS

b) Ground current trend of floated discharge model in GIS

Figure 6.   Partial discharge volume and ground current trend of floated discharge in GIS.

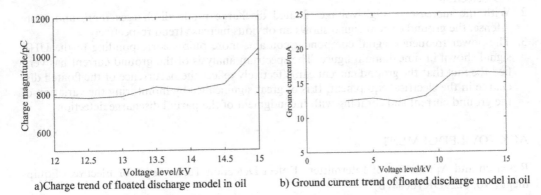

a)Charge trend of floated discharge model in oil

b) Ground current trend of floated discharge model in oil

Figure 7.   Partial discharge volume and ground current trend of floated discharge in transformer.

4.4   *Ground current signal spectrum analysis*

Read ground current signal under the two test conditions by MATLAB and analyze the characteristic of signal spectrum as shown in Figure8. The signal spectrum is composed of power

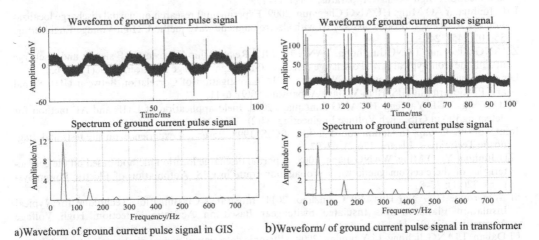

a)Waveform of ground current pulse signal in GIS

b)Waveform/ of ground current pulse signal in transformer

Figure 8.   The spectrum of ground current for floated discharge in GIS and transformer.

frequency signal and harmonic components in high frequency. The power frequency signal contains a certain number of single pulses which coupled by a ground line when the partial discharge occurs. The spectrum analysis of the ground current has verified the fact that the ground current can effectively reflect the occurrence of the floated discharge in the electrical equipment.

5 CONCLUSION

Experiments and research are carried out by building floated discharge model and the test platforms simulating the GIS and oil-immersed transformer. The variation of partial discharge signal and ground current signal were observed by increasing the high voltage under the two test conditions. By the analysis of the signal characteristic, conclusions can be obtained as follow.

1. The pulse rising edges exist exactly corresponding relationship between the PD signal and the ground current signal in time domain explains the validity of ground current method in PD detection.
2. With the increase of high-voltage, floated discharge mode discharges more and more intense, the ground current signal shows an obvious increase trend respectively.
3. The power frequency signal component contains more pulses corresponding to the UHF signal shown in time domain figure. The spectrum analysis of the ground current has verified the fact that the ground current can effectively reflect the occurrence of the floated discharge in the electrical equipment. It is of great significance by monitoring the variation of the ground current and combing with the judgment of the partial discharge detection

ACKNOWLEDGEMENT

Research and Application of Intermittent Defect Detection Technology for Electrical Equipment (No.:5229DK160001B)

REFERENCES

QIU Yuchang. 1994. GIS device and its insulation technology. Beijing Water Resources and Electric Power Press.
LI Junhao. HAN Xutao. LIU Zehui. 2015. Review on partial discharge measurement technology of electrical equipment. High Voltage Engineering. 41(8): 2583–2601.
LIU Weidong. HUANG Yulong. WANG Jianfeng. 1999. GIS online partial discharge UHF detection and location. High Voltage Apparatus. 35(1): 11–15.
LIU Junhua. YAO Ming. HUANG Chengjun. 2009. Experimental research on partial discharge localization in GIS using ultrasonic associated with electromagnetic wave method. High Voltage Engineering. 35(10): 2458–2463.
ZHANG Xiaoxing. TANG Junzhong. TANG Ju. 2012. Relationship between UHF signals and discharge magnitude from typical partial discharge defects in GIS. High Volt-age Engineering. 38(1): 59–65.
Ding Dengwei, Gao Wensheng, Liu Weidong. 2014. Investigation of Correlation Between UHF Signal and PD Severity in GIS. High Voltage Apparatus. 50(9): 6–11.
YAO Yong. YUE Yanfeng. HUANG Xingquan. 2008. Field application of UHF and AE method for detecting PD in GIS. High Voltage Engineering. 34(2): 422–424.
Joint working Group. Insulation coordination of GIS. 1998. Return of experience, on site tests and diagnostic Techniques. Electra. 176(2): 67–97.
LIU Junhua. YAO Ming. WANG Jiang. 2008. A partial discharge location method based on the characteristics of the electromagnetic wave propagation route in GIS. Automation of Electric Power Systems. 32(21): 77–81.
SI Wen-rong. HUANG Hua. Fu Chenzhao. 2011. Integrated Classification Technique for Typical Insulation Flaws in Gas Insulated Switchgear Based on Acoustic Detection. High Voltage Apparatus. 47(12): 11–17.
LI Dajian. LIANG Jichong. Li Yanming. 2009. Ultrsonic detection of partial discharge on typical defects in GIS. High Voltage Apparatus. 45(1): 72–75.

Emerging Developments in the Power and Energy Industry – Dufo-López, Krzywanski & Singh (eds)
© 2020 Taylor & Francis Group, London, ISBN 978-0-367-27169-5

A coordinated damping control for ultralow-frequency oscillation in a DC isolated system

Xin Zhou & Peng He
Yunnan Electrical Power Experiment Institute, Kunming, China

Ningning Ma*, Wei Deng & Zhenpeng Li
School of Electrical Engineering, Southwest Jiaotong University, Chengdu, China

Shuijun Wu
Yunnan Electric Power TestResearch Institute, Kunming, China

ABSTRACT: After asynchronous interconnection, an ultralow-frequency oscillation of long duration occurred in some isolated systems, which seriously affected the safe and stable operation of the power system. In this article, the damping characteristics of the governor system are analyzed. A proportional–integral–derivative (PID) controller parameter optimization simulation experiment is performed in the actual DC isolated system, and the frequency stability is compared under different PID parameters. A coordinated control method between the speed control system and the DC frequency controller is proposed in this article. Finally, an actual DC isolated system is used as an example to verify the effectiveness of this control method. The simulation results indicate that the time constant of the water hammer effect of the hydraulic turbine in the power grid becomes larger, the damping ratio will decrease, and the frequency stability of the system will deteriorate.

Keywords: damping characteristic, DC frequency controller speed control system, ultralow-frequency oscillation, water hammer effect

1 INTRODUCTION

In 2016, a back-to-back direct current (DC)-based asynchronous interconnection between the Yunnan Power Grid (YNPG) and the China Southern Power Grid (CSPG) was achieved. An experiment was performed to test the stability of asynchronous networked systems. An oscillation of system frequency with a period of 20 s was present in YNPG. But the system frequency begins to normalize when the frequency regulation equipment in Xiaowan and Nuozhadu power plants are out of service (Liu et al. 2016; Chen et al. 2017). Studies have shown that the ultralow-frequency oscillation is caused mainly by the water hammer effect and the hydro-turbine governors. Therefore, tuning the proportional–integral–derivative (PID) parameters of the governors to damp the ultralow-frequency oscillation will be an effective method (Zhang et al. 2016).

The islanding operation realized by UH VDC can effectively eliminate transient instability and improve the transmission limit. However, some problems exist in the isolated system, such as frequency stability, voltage stability, overvoltage and subsynchronous oscillation, and so on. Relevant research about the problems has been carried out in a DC transmission system in an islanding operation (Dhaliwal et al. 1996; Huang and Li 2013). When the system is in an islanding state, the governor setting may lead to

*Corresponding author: mnsdxt@163.com

frequency instability of the system under certain conditions. In Huang and Li (2013), the influence of load on frequency stability was analyzed via the detailed model of an isolated system. However, because of the nonlinear characteristics of an isolated system, important information may be omitted from the linear analysis method. According to Gong et al. (2014), when the disturbance occurs in the isolated power system, the frequency fluctuation will bring a great security risk to the isolated system. The analysis of the small signal eigenvalue shows that the ultralow-frequency oscillation has a strong correlation with the hydraulic turbine governor, and the operating model and parameters of the governor can significantly change the position of the eigenvalue on the complex plane, thus affecting the damping ratio. The coordinated optimization of governor parameters can effectively improve the stability of the power system and suppress the phenomenon of ultralow-frequency oscillation. Zheng et al. (2018) report that the phenomenon of ultralow-frequency oscillation has occurred many times in the Southwest China Power Grid under some fault conditions, and the energy characteristics of the power plants participating in the ultralow-frequency oscillation are revealed through analysis of the oscillatory energy flow of ultralow-frequency oscillation. In addition, by analyzing the damping characteristics of the main hydropower plants, the causes of ultralow-frequency oscillation are revealed, that is, the insufficient damping of hydropower stations. In an isolated power system with concentrated hydropower, the negative damping effect provided by the speed regulation system of hydropower units is one of the important factors leading to ultralow-frequency oscillation.

In this article a novel method is used to optimize the PID parameters. Then, the influence of primary frequency regulation on system frequency and DC power is studied. Next, a coordinated control method of frequency and DC frequency controller is proposed. Finally, a practical isolated DC system example of YNPG of China is used to verify the effectiveness of the control method and the DC frequency controller presented in this article.

2 DC ISOLATED SYSTEM

The hydropower station N is located in YNPG, which is one part of CSPG. Under normal operation, this station connects to power grids in other areas through an AC line and a DC line. The AC line is from A to B, and the DC line is from A to C. For the DC line, the voltage level is ±800 kV, and the transmission capacity is 5000 MW. Hydropower station N contains nine generator sets, each of which is a hydroelectric generator with a rated power of 650 MW. However, when the AC lines from A to B are disconnected, the link between hydropower station N and the main network of YNPG will disappear. As a result, a DC isolated system will form, as shown in Figure 1. After the formation of an isolated DC system, six generators operate normally and three other generators disconnect from the power grid. When a power load of 75 MW is rejected, there will be a frequency fluctuation with a period of about 20 s and an amplitude of about 49.80–50.28 Hz. The frequency and active power of one generator are shown in Figure 2. This oscillation belongs to the scope of ultralow-frequency oscillation.

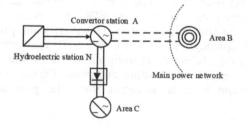

Figure 1. Geographical wiring diagram of hydropower station N.

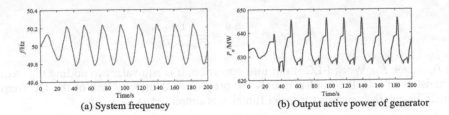

(a) System frequency   (b) Output active power of generator

Figure 2. Ultralow-frequency oscillation in the hydropower station N.

3 BASIC MODELS

3.1 *Generator and turbine governor*

The generators in hydropower plant N are all hydroelectric, so mainly the hydro-turbine model is discussed in this article. The swing equation in the incremental form is

$$T_M \frac{d\Delta\omega}{dt} = \Delta P_m - \Delta P_e - D\Delta\omega \tag{1}$$

where T_M is the inertia time constant, ΔP_m and ΔP_e are the mechanical and electromagnetic power deviations, and D is the damping coefficient.

The turbine model is the dynamic relationship between the turbine steering opening and the output mechanical power P_m (Kundur 1994). In this article, a simplified dynamic model of hydraulic turbines and their water diversion pipelines is used. Only the water hammer effect due to water flow inertia is considered, and the transfer function is

$$G_w(s) = \frac{1 - sT_W}{1 + 0.5sT_W} \tag{2}$$

where T_W is the time constant of water hammer effect and is generally 0.5–4.0s under the full load condition.

The hydraulic turbine governor mainly adopts the electric adjustable speed control, which consists of a regulating control system, an electrohydraulic servo system, and a prime mover. The adjustment system model adopts a PID-type governor (Kundur 1994). The transfer function is generally expressed as

$$G_{GOV}(s) = \frac{K_D s^2 + K_P s + K_I}{B_P K_I + s} \frac{1}{1 + T_G s} \tag{3}$$

where K_P, K_I, and K_D are the ratio, integral and differential coefficients of the governor, B_P is the coefficient of adjustment, and T_G is the time constant of the actuator.

3.2 *DC frequency controller models*

When the DC frequency limiting controller (FLC) is not used, the DC power will remain constant and not be affected by frequency change. So $\Delta P_e = K_L \Delta\omega$, and K_L is the load frequency adjustment effect coefficient, and the generator transfer function can be written as (Wang et al. 2016)

$$G_{d1}(s) = \frac{\Delta\omega}{\Delta P_e} = \frac{1}{T_M s + D_s} \tag{4}$$

where $D_s = D + K_L$. When FLC is put into operation, it is equivalent to adding the frequency characteristics of DC. If considering only the proportional–integral link of the DC frequency controller model, the generator transfer function is simplified as

$$G_{d2}(s) = \frac{1}{T_M s + D'_s + \frac{K_{HI}}{s}} \tag{5}$$

where $D'_s = D + K_L + K_{HP}$, K_{HP} and K_{HI} are the proportional and integral factors.

4 DAMPING ANALYSIS OF A DC ISOLATED SYSTEM

4.1 Damping characteristics analysis of the turbine governor

The governor model is now simplified, and the transfer function of the prime mover adjustment system is

$$G_{GW}(s) = \frac{1}{B_P} \frac{1}{1 + T_G s} \frac{1 - s T_W}{1 + 0.5 s T_W} \tag{6}$$

According to the Phillips–Heffron model, the relationship between mechanical power change ΔP_m and the speed deviation $\Delta\omega$ generated by the prime mover governor is

$$\Delta P_m = G_{GW}(s)(-\Delta\omega) \tag{7}$$

Substituting $s = j\omega_d$ into Equation (6),

$$\Delta P_m = (D_m + j T_m)(-\Delta\omega) \tag{8}$$

where $T_m = \mathrm{Im}(G_{GW}(j\omega))$ is the synchronous torque coefficient, $D_m = \mathrm{Re}(G_{GW}(j\omega))$ is the damping torque coefficient, and

$$D_m = \frac{1}{B_P} \frac{1 - (1.5 T_W T_G + 0.5 T_W^2)\omega_d^2}{(1 + T_G^2 \omega_d^2)(1 + 0.25 T_W^2 \omega_d^2)} \tag{9}$$

and

$$\omega_d = \sqrt{\frac{1}{1.5 T_W T_G + 0.5 T_W^2}} \tag{10}$$

The frequency corresponding to $D_m = 0$ is called the demarcation frequency. When the system oscillation frequency is higher than the demarcation frequency, the hydraulic turbine speed control system produces negative damping; when the oscillation frequency is lower than the demarcation frequency, the hydraulic turbine speed control system produces positive damping.

The damping characteristics of the control system under two different parameters are shown in Table 1, and the damping torque coefficient and phase of the control system are calculated. Calculated from Equation (8), the demarcation frequency of parameter group 1 is 0.061 Hz, and the demarcation frequency of parameter group 2 is 0.089 Hz. −

Table 1. Damping characteristics of turbine governor with different parameters.

Frequency/ Hz	Parameter group 1 $T_W = 3$, $B_P = 0.04$, $T_G = 0.5$		Parameter group 2 $T_W = 2$, $B_P = 0.02$, $T_G = 0.4$	
	Phase/(°)	D_m	Phase/(°)	D_m
0.050	−77.46	6.66	−56.74	30.65
0.061	−89.73	−0.15	−67.16	22.57
0.070	−98.66	−5.08	−75.06	15.48
0.080	−107.57	−10.58	−83.21	7.35
0.089	−114.81	−15.12	−89.98	−0.02
0.100	−122.80	−20.06	−97.74	−8.88

4.2 Governor PID parameter control

Considering the PID parameters of the speed control system, the transfer function of prime mover adjustment system is

$$G_{GW}(s) = G_{GOV}(s) G_w(s) \tag{11}$$

Substituting $s = j\omega_d$ into Equation (11), we obtain

$$
\begin{cases}
D_G = \dfrac{K_1 \omega_d^6 + K_2 \omega_d^4 + K_3 \omega_d^2 + K_4}{(B_P^2 K_I^2 + \omega_d^2)(1 + T_G^2 \omega_d^2)(1 + 0.25 T_W^2 \omega_d^2)} \\
K_1 = -0.5 T_W^2 T_G K_D \ , K_4 = B_P K_I^2 \\
K_2 = T_G K_D + 0.5 T_W^2 K_D K_I B_P - 0.5 T_W^2 K_P - 0.5 T_W^2 T_G K_P B_P K_I \\
\quad\quad + 0.5 T_W^2 T_G K_I + 1.5 T_W T_G K_D K_I B_P - 1.5 T_W T_G K_P + 1.5 T_W K_D \\
K_3 = -K_D K_I B_P + K_P + T_G K_P B_P K_I - T_G K_I - 0.5 T_W^2 B_P K_I^2 \\
\quad\quad - 1.5 T_W T_G B_P K_I^2 + 1.5 T_W K_P B_P K_I - 1.5 T_W K_I
\end{cases} \tag{12}
$$

The damping torque characteristics influenced by the governor control parameters are analyzed, and the parameters of the K_P, K_I, K_D, and B_P parameters are changed. The damping torque coefficient of the turbine governor is analyzed as a function of frequency, as shown in Figures 3 and 4.

In Figure 3, the ultralow-frequency band is between 0.01 and 0.1 Hz; when K_P increases, its damping gradually weakens, and even negative damping occurs. The negative damping characteristic influenced by K_I within the ultralow-frequency band is significant, and the larger the K_I is, the larger the negative damping amplitude is. Figure 4 shows that in the ultralow-

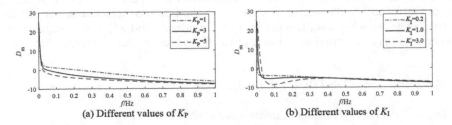

(a) Different values of K_P (b) Different values of K_I

Figure 3. Damping characteristic with different values of K_P and K_I.

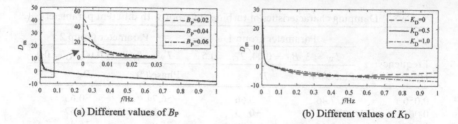

(a) Different values of B_P (b) Different values of K_D

Figure 4.   Damping characteristics with different values of B_P and K_D.

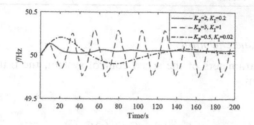

Figure 5.   System frequency with different PI parameters.

frequency band, B_P decreases and damping slightly decreases; and K_D basically provides positive damping.

The foregoing analysis shows that the magnitude of the turbine governor control parameters can easily affect the damping characteristics of the system. Therefore, in order to generate a stable frequency response and improve the system damping, it is necessary to properly decrease K_P, K_I so that the speed regulation system exhibits positive damping characteristics, as it is more conducive to improving the stability of governor.

Simulation experiments are carried out in the DC isolated system of hydropower station N. The parameters of governor in this station are: $K_P = 3$, $K_I = 1$, $K_D = 1$, $B_P = 0.04$. After decreasing K_P and K_I parameters, the frequency of the isolated system changes as shown in Figure 5. In the figure, $K_P = 3$ and $K_I = 1$, and ultralow-frequency oscillations occur. The primary motor speed control system generates negative damping torque, and the turbine adjustment system provides negative damping. When $K_P = 2$ and $K_I = 0.2$, the system oscillates, the positive-moment governing system produces positive damping torque, and the governor provides positive damping. When $K_P = 0.5$ and $K_I = 0.02$, the system oscillation attenuates, but when $K_P = 2$, the oscillating peak is significantly greater than when $K_I = 0.2$, and the governor response time is also longer.

The simulation results show that decreasing K_P and K_I parameters may increase the positive damping provided by the speed control system and help to suppress the ultralow-frequency oscillation. This is consistent with the results obtained in Figures 3 and 4. However, if the values of the K_P and K_I parameters are too small, the governor response speed will be slow. Therefore, in an actual system, it is necessary to coordinate and adjust the magnitudes of K_P and K_I according to the actual situation in order to produce an ideal damping effect.

5   COORDINATED CONTROL METHOD

5.1   *Test under primary frequency regulation out of service*

In an actual DC isolated system, the proportional coefficient of the DC frequency controller is usually $K_{HP} = 30$, the integral coefficient $K_{HI} = 22.2$, the dead zone $D_F = \pm 0.2$ Hz, and the

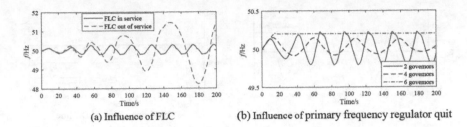

| (a) Influence of FLC | (b) Influence of primary frequency regulator quit |

Figure 6.　Frequency curve.

governor dead zone $D_G = \pm 0.05$ Hz. Simulation of the frequency dynamics when the FLC is put into operation and withdrawn from the operation in the DC isolated system is shown in Figure 6a. After the DC frequency controller exits, the system becomes unstable, and after losing the function of the FLC of DC, the frequency recovery effect of primary frequency regulation is limited, resulting in a large oscillation frequency of the DC isolated system. After the FLC is put into operation, the frequency of isolated system has an equiamplitude oscillation mode with an oscillation scope between 49.73 and 50.25 Hz and an oscillation amplitude of more than 0.2 Hz. After the FLC action, under the joint effect of the speed control system and FLC, the system does not experience destabilization, but there is a slight oscillation. The frequency changes of the isolated system after the FLC was put into operation, with the primary frequency regulation withdrawn in turn, are shown in Figure 6b.

In Figure 6b, when two governors are out of service, there will be a negative damping, and the maximum amplitude does not exceed 50.2 Hz, which is slightly decreased compared to the frequency before the governor exits. After the four governors exit, the oscillation amplitude of the system frequency is significantly decreased, the maximum amplitude is 50.17 Hz, the oscillation is in the attenuation state, the negative damping characteristic is weakened, and the FLC does not participate in frequency regulation. After the six governors exit, the system frequency is stable at 50.2 Hz, and the unit frequency is restored only under FLC action.

The Prony method is used to analyze the time-domain simulation results, and the results are shown in Table 2. The greater the number of exiting governors is, the larger the damping ratio is. The attenuation becomes larger gradually, and the system also stabilizes gradually. According to the foregoing simulation and calculation results, all the units in the isolated system are involved in the ultralow-frequency oscillation, and the hydro-generator unit speed control system provides negative damping for the oscillation. The more governors participate in the unit, the greater the negative damping provided, and the more prominent the ultralow-frequency oscillation is.

5.2　FLC of DC and primary frequency regulation control

5.2.1　Influence of FLC dead zone of DC
Without quitting the speed control system, the dead zone D_G of the governor is ± 0.05 Hz, the proportional coefficient of the FLC is $K_{HI} = 22.2$, and the integral coefficient $K_{HP} = 30$. Change the dead zone D_F of the DC frequency controller to ± 0.02 Hz and ± 0.1 Hz,

Table 2.　Prony calculation results.

Speed control system	Amplitude	Attenuation coefficient	Damping ratio
No exit	0.1047	−0.0001	0.0002
Two exits	0.1326	−0.0200	0.0089
Four exits	0.2949	−0.0392	0.1279
Six exits	0.7713	−0.1204	0.2740

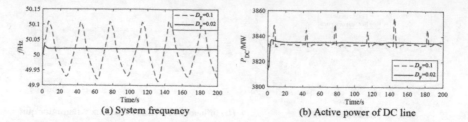

(a) System frequency

(b) Active power of DC line

Figure 7.   Influence of different dead band of FLC.

respectively. The initial disturbance is the same as that in Section 5.2, the simulation results of the frequency change, and the transmission power on the DC line are shown in Figure 7.

In Figure 7, when $D_F = \pm 0.1$ Hz, the amplitude of frequency fluctuation is significantly reduced with respect to $D_F = \pm 0.2$ Hz (Figure2). The negative damping effect produced by the governor still exists. Due to the FLC limitation, the system experiences small-amplitude oscillations. When $D_F = \pm 0.02$ Hz, the FLC dead zone is smaller than that of the governor, which causes the FLC to operate before the governor. The system frequency has been stabilized due to FLC before the governor acts, and the frequency oscillation amplitude does not exceed 50.03 Hz. The negative damping provided by the speed control system loses its effect and thus achieves the purpose of suppressing the ultralow-frequency oscillation.

5.2.2   *PI parameter control of FLC*

For the integral part of the DC frequency controller, during the actual operation of YNPG, due to the large inertia of the system, the FLC may cause frequent, rapid, and wide-ranging action of the DC power when fast and zero-difference adjustment is performed. K_{HI} is generally set to zero or a smaller value. The following is a simulation analysis on the proportional coefficient K_{HP}.

In the DC isolated system of the hydropower station N, $K_{HI} = 22.2$, $D_F = \pm 0.02$ Hz, $D_G = \pm 0.05$ Hz, K_{HP} is taken as 30 and 100 respectively, and the change of frequency and active power are shown in Figure 8. When K_{HP} becomes larger, the peak frequency of the system frequency decreases obviously, and the frequency oscillation amplitude decreases. The change trend of the active power is consistent with the change of the frequency of isolated system, which indicates that the K_{HP} is more favorable for suppressing the ultralow-frequency oscillation phenomenon.

Through the foregoing test, based on the coordinated control of the speed control system and the FLC of DC, in order to suppress the ultralow-frequency oscillation in the DC isolated system, it is recommended to amplify the dead zone of the speed control system, and at the same time reduce the dead zone of the DC frequency controller. When $D_G > D_F$ the role of FLC regulation can fully come into play, and it is conducive to the stability of the governor. In addition, the proportional coefficient K_{HP} of FLC should be increased so that the FLC of DC adjusts linearly as much as possible when the frequency fluctuates. In addition, the

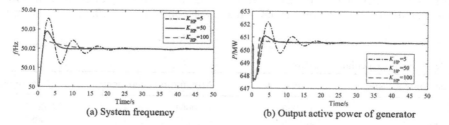

(a) System frequency

(b) Output active power of generator

Figure 8.   Frequency variation with different K_{HP}.

integral coefficient K_{HI} should be reduced. In order to avoid frequent and rapid large-scale adjustment of DC power, it is better to set K_{HI} to zero or a smaller value.

6 CONCLUSION

In actual DC isolated system, ultralow-frequency oscillations are apt to occur. This article analyzes this phenomenon and draws the following conclusions:

(1) Based on the frequency-domain response, the influence of FLC of DC on isolated system is analyzed. It is beneficial to improve the stability of the system when the FLC of DC is put into use.

(2) The damping torque analysis of the speed control system shows that appropriate reduction of K_P and K_I parameters is beneficial to suppressing ultralow-frequency oscillation. In the isolated system simulation, the system governor provides positive damping when $K_P = 2$, $K_I = 0.2$, and the system frequency can stabilize gradually.

(3) The simulation study was conducted on the coordinated control between a primary frequency controller and a DC frequency controller. The results show that reducing the dead zone of FLC of DC and increasing the proportional coefficient of the DC frequency controller can effectively suppress the ultralow-frequency oscillation.

REFERENCES

Chen, Y., Liu, Y., Tang, Z., et al. 2017. Analysis of ultralow frequency oscillation in Yunnan asynchronous sending system. In IEEE Power & Energy Society General Meeting, Chicago, IL, USA.

CIGRE WG 14. 07, IEEE WG 15. 05. Guide for planning DC links terminating at AC system locations having low short circuit capacities, part I: AC/DC system interaction phenomena. *Electra*, 1992, 142(1): 119–123.

CIGRE WG 14. 07, IEEE WG 15. 05. Guide for planning DC links terminating at AC system locations having low short circuit capacities, part II: planning guidelines. *Electra*, 1992, 142(1): 56–62.

CIGRE WG 14. 05. Report on voltage and power stability in AC/DC systems. *Electra*, 2003, 153(1): 81–85.

Dhaliwal, N.S., Recksiedler, L.D., & Tang, D.T.Y. 1996. Operating experiences of Nelson River HVDC system. In Transmission and Distribution Conference, Los Angeles, 174–180.

Gong, T., Wang, G., & Li, T. 2014. Analysis and control on ultralow frequency oscillation at seeding end of UHVDC power system. In International Conference on Power System Technology, Chengdu, China.

Huang, H. & Li, F. 2013. Sensitivity analysis of load-damping characteristic in power system frequency regulation. *IEEE Transactions on Power Systems* 28(2):1324–1335.

Kundur, P. 1994. *Power System Stability and Control*. New York: McGraw-Hill.

Liu, C., Zhang, J., Chen, Y., et al. (2016). Mechanism analysis and simulation on ultralow frequency oscillation of YNPG in asynchronous interconnection mode. *Southern Power System Technology 10* (7):29–34 (in Chinese).

Wang, G., Yu, Z., Zhang Y., et al. 2016. Troubleshooting and analysis of ultralow frequency oscillation mode in power system. *Power System Technology*, 40(8): 2324–2330 (in Chinese).

Wichert, H. & Dhaliwal, N. (1978). Analysis of PID governors in multimachine system. *IEEE Transactions on Power Apparatus and Systems* no. 2: 456-463.

Zhang, J., Liu, C., Chen, Y., et al. (2016). Countermeasures and experiments on ultralow frequency oscillation of YNPG in asynchronous interconnection mode. *Southern Power System Technology* 10(7): 35 –39 (in Chinese).

Zheng, C., Ding, G., Liu B., et al. (2018). Analysis and control to the ultralow frequency oscillation in southwest power grid of China: a case study. In Chinese Control and Decision Conference (CCDC), Shenyang, China.

A method for testing the sensitivity of equipment to voltage sag

Junxia Meng & Limin Jiang
China Electric Power Research Institute, Beijing, China

Kai Ding & Wei Li
State Grid Hubei Electric Power Research Institute, Wuhan, China

ABSTRACT: Testing the voltage tolerance curve (VTC) of equipment is the basis for cognizing and understanding the voltage sag. This paper compares the existing five test methods and proposes a new test method. This method is based on the idea of curve classification, and can measure the VTC of various sensitive equipment efficiently and accurately. This paper proves the superiority of the method by building a test platform and comparing the test accuracy and workload of each method through the measured data.

1 INTRODUCTION

In the existing defined power disturbances, voltage sag is the most common power quality problem complained by users, which results in user production interruption and huge economic loss (PELEGRINI M A et al. 2012, CEBRIAN J C et al. 2015). Due to the wide range of sensitive equipment in the industry, there is an urgent need for an accurate and efficient method for the tolerance test of the equipment to voltage sag.

The tests of sensitive equipment to voltage sags can be divided into compliance test and tolerance test. Compliance test refers to the test of the compliance of the sensitive equipment's immunity to voltage sag by a professionally certified test laboratory. IEC has promulgated standards such as IEC 61000-4-30/11/34 and IEC 61000-2-1 . These standards only specify several specific test points, and the results do not accurately reflect the overall perspective of the tolerance capacity of the equipment. The tolerance test is to obtain more detailed information about the performance of the equipment under voltage sag conditions. Compared with the compliance test, the tolerance test needs to be tested multiple times at more test points, but the requirements for the specific details of each test are relatively loose. The voltage tolerance curve (VTC) of the sensitive equipment can be gained through the tolerance test. VTC is used to describe the tolerance of sensitive load to voltage in the two dimensions of sag magnitude and sag time. It can describe the electrical properties of the equipment during voltage sag, and reflect whether the equipment can operate normally.

The selection of test methods is the key to improving the accuracy and efficiency of the test. The existing test methods are based on IEEE Std. 1668-2017 and the step-by-step test method. IEEE Std. 1668-2017 proposes complete test methods for voltage sags tolerance characteristics of low-voltage equipment . The test methods mainly include top-down, left-to-right and box-in methods. However, the number of the test points recommended by these three methods is small, so the test accuracy is relatively low. In the step-by-step test method, the step size of the sag magnitude is 0% to 1%, and the duration of sag starts from 0ms. The step of duration depends on the actual situation and can be 5ms or 10ms. This test method can obtain more accurate test results, but there are a lot of unnecessary test points, which is not conducive to the measurement of multiple equipment (S. Z. DJOKIC et al. 2004, S. Z. DJOKIC et al. 2005). In addition, some scholars have proposed a dichotomy test method (Wang, Y et al. 2018), which can measure all types of tolerance curves and improve the accuracy of the vertical part of the tolerance curve to some extent, but the accuracy of the tolerance curve is still low. For the equipment with a rectangular tolerance curve, the test workload is heavy, but the accuracy of the test results is not greatly improved.

Based on the existing test methods, this paper proposes a new test method for tolerance test, which comprehensively considers the characteristics of various types of equipment tolerance curves. The method can adjust the test strategy according to the shape of the tolerance curve, greatly reduce the test procedure under the condition of ensuring the accuracy of the experiment. The superiority of the method is proved by actual measurement.

2   CURRENT METHODS

At present, the main methods for testing the tolerance capacity of equipment to voltage sag are proposed by IEEE Std. 1668-2018, including top-down, left-to-right and box-in test methods. In addition, relevant scholars mention step-by-step test method and dichotomy test method.

2.1   *The methods are proposed by IEEE Std. 1668-2018*

The top-down method is a commonly used test method. In each test cycle, the sag magnitude is gradually reduced from 85% by 5% intervals until the test point is at the trigger point of the failure criterion. Thereafter, the duration of the voltage sag is reduced to the next predetermined value until all the preset duration is measured.

The left-right test method is to keep the voltage sag magnitude constant, and the voltage sag duration is from the set minimum value to the maximum value until the duration reaches maximum or the equipment under test has failed. Thereafter, the voltage sag magnitude is reduced to the next predetermined value until all of the preset magnitudes have been measured.

The box-in test method enables researchers to quickly obtain a voltage tolerance curve. First, longest tolerated time of the equipment under test for short-term voltage interruption can be found. Then, set the test maximum sag duration, and the sag magnitude gradually decreases by 5% intervals until the critical point is found from 85%. Continue to reduce the duration, enter the next cycle, the initial point of the sag magnitude is no longer 85% but from the point greater than 5% of the previous critical magnitude, still decreases to the critical point by 5% interval. This process will be repeated several times until the tolerance curve of the equipment is obtained.

These three test methods recommend few test points, so the test accuracy is relatively low. Even if a certain test point is added, the test result is not accurate and it is difficult to meet the requirements. The top-down method and box-in method can not measure the voltage tolerance curve of all equipment, but can only measure the monotonic curve.

2.2   *Step-by-step test method*

The test method starts from 0V (the minimum test plan magnitude), keeps the magnitude constant, and the voltage sag duration is changed from the set minimum value (specified in the test plan) to the maximum value (also as specified in the test plan), but the step size is 5ms until the voltage sag duration reaches the maximum value or the equipment under test fails. Thereafter, the voltage sag magnitude is increased to the next test point with the step of 1%, and the duration is repeated. The duration in the sag magnitude loop becomes larger until all preset magnitudes have been measured.

Although the method can obtain the tolerance curve of any equipment and the test accuracy is high, but the step size is small and the workload is greatly large, which is not conducive to the measurement of multiple equipment.

2.3 Dichotomy test method

This test method is to keep the magnitude constant, and determines the duration by dichotomy method. Based on the results of the last two tests, the next test point is determined by the dichotomy, until a critical test point between the device tripping and normal operation is found.

Dichotomy test method can measure all types of tolerance curves, and improve the accuracy of the vertical part of the tolerance curve to a certain extent, but the accuracy of the horizontal part of the tolerance curve is still low. For the equipment whose tolerance curve is rectangular, the test workload is large, but the accuracy of the test results is not improved greatly.

The existing test methods are difficult to guarantee the accuracy and speed of the test at the same time. A new test scheme is proposed for this problem. The tolerance curves are divided into two categories (i.e. Type I and Type II) for different tolerance curves of equipment. According to the characteristics of different types of tolerance curves, different test methods are applied to improve the test speed and accuracy of the test plan.

3 THE PROPOSED METHOD

In order to improve the test speed and obtain more accurate tolerance curve, this paper proposes an adaptive test method. The basic idea of the method is to divide the tolerance curve into two categories according to the difference of tolerance curves (i.e. Type I and Type II). Based on the characteristics of different types of tolerance curves, different test methods are applied to improve the test speed and accuracy of the test plan. The test method is as follows:

Step 1: Set the voltage magnitude to 0, and test the fault condition of the equipment under the maximum duration T_{max}. If the equipment does not malfunction under the maximum duration, according to the IEEE Std. 1668-2017, the equipment can be considered to be able to withstand voltage interruption under this duration. The equipment has a strong tolerance, and the test is stopped.

Step 2: If the equipment fails under the maximum duration T_{max}, the magnitude remains unchanged, and the fault duration of the next test point is $T_1 = (T_{max} + T_{min})/2$. If the equipment fails under time T_1, let $T_{max} = T_1$, the next duration test point is $T_1 = (T_{max} + T_{min})/2$. If the equipment does not fail under time T_1, let $T_{min} = T_1$, the next duration test point is $T_1 = (T_{max} + T_{min})/2$, and so on until the difference between T_{max} and T_{min} is less than 2ms, the longest withstand time of voltage short interruption is $T_{0max} = (T_{max} + T_{min})/2$. After the test is completed, the values of T_{max} and T_{min} are restored to the initial values.

Step 3: Set the voltage sag time to: $t = T_{0max}-\Delta t$ (Δt recommended value is 5ms, can be determined according to the actual situation), and keep the voltage sag time unchanged, the sag voltage magnitude changes from the set minimum value V_{min} to the maximum value V_{max}, until the sag the voltage has reached the maximum value or the equipment under test has failed. If the fault occurs, the tolerance curve is Type I, otherwise the tolerance curve is Type II.

Step 4: Test the maximum magnitude and maximum duration point in the plan, i.e. point (T_{max}, V_{max}). If the equipment fails,it indicates that the equipment is weak in tolerance, stop testing.

Step 5: If the equipment does not fail, keep the duration unchanged, and the next test magnitude is $V_1 = (V_{max} + V_{min})/2$. If the equipment under the magnitude V_1 fails, let $V_{min} = V_1$, the next magnitude test point is $V_1 = (V_{max} + V_{min})/2$. If the equipment is not faulty, let $V_{max} = V_1$, the next magnitude test point is $V_1 = (V_{max} + V_{min})/2$, and so on until the difference between V_{max} and V_{min} is less than 2%, then the critical magnitude of withstanding voltage for duration of T_{max} is $V_{0max} = (V_{max} + V_{min})/2$. Type I tolerance curve jumps to step 6, and Type II tolerance curve jumps to step 7.

Step 6: Reduce the duration t to $(T_{max} + t_{0max})/2$, and let $T_{max} = t$. Keep the duration a constant, the initial point of the sag magnitude is 1% greater than the previous critical magnitude, and decreases to the critical point by 1% interval. Reduce the duration and execute the next cycle, and repeat the aforementioned cycle for eight times (which can be determined according to the accuracy requirements). Then let $t = t_{0max}$, and measure the critical point when sag duration is t_{0max}. The test terminates.

Step 7: Set the duration to t_{0max}, and keep the duration a constant. The initial point of the sag magnitude is 1% greater than the previous critical magnitude, and decreases to the critical point by 1% interval. Find the ordinate U_2 of the critical test point corresponding to t_{0max}. Set the maximum test time that is t_{0max}, step size of 5% magnitude, and magnitude range from 0 to U_2. Set the voltage magnitude to the maximum value to be tested, and use the dichotomy method to determine the critical time point of the voltage sag test (test step is the same as Step 2). After determining the critical time point under the magnitude, reduce the magnitude and enter the next cycle until the voltage is 0.Eventually the test ends.

Figure 1 shows the flow chart of the method. Figure 2 is the diagram of the method. As shown in the figure, we can classify the type of the tolerance curve according to cycle 2. According to the characteristics of the shape of the tolerance curve, the method divides the tolerance curve into two types, Type I and Type II. The Type I tolerance curve has the minimum critical tolerance time when the sag magnitude is 0, while the minimum critical tolerance time of Type II exists when the sag magnitude is not 0. The advantage of this method is that the test steps can be changed according to the shape of the curve, and the best test steps are selected according to the characteristics of the curve, thereby reducing the test workload.

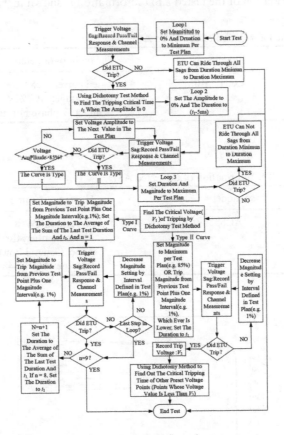

Figure 1.   New voltage-sag characterization test method flow chart.

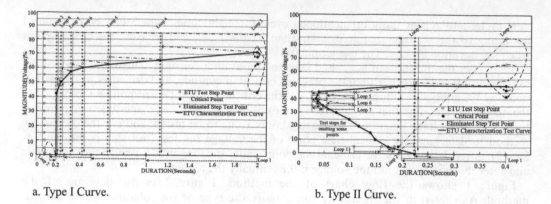

| a. Type I Curve. | b. Type II Curve. |

Figure 2. Illustration of the new test method.

4 TEST

In order to verify the applicability of the method to Type I and Type II curves, the paper do the test on an ACC and an ASD respectively. The accuracy and the workload of the method were compared with the four published methods. The information of the tested ACC is shown in Table 1. The information of the tested ASD information is shown in Table 2.

Table 1. The information of the tested ACC.

Brand	model	The main parameters
Schneider	LC1E2510M5N	Rated current: 25A Coil voltage: AC220V Contact: 3-phase normally open type

Table 2. The information of the tested ASD.

Brand	model	The main parameters
ABB	ACS550-01-05A4-4	Power: 2.2KW Phase number: three phase Type: General Power supply Voltage: AC380-480V

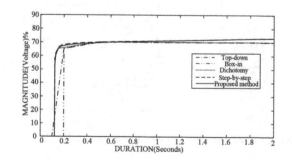

Figure 3. Test results of the ASD.

4.1 The test results of the ASD

From the above test results, it can be seen that the error of the methods proposed by IEEE Std. 1668-2017 is large. Because there are few test points in the test plan, the tolerance curve differs greatly from the curve measured by step-by-step test method. If you want to improve the accuracy, you need to increase the test points substantially. It can be seen that these two methods have certain defects. In this test, the time error of the test results of the two methods proposed by IEEE Std. 1668-2017 reached 75ms. Obviously, for sensitive equipment such as ASD, the test results are not satisfactory. The step-by-step test method has higher accuracy, but there are 7716 test points, which take a long time. The accuracy of the vertical part of the tolerance curve measured by the dichotomy test method is high and there are only 168 test points, but there are large errors in the corners and horizontal parts of the curve. To reduce the error, the amplitude step size needs to be reduced, but this way will greatly increase the workload. Compared the method proposed in this paper(adaptive method) with step-by-step test method, we found that the tolerance curves measured by the two methods are almost the same, and there are only 62 test points. It can be seen that the method can greatly reduce the workload while ensuring the accuracy of the tolerance curve. Table 3 shows the workload and accuracy of different methods.

4.2 The test results of the ACC

Since the methods proposed by IEEE Std. 1668-2017 cannot measure the ACC tolerance curve, only the dichotomy test method, the step-by-step test method, and the test method proposed in this paper are used for testing the ACC. As with the test results of the ASD, the accuracy of the vertical part of the ACC tolerance curve measured by the dichotomy method is higher, but there is a large error in the corner and horizontal part of the curve. The main advantage is that the test workload is small, which are only 81 test points. The step-by-step test method requires 8881 points to be tested, and the test workload is further increased than ASD. Comparing the method proposed in this paper with the step-by-step test method, it is found that there are some errors in the corner of the curve measured by the method, but the accuracy is still higher than the dichotomy test method. And the workload of the method is smaller than the dichotomy test method, which are only 67 test points. Table 4 shows the workload and accuracy of different methods.

Table 3. The workload and accuracy of different methods (ASD).

Methods	Top-down	Box-in	Dichotomy	Step-by-step	Proposed method
Workload	82 times	15 times	168 times	7716 times	62 times
Duration error	75ms	75ms	1ms	5ms	1ms
Magnitude error	< 5%	< 5%	< 5%	< 1%	< 1%

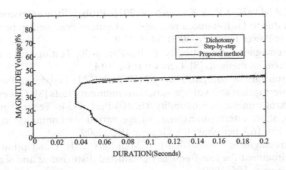

Figure 4. Test results of the ACC.

Table 4. The workload and accuracy of different methods (ACC).

Methods	Dichotomy	Step-by-step	Proposed method
Workload	81 times	8881 times	67 times
Duration error	1ms	5ms	1ms
Magnitude error	< 5%	< 1%	< 5%

4.3 *Comparison and summary of test methods*

After testing the ASD and ACC, comparing the workload and accuracy in Table 3 and Table 4, the methods proposed by IEEE Std. 1668-2017 have a small workload, but the error is large. It is absolutely unacceptable for sensitive equipment. The accuracy of the step-by-step test method is high, but the test workload is too large. The dichotomy test method can meet the accuracy requirements to a certain extent, and the test points are few, which is better than the previous methods. The method proposed in this paper can greatly reduce the workload of the test while ensuring the accuracy, and through experiment comparison, it is found that the method is superior to the dichotomy test method.

5 CONCLUSION

This paper proposes a new test method for testing the voltage tolerance curve. According to the difference of tolerance curves, the tolerance curves were divided into two categories (i.e. Type I and Type II). The best test method is applied according to the characteristics of different types of tolerance curves to improve the test speed and accuracy of the test plan. The experimental results show that compared with the traditional box-in test method, the test accuracy is greatly improved, and the test workload is not increased much. In addition, the method can measure the tolerance curve of any sensitive equipment, and solves the problem that the traditional test method can only measure the monotonic curve.

ACKNOWLEDGEMENT

This research is funded by [State Grid Science and Technology Project] project number [52153218000B]

REFERENCES

PELEGRINI M A. & ALMEIDA C F M. & KONDO D V. 2012. Survey and applications of interruption costs in large customers [C]//2012 IEEE 15th International Conference on Harmonics and Quality of Power, (4): 860–864.

CEBRIAN J C. & MILANOVIC J V. & KAGAN N. 2015. Probabilistic assessment of financial losses in distribution network due to fault-induced process interruptions considering process immunity time [J]. IEEE Transactions on Power Delivery, 30(3): 1478–1486.

IEC 61000–4–30, Electromagnetic compatibility (EMC)-Part 4–30: Testing and measurement techniques-power quality measurement methods [S]. Geneva: IEC, 2014.

IEC 61000–4–11, Electromagnetic compatibility (EMC)-Part 4–11: Testing and measurement techniques voltage dips, short interruption and voltage variations immunity tests [S]. Geneva: IEC, 2001.

IEC 61000–4–34, Electromagnetic compatibility (EMC)-Part 4–34: Testing and measurement techniques–Voltage dips, short interruptions and voltage variations immunity tests for equipment with mains current more than 16A per phase [S]. Geneva: IEC, 2009.

IEC 61000–2–1, Electromagnetic compatibility (EMC)-Part 2–1: Description of the environment-Electromagnetic environment for low-frequency conducted disturbance and signaling in public power supply systems [S]. Geneva: IEC,1990.

IEEE Approved Draft Recommended Practice for Voltage Sag and Short Interruption Ride-Through Testing for End-Use Electrical Equipment Rated Less than 1000 V, IEEE P1668/D3, March 2017, vol., no., pp.1–90, 1 Jan. 2017.

S. Z. DJOKIC. & J. V. MILANOVIC. & D. S. KIRSCHON. 2004."Sensitivity of AC coil contactors to voltage sags, short interruptions and undervoltage transients", IEEE Trans. Power Delivery, vol.19, no.3, pp.1299–1307.

S. Z. DJOKIC. & K. STOCKMAN. & J. V. MILANOVIC. et al. 2005. "Sensitivity of AC adjustable speed drives to voltage sags and short interruptions", IEEE Trans. Power Del, vol.20, no. 1, pp.494–505.

S. Z. DJOKIC. & J. DESMET. & G. VANALME, et al. 2005. "Sensitivity of Personal computers to voltage sags and short interruptions", IEEE Trans. Power Del, vol.20, no.1, pp.375–383.

Wang, Y. & MA Z.Y. & XU, Z. et al. 2018. "A new method for sensitive single-phase equipment testing to voltage sag". ICHQP2018, 18th International Conference on Harmonics and Quality of Power, Ljubljana, Slovenia.

Emerging Developments in the Power and Energy Industry – Dufo-López, Krzywanski & Singh (eds)
© 2020 Taylor & Francis Group, London, ISBN 978-0-367-27169-5

PIT analysis of typical industrial production processes under voltage sag

Z. Xu, Z.Y. Ma, K. Zhou & Q.W. Guo
Guangzhou Power Supply Bureau Electric Power Test Research Institute, Guangzhou, China

ABSTRACT: Voltage sag is a hot and difficult issue in power quality problems. Voltage sags lead to industrial process malfunction and huge economic losses. The voltage tolerance curve(VTC) describes the electrical characteristics of the equipment, and the process immunity time(PIT) describes the physical characteristics of the equipment. This paper proposes a new and practical method for analyzing the voltage sag immunity of industrial processes. Finally, the automobile production process is analyzed in detail

1 INTRODUCTION

With the wide application of digital and informative technology in various fields, more and more sensitive equipment is connected to the power system. Sensitive equipment has increased the production level of all walks of life, but the probability of failure of industrial equipment and production processes has increased, which has brought huge economic losses to industrial users and society. voltage sag has become an issue of increasing concern to the power industry and the whole society, and it is considered by industry and academia to be the most serious power quality problem[1-2]. Voltage sag problem includes the internal characteristics of the power system and the power consumption characteristics of the user-side sensitive equipment. It is the most prominent, influential and most harmful transient power quality problem in the power system. The contradiction between voltage sag events and sensitive equipment is deeper. Therefore, voltage sag has become the most concerned and most urgently power quality problem in industry and academia in recent decades.

In recent years, in order to investigate the sensitivity of equipment to voltage sag, a large number of test experiments were carried out. Test experiments include setting influencing factors, selecting malfunction criterion of equipment under test(EUT), experimental setup, the test method and the test procedure[3-5]. EUT includes AC coil contactors(ACC)[6-9], personal computers(PC)[10-11], adjustable speed drivers(ASD)[12-14], programmable logic controllers(PLC)[15] and so on. In practical applications,engineers are usually concerned with the tolerance of individual sensitive equipment and believe that the impact of individual component or equipment can be affected by the entire industrial process, due to the limitations of cognitive means and methods.

In fact, the influence of voltage sag on industrial production process or equipment is not only related to magnitude and duration, but also to self design, selecting equipment, industrial process, equipment properties and the state of user accept. In other words, to evaluate the voltage sag immunity of industrial processes, we need to consider not only the basic characteristics of sag, but also the structural and functional characteristics of industrial processes.

Therefore, in order to assess the impact of voltage sag on industrial production processes, it is necessary to find a bridge that links the voltage on the supply side to the state of the process consequences.

This paper evaluates the voltage sag immunity process of a typical industrial production process. According to the PIT proposed in the C4.110 Working Group report, combined with the voltage sag tolerance of sensitive equipment and its role in industrial production, key components are identified. The immunity of the production process, equipment and components of the industrial process is then ranked and ranked according to the role of the component throughout the process and the state of acceptable consequences for the user. Finally, a hierarchical evaluation method is used to evaluate the voltage sag immunity of a typical industrial production process, and to find the key links and equipment in the industrial production process.

2 PROCESS PARAMETER CURVE

In actual production, it is not electrical engineers but process or instrumentation engineers who can directly perceive the impact of voltage sags. This perception represents the physical parameters in the production process, such as wind speed, pressure, temperature and flow rate. JWG C4.110, a joint working group composed of CIGRE, CIRED and UIE, which proposes the concept of PIT, as shown in Figure 1. It is also recommended to use PIT to identify sensitive elements in industrial processes.

In the Figure 1, the process parameters are the implicit physical parameters, such as wind speed, flow rate, etc. P_{nom} and P_{limit} are the normal and acceptable limit values of process parameters. t_1, Δt, t_2 are the occurrence, response delay and critical time of sags.

3 ANALYSIS OF TYPICAL PRODUCTION PROCESSES

3.1 *User perceived loss based on VTC and PIT*

When voltage sag causes the equipment to malfunction and the structure or function of the production process is missing, it will cause a process interruption loss similar to that caused by the unplanned power outage. When the voltage sag does not cause equipment malfunction, the physical parameters in the process may be affected to varying degrees. According to the degree of influence of the physical parameters, the user will suffer different degrees of loss. In this paper, a loss event with no equipment failure and uninterrupted process is defined as a user-aware loss event. Therefore, the voltage sags loss event includes a process interruption loss event and a user perceptual loss event, as shown in Figure 2.

In order to assess the user's voltage sag loss more accurately, the voltage sag loss event can be divided into three parts: equipment electrical characteristics, physical properties and perceived loss, as shown in Figure 3. When a voltage sag occurs, the device VTC can be used to determine the equipment malfunction state that causes significant loss on the voltage magnitude (V)-duration (T) plane, while CIGRE C4.110 The recommended PIT curve makes it easier to identify the user's perceived loss when the electrical state of the device is difficult to determine. Therefore, it is more reasonable to combine the VTC and PIT curves and evaluate the user's perceived loss according to the mapping law of VTC to PIT in different production processes.

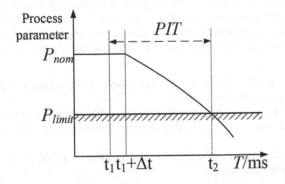

Figure 1.   Definition of the Process Immunity Time (PIT) for an equipment within the process.

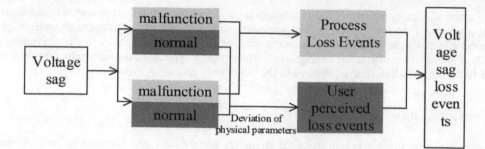

Figure 2. Voltage sag loss event of production process.

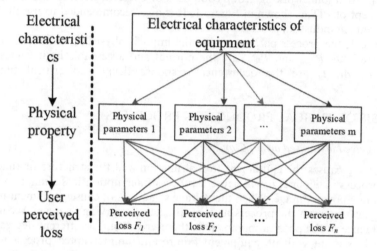

Figure 3. Electrical Characteristic-Physical Attribute-Perceived Losses.

3.2 VTC and PIT in production process

The sensitive equipment commonly used in the production process is ACC, PLC, ASD, etc. Only ASD is analyzed here. Typically, ASD is used to drive and control motors at different load rates during production. Engineers can understand the working status of the ASD drive system by monitoring physical parameters such as speed, torque, pressure and pressure in real time.

The VTC curve and the PIT curve reflect electrical and physical properties, respectively. When the load or process is different, the mapping relationship between the two is different. When the electrical state of the industrial process is difficult to determine, the operating state of the ASD drive system can be reflected by physical parameters. The mapping method of the mapping between electrical characteristics and physical attributes of sensitive equipment is shown in Figure 4:

1) When the voltage sag does not affect the operation of the equipment, the physical parameters maintain the normal value of PP_N.
2) When the voltage sag occurs in uncertain regions, different sags correspond to different PIT curves. $V_0 \rightarrow V_1 \rightarrow V_2 \rightarrow V_3$ corresponding process immunization time is $PIT_0 \rightarrow PIT_1 \rightarrow PIT_2 \rightarrow PIT_3$
3) When the voltage sag occurs in malfunction area, the physical parameters exceed PP_{limit}, and the equipment malfunction.

Therefore, the production status of industrial process can be evaluated by PIT curve

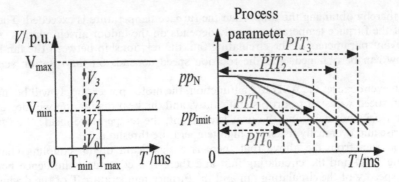

Figure 4.　Mapping relationship between Electrical and physical performances.

This paper is based on an automobile manufacturing plant in Guangzhou. The automotive industry mainly includes stamping, welding, painting, final assembly process, and engine parts production. On-site engineer feedback, affected by voltage sag or short-term interruption accidents, prone to failure and make the production recovery time longer, there are electrophoretic coating drying, sealing and drying, the equipment involved are electrophoresis drying furnace, sealing Glue drying oven, control system, etc. The recovery time after such interruption is long, which seriously affects the production recovery time. The survey analysis found that the sensitive part of the automotive industry is mainly concentrated in the process of painting workshop. This analysis discusses the coating process in detail.

4 ANALYSIS OF CASE

4.1 *Describe the problem*

The drying process of the automobile coating production line is easily interrupted due to the tripping of power quality problems. After analysis, it is found that the reason for the drying furnace tripping is that the heat dissipation system in the furnace is out of adjustment, causing the furnace temperature to be too high, which triggers the furnace overheating protection, resulting in the furnace burner is off.

According to the relevant safety guidelines, after the burner is closed, it needs to be completely inspected and cleaned before it can be used for secondary ignition. The pre- and post-processing time is about 30-45 minutes, which is a key factor limiting the restart time of the link, which brings serious problems to the enterprise. Loss of capacity. The furnace heat dissipation system is mainly composed of a circulation fan (fan fan), and the device driver (ASD) is easy to stop under a sag event, thereby causing the circulation fan to fail. During the investigation, it was found that there were two sets of circulating fans in the drying furnace of the plant, and each group of circulating fans was supplied by the same factory for transformers.

4.2 *Analysis problem*

According to the analysis method proposed in C4.110, the drying furnace system is regarded as the industrial process of concern, and the process side is combed to the power supply end, and the process includes the drying process and the heat dissipation link. The heat dissipation link is mainly the circulating fan and its driving and control equipment. The key parameter of the process is the furnace temperature T, which can be known according to the relevant engineering safety guidelines. The stability of T depends mainly on the balance between the heat supply of the burner and the heat dissipation of the heat dissipation system. According to the heat dissipation reduced by the fan, the trend of the furnace temperature change can be obtained, thereby obtaining the time when the furnace temperature is exceeded. The heat dissipation of the furnace temperature mainly depends on the indoor air circulation. According to the relevant parameters of the circulating fan, the relationship between the fan speed and the air flow can be obtained, and the rotation speed depends on the rotation speed of the motor.

When an event causes the ASD to malfunction, the motor power supply will be interrupted, and the fan speed will start to drop significantly, and the heat dissipation will decrease. Since the burner works normally and the heat is constant, the temperature inside the furnace will gradually rise and eventually reach the high temperature threshold.

It can be seen from the above that there is a significant digital coupling characteristic between the ASD and the circulating fan, and there is a continuous influence between the rotational speed v of the circulating fan and the furnace temperature T of the drying furnace, so that the above industrial micro-process can be roughly divided into two. The level, that is, the level of the power supply link (composed of ACC and ASD for sensitive equipment) and the temperature system level. However, considering the feasibility of the research scheme, the circulating fan is divided into one part of the furnace temperature system, and the power supply link belongs to the circulating fan link, which is divided into three levels, as shown in Figure 5.

In Table 1, PIT1 is the PIT value of each specific sensitive device, PIT2 is the PIT value of each fan's individual fault, and PIT3 is the PIT value when both sets of fans are faulty.

Based on the above theoretical analysis, K_3 is the PIT value of two fans in case of simultaneous malfunction and the time when the furnace temperature T exceeds the limit after the fans malfunction. The PIT value K_{2-1}/K_{2-2} of one fan malfunction is the time needed for the fan malfunction and the time of temperature T exceeding the limit in the case of one fan malfunction.

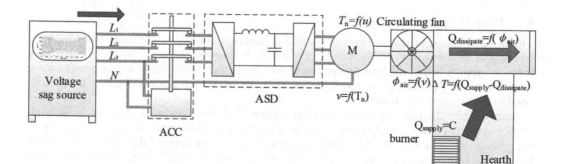

a. Process structure

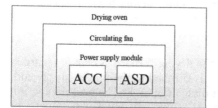

b.Equivalent structure

Figure 5.   Sensitive process of drying oven. a.Process structure. b.Equivalent structure.

Table 1. Record table of analysis process.

Level I	Level II	Level III	Process parameter	PIT1	PIT2	PIT3
Drying hearth	Circulating fan 1	ACC	ΔT	K_{1-1}	K_{2-1}	K_3
		ASD	ΔT	K_{1-2}		
	Circulating fan 2	ACC	ΔT	K_{1-3}	K_{2-2}	
		ASD	ΔT	K_{1-4}		

Therefore, it can be found that the lower the level and the smaller the scale, the longer the process fails, but at the same time, it is necessary to note that the probability that the link will suffer a sag is greater, and the risk of process failure may be higher. high. Since the two sets of fans are powered by the same transformer, when the sensitive event occurs, all the fans are tripped due to the action of the Class III sensitive equipment, so that the furnace temperature rises rapidly and the PIT value is short.

4.3 *Analysis and summary of case*

This survey can be divided into the following steps:
Step 1: Learn from the field engineers about the power supply of the factory;
Step 2: Conduct relevant analysis on the production process;
Step 3: Hierarchical grading of production links
Step 4: Get the PIT for each process

The conclusion that can be obtained from the analysis of this link is that the failure of this link can lead to the failure of the entire industrial process, and the recovery time is long, which brings huge economic losses to the enterprise. Through the analysis of the power grid around the plant, the power supply mode of the plant, and the situation of the affected equipment. The mitigation measures can be proposed through the following aspects:
1) Chang the power supply mode
2) Add DVR.

5 CONCLUSION

Using the proposed method in this paper to analyze the industrial production line, the voltage sag is effectively combined with the process parameters in the industrial production line, and the immune time of the industrial production line to the voltage sag is accurately evaluated. It provides a scientific basis for industrial users' production process, equipment selection, and operation and maintenance of industrial processes. Effectively avoid unnecessary compensation investment and possible damage caused by equipment after operation.

REFERENCES

[1]GALLO D, LANDI C, LUISO M, FIORUCCI E. 2018. Survey on voltage dip measurements in standard framework [J]. IEEE Trans. Instrument. Meas, 63(2): 374–387.
[2]CEBRIAN J C, MILANOVIC J V, KAGAN N. 2015. Probabilistic assessment of financial losses in distribution network due to fault-induced process interruptions considering process immunity time [J]. IEEE Transactions on Power Delivery, 30(3): 1478–1486.
[3]IEEE Std 1668-2017 IEEE Recommended Practice for Voltage Sag and Short Interruption Ride-Through Testing for End-Use Electrical Equipment Rated Less than 1000 V. USA, 2017.
[4]IEC 61000-4-11, Electromagnetic compatibility (EMC)-Part 4-11 : Testing and measurement techniques-Voltage dips, short interruption and voltage variations immunity tests. IEC, press, 2001.
[5]IEC 61000-4-34: Testing and measurement techniques–Voltage dips, short interruptions and voltage variations immunity tests for equipment with input current more than 16A per phase. IEC, Elektron press, 2009.

[6] DJOKIC S, MILANOVIC J V, KIRSCHEN D. 2004. Sensitivity of AC Coil Contactors to voltage sags, short interruptions, and under-voltage transients. IEEE Trans Power Delivery, 19 (3):1299–1307.

[7] ELIPHICK S, SMITH V, GOSBELL V.2013. Voltage sag susceptibility of 230V equipment [J]. IET Generation Transmission & Distribution, 77 (6):576–583.

[8] MC G M, MUELLER D, SAMOTYI M. 1993. Voltage sags in industrial systems [J]. IEEE Transactions on industry applications. 29 (2):397–403.

[9] HONRUBIA E A, GOMEZ L E, MOLINA G A. 2014l. Load influence on the response of AC Coli Contactors under power quality disturbances [J]. International Journal of Electrical Power and Energy Systems, 63 (4):846–854.

[10] DJOICK S Z, STOCKMAN K, MILANOVIC J V. 2005. Sensitivity of AC Adjustable Speed Drives to voltage sags and short interruptions [J]. IEEE Transactions on Power Delivery, 20 (1):494–505.

[11] BOLLEN M H, LIDONG D. 2000. Analysis of Voltage Tolerance of AC Adjustable-Speed Drives for Three-Phase Balanced and Unbalanced Sags [J]. IEEE Transaction on Industry and Application, 36 (3):904–910.

[12] DJOICK S Z, DESMET J, VANALME G. 2005. Sensitivity of Personal Computers to Voltage Sags and Short Interruptions [J]. IEEE Transactions on Power Delivery, 20 (1):375–383.

[13] YANG D, XIAO X Y, WANG Y. 2013. Evaluation of equipment voltage sags with failure and severity uncertainty [J]. Electric Power Automation Equipment. 33 (10):107–117.

[14] GUPA C P, MILANOVIC J V. 2006. Probabilistic assessment of equipment trips due to voltage sags [J]. IEEE Transactions on Power Delivery. 21 (2):711–718.

[15] SURYA H, DAUT I. Sensitivity of Low Voltage Consumer Equipment to Voltage Sags [C]. The 4th International Power Engineering and Optimization, Malaysia, Shah, 2010.

[16] CIGRE/CIRED/UIE Joint Working Group C4.110. Voltage dip immunity of equipment and installations [R]. London: Working Group C4.110, 2010.

Calculation of formation pressure during drilling and shut-in of oilfields with multiwell and multilayer conditions

Nannan Zhang

Oil Production Plant, Daqing Oil Field Company Limited, Daqing, Heilongjiang

ABSTRACT: At present, multiwell and multilayer conditions have been adopted for most oilfields. In this article, a formation pressure calculation model is established for multiwell, multilayer, variable production and unstable seepage oilfields. An example of the drilling and shut-in area in eastern North Block 3 is calculated. Based on the actual parameters of the basic well pattern in the drilling and shut-in area, a physical model of formation pressure variation is established. When the physical properties of vertical reservoirs are quite different, the liquid-producing capacity of each layer is divided according to the *Kh* value of each layer. Combined with the principle of pressure superposition, the suitable definite solution conditions are given. A mathematical model for calculating formation pressure with multiwell and multilayer variable flow rate is established. Using the ratio X1: X2: X3: X4: X5 = 2:3:4:5:6, giving the permeability of each layer, and a schematic diagram of the formation pressure distribution field a mathematical model is constructed. Six different drilling points are selected and the formation pressure is calculated when the composite water cut reaches 96%. The results show that the pressure of the high-permeability layer is low and the pressure of the low-permeability layer is high. It can be seen that the plane formation pressure distribution of the mathematical model presents a regular distribution pattern of a dumbbell–spindle type. In the longitudinal direction, the positive rhythm oil layer presents a scheme in which the formation pressure gradually decreases.

1 INTRODUCTION

Formation pressure distribution is a very important parameter for oilfield dynamic prediction and calculation of reserves. It is particularly important to obtain the accuracy of interwell formation pressure distribution in oilfield development. On the basis of theoretical studies, Wang et al. (1999) provided an empirical formula for reasonably delimiting the bottom-hole pressure (BHP) of oil wells according to the actual data of different oilfields. Based on the theory of seepage mechanics, Zhang et al. (2005) established a reservoir formation pressure prediction model that satisfies the field application. Cang (2014) and others proposed the idea that in the drilling and shut-in process, for water injection wells with large interlayer contradictions, first depressurizing a high-pressure zone and then keeping up pressure and water injection of the weak zone are beneficial to a balanced decline in formation pressure. Zhou (2015) drew up a reasonable drilling and shut-in recovery plan in order to balance the formation pressure, control ascending velocity of water content, and reduce the impact of drilling and shut-in on production. Ai et al. (2016) emphasized that predicting the time of shut-in of water injection wells can effectively reduce the incidence of complex drilling accidents and drilling costs.

Based on the actual data of oil layers in the drilling and shut-in area in eastern North Block 3, this article applies the pressure field superposition principle to establish a mathematical model of formation pressure changes. The distribution law of the formation pressure in planar and longitudinal directions is deduced, which provides a theoretical basis for dynamical analysis, development plan adjustment, and implementation of a water-flooding oilfield in the future.

2 PHYSICAL MODEL

Based on the actual dynamic and static parameters of the basic well pattern in eastern North Block 3, a physical model of formation pressure changes during drilling and shut-in is established. In the longitudinal direction, the physical model has numerous oil layers. The well network deployment method of line water drive is adopted, the well spacing for producing wells is 500 m, and the distance of the well array for producing wells is 500 m. Well spacing for water injection wells is 300 m, and well locations of producing wells are mutually staggered. It is assumed that there are 77 wells in the line water drive pattern in the model, including 27 injection wells and 50 production wells.

In the case of differences in water intake capacity (liquid-producing capacity) between the oil layers in the longitudinal direction, there is a large difference in formation pressure. Therefore, for the oil layers in the longitudinal direction, especially when the physical properties of vertical reservoirs are quite different, the liquid-producing capacity of each layer is divided according to the Kh value of each layer. By establishing physical and mathematical models for liquid flow volume splitting and pressure distribution in a multilayer commingled production reservoir, it is possible to solve the pressure distribution of the reservoir in the longitudinal direction. A vertical schematic of the physical model of formation pressure distribution is shown in Figure 2.

For computational simplifications, we make the following assumptions about the physical model: (1) Each layer has uniform thickness and homogeneous formation. The same layer has the same permeability, porosity, fluid viscosity, thickness, and applicable relative permeability curve. (2) Different layers may have different permeability, relative permeability curve, thickness, porosity, viscosity, initial formation pressure, start-up pressure gradient, and test work system. (3) The production wells are produced with a fixed liquid production rate (fixed water injection rate for water injection wells) and have the same production time. (4) Each layer has an identical area. All these layers are controlled within the control area in all oil and water wells. (5) For the same layer, the stratigraphic supply radius within the control range of all oil and water wells is half of the injector producer distance. (6) There is a good barrier between the layers, with no channeling. (7) The flow of fluid in each layer conforms to Darcy's law and runs radially in the stratum. (8) The force of gravity and the capillary force can be omitted.

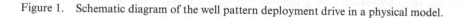

Figure 1.   Schematic diagram of the well pattern deployment drive in a physical model.

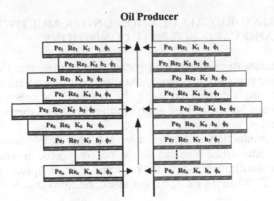

Oil Producer

Figure 2. Schematic diagram of oil well production.

The average daily liquid production rate is determined according to the current average daily liquid production rate of single wells in the basic well pattern. Other formation parameters are selected according to the average static parameters of the strata controlled by the basic well pattern.

In order to determine the planar and longitudinal pressure distributions at a point, we first need to determine which production well/injection well controls this point. Then the filtrational resistance and formation pressure of each layer within the control range of this well are dynamically computed. Based on the calculated resistance, the liquid volume is dynamically split. Finally, the formation pressure distribution of this point in different layers is determined through calculation.

As shown in Figure 2, to calculate the formation pressure of point P in the studied area, we should first confirm that point P is controlled by well N1. Then according to the daily water injection rate in well N1, the liquid volume of each layer within the control range is dynamically split. The distribution of formation pressure of each layer can be computed according to the formation pressure calculation model. By calculating the formation pressure of different points, the formation pressure distribution in the studied area can be determined, namely the formation pressure distribution field. Furthermore, by calculating the distribution fields of formation pressure at different times, it is possible to determine the changes of the formation pressure distribution field during drilling and shut-in.

When the oil/water well working system remains unchanged for some time, it can be considered that the entire production well is produced with a fixed liquid production rate and the entire injection well is operated with a fixed water injection rate. By solving the variation of formation pressure and the variation of bottom-hole flowing pressure (BHFP) to the different seepage flow stages, we have looked at the liquid volume splitting in different oil layers at the to-be-drilled point at different times.

Table 1. Mean parameters of the basic well pattern.

	Daily liquid production (m³/d)	Formation pressure (MPa)	Effective thickness (m)	Crude oil viscosity (mD)	Permeability (mD)	Spacing of production wells (m)	Spacing of injection wells (m)	Row spacing (m)
Basic well Pattern	160	11.35	9.5	9.6	339.3	500	300	500

3 FORMATION PRESSURE CALCULATION UNDER MULTIWELL, MULTILAYER, AND VARIABLE-RATE CONDITIONS

For a real oilfield, the oil/water well working system may change over time, which would lead to a change in formation pressure accordingly. At this point, the injection–production relation in this area cannot be simply considered stable. In this case, the pressure calculation at variable production rates can be realized by applying the pressure superposition principle commonly used in seepage flow mechanics.

Assuming that the production system of a well in the oilfield area changes at time T, a coordinate system is established centered on the borehole axis, in which $P_1(r, t)$ represents the formation pressure distribution in time $0 \le t \le T$ and $P_2(r, t)$ represents the formation pressure distribution at $t \ge T$. When $0 \le t \le T$, the definite conditions suitable for $P_1(r, t)$ are

$$
\begin{cases}
\frac{\partial^2 P_1}{\partial r^2} + \frac{1}{r}\frac{\partial^2 P_1}{\partial r} = \frac{1}{\eta}\frac{\partial P_1}{\partial t} \\
P_1(r, 0) = P_0 \\
P_1(\infty, T) = P_0 \\
\lim_{r \to 0} r\frac{\partial P_1}{\partial r}\big|_{r \to 0} = \frac{q_1\mu}{2\pi Kh}
\end{cases}
\tag{1}
$$

Through solving this equation, the result obtained is

$$
P_1(r, t) = P_0 + \frac{q_1\mu}{4\pi Kh} E_i\left(-\frac{r^2}{4\eta t}\right)
\tag{2}
$$

At $t \ge T$, the problem of determining solutions suitable for $P_2(r, t)$ is

$$
\begin{cases}
\frac{\partial^2 P_2}{\partial r^2} + \frac{1}{r}\frac{\partial^2 P_2}{\partial r} = \frac{1}{\eta}\frac{\partial P_2}{\partial t} \\
P_2(r, t) = P_1(r, T) \\
P_2(\infty, t) = P_0 \\
\lim_{r \to 0} r\frac{\partial P_2}{\partial r}\big|_{r \to 0} = \frac{q_2\mu}{2\pi Kh}
\end{cases}
\tag{3}
$$

Assuming that the oil well is still producing at a rate q_1 at $t = T$, the formation pressure distribution is obtained by substituting the rate into Equation (3). However, it will create a pressure difference $P_3(r, t)$ from the actual pressure distribution, namely:

$$
P_3(r, t) = P_2(r, t) - P_1(r, t)
\tag{4}
$$

This is easily verified. At $t > t_0$, $P_3(r, t)$ is the solution to the following definite conditions problem:

$$
\begin{cases}
\frac{\partial^2 P_3}{\partial r^2} + \frac{1}{r}\frac{\partial^2 P_3}{\partial r} = \frac{1}{\eta}\frac{\partial P_3}{\partial t} \\
P_3(r, t_0) = 0 \\
P_3(\infty, t) = 0 \\
r\frac{\partial P_3}{\partial r}\big|_{r \to 0} = \frac{(q_2 - q_1)\mu}{2\pi Kh}
\end{cases}
\tag{5}
$$

The result is

$$
P_3(r, t) = \frac{(q_2 - q_1)\mu}{4\pi Kh}\left[-E_i\left(-\frac{r^2}{4\eta(t - t_0)}\right)\right]
\tag{6}
$$

Thus at $t > t_0$, the result is as follows:

$$P_2(r,t) = P_0 + \frac{q_1\mu}{4\pi Kh}\left[E_i\left(-\frac{r^2}{4\eta t}\right)\right] + \frac{(q_2-q_1)\mu}{4\pi Kh}\left[-E_i\left(-\frac{r^2}{4\eta(t-T)}\right)\right] \tag{7}$$

Equation (7) indicates that the formation pressure field from $t = T$ can be seen as a superposition of two pressure fields, of which one is formed due to the maintenance of productivity q_1 in the production well and the other is formed due to the increment from $t = T$. In other words, from $t = T$, it is as if the oil well is still producing at the original rate q_1, and another production well at a rate $(q_2 - q_1)$ is added at the same place.

By analogy, when a production well in infinite reservoir is producing at variable rates (as shown in Figure 3), according to the superposition principle, we can consider in this situation that there are n wells that are put into production at different times but at the same location. Specifically, the first well is produced at a rate $(q_1 - q_0 = q_1$, $q_0 = 0)$ from $t_1 = 0$ to t, with a pressure drop Δp_1... and the nth well is produced at a rate $(q_n - q_{n-1})$ from t_n to t, with a pressure drop Δp_n. Therefore, the total pressure drop is

$$\Delta p = \Delta p_1 + \Delta p_2 + \Delta p_3 + \cdots \Delta p_n = \sum_{i=1}^{n}\Delta p_i = \frac{\mu}{4\pi Kh}\left\{\sum_{i=1}^{n}(q_i - q_{i-1})\left[-E_i\left(-\frac{r^2}{4\eta(t-t_i)}\right)\right]\right\} \tag{8}$$

If each item can be calculated using an approximation formula, Equation (8) can be written as

$$\Delta p = p_0 - p(r,t) = \sum_{i=1}^{n}\Delta p_i = \frac{\mu}{4\pi Kh}\left\{\sum_{i=1}^{n}(q_i - q_{i-1})\ln\frac{2.25\eta(t-t_i)}{r^2}\right\} \tag{9}$$

Then the formation pressure at point P is calculated according to

$$p(r,t) = p_0 + \frac{\mu\xi}{4\pi}\left\{\sum_{i=1}^{n}\frac{(q_i - q_{i-1})}{(Kh)_i}\ln\frac{2.25\eta(t-t_i)}{r^2}\right\} \tag{10}$$

where t = well production time, s; μ = fluid viscosity, mPa·s; ξ = fit coefficient, obtained by fitting to the actual formation pressure. The formation fluid viscosity μ can be derived from the relative permeability curve interpolation according to the water saturation at different formation stages. Considering multiple well interference effects, the formation pressure at point P is calculated according to

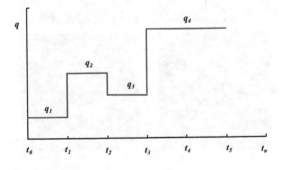

Figure 3.   Single-well production change curve.

194

$$p = p_0 + \sum_{i=1}^{m} \Delta p_i \qquad (11)$$

where m = the number of oil and water wells.

4 EXAMPLE CALCULATION

The oil layer controlled by the current basic well pattern is 9.5 m in effective thickness, 339.3 mD in average permeability, and 27% in porosity, and the spacing of production wells and the row spacing are both 500 m while the spacing of injection wells is 300 m. When point P lies within the supply radius of a well, we believe that point P is controlled by this well. Based on the actual data of the basic well pattern in eastern and western North Block 3, we assume that there are longitudinally five oil layers with relatively low, low, medium, relatively high, and high permeabilities that are given respectively according to the ratio X1:X2:X3:X4:X5 = 2:3:4:5:6.

In the test area, production is kept constant for a quite long time. Therefore, it can be assumed that this is a staged variable-rate production system with unstable seepage flows. The average single-well daily liquid production rate of the basic well pattern is 160 m³/d, and production is maintained in a balanced state of injection and production.

Table 2. Basic parameters of strata controlled by a basic well pattern.

Parameter oil layer	Effective thickness (m)	Permeability (mD)	Porosity (%)	Supply radius (m)	Initial formation pressure (MPa)	Oil viscosity (mD)	Water viscosity (mD)
1	1.9	169.65	27	250	12	9.5	0.5
2	1.9	254.48	27	250	12	9.5	0.5
3	1.9	339.3	27	250	12	9.5	0.5
4	1.9	424.13	27	250	12	9.5	0.5
5	1.9	508.95	27	250	12	9.5	0.5

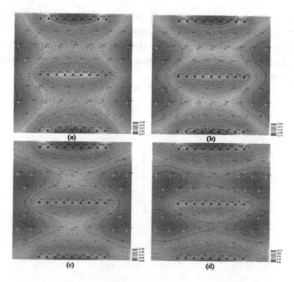

Figure 4. Formation pressure distribution fields in each layer.

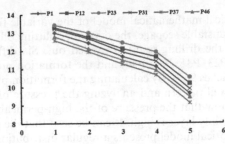

Figure 5.   Schematic diagram of calculation results of formation pressure distribution in five layers.

Table 3.   Calculation results of formation pressure distribution in five layers of a basic well pattern.

No.	Formation Pressure in layer 1 (MPa)	Formation Pressure in layer 2 (MPa)	Formation Pressure in layer 3 (MPa)	Formation Pressure in layer 4 (MPa)	Formation Pressure in layer 5 (MPa)
P1	12.984	12.418	11.763	11.488	9.823
P12	13.277	12.739	12.098	11.842	10.177
P23	13.314	12.777	12.237	11.881	10.214
P31	13.203	12.662	12.021	11.763	10.104
P37	12.557	11.939	11.255	10.947	9.267
P46	12.770	12.101	11.364	11.030	9.249

According to the theoretical mathematical model of a variable flow rate production pressure system in the stage of unstable seepage, the dynamic splitting of the liquid flow volume in different aquifer stages at the drilling point is carried out. Six different drilling points are selected (such as P1, P12, P23, P31, P37, P46) and formation pressure is calculated when the composite water cut reaches 96% (Table 3).

Because there is good connectivity among oil and water wells in the basic well pattern, the adjacent wells are basically exploited in the same horizon, with good pressure distribution continuity and little interlayer interference. Therefore, by calculating the formation pressure distribution fields of five layers in the basic well pattern and analyzing the pressure of oil layers with different permeability, the results show that the pressure of the high-permeability layer is low and the pressure of the low-permeability layer is high. It can be seen that the plane formation pressure distribution of the mathematical model presents a regular distribution pattern of a dumbbell–spindle type. In the longitudinal direction, the positive rhythm oil layer presents a scheme in which the formation pressure gradually decreases.

5   CONCLUSION

Based on the actual parameters of the basic well pattern in the drilling and shut-in area, a physical model of formation pressure variation is established. The well network deployment method of the line water drive is adopted, well spacing for producing wells is 500 m, and the distance of the well array for producing wells is 500 m. Well spacing for water injection wells is 300 m, and well locations of producing wells are mutually staggered. When the physical properties of vertical reservoirs are quite different, the liquid-producing capacity of each layer is divided according to the Kh value of each layer. In the longitudinal direction, the formation pressure distribution of reservoirs is computed. Using the ratio X1: X2: X3: X4: X5 = 2:3:4:5:6, giving the permeability of each layer, and a schematic diagram of the formation pressure distribution field a mathematical model is constructed.

According to the theoretical mathematical model of the variable flow rate production pressure system in the stage of unstable seepage, the dynamic splitting of the liquid flow volume in different aquifer stages at the drilling point is carried out. Six different drilling points are selected (such as P1, P12, P23, P31, P37, P46) and the formation pressure is calculated when the composite water cut reaches 96%. By calculating the formation pressure distribution fields of five layers in the basic well pattern and analyzing the pressure of oil layers with different permeability, The results show that the pressure of the high-permeability layer is low and the pressure of the low-permeability layer is high. It can be seen that the plane formation pressure distribution of the mathematical model presents a regular distribution pattern of a dumbbell–spindle type. In the longitudinal direction, the positive rhythm oil layer presents a scheme in which the formation pressure gradually decreases.

ACKNOWLEDGMENTS

Funding was provided by the Natural Science Foundation of Heilongjiang Province, China (E201407).

REFERENCES

Ai, C., Gao, J., Feng, F., et al. 2016. Pressure drop simulation of water injection well in low permeability fractured reservoir. *Contemporary Chemical Industry* 45(12): 2845–2847.

Cang, J. 2014. Improve drilling method and balance formation pressure. *Inner Mongolia Petrochemical Industry* 40(8): 63–64.

Gao, Z., Hou, D., & Tang, Li. 2005. Research on prediction method of adjusting well formation pressure. *Petroleum Geology & Oilfield Development in Daqing* 24(3): 57–58.

Li, G. 2009. Application and research on the technology of formation pressure predication. Chengdu University of Technology.

Wang, J., Li, Y., & Zhao, G. 1999a. Determination of oil well inflow dynamic curve and reasonable bottom hole pressure. *Xinjiang Petroleum Geology* no. 5: 414–417, 450.

Wang, J., Wang, C., & Fang, L. 1999b. Method for determining reasonable bottom hole pressure limit of oil production well. *Petroleum Geology & Oilfield Development in Daqing* no. 5: 21–22, 53.

Wang, Y., Chen, Z., Yu, W., et al. 2005. Research and application of formation pressure prediction technology in Qingxi oilfield. *Petroleum Drilling Techniques* 33(4): 29–31.

Yao, J., Yuan, Li, Wang, C., et al. 2012. Prediction method and application of inter-well formation pressure distribution. *Science Technology and Engineering* 12(30): 7854–7858, 7863.

Zhang, J., Zhang, T., Chen, P., et al. 2005. Adjustment well formation pressure prediction method based on seepage theory. *Drilling & Production Technology* 28(3): 7–9.

Zhou, B. 2015. Optimizing drilling and recovery plan to reduce drilling yield impact. *Inner Mongolia Petrochemical Industry* 41(1): 103–105.

Emerging Developments in the Power and Energy Industry – Dufo-López, Krzywanski & Singh (eds)
© 2020 Taylor & Francis Group, London, ISBN 978-0-367-27169-5

Multi-index program decision model for the development of a condensate gas reservoir group and its application

Jing Xia & Aifang Bie
Institute of Oilfield Development, PetroChina Research Institute of Petroleum Exploration and Development, Beijing, China

Bingyu Lv, Jiale Fan & Jicheng Zhang
School of Petroleum Engineering, Northeast Petroleum University, Daqing, Heilongjiang, China

ABSTRACT: In order to determine the relationship between annual gas production and number of wells, a multi-index program decision model for the development of a condensate gas reservoir group is established. According to the different numbers of production wells and gas production scales in old blocks and new blocks, several sets of development plans are compiled. The reservoir engineering method is used to predict the number of wells, annual gas production, net present value, and internal rate of return of a condensate gas reservoir. Based on the gas production and well number data of an actual gas reservoir, according to the multi-index fuzzy evaluation method, the optimum development scheme of condensate gas reservoirs in old areas and new areas is obtained. The results show that the optimum scheme for a condensate gas reservoir (old area) are scheme 1(1) and scheme 1(3). The parameters of scheme 1(1) are 29 wells, annual gas production is $150.58 \times 10^8 \, m^3$, net present value is 6658.00 million RMB, and internal rate of return is 13.28%. The parameters of scheme 1(3) are 31 wells, annual gas production is $160.82 \times 10^8 \, m^3$, net present value is 7058.00 million RMB, and internal rate of return is 9.77%. The optimum scheme for a condensate gas reservoir (new area) is scheme 20. The parameters of scheme 20 are 3 wells, annual gas production is $15.8 \times 10^8 \, m^3$, net present value is 374.27 million RMB, and internal rate of return is 2.02%. The optimal scheme can provide a theoretical basis for decision-makers.

1 INTRODUCTION

Development of a condensate gas reservoir group is a large- system decision. There are many factors affecting the condensate gas reservoir group. It is necessary to adopt the means and methods of a large-system decision to deal with a condensate gas reservoir group. Therefore, during the optimization of multiple development plans, a technical and economic indicator system that can reflect these influencing factors shall be established. Then, the decision-makers can make a corresponding decision according to a specified criterion or method.

Zhou (2011) tried to establish a comprehensive mathematical model by using an analytic hierarchy process and fuzzy comprehensive judgment method. Based on fuzzy mathematics and fuzzy decision matrix, the evaluation method for solving the problem concerning archive information utilization services was discussed. He analyzed the development mode from the technical and economic points of view. Comparing these two indicators and combining the geological characteristics of gas reservoirs, he found the way to maximize economic benefits was through a depletion type of development. With a view toward realizing growth of benefits, the concept of deploying two kinds of different well patterns was proposed to improve the condensate oil recovery ratio (Li & Zhong 2015). Wei (2017) studied the reserves of a gas field by using gas reservoir engineering and a numerical simulation method. The recovery factor of the gas reservoir was judged by the material balance method.

2 MULTI-INDEX PROGRAM DECISION MODEL

2.1 Program design

There must be at least two alternative plans for any decision. For a condensate gas reservoir group, assuming different well numbers, multiple development schemes are formed. Therefore, by setting different wells, we can obtain several feasible development plans and then calculate the corresponding gas production.

2.1.1 Calculation of the number of production wells

For the old blocks of a condensate gas reservoir group in a certain region, the annual gas production results of the condensate gas reservoir group can be obtained according to the dynamic programming for the development of the condensate gas reservoir group. The gas production speed of the condensate gas reservoir group can be calculated according to Equation (1). The gas production speed is an extremely important development indicator in gas field development. It is the ratio between the annual gas production and the geological reserves of a gas reservoir. Then the number of gas wells and total number of wells needed to meet reasonable gas production speed requirements under a certain production pressure difference can be calculated. The advantage of the method lies in there being no limitation of regions and development phases.

$$v_g = \frac{Q_g}{G_{remaining}} \tag{1}$$

where $G_{remaining}$ = remaining geological reserves of the gas reservoir, $10^8\,m^3$; v_g = gas production speed, %; and Q_g = planned annual gas production, $10^8\,m^3$.

The number of gas wells for old block development is as follows:

$$N_{gw} = \frac{G_{remaining} \times v_g}{330 \times \eta \times \frac{Kh}{\mu}\Delta p} \times 10^4 \tag{2}$$

where Δp = production pressure difference, MPa; η = comprehensive utilization rate of a gas well, 80%–98%, 90% here; k = gas reservoir permeability, μm^2; h = gas reservoir thickness, m; kh/μ = formation flow coefficient, $\mu m^2 \cdot m/mPa \cdot s$; and v_g = gas production speed, %.

2.1.2 Calculation of the number of production wells in new blocks

For the new blocks of a condensate gas reservoir group in a certain region, the annual gas production results of a condensate gas reservoir group can be obtained according to the dynamic programming for the development of a condensate gas reservoir group. The number of wells that can satisfy the planned gas production can be calculated according to Equation (3). It is assumed that the new blocks of the condensate gas reservoir group have a daily gas production of $30 \times 10^4\,m^3/d$ and annual gas production of $0.9 \times 10^8\,m^3/a$.

$$N_{gw} = \frac{Q_g}{q_g} \tag{3}$$

where q_g = average productivity of a single well, $10^8\,m^3$ and N_{gw} = number of wells.

2.1.3 Program compilation

According to the planned annual gas production, the total number of wells in old blocks and new blocks is determined. Information about the damage to the production wells is considered, and it is guaranteed that the gas production will be no less than the planned gas production. The number of wells is changed and multiple development plans are compiled according to the actual production of a gas reservoir.

2.2 The multi-index system

2.2.1 Calculation of technical indicators

(1) Annual gas production of old wells in old blocks

According to the basic plan, information on gas field production is collected and analyzed. The annual gas production in more than 10 old blocks of a condensate gas reservoir group from the year 2010 to 2030 is predicted. The single well (average) annual gas production of old wells can be calculated according to Equation (4).

$$q_g = \frac{Q_g}{\eta \times N_{original}} \tag{4}$$

where q_g = single well average annual productivity of old wells in old blocks, 10^8 m^3; $N_{original}$ = original number of wells in old blocks; and Q_g = predicted gas reservoir production, 10^8 m^3. According to the calculation, the average annual productivity of a single well in an old well is obtained. By multiplying the number of old wells in various schemes, the annual gas production of old wells in old areas can be obtained.

$$Q_{oldwell} = q_g \times N_{oldwell} \times \eta \tag{5}$$

where $N_{old\ well}$ = number of old wells in the plans and $Q_{old\ well}$ = annual gas production of old wells in old blocks in the plans, 10^8 m^3.

(2) Annual gas production of new wells in old blocks

According to the remaining geological reserves of a condensate gas reservoir group, the planned gas production speed results and number of planned wells can be calculated. The annual gas production of new wells in old blocks can be calculated according to Equation (6).

$$q_{g\ new\ (spudded)\ well} = \frac{Q_g - Q_{oldwell}}{\eta \times N_{gw}} \tag{6}$$

where $q_{g\ new\ (spudded)\ wells}$ = single well (new well) average annual productivity in old blocks, 10^8 m^3; N_{gw} = number of new (spudded) wells in the plans; and Q_g = planned annual gas production, 10^8 m^3. According to the calculation, the average annual productivity of a single well in a new well is obtained. By multiplying the number of new wells in various schemes, the annual gas production of new wells in old areas can be obtained.

$$Q_{new\ (spudded)\ wells} = q_{g\ new\ (spudded)\ wells} \times N_{new\ (spudded)\ wells} \times \eta \tag{7}$$

where $N_{new\ (spudded)\ wells}$ = number of new (spudded) wells in the plans and $Q_{new\ (spudded)\ wells}$ = annual gas production of new (spudded) wells in old blocks in the plans, 10^8 m^3.

(3) Annual gas production of old wells in new blocks

According to the basic plan, gas field production is collected and analyzed. The annual gas production in more than 10 old blocks of a condensate gas reservoir group from the year 2010 to 2030 is predicted. The single well (average) annual gas production of old wells can be calculated according to Equation (4). According to Equation (5), single well average annual productivity of old wells can be calculated. It is multiplied by the number of old wells in the several compiled plans, and then the annual gas production of old wells in new blocks can be obtained.

(4) Annual gas production of new wells in new blocks

As the basic parameters in new blocks are uncertain, the new blocks have daily gas production of 30×10^4 m^3/d and annual gas production of 0.9×10^8 m^3/a. The annual gas production of new wells in new blocks can be calculated according to Equation (8).

200

$$Q_{\text{g new blocks}} = q_{\text{g new blocks}} \times \eta \times N_{\text{new wells}} \tag{8}$$

where $N_{\text{new well}}$ = the number of new wells in new blocks; $q_{\text{g new block}}$ = average single well productivity of new wells in new blocks, 10^8 m^3; $Q_{\text{g new block}}$ = annual productivity of new wells in new blocks, 10^8 m^3; and G = geological reserves of a gas reservoir, 10^8 m^3.

2.2.2 Calculation of economic indicators

(1) Net present value

Net present value (NPV) is an indicator commonly used in practice to evaluate the economic effects of project plans. NPV can reflect the profitability of the project in the economic life cycle. The indicator expresses the net cash flow during the project's economic life cycle as the present cumulative value in some base year (generally in the early stages of investment) as per a certain benchmark discount rate, which can be expressed with the following formula:

$$\text{NPV} = \sum_{t=0}^{T} C_t \cdot (1 + i_o)^{-t} \tag{9}$$

$$C_t = (C_I - C_O)_t \tag{10}$$

where C_I = cash inflow in the tth year, $\times 10^4$ RMB; C_t = net cash flow in the tth year, $\times 10^4$ RMB; C_O = cash outflow in the tth year, $\times 10^4$ RMB; NPV = net present value, $\times 10^4$ RMB; T = mining period, years; i_o = benchmark rate of return, 12%.

(2) Internal rate of return

The internal rate of return (IRR) is used to evaluate the project's profitability, with the mathematical expression as follows:

$$\sum_{t=0}^{T} C_t \bullet (1 + \text{IRR})^{-t} = 0 \tag{11}$$

$$C_t = (C_I - C_O)_t \tag{12}$$

where IRR = internal rate of return, %.

2.3 The multi-index fuzzy judgment method

(1) Determination indicator

The technical indicators include gas production speed and annual gas production. The economic indicators include net present value and internal rate of return.

(2) Determining weight set with the analytic hierarchy process

Take n matters as an example, with A as the target and u as the factor. Take two factors and compare them with each other to form an $n \times n$ matrix:

$$A = \begin{bmatrix} \frac{u_1}{u_1} & \frac{u_1}{u_2} & \frac{u_1}{u_3} & \cdots & \frac{u_1}{u_n} \\ \frac{u_2}{u_1} & \frac{u_2}{u_2} & \frac{u_2}{u_3} & \cdots & \frac{u_2}{u_n} \\ \cdots & \cdots & \cdots & \cdots & \\ \frac{u_n}{u_1} & \frac{u_n}{u_2} & \frac{u_n}{u_3} & \cdots & \frac{u_n}{u_n} \end{bmatrix} = \begin{bmatrix} 1 & a_{12} & a_{13} & \cdots & a_{1n} \\ a_{21} & 1 & a_{23} & \cdots & a_{2n} \\ \cdots & \cdots & \cdots & \cdots & \cdots \\ a_{n1} & a_{n2} & a_{n3} & \cdots & a_{nn} \end{bmatrix} \tag{13}$$

This matrix names the judgment matrix, with the following characteristics: (1) $a_{ij} > 0$; ② $a_{ij} = 1$; ③ $a_{ij} = 1/a_{ji}$. This kind of matrix is called an antisymmetric matrix.

Matrix A is multiplied by vector $W = [w_1 \; w_2 \; \ldots \; w_n]$ from the right; then

$$A \times W = \begin{bmatrix} 1 & a_{12} & a_{13} & \cdots & a_{1n} \\ a_{21} & 1 & a_{23} & \cdots & a_{2n} \\ \cdots & \cdots & \cdots & \cdots & \cdots \\ a_{n1} & a_{n2} & a_{n3} & \cdots & a_{nn} \end{bmatrix} \times \begin{bmatrix} w_1 \\ w_2 \\ \cdots \\ w_n \end{bmatrix} = nW \tag{14}$$

According to the matrix principle, the maximum eigenvalue of the antisymmetric matrix is equal to its order, namely $\lambda_{\max} = n$. Therefore, Equation (14) translates into

$$A \times W = \lambda_{\max} W \tag{15}$$

where the vector the W is the vector value of the weight to be calculated. This indicates that, if the judgment matrix has a reasonable structure, then it can approximately satisfy Equation (15). Therefore, calculating the weight vector value with this method is converted to a simple matrix operation. If the variable W is multiplied by a constant k, then the formula will still remain unchanged; namely, $k \cdot W$ is still a feature vector and W is a basic feature vector.

i. Construct a judgment matrix

Compare two factors u_i and u_j according to a certain criterion. Determine which one is more important and how much more importance it has than the other one; this extra importance shall be given a certain value. As for n factors, the judgment matrix A comparing two factors with each other can be obtained.

ii. Calculate the weight vector

Considering the particularity of the judgment matrix, the square root method can be used to calculate the weight vector, and then the maximum eigenvalue $\lambda_{\max}$ can be calculated:

$$\lambda_{\max} = \sum_{i=1}^{n} \frac{(AW)_i}{nW_i} \tag{16}$$

iii. Make a consistency check

The indicator for a consistency check is the consistency ratio CR, which is defined as

$$CR = \left(\frac{\lambda_{\max} - n}{n - 1} \right) / (RI) \tag{17}$$

where RI is the average random consistency indicator, which is related to matrix order and can be calculated according to the values listed in Table 1. The standard for inspection is $CR < 0.1$. Therefore, it is recognized that the judgment matrix is acceptable and the calculated weight is reasonable.

(3) Method for membership determination

According to the analysis results for single factors, if the single factor takes a larger value, there is a greater possibility for low-efficiency circulation. Then, an ascending semitrapezoidal distribution will be used to calculate the membership, as shown in Equation (18). If the single factor takes a smaller value, there is a greater possibility for low-efficiency circulation. Then, a descending semi-trapezoidal distribution will be used to calculate the membership, as shown in Equation (19).

$$R_{kj} = \frac{r_{kj} - (r_{kj})_{\min}}{(r_{kj})_{\max} - (r_{kj})_{\min}} \tag{18}$$

Table 1. Relationship between RI value and matrix order.

Matrix order	1	2	3	4	5	6	7	8
RI	0	0	0.52	0.89	1.12	1.26	1.36	1.41
Matrix order	9	10	11	12	13	14	15	
RI	1.46	1.49	1.52	1.524	1.56	1.58	1.59	

$$R_{kj} = \frac{(r_{kj})_{max} - r_{kj}}{(r_{kj})_{max} - (r_{kj})_{min}} \tag{19}$$

where R_{kj} is the membership for the kth factor of the jth well; r_{kj} is the value for the kth factor of the jth well; (r_{kj})min is the minimum value of the kth factor in n wells; and (r_{kj})max is the maximum value of the kth factor in n wells.

(4) Fuzzy comprehensive judgment

During the evaluation of some complex projects, various factors affect the judgment results at different levels. Therefore, it is necessary to summarize the different levels of factors into a total judgment result on a level-by-level basis.

Let the importance fuzzy subset of the second level of factors be W_i, and the total evaluation matrix of k_i factors will be R_i.

Table 2. Optimal results of the plans of old blocks.

Condensate gas reservoir (old block)	Plan	Number of wells	Production	Net present value	Internal rate of return	Fuzzy judgment results
1	1	29	150.58	665,080	13.28	0.50
	3	31	160.82	705,800	9.77	0.50
2	3	11	64.02	239,540	4.45	0.50
3	3	5	50.61	208,790	5.66	0.69
4	3	27	432.37	176,700	24.07	0.56
5	3	7	24.57	95,962	1.19	0.51
6	1	4	1.57	−245.264	0.09	0.86
7	1	9	1.92	−1007.6	0.05	0.86
8	3	21	186.11	731,360	9.81	0.73
9	3	4	180.13	630,860	8.50	0.80
10	2	14	146.11	56,720	4.4	0.64
11	3	21	92.10	344,350	4.42	0.74
12	3	74	154.94	584,240	23.69	0.62
13	3	5	7.80	25,314	1.43	0.61
14	3	2	5.40	25,042	1.94	0.86

Table 3. Optimal results of the plans of new blocks.

Condensate gas reservoir (new block)	Plan	Number of wells	Production	Net present value	Internal rate of return	Fuzzy judgment results
15	1	4	18.21	42,525	2.15	0.50
16	3	9	49.14	140,540	2.06	0.64
17	1	9	24.91	53,670	2.03	0.86
18	2	2	3.20	14,427	2.11	0.78
19	3	7	19.7	41,598	2.12	0.65
20	2	3	15.8	37,427	2.02	0.58

Let the importance fuzzy subset of the first level of factors be W and $W = (W_1, W_2, \ldots, W_N)$. Then, the total evaluation matrix R will be

$$
R = \begin{pmatrix} Y_1 \\ Y_2 \\ \cdots \\ Y_N \end{pmatrix} = \begin{pmatrix} y_{11}, & y_{12}, & \cdots & y_{1m} \\ y_{21}, & y_{22}, & \cdots & y_{2m} \\ \cdots & \cdots & \cdots & \cdots \\ y_{N1}, & y_{N2}, & \cdots & y_{Nm} \end{pmatrix} \tag{20}
$$

Then the total (second level) comprehensive judgment result will be

$$
Y = W \otimes R = \begin{pmatrix} W_1 \otimes R_1 \\ W_2 \otimes R_2 \\ \cdots \\ W_N \otimes R_N \end{pmatrix} \tag{21}
$$

Based on this result, according to the maximum membership principle, the calculated result will be the final judgment result.

3 EXAMPLE CALCULATION ON A MULTI-INDEX PROGRAM DECISION MODEL FOR THE DEVELOPMENT OF A CONDENSATE GAS RESERVOIR GROUP

3.1 *Compile plan for old blocks*

According to the results calculated from the dynamic planning model for the development of a condensate gas reservoir group, based on the parameters of the condensate gas reservoirs in old blocks, the planned gas production speed and number of planned wells in old blocks can be calculated.

The plan shall be compiled according to the principle, but the number of planned wells shall be decided according to the actual gas production in old blocks. The number of wells shall be changed according to the actual production of the gas reservoir, and multiple sets of development plans can be compiled.

3.2 *Compile plan for new blocks*

According to the results calculated from the dynamic planning model for the development of a condensate gas reservoir group and the annual gas production (as shown in the following table) of the condensate gas reservoirs in new blocks, and based on Equation (11), the number of planned wells in new blocks can be calculated.

The plan shall be compiled according to the principle, but the number of planned wells shall be decided according to the actual gas production in old blocks. The number of wells shall be changed according to the actual production of the gas reservoir, and multiple sets of development plans can be compiled.

3.3 *Development plan optimization*

1. According to Equation (14), the average single well production of old wells in the old area is obtained.
2. According to Equation (6), the average single well production of new wells in old area is obtained.
3. According to Equation (16), the average single well production of old wells in new area is obtained.

4. Considering that the basic parameters in new blocks are uncertain, it is assumed that the new blocks have daily gas production of $30 \times 10^4 \ \mathrm{m}^3/\mathrm{d}$ and annual gas production of $0.9 \times 10^8 \ \mathrm{m}^3/\mathrm{a}$.

5. The inflow, outflow, and net cash flow of a condensate gas reservoir group in the tth year can be calculated according to Equation (20). It is assumed that, as for new wells, the costs such as drilling cost, etc. are 25.18 million RMB; for old wells, the costs such as maintenance cost, etc. are 0.50 million RMB; and the natural gas price is $0.7 \times 10^8 \ \mathrm{RMB}/10^8 \ \mathrm{m}^3$.

6. Fuzzy judgment
   i. Calculate the weight with the analytic hierarchy process
   a. Determine the factor set and judgment set: the number of wells, gas production, net present value, and internal rate of return.
   b. Determine all weight vector W for all indicators: $W = (0.139, 0.139, 0.361, 0.361)^T$
   Calculate the weight of the indicator system with the foregoing methods; i.e., the weight is 0.139 for the number of wells, 0.139 for the gas production, 0.361 for the net present value, and 0.361 for the internal rate of return.
   ii. Fuzzy judgment
   Calculate various weight results of the indicator system with the analytic hierarchy process, and then perform the judgment calculation level by level with the fuzzy judgment method. The calculation results of fuzzy judgment are shown in the following table.
   Choose the optimal plans according to the foregoing results, namely, choose plans 1 and 3 for condensate gas reservoir 1; choose plan 3 for condensate gas reservoir 2; choose plan 3 for condensate gas reservoir 3; choose plan 3 for condensate gas reservoir 4; and choose plan 2 for condensate gas reservoir 20.

4 CONCLUSIONS

1. A multi-index program decision model for the development of a condensate gas reservoir group is established. Based on the different number of production wells and gas production scales in old blocks and new blocks, several sets of development plans are compiled. The reservoir engineering method is used to predict the number of wells, annual gas production, net present value, and internal rate of return of a condensate gas reservoir. Based on the gas production and well number data of an actual gas reservoir, according to the multi-index fuzzy evaluation method, the optimum development scheme of condensate gas reservoirs in old areas and new areas is obtained.

2. The optimum schemes for a condensate gas reservoir (old area) are scheme 1(1) and scheme 1(3). The parameters of scheme 1(1) are 29 wells, annual gas production is $150.58 \times 10^8 \ \mathrm{m}^3$, net present value is 6658.00 million RMB, and internal rate of return is 13.28%. The parameters of scheme 1(3) are 31 wells, annual gas production is $160.82 \times 10^8 \ \mathrm{m}^3$, net present value is 7058.00 million RMB, and internal rate of return is 9.77%. The optimum scheme for a condensate gas reservoir (new area) is scheme 20. The parameters of scheme 20 are 3 wells, annual gas production is $15.8 \times 10^8 \mathrm{m}^3$, net present value is 374.27 million RMB, and internal rate of return is 2.02%. The optimal scheme can provide a theoretical basis for decision-makers.

ACKNOWLEDGMENTS

This work is supported by the National Science and Technology Major Project of the Ministry of Science and Technology of China (2016ZX05015-002).

REFERENCES

Gao, Y. 2010. A study on the method of multiple attributes decision making based on fuzzy decision making matrix. Nanjing University of Aeronautics and Astronautics.

Li, Y. & Zhong, G. 2015. The discussion on the development mode of Longfengshan condensate gas field. *Neimenggu Petroleum and Chemical* 41(12): 50–51.

Li, Z. 2016. Productivity analysis and development effect evaluation & study on YaHa recycling gas-injection condensate gas reservoir. Southwest Petroleum University.

Lu, J. & Li, L. 2001. Comparison and analysis between internal rate of return method and net present value method. *Journal of Xinjiang University (Science and Technology Edition)* no. 2: 200–203, 207.

Ma, K. 2017. The study on development potential evaluation & countermeasure research in Bai-55 Fault-block Condensate Gas Reservoir. Chongqing University of Science and Technology.

Wang, Z. 1994. An important indicator for investigating the profitability of construction projects internal rate of return. *East China School of Economics and Management* no. 3: 50–53.

Wei, F. 2017. Overall development plan of LW 3-1 gas field–gas reservoir engineering research. Southwest Petroleum University.

Wu, J. 2016. A study on recoverable reserves calculation of KeKeYa X_7–2 condensate gas reservoir. Southwest Petroleum University.

Xu, Z. 2016. A study on the reasonable gas production speed of X water-driven gas reservoir. Southwest Petroleum University.

Zhou, C. 2011. An evaluation on the archiving information utilization services based on AHP and fuzzy comprehensive judgment. *Archives Science Bulletin* no. 3: 88–91.

Emerging Developments in the Power and Energy Industry – Dufo-López, Krzywanski & Singh (eds)
© 2020 Taylor & Francis Group, London, ISBN 978-0-367-27169-5

An analytical method for voltage sag profile estimation of a power system with STATCOMs

Qingbin Wang, Ri Cai, Yun Yang, Hui Huang, Xiongkeng Zhan, Chuangdao Li, Gai Li & Rongqin Huang
Yunfu Power Supply Company, Guangdong Power Grid Corporation, Yunfu, Guangdong, China

ABSTRACT: A static synchronous compensator (STATCOM) is a an electronic-based power device that is widely used for providing fast reactive power for a power system. With a STATCOM connected in the system, the voltage sag profile during a fault will be improved. In order to estimate this voltage sag profile improvement, a traditional voltage sag profile estimation method based on a short-circuit calculation needed to be amended to incorporate the impact of STATCOM on the system bus voltage. This article presents a current source-based STATCOM model for voltage sag profile estimation, and by incorporating the operation and control constraints into the short-circuit calculation model, a mathematical voltage sag profile estimation model for a power system with STATCOMs is proposed. The proposed model is solved via a nonlinear programming–based method. Case studies based on an IEEE 30-bus system with STATCOMs are used to illustrate the feasibility and effectiveness of the proposed method.

Keywords: voltage sag, STATCOM, voltage sag profile estimation, nonlinear programming method

1 INTRODUCTION

Power quality has received increasing attentions both from customers and electricity suppliers as a result of the deregulation of the power industry and wider application of sensitive equipment (Xiao et al. 2016). Among different types of power disturbances, voltage sag is considered the most severe issue. Enormous losses are caused by voltage sags at the customer side, consequently increasing complaints to electricity suppliers (CIGRE/CIRED Joint Working Group C4.107 2011). Thus, various mitigation solutions for voltage sag have been proposed including reducing the number of faults in a power system, improving the protection of sensitive equipment, and installing compensation devices for voltage sags (Chang & Yu 2004).

A static synchronous compensator (STATCOM) is a typical electronic-based power device that can maintain connected bus voltage by injecting reactive power when system disturbances occur, such as faults and large load variation. Thus, it can also improve the voltage sag performance of a power system (Hingorani et al. 2000). With the rapid technical development and decreasing costs of electronic devices, STATCOM could be an effective and economical compensation device for voltage sag, especially as a system-level solution (Chan & Milanović 2015).

Voltage sag is defined as a short-duration reduction of the root mean square value of the voltage that is mostly caused by faults in the power system (Bollen 2000). The voltage sag profile is defined as the voltage sag performance of all system buses during a given fault event, and short-circuit calculation is commonly used for voltage sag profile estimation (Cruz et al. 2015). In order to assess the impact of STATCOM installation on voltage sag and allocate STATCOM reasonably, including a STATCOM model in the voltage sag profile estimation is essential. Many studies has been conducted for power flow (Kazemtabrizi & Acha 2014), optimal power flow (Zhang et al. 2001), and stability analysis of a power system with STATCOMs (Zhang et al. 2006), but few studies addressed the voltage sag profile estimation problem. In the past, electromagnetic transient model simulation was often used to assess the voltage sag in the system with STATCOM (Ramamurthy et al. 2018). Although a detailed electromagnetic transient model can provide an accurate voltage waveform during both the transient and steady state of power system, it is unnecessary and impractical for voltage sag profile estimation of a large system because of the time-consuming process of build electromagnetic models Milanovic and Zhang (2010) is a pioneer work that modeled STATCOM and included it in short-circuit calculation equations. However, the method may not be feasible in the case of multiple STATCOMs connected in the system situation because the injected current of STATCOM needs to be calculated beforehand.

This article analyzes the operation characteristics of STATCOM and proposes a current source model of STATCOM for voltage sag profile estimation. Based on the proposed model and considering STATCOM operation and control constraints, a nonlinear voltage sag profile estimation model of a power system with STATCOMs is presented. To estimate voltage sag profile efficiently, a nonlinear programming-based method is proposed and the formulated nonlinear programming model is solved by a branch and bound algorithm using the commercial optimization solver SCIP. By comparing the numerical results of the proposed method and PSCAD/EMTDC simulation results based on IEEE30 bus system with STATCOMs, the feasibility and correctness of the proposed method are verified.

2   MODELING OF STATCOM FOR VOLTAGE SAG PROFILE ESTIMATION

STATCOM is a shunt-connected device as shown in Figure 1. The basic structure of STATCOM consists of a coupling transformer, a voltage source converter, and a DC capacitor (Hingorani et al. 2000). The reactive power exchange of STATCOM with the system is controlled by regulating the magnitude of the voltage source converter (VSC) output voltage V_{sh}. When V_{sh} is larger than bus voltage Vs, the current flows through the transformer from the STATCOM to the system, and the device generates reactive power.

2.1   *Operation constraints*

Because usually STATCOM is not capable of active power exchange with a power system, the operating constraint of STATCOM can be described as (Zhang et al. 2012)

$$P_{stat} = Re\left(\vec{V}_s \vec{I}_{sh}^*\right) = 0 \tag{1}$$

where $\vec{V}_s$ = the bus voltage of the STATCOM connected bus; $\vec{I}_{sh}$ = the injected current of STATCOM; and P_{stat} = the active power injected from STATCOM to the system.

Due to the physical limitation of semiconductors in the STATCOM, the RMS values of the output current I_{sh} should not exceed its rated value during the STATCOM operation. The rated valued of the output current equals the STATCOM rated capacity, so this operation constraint can be described as

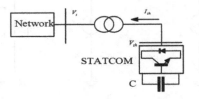

Figure 1.   Structure of STATCOM.

$$I_{sh} \le I_{max} \tag{2}$$

where I_{max} is the current limit of STATCOM.

2.2 *Control constraints*

Voltage regulation is the most common control function of STATCOM application in the power system (Xu & Li 2014). When the voltage of a STATCOM connected bus deviates from the reference value, STATCOM will inject a capacitive current into the system to maintain the voltage. If the voltage cannot be maintained because of the current rating limitations, STATCOM injected current will be kept at its maximum value. This switching logic control constraint of STATCOM can be mathematically described as (Zhang et al. 2012)

$$\begin{cases} V - V_{ref} = 0 & if \ I_{sh} < I_{max} \\ I - I_{max} = 0 & if \ V < V_{ref} \end{cases} \tag{3}$$

where V_{ref} is the predefined reference value of a STATCOM connected bus voltage.

2.3 *Modeling of STATCOM*

When a fault occurred in the power system with STATCOMs and caused voltage sags at all buses, based on the analyzed STATCOM operation principle, STATCOM will inject a capacitive current into the system to maintain the voltage of the connected bus. So STATCOM can be modeled as a current source (Milanovic & Zhang 2010). The equivalent network of a power system with STATCOM under fault is shown in Figure 2.

3 VOLTAGE PROFILE ESTIMATION MODEL FOR A POWER SYSTEM WITH STATCOMS

3.1 *Traditional voltage sag profile estimation model*

For a bus power system with no STATCOM, a voltage sag profile can be directly estimated based on the traditional steady-state short-circuit calculation (Zhang et al. 2012). For a given symmetrical bus fault in a large meshed network, the impedance-based short-circuit equations (4) and fault condition boundary equation (5) are built, and then the voltage sag magnitude and phase of each bus during the fault can be calculated.

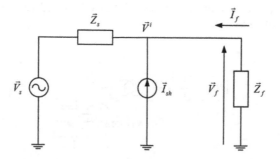

Figure 2. Equivalent network of a power system with STATCOM under fault.

$$\begin{pmatrix} \vec{V}_1 \\ \vec{V}_i \\ \vdots \\ \vec{V}_f \\ \vec{V}_n \end{pmatrix} = [\vec{V}^{\text{prefault}}] - [\vec{Z}]^k * [\vec{I}_f] = \begin{pmatrix} \vec{V}_1^{\text{prefault}} \\ \vec{V}_i^{\text{prefault}} \\ \vdots \\ \vec{V}_f^{\text{prefault}} \\ \vec{V}_n^{\text{prefault}} \end{pmatrix} - \begin{pmatrix} \vec{Z}_{11} & \cdots & \vec{Z}_{1f} & \vec{Z}_{1n} \\ \vec{Z}_{i1} & \cdots & \vec{Z}_{if} & \vec{Z}_{in} \\ \vdots & \ddots & \vdots & \vdots \\ \vec{Z}_{f1} & \cdots & \vec{Z}_{ff} & \vec{Z}_{fn} \\ \vec{Z}_{n1} & \cdots & \vec{Z}_{nf} & \vec{Z}_{nn} \end{pmatrix}^k \times \begin{pmatrix} 0 \\ \vec{I}_f \\ 0 \end{pmatrix}$$

(4)

$$\vec{V}_f - \vec{I}_f \vec{Z}_f = 0 \tag{5}$$

where $\vec{V}_i$ is the voltage sag phasor of bus i, $|\vec{Z}|$ is the bus impedance matrix under fault k, and $\vec{V}_i^{\text{prefault}}$ is the pre-fault voltage of bus i and fault point f, where $i = 1\ldots n$. $\vec{V}_f$ is the fault point voltage, $\vec{I}_f$ is the fault current, and $\vec{Z}_f$ is the fault impedance.

3.2 *Voltage sag profile estimation model for a power system with STATCOMs*

With STATCOM connected in the system, STATCOM is modeled as a current source during the fault. Based on the analysis of STATCOM operation characteristics, the direction of the injected current is opposite that of the fault current. So incorporating the STATCOM model in the short-circuit calculation equations, the short-circuit equation of a power system with STATCOM is

$$\begin{pmatrix} \vec{V}_1 \\ \vec{V}_i \\ \vdots \\ \vec{V}_f \\ \vec{V}_n \end{pmatrix} = \begin{pmatrix} \vec{V}_1^{\text{prefault}} \\ \vec{V}_i^{\text{prefault}} \\ \vdots \\ \vec{V}_f^{\text{prefault}} \\ \vec{V}_n^{\text{prefault}} \end{pmatrix} - \begin{pmatrix} \vec{Z}_{11} & \cdots & \vec{Z}_{1f} & \vec{Z}_{1n} \\ \vec{Z}_{i1} & \cdots & \vec{Z}_{if} & \vec{Z}_{in} \\ \vdots & \ddots & \vdots & \vdots \\ \vec{Z}_{f1} & \cdots & \vec{Z}_{ff} & \vec{Z}_{fn} \\ \vec{Z}_{n1} & \cdots & \vec{Z}_{nf} & \vec{Z}_{nn} \end{pmatrix}^k \times \begin{pmatrix} 0 \\ -\vec{I}_{sh} \\ \vec{I}_f \\ 0 \end{pmatrix}$$

(6)

where $\vec{I}_{sh}^i$ is the injected current of STATCOM connected at bus i.

During the operation of STATCOM under system faults, the injected current in the short-circuit equation should satisfy the operation constraints and control constraints. So, Equations (1)–(3) should be incorporated in the short-circuit calculation model. The mathematical voltage sag profile estimation model is composed of Equations (1)–(3) and (5)–(6).

For multiple STATCOMs connected in the system, the complete voltage sag profile estimation model is as follows:

$$\begin{pmatrix} \vec{V}_1 \\ \vec{V}_i \\ \vdots \\ \vec{V}_f \\ \vec{V}_n \end{pmatrix} = \begin{pmatrix} \vec{V}_1^{\text{prefault}} \\ \vec{V}_i^{\text{prefault}} \\ \vdots \\ \vec{V}_f^{\text{prefault}} \\ \vec{V}_n^{\text{prefault}} \end{pmatrix} - \begin{pmatrix} \vec{Z}_{11} & \cdots & \vec{Z}_{1f} & \vec{Z}_{1n} \\ \vec{Z}_{i1} & \cdots & \vec{Z}_{if} & \vec{Z}_{in} \\ \vdots & \ddots & \vdots & \vdots \\ \vec{Z}_{f1} & \cdots & \vec{Z}_{ff} & \vec{Z}_{fn} \\ \vec{Z}_{n1} & \cdots & \vec{Z}_{nf} & \vec{Z}_{nn} \end{pmatrix}^k \times \begin{pmatrix} 0 \\ -\vec{I}_{sh}^i \\ \vdots \\ -\vec{I}_{sh}^j \\ \vec{I}_f \\ 0 \end{pmatrix}$$

(7)

$$\vec{V}_f - \vec{I}_f \vec{Z}_f = 0 \tag{8}$$

210

$$
\begin{cases}
P^i_{stat} = Re\left(\vec{V}_i \vec{I}^{i}_{sh}{}^{*}\right) = 0 \\
I^i_{sh} \leq I^i_{max} \\
V^i - V^i_{ref} = 0 \quad \text{if } I^i_{sh} \leq I^i_{max} \\
I^i - I^i_{max} = 0 \quad \text{if } V^i \leq V^i_{ref}
\end{cases}
\cdots
\begin{cases}
P^j_{stat} = Re\left(\vec{V}_j \vec{I}^{j}_{sh}{}^{*}\right) = 0 \\
I^j_{sh} \leq I^j_{max} \\
V^j - V^j_{ref} = 0 \quad \text{if } I^j_{sh} \leq I^j_{max} \\
I^j - I^j_{max} = 0 \quad \text{if } V^j \leq V^j_{ref}
\end{cases}
\tag{9}
$$

where Equation (7) are short-circuit equations of a system with multiple STATCOMs, (8) is a fault condition boundary equation, and Equation (9) are operation and control constraint functions of STATCOM connected at bus i and bus j. The unknown variables in this model are STATCOM injected current $I^m_{sh}, m = 1 \ldots j - i$, and bus voltage $\vec{V}_i$, $i = 1 \ldots n$.

4 SOLUTION METHOD FOR A VOLTAGE PROFILE ESTIMATION MODEL FOR A POWER SYSTEM WITH STATCOMS

4.1 Nonlinear programming formulation

It can be noted that due to the operation, control constraints, and the coupling of STATCOMs with the system, the proposed model is a combination of linear and non-linear equations. The voltage sag profile estimation is based on calculating the voltage by solving this nonlinear system. The iteration-based method has been widely used for solving nonlinear equations. But for this model, due to the switching logic of control constraints, a numerical oscillation may occur in the iteration and lead to divergence (Haque 2003). A nonlinear programming based method has been widely used for solving nonlinear equations (Yude & Zhijun 2008). To solve the proposed model efficiently, we formulated a nonlinear programming model to estimate the voltage sag profile in the system with STATCOMs. The voltage sag profile model was changed to a nonlinear programming model to minimize the sum of squares of pre-fault voltage mismatch α. The formulation of this nonlinear programming model is:

$$
\text{Min } F = \alpha \alpha^T
\tag{10}
$$

$$
\text{Subject to : } \left([\vec{V}] - [\vec{Z}]^k \begin{pmatrix} 0 \\ -\vec{I}^i_{sh} \\ \vdots \\ -\vec{I}^j_{sh} \\ \vec{I}_f \\ 0 \end{pmatrix} \right) - [\vec{V}]^{\text{prefault}} = \alpha
\tag{11}
$$

short-circuit equations (Equation 7)
fault boundary condition equations (Equation 8)
operation and control constraints of STATCOMs (Equation 9)
where decision variables are the unknown variables $I^m_{sh}, m = 1 \ldots j - i$, and $\vec{V}_i$, $i = 1 \ldots n$.

4.2 Solution method

To reduce the complexity of the proposed model caused by the switch logic of control constraints, the intermediate variable $d^i_k = V_{ref} - V_i$ is introduced and the control constraint of STATCOM connected with bus i is changed as follows:

$$\begin{cases} P_{stat}^i = Re\left(\vec{V}_i \vec{I}_{sh}^{i\,*}\right) = 0 \\ I_{sh}^i \le I_{max}^i \\ d_k^i = 0 \qquad\qquad \text{if } I_{sh}^i \le I_{max} \\ 0 \le d_k^i \le V_{ref}^i \quad \text{if } I_{sh}^i = I_{max} \end{cases} \tag{12}$$

The Big-M reformulation (Ding et al. 2014) is used to deal with the switching logic in Equation (12), which is converted as follows:

$$\begin{cases} 0 \le d_k^i \le (1 - x_1^i) \times M_1^i \\ 0 \le I_{sh}^i \le (1 - x_1^i) \times M_2^i + I_{max}^i \\ 0 \le d_k^i \le V_{ref}^i - (1 - x_2^i) \times M_1^i \\ -(1 - x_2^i) \times M_2^i + I_{max}^i \le I_{sh}^i \le (1 - x_2^i) \times M_2^i + I_{max}^i \\ 0 \le d_k^i \le (1 - x_3^i) \times M_3^i \\ -(1 - x_3^i) \times M_3^i + I_{max}^i \le I_{sh}^i \le (1 - x_3^i) \times M_3^i + I_{max}^i \\ x_1^i + x_2^i + x_3^i = 1 \end{cases} \tag{13}$$

where x_1^i, x_2^i, x_3^i are binary variables to identify which control constraint is enforced. M_3^i, M_2^i, M_3^i are selected as $V_{ref}^i + 1$, $I_{max}^i + 1$ and $\min\left\{V_{ref}^i + 1, I_{max}^i + 1\right\}$ conservatively because an extremely large M will cause numerical instability (Ding et al. 2014). The same process can be applied to STATCOM connected at bus j. After this modification, the proposed model is a mixed-integer nonlinear programming problem that can be solved using a branch and bound algorithm using a mature commercial optimization solver.

5 CASE STUDIES

Case studies were conducted in order to verify the STATCOM model and correctness of the proposed voltage sag profile estimation method based on nonlinear programming in this article. The IEEE30 bus system with the STATCOM model under fault is established in PSCAD/ EMTDC as shown in Figure 3. The system parameters are obtained from (http://www.ee.wash ington.edu/research/pstca/pf30/pg_tca30bus.htm) and STATCOM is modeled based on Anaya-Lara and Acha (2007). Two STATCOMs are connected with bus 4 and bus 6 respectively. The capacity of both STATCOMs is set at 100 MVA. Numerical studies of the proposed nonlinear programming method for voltage sag profile estimation also have been carried out on the IEEE30 bus system with STATCOM using the commercial optimization solver SCIP through the YALMIP platform (Löfberg 2005.).

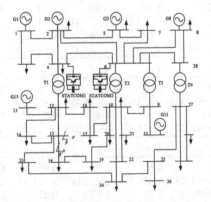

Figure 3.   IEEE30 bus system with STATCOMs under different faults.

5.1 Voltage sag profile of a power system with STATCOMs under bus fault

The voltage sag profile estimation results are shown in Table 1 for a symmetrical fault that occurred at bus 15, assuming the fault impedance is 0,

Table 1. Voltage sag profile under bus 15 fault.

Bus voltage	V_1	V_2	V_3	V_4	V_5	V_6	V_7	V_8
Simulation result	1.06 ∠0°	1.04 ∠−5.35°	1.01 ∠−8.19°	1.00 ∠−10.11°	1.01 ∠−14.16°	1.00 ∠−11.66°	1.00 ∠−13.22°	1.01 ∠−11.81°
Numerical result	1.05 ∠0°	1.02 ∠−5.61°	0.96 ∠−8.31°	0.94 ∠−10.11°	1.00 ∠−14.62°	0.95 ∠−12.10°	0.96 ∠−13.68°	0.97 ∠−13.31°

Bus voltage	V_9	V_{10}	V_{11}	V_{12}	V_{13}	V_{14}	V_{15}	V_{16}
Simulation result	0.90 ∠−17.17°	0.74 ∠−23.54°	1.08 ∠−14.10°	0.63 ∠−25.61°	1.07 ∠−14.94°	0.33 ∠−36.98°	0.00 ∠−8.43°	0.67 ∠−25.47°
Numerical result	0.86 ∠−17.91°	0.71 ∠−24.57°	1.02 ∠−14.91°	0.56 ∠−27.06°	0.88 ∠−17.10°	0.29 ∠−29.51°	0.00 ∠0.13°	0.62 ∠−26.93°

Bus voltage	V_{17}	V_{18}	V_{19}	V_{20}	V_{21}	V_{22}	V_{23}	V_{24}
Simulation result	0.71 ∠−24.35°	0.25 ∠−26.01°	0.41 ∠−25.46°	0.49 ∠−25.13°	0.70 ∠−24.20°	0.69 ∠−24.20°	0.23 ∠−23.42°	0.54 ∠−23.01°
Numerical result	0.68 ∠−25.48°	0.24 ∠−28.57°	0.38 ∠−28.03°	0.46 ∠−27.17°	0.67 ∠−25.37°	0.66 ∠−25.31°	0.22 ∠−25.30°	0.52 ∠−24.28°

Bus voltage	V_{25}	V_{26}	V_{27}	V_{28}	V_{29}	V_{30}
Simulation result	0.71 ∠−22.45°	0.69 ∠−23.10°	0.82 ∠−21.60°	0.97 ∠−12.22°	0.79 ∠−23.46°	0.78 ∠−24.81°
Numerical result	0.69 ∠−23.11°	0.67 ∠−23.90°	0.80 ∠−21.95°	0.93 ∠−12.83°	0.78 ∠−23.70°	0.78 ∠−24.90°

Table 2. Voltage sag profile under the midpoint of transmission line 15–18 fault.

Bus voltage	V_1	V_2	V_3	V_4	V_5	V_6	V_7	V_8
Simulation result	1.06 ∠0°	1.04 ∠−5.35°	1.01 ∠−8.06°	1.00 ∠−9.95°	1.01 ∠−14.16°	1.00 ∠−11.55°	1.00 ∠−13.15°	1.01 ∠−11.81°
Numerical result	1.06 ∠0°	1.04 ∠−5.61°	1.00 ∠−8.62°	0.98 ∠−10.47°	1.01 ∠−14.50°	0.98 ∠−12.16°	0.98 ∠−13.67°	0.99 ∠−13.22°

Bus voltage	V_9	V_{10}	V_{11}	V_{12}	V_{13}	V_{14}	V_{15}	V_{16}
Simulation result	0.92 ∠−16.65°	0.78 ∠−22.01°	1.08 ∠−14.10°	0.80 ∠−21.41°	1.07 ∠−14.94°	0.61 ∠−25.26°	0.43 ∠−21.27°	0.78 ∠−22.12°
Numerical result	0.90 ∠−17.28°	0.78 ∠−22.66°	1.04 ∠−14.81°	0.76 ∠−22.35°	0.96 ∠−16.74°	0.58 ∠−26.24°	0.41 ∠−22.14°	0.76 ∠−23.01°

Bus voltage	V_{17}	V_{18}	V_{19}	V_{20}	V_{21}	V_{22}	V_{23}	V_{24}
Simulation result	0.78 ∠−22.41°	0.16 ∠−24.32°	0.36 ∠−23.67°	0.47 ∠−23.48°	0.76 ∠−22.62°	0.75 ∠−22.57°	0.54 ∠−22.04°	0.69 ∠−22.15°
Numerical result	0.77 ∠−23.08°	0.16 ∠−26.42°	0.35 ∠−25.87°	0.46 ∠−25.07°	0.75 ∠−23.27°	0.75 ∠−23.21°	0.53 ∠−22.90°	0.69 ∠−22.82°

Bus voltage	V_{25}	V_{26}	V_{27}	V_{28}	V_{29}	V_{30}
Simulation result	0.80 ∠−22.24°	0.78 ∠−21.89°	0.88 ∠−20.25°	0.98 ∠−12.18°	0.85 ∠−21.94°	0.84 ∠−23.16°
Numerical result	0.80 ∠−21.61°	0.79 ∠−22.26°	0.88 ∠−20.49°	0.97 ∠−12.87°	0.87 ∠−22.04°	0.86 ∠−23.12°

The voltage sag magnitude and phase numerical results obtained by the proposed method are both consistent with the electromagnetic transient model simulation results. The average error of voltage sag magnitude is 0.01 p.u. and the average error of voltage sag phase is 0.76°, which verified the correctness of the proposed model.

5.2 *Voltage sag profile of a power system with STATCOMs under a transmission line fault*

The voltage sag profile estimation results using the proposed method for a symmetrical fault that occurred at the mid-point of transmission line 15–18, assuming the fault impedance is 0, are shown in Table 2. The average error of the voltage sag magnitude is 0.03 p.u. and the average error of the voltage sag phase is 0.71°, which shows the feasibility of the proposed method under a line fault.

6 CONCLUSIONS

To assess the impact of multiple STATCOMs on the voltage sag profile quickly and accurately, this article modeled the STATCOM as a current source considering its operation characteristics. Then incorporating the STATCOM model into the short-circuit calculation, the mathematical model of voltage profile estimation in networks with STATCOMs is presented. A nonlinear programming–based method is used to solve the model and results of different cases show that it is correct. It can also meet the requirement of optimal STATCOM allocation for voltage sag mitigation. The analytical method is both accurate enough for planning purposes and also easy to program, which gives it a high application value.

REFERENCES

Anaya-Lara, O. & Acha, E. 2007. Modeling and analysis of custom power systems by PSCAD/EMTDC[J]. *IEEE Power Engineering Review* 21(11): 56.

Bollen, M. H. J. 2000. *Understanding Power Quality Problems: Voltage Sags and Interruptions*. IEEE Press Series on Power Engineering. Hoboken, NJ: Wiley.

Chan, J. Y. & Milanović, J. V. 2015. Assessment of the economic value of voltage sag mitigation devices to sensitive industrial plants. *IEEE Transactions on Power Delivery* 30(6): 2374–2382.

Chang, C. S. & Yu, Z. 2004. Distributed mitigation of voltage sag by optimal placement of series compensation devices based on stochastic assessment[J]. *IEEE Transactions on Power Systems* 19(2): 788–795.

CIGRE/CIRED Joint Working Group C4.107. 2011. Economic framework for power quality. Paris: CIGRE/CIRED.

Cruz, I. B. N. C., Lavega, A. P., & Orillaza, J. R. C. 2015. Overview of voltage sag profile estimation. In TENCON 2015–2015 IEEE Region 10 Conference. *IEEE*: 1–6.

Ding, T., Bo, R., Gu, W., et al. 2014. Big-M based MIQP method for economic dispatch with disjoint prohibited zones. *IEEE Transactions on Power Systems* 29(2): 976–977. http://www.ee.washington.edu/research/pstca/pf30/pg_tca30bus.htm

Hingorani, N. G., Gyugyi, L., & El-Hawary, M. 2000. *Understanding FACTS: Concepts and Technology of Flexible AC Transmission Systems*. New York: IEEE Press.

Haque, M. H. 2003. On-line monitoring of maximum permissible loading of a power system within voltage stability limits. *IEE Proceedings: Generation, Transmission and Distribution* 150(1): 107–112.

Kazemtabrizi, B. & Acha, E. 2014. An advanced STATCOM model for optimal power flows using Newton's method. *IEEE Transactions on Power Systems* 29(2): 514–525.

Löfberg, J. 2005. YALMIP: a toolbox for modeling and optimization in MATLAB. In IEEE International Symposium on Computer Aided Control Systems Design.

Milanovic, J. V. & Zhang, Y. 2010. Modeling of FACTS devices for voltage sag mitigation studies in large power systems. *IEEE Transactions on Power Delivery* 25(4): 3044–3052.

Ramamurthy, J. R., Kolluri, S., Mader, D. J., et al. 2018. Mitigation of motor starting voltage sags using distribution-class STATCOM. In 2018 IEEE/PES Transmission and Distribution Conference and Exposition (T&D). IEEE: 1–9.

Xiao, X. Y., Ma, Y. Q., Zhang, Y., et al. 2016. Premium power valuation method based on customer perception of utility for high-technology manufacturing customers. *IEEE Transactions on Power Delivery* 31(4): 1655–1662.

Xu, Y. & Li, F. 2014. Adaptive PI control of STATCOM for voltage regulation. *IEEE Transactions on Power Delivery* 29(3): 1002–1011.

Yude, Y. & Zhijun, Q. 2008. Vectorial power flow calculation based on nonlinear programming. In Power & Energy Society General Meeting: Conversion & Delivery of Electrical Energy in the Century.

Zhang, X. P., Handschin, E., & Yao, M. 2001. Modeling of the generalized unified power flow controller (GUPFC) in a nonlinear interior point OPF. *IEEE Transactions on Power Systems* 16(3): 367–373.

Zhang, Y., Zhang, Y., Wu, B., et al. 2006. Power injection model of STATCOM with control and operating limit for power flow and voltage stability analysis. *Electric Power Systems Research* 76(12): 1003–1010.

Zhang, X. P., Rehtanz, C., & Pal, B. 2012. *Flexible AC Transmission Systems: Modelling and Control.* Berlin and Heidelberg: Springer Science+Business Media.

Emerging Developments in the Power and Energy Industry – Dufo-López, Krzywanski & Singh (eds)
© *2020 Taylor & Francis Group, London, ISBN 978-0-367-27169-5*

Evaluation of development effects of SZ36-1 oilfield after overall infilling

Jicheng Zhang, Jiale Fan, Jun Zhang & Qingqing Li
College of Petroleum Engineering, Northeast Petroleum University

ABSTRACT: SZ36-1 Oilfield was put into production in 1993 and underwent well pattern infilling in 2009-2010. After that, the well pattern type was changed from inverted nine-spot pattern to line drive pattern. Based on the dynamic data as of June 2016, eight development indexes were calculated as follows: recovery factor 43.04%, degree of control of waterflooding reserves 93.08%, producing degree of waterflooding reserves 90.06%, water cut rising rate 0.8%, decline rate 1.01%/a, stage water storage rate evaluation coefficient 0.654, stage water drive index 1.721, average formation pressure 10.11MPa, and pressure maintenance level 73.3%. In accordance with CNOOC's industrial standards, the recovery factor, degree of control of waterflooding reserves, producing degree of waterflooding reserves, water cut rising rate, decline rate, and stage water storage rate after infilling belong to Class I. Determined by the industrial standard method, the development effect of SZ36-1 Oilfield after overall infilling also belongs to Class I. In SZ36-1 Oilfield, the water cut is 74.04%, and the recovery percent is 23.11%, but the forecasted recovery factor is 43.04%, therefore, there is still a large residual oil potential.

1 INTRODUCTION

From the 1940s through the 1950s, the United States began to consider the rationality of water-flooding development and formed a lot of empirical prediction formulas for static and dynamic recoverable reserves. Since the 1950s, China also began the research on water-flooding development effects and has formed several methods such as state contrast, recoverable reserves evaluation, gray system theory and numerical simulation evaluation after several decades of development.

They applied fuzzy comprehensive evaluation method to the evaluation of waterflooding development effects and obtained an ideal effect, of which the key is the determination of single factor evaluation vector (WANG Nutao, LUO Xingwang & ZHANG Yanmei, 2008). They applied gray fuzzy theory to the comprehensive evaluation of such indicators as water drive recovery factor, control extent of water drive reserves, stage water storage rate, water drive index, water cut rising rate and natural decline rate (LUO Erhui & WANG Xiaodong, 2010). They proposed predicting the waterflooding recoverable reserves in water-flooded oilfields by using a variety of displacement characteristic curves, of which Type A and Type C water-drive law curves are most widely used and have the highest precision (CHEN Yuanqian, ZOU Cunyou & ZHANG Feng, 2011). He based on the relative permeability curves of oilfields, derived the expression of water cut rising rate corresponding to the recovery degree of geological reserves, and demonstrated it by practice (MIAO Feifei, 2014).

2 EVALUATION OF SINGLE-INDEX WATERFLOODING DEVELOPMENT EFFECTS

2.1 *Recovery factor*

The recovery factor of SZ36-1 Oilfield was calculated by using Type A and Type C water-drive law curve methods and then rated according to the target recovery factor obtained by the empirical formula of Bohai Oilfield.

2.1.1 *Type A water-drive law curve method*

The Type A water-drive law curve is mathematically expressed as follows:

$$\lg W_p = A_1 + B_1 N_p \tag{1}$$

Where: W_p - cumulative water production, $10^4 m^3$; N_p - cumulative oil production, $10^4 m^3$.

According to the intercept and slope in the above Type A water-drive law curve fitting formula, the oilfield's recovery factor can be calculated (see Table 1):

$$R = \frac{1}{NB_1}\left(\lg\frac{f_w}{2.303B(1-f_w)} - A_1\right) \tag{2}$$

2.1.2 *Type C water-drive law curve method*

The Type C water-drive law curve is mathematically expressed as follows:

$$\frac{L_p}{N_p} = A_2 + B_2 N_p \tag{3}$$

Where: L_p - cumulative liquid production, $10^4 m^3$; N_p - cumulative oil production, $10^4 m^3$.

According to the intercept and slope in the above Type C water-drive law curve fitting formula, the oilfield's recovery factor can be calculated (see Table 1):

$$N_p = \frac{1-\sqrt{A(1-f_w)}}{NB} \tag{4}$$

In SZ36-1 Oilfield, the water cut and recovery percent are now 74.04% and 23.11% respectively. It is predicted by two water-drive law curve methods that the recovery factor is 43.04%, so there is still a large residual oil potential.

2.1.3 *Target recovery factor*

According to the empirical formula of Bohai Oilfield, the target recovery factor is obtained.

$$\text{Ultimate Well} - \text{controlled Reserves} = (-3.4525 + 1.7988\log_{10}K - 0.5412\log_{10}\mu_o - 0.03V_K)h \tag{5}$$

Where: K - reservoir permeability; μ_o - crude oil viscosity; V_K - variation coefficient of permeability; h - reservoir thickness

Through calculation, the recovery factor evaluation coefficient after infilling is A = 43.04%/47.37% = 90.86% ≥ 85%. According to the evaluation criteria, the recovery factor is rated as Class I.

2.2 *Degree of control of water drive reserves*

Separate oil sand body method can be used to analyze the extent to which the degree of water-flooding control is affected by well spacing density. The degree of waterflooding control under

Table 1. Results of calculation by water-drive law curves.

	Type A Water-drive law curve	Type C Water-drive law curve	Average Recovery Factor
Recovery Factor	41.83%	44.24%	43.04%

Table 2. SZ36-1 oilfield's target recovery factor calculated by empirical formula method.

Ultimate Well-controlled Reserves	Ultimate Well Spacing Density (wells/km^2)	Well-controlled Area (ha/well)	Target Recovery Factor
77.28	9.35	0.107	47.37%

Table 3. Target recovery evaluation criteria.

Evaluation Index	Description	Evaluation Coefficient	Evaluation Criteria		
			Class I	Class II	Class III
Recovery Factor	Calibrated recovery factor or target recovery factor	Coefficient A: the ratio of the calibrated recovery factor to the target recovery factor	≥85%	85% ~ 75%	≤75%

different well spacing conditions can be calculated by sorting out the perimeter and area of each sand body in reservoirs.

The degree of waterflooding control over oil sands is as follows:

$$M_i = 1 - 0.470698 \cdot D \cdot L_i^{0.5}/A_i^{0.75} \tag{6}$$

The degree of waterflooding control over development block is as follows:

$$M = \sum_{i=1}^{n} (M_i \cdot N_i)/\sum_{i=1}^{n} N_i \tag{7}$$

Where: L_i - oil sand perimeter, km; A_i - oil sand area, km^2; N_i - geological reserves in each oil sand body, t; D - well spacing, km; M_i - degree of waterflooding control over oil sands, decimal; M - degree of waterflooding control over development block, decimal.

Through calculation, the degree of control over water drive reserves after infilling is 92.8%. According to the evaluation criteria in Table 4, SZ36-1 Oilfield belongs to Class I Oilfield with delta facies sediments.

2.3 *Producing degree of water drive reserves*

Mine field practices indicate that, due to the prediction based on Type C water drive characteristic curves, the computational process becomes simple and easy, and the calculation results are also relatively objective and accurate. The mathematical expression is:

Table 4.   Evaluation criteria for degree of control over water drive reserves.

Item		Category		
		Class I	Class II	Class III
Degree of Control over Water Drive Reserves	Delta	≥85%	85% ~ 75%	≤75%
	Fluvial facies	≥80%	80% ~ 70%	≤70%

$$\frac{L_p}{N_p} = A_2 + B_2 N_p \tag{8}$$

$$R_{OM} = \frac{1}{B_2} \frac{N_{OM}}{N \times R_{gm}} \tag{9}$$

Where: B_2 - a constant of Type C water-drive law curve; N_{OM} - waterflooding controllable reserves, $10^4 m^3$; N - geological reserves, $10^4 m^3$; R_{gm} - the final recovery percent of reservoir evaluated according to the reservoir's geologic characteristic parameters, decimal; R_{OM} - the producing degree of reserves, decimal.

Calculated by the Type C water-drive law curve method, the producing degree of water drive reserves after infilling is 90.06%. Since the sedimentary facies type of SZ36-1 Oilfield belongs to delta facies sediments, the producing degree of water drive reserves after infilling is rated as Class I.

2.4   Water cut rising rate

The actual water cut rising rate of the oilfield is calculated according to the actual water cut after infilling. The water cut rising rate is evaluated by using theoretical and actual water cut rising rates and with reference to the evaluation methods and criteria in Table 6.

After pattern infilling was applied to SZ36-1 Oilfield, the water cut rising rate was actually 0.8% but 3.19% in theory. In 2016, the coefficient α for water cut rising rate after infilling was 0.25, so the evaluation result is Class I Oilfield.

2.5   Decline rate

By means of reservoir engineering, and based on the definition of decline rate, the relative permeability curve and fractional flow equation as well as the actual oilfield production conditions, it can be concluded that the water cut rising rate is directly proportional to the production decline rate under fixed-rate production conditions, and thus the theoretical decline rate in 2016 was 0.1621/a.

Calculated by the annual decline rate formula (Formula 10), the annual decline rate after infilling was A = 1.01%/a in 2016.

Table 5.   Evaluation criteria for producing degree of water drive reserves.

Item		Category		
		Class I	Class II	Class III
Producing Degree of Water Drive Reserves	Delta	≥ 80%	80% ~ 70%	≤ 70%
	Fluvial facies	≥ 75%	75% ~ 65%	≤ 65%

Table 6. Classification criteria for evaluation coefficient of water cut rising rate.

Evaluation Index	Evaluation Coefficient and Method	Evaluation Criteria		
		Class I	Class II	Class III
Water Cut Rising Rate	Coefficient α: the ratio of the actual water cut rising rate to the theoretical water cut rising rate	≤ 1.2	1.2 ~ 2.0	≥2.0

$$\lg q_t = \lg q_i - D_i t \lg e \qquad (10)$$

Taking the relation curve for theoretical decline rate vs. water cut of reservoirs as the basis of comparative analysis, and considering the different development stages, the relative index β was introduced for the evaluation. The evaluation criteria are shown in Table 7.

Through calculation, the actual decline rate after infilling was 1.01%/a in 2016, with the coefficient β of 0.062, so the evaluation result is Class I Oilfield.

2.6 Stage water storage rate

Broadly, the relationship between water storage rate and water content is expressed as follows:

$$C_{PG} = 1 - \frac{1}{Z_G(1 + B_o \frac{1-f_w}{f_w})} \qquad (11)$$

According to the stage injection-production ratio calculated under real oilfield conditions, the theoretical stage water storage rate curve is plotted, and the theoretical curve slope is obtained accordingly. The water storage rate in the actual oilfield evaluation stage is put on the theoretical curve graph, and then connected to the point where the water cut is zero and the water storage rate is 100%. In this way, the actual slope is calculated as shown in Figure 1, and the evaluation criteria are shown in Table 8.

Through calculation, the theoretical rate slope after infilling is -1.2502, while the actual water storage rate slope is -0.8182 and the stage water storage rate (λ coefficient) is 0.654. So the water storage rate after infilling is rated as Class I.

2.7 Stage water drive index

The relationship between stage water drive index and water cut is expressed as follows:

$$S_{PG} = (Z_G - 1)\frac{f_w}{B_o(1-f_w)} + Z_G \qquad (12)$$

Where: Z_G - injection-production ratio; f_w - water cut; S_{PG} - stage water drive index.

Table 7. Decline rate evaluation criteria.

Evaluation Index	Evaluation Coefficient	Evaluation Criteria		
Decline Rate	Coefficient β: the ratio of the actual decline rate to the theoretical decline rate	Class I ≤ 1.0	Class II 1.0 ~ 2.0	Class III ≥ 2.0

Table 8. Stage water storage rate evaluation criteria.

Remark	Class I	Class II	Class III
Stage Water Storage Rate (λ Coefficient)	≤ 1.0	1.0 ~ 2.0	≥ 2.0

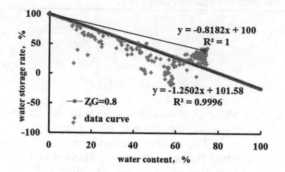

Figure 1. Schematic diagram of stage water storage rate slope.

According to the above formula, the theoretical curve for stage water drive index vs. water cut is made. By virtue of the comparison between the oilfield's actual production curve and the theoretical curve, the oilfield's waterflooding development effect can be evaluated accordingly. The relative index of coefficient γ was introduced for the evaluation (see Table 9):

In SZ36-1 Oilfield, the water drive index in 2016 was theoretically 1.814 but actually 1.721, with the evaluation coefficient (γ coefficient) of 0.949. Since the water drive index is greater than 0, the evaluation result is Class II Oilfield.

2.8 *Formation pressure maintenance level*

By sorting out the formation pressure of typical oil wells in this block, the formation pressure level of the block can be obtained. The pressure maintenance level is divided into the following three categories, as shown below.

Based on the measured data, the average oilfield-wide formation pressure is 10.11MPa, and the pressure maintenance level is 0.733, belonging to Class II.

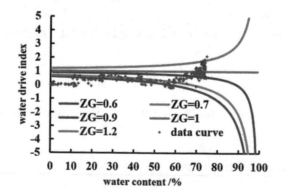

Figure 2. Water cut vs. water drive index curve.

Table 9. Stage water drive index evaluation criteria.

Stage Water Drive Index	Remark	Class I	Class II	Class III
<0	Stage Water Drive Index (γ coefficient)	≤ 1.0	1.0 ~ 2.0	≥ 2.0
>0		≥ 1	0.5 ~ 1.0	≤ 0.5

Table 10. Evaluation criteria for formation pressure maintenance level.

Evaluation Index	Category		
Pressure Maintenance Level	Class I The formation pressure is above 85% of the initial formation pressure	Class II The formation pressure is 60-85% of the initial formation pressure	Class III The formation pressure is below 60% of the initial formation pressure

3 COMPREHENSIVE EVALUATION OF WATERFLOODING DEVELOPMENT EFFECTS

According to the evaluation results of single-factor indexes after infilling as previously mentioned, the evaluation result before infilling can be obtained by the same calculation method. Combined with the criteria stated in the *Evaluation on the Development Effects Water Drive Reservoirs with Middle & High Porosity and Permeability in Bohai Sea*, the final evaluation results before and after infilling are listed as follows:

4 CONCLUSIONS

(1) Based on the dynamic data as of June 2016, eight development indexes were calculated, including recovery factor, degree of control of waterflooding reserves, producing degree of waterflooding reserves, water cut rising rate, decline rate, stage water storage rate, stage water drive index and formation pressure maintenance level. Results indicate that the recovery factor

Table 11. Evaluation results of development effect evaluation indexes of SZ36-1 oilfield.

			Before Infilling		After Infilling	
No.	Evaluation Index	Evaluation Coefficient	Evaluation Result	Conclusion	Evaluation Result	Conclusion
1	Recovery factor	A	47.52%	Class III	90.86%	Class I
2	Control extent of water drive reserves		85.17%	Class I	93.08%	Class I
3	Producing degree of water drive reserves		42.85%	Class III	90.06%	Class I
4	Water cut rising rate	α	1.45	Class II	0.25	Class I
5	Decline rate	β	1.117	Class II	0.062	Class I
6	Stage water storage rate	λ	0.829	Class I	0.654	Class I
7	Stage water drive index	γ	1.058	Class I	0.949	Class II
8	Formation pressure maintenance level	B	0.685	Class II	0.733	Class II
9	Comprehensive Evaluation Result of SZ36-1 Oilfield			Class II		Class I

is 43.04%, the degree of control of waterflooding reserves is 93.08%, the producing degree of waterflooding reserves is 90.06%, the water cut rising rate is 0.8%, the decline rate is 1.01%/a, the stage water storage rate evaluation coefficient is 0.654, the stage water drive index is 1.721, the average formation pressure is 10.11MPa, and the pressure maintenance level is 73.3%.

(2) These eight indexes were comprehensively evaluated according to CNOOC's industrial standards. Results indicate that the recovery factor, degree of control of waterflooding reserves, producing degree of waterflooding reserves, water cut rising rate, decline rate and stage water storage rate after infilling belong to Class I. Determined by the industrial standard method, the development effect of SZ36-1 Oilfield after overall infilling belongs to Class I; however, it belongs to Class II before infilling. This proves that the overall infilling achieved good results.

ACKNOWLEDGEMENT

"Study on the Development Model of Offshore Heavy Oilfields" in the National Major Science and Technology Project during the 13th Five-Year Plan Period (2016ZX05025-001).

REFERENCES

CHEN Yuanqian, ZOU Cunyou and ZHANG Feng. 2011. Application of Water-drive Law Curve Method in Oilfield Development Evaluation [J]. *Fault Oil & Gas Fields*, 18 (06): 769–771+779.

Guthrie R K, Gerenbegrer M H. 1955. Use of Multiple Correlation Analysis for Interpreting Petroleum Engineering Data [C]. *Drilling and Production Practice*, API, 130–137.

Hongzhao T. 2011. Comprehensive Forecasting of Development Indexes in Water Drive Oilfield [J]. *Fault Block Oil & Gas Filed*.

LUO Erhui, WANG Xiaodong and WANG Jiqiang. 2010. Comprehensive Evaluation of Waterflooding Development Effects Based on Gray Fuzzy Theory [J]. *Xinjiang Oil & Gas*, 6(02): 30–34+109.

MIAO Feifei, LIU Xiaohong and ZHANG Hongyou. 2014. Comprehensive Study and Practices on Water Cut Rising Rules of Waterflooding Oilfields [J]. *Journal of Chongqing University of Science and Technology (Natural Science)*, 16(02): 71–73+153.

WANG Feiqiong, CHENG Mingjia and CHENG Zili. 2011. Bohai Offshore Oilfield Development Adjustment Strategy and Effect [J]. *Journal of Oil and Gas Technology*, 33(12): 148–151+176.

WAN Jiye. 1982. Displacement Series and Application of Waterflooding Oilfield [J]. *Petroleum Exploration and Development*, (06): 65–73.

WANG Nutao, LUO Xingwang and ZHANG Yanmei. A New Method for Determination of Single Factor Evaluation Vector in Waterflooding Development Effect Evaluation [J]. *Daqing Petroleum Geology and Development*, (01): 61–62+66.

WANG Yingying. 2013. Evaluation Research of Waterflooding Development Effects in SZ36-1 Oilfield [D]. *Southwest Petroleum University*.

ZHANG Jicheng and SONG Kaoping. 2007. Relative Permeability Characteristic Curve and Its Application [J]. *ACTA PETROLEI SINICA*, (04): 104–107.

ZHANG Jifeng. 2012. Overview on Effect Evaluation Methods and Development Trends of Waterflooding Oilfield Development [J]. *Lithologic Reservoir*, 24(03): 118–122.

Emerging Developments in the Power and Energy Industry – Dufo-López, Krzywanski & Singh (eds)
© 2020 Taylor & Francis Group, London, ISBN 978-0-367-27169-5

Cyberphysical paradigm of multiagent systems for smart microgrids

Jiahong Dai, Yu Wang & Yan Xu*
School of Electrical and Electronic Engineering, Nanyang Technological University, Singapore

ABSTRACT: Communication infrastructures play a key role in a smart grid. In particular, the multiagent system (MAS) will have a wide application in microgrids, which are the basic subsystems of modern power systems. In this article, the cyberphysical paradigm for a communication MAS in a smart grid application is introduced. The state of the-art in this field is reviewed. First, the structures and concepts of MAS for smart microgrids are presented. Then, the technologies of communication and computation infrastructure in the physical layer, as the hardware of communication systems, are reviewed. Finally, the methodologies for data transmission, processing, decision, and optimization in the cyber layer are discussed. A feasibility study of Internet of things devices and technologies including cloud and edge computation for MAS in microgrids has been conducted.

Keywords: MAS, Microgrid, Communication Infrastructure, Cyberphysical Paradigm

1   INTRODUCTION

In recent years, the smart grid has been a new trend for digitalization of power systems, integrating advanced sensing, communication, and control functionalities in the power grid's operation. The motivation of smart grid technology is to improve efficiency, reliability, and security and to reduce emissions, which is difficult for the legacy grid to achieve considering the prevailing and increasing distributed energy resources (Hurtado et al. 2018). The microgrids, as the fundamental subsystems in future smart grids, cover both energy consumers and producers in a small-range power distribution system.

With its significant advantages, flexibility, fault tolerance, autonomy, responsiveness, proactiveness, and scalability in distributed systems, the multiagent system (MAS) has become a hot research pursuit for distributed generations, microgrids, and smart grids in the future (Coelho et al. 2017). A MAS equipped with data-driven methods, such as machine learning and artificial intelligence, is a highly competitive direction to realize the concepts of he smart grid. There has been substantial progress in related fields, such as energy management, service restoral, and demand response (DR) (Meskina et al. 2017; Wang & Paranjape 2017; Hurtado et al. 2018; Netto et al. 2018; Wang et al. 2018).

As a bottleneck, the communication technologies are limiting the current development of smart grids. Latency sensitivity is an important feature of many industrial control networks, as it is in the smart grid (Coelho et al. 2017). MAS addresses the challenges that traditional power grids cannot cope with through the coordination of agents. However, every decision and action of the agent requires high-speed and reliable communication between the agents, and failures may occur in these communication processes. Considering the widespread adoption of MAS over the smart grid, the existing communication and computation systems are open to doubt, especially confronted with concerns regarding security of implementation and scalability in large-scale real problems.

The smart grid is a typical cyberphysical system (CPS) running in a mission-critical industrial environment, which requires a set of harsh conditions to satisfy the supply and management of high-quality energy (Yu & Xue 2016). From the perspective of a CPS, both physical and cyber layers would be potential bottlenecks to constrain the performance of the power grid or could even cause a catastrophic accident. To avoid a potential but fatal accident, it is necessary to clear up the related work and discuss the research direction of the communication system on MAS in the smart grid and microgrids.

The rest of the article is organized as follows. Section 2 introduces the cyberphysical paradigm of a communication system in MAS. Section 3 describes the communication and computation infrastructure in the physical layer. Section 4 discusses the transmission, processing, decision, and optimization in the cyber layer. Finally, Section 5 draws conclusions.

2 COMMUNICATION SYSTEM OF MAS IN A MICROGRID

2.1 MAS in a microgrid

A microgrid can be operated either in grid-connected mode or in an islanded mode. In the grid-connected mode, the main grid can either inject power into the microgrid for the power shortage or absorb surplus power from the microgrid. Hence, the main grid can maintain the supply–demand balance in the grid-connected mode. However, in the islanded mode, the microgrid must maintain its own supply–demand balance of power, which is achieved by load sharing control.

MAS is an application-oriented system design. The difference between it and the traditional control system lies in its group intelligence, which is a wide concept and plays various roles in different scenarios. The application of MAS in the field of smart grids can be traced back to the work of McArthur, which advocated investing in agent technology for power grids (McArthur et al. 2007a, 2007b). They concluded that a MAS could be used either as a way of building robust and flexible hardware/software systems or as a modeling approach. They also emphasized techniques and tools that could favor and support engineers to use MAS, such as the Java Agent Development Framework (JADE), which is a Foundation for Intelligent Physical Agents (FIPA) standard-based MAS framework supporting multiagent development as an important agent platform implementation.

However, the information in MAS is exchanged periodically and the agent needs to broadcast its state to its neighbors at every iteration. Because of its ease and simplicity, this polling communication has been widely used in MAS. Unfortunately, the useless data flow caused by polling is a heavy burden considering the low communication and computation resources of most processors in MAS. It is necessary to redesign the communication mode to alleviate the load of both the network and agent.

Combined with edge and cloud services, the MAS can scale down the data throughput and the resource cost. Considering the tenancy sensitivity, MAS tasks are classified as cloud based and edge based. When the decision-makers and actuators need to respond quickly, the edge computation embedded in the agent will act by itself. When the tasks require a global vision to optimize the whole performance of the grid, the cloud will take over the work.

2.2 Cyberphysical paradigm of communication in MAS

From the perspective of a CPS to reconsider the communication system of MAS in microgrids, the highly integrated grid can be divided into a physical layer and a cyber layer. Decoupling of the two dimensions enables clarification of common design goals and makes it possible to sort out the priorities of reliability, manageability, efficiency, and smart operation. Figure 1 shows the cyberphysical paradigm of communication in MAS.

It is not certain sensors or actuators but the communication system that is the foundation for maintaining reliable power grid operation in the MAS environment in a smart grid. If the communication can be reliably maintained, even if the individual components fail, the features of MAS, a distributed system, can realize service restoral and self-correction. This means

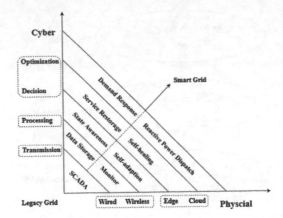

Figure 1. Cyberphysical paradigm of communication in MAS.

a huge shift in the design purpose: the reliability of the overall network service and the integrity of network functionality replace those of individual components.

In a successful MAS design in a smart grid, customers do not observe the failures of the power grid and enjoy the high-quality energy even if several components have failed in the related system. In other words, MAS serves as a hybrid of platform as a service (PaaS) and software as a service (SaaS), which means it provides consumers with the service of power but the source of the service is not any specific generators associated with geographic locations and power companies but the joint work of MAS. In this case, the agent and the information communication technology devices connected to it serve as a virtual machine and container in the cloud architecture, whereby physical resources are separated from computation services to make the product and consumer of power separate entities, which is impossible in a legacy grid. In short, failure is no longer a non-existent thing. On the contrary, it will appear commonly, but the service will no longer be suspended due to the occurrence of faults. Even if an accident occurs, its scale will be as low as possible without affecting customers.

In a legacy power grid, it is common practice to use a supervisory control and data acquisition system (SCADA) to monitor the operation state of the overall grid (Boyer 2009). Although the auto-decision of a power grid dispatch and automation of protection have been widely applied in a production environment, the legacy grid is still highly dependent on human orders. It requires data transmission and a related legacy communication infrastructure. However, it can only passively get the data of sensors and barely change anything unless it contains components out of the communication system, SCADA.

With the uniformed interfaces of wired and wireless devices and broadened bandwidth, such as the standard sets of a Wide Area Measurement System (WAMS), DR, and cognitive radio network (CRN) technology, the legacy grid is enabled to monitor and store the real-time big data produced from massive devices and formulate the knowledge base.

Equipping the power grid with mature data processing technology such as software-defined networking (SDN) and agents connected to massive devices, especially smart meters and actuators, give the power grid the ability to be aware of its operation state and self-adapt to the changes in the environment such as emergency islanding and a blackout at several points.

Massive edge computation devices and decision-making based on a formulated knowledge base make it possible to realize service restoral by an automatic change of dispatch mode and self-correction based on the physical actors. When the cloud devices are ready for global optimization, the DR and reactive power dispatch are within reach. Figure 2 demonstrates MAS in microgrids combined with edge and cloud services.

In the following sections, the details of the physical layer and cyber layer of a MAS communication system for smart grid applications are introduced and reviewed.

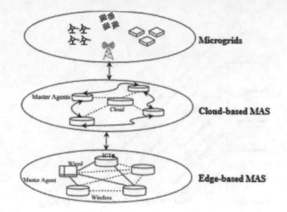

Figure 2. MAS in microgrids combined with edge and cloud services.

3 PHYSICAL LAYER

In CPS, communication involves data transmission, network maintenance, and actuator control. Implementing these features requires the physical layer to have the infrastructure for communication and computation. Actuators can be abstracted into artificial changes in a state in the network caused by parameters passing by communication networks. The focus of this article is the communication design of MAS. The communication system does not cover the local functional facilities such as circuit breakers, transformers, and power electronics but rather the computation result that passes to these, which is classified into the computation infrastructure of the physical layer.

3.1 *Communication infrastructure*

The mainstream communication technologies for a smart grid can be classified into two sets: wired or wireless. Traditional communication devices are commonly based on wired communication for the sake of high reliability and low latency and are relatively economical. However, wireless communication shows more advantages considering the requests of fast-growing smart grids and is more suitable for seamless integration with other components in the smart grids. Although the applications are service constrained, wireless communication has proven to be necessary in smart grids, especially in the microgrid and distributed grid environments.

In general, wired communication devices are more reliable and are commonly deployed in the critical components of the grid. The representatives are fiber-optic communications and power line comunications (PLCs) (Latchman & Katar 2002; Keiser 2003). Fiber-optic communication offers a high data rate up to 50 Gb/s and promises an ultra-low latency, which often serves as the backbone to cover long-distance and real-time information. However, the deployment and maintenance are costly and difficult, which constrains the application environment to the backbones of the grid. PLC utilizes the existing power cables for data transmission, which reduces the installation cost of the communications infrastructure. Since the signal propagation environment of a PLC is harsh and noisy, the channel is difficult to model and the data transmissions via power lines may not be reliable. Nevertheless, thanks to the numerous emerging achievements from both academia and industry, the transmission problems are expected to be solved with the rapid development of PLCs.

Compared with wired communication, wireless communication provides a more flexible and scalable method to organize and manage the subnetwork of the grid although it works properly only in if a specific application or component has enough margin for occasional traffic jams and failures. The array of wireless communication is growing quickly, ranging from traditional techniques such as Zigbee and WLAN to emerging technology, such as Long Range Radio (Lora), and Narrowband IoT (NB-IoT).

Lora is also an emerging representative unlicensed Low-Power Wide-Area Network (LPWAN) technology that operates on the 433–868 MHz, or 915 MHz Industrial Scientific Medical (ISM) bands. It was proposed by Sentech and further promoted by the Lora Alliance and is a highly customized system offering specification of standards and adaptation to legacy devices based on the developers' needs, all of which is free of charge (Georgiou & Raza 2017). The modulation of Lora is based on a chirp spread spectrum scheme that helps fight against heavy multipath fading. The coverage of Lora scales up to 8 km in urban areas and 22 km in rural areas in theory. It offers an adaptive data rate range from 0.3 to 50 kb/s that is determined by six orthogonal spreading factors. The low-data-rate characteristic of Lora signifies that it is applicable only to applications with small payloads. However, under a harsh electromagnetic environment such as the neighborhood of a power line, the communication coverage will drop and cannot maintain the theoretical indicator.

NB-IoT is a new Third Generation Partnership Project (3GPP) radio-access technology and designed based on the existing long-term evolution (LTE) facilities. 3GPP has just released the standard of this technology and it can be compatible with existing Global System for Mobile (GSM) communications, general packet radio service, and LTE standard. It is also a low-rate wireless technology and its peak speed of downlink and uplink is 230 and 250 kb/s respectively, which is suitable for some event-based monitors and special signaling (Wang et al. 2017). However, just like in Lora, the transmission speed will decrease as the transmitting data increase while NB-IoT has a better performance as a hub node to provide local devices access to the network service. Moreover, it requires the transformation of communication base station while most Internet service providers (ISPs) are still observing the development of 5G and NB-IoT and do not want to make a premature decision. More importantly, NB-IoT does not promise a low latency of communication and compromises its quality of service (QoS) to maintain a larger coverage, which constrains its application in a smart grid.

3.2 Computation infrastructure

The high-reliability requirements of the power grid make the reliability of communication the core target in many design indicators of system design. In recent years, in order to improve the reliability and service quality of communication systems under existing equipment technology conditions, a paradigm shift occurred: from cloud computing to the edge computing, or the hybrid mode of cloud computing and edge computing (Armbrust et al. 2010; Bonomi et al., n.d.). The main feature of edge computation is to push light computation, network control, and data storage to the network edges such as base stations and ISP access points, to enable computation-intensive and latency-critical applications at resource-limited devices. This section focuses on the cloud and edge computing model from the perspective of computation infrastructure.

Virtualization and container technology free the physical resources of computers from the limitations of time and space. Users do not need to be equipped with advanced computing devices on site if they can connect to the network and hand over their local work to a remote cloud service. The computation power is no longer the specific performance of the local device but an abstract infrastructure, platform, or service, which is also called IaaS, PaaS, and SaaS. From the perspective of the physical layer, the cloud comprises the server clusters on the core network, while cloud computing organizes and schedules the clusters of these core networks in an efficient, economical, and reliable way through virtualization and container technology such as Docker and Kubernetes.

Cloud facilities can be divided into public and private clouds. Currently, mainstream cloud providers include Google, Amazon, and Microsoft. Many companies have completely transferred their technology stacks onto these platforms to reduce development, operation, and maintenance costs, but cloud services are billed according to the flow, which imposes many restrictions on the cost of some high data traffic scenarios in industries. MAS is an example of machine to machine (M2M) communication between the intelligent devices and agents and the coordination and operation scheduling in the whole network view consuming huge network traffic all the time. Therefore, private clouds are more popular with these industrial scene applications. With the support of the open source community, the technology stack represented by OpenStack and Docker makes private clouds popular in a low-cost way. Without this, many industries and business specificities and corresponding laws and regulations determine that their data cannot be handed over to third-party platforms, although in terms of data security, these third-party platforms may have a better performance.

Owing to the rapid increase in the number of mobile devices, conventional centralized cloud computing is struggling to satisfy the QoS for many applications. With 5G network technology on the horizon, edge computing will become the key solution to solving this issue. One of the major challenges associated with 5G technology is the radio-access network (RAN). In RAN, mobile edge computing provides real-time RAN information. By using the real-time RAN information, the network providers can improve the quality of experience (QoE) for end users, because real-time RAN will offer context-aware services.

These edges will not be very powerful nodes; they would just complement the cloud. Essentially, edge computing is about knowing which data must be analyzed at which point. The function of edge nodes is to broaden access for devices and realize data preprocessing to alleviate the load of the cloud and improve the local performance that doesn't require rich hardware resources. Even embedded systems such as Raspberry P can act as edge nodes.

At present, there are no enterprises in the industry that focus on edge products, and most of them are just sinking mature cloud services to the edge by a small cloud size or mobile embedded systems, thus achieving complementarity between the edge and the cloud. The main players are still Google, Amazon, and Microsoft. However, as the market continues to be subdivided, new players have emerged in the market for industrial scenes, such as Predix of GE and MindSphere of Siemens.

4 CYBER LAYER

After the MAS communication is decoupled from the physical layer, the role of the cyber layer is to achieve access and transmission of massive devices, process the data in access and transmission, and provide a low-latency and high-reliability data link for edge and cloud computation.

Existing power communication systems are designed based on the legacy grid with a one-way flow of information and power. With the increasing adoption of distributed energy and smart devices, its transmission capacity can no longer satisfy the needs of the smart grid. Considering the trend, there are two major tasks: the first is the unification of the interface. Existing communication protocols and standards are mostly based on wired communication systems. Smart devices, especially wireless sensors and edge nodes in the power grid, are not compatible with wired communication standards and protocols due to resource constraints, followed by bandwidth expansion which is necessary for maintenance of a low-latency and high-quality data link of edge and cloud computation.

4.1 *Data transmission*

Transformation from wired to wireless communication is an inevitable trend in the development of power communication in which WMAS and DR are always the core issues.

IEC 61970 is the common guild line in the area of electric transmission. IEC 61850 is widely considered to the best choice for a smart substation involving performance protection, monitoring, control, and automation functions. With the increasing applications associated with interoperability among intelligent electronic devices (IEDs) and remote terminal units (RTUs), IEEE 1379 has been the popular standard.

The demand response business network (DRBizNet) is a highly flexible, reliable, and scalable platform to support DR applications, which enables market operators and utilities to efficiently, reliably, and securely manage DR processes supporting management through web browsers.

Compared with increasing the bandwidth of wired communications, which requires updating many devices at an alarming cost, the communication bandwidth of wireless devices depends on the utilization of the spectrum, and the cost is relatively low via the adoption of CRN (Letaief & Zhang 2009).

Dynamic spectrum access using CRN is an important technology to improve the spectrum utilization of SG communications (Khan et al. 2016). To increase the efficiency of spectrum utilization, CRN permits unlicensed users to opportunistically operate in vacant licensed spectrum bands,

However, the CRN also brings new problems. A secondary user (SU) in CR must continuously monitor radio spectrum usage to give precedence to the primary user (PU). As such, if a PU starts to transmit signals, then SUs must switch to another spectral hole immediately, which may occur in a random fashion. For this reason, the random interruptions of SU traffic will unavoidably cause packet losses and delays for SU data delivery. Moreover, the loss of data packets may have considerable effects on the control and management of the smart grids, particularly in emergency situations.

In real production environments, communication failures in communication channels can significantly degrade the reliability of the cooperative control of distributed energy resources in distribution networks, which may cause very serious problems for both system operation and control in a power grid and can interrupt the wide area damping control of power systems.

4.2 *Data processing*

Software-defined networking (SDN) provides an open architecture for enabling centralized control and automatic management of networks through the decoupling of the control plane and data plane, and the incorporation of network programmable capability. The design, deployment, management, and maintenance of networks can easily be implemented on an open-standard-based centralized controller rather than directly configuring a massive number of heterogeneous devices.

The key principle of SDN is to provide logically centralized control through an SDN centralized controller. SDN's centralized control architecture enables it to maintain a global view of the network states and information. Therefore, by using its centralized controller SDN can take forwarding and routing decisions much faster and better (Rehmani et al. 2018).

SDN leverages a multitude of functionalities and access to fine-grained packet-related information through SDN controllers, such as OpenFlow and OpenDayLight. This information helps the SDN controllers to access packet collision related information, port information, hardware description, and the type of connection used. Moreover, SDN controllers are capable of dynamically configuring the flow entries on switches and routers. In addition, SDN controllers can also identify errors in data paths, and unidentified packets and may remove or alter the data flow path entries. All these capabilities make the SDN controller very powerful and thus highly suitable for SG communication.

4.3 *Decision and optimization*

As the cloud servers are deployed farther away from the end devices, the transmission latency is significant in the networks. Nonetheless, the cloud servers in the far-end environment can provide more computing power and more data storage. For example, cloud servers can provide massive parallel data processing, big data mining, big data management, and machine learning.

To decrease latency and improve reliability, more and more functional units and privileges are pushed toward the edge nodes, which are closer to the real actors and sensors. It is obvious that

Table 1. Comparison between cloud and edge computation.

Functionalities	Cloud	Edge
Location	Centralized	Distributed
Mobility	Limited	Supported
Service distance	Multiple hops	Single hop
Latency	Media	Low
Scalability	Low	High
Response time	High	Very low
Location	No	Yes
Point of service	Within network	Edge of network
Security	Weak	Strong

the decrease of physical distance could slash the latency and improve local reliability but the massive and heterogeneous processing also brings new challenges, such as manageability and complexity.

In the IIoT case, the limitations in terms of processing and computing capabilities are more pronounced. Rapid mobility patterns, high throughput, reliable sensing, reliable control and actuation, very low latency, big data management, different levels of real-time analytics, and data aggregation are the main IIoT requirements that cannot be met concurrently by cloud technology. Low latency, low jitter, mobility support, location awareness, augmented reality, geo-distributiveness, and multitenancy application support are common characteristics provided by edge computing in its different flavors, such as fog computing. Proposed by Cisco in 2012, it is an extension of the cloud to the network edge (Salman et al. 2018), which can be considered as a hybrid type of cloud and edge computation. Table 1 shows a comparison between edge and cloud computation.

Fog and cloud are complementary technologies and none of them replaces the other (Chiang 2016). The differentiation between fog and cloud is meant to be in the type of required data and the speed with which data must be processed. Local information can be served by fog nodes and global information can be served by the cloud. The short distance to end users makes the fog distributed platform more suitable for IIoT applications while the cloud is relatively farther away and requires quick action of local devices.

5 CONCLUSION

Communication infrastructures play a key role in a smart grid. In particular, the MAS will have a wide application in microgrids, which are the basic subsystems of modern power systems. This article reviews the state-of-the-art MAS communication from the perspective of a cyberphysical system. First, the structures and concepts of MAS for smart microgrids are presented. Then, the infrastructure of communication and computation in the physical layer are reviewed, especially the hardware of emerging completive IoT technologies such as Lora and NB-IoT, and widely used edge and cloud devices. Finally, the methodologies for data transmission, processing, decision, and optimization in the cyber layer are discussed, ranging from the mainly used standard sets of communication and spectrum access technology CRN to advanced cloud and edge computation.

ACKNOWLEDGMENTS

The work in this paper is supported by Ministry of Education, Republic of Singapore, under grant AcRF TIER 1 2017-T1-001-228 (RG92/17). Y. Xu's work is supported by a Nanyang Assistant Professorship from Nanyang Technological University, Singapore.

REFERENCES

Armbrust, M., Stoica, I., Zaharia, M., et al.2010. A view of cloud computing. *Communications of the ACM* 53(4): 50.

Bonomi, F., Milito, R., Zhu, J., & Addepalli, S. (2012, August). Fog computing and its role in the internet of things. In *Proceedings of the first edition of the MCC workshop on Mobile cloud computing* (13-16). ACM.

Boyer, S.A. 2009. *Scada: Supervisory Control and Data Acquisition*, 4th ed. Research Triangle Park, NC: International Society of Automation.

Chiang, M. 2016. Fog networking: an overview on research opportunities. *ArXiv:1601.00835 [Cs]*, January.

Coelho, V.N., Weiss Cohen, M., Coelho, I.M., Liu, N., & Gadelha Guimarães, F. 2017. Multi-agent systems applied for energy systems integration: state-of-the-art applications and trends in microgrids. *Applied Energy* 187 (February): 820–832.

Georgiou, O. & Raza, U. 2017. Low power wide area network analysis: can LoRa scale? *IEEE Wireless Communications Letters* 6(2): 162–165.

Hurtado, L.A., Mocanu, E., Nguyen, P.H., Gibescu, M., & Kamphuis, R.I.G. 2018. Enabling cooperative behavior for building demand response based on extended joint action learning. *IEEE Transactions on Industrial Informatics* 14(1): 127–136.

Keiser, G. 2003. Optical fiber communications. In *Wiley Encyclopedia of Telecommunications*. Hoboken, NJ: Wiley.

Khan, A.A., Rehmani, M.H., & Reisslein, M. 2016. Cognitive radio for smart grids: survey of architectures, spectrum sensing mechanisms, and networking protocols. *IEEE Communications Surveys Tutorials* 18(1): 860–898.

Latchman, H.A. & Katar, S. 2002. A power line communication network infrastructure for the smart home. *IEEE Wireless Communications* 9 (6): 104–111.

Letaief, K.B. & Zhang, W. 2009. Cooperative communications for cognitive radio networks. *Proceedings of the IEEE* 97(5): 878–893.

McArthur, S.D.J., Davidson, E.M., Catterson, V.M., et al. 2007a. Multi-agent systems for power engineering applications–Part 2: Technologies, standards and tools for building multi-agent systems. *IEEE Transactions on Power Systems* 22 (November): 1753–1759.

McArthur, S.D.J., Davidson, E.M., Catterson, V.M., et al. 2007b. Multi-agent systems for power engineering applications—Part I: Concepts, approaches, and technical challenges. *IEEE Transactions on Power Systems* 22(4): 1743–1752.

Meskina, S.B., Doggaz, N., Khalgui, M., & Li, Z. 2017. Multiagent framework for smart grids recovery. *IEEE Transactions on Systems, Man, and Cybernetics: Systems* 47(7): 1284–1300.

Netto, R.S., Ramalho, G.R., Bonatto, B.D., et al. 2018. Real-time framework for energy management system of a smart microgrid using multiagent systems. *Energies* 11(3): 656.

Rehmani, M.H., Davy, A., Jennings, B., & Assi, C. 2018. Software defined networks based smart grid communication: a comprehensive survey. *ArXiv:1801.04613*, January.

Salman, O., Elhajj, I., Chehab, A., & Kayssi, A. 2018. IoT survey: an SDN and fog computing perspective. *Computer Networks* 143 (October): 221–246.

Wang, Y.E., Lin, X., Adhikary, A., et al. 2017. A primer on 3GPP narrowband Internet of Things. *IEEE Communications Magazine* 55(3): 117–123.

Wang, Y., Xu, Y., Tang, Y., et al. 2018. Aggregated energy storage for power system frequency control: a finite-time consensus approach. *IEEE Transactions on Smart Grid* 10(4): 3675-3686.

Wang, Z. & Paranjape, R. 2017. Optimal residential demand response for multiple heterogeneous homes with real-time price prediction in a multiagent framework. *IEEE Transactions on Smart Grid* 8(3): 1173–1184.

Yu, X. & Xue, Y. 2016. Smart grids: a cyber–physical systems perspective. *Proceedings of the IEEE* 104(5): 1058–1070.

Emerging Developments in the Power and Energy Industry – Dufo-López, Krzywanski & Singh (eds)
© *2020 Taylor & Francis Group, London, ISBN 978-0-367-27169-5*

Analysis of the error of harmonic current detection based on the instantaneous reactive power theory

Xingyong Zhao*
ShanXi University, Taiyuan, China

Jiawei Wang
Economic and Technical Research Institute of State Grid Shanxi Electric Power Company, Taiyuan, China

Zhilan Wang & Fang Yao
ShanXi University, Taiyuan, China

ABSTRACT: The detection unit of a harmonic current is an important part of harmonic control. The mathematical model is built according to the worst operating state of the system. Through theoretical derivation, two harmonic current detection methods based on instantaneous reactive power theory, the p–q method and the i_p–i_q method, and their harmonic current detection errors are analyzed. An improved i_p–i_q method developed by removing the phase-locked loop is proposed and a detailed control block diagram is constructed. According to the different operating conditions of the system voltage, PSCAD is used to build a system simulation model with nonlinear and linear loads. Through simulation analysis, the correctness of the theoretical analysis and the effectiveness of the improved scheme are verified.

Keywords: *Harmonic current detection, Instantaneous Reactive Power Theory, error analysis, Voltage distortion*

1 INTRODUCTION

With a large nonlinear load connected to the grid, the problem of harmonic pollution goes from bad to worse. The excessive harmonic current does great harm to the safe operation of the system and equipment life (Zhao 2015; Wang 2016). Therefore, research on system harmonic control is very meaningful.

Harmonic control issues are concentrated in two areas: harmonic current detection and compensation current generation. At present, most of the research on active power filter (APF) harmonic detection is based on Instantaneous Reactive Power Theory (IRPT) (Chen 2013). However, there is a certain error in the actual detection process. Many scholars have proposed new detection schemes for the influence of errors. Literature (Lucian 2006) proposes a Fryze–Buchholz–Dpenbrock (FBD) algorithm for time-domain analysis. The literature (Tenti & Mattavelli 2003) proposes another current time-domain decomposition method: Conservative Power Theory (CPT) algorithm. However, the two aforementioned methods are only in the research stage and have not been put into practical use.

For harmonic control schemes based on IRPT, more experts tend to improve the detection accuracy on the basis of not increasing economic and structural complexity. Studies reported in the literature (Tang & Chen 2005) have achieved harmonic current detection based on

---

*Corresponding author: E-mail:13546725086@163.com

coordinate transformation, and analyzed the detection error under different voltage environments. In Yan and Ge (2007), the i_p–i_q decomposition method is studied, and the mean filter is proposed to improve the dynamic response characteristics of the low-pass filter for the case of a small amount of harmonic current in the grid. Zhou (2007) analyzed the influence of the phase-locked loop and the working characteristics of the low-pass filter on the detection accuracy in the harmonic detection scheme based on IRPT. Although these articles analyze the error of the harmonic detection scheme based on IRPT, they do not establish a unified analysis model for various operating conditions of the system, and there is no in-depth discussion of detection error issues.

In the actual operating system, the asymmetry and nonlinear characteristics of the load will cause a three-phase imbalance of the power grid and a distortion of the electrical quantity. In this article, a unified mathematical model is established based on the worst operating case of an actual system. It studies detection errors of the p–q method and the i_p–i_q method, which are both based on IRPT, in different operating conditions of the system, and the corresponding inhibition measures are proposed. The PSCAD simulation results verify the correctness of the theoretical analysis and the effectiveness of the proposed improvement.

2 ESTABLISHMENT OF A SYSTEM OPERATION MODEL

Considering the harmonic pollution brought by asymmetry and nonlinearity of the load, the mathematical model of the system is established as follows.

$$\begin{cases} u_a = \sqrt{2} \sum_{m=1}^{\infty} [U_{1m} \sin(m\omega t + \varphi_{1m}) + U_{2m} \sin(m\omega t + \varphi_{2m}) + u_0] \\ u_b = \sqrt{2} \sum_{m=1}^{\infty} [U_{1m} \sin(m\omega t + \varphi_{1m} - \frac{2\pi}{3}) + U_{2m} \sin(m\omega t + \varphi_{2m} + \frac{2\pi}{3}) + u_0] \\ u_c = \sqrt{2} \sum_{m=1}^{\infty} [U_{1n} \sin(m\omega t + \varphi_{1m} + \frac{2\pi}{3}) + U_{2m} \sin(m\omega t + \varphi_{2m} - \frac{2\pi}{3}) + u_0] \end{cases}$$

$$\begin{cases} i_a = \sqrt{2} \sum_{n=1}^{\infty} [I_{1n} \sin(n\omega t + \phi_{1n}) + I_{2n} \sin(n\omega t + \phi_{2n}) + i_0] \\ i_b = \sqrt{2} \sum_{n=1}^{\infty} [I_{1n} \sin(n\omega t + \phi_{1n} - \frac{2\pi}{3}) + I_{2n} \sin(n\omega t + \phi_{2n} + \frac{2\pi}{3}) + i_0] \\ i_c = \sqrt{2} \sum_{n=1}^{\infty} [I_{1n} \sin(n\omega t + \phi_{1n} + \frac{2\pi}{3}) + I_{2n} \sin(n\omega t + \phi_{2n} - \frac{2\pi}{3}) + i_0] \end{cases}$$

In the two formulas above, U_{1m} and U_{2m} represent the effective value of the mth harmonic positive and negative sequence voltage of the system, respectively; I_{1n} and I_{2n} are the effective value of the nth harmonic positive and negative sequence current of the system, respectively; u_0 and i_0 are the zero sequence voltage and current of the system, respectively. ω is the angular velocity of the system. φ and ϕ represent the initial phase angle of each harmonic voltage and current, respectively. m and n are the harmonic order of the system voltage and current.

The above model is a three-phase four-wire system model. The method based on IRPT is mostly used in the three-phase three-wire system. So it is necessary to add an elimination link of the zero-sequence component.

The zero sequence component is calculated as follows:

$$x_0 = \frac{1}{3}(x_a + x_b + x_c)$$

where x represents the corresponding voltage or current component. So the mathematical model after eliminating the zero sequence component is established as follows:

$$
\begin{cases}
\begin{aligned}
u'_a &= u_a - u_0 \\
&= \sqrt{2} \sum_{m=1}^{\infty} [U_{1m} \sin(m\omega t + \varphi_{1m}) + U_{2m} \sin(m\omega t + \varphi_{2m})] \\
u'_b &= u_b - u_0 \\
&= \sqrt{2} \sum_{m=1}^{\infty} [U_{1m} \sin(m\omega t + \varphi_{1m} - \tfrac{2\pi}{3}) + U_{2m} \sin(m\omega t + \varphi_{2m} + \tfrac{2\pi}{3})] \\
u'_c &= u_c - u_0 \\
&= \sqrt{2} \sum_{m=1}^{\infty} [U_{1m} \sin(m\omega t + \varphi_{1m} + \tfrac{2\pi}{3}) + U_{2m} \sin(m\omega t + \varphi_{2m} - \tfrac{2\pi}{3})]
\end{aligned}
\end{cases}
\tag{1}
$$

$$
\begin{cases}
\begin{aligned}
i'_a &= i_a - i_0 \\
&= \sqrt{2} \sum_{n=1}^{\infty} [I_{1n} \sin(n\omega t + \phi_{1n}) + I_{2n} \sin(n\omega t + \phi_{2n})] \\
i'_b &= i_b - i_0 \\
&= \sqrt{2} \sum_{n=1}^{\infty} [I_{1n} \sin(n\omega t + \phi_{1n} - \tfrac{2\pi}{3}) + I_{2n} \sin(n\omega t + \phi_{2n} + \tfrac{2\pi}{3})] \\
i'_c &= i_c - i_0 \\
&= \sqrt{2} \sum_{n=1}^{\infty} [I_{1n} \sin(n\omega t + \phi_{1n} + \tfrac{2\pi}{3}) + I_{2n} \sin(n\omega t + \phi_{2n} - \tfrac{2\pi}{3})]
\end{aligned}
\end{cases}
\tag{2}
$$

3 HARMONIC CURRENT DETECTION BASED ON IRPT

IRPT converts the three-phase voltage and current of the system from the abc coordinate system to the $\alpha\beta$ coordinate system for discussion. The harmonic current detection algorithm based on IRPT has two calculation methods, p–q and i_p–i_q. The principle block diagram is shown in Figures 1 and 2, respectively.

In Figure 1, $\bar{p}$ and $\bar{q}$ refer to the DC component of active power p and reactive power q, respectively. i_a^*, i_b^*, i_c^* are the compensation current values of three phases. In Figure 1, the values of C_{32} and C_{pq} are as follows:

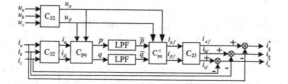

Figure 1.   Simulation diagram of the p–q method.

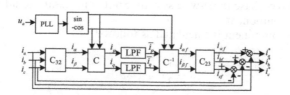

Figure 2.   Simulation diagram of i_p–i_q method.

$$C_{32} = \sqrt{2/3}\begin{bmatrix} 1 & -1/2 & -1/2 \\ 0 & \sqrt{3}/2 & -\sqrt{3}/2 \end{bmatrix} \quad C_{pq} = \begin{bmatrix} u_\alpha & u_\beta \\ u_\beta & -u_\alpha \end{bmatrix}$$

In Figure 2, PLL represents the phase-locked loop. The real-time phase of system voltage u_a is obtained by PLL.

Both $p\text{–}q$ and $i_p\text{–}i_q$ methods need to convert from the *abc* coordinate system to the *αβ* coordinate system. The conversion results corresponding to Equations (1) and (2) are as shown in Equations (3) and (4).

$$u_{\alpha\beta} = C_{32}u_{abc} = \sqrt{3}\begin{bmatrix} \displaystyle\sum_{m=1}^{\infty} U_{1m}\sin(m\omega t + \varphi_{1m}) + \sum_{m=1}^{\infty} U_{2m}\sin(m\omega t + \varphi_{2m}) \\ \displaystyle\sum_{m=1}^{\infty} -U_{1m}\cos(m\omega t + \varphi_{1m}) + \sum_{m=1}^{\infty} U_{2m}\cos(m\omega t + \varphi_{2m}) \end{bmatrix} \tag{3}$$

$$i_{\alpha\beta} = C_{32}i_{abc} = \sqrt{3}\begin{bmatrix} \displaystyle\sum_{n=1}^{\infty} I_{1n}\sin(n\omega t + \phi_{1n}) + \sum_{n=1}^{\infty} I_{2n}\sin(n\omega t + \phi_{2n}) \\ \displaystyle\sum_{n=1}^{\infty} -I_{1n}\cos(n\omega t + \phi_{1n}) + \sum_{n=1}^{\infty} I_{2n}\cos(n\omega t + \phi_{2n}) \end{bmatrix} \tag{4}$$

3.1 *Fundamental current error under the p–q method*

Using the $p\text{–}q$ method, the active power p and reactive power q are as shown in Equation (5).

$$\begin{bmatrix} p \\ q \end{bmatrix} = \begin{bmatrix} u_\alpha & u_\beta \\ u_\beta & -u_\alpha \end{bmatrix}\begin{bmatrix} i_\alpha \\ i_\beta \end{bmatrix}$$

$$= 3\begin{bmatrix} \displaystyle\sum_{m=1}^{\infty}\sum_{n=1}^{\infty}\{U_{1m}I_{1n}\cos[(m-n)\omega t + \varphi_{1m} - \phi_{1n}] + U_{1m}I_{2n}\cos[(m+n)\omega t + \varphi_{1m} + \phi_{2n}]\}+ \\ \displaystyle\sum_{m=1}^{\infty}\sum_{n=1}^{\infty}\{U_{2m}I_{2n}\cos[(m-n)\omega t + \varphi_{2m} - \phi_{2n}] + U_{2m}I_{1n}\cos[(m+n)\omega t + \varphi_{2m} + \phi_{1n}]\} \\ \displaystyle\sum_{m=1}^{\infty}\sum_{n=1}^{\infty}\{U_{1m}I_{1n}\sin[(m-n)\omega t + \varphi_{1m} - \phi_{1n}] - U_{1m}I_{2n}\sin[(m+n)\omega t + \varphi_{1m} + \phi_{2n}]\}+ \\ \displaystyle\sum_{m=1}^{\infty}\sum_{n=1}^{\infty}\{U_{2m}I_{2n}\sin[(n-m)\omega t + \phi_{2n} - \varphi_{2m}] + U_{2m}I_{1n}\sin[(m+n)\omega t + \varphi_{2m} + \phi_{1n}]\} \end{bmatrix} \tag{5}$$

The DC components of p and q are as follows:

$$\begin{bmatrix} \bar{p} \\ \bar{q} \end{bmatrix} = 3\begin{bmatrix} \displaystyle\sum_{n=1}^{\infty} U_{1n}I_{1n}\cos(\varphi_{1n} - \phi_{1n}) + U_{2n}I_{2n}\cos(\varphi_{2n} - \phi_{2n}) \\ \displaystyle\sum_{n=1}^{\infty} U_{1n}I_{1n}\sin(\varphi_{1n} - \phi_{1n}) - U_{2n}I_{2n}\sin(\varphi_{2n} - \phi_{2n}) \end{bmatrix} \tag{6}$$

The three-phase corresponding fundamental current is shown in Equation (7).

$$\begin{bmatrix} i_{af} \\ i_{bf} \\ i_{cf} \end{bmatrix} = C_{32}\frac{1}{u^2}\begin{bmatrix} u_\alpha & u_\beta \\ u_\beta & -u_\alpha \end{bmatrix}\begin{bmatrix} \bar{p} \\ \bar{q} \end{bmatrix} \tag{7}$$

In Equation (7), $u^2 = 3\left(\displaystyle\sum_{n=1}^{\infty} u_{1n}^2 + \sum_{n=1}^{\infty} u_{2n}^2\right)$, which is obtained from the values of u_α and u_β.

When the system voltage is distorted and asymmetrical, the error of the fundamental current is as follows in Equation (8).

$$
\begin{bmatrix} \Delta i_{af} \\ \Delta i_{bf} \\ \Delta i_{cf} \end{bmatrix} = C_{32} \begin{bmatrix} u_{ah}\frac{\bar{p}}{u^2} + u_{\beta h}\frac{\bar{q}}{u^2} \\ u_{\beta h}\frac{\bar{p}}{u^2} - u_{ah}\frac{\bar{q}}{u^2} \end{bmatrix}
$$

$$
+ C_{32} \begin{bmatrix} u_{af}\left(\frac{\bar{p}}{u^2} - \frac{I_{11}\cos\phi_{11}}{U_{11}}\right) + u_{\beta f}\left[\frac{\bar{q}}{u^2} - \frac{I_{11}\sin(-\phi_{11})}{U_{11}}\right] \\ u_{\beta f}\left(\frac{\bar{p}}{u^2} - \frac{I_{11}\cos\phi_1}{U_{11}}\right) + u_{af}\left[\frac{\bar{q}}{u^2} - \frac{I_{11}\sin(-\phi_{11})}{U_{11}}\right] \end{bmatrix} \tag{8}
$$

The reason for the error is as follows:

(1) When the voltage is distorted and asymmetrical, the distortion components u_{ah} and $u_{\beta h}$ of the voltage exist in the $\alpha\beta$ coordinate system, resulting in partial components of the current error $\Delta i_{af}, \Delta i_{bf}, \Delta i_{cf}$.

(2) When the value of u^2 is obtained, if the voltage is not distorted, it is only three times the square of the effective value of the fundamental voltage. If the voltage is distorted, u^2 is three times the sum of the squares of the effective values of the positive and negative sequence voltages of each frequency.

3.2 *Fundamental current error under the i_p–i_q method*

Using the i_p–i_q method, when the system voltage is asymmetrical, the unbalanced sequence components appear, and the phase measured by PLL will be the sum of the phases of the three orders, which has a phase difference θ with the phase of the desired positive sequence component. At the same time, the measured frequency is also different from the fundamental frequency, and the voltage signals obtained through the PLL would be $\sin(\lambda\omega t + \theta)$ and $-\cos(\lambda\omega t + \theta)$, where λ is a positive real number. So, the conversion matrix C is adjusted to

$$
C = \begin{bmatrix} \sin(\lambda wt + \theta) & -\cos(\lambda wt + \theta) \\ -\cos(\lambda wt + \theta) & -\sin(\lambda wt + \theta) \end{bmatrix}
$$

Then, the active and reactive components of the current are obtained as follows:

$$
\begin{bmatrix} i_p \\ i_q \end{bmatrix} = \begin{bmatrix} \sin(\lambda\omega t + \theta) & -\cos(\lambda\omega t + \theta) \\ -\cos(\lambda\omega t + \theta) & -\sin(\lambda\omega t + \theta) \end{bmatrix} \begin{bmatrix} i_a \\ i_\beta \end{bmatrix}
$$

$$
= \begin{bmatrix} \sqrt{3}\sum_{\substack{n=1 \\ m=1}}^{\infty} I_{1n}\cos[(n-\lambda)\omega t + \phi_{1n} - \theta] - \sqrt{3}\sum_{\substack{n=1 \\ m=1}}^{\infty} I_{2n}\cos[(n+\lambda)\omega t + \phi_{2n} + \theta] \\ -\sqrt{3}\sum_{\substack{n=1 \\ m=1}}^{\infty} I_{1n}\sin[(n-\lambda)\omega t + \phi_{1n} - \theta] - \sqrt{3}\sum_{\substack{n=1 \\ m=1}}^{\infty} I_{2n}\sin[(n+\lambda)\omega t + \phi_{2n} + \theta] \end{bmatrix}
$$

The DC power components are shown in Equation (9).

$$\begin{bmatrix} \bar{i}_p \\ \bar{i}_q \end{bmatrix} = \sqrt{3} \begin{bmatrix} \sum_{\lambda=1}^{\infty} I_{1\lambda}\cos(\phi_{1\lambda} - \theta) \\ -\sum_{\lambda=1}^{\infty} I_{1\lambda}\sin(\phi_{1\lambda} - \theta) \end{bmatrix} \tag{9}$$

Then, the fundamental current is obtained, which is shown in Equation (10).

According to Equation (10), there is an error in the fundamental wave current obtained when the voltage is distorted. When the voltage is not distorted, the frequency detected by the PLL is the fundamental frequency, that is, $\lambda = 1$. So, the current component is obtained as the true fundamental current according to the above steps.

In addition, it can be seen from Equation (10) that the fundamental current obtained is related to the initial phase of the current and the frequency of the system voltage. The phase difference θ caused by the system voltage asymmetry has almost no effect on the decomposition result.

$$\begin{bmatrix} i_{af} \\ i_{bf} \\ i_{cf} \end{bmatrix} = C_{23} \begin{bmatrix} \sin(\lambda\omega t + \theta) & -\cos(\lambda\omega t + \theta) \\ -\cos(\lambda\omega t + \theta) & -\sin(\lambda\omega t + \theta) \end{bmatrix} \begin{bmatrix} \bar{i}_p \\ \bar{i}_q \end{bmatrix}$$

$$- \sqrt{2} \begin{bmatrix} \sum_{\lambda=1}^{\infty} I_{1\lambda}\sin(\lambda\omega t + \phi_{1\lambda}) \\ \sum_{\lambda=1}^{\infty} I_{1\lambda}\sin(\lambda\omega t - \frac{2\pi}{3} + \phi_{1\lambda}) \\ \sum_{\lambda=1}^{\infty} I_{1\lambda}\sin(\lambda\omega t + \frac{2\pi}{3} + \phi_{1\lambda}) \end{bmatrix}$$

Therefore, the i_p–i_q method can be applied to the harsh conditions of system voltage and current asymmetry and current distortion, and also applies to system current distortion or asymmetry. However, when the voltage is also distorted, there will be an error.

4   IMPROVEMENT MEASURES

According to the comparison of the p–q and i_p–i_q methods, in the system with current asymmetry and distortion, the fundamental current decomposed is almost not affected by voltage

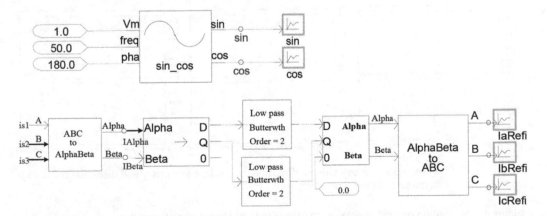

Figure 3.   Simulation diagram of the improved i_p–i_q method.

asymmetry, and the detection error is smaller. However, when the voltage is distorted, the PLL detects a voltage frequency other than the fundamental frequency, which will cause an error in the final result. When the voltage and current are both asymmetrical and distorted, the PLL can only detect the frequency and phase of the phase A voltage deviating from the fundamental positive sequence voltage, and an error occurs during the current decomposition process.

Therefore, this article proposes removing the PLL and connecting the sine and cosine components involved in the coordinate transformation as a given reference value. The actual simulation is shown in Figure 3. In the analysis, it is equivalent to giving the voltage frequency collected by the PLL as $\lambda = 1$. According to the analysis, the phase θ has almost no effect on the current detection error.

The low-pass filter (LPF) is an important component of the harmonic detection scheme based IRPT. According to the design concept of LPF, it can be divided into the Butterworth type, Chebyshev type, and elliptic function type LPF. Among them, the Butterworth type has the best harmonic detection accuracy. The increase of the filtering order can improve the filtering accuracy of LPF, but it will reduce the dynamic response characteristics (Tang & Chen 2008). Therefore, in order to meet the accuracy and sensitivity of the harmonic detection unit, the second-order Butterworth type LPF is used in this article, as shown in Figure 3.

5 SIMULATION ANALYSIS

Four system operation models are built by PSCAD: voltage symmetry without distortion, voltage symmetry and distortion, voltage asymmetry without distortion, and voltage asymmetry distortion. The simulation system consists of three uncontrollable rectification chargers simulating nonlinear loads (He 2017), and a certain linear load is set according to different operating conditions of the system.

In this article, the feasibility of the harmonic detection scheme is analyzed by comparing the waveform of the fundamental wave current and the total distortion rate. The two methods are

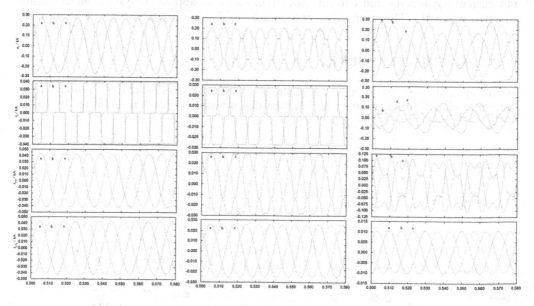

(a) voltage symmetry without distortion (b) voltage symmetry and distortion (c) voltage asymmetry without distortion

Figure 4.   Simulation on fundamental-wave current in three system operating states.

used to filter the fundamental current of the system. The simulation results are shown in Figure 4.

It can be seen from Figure 4a that when the system voltage is symmetric and distortion-free, the fundamental wave filtering effects of the p–q method and the i_p–i_q method are basically the same. According to FFT, the voltage total harmonic distortion rate THD_{u_s} is 3.80%, the current total harmonic distortion rate THD_{i_s} is 23.58%, and the fundamental current distortion rates THD_i corresponding to the p–q method and i_p–i_q method are 0.69% and 1.17%, respectively. Therefore, both methods are applicable when the system voltage is symmetric and distortion-free, but the p–q method is more accurate.

When the simulated system voltage is symmetrically distorted, THD_{u_s} is 12.16% and THD_{i_s} is 19.51%. From the waveform comparison of Figure 4b, it is obvious that the i_p–i_q method is better than p–q method for filtering the fundamental wave, and THD_i of the two are 0.99% and 8.46%, respectively.

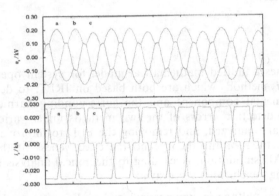

Figure 5. Waveform when the voltage is asymmetrically distorted.

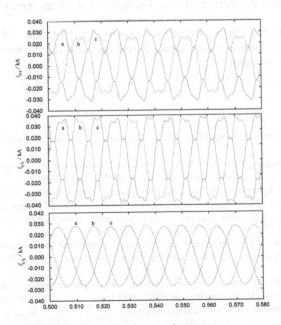

Figure 6. Simulated waveform of the filtered fundamental current.

240

According to Figure 4c, when the system is asymmetrical and undistorted, the simulation results show that THD_{u_s} is 3.18% and THD_{i_s} is 22.53%. The filtering error of the p–q method is extremely large, and the ip-iq method can still filter out a good fundamental current, and THD_i of the two are 36.06% and 0.49%, respectively. The foregoing results verify that the detection phase difference θ caused by the system voltage asymmetry has little effect on the i_p–i_q method. For this type of operation, the i_p–i_q method is more applicable.

Figure 5 shows the waveform of the system when the voltage is asymmetrically distorted, with THD_{u_s} reaching 11.33% and THD_{i_s} reaching 18.78%. This is the worst case for system operation. Figure 6 shows the fundamental current waveform obtained by the two methods. It can be seen from Figure 6 that the filtering effects of the two methods are not good, and the harmonic distortion rates of the fundamental current are 11.93% and 6.79%, respectively. In particular, the fundamental wave current simulation waveform of the improved i_p–i_q method after removing the phase-locked loop is good, and its THD_i is only 0.96%. Therefore, the improved i_p–i_q method after removing the PLL has a better detection effect.

6 CONCLUSION

This article establishes a unified mathematical model for system operation, and uses the p–q method and i_p–i_q method, which are both based on IRPT, to decompose the system current. The fundamental component and the corresponding harmonic component are filtered out, and the detection errors of the two methods under various operating conditions of the system are analyzed, and removing the PLL to improve the i_p–i_q method is proposed. The correctness of the theory and the effectiveness of the proposed method are verified by PSCAD simulation. According to theoretical analysis and simulation verification, the following conclusions are drawn:

(1) When the system voltage is symmetrical and undistorted, and only the current is distorted, the effects of the p–q method and the i_p–i_q method are similar.
(2) When the system voltage is asymmetrical or distorted, there is a large error in the fundamental current filtering of the p–q method, and the error of the i_p–i_q method is small.
(3) After the i_p–i_q method is improved by removing the PLL, the influence of the voltage distortion on the decomposition effect is reduced, and the filtering error of the fundamental current is reduced.

REFERENCES

Chen, Z. L. 2013. Harmonic Current Detection and Tracking Control of Active Power Filter. Jiangsu University.

He, T. Y., et al. 2017. Harmonic amplification effect and restrain measures based on APF for electric vehicle charging station. *Renewable Energy Resources* 35(11): 1647–1654.

Lucian, A., et al. 2006. A new control structure for hybrid power filter to reduce the inverter power rating. IECON 2006. In 32nd Annual Conference on IEEE Industrial Electronics. 1257–1262.

Tang, L., & Chen, W. R. 2005. Detection error analysis of active power filter. *Electric Power Automation Equipment* (10): 29–32.

Tang, L., & Chen, W. R. 2008. Deduction of coordinate transform for instantaneous reactive power theory and analysis on the principle of harmonic current detection method. *Power System Technology* 5: 66–69, 86.

Tenti, P., & Mattavelli, P. 2003. A time-domain approach to power term definitions under non-sinusoidal conditions. In Proceedings of 2003 Sixth International Workshop on Power Definitions and Measurements under Non-Sinusoidal Conditions.

Wang, Z. A., et al. 2016. *Harmonic Suppression and Reactive Power Compensation*. Beijing: Mechanical Industry Press.

Yan, Q. R., & Ge, J. W. 2007. Harmonics detection error analysis based on the instantaneous reactive power theory. *Electric Drive* 5: 57–60.

Zhao, X.Y. 2015. *Grid-Connected Distributed Generation and Voltage Stability*. Beijing: China Electric Power Press.

Zhou, K., et al. 2007. An improved i_p–i_q harmonic current detecting method and digital low-pass filter's optimized design. *Proceedings of the CSEE* 34: 96–101.

Emerging Developments in the Power and Energy Industry – Dufo-López, Krzywanski & Singh (eds)
© 2020 Taylor & Francis Group, London, ISBN 978-0-367-27169-5

Method for evaluating the validity of synchronous generator model parameters considering disturbance scene classification

Zhenguo Jiang & Jun An
School of Electrical Engineering, Northeast Electric Power University, Jilin, Jilin Province, China

Dexin Li, Tong Li & Jiarui Wang
Electric Power Research Institute of State Grid Jilin Electric Power Supply Company, Changchun, Jilin Province, China

ABSTRACT: Aimed at the problem of generator parameters dispersed under different disturbance scenarios, this article proposes a method to verify generator model parameters, which has taken the classification of disturbance scenes into account by combining probability and statistics theory. First, the obtained disturbance scenes are clustered and analyzed, and the target unit model parameters are checked via different disturbance scenarios under various disturbances. Then, based on the recognition dominant parameter and the gray distance measure, an interval for its value is given at a certain confidence level. Finally, because there are different disturbance types with their corresponding error intervals, different modes of generator parameters that provide a reasonable and sufficiently auxiliary basis for the selection and verification of simulation parameters are given. The method described in this article is verified by a large amount of measured information derived from random disturbances in the provincial power grid.

Keywords: disturbance classification, dynamic simulation, parameter correction, trajectory sensitivity, Wide Area Measurement System (WAMS)

1   INTRODUCTION

Power system dynamic simulation, which lays the foundation for guiding planning, design, and operation, is an important tool for mastering the dynamic behavior in power systems. The credibility of simulation results is of great significance to the safe and economic operation of a power system (Hauer et al. 2000). In addition, the inaccuracy of model parameters is considered the main factor causing simulation error. The checking and verification of the parameters has therefore been a fundamentally important work (Liang et al. 2010). As the most important dynamic component in a power system, the accuracy of the parameters is particularly important for simulation results. However, most of the parameters used in current dynamic simulation do not match the values actually. To improve simulation accuracy, using the measured data that contain system disturbance information is an effective method to check the parameters. In recent years, based on the Global Positioning System (GPS), the Wide Area Measurement System (WAMS) can capture every random disturbance event in a power grid, and the measured information covering a large number of random disturbance events can provide important data supporting the model parameters under verification (Allen et al. 2010). The vast power system, consisting of many components and parameters, has great difficulties in backtracking and may check the wrong parameters because of the close coupling relationship among components (Huang et al. 2006). Based on measured data, An et al. (2013) proposed a simulation and verification strategy for power system stratification and segmentation, greatly reducing the scope of error traceability.

In this article, the dispersed problem of generator parameters under different disturbance scenarios is presented. Based on the estimation of a small sample data interval, a reasonably evaluated method of generator model parameters, considering disturbance scene classification, is proposed. Cluster analysis of the disturbance scenes reveals three typical disturbance scenarios. The dominant parameters are identified and given based on the gray distance measure at a certain confidence level. The confidence interval of the value captures different generator parameters representing different disturbance types and guides the grid operator to select effective simulation parameters that improve the accuracy of the dynamic simulation.

2 DISTURBANCE SCENE CLASSIFICATION

During the actual operation of a power grid, some parameter values will be changed because there are different responses of components in different disturbance scenarios. The index established by the clusters of disturbance depth and the running state of the generator, which indicates the dynamic response in different disturbance scenarios, is significant for checking the model parameters for dynamic components.

The power flow in components will fluctuate to different extents when the disturbances to the power system occur. Voltage amplitude and power changes can directly reflect the extent to which the model is disturbed. So, the fluctuation of voltage amplitude is regarded as the depth of disturbance in this article.

For the case in which the component port is configured with a phasor measurement unit (PMU), the disturbance depth can be defined as follows:

$$\eta = \max\left(\left|\frac{U_i - U_{\text{sta}}}{U_{\text{sta}}}\right|\right) \tag{1}$$

where η is the disturbance depth, U_i is the positive sequence component of the voltage recorded at the time of i by the PMU at the node where the component is located, and U_{sta} is the steady-state voltage value of the system before the disturbance occurs.

When large amounts of disturbed samples are clustered, the disturbed depth and generator running state could be clustered as the attributes corresponding to the cluster samples. The k-means method, with its advantages of simplicity and efficiency, is chosen in this article.

3 GENERATOR SIMULATION VERIFICATION IN MULTIDISTURBANCE SCENARIOS

3.1 *Error indicator system*

Quantifying dynamic simulation error is the basis for adjusting simulation parameters in a power system. By evaluating the error of simulation, the main cause of the error can be found. Experts with a knowledge of error calculation can understand the simulation situation clearly and judge it correctly. The error indicators are defined as the following in this article.

(1) Active power first swing amplitude error index:

$$\Delta P = \frac{\max(P_{\text{meas}}(t)) - \max(P_{\text{simu}}(t))}{\max(P_{\text{meas}}(t))} \tag{2}$$

(2) Voltage overshoot error indicator:

$$\Delta U_1 = \frac{\max(U_{\text{meas}}(t)) - \max(U_{\text{simu}}(t))}{U_{\text{meas}}(0)} \tag{3}$$

(3) Voltage lowest point error index:

$$\Delta U_2 = \frac{\min(U_{\text{meas}}(t)) - \min(U_{\text{simu}}(t))}{U_{\text{meas}}(0)} \tag{4}$$

(4) Error energy ratio indicator:

$$\text{REE} = \frac{\sum\limits_{i=1}^{n} \left(y_{\text{meas}}(t) - y_{\text{simu}}(t)\right)^2}{\sum\limits_{i=1}^{n} \left(y_{\text{meas}}(t) - y_{\text{stab}}\right)^2} \tag{5}$$

3.2 *Dominant parameter identification method based on trajectory sensitivity*

The simulation error can be improved by correcting the dominant parameters, which are determined in the process of checking model parameters. The degree of change can be reflected sensitively by the trajectory sensitivity when a change of parameter or structure occurs in a power system.

Calculation of trajectory sensitivity mainly includes an analytic and a perturbed method. The calculation cost, which is related to the physical model and parameters, is higher than the cost of the perturbed method to obtain the trajectory sensitivity of multiple parameters in a large-scale power system by the analytic method. In addition, the accuracy of the simulation model and parameters has a great influence on the calculation results. Therefore, the trajectory sensitivity has been calculated by the perturbed method in this article. The principle is as follows.

For a multimachine system, the trajectory sensitivity is usually described by differential algebraic equations:

$$\begin{cases} x_\theta = \frac{x(\theta + \Delta\theta, t) - x(\theta, t)}{\Delta\theta} \\ y_\theta = \frac{y(\theta + \Delta\theta, t) - y(\theta, t)}{\Delta\theta} \end{cases} \tag{6}$$

In the parameter calibration process of multidisturbance scenes, evaluating the importance of parameters requires integration of the disturbance depth corresponding to each disturbance scene and the calculated trajectory sensitivity results for the parameters. The four most important parameters are identified as the dominant parameters for parameter checking in multidisturbance scenarios.

4 CONFIDENCE INTERVAL ESTIMATION OF PARAMETER CHECK RESULTS

Confidence interval estimation, using existing samples, can effectively describe the dispersion problem of generator model parameters under different disturbance scenarios. It can also be used to guide the system operator to grasp the dynamic characteristics of generators, select simulation parameters, and evaluate the validity of the parameters. Therefore, we can regard the parameter calibration as an independent test based on measured information for different disturbances, and finally form a model parameter library through several independent experiments. $(X_1, X_2..., X_n)$ is considered an independent and identically distributed sample of X in several courses of parameter checking. For convenience, it can be regarded as X, and the distribution function of X_i is an unknown distribution $F(X)$.

The bootstrap method is a nonparametric statistical method. It doesn't make any assumptions about the sampling distribution function. The simulated large sample can be directly sampled by the empirical distribution function of the observed data to estimate the relevant statistics. Based on the Grey System Theory, this articlepaper uses the small sample confidence interval calculation method, which combines the bootstrap resampling method and the small sample interval estimation method (Wu et al. 2016) to define the confidence interval (value range) of the dominant parameter at a certain confidence level.

5 CASE ANALYSIS

5.1 *Introduction to simulation*

By a preliminary analysis of the disturbance information, the 1# machine of a thermal power plant in Song Bai area is selected. The simulation uses the generator parameters with 5th-order and voltage regulator parameters with 12-type provided by the Northeast Power Grid Standard Database. The data are the recorded information measured by WAMS in a provincial power grid from 2016 to 2017.

5.2 *Disturbed scene clustering analysis*

Set the number of clusters $K = 3$ by calculating the Euclidean metric among the cluster samples and using the k-means algorithm for cluster analysis. The clustering results are shown in Figure 1 and Table 1.

Based on Figure 1, in Table 1, the three types of disturbance scenes are defined as deep disturbance, middle disturbance, and shallow disturbance.

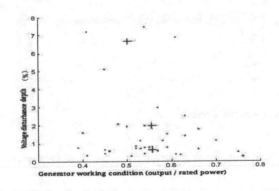

Figure 1. Disturbing scene clustering result.

Table 1. Disturbing scene clustering result.

Disturbing scene category	Number of disturbance events	Disturbance event number
Deep disturbance	4	1, 2, 12, 21
Middle disturbance	9	5, 7, 16, 0, 23, 26, 27, 31, 36
Shallow disturbance	23	3, 4, 6, 8–10, 11, 13, 14, 15, 17–19, 22, 24, 25, 28, 29, 30, 32, 33–35

5.3 Parameter checking process in a specific disturbance scenario

Owing to space limitations, the calibration of the target generator is based on the actually measured information of the B-phase short-circuit disturbance event on a 220-kV line in Song Bai District on August 6, 2015.

The measured information of the power is directly injected into the generator to realize the decouple in the system and perform a mixed dynamic simulation. The perturbed method is used to calculate the trajectory sensitivity of the model parameters. The calculation results are shown in Figure 2.

The four parameters, which have a great influence on the trajectory, are selected by sensitivity calculation. After parameter adjustment, the simulation results can be close to the measured trajectory. The parameter correction results are shown in Table 2 (sorted by sensitivity results order).

The error indicators between simulated trajectory and measured trajectory are shown in Table 3. It's known that the simulation results deviate significantly from the measured trajectory according to the parameters supported by the Northeast Power Grid Standard Database. After adjustment of the parameters, the error is greatly reduced and the simulated results are obviously close to the measured trajectories.

5.4 Generator parameter verification method considering the classification of disturbance scenes

The importance of each parameter in the three parameter sets is evaluated by the method described in the article, and three dominant parameter sets corresponding to various disturbance scenarios are identified. Tables 4 and 5 show the evaluation results of the importance of each parameter in the deep layer. Owing to space limitations, the results of the moderate and shallow disturbance evaluations will not be listed.

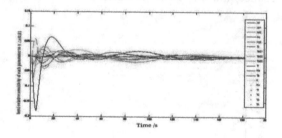

Figure 2. Parameter track sensitivity result.

Table 2. Generator original parameters and corrected parameters.

Parameter	x_q	T'_{d0}	T_4	T_2
Before correction	2.42985	9.84	0.1	10
After correction	1.24	6.1	0.12	14.2

Table 3. Parameter correction and error indicators.

Error index	Voltage overshoot error (%)	Voltage lowest point error (%)	Voltage overall error energy (%)	First swing amplitude error (%)	Active power total error energy (%)
Before correction	0.0763	−0.0242	1.5975	−0.052	0.3189
After correction	0.016	−0.0039	0.2	−0.0197	0.0445

Table 4. Deep disturbance corresponding to parameter importance.

Parameter	x_q	T''_{q0}	T_4	T_2	T'_{d0}
Importance	3.73	2.94	2.07	0.8	0.5

Table 5. Range of parameters corresponding to depth disturbance.

Parameter	x_q	T''_{q0}	T_4	T_2
Initial value	2.42985	0.067	0.1	10
Ranges	[1.35,1.46]	[0.09,0.12]	[0.1,0.12]	[9.92,12.6]

According to the results shown in Tables 4 and 5, the first four parameters in each group are taken as the dominant parameters of each group. According to small-sample confidence interval calculation, the value ranges of the dominant parameters are listed when the confidence is 95%. The error interval can be listed at the same time.

By analyzing the parameter calibration results based on large amounts of measured disturbance data, three sets of generator parameters can be obtained. In addition, the correction values of each parameter are mostly concentrated in a close interval with small dispersion, and the corresponding error range is also much smaller than the error level of the original simulation results.

6 CONCLUSION

(1) Based on the simulation results calculated in the existing parameter library, the error between simulated results and measured data is large and cannot objectively describe the dynamic behavior of a power system.

(2) According to theoretical analysis and simulation results, it can be concluded that the parameter correction results tend to be dispersive because of the difference between the disturbed depths and operating situations.

(3) A generator parameter verification method is proposed and tested. By checking large amounts of different disturbance scenarios in the provincial grid, three parameter modes can be obtained based on clustering of the disturbance scenes. The range of the dominant parameters can be calculated in a certain confidence level under different parameter models, which provides a powerful capability for grid operators to select the simulation parameters.

ACKNOWLEDGMENTS

This work was supported by the National Key Research and Development Plan "Smart Grid Technology and Equipment" on special high-performance analysis and situational awareness technology for interconnected large power grids (2018YFB0904500). The authors also thank the State Grid Corporation for science and technology funding (18-GW-05).

REFERENCES

Allen, E., et al. 2010. Validation of power system models. In Power and Energy Society General Meeting. IEEE: 1–7.
An, J., et al. 2013. Tracing and credibility verification method for dynamic simulation of power system based on WAMS measurement and Thevenin equivalent. *Power System Technology* 37(5): 1389–1394.

Fang, Y. 2012. Enlightenment of the "9·8" blackout in America on cascading failure prevention and control technology. *Automation of Electric Power Systems* 36(15): 1–7.

Fang, Z., et al. 2007. Influence of generator excitation and governor model parameters on simulation calculation of large disturbance test in Northeast power grid. *Power System Technology* 31(4): 69–74.

Hauer, J. F., et al. 2000. Dynamic performance validation in the Western power system. Association of Power Exchanges.

Huang, Z., et al. 2004. Large-scale hybrid dynamic simulation employing field measurements. In Power Engineering Society General Meeting. IEEE: 1570–1576.

Huang, Z., et al. 2006. Model validation of power system components using hybrid dynamic simulation. In Transmission and Distribution Conference and Exhibition, 2005/2006 IEEE PES. IEEE Xplore: 153–160.

Kosterev, D. N., et al. 1999. Model validation for the August 10, 1996 WSCC system outage. *IEEE Transactions on Power Systems* 14(3): 967–979.

Liang, Y, et al. 2010. Load calibration and model validation methodologies for power distribution systems. *IEEE Transactions on Power Systems* 25(3): 1393–1401.

Ma, Jin, et al. 2007. Power system dynamic simulation verification strategy based on wide area measurement system. *Automation of Electric Power Systems* 31(18): 11–15.

Wang, Y., & Tian, B. 2005. *Probability Theory and Mathematical Statistics*. Beijing: Science Press, 154–161.

Wu, S., et al. 2010. Hybrid dynamic simulation verification strategy for setting V-θ nodes using PMU measurement. *Automation of Electric Power Systems* 34(17): 12–16.

Wu, Y., et al. 2016. Evaluation method of credibility of test system based on small sample. *Journal of Beijing University of Aeronautics and Astronautics* 42(9): 1911–1917.

Emerging Developments in the Power and Energy Industry – Dufo-López, Krzywanski & Singh (eds)
© 2020 Taylor & Francis Group, London, ISBN 978-0-367-27169-5

Method for judging the severity of atmospheric corrosion of power equipment metal based on colorimetric and topographical texture features

Y. Zhong, J. Hao & X.P. Wang
State Key Laboratory of Power Transmission Equipment & System Security and New Technology, Chongqing University, Chongqing, China

G.L. Wu, X. Ren & Q. Wang
State Grid Chongqing Electric Power Company Chongqing Electric Power Research Institute, Chongqing, China

ABSTRACT: This paper examines the atmospheric corrosion characteristics of power metal frame equipment in Chongqing Power Grid. The "exposure" experiment of Q_{235} steel was carried out in the substation, and the on-site corrosion data of Q_{235} steel with different corrosion degrees were obtained. For metal corrosion morphology at different exposure time, different data were measured, including the color, hue and saturation related features which were based on HSI and RGB image models: M_R, M_G, M_B, σ_R, σ_G, σ_B, M_H, M_S, σ_H, σ_S; statistical characteristics: corrosion energy E, corrosion entropy S; binary characteristic: corrosion rate k; fractal feature: fractal dimension FD. A total of 14 features were selected as the evaluation system of corrosion feature vector. We use the digital image processing technology and the BP neural network algorithm to qualitatively evaluate the corrosion state of equipment metal. By testing the on-site samples of the other 3 substations, the corrosion state values of the samples were 1.0293, 0.9967 and 0.8508 respectively, which were basically consistent with the actual corrosion degree. The system had a good evaluation effect.

1 INTRODUCTION

The corrosion laws and characteristics of power grid equipment metal materials in natural environment have obvious regional differences, and the service life of power grid equipment metal materials is obviously different due to different environment. Because of the lack of corrosive data of related materials in natural environment, it is very difficult to establish standards for corrosion selection and anti-corrosion design of power transmission and transformation materials. As a result, the service life of equipment or materials can not meet the design requirements, and corrosion problems are prominent in operation, which greatly increases the maintenance work and cost of corrosion protection in the later period.

In an atmospheric environment, the propagation of corrosion is always accompanied by changes in corrosion morphology and color (Feliciano, F.F. et al. 2015). Using the different information such as grayscale and texture of the corrosion image, it is possible to judge the type of corrosion, analyze the degree of corrosion, and study the corrosion law. D.Itzhak et al. obtained the corrosion image of AISI 304 stainless steel immersed in 10% FeCl3 at 50 °C for 20 minutes with a scanner (Itzhak, D. et al. 1981). It was found that the number of etched holes of different sizes on the surface of the whole sample can be counted by binarization.

The transformation of corrosion products on the metal surface will affect the irregularity of the surface. The irregularity of the extraction surface is an important method to evaluate the corrosion degree of the metal. Wang, S. calculated the fractal dimension, area factor and gap feature of the corrosion pattern based on the gray value of the image, and established the corrosion image analysis and diagnosis system by fuzzy mathematics (Wang, S. et al. 2001). Ghanta, S. used wavelet transformation to detect rust images (Ghanta, S. et al. 2011).

In this paper, firstly, the on-site hanging test was carried out under the corrosive environment conditions of the Chongqing power grid to obtain the scene image. Then the graphic processing method was adopted to obtain image features of metals in different corrosion states.

2 IMAGE ACQUISITION

This paper tested the standard reference samples of Q_{235} carbon steel for transmission and transformation engineering structural materials in substation A_0, which is located in northern Chongqing. The substation has a voltage rating of 220kV.

The size of the carbon steel coupon was 100mm × 50mm × 3mm. Before the experiment, the samples were polished with 2# to 8# sandpaper, and the surface of the hanger was scrubbed with alcohol and acetone to remove the oil stain. The hanger was fixed with plastic bolts. The bracket had a tilt angle of 45° and faces south. The hanging scene is shown in Figure 1, left.

The live image of metal was acquired respectively before the hanging, at 2 weeks, 1.5 months and 4 months. Figure 1, right shows the on-site corrosion patterns of samples at different corrosion stages. The high temperature and high humidity atmosphere in Chongqing is the reason why the corrosion rate of the sample is so fast.

3 EXTRACTION AND CALCULATION OF CORRISION IMAGE TOPOGRAPHY

Corroded images record a large amount of information, and image processing techniques can be used to extract color, texture, and topographic features.

3.1 *Extraction of colorimetric features of corrosion images based on RGB-HSI space*

As the corrosion process progressed, the corrosion products on the metal surface would gradually increase, and the color and morphology would gradually change as well. The RGB

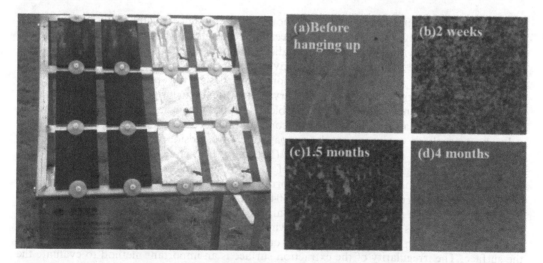

Figure 1. "Exposure" method-a live image of atmospheric metal corrosion coupon test and corrosion images of metal corrosion samples taken at different corrosion stages (a)(b)(c)(d).

model can fully reflect the original color information of the image, and the HSI model can simultaneously extract information such as hue and saturation of the image. Therefore, this paper selected the RGB model to extract the color features, at the same time, converted the RGB model into an HSI model to extract the hue and saturation of the image.

The conversion models for the H (hue), S (saturation) and I (intensity, brightness) values of the R (red), G (green) and B (blue) components of the original image are as follows:

$$I = \frac{1}{3}(R + G + B) \tag{1}$$

$$S = 1 - \frac{3}{R + G + B}[\min(R, G, B)] \tag{2}$$

$$H = \cos^{-1} \frac{\frac{1}{2}[(R - G) + (R - B)]}{\sqrt[2]{(R - G)^2 + (R - B)(G - B)}} \tag{3}$$

The mean (4) and standard deviation (5) of each corresponding variable of the RGB-HSI model of the corroded image were selected as the feature quantity and calculated. The result can be obtained as shown by Table 1.

$$\overline{M} = \frac{1}{n}\sum_i xi, x = \{R, G, B, H, S\} \tag{4}$$

$$\sigma = \sqrt{\frac{1}{N}\sum_{i=1}^{N}(xi - \mu)}, x = \{R, G, B, H, S\} \tag{5}$$

3.2 Statistical and binary characteristics of corroded images

Although color images contain more useful information, there is also a large amount of redundant information. Converting a color image into a grayscale image does not cause an incorrect discrimination for the formed component, which can reduce the amount of calculation, and at the same time, suppress the noise well, make the image clear and dark, clear outline, and facilitate subsequent processing.

The energy E of the etched image can be calculated according to the formula (6), which can reflect the gradation distribution of different regions, and $p(m)$ is the ratio of the point of the pixel value m to the total pixel point.

Table 1. Partially corroded image RGB-HSI model eigenvalues at different corrosion stages.

Feature	0(d)	14(d)	45(d)	120(d)
M_R	146.139	124.499	117.194	189.250
M_G	148.162	110.215	91.658	109.207
M_B	143.160	100.611	80.878	51.579
σ_R	9.181	21.632	22.175	14.832
σ_G	9.359	24.607	23.195	14.027
σ_B	9.436	25.181	34.319	14.776
M_H	0.243	0.069	0.213	0.070
M_S	0.036	0.202	0.359	0.727
σ_H	0.071	0.016	0.287	0.007
σ_S	0.012	0.074	0.179	0.072

The entropy S of the corrosion topographic image can reflect the randomness of the image content, which can be expressed as formula (7).

$$E = \sum_{m=0}^{255} p_m^2 \tag{6}$$

$$S = -\sum_{i=0}^{255} p(m) \log 2[\, p(m)] \tag{7}$$

$$k = \frac{S_{co}}{S_0} \tag{8}$$

The shape of the corroded area has significant irregularity. The image was binarized, and the area of the corroded hole in the corroded topography image was measured by the number of pixels to calculate the corrosion ratio to evaluate the degree of corrosion of the material. The corrosion rate calculation method is as shown in the formula (8), where S_{co} is the area of the corrosion area, and S_0 is the total area of the original sample image.

The white balance correction, gradation transformation, filtering and enhanced corrosion images of different corrosion stages were combined with the formulas (6)~(8) to calculate and extract the eigenvalues, and the results could be obtained as shown by Table 2.

3.3 Fractal characteristic of corroded images

The metal corrosion image contains a large difference in the degree of irregularity in different corrosion states due to its irregularity and complexity. The fractal dimension can well characterize the irregularity of the corrosion image. Chaudhuir et al. proposed a simple, fast and high-precision method called DBC "box" dimension to obtain the fractal dimension, as shown in Figure 2.

The reasonable interval of the fractal dimension FD is [2, 3], where 2 corresponds to a pure plane. The larger the FD, the more complicated the grayscale surface is.

In this paper, the fractal dimension FD was selected as the corrosion feature quantity, and the fractal dimension FD calculation result of the partial corrosion image is shown as Table 3.

4 CORROSION STATE EVALUTION SYSTEM BASED ON BP NETURAL NETWORK

4.1 Establishment of corrosion state evaluation system

BP neural network has good classification effect and high computational efficiency. In this paper, BP neural network was used as the evaluation method to diagnose the corrosion state of Q_{235} steel based on the morphology of the corrosion image.

The color, hue and saturation related features based on HSI and RGB image models: M_R, M_G, M_B, σ_R, σ_G, σ_B, M_H, M_S, σ_H, σ_S; statistical characteristics: corrosion energy E, corrosion entropy S; binary characteristic: corrosion rate k; fractal feature: fractal dimension FD

Table 2. Statistical and binary eigenvalues of partially corroded images at different corrosion stages.

Feature	0(d)	14(d)	45(d)	120(d)
E	0.037	0.013	0.019	0.022
S	5.128	6.471	6.105	5.789
k	89.182	23.446	10.050	18.353

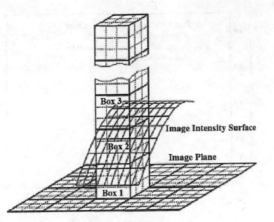

Figure 2. DBC "box" dimension FD calculation model.

Table 3. Fractal dimension eigenvalues of partial corrosion Images at different corrosion stages.

Feature	0(d)	14(d)	45(d)	120(d)
FD	2.662	2.866	2.826	2.723

totaled 14 features were selected as corrosion feature vector, ie the input layer of the neural network. The input factor was 14. So the input feature vector C_i could be characterized as:

$$C_i = [M_R, M_G, M_B, \sigma_R, \sigma_G, \sigma_B, M_H, M_S, \sigma_H, \sigma_S, E, S, k, FD] \tag{9}$$

Since the sampling time is 4 time nodes, there are 4 different corrosion states corresponding to the corrosion image. Therefore, the output layer node no is set to 1 here, and the expected corrosion state output value is C_o, and the typical output values are 0.25, 0.5, 0.75, and 1. Among them, 1 represents severe corrosion, 0.75 represents general corrosion, 0.5 represents slight corrosion, and 0.25 represents non-corrosion, corresponding to corrosion time of 4 months, 1.5 months, 2 weeks and before corrosion. According to the empirical formula, the hidden layer nodes were selected as 9.

Based on 14 feature quantities such as fractal dimension FD, this paper could establish a corrosion state evaluation system as shown in Figure 3:

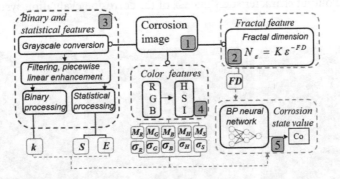

Figure 3. Corrosion state evaluation system based on corrosion morphology.

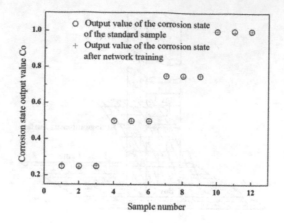

Figure 4. Comparison of the output results of the original sample of the base network and the corrosion state value after training.

4.2 Calculation of corrosion sample eigenvalues and operation and inspection of the evaluation system

In this paper, the Maltab2015 image processing toolbox is used to visually program the corrosion state evaluation system with reference to Figure 3.

The BP neural network is trained by using the eigenvalue vectors typical corrosion samples at different stages of the A_0 substation, and the eigenvalues based on the corrosion image can be obtained.It can be seen from Figure 4. that after a large amount of training, the network test output value is very close to the actual value, and the error is small. Explain that the network is an ideal evaluation network.

The four-month coupon samples from the A_1 substation in Chongqing and the A_2 substation in the northeast and the A_3 substation in the southeastern part of the country were selected for operational calculation and verification.

The corrosion morphology of the coupon samples are shown in Figure 5. and the results are shown in the Table 4.

It can be seen from the Table 4 evaluation results that the corrosion state output value of the A_1 and A_2 substation coupons are 1.0293, 0.9927, which are a serious state. The corrosion state output value of the A_3 substation coupon is 0.8508, which is a relatively serious state.

Since the atmospheric environment of the A_1 substation is the urban atmosphere, the A_2 substation is the industrial atmosphere, and the atmospheric environment of the A_3 substation is the rural atmosphere, the A_1 and A_2 substations are more severely corroded than the A_3 substation. The output characteristic value C_o of the corrosion evaluation system are basically

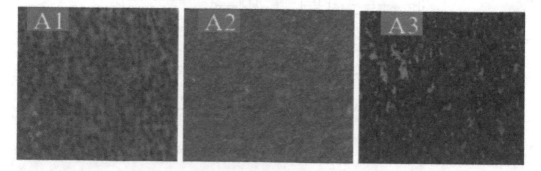

Figure 5. The corrosion morphology of the coupon sample A_1, A_2, A_3.

Table 4. Corrosion characteristics value and evaluation result.

Substation	A_1	A_2	A_3
M_R	193.8791	176.4362	128.9333
M_G	107.4294	93.8853	97.6994
......	......	......	......
k	17.3887	10.5044	4.2496
FD	2.5218	2.4812	2.4729
C_o	1.0293	0.9967	0.8508

consistent with the corresponding corrosion states, indicating that the evaluation system is reliable and basically realizes the qualitative evaluation of the corrosion state.

5 CONCLUSIONS

Combining the color feature quantity of the three-dimensional color space, the statistical features of the gray image, the binary feature and the fractal feature can reflect the corrosion information of the corrosion image more comprehensively.

The BP neural network method is suitable for judging the severity of atmospheric corrosion of electrical equipment metal.

ACKNOWLEDGMENTS

The authors thank the Project Supported by the State Grid Chongqing Electric Power Company Chongqing Electric Power Research Institute (2018 Yudian Keji 9#).

REFERENCES

Feliciano, F.F. Leta, F.R. & Mainier, F.B. 2015. Texture Digital analysis for corrosion monitoring. *Corrosion Science*. 2:138-147.

Itzhak, D. Dinstein, I. & Zilberberg, T. 1981. Pitting corrosion evaluation by computer image processing. *Corrosion Science*. 21:17-22.

Wang, S. Kong, D. & Song, S. 2001. Diagnosing corrosion modality system of metallic material in seawater based on fuzzy pattern recognition. Acta Metallrugica Sinica. *Acta Metallrugica Sinica*. 37 (5):517-521.

Ghanta, S. Karp, T. & Lee, S. 2011. Wavelet domain detection of rust in steel bridge images. *IEEE Xplore*.

Emerging Developments in the Power and Energy Industry – Dufo-López, Krzywanski & Singh (eds)
© 2020 Taylor & Francis Group, London, ISBN 978-0-367-27169-5

Safety analysis of entering equipotential on UHV transmission lines

Xinglie Lie & Yong Peng
China Electrical Power Research Institute & State Key Laboratory of Power Grid Environmental Protection, Wuhan, China

Yuqun Fang
Jinhua Power Supply Company of State Grid Zhejiang Electric Power Company, Jinhua, China

Fanglin Xu, Ling Meng & Jiejin Yan
Anhui transmission and transformation engineering company of national power grid, Hefei, China

ABSTRACT: Ultrahigh-voltage (UHV) transmission lines have high operating parameters and electric field intensity around the lines. It is necessary to study the electric field distribution characteristics according to different tower types and find a safe method of entering equipotential. In this article, based on the typical tower structure of UHV transmission lines, we analyzed a possible method of entering equipotential. The electric field distributions on the worker's body under different conditions were analyzed by simulation, and safety protection measures were put forward. The study shows that entering equipotential is safe by means of a basket, insulating ladder, or climbing through a tension insulator-string. Then safety precautions were proposed for each method, and the precautions were proved safe and reliable through application, which can guide live working on UHV transmission lines.

Keywords: entering equipotential, safety, UHV

1 INTRODUCTION

With the development of ultrahigh voltage (UHV) technology in China, the number of UHV transmission projects is increasing and the scale is expanding daily. The reliability of a UHV power grid is of great significance to China's economy and society (Liang & Chen 2017; Peng et al. 2017; Zhang 2017; Zhao 2017). Live working is an important means of maintenance for a UHV power grid to ensure its continuous operation.

UHV transmission lines have the characteristics of high structure, large operating parameters, and long distance across regions, which pose a series of challenges to live working (Du et al. 2015; Fang et al. 2015; Mengsuimin et al. 2016; Xiao et al. 2016). Entering equipotential is an important link in live working. Because the operation parameters are large and the electric field intensity is high, the requirements for safety protection and prevention of injury to workers in the process of entering equipotential are high (Li 2017, Liu et al. 2017; Wang et al. 2017; Wei et al. 2017). It is therefore necessary to study and analyze the equipotential method so as to ensure the safety of live working.

The study described in this article carried out a safety analysis of entering equipotential on straight and tension towers and calculated the electric field intensity distribution of a live worker's body surface by means of ANASYS. Combined with practical application and measurement, we verified the safety of the proposed method, which can guide the entering and exiting equipotential on the UHV lines. The results of this study are of practical significance to guarantee the safety of workers when performing live working.

2 METHOD OF ENTERING EQUIPOTENTIAL

This article takes a UHV double-circuit transmission line as an example for analysis. Carrying out live working on UHV lines includes two ways of entering equipotential: from a straight tower or a tension tower.

(1) Entering equipotential from a straight tower

According to the experience of live working on UHV transmission lines, two methods can be adopted to enter equipotential from a straight tower: by a hanging basket or by an insulating flexible ladder.

Figure 1 shows the arrangement for entering equipotential by a hanging basket. First, the hanging basket and pulley block should be installed at the appropriate fixed point on the upper cross arm, and an appropriate fixed point should be selected to fix the worker's personal protection rope. The equipotential worker shall take corresponding protective measures before entering the basket. The worker on the tower shall release the control end of the hanging rope at a constant speed, only after getting to an appropriate position should the equipotential worker start to get into equipotential.

Figure 2 shows the arrangement for entering equipotential by an insulating flexible ladder. First, the insulating flexible ladder and its control rope should be fixed at the appropriate position on the cross arm. The equipotential worker shall take corresponding protection measures before climbing down along the insulating flexible ladder to a specified length. Then, the worker on the tower should release the control rope to make the equipotential operators approach the conductor at a constant speed, and the equipotential worker shall be equipotential at an appropriate position.

(2) Entering equipotential from tension tower

On the tension tower, workers generally move into the high electric field area along the tension insulator string. The method and trajectory are shown in Figure 3.

While entering equipotential, the workers will be in a very strong electric field environment, and the electric field strength on the worker's body will be different under each path, so it's necessary

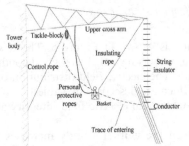

Figure 1.   Entering equipotential by basket.

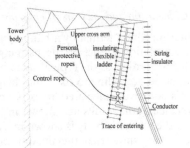

Figure 2.   Entering equipotential by insulating flexible ladder.

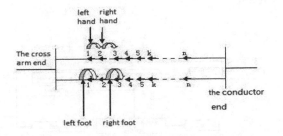

Figure 3.   Entering high potential through a tension insulator-string.

to study and determine the electric field distribution under different methods in order to ensure the safety of the workers and to choose the proper security method to enter equipotential.

3 SIMULATION ANALYSIS OF ELECTRIC FIELD DISTRIBUTION CHARACTERISTICS

3.1 *Finite element calculation and ANSYS analysis*

In this article, the electric field distribution is analyzed and calculated by using the finite element method and ANSYS. The relation between the unknown potential function in the whole region and the excitation source can be expressed by the overall matrix equation, as shown in Equation (1):

$$[K][\phi] = [f] \tag{1}$$

where $[K]$ is an $N \times N$ order coefficient matrix, $[\phi]$ is an Nth order node potential function matrix, $[f]$ is an Nth order incentive matrix. In the matrix

$$K_{ij} = K_{ji} = \int_\Omega \nabla \psi_i \nabla \psi_j d\Omega \tag{2}$$

$$f_i = \int_\Omega q \psi_i d\Omega \tag{3}$$

Because the whole region is divided into many finite elements, the calculation of any element of the coefficient matrix can be carried out for each element separately, and then the integral results of each element can be added up. The basic flow of finite element analysis using ANSYS software is shown in Figure 4.

3.2 *Calculation and analysis of entering equipotential from a straight tower*

(1) Entering equipotential by a hanging basket

The simulation model of the hanging basket method is shown in Figure 5. The tower is modeled according to the typical structure. The human body is regarded as a good conductor and modeled according to its actual height and posture. The electric field distribution on the worker's body was calculated at six typical positions, which are described in Table 1.

The calculation results are shown in Figure 6. The maximum electric field intensity at key parts of the worker's body is shown in Table 2.

The results show that the maximum electric field on the worker's body increases from 85.2 kV/m to 1230 kV/m while getting closer to the conductor. The electric field on knees and hands is stronger than that on other parts, and the maximum electric field appears on hands. At the moment of equal potential (position 5), the operator needs to use the potential transfer bar. At this time, the arm is in the extended state and the electric field distortion is severe.

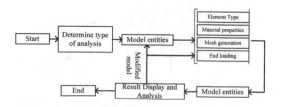

Figure 4. Finite element analysis using ANASYS.

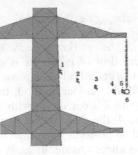

Figure 5. Simulation model of entering high potential by basket.

Table 1. Description of simulation position.

Position	1	2	3	4	5	6
Horizontal distance between worker and the tower (m)	0	4	7	12	0.5 m away from the conductor	on the conductor

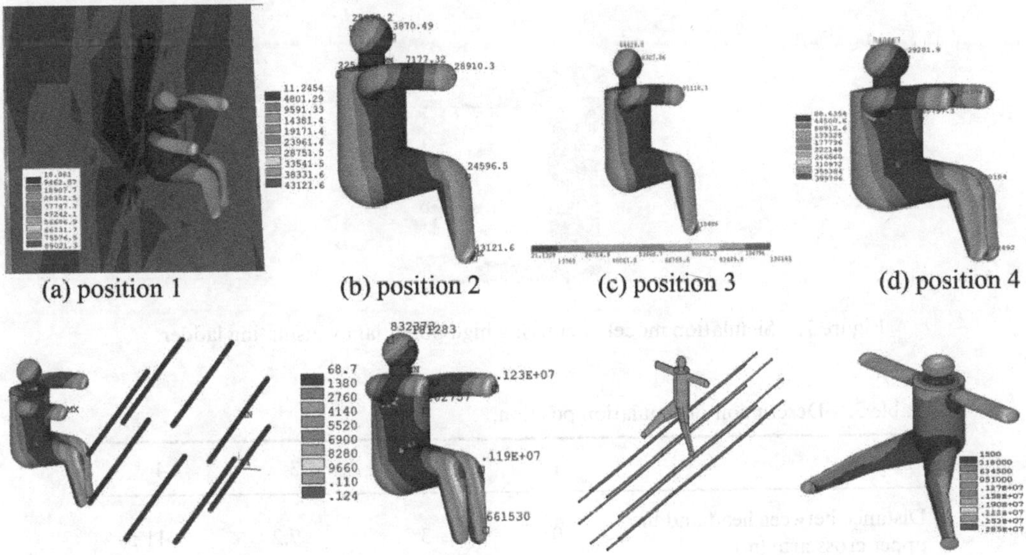

(a) position 1 (b) position 2 (c) position 3 (d) position 4

(e) position 5 (f) position 6

Figure 6. Electric field distribution of entering high potential by basket.

Table 2. Electric potential of the human body (kV/m).

Part of worker's body	Position 1	Position 2	Position 3	Position 4	Position 5	Position 6
Head	12.7	26.0	64.4	240.7	832.4	2291.7
Face	25.6	3.9	8.3	29.2	181.3	1066.6
Chest	18.1	7.2	10.3	65.8	202.8	506.2
Knee	35.8	24.6	69.8	290.2	1190.0	523.3
Hand	69.5	28.9	81.1	399.8	1230.0	2857.4
Feet	85.2	43.1	118.5	382.5	661.5	1062.4

(2) Entering equipotential by insulating flexible ladder

The simulation model of the insulating flexible ladder method is shown in Figure 7. The electric field distribution on the worker's body was calculated at four positions during the process, and the description of positions is shown in Table 3. The maximum electric field at key parts of the worker's body is shown in Table 4.

When adopting the insulating flexible ladder method, the electric field distribution on the worker at positions 5 and 6 in Figure 7 is consistent with the hanging basket method. In the process from position 1 to position 4, the electric field intensity on the worker's feet was the highest. When the worker climbed to position 4 (as high as the conductor), the electric field intensity on the worker's hands and knees was significantly enhanced.

3.3   Calculation and analysis of entering equipotential from a straight tower

The simulation model of moving along the tension insulator string method is shown in Figure 8. The tension string is a porcelain insulator of U550B/240 (51 pieces in total). The electric field distribution on the worker's body at five typical positions is calculated, and the position description is shown in Table 5. The maximum electric field at key parts of the worker's body is shown in Table 6.

It can be seen from the results that the maximum electric field intensity appears at the top of the head and shoulder during entering equipotential along the tension string, and after equal potential, the maximum electric field intensity appears on the head.

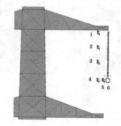

Figure 7.   Simulation model of entering high potential by insulating ladder.

Table 3.   Description of simulation position.

Position	1	2	3	4
Distance between head and the upper cross arm (m)	0	3.7	7.2	11.5

Table 4.   Electric potential of the human body (kV/m).

Part of worker's body	Position 1	Position 2	Position 3	Position 4
Head	10.2	97.9	204.5	331.2
Face	9.5	63.4	109.9	42.9
Chest	12.0	2.5	2.1	78.8
Knee	72.9	20.8	80.0	361.6
Hand	70.9	44.2	62.5	452.4
Feet	289.5	166.3	422.3	397.3

261

Figure 8. Simulation model of entering high potential through a tension insulator-string.

Table 5. Description of simulation position.

Position	Description
1	Feet are located on the first and third insulators respectively.
2	Feet are located on the 16th and 18th insulators respectively.
3	Feet are located on the 33th and 35th insulators respectively.
4	Feet are located on the 47th and 49th insulators respectively.
5	The left foot is on the 51st insulator and the right foot is on the conductor.

Table 6. Maximum electric field intensity of the human body.

Position	Maximum electric field intensity (kV/m)	Description
1	179.8	Appear on head
2	87.2	Appear on shoulder
3	116.6	Appear on shoulder
4	195.4	Appear on head
5	226.8	Appear on head

Table 7. Maximum electric potential of the human body (kV/m).

Part of the worker's body	Straight tower		Tension tower	
	Outer part of the screen clothes	Inner part of the screen clothes	Outer part of the screen clothes	Inner part of the screen clothes
Head	1896	0.9	194	0.7
Knee	452	1.0	-	-
Hand	2399	1.1	-	-
Feet	954	1.0	-	-
Face	908	15.1	147	13.2

3.4 *Analysis of simulation results*

It can be seen from the results that the electric field intensity of the worker's hands and knees is relatively high during entering equipotential by the hanging basket method, and the electric field intensity of the worker's hands is relatively high at the moment of potential transfer. Therefore, if using this method, workers should not extend their hands, feet, knees, and other parts out of the hanging basket to avoid overprotruding. At the moment of potential transfer, the action should be stable and rapid, and it is not appropriate to extend the hand out of the hanging basket too early to avoid electric discharge caused by electric field distortion.

In the process of entering equipotential by the insulating flexible ladder method, the electric field intensity of the worker's feet is relatively high, and the electric field distribution is similar to that of the hanging basket method when workers reach the same height as the conductor and during the potential transfer. Therefore, when adopting this method, the suspension point of the insulating flexible ladder should be as far away from the insulator string as possible. At the moment of potential transfer, the action should be stable and rapid, and it is not recommended to extend the hand too early to avoid the electric discharge caused by electric field distortion.

In the process of entering the equipotential along the tension string, the electric field intensity of the head and shoulder of the workers was relatively high. At the moment of equipotential, the operator extended his hand laterally to use the potential transfer bar for potential transfer. At this point, the hand and head were similar to the tip, and the electric field intensity was seriously distorted. Therefore, when using this method, excessive movement should be avoided. In the moment of potential transfer, the action should be stable and rapid.

4 APPLICATION

According to the provisions of GB/T 25726-2008, the shielding efficiency of screen clothes for live working on UHV transmission lines shall not be less than 60 dB, and that of the mask shall not be less than 20 dB. According to the simulation results, after wearing a full set of screen clothes, the maximum electric field intensity on the body surface shall not exceed 3 kV/m, and the maximum facial electric field intensity shall not exceed 100 kV/m, which meets the provisions of GB/T 6568-2008.

Based on the simulation results, we performed live working on a 1000 kV UHV transmission line. The workers completed the safety protection in accordance with the management regulations and operation instructions. The equipotential workers wore a full set of screen clothes, and then entered equipotential through the hanging basket method, insulating flexible ladder method, and tension insulator string respectively, and the distribution of electric field intensity on the body surface was measured. The measurement results are shown in Table 7.

The measured results for the straight tower are 11%–17% smaller than the calculated results, and the maximum value for the tension tower is 14% smaller than the calculated results. The measured electric field distribution value is lower than the calculated results because the model consideration and boundary conditions are more rigorous than the actual situation. The measurement results show that the maximum electric field intensity and distribution characteristics on the surface of the worker's body are consistent with the simulation results. Corresponding protective measures and methods optimized according to the simulation results can ensure the safety of workers.

5 CONCLUSION

Based on the simulation calculation and analysis of the safety of entering equipotential on UHV transmission lines, combined with the field application, the following conclusions can be drawn:

(1) It is safe to enter the straight line tower by the hanging basket method, insulating flexible ladder method, and climbing along the tension insulator string method.

(2) In the process of entering equipotential, the worker should avoid movement that could cause overprotruding.

(3) At the moment of potential transfer, the action should be smooth and rapid, and it is not advisable to extend the hand out too early.

ACKNOWLEDGMENTS

This work was supported by a grant from the Programs of STATE GRID Corporation of China (No. SGZJ0000KXXJS1700312).

REFERENCES

Du, Y., et al. 2015. Experimental study on safety clearance distance of live operation of UHV AC transmission line platform helicopter. *High Voltage Technology* 41(4): 1292–1298.

Fang, Y., et al. 2015.Simulation calculation of body surface electric field of live workers on UHV compact lines. *China Electric Power* 48(10): 77–83.

Li, M. 2017. Electric field calculation and corona effect research on UHV AC transmission lines. Xi 'an University of Science and Technology.

Liang, P., & Chen, X. 2017. Analysis on the current situation of ultra-high voltage transmission line operation and maintenance technology. *Low-Carbon World* 35: 120–121.

Liu, S., et al. 2017. Calculation of surface electric field of valve hall of uhvdc converter station using galerkin boundary element method. *High Voltage Technology* 43(6): 2049–2054.

Meng, S., et al. 2016. Live operation safety evaluation of uhvdc transmission lines based on fuzzy analytic hierarchy process. *High Voltage Electrical Appliances* 52(11): 42–46.

Peng, L., et al. 2017. Reliability and economic comparison of UHV AC and HVDC transmission systems. *Power Grid Technology* 41(4): 1098–1107.

Wang, S., et al. 2017. Characteristics and application of UHV DC transmission technology. *China Equipment Engineering* 17: 166–167.

Wei, M., et al. 2017. Electrostatic field distribution of y-type composite insulator string for UHV AC lines. *Journal of Electrotechnics* 32(21): 203–210.

Xiao, B., et al. 2016. Experimental study on live operation of 1000 kV UHV substation. *High Voltage Electrical Appliances* 52(1): 21–26.

Zhang, J. 2017. Influence of ultra-high voltage transmission lines on power system operation. Communications World 11: 210–211.

Zhao, Gu. 2017. Research on the development of ultra-high voltage transmission technology. *Electronic Technology and Software Engineering* 4: 233–234.

Emerging Developments in the Power and Energy Industry – Dufo-López, Krzywanski & Singh (eds)
© 2020 Taylor & Francis Group, London, ISBN 978-0-367-27169-5

Application of single-phase three-leg inverter for grid connected transformerless photovoltaic system

Na Su, Pan Mei, Lingwei Wu & Yue Chen
School of Electronics & Information Engineering, Taizhou University, China

ABSTRACT: The full-bridge inverter with hybrid PWM (HPWM) has higher efficiency and lower filtering requirement. However, when it is used for grid-connected transformerless photovoltaic system, the ground current may be high due to the parasitic capacitors and high frequency common-mode voltage. In this paper, the single-phase three-leg inverter is proposed to suppress the ground current. The additional leg is parallelled with the full-bridge inverter. The analytical method is established to calculate the ground current. The common mode voltage of the inverter can be reduced to suppress the ground current. Finally, the validity of the proposed method is verified by simulation results.

1 INTRODUCTION

Grid connected photovoltaic (PV) systems have an important role in distributed power generation. Many grid-connected PV systems have a transformer for the electrical isolation between the conversion stage and the grid for safety reasons. The transformer increases the weight, size, cost, and reduces the overall efficiency (Gonzalez, R et al. 2008). Transformerless structure seems to be more advantageous. However, without the isolation transformer, the grid connected transformerless PV system has a galvanic connection between the PV arrays and the grid, due to the existence of parasitic capacitors. The capacitors, which are marked with dashed lines in Figure 1, are formed between the solar cell and the normally ground frame (Calais, M. & Agelidis, V. G. 1998). The parasitic capacitors, the PV inverter, the filter and the grid constitute a common-mode loop where the ground current flows through (Ó. López, R. et al. 2007).

The circulation of the ground current leads to increased electromagnetic emissions, grid current distortion, additional power losses in the system, and especially safety hazards in the event of ground faults (Myrzik, J.M.A. & M. Calais 2003). Therefore, it needs to be limited. The ground current is dependent on the amplitude and frequency content of the voltage fluctuations which present at the PV panel terminals, and the parasitic capacitors (Kerekes, T. et al. 2009). The parasitic capacitors are determined by many factors, such as physical structure of the array and the weather conditions. The voltage fluctuations are determined by the topologies and the inverter's modulation technique.

Using bipolar PWM (BPWM), the diagonal switch-pairs of the full-bridge inverter are switched alternatively. The common-mode voltage is kept constant. Thus no ground current would appear. Using hybrid PWM (HPWM), the full-bridge inverter has higher efficiency and lower filtering requirements. But this modulation technique is not suitable for transformerless PV systems. The common-mode voltage has a content of square waves, which leads to a large ground current (Gonzalez, R. et al. 2007).

The ground current suppression methods can be normally divided into three types. The first method is to use additional capacitors which are paralleled with the parasitic capacitors, such as 3FB-SC and 3xNPC using the DC-link capacitor (A. Nabae et al. 1981). Then, the ground current is diverted to the paralleled capacitors. The second method is to increase the common mode loop impedance, such as the common mode filter and the common mode transformer (Von Jouanne, A. et al. 1998). The third method is to reduce the common mode voltage by

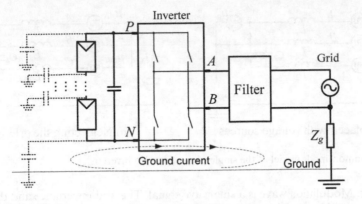

Figure 1.   Scheme of ground current in grid connected transformerless PV system.

decoupling the AC side and DC side, such as HERIC topology (Schmidt Heribert et al. 2005) and H5 topology (Matthias Victor et al. 2005). Or superimposing a compensating voltage on the inverter output to cancel the common-mode voltage, such as ACC (the active common-mode noise canceller) (Ogasawara, S. & Akagi, H. 1996). Another method for reducing the common mode voltage is to connect additional leg paralleled with the inverter, such as four-leg inverter using for the three-phase inverter (Eugenio Gubı́a et al. 2010).

In this paper, a ground current suppression method is proposed by using additional leg for the non-isolated full-bridge inverter. The rest of the paper is organized as follows. Section 2 gives the common-mode model to estimate the ground current for the proposed single-phase three-leg inverter using proposed modulation technique. Simulation results are given in section 3 to verify the theoretical analysis.

2 GROUND CURRENT ANALYSIS

2.1 *Model of single-phase three-leg inverter*

Figure 2 shows the grid connected transformerless single-phase three-leg inverter. The parasitic capacitors C_{pv1} and C_{pv2}, which is shown as dashed lines in Figure 2, present between the PV terminals and ground connection points of the inverter (0). L_1, L_2 and L_3 are the output filter inductors which are used to filter out the switching component of the output current. C_1 and C_2 are the output filter capacitors. Z_g is the grounding resistance.

In order to simplify the analysis, the following conditions are assumed. The inverter has reached steady state. All power devices are assumed as ideal switches. The dc-link voltage is

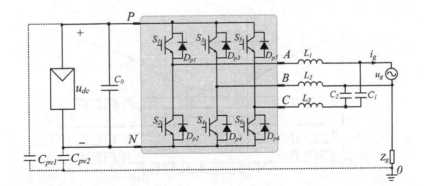

Figure 2.   Scheme of the single-phase three-leg inverter.

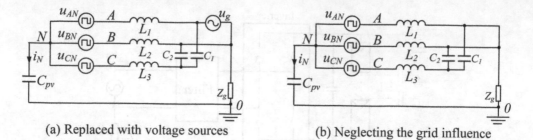

(a) Replaced with voltage sources (b) Neglecting the grid influence

Figure 3. Common-mode model of the single-phase three-leg inverter.

kept constant. Modulation wave is a sine-wave signal. The grid impedance and the line imped-ance are not considered. The mismatch of the power devices and the dead-time are not considered.

The model of the single-phase three-leg inverter is shown in Figure 3(a), where C_{pv} is the parasitic capacitor, and $C_{pv} = C_{pv1} + C_{pv2}$; u_{AN}, u_{BN} and u_{cN} is the converter output voltage and N is chosen as the reference point. The impact of the grid voltage u_g on the common mode current is small due to its low frequency of 50 or 60 Hz, the model can be simplified and it is shown in Figure 3(b). Based on the assumptions of the output filter inductors $L_1 = L_2 = L_3 = L$ and the output filter capacitors $C_1 = C_2 = C$, the ground current can be derived as,

$$i_N = H(s)\,u_\Sigma = -\frac{1}{sL + (Z_g + 1/sC_{pv})\left(2 + \frac{1}{1+1/s^2 2LC}\right)} \times \left(u_{AN} + u_{BN} + \frac{1}{1+1/s^2 2LC}u_{CL}\right)$$

(1)

It can be seen that the ground current is determined by the parasitic capacitor C_{pv}, the output filter inductor L, the output filter capacitor C and the voltage u_{AN}, u_{BN}, u_{CN}. The volt-age u_{AN}, u_{BN}, u_{CN} is determined by the dc-link voltage and the modulation technique.

2.2 Modulation technique

The modulation technique determines the switch states of each phase in the inverter. Once the switch states and the dc-link voltage are known, the magnitude and the frequency of the common mode voltage are decided.

The schematic of modulation technique for the single-phase three-leg inverter is shown in Figure 4, where u_{AB} is the output voltage of inverter. Using HPWM, Leg A is switched with high PWM frequency and S_3 of leg B is switched off during the positive half cycle. During the negative half cycle Leg B is switched with high PWM frequency and S_1 of leg A is switched

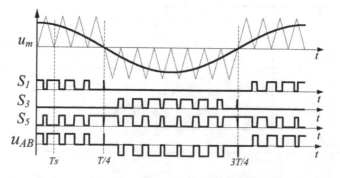

Figure 4. Schematic of modulation technique for the single-phase three-leg inverter.

off. Conduction states for S_5 are the complementary states of S_1 in the positive half cycle and S_3 in the negative half cycle. S_1, S_3, S_5 and S_2, S_4, S_6 are operated complementarily, respectively.

The double Fourier series of u_Σ for HPWM is deduced as follows.

$$
u_\Sigma = H_1(s) \left\{ \frac{2m_a U_{pv}}{\pi} - \frac{4}{\pi} m_a U_{pv} \sum_{n=2,3,\ldots}^{\infty} \frac{\cos(n\pi/2)}{(n+1)(n-1)} \cos(n\omega_s t + \varphi_n) + \frac{8}{\pi^2} U_{pv} \times \right.
$$

$$
\sum_{m=1,2,\ldots}^{\infty} \frac{1}{m} \sum_{k=1,2,\ldots}^{\infty} \frac{J_{2k-1}(mm_a\pi)}{2k-1} \cos(mm_f\omega_s t + \varphi_{m0}) - \frac{8U_{pv}}{\pi^2} \times \tag{2}
$$

$$
\left. \sum_{m=1}^{\infty} \sum_{n=\pm1}^{\pm\infty} \sum_{k=1,2,\ldots}^{\infty} \left. \frac{(2k-1)\cos(n\pi/2)J_{2k-1}(mm_a\pi)}{(n+2k-1)(n-2k+1)} \right|_{n \neq 2k-1} \cos\left[(mm_f + n)\omega_s t + \varphi_{mn}\right] \right\}
$$

where the transfer function $H_1(s) = 1/(1+2s^2LC)$, m_a is the amplitude modulation ratio, m_f is the frequency modulation ratio; ω_s is the fundamental angular frequency; φ_n is the phase angle of the band harmonics for the fundamental wave; φ_{m0} is the phase angle of the carrier wave harmonic and m order harmonics of the carrier wave harmonics; φ_{mn} is the phase angle of their corresponding upper and lower side band harmonics.

2.3 Ground current analysis

Figure 5 shows the harmonic spectrum of the ground current i_N for the single-phase three-leg inverter with the proposed modulation method, where h is the harmonic order, I_{Nhm} is the amplitude of hth order harmonic. In this study, the parameters of the PV systems are listed in Table 1.

The calculation result of the single-phase three-leg inverter is shown in Figure 5(a). It is seen that the sideband harmonics of the fundamental component have higher amplitude due to the resonance of $H_1(s)$. The relationship between the resonant frequency of $H_1(s)$ and the output filter capacitor C is shown in Figure 5(b). The common voltage can't be reduced very well around the resonant frequency. As the output filter capacitors change, the highest amplitude of harmonic varies between the fundamental frequency and the switching frequency. Therefore, it is better to choose smaller output filter capacitor for filtering consideration.

Figure 6 shows the ground current rms value I_{Nrms} varies with the output filter capacitors. The expression of ground current rms value is as follows with the harmonics above 50 kHz ignored,

$$
I_{N1rms} = \sqrt{\sum_{h=1}^{1000} (I_{N1hm})^2} \tag{3}
$$

Table 1. Converter circuit parameter values.

Parameters	Symbols	Values
System power	P	3.6 kW
PV array output voltage	U_{pv}	360 V
Amplitude modulation ratio	m_a	0.87
Frequency modulation ratio	m_f	400
Switching frequency	f_s	20 kHz
Output filter inductor	$L_1 \& L_2 \& L_3$	0.86 mH
Grounding impedance	Z_g	0.5 Ω
Parasitic capacitor	$C_{pv1} \& C_{pv2}$	0.3 μF
Output filter capacitor	$C_1 \& C_2$	11 μF

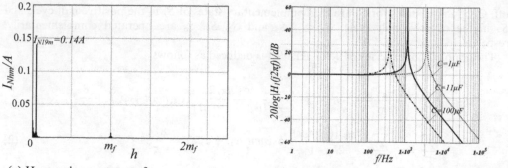

(a) Harmonic spectrum of i_N (b) Amplitude frequency characteristic of $H_1(s)$

Figure 5. i_N for the single-phase three-leg inverter.

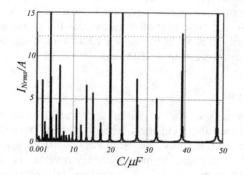

Figure 6. i_{Nrms} varies with output filter capacitor.

It can be seen that when the output filter capacitor increases, the ground current rms value reduces despite some resonance exists. In practice, the resonance will be restricted due to the impedance in the inductance, switches and line. It seems that the larger output filter capacitance has better effective on ground current suppression. But it will lead to large inter phase current circulation for its low impedance in the inter phase loop. Therefore, the output filter capacitors selection needs to consider the aspects of the ground current suppression in Figure 6, the filtering effect in Figure 5(b) and the inter phase current circulation.

3   SIMULATION RESULTS

The simulation is implemented in Matlab\Simulink. The grid is modeled as a f_g=50Hz grid with AC line voltage 220Vrms. The other parameters for simulation are the same with that used for calculation in the above.

Figure 7(a) shows the simulation results of full-bridge inverter with HPWM. The waveforms are the grid voltage u_g, the grid current i_g and the ground current i_N. A Fast Fourier Transform (FFT) of the ground current i_N is also presented. Figure 7(b) (c) show the simulation results of the single-phase three-leg inverter with the output filter capacitance 11μF and 45μF respectively. It can be seen that the ground current has been significantly reduced in the single-phase three-leg inverter. The ground current of the inverter with the output filter capacitance 45μF has lower amplitude and lower frequency. The simulation results in Figure 7(b) and the calculation results in Figure 5(a) are nearly the same. The highest harmonic is the 19th harmonic with the amplitude 0.128A, which is close to that in Figure 5(a).

269

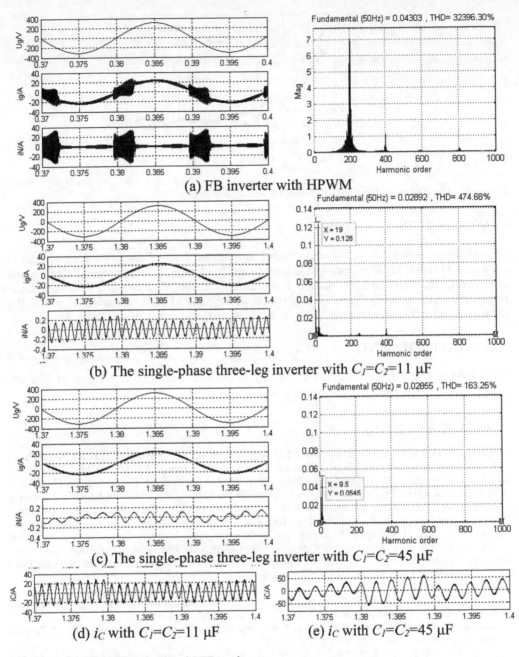

(a) FB inverter with HPWM

(b) The single-phase three-leg inverter with $C_1=C_2=11$ μF

(c) The single-phase three-leg inverter with $C_1=C_2=45$ μF

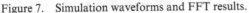

(d) i_C with $C_1=C_2=11$ μF   (e) i_C with $C_1=C_2=45$ μF

Figure 7.   Simulation waveforms and FFT results.

Figure 7(d) (e) show the current of leg C i_c with the output filter capacitance 11μF and 45μF respectively. The inter-phase circulation current with 45μF capacitance has higher amplitude and lower frequency. The simulation results are consistent with the theoretical analysis.

Figure 8 shows the rms ground current by simulation and calculation. It is observed that the ground current reduces when the output capacitance varies from 5μF to 45μF. The theoretical analysis results are often larger than the simulation results due to the circuit impedance of the simulation model.

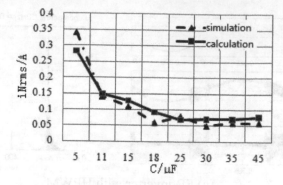

Figure 8. The rms ground current by simulation and calculation.

4 CONCLUSION

This paper proposed a novel ground current suppression method for full-bridge inverter. It adds an additional leg to suppress the common mode voltage. The suppression effect is analyzed using the proposed modulation methods by theoretical calculation. The simulation results are given to verify the validity of the proposed method and calculation. It proves that the single-phase three-leg inverter is effective for the ground current suppression.

ACKNOWLEDGEMENT

The authors would like to thank the support of Natural Science Foundation of Zhejiang under Grant Q15E070001.

REFERENCES

Gonzalez, R. Gubia, E. Lopez, J. Marroyo, L., "Transformerless Single-Phase Multilevel-Based Photovoltaic Inverter", Industrial Electronics, IEEE Transactions on, July 2008, 55–7, pp 2694–2702.

Calais, M. & Agelidis, V. G. 1998. Multilevel converters for single-phase grid connected Photovoltaic systems-an overview. Proc IEEE ISIE 1: 224–229. IEEE International Symposium on Industrial Electronics.

Ó. López, R. Teodorescu, F. Freijedo & J. Doval-Gandoy 2007. Eliminating ground current in a transformerless photovoltaic application, IEEE Power Engineering Society General Meeting, June:1–5.

Myrzik, J.M.A. & M. Calais. 2003. String and module integrated inverters for single-phase grid connected photovoltaic systems - a review. In Power Tech Conference Proceedings, IEEE Bologna.

Kerekes, T.; Liserre, M.; Teodorescu, R.; Klumpner, C.; Sumner, M. 2009. Evaluation of Three-Phase Transformerless Photovoltaic Inverter Topologies, Power Electronics, IEEE Transactions on, 24(9):2202–22211.

Gonzalez, R.; Lopez, J.; Sanchis, P.; Marroyo, L. 2007. Transformerless Inverter for Single-Phase Photovoltaic Systems, Power Electronics, IEEE Transactions on, 22(2): 693–6697.

Nabae, A., Takahashi, I., & Akagi, H.. 1981. A new neutral-point-clamped pwm inverter. IEEE Transactions on Industry Applications, IA-17(5), 518–523.

Jouanne, A. V., Zhang, H., & Wallace, A. K.. 1998. An evaluation of mitigation techniques for bearing currents, emi and overvoltages in asd applications. IEEE Transactions on Industry Applications, 34(5), 1113–1122.

Schmidt Heribert, Siedle, Christoph, Ketterer, Jürgen, 2005. US0174817A1, DC/AC converter to convert direct electric voltage into alternating voltage or into alternating current.

Matthias Victor, Frank Greizer, Sven Bremicker, Uwe Hubler, 12.29.2005. US0286281A1, Method of converting a direct current voltage from a source of direct current voltage, more specifically from a photovoltage couse of direct current voltage, into a alternating current voltage.

Ogasawara, S. & Akagi, H. 1996. Modeling and damping of high-frequency leakage currents in pwm inverter-fed ac motor drive systems. IEEE Transactions on Industry Applications, 32(5), 1105–1114.

Eugenio Gubía, Sanchis, P., Alfredo Ursúa, Jesús López, & Marroyo, L. 2010. Ground currents in single-hase transformerless photovoltaic systems. Progress in Photovoltaics Research & Applications, 15(7), 629–650.

Emerging Developments in the Power and Energy Industry – Dufo-López, Krzywanski & Singh (eds)
© 2020 Taylor & Francis Group, London, ISBN 978-0-367-27169-5

Sustainable development in green energies and the environment

A.M. Omer*
Energy Research Institute (ERI), Nottingham, UK

ABSTRACT: The move towards a de-carbonised world, driven partly by climate science and partly by the business opportunities it offers, will need the promotion of environmentally friendly alternatives, if an acceptable stabilisation level of atmospheric carbon dioxide is to be achieved. This requires the harnessing and use of natural resources that produce no air pollution or greenhouse gases and provides comfortable coexistence of human, livestock, and plants. This article presents a comprehensive review of energy sources, and the development of sustainable technologies to explore these energy sources. It also includes potential renewable energy technologies, efficient energy systems, energy savings techniques and other mitigation measures necessary to reduce climate changes. The article concludes with the technical status of the ground source heat pumps (GSHP) technologies.

Keywords: Renewable energy resources, technologies, sustainable development, environment

1 INTRODUCTION

Over millions of years ago, plants have covered the earth converting the energy of sunlight into living plants and animals, some of which was buried in the depths of the earth to produce deposits of coal, oil and natural gas Cantrell and Wepfer (1984), ASHRAE (1995), and Kavanaugh, and Rafferty (1997). The past few decades, however, have experienced many valuable uses for these complex chemical substances and manufacturing from them plastics, textiles, fertiliser and the various end products of the petrochemical industry. Indeed, each decade sees increasing uses for these products. Coal, oil and gas, which will certainly be of great value to future generations, as they are to ours, are however non-renewable natural resources. The rapid depletion of these non-renewable fossil resources need not continue. This is particularly true now as it is, or soon will be, technically and economically feasible to supply all of man's needs from the most abundant energy source of all, the sun. The sunlight is not only inexhaustible, but, moreover, it is the only energy source, which is completely non-polluting United Nations (2003).

Industry's use of fossil fuels has been largely blamed for warming the climate. When coal, gas and oil are burnt, they release harmful gases, which trap heat in the atmosphere and cause global warming. However, there had been an ongoing debate on this subject, as scientists have struggled to distinguish between changes, which are human induced, and those, which could be put down to natural climate variability. Notably, human activities that emit carbon dioxide (CO_2), the most significant contributor to potential climate change, occur primarily from fossil fuel production. Consequently, efforts to control CO_2 emissions could have serious, negative consequences for economic growth, employment, investment, trade and the standard of living of individuals everywhere.

*Corresponding author: e-mail: abdeenomer2@yahoo.co.uk

2 ENERGY SOURCES AND USE

2.1 *Energy sources*

Scientifically, it is difficult to predict the relationship between global temperature and greenhouse gas (GHG) concentrations. The climate system contains many processes that will change if warming occurs. Critical processes include heat transfer by winds and tides, the hydrological cycle involving evaporation, precipitation, runoff and groundwater and the formation of clouds, snow, and ice, all of which display enormous natural variability. The equipment and infrastructure for energy supply and use are designed with long lifetimes, and the premature turnover of capital stock involves significant costs. Economic benefits occur if capital stock is replaced with more efficient equipment in step with its normal replacement cycle. Likewise, if opportunities to reduce future emissions are taken in a timely manner, they should be less costly. Such a flexible approach would allow society to take account of evolving scientific and technological knowledge, while gaining experience in designing policies to address climate change United Nations (2003).

The World Summit on Sustainable Development in Johannesburg in 2002 United Nations (2003) committed itself to "encourage and promote the development of renewable energy sources to accelerate the shift towards sustainable consumption and production". Accordingly, it aimed at breaking the link between resource use and productivity. This can be achieved by the following:

- Trying to ensure economic growth does not cause environmental pollution.
- Improving resource efficiency.
- Examining the whole life-cycle of a product.
- Enabling consumers to receive more information on products and services.
- Examining how taxes, voluntary agreements, subsidies, regulation and information campaigns, can best stimulate innovation and investment to provide cleaner technology.

The energy conservation scenarios include rational use of energy policies in all economy sectors and the use of combined heat and power systems, which are able to add to energy savings from the autonomous power plants. Electricity from renewable energy sources is by definition the environmental green product. Hence, a renewable energy certificate system, as recommended by the World Summit, is an essential basis for all policy systems, independent of the renewable energy support scheme. It is, therefore, important that all parties involved support the renewable energy certificate system in place if it is to work as planned. Moreover, existing renewable energy technologies (RETs) could play a significant mitigating role, but the economic and political climate will have to change first. It is now universally accepted that climate change is real. It is happening now, and GHGs produced by human activities are significantly contributing to it. The predicted global temperature increase of between 1.5 and 4.5°C could lead to potentially catastrophic environmental impacts UNFCCC (2009). These include sea level rise, increased frequency of extreme weather events, floods, droughts, disease migration from various places and possible stalling of the Gulf Stream. This has led scientists to argue that climate change issues are not ones that politicians can afford to ignore, and policy makers tend to agree UNFCCC (2009). However, reaching international agreements on climate change policies is no trivial task as the difficulty in ratifying the Kyoto Protocol and reaching agreement at Copenhagen have proved.

Therefore, the use of renewable energy sources and the rational use of energy, in general, are the fundamental inputs for any responsible energy policy. However, the energy sector is encountering difficulties because increased production and consumption levels entail higher levels of pollution and eventually climate change, with possibly disastrous consequences. At the same time, it is important to secure energy at an acceptable cost in order to avoid negative impacts on economic growth. To date, renewable energy contributes only as much as 20% of the global energy supplies worldwide UNFCCC (2009). Over two thirds of this comes from biomass use, mostly in developing countries, and some of this is unsustainable. However, the potential for energy from sustainable technologies is huge. On the technological side, renewables have an obvious role to play. In general, there is no problem in terms of the technical potential of renewables to deliver energy. Moreover, there are very good opportunities for RETs to play an important role in reducing emissions of GHGs into the atmosphere, certainly far more than have been exploited so far. However, there are still some technical issues to address in order to cope with the intermittency of some renewables, particularly wind and solar. Nevertheless, the biggest problem with relying on renewables to deliver the necessary cuts in GHG emissions is more to do with politics and policy issues than with technical ones Rees (1999). For example, the single most important step governments could take to promote and increase the use of renewables is to improve access for renewables to the energy market. This access to the market needs to be under favourable conditions and, possibly, under favourable economic rates as well. One move that could help, or at least justify, better market access would be to acknowledge that there are environmental costs associated with other energy supply options and that these costs are not currently internalised within the market price of electricity or fuels. This could make a significant difference, particularly if appropriate subsidies were applied to renewable energy in recognition of the environmental benefits it offers. Similarly, cutting energy consumption through end-use efficiency is absolutely essential. This suggests that issues of end-use consumption of energy will have to come into the discussion in the foreseeable future Bos, My, Vu and Bulatao (1994).

However, RETs have the benefit of being environmentally benign when developed in a sensitive and appropriate way with the full involvement of local communities. In addition, they are diverse, secure, locally based and abundant. In spite of the enormous potential and the multiple benefits, the contribution from renewable energy still lags behind the ambitious claims for it due to the initially high development costs, concerns about local impacts, lack of research funding and poor institutional and economic arrangements Duchin (1995). Hence, an approach is needed to integrate renewable energies in a way that meets the rising demand in a cost-effective way.

2.2   Role of energy efficiency systems

The prospects for development in power engineering are, at present, closely related to ecological problems. Power engineering has harmful effects on the environment, as it discharges toxic gases into atmosphere and also oil-contaminated and saline waters into rivers, as well as polluting the soil with ash and slag and having adverse effects on living things on account of electromagnetic fields and so on. Thus there is an urgent need for new approaches to provide an ecologically safe strategy. Substantial economic and ecological effects for thermal power projects (TPPs) can be achieved by improvement, upgrading the efficiency of the existing equipment, reduction of electricity loss, saving of fuel, and optimisation of its operating conditions and service life leading to improved access for rural and urban low-income areas in developing countries through energy efficiency and renewable energies.

Sustainable energy is a prerequisite for development. Energy-based living standards in developing countries, however, are clearly below standards in developed countries. Low levels of access to affordable and environmentally sound energy in both rural and urban low-income areas are therefore a predominant issue in developing countries. In recent years many programmes for development aid or technical assistance have been focusing on improving access to sustainable energy, many of them with impressive results. Apart from success stories, however, experience also shows that positive appraisals of many projects evaporate after completion and vanishing of the implementation expert team. Altogether, the diffusion of sustainable technologies such as energy efficiency and renewable energy for cooking, heating, lighting, electrical appliances and building insulation in developing countries has been slow. Energy efficiency and renewable energy programmes could be more sustainable and pilot studies more effective and pulse releasing if the entire policy and implementation process was considered and redesigned from the outset Givoni (1998). New financing and implementation processes, which allow reallocating financial resources and thus enabling countries themselves to achieve a sustainable energy infrastructure, are also needed. The links between the energy policy framework, financing and implementation of renewable energy and energy efficiency projects have to be strengthened and as well as efforts made to increase people's knowledge through training.

3   RENEWABLE ENERGY TECHNOLOGIES

Buildings consume energy mainly for cooling, heating and lighting. The energy consumption was based on the assumption that the building operates within American Society of heating, Refrigeration and Air-conditioning Engineers (ASHRAE) - thermal comfort zone during the cooling and heating periods ASHRAE (1993). Most of the buildings incorporate energy efficient passive cooling, solar control, photovoltaic, lighting and day lighting, and integrated energy systems. It is well known that thermal mass with night ventilation can reduce the maximum indoor temperature in buildings in summer Kammerud, Ceballos, Curtis, Place and Anderson (1984). Hence, comfort temperatures may be achieved by proper application of passive cooling systems. However, energy can also be saved if an air conditioning unit is used Shaviv (1989). The reason for this is that in summer, heavy external walls delay the heat transfer from the outside into the inside spaces. Moreover, if the building has a lot of internal mass the increase in the air temperature is slow. This is because the penetrating heat raises the air temperature as well as the temperature of the heavy thermal mass. The result is a slow heating of the building in summer as the maximal inside temperature is reached only during the late hours when the outside air temperature is already low. The heat flowing from the inside heavy walls could be reduced with good ventilation in the evening and night. The capacity to store energy also helps in winter, since energy can be stored in walls from one sunny winter day to the next cloudy one. However, the admission of daylight into buildings alone does not guarantee that the design will be energy efficient in terms of lighting. In fact, the design for increased daylight can often raise concerns relating to visual comfort (glare) and thermal comfort (increased solar gain in the summer and heat losses in the winter from larger apertures). Such issues will clearly need to be addressed in the design of the window openings, blinds, shading devices, heating system, etc. In order for a building to benefit from daylight energy terms, it is a prerequisite that lights are switched off when sufficient daylight is available. The nature of the switching regime; manual or automated, centralised or local, switched, stepped or dimmed, will determine the energy performance. Simple techniques can be implemented to increase the probability that lights are switched off Shaviv (1989). These include:

- Making switches conspicuous and switching banks of lights independently.
- Loading switches appropriately in relation to the lights.
- Switching banks of lights parallel to the main window wall.

There are also a number of methods, which help reduce the lighting energy use, which, in turn, relate to the type of occupancy pattern of the building Shaviv (1989). The light switching options include:

- Centralised timed off (or stepped)/manual on.
- Photoelectric off (or stepped)/manual on.
- Photoelectric and on (or stepped), and photoelectric dimming.
- Occupant sensor (stepped) on/off (movement or noise sensor).

Likewise, energy savings from the avoidance of air conditioning can be very substantial. Whilst day-lighting strategies need to be integrated with artificial lighting systems in order to become beneficial in terms of energy use, reductions in overall energy consumption levels by employment of a sustained programme of energy consumption strategies and measures would have considerable benefits within the buildings sector. The perception is often given however is that rigorous energy conservation as an end in itself imposes a style on building design resulting in a restricted aesthetic solution. It would perhaps be better to support a climate sensitive design approach that encompasses some elements of the pure conservation strategy together with strategies, which work with the local ambient conditions making use of energy technology systems, such as solar energy, where feasible. In practice, low energy environments are achieved through a combination of measures that include:

- The application of environmental regulations and policy.
- The application of environmental science and best practice.
- Mathematical modelling and simulation.
- Environmental design and engineering.
- Construction and commissioning.
- Management and modifications of environments in use.

While the overriding intention of passive solar energy design of buildings is to achieve a reduction in purchased energy consumption, the attainment of significant savings is in doubt. The non-realisation of potential energy benefits is mainly due to the neglect of the consideration of post-occupancy user and management behaviour by energy scientists and designers alike. Calculating energy inputs in agricultural production is more difficult in comparison to the industry sector due to the high number of factors affecting agricultural production. However, considerable studies have been conducted in different countries on energy use in agriculture Singh (2000), CAEEDAC (2000), Yaldiz, Ozturk, Zeren (1993), Dutt (1982), Baruah (1995), Thakur, Mistra (1993) in order to quantify the influence of these factors.

3.1 *Sustainable development*

Sustainable energy is the energy that, in its production or consumption, has minimal negative impacts on human health and the healthy functioning of vital ecological systems, including the global environment. It is an accepted fact that renewable energy is a sustainable form of energy, which has attracted more attention during recent years. Increasing environmental interest, as well as economic consideration of fossil fuel consumption and high emphasis of sustainable development for the future helped to bring the great potential of renewable energy into focus Wu, and Boggess (1999). Nearly a fifth of all global power is generated by renewable energy sources, according to a new book published by the Organisation for Economic Co-operation and Development (OECD)/International Energy Association (IEA) OECD/IEA (2004). "Renewables for power generation: status and prospects" claims that, at approximately 20%, renewables are the second largest power source after coal (39%) and ahead of nuclear (17%), natural gas (17%) and oil (8%) respectively. From 1973-2000 renewables grew at 9.3% a year and it is predicted that this will increase by 10.4% a year to 2010. Wind power grew fastest at 52% and will multiply seven times by 2010, overtaking biopower and hence help reducing green house gases, GHGs, emissions to the environment.

4 RESULTS AND DISCUSSIONS

The availability of data on solar radiation is a critical problem. Even in developed countries, very few weather stations have been recording detailed solar radiation data for a period of time long enough to have statistical significance. Solar radiation arriving on earth is the most fundamental renewable energy source in nature. It powers the bio-system, the ocean and atmospheric current system and affects the global climate. Reliable radiation information is needed to provide input data in modelling solar energy devices and a good database is required in the work of energy planners, engineers, and agricultural scientists. In general, it is not easy to design solar energy conversion systems when they have to be installed in remote locations. First, in most cases, solar radiation measurements are not available for these sites. Second, the radiation nature of solar radiation makes the computation of the size of such systems difficult. While solar energy data are recognised as very important, their acquisition is by no means straightforward. The measurement of solar radiation requires the use of costly equipment such as pyrheliometers and pyranometers. Consequently, adequate facilities are often not available in developing countries to mount viable monitoring programmes. This is partly due to the equipment cost as well as the cost of technical manpower. Several attempts have, however, been made to estimate solar radiation through the use of meteorological and other physical parameter in order to avoid the use of expensive network of measuring instruments.

A sustainable energy system includes energy efficiency, energy reliability, energy flexibility, fuel poverty, and environmental impacts. A sustainable biofuel has two favourable properties, which are availability from renewable raw material, and its lower negative environmental impact than that of fossil fuels. Global warming, caused by CO_2 and other substances, has become an international concern in recent years. To protect forestry resources, which act as major absorbers of CO_2, by controlling the ever-increasing deforestation and the increase in the consumption of wood fuels, such as firewood and charcoal, is therefore an urgent issue. Given this, the development of a substitute fuel for charcoal is necessary. Briquette production technology, a type of clean coal technology, can help prevent flooding and serve as a global warming countermeasure by conserving forestry resources through the provision of a stable supply of briquettes as a substitute for charcoal and firewood.

District Heating (DH), also known as community heating can be a key factor to achieve energy savings, reduce CO_2 emissions and at the same time provide consumers with a high quality heat supply at a competitive price. Generally, DH should only be considered for areas where the heat density is sufficiently high to make DH economical. In countries like Denmark for example, DH may today be economical even to new developments with lower density areas, due to the high level of taxation on oil and gas fuels combined with the efficient production of DH.

Platinum is a catalyst for fuel cells and hydrogen-fuelled cars presently use about two ounces of the metal. There is currently no practicable alternative. Reserves are in South Africa (70%), and Russia (22%). Although there are sufficient accessible reserves in South Africa to increase supply by up to 5% per year for the next 50 years, there are significant environmental impacts associated with its mining and refining, such as groundwater pollution and atmospheric emissions of sulphur dioxide ammonia, chlorine and hydrogen chloride

Hydrogen is now beginning to be accepted as a useful form for storing energy for reuse on, or for export off, the grid. Clean electrical power harvested from wind and wave power projects can be used to produce hydrogen by electrolysis of water. Electrolysers split water molecules into its constituent parts: hydrogen and oxygen. These are collected as gases; hydrogen at the cathode and oxygen at the anode. The process is quite simple. Direct current is applied to the electrodes to initiate the electrolysis process. Production of hydrogen is an elegant environmental solution. Hydrogen is the most abundant element on the planet, it cannot be destroyed (unlike hydrocarbons) it simply changes state (water to hydrogen and back to water) during consumption. There is no CO or CO_2 generation in its production and consumption and, depending upon methods of consumption, even the production of oxides of nitrogen can be avoided too. However, the transition will be very messy, and will take many technological paths to convert fossil fuels and methanol to hydrogen, building hybrid engines and so on. Nevertheless, the future of hydrogen fuel cells is promising. Hydrogen can be used in internal combustion engines, fuel cells, turbines, cookers gas boilers, road-side emergency lighting, traffic lights or signalling where noise and pollution can be a considerable nuisance, but where traffic and pedestrian safety cannot be compromised.

Water is the most natural commodity for the existence of life in the remote desert areas. However, as a condition for settling and growing, the supply of energy is the close second priority. The high cost and the difficulties of mains power line extensions, especially to a low populated region can focus attention on the utilisation of different and more reliable and independent sources of energy like renewable wind energy. Accordingly, the utilisation of wind energy, as a form of energy, is becoming increasingly attractive and is being widely used for the substitution of oil-produced energy, and eventually to minimise atmospheric degradation, particularly in remote areas. Indeed, utilisation of renewables, such as wind energy, has gained considerable momentum since the oil crises of the 1970s. Wind energy, though site-dependent, is non-depleting, non-polluting, and a potential option of the alternative energy source. Wind power could supply 12% of global electricity demand by 2020, according to a report by the European Wind Energy Association and Greenpeace.

The challenge is to match leadership in GHG reduction and production of renewable energy with developing a major research and manufacturing capacity in environmental technologies (wind, solar, fuel cells, etc.). More than 50% of the world's area is classified as arid, representing the rural and desert part, which lack electricity and water networks. The inhabitants of such areas obtain water from borehole wells by means of water pumps, which are mostly driven by diesel engines. The diesel motors are associated with maintenance problems, high running cost, and environmental pollution. Alternative methods are pumping by photovoltaic (PV) or wind systems. At present, renewable sources of energy are regional and site specific. It has to be integrated in the regional development plans.

5 RECOMMENDATIONS AND CONCLUSIONS

5.1 *Conclusions*

There is strong scientific evidence that the average temperature of the earth's surface is rising. This is a result of the increased concentration of carbon dioxide and other GHGs in the atmosphere as released by burning fossil fuels. This global warming will eventually lead to substantial changes in the world's climate, which will, in turn, have a major impact on human life and the built environment. Therefore, effort has to be made to reduce fossil energy use and to promote green energy, particularly in the building sector. Energy use reductions can be achieved by minimising the energy demand, rational energy use, recovering heat and the use of more green energy. This study was a step towards achieving this goal. The adoption of green or sustainable approaches to the way in which society is run is seen as an important strategy in finding a solution to the energy problem. The key factors to reducing and controlling CO_2, which is the major contributor to global warming, are the use of alternative approaches to energy generation and the exploration of how these alternatives are used today and may be used in the future as green energy sources. Even with modest assumptions about the availability of land, comprehensive fuel-wood farming programmes offer significant energy, economic and environmental benefits. These benefits would be dispersed in rural areas where they are greatly needed and can serve as linkages for further rural economic development. The nations as a whole would benefit from savings in foreign exchange, improved energy security, and socio-economic improvements. With a nine-fold increase in forest – plantation cover, a nation's resource base would be greatly improved. The international community would benefit from pollution reduction, climate mitigation, and the increased trading opportunities that arise from new income sources. The non-technical issues, which have recently gained attention, include: (1) Environmental and ecological factors, e.g., carbon sequestration, reforestation and revegetation. (2) Renewables as a CO_2 neutral replacement for fossil fuels. (3) Greater recognition of the importance of renewable energy, particularly modern biomass energy carriers, at the policy and planning levels. (4) Greater recognition of the difficulties of gathering good and reliable renewable energy data, and efforts to improve it. (5) Studies on the detrimental health efforts of biomass energy particularly from traditional energy users.

5.2 *Recommendations*

The following are recommended:

- Launching of public awareness campaigns among local investors' particularly small-scale entrepreneurs and end users of RETs to highlight the importance and benefits of renewable, particularly solar, wind, and biomass energies.

- Amendment of the encouragement of investment act, to include furthers concessions, facilities, tax holidays, and preferential treatment to attract national and foreign capital investment.
- Allocation of a specific percentage of soft loans and grants obtained by governments to augment budgets of (R & D) related to manufacturing and commercialisation of RETs.
- Governments should give incentives to encourage the household sector to use renewable energy instead of conventional energy.
- Execute joint investments between the private sector and the financing entities to disseminate the renewable with technical support from the research and development entities.
- Availing of training opportunities to personnel at different levels in donor countries and other developing countries to make use of their wide experience in application and commercialisation of RETs particularly renewable energy.
- The governments should play a leading role in adopting renewable energy devices in public institutions, e.g., schools, hospitals, government departments, police stations, etc., for lighting, water pumping, water heating, communication and refrigeration.
- Encouraging the private sector to assemble, install, repair and manufacture renewable energy devices via investment encouragement and more flexible licensing procedures.

ACKNOWLEDGMENT

It is a pleasure to acknowledge, with gratitude, all those who, at different times and in different ways, have supported Cleaner and Greener Energy Technologies, Sustainable Development and Environment. This article would not have been possible without the contributions of several people. Also, I wish to express my gratitude. Thanks to my wife Kawthar Abdelhai Ali for her warmth and love. Her unwavering faith in me, her intelligence, humour, spontaneity, curiosity and wisdom added to this book and to my life.

REFERENCES

Cantrell, J., and Wepfer, W. (1984) 'Shallow Ponds for Dissipation of Building Heat: A case Study', ASHRAE Transactions 90 (1): 239–246. l.

ASHRAE. (1995) 'Commercial/Institutional Ground Source Heat Pump Engineering Manual', American Society of heating, Refrigeration and Air-conditioning Engineers, Inc. Atlanta, GA: USA.

Kavanaugh, S., Rafferty, K. (1997) 'Ground source heat pumps. Design of Geothermal Systems for Commercial and Institutional Buildings', American Society of heating, Refrigeration and Air-conditioning Engineers, Inc. Atlanta, GA: USA.

United Nations. (2003) 'World urbanisation project: the 2002 revision', New York: The United Nations Population Division.

The United Nations Framework Convention on Climate Change (UNFCCC). (2009) 'The draft of the Copenhagen Climate Change Treaty', p. 3–181.

Rees, W.E. (1999) 'The built environment and the ecosphere: a global perspective', Building Research and information 1999; 27(4): 206–220.

Bos, E., My, T., Vu, E. and Bulatao, R. (1994) 'World population projection: 1994-95', Baltimore and London: World Bank by the John Hopkins University Press; 1994.

Duchin, F. (1995) 'Global scenarios about lifestyle and technology, the sustainable future of the global system', Tokyo: United Nations University; 1995.

Givoni, B. (1998) 'Climate consideration in building and urban design', New York: Van Nostrand Reinhold; 1998.

ASHRAE. (1993) 'Energy efficient design of new building except new low-rise residential buildings', BSRIASHRAE proposed standards 90-2P-1993, alternative GA. American Society of Heating, Refrigerating, and Air Conditioning Engineers Inc., USA. 1993.

Kammerud, R., Ceballos, E., Curtis, B., Place, W., and Anderson, B. (1984) 'Ventilation cooling of residential buildings', ASHRAE Trans: 90 Part 1B, 1984.

279

Shaviv, E. (1989) **'The influence of the thermal mass on the thermal performance of buildings in summer and winter'**, In: Steemers TC, Palz W., editors. Science and Technology at the service of architecture. Dordrecht: Kluwer Academic Publishers, 1989. p. 470–472.

Singh, J. (2000) **'On farm energy use pattern in different cropping systems in Haryana, India'**, Germany: International Institute of Management-University of Flensburg, Sustainable Energy Systems and Management, Master of Science; 2000.

CAEEDAC. (2000) **'A descriptive analysis of energy consumption in agriculture and food sector in Canada'**, Final Report, February 2000.

Yaldiz, O., Ozturk, H., Zeren, Y. (1993) **'Energy usage in production of field crops in Turkey'**, In: 5th International Congress on Mechanisation and Energy Use in Agriculture. Turkey: Kusadasi; 11-14 October 1993.

Dutt, B. (1982) **'Comparative efficiency of energy use in rice production'**, Energy 1982; 6:25.

Baruah, D. (1995) **'Utilisation pattern of human and fuel energy in the plantation'**, Journal of Agriculture and Soil Science 1995; 8(2): 189–192.

Thakur, C. Mistra, B. (1993) **'Energy requirements and energy gaps for production of major crops in India'**, Agricultural Situation of India 1993; 48: 665–689.

Wu, J. and Boggess, W. (1999) **'The optimal allocation of conservation funds'**, Journal Environmental Economic Management 1999; 38.

OECD/IEA. (2004) **'Renewables for power generation: status and prospect'**, UK, 2004.

Emerging Developments in the Power and Energy Industry – Dufo-López, Krzywanski & Singh (eds)
© 2020 Taylor & Francis Group, London, ISBN 978-0-367-27169-5

Study on radiated electromagnetic disturbance characteristics of a ±800 kV VSC-HVDC converter valve

Yang Xiang & Lei Liu
China Southern Power Grid Research Institute Co. Ltd., Guang zhou, Guang dong, China

ABSTRACT: In order to accurately predict the radiated electromagnetic disturbance generated during the normal operation of an extra-high voltage (EHV) large-capacity VSC-HVDC transmission system, this article studies the electromagnetic radiation problem based on a certain ±800 kV modular multilevel converter (MMC) valve in China. First, the disturbance source of the VSC-HVDC converter valve is analyzed. A new calculation method based on the complex geometric structure of the converter is proposed that uses DC bus current as the disturbance source. Then the antenna model with reference to the actual structure of the converter valve is built in FEKO. The radiated electric field and magnetic field strength generated in a valve hall of the converter are calculated, and the variation of electromagnetic disturbance with distance at different frequencies is analyzed. The calculation method and results can be used to guide electromagnetic disturbance research of UHV large-capacity VSC-HVDC transmission systems.

Keywords: VSC-HVDC, radiated electromagnetic disturbance, disturbance source, calculation method

1 INTRODUCTION

Voltage-sourced converter–high-voltage direct current (VSC-HVDC) technology is a new high-voltage direct current transmission technology based on a voltage source converter developed on the basis of insulated-gate bipolar transistor (IGBT) and pulse width modulation (PWM) technology. Compared with traditional HVDC, VSC-HVDC is especially suitable for asynchronous networking, offshore wind farm and distributed power supply accessing to a power grid, remote area power supply, urban DC distribution network, etc. (Xu et al. 2014). The modular multilevel converter (MMC) is a new voltage source converter topology proposed by A. Lesnicar and R. Marquardt in 2002, which has the advantages of no need for large number of IGBTs series, low voltage variation rate of devices, and low harmonic content of the output waveform (Tang et al. 2014). Nowadays, with the increasing protection of the environment, the electromagnetic disturbance of VSC-HVDC has attracted much attention.

The VSC-HVDC system has the same electromagnetic compatibility problem as the traditional converter valve, and also has its particularity. The voltage and current passing through the valve change rapidly due to the high-frequency action of IGBTs, resulting in a very steep pulse. This high-amplitude, fast-changing voltage and current create spatial radiated electromagnetic disturbances through the valve, rectifier and inverter circuit, and DC and AC side conductors. North China Electric Power University is conducting research on radiated electromagnetic disturbances of VSC-HVDC systems. The antenna model of the two-phase three-level converter is established in accordance with the size and connection mode of the actual converter system (Sun et al. 2015). The electromagnetic disturbance intensity at different positions is calculated, and the actual radiated electromagnetic disturbance is measured by the loop antenna to verify the correctness of the model. Based on the same modeling method, the calculation model of the 49-level MMC converter valve is established (Sun et al. 2016). The voltage waveform of the normal switching process of the submodule is used as the electromagnetic disturbance source, and the electromagnetic field of the near field and far field of the converter valve is calculated. But its voltage level is only 35 kV and cannot reflect the disturbance characteristics of the higher voltage level VSC-HVDC.

In this article, the calculation model of a ±800 kV UHV large-capacity VSC-HVDC converter valve is established, and the disturbance source characteristics and disturbance path are introduced in detail. Then the direct time domain method and direct frequency domain method commonly used in electromagnetic disturbance calculation are analyzed. Then, a new calculation method based on the complex geometric structure of the converter is proposed, which uses a DC bus current as the disturbance source. Finally, the antenna model of the converter is built, and the radiated electromagnetic disturbance characteristics of the VSC-HVDC converter valve are analyzed.

2 DISTURBANCE SOURCE ANALYSIS

Different from the generation mechanism, the disturbance source of a VSC-HVDC converter valve can be divided into the following three types (Yu et al. 2008).

- Continuous electromagnetic disturbance caused by the operation of the converter valve
- Electromagnetic disturbance caused by the corona of high-voltage equipment
- Electromagnetic disturbance caused by discharge of high-voltage equipment

For electromagnetic disturbance caused by the corona and discharge of equipment, it is generally assumed that the converter station design can meet the immunity requirements. When the above phenomenon occurs, the equipment can be guaranteed to be unaffected and operate normally. Electromagnetic disturbances caused by the corona and discharge are no longer considered.

The electromagnetic disturbance is mainly due to the transient voltage and current generated during the turn-on and turn-off of IGBTs inside the submodule, which forms a continuous disturbance noise that is the main disturbance source for the VSC-HVDC converter valve. In this case, the transmission line, the metal frame, and the wire structure can be regarded as the antenna structure, which can generate an the electromagnetic disturbance around under the excitation of transient voltage and current.

The propagation path of an electromagnetic disturbance is shown in Figure 1. The disturbance source in the valve hall can be regarded as composed of two parts. One part is the transient quantity during the turn-on and turn-off process of IGBT. Due to the shield of the MMC metal case, this part of the disturbance is hardly radiated to the external space. The other part is the transient quantity along the connecting metal plate between different MMCs, which is the main factor in electromagnetic disturbance in the valve hall.

The disturbance source outside the valve hall can also be regarded as consisting of two parts. One part is the electromagnetic disturbance through the valve hall, thus causing the disturbance in the external space. This part of the disturbance is closely related to the shielding effectiveness of the valve hall and needs to be suppressed. The other part is the transient

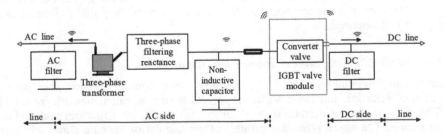

Figure 1. Disturbance path of converter system.

quantity in the valve hall along the main loop of the DC system. This part of the disturbance is mixed with the radio interference generated by the corona and ~~becoming~~becomes the main component of the electromagnetic environment near the station.

3 CALCULATION METHOD OF ELECTROMAGNETIC DISTURBANCE

3.1 *The direct time domain method*

Since the scatterers of the converter valve are all wires or surface conductors, the method of moments is a simple and feasible method to calculate the electromagnetic radiation. The conductor current of each section of the converter valve can be directly used to calculate disturbance by the radiation field formula. It is called the direct time domain calculation method (Zhao et al. 2010). This method takes the time domain current on the connecting conductors between the MMC modules as the excitation source, respectively calculates the radiation field of each segment conductor, and then obtains the electromagnetic disturbance level of the valve hall by superimposing. Specific steps are as follows:

- Currents of all wires are obtained by measurement or calculation.
- For the spatial point where the electromagnetic field needs to be calculated, the field strength of each segment current at the same time is calculated and superimposed.

In general, the Fourier transform of the current in the time domain is used to obtain its frequency domain response, and then the disturbance is calculated at different frequency points.

3.2 *The direct frequency domain method*

Although the time domain method is simple, it can only consider the radiation field generated by the current flowing through the wire, and it is not easy to establish the scattering field of other wires or metal structures in the valve hall, so the prediction of ~~a~~the high-frequency electromagnetic disturbance may have a large deviation. Therefore, it is necessary to find a more accurate calculation method from the perspective of the frequency domain.

Since the electromagnetic disturbance of the converter valve is generated by the IGBT in the submodule, the port voltage of the submodule can be used as an excitation source for electromagnetic field analysis. The specific steps are as follows:

- The voltage time domain $u(t)$ of the submodule port during normal operation is obtained through testing or simulation.
- Obtain the voltage $u_L(t)$ and current $i_L(t)$ on the outgoing side of AC and DC bussing of the valve hall.
- Transform each voltage and current time domain into the frequency domain. The submodule port voltage frequency domain result is $U(f)$. The model terminal voltage and current are $U_L(f)$ and $I_L(f)$ respectively, and the terminal load can be obtained as $Z_L(f)$.

Confirm the number and position of submodules to be switched, add the disturbance source $U(f)$ at both ends, and add the terminal load $Z_L(f)$. Then calculate the spatial radiated electromagnetic fields at each frequency point, and the electromagnetic field distribution at each frequency point f_0 is obtained.

3.3 *A new method for calculating electromagnetic radiation*

Comparing the foregoing two methods, the direct time domain method uses the current source as the excitation and the calculation process is simple, but it has a large calculation error. The direct frequency domain method can consider complex structures such as lines and metal structures. The model is more accurate, but the calculation speed is slow and the modeling process is more complicated. Considering these two calculation methods, this article establishes the complex geometry model of the converter valve by the software FEKO, and uses the DC bus current as the disturbance source to calculate the electromagnetic radiation. The calculation steps are as follows:

- Establish the model of the converter valve. To simplify the model, the part that enters the submodule is no longer considered, and all are modeled by a straight-through metal plate.
- Connect the DC bus with the unit current as the disturbance source at the focus frequency. The current injects from one end and extracts from the other end.
- Specify the calculation point or the calculation area, and carry out the calculation to obtain the radiation field generated by unit excitation considering the metal scatterer in the valve hall.
- Fourier transform the DC bus current obtained by measurement or simulation calculation to obtain the current values of different frequencies. Multiply the current value of the focus frequency with the radiation field results obtained in step 4 to obtain the final electromagnetic radiation.

4 MODELING AND ELECTROMAGNETIC DISTURBANCE CHARACTERISTIC OF A ±800 KV VSC-HVDC CONVERTER VALVE

4.1 Model of converter valve

In this article, a model is established in FEKO with reference to the valve tower geometry of a ± 800 kV VSC-HVDC project in China. The modeling process adopts the following principles: (a) the valve arm adopts the thin wire model with series impedance; (b) the metal frame adopts the pure conductor plate model; (c) the connecting wire adopts the thin wire model; and (d) AC and DC outgoing lines adopt the voltage source model of series impedance. In order to simplify the analysis, the wire radius in the model is always 0.01 m, and the metal frame is replaced by a metal surface conductor, regardless of the thickness of the metal frame. The calculation models for a single valve tower and valve hall are shown in Figures 2 and 3.

4.2 Radiation electromagnetic disturbance characteristic

Based on the calculation method in Section 3.3 and the model built in Section 4.1, this section uses the unit current as the radiated disturbance source, and calculates the electromagnetic radiation characteristics of the ±800 kV UHV large-capacity VSC-HVDC converter valve. The result sets up the preliminary work foundation for the radiated electromagnetic disturbance calculation after obtaining the actual current.

Accordingly, the electromagnetic disturbance of all valve towers in the valve hall shown in Figure 3 is calculated. The spatial electric field and magnetic field distribution at a height of

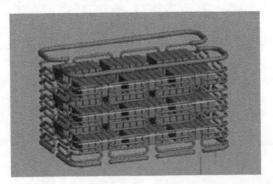

Figure 2.   Calculation model of a single valve tower.

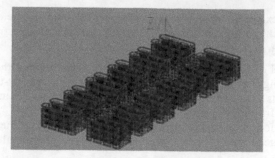

Figure 3.   Calculation model of all valve towers in a valve hall.

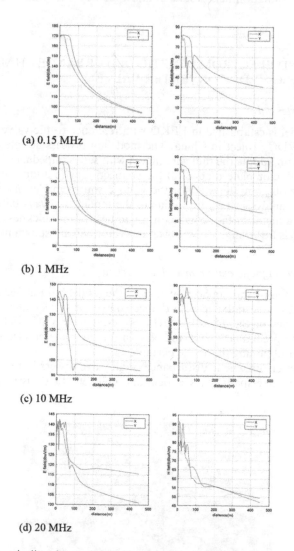

(a) 0.15 MHz

(b) 1 MHz

(c) 10 MHz

(d) 20 MHz

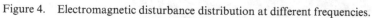

Figure 4.   Electromagnetic disturbance distribution at different frequencies.

1.5 m from the ground are given in the mutually orthogonal x and y directions, within a range of 450 m. As shown in Figure 4, the left side is the distribution of the radiated electric field and the right side is the distribution of the radiated magnetic field.

As can be seen from Figure 4, the electric field strength is numerically much larger than the magnetic field strength. In fact, various electromagnetic compatibility standards are also concerned with the limits of the radiated electric field. At low frequencies, the distribution of the radiated electric field in different directions is almost the same, but above 10 MHz, the radiation electric field in different directions appears to be quite different. At the same time, both the radiated electric field and the magnetic field strength at low frequencies decrease with increasing distance. But as the frequency increases, obvious fluctuation of the distribution appears in the area closer to the valve tower. The reason is that with the increase of the number of valve towers, there will be mutual interference of electromagnetic radiation between different valve towers.

5 CONCLUSION

In this article, the electromagnetic radiation model of a ±800 kV VSC-HVDC converter valve is constructed, and the main disturbance source characteristics are analyzed. Then, a method to calculate the electromagnetic disturbance using a DC bus current as the excitation is proposed, and the relationship of disturbance to distance at different frequencies is analyzed.

(1) The transient quantity in the IGBT turn-on and turn-off process propagates along the connecting metal plate between different MMCS and the DC main loop system, which is the main disturbance source.

(2) The frequency domain calculation method combined with the DC bus current as the source is applied to calculate the radiation electromagnetic disturbance.

(3) Mutual interference between different valve towers will cause fluctuations in electromagnetic radiation near the valve tower at low frequencies.

REFERENCES

Konstantinou, G. S., Ciobotaru, M., & Agelidis, V. G. 2011. Operation of a modular multilevel converter with selective harmonic elimination PWM. *International Conference on Power Electronics and Ecce Asia*. IEEE, 999–1004.

Liu, B., Ruan, J., Han, H., et al. 2009. Analysis of electromagnetic environment of VSC-HVDC. *High Voltage Engineering* 35(11): 2747–2752.

Liu, J., He, W, & Bao, H. 2008. Study on the technology of HVDC based voltage source converter and its application prospect. *Distribution & Utilization* 25(1): 6–9.

Liu, Y., Sebo, S. A., & Caldecott, R. 2002. Modeling of converter transformers using frequency domain terminal impedance measurements. *IEEE Transactions on Power Delivery* 8(1): 66–72.

Sun, H., Du, L., & Liang, G. 2015. Calculation of radiated electromagnetic disturbance of VSC-HVDC converter system. *High Voltage Engineering* 41(12): 4176–4183.

Sun, H., Du, L., & Liang, G. 2016. Antenna model of MMC-HVDC converter valve system and its radiated electromagnetic disturbance analysis. *Proceedings of the CSEE* 36(3): 879–888.

Tang, X., Zhang, W., & Zeng, X. 2014. Resonance problem analysis and active suppression method for VSC-HVDC transmission system. *Proceedings of the CSEE* 34(30): 5352–5359.

Xu, Z., Xue, Y., & Zhang, Z. 2014. VSC-HVDC technology suitable for bulk power overhead line transmission. *Proceedings of the CSEE* 34(29): 5051–5062.

Yu, Z., He, J., Zeng, R., et al. 2008. Characteristics of major EMD sources in HVDC converter stations. *High Voltage Engineering* 34(5): 898–902.

Zhao, Z., Cui, X., & Wang, Q. 2010. Analysis of electromagnetic disturbance from valve hall in converter station. *High Voltage Engineering* 36(3): 90726–90731.

Emerging Developments in the Power and Energy Industry – Dufo-López, Krzywanski & Singh (eds)
© *2020 Taylor & Francis Group, London, ISBN 978-0-367-27169-5*

Research on multi-resource co-control method and defense system for bundled wind-thermal power transmission

Shiying Ma, Xuetao Yang, Dunwen Song, Qing Wang, Dezhi Chen & Jian Zhang
China Electric Power Research Institute, Haidian District, Beijing, China

Ningbo Wang
Wind Power Technology Center of State Grid Gansu Electric Power Corporation, Lanzhou, Gansu Province, China

ABSTRACT: With large-scale new energy centralized access, power grid is in the difficulty of new energy consumption and stable operation. Firstly, the stability characteristics and risk of terminal bundled wind-thermal power grid are analyzed. A system frame of security defense system for bundled wind-thermal power transmission is proposed. The design objectives, the four-tier deployment architecture of the defense system and supporting strategies including the frequency coordination control and voltage coordination control are emphasized. The system aims to use the whole network controllable resources to reduce the frequency and voltage fluctuations caused by large-scale new energy, and to block the path of cascading failures, then hence to ensure the safe operation of power network.

Keywords: bundled wind-thermal, multi-resource co-control, security defense system

1 INTRODUCTION

In recent years, China has carried out the construction of several tens of kilowatt-class wind power bases in the wind energy resource-rich areas such as Gausu, Xinjiang, Ningxia, and a high proportion of new energy delivery systems have gradually been formed. In the coming years, large-scale new energy development will still be in a stage of rapid development.

The above large-scale new energy bases are located at the end of the power system with less local load. New energy can't be fully absorbed locally, and needs to be sent out from a long distance (XIAO Ch. & WANG N. 2010, TANG Y. & ZHAO L. 2013). However, the randomness and fluctuation of wind power make it difficult to control wind power delivery. In addition, the annual utilization hours of wind power are low, so the economy of long-distance transmission alone is relatively poor. These problems seriously affect the stable operation of the power system (TU J. & ZHANG J. 2015, WU P. & CHEN H. 2016, GUO X. & MA S. 2012). For the above reasons, wind power and nearby thermal power bundled through UHVDC transmission can achieve smooth output, which greatly improves the utilization ratio of transmission channels.

However, in the case of large-scale transmission of new energy, on the one hand, the uncertainty of wind power may lead to millions of kilowatts of random power shortage. The failure of DC commutation easily leads to overvoltage of the system and triggers large-scale off-grid of new energy. DC blocking will cause a large-scale shift of power flow, which is likely to cause the power of critical channels to exceed the limit. On the other hand, the inertia level of new energy is relatively low, the capacity of frequency/voltage regulation is insufficient, and the performance of network source coordination has not been significantly improved.

Therefore, frequency/voltage fluctuation and source-network coordination caused by multiple kinds of impulses constitute the main contradiction of security and stability of large-scale new energy transmission grid (ZHENG Ch. & TANG Y. 2014, CHEN Zh. & CHEN Y. 2011).

In view of the above problems, this paper analyses the operation characteristics and risks of wind-thermal bundled AC/DC transmission system. A security defense system for wind-fire bundled energy base based on multi-resource cooperation is proposed. The key technologies of frequency coordination and voltage coordination strategies are elaborated. The system aims to coordinate the dispatching of wind, fire, FACTS and other system resources, and exploit the potential available measures, resources, information or improvement methods of the system to comprehensively improve the safe and stable operation level and comprehensive defense capability of the power grid.

2 CHARACTERISTICS OF HIGH PROPORTION NEW ENERGY DELIVERY GRID

2.1 Frequency fluctuation/overvoltage caused by large-scale wind power disconnection

Fault disturbances such as DC blocking, DC commutation failure, and short circuit are easy to cause large-scale off-grid of new energy. At this time, the system's active power imbalance is intensified, causing a large fluctuation and decrease of the system frequency. During the low-voltage process of DC commutation failure, the fan output drops, which easily causes the voltage around the wind farm to climb. Taking a typical mode of Gansu power grid as an example, the simulation shows that if there is a fault of 3629MW wind turbine off-grid in the wind farm, the voltage fluctuation of the 800kV bus of Liuyuan, Qiaowan and Dunhuang substations is shown in Figure 1. If there is no timely and effective adjustment measure, it will affect the safe operation of the system.

2.2 The impact of new energy power on transient stability

When DC blocking or N-2 serious AC fault occurs, the power will be transferred over a wide range of channels, and unbalanced energy will result in transient stability problems. When the wind turbine is close to the fault oscillation center, the wind turbine output change has a great influence on the transient stability, that is, the system transient stability deteriorates with the increase of the wind power output ratio.

Taking a typical mode of the Northwest power grid of China as an example, the simulation shows that the system is unstable when the N-2 fault occurs in the "Shazhou-Yuka" line under the four modes. However, with the reduction of the proportion of wind power and wind power output (that is, from mode one to mode four), the degree of power swing of the Xinjiang unit is gradually weakened, as shown in Figure 2.

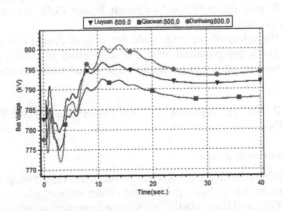

Figure 1.   Voltage fluctuation under the fault of fan off the grid.

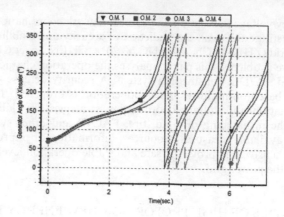

Figure 2.   Transient power angle instability of Xinsaier.

2.3   *Increased risk of cascading failure*

The main reason for the serious fault development process comes from two aspects. First, the failure of the large-capacity DC commutation failure causes the frequency and synchronization stability problems. Second, during the AC-DC fault process, the transient reactive power shock caused by the sudden drop of DC power leads to voltage stability and fan disconnection. Coupled with the weak anti-disturbance ability of new energy, if there are no timely and reasonable joint prevention measures, it will bring the risk of cascading failure.

3   PROBLEMS WITH TRADITIONAL METHODS OF STABILITY CONTROL

In the face of the continuous growth of new energy capacity, DC failure and other factors, traditional stability control methods have begun to show its limitations, mainly reflected in three aspects.

(1) The existing stability control system can only solve the local stability problem, and lacks inter-regional or overall considerations. It is difficult to control and block the fault under the large-capacity impact.

(2) The types of existing stable control measures are mainly limited to cutting machines and a small amount of load shedding. It is difficult to fully and effectively mobilize and coordinate potential supporting control resources to jointly improve the grid's ability to resist failures.

(3) There is a lack of coordination between the existing stability control systems, and multiple stability control systems have the risk of disordered actions. At present, the stability control devices of 220 kV and above in the Northwest Power Grid have reached 786 sets, but there is less collaborative control information.

In order to ensure the safety of the large-scale power grid during the transition period, it is necessary to fully dispatch the potential available supporting resources, and build a security defense system for wind-thermal power transmission based on multi resource collaboration.

4   MULTI-RESOURCE COLLABORATIVE DEFENSE SYSTEM FRAMEWORK CONSTRUCTION

4.1   *System deployment architecture*

The multi-resource collaborative defense system implements closed-loop control according to the 4-layer architecture, which includes the control center station, the co-control main station, the co-control sub-station, and the execution station (information acquisition layer). The overall structure of the system and the information interaction relationship are shown in Figure 3.

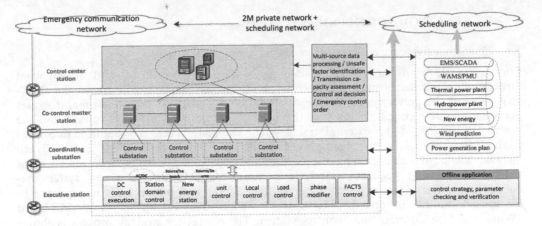

Figure 3. The framework wind & thermal energy transmission security defense system.

The system makes full use of the existing dispatch automation system software and hardware resources and information resources. The networking mode between the internal control layers is composed of a scheduling data network and a 2M emergency communication private network. The sending of the real-time data collection and the non-emergency control command are sent based on the scheduling data network. The control center station communicates directly with the coordinated control master station, the control substation, and the execution station. The emergency control command is sent to the delivery and is quickly completed through the private network.

4.2 Main function of the system

(1) Description of station layer function

The control center station is located in the provincial dispatch center and interfaces with information such as grid dispatching support system, wind forecasting system, and stability control device. The control center station performs security and stability analysis and global coordinated control, implements auxiliary decision-making, and sends the online control strategy to the coordinated control primary station.

The main control station is located in the 750kV substation of the network. The main station accepts the online control strategy of the control center station, summarizes and uploads the measured information, and combines the information of other coordinated control master stations to perform regional control and coordinated control by adding control strategies.

The control substation is located in the main 330kV substation of the grid. The substation uploads the measured information of the local station and the execution station, receives and forwards the control command of the coordinated control master station, calculates the control amount according to the control object, controls the own station and the associated execution station, and implements local control.

The implementation station is located in the wind farm/photovoltaic power plant collection booster station, thermal power plant and DC converter station. The implementation station issues a control command and uploads the measured information of the station.

(2) Communication network

A 2M dedicated channel is used between the control units. The uplink information mainly includes the real-time status of the grid related to the master station strategy. The downlink information mainly includes emergency control commands. The control center station sends control commands in the manner of Coordinating Master Station -> Control Substation -> Execution Station.

5 KEY STRATEGIES OF DEFENSE SYSTEM

5.1 *Emergency frequency coordinated control for wind power off-grid*

For large-scale wind-thermal bundle delivery systems, large-scale off-grid of the wind turbine will cause the frequency to decrease, and in severe cases, it will exceed the allowable value of the power system frequency deviation ± 0.2Hz. At the same time, it causes many unfavorable conditions such as the increase of the main network voltage and the power fluctuation of the external transmission channel. When the traditional control method is applied, the dispatching center will notify the power plant artificially to increase the power after finding the fault. This method has a large delay and cannot reduce the transient frequency offset. In response to this situation, a new system frequency control strategy for emergency boosting of thermal power units in response to fan off-grid failures is applied. This strategy is achieved by establishing a WAMS-based "Wind and Fire Coordination" control system. The emergency strategy process is shown in the following Figure 4.

When a group fan blowout failure occurs, the method performs the following steps.

1. Determining the amount of fan off-network.
2. Determining the available spin reserve of the thermal power unit.
3. Judging the start condition of the emergency power boosting allocation strategy of the thermal power unit.
4. Determining the emergency power boost of the thermal power unit.
5. Issuing the emergency power boost command of the thermal power unit.

Therefore, the task of responding to the off-grid of the fan for emergency power generation of thermal power can be realized.

5.2 *Voltage coordinated control based on multiple FACTS coordination*

Firstly, the HSVAC method is proposed. The method adds the voltage deviation signal of the high voltage side of main transformer and then controls the excitation of the generator. Broadly speaking, the voltage signal may also be taken from any "virtual" point between the exiting end and the high side without being limited to the high side bus voltage. According to the HSVAC control principle, each unit operates at approximately the unit's adjustment rate, significantly increasing the static stability limit of the power system and the excitation system's support for the grid voltage.

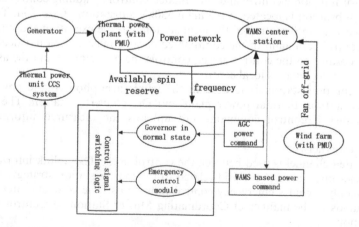

Figure 4. Emergency control architecture diagram of thermal power unit to deal with fan off grid.

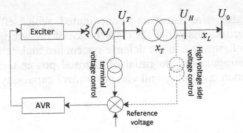

Figure 5. HSVAC wiring diagram.

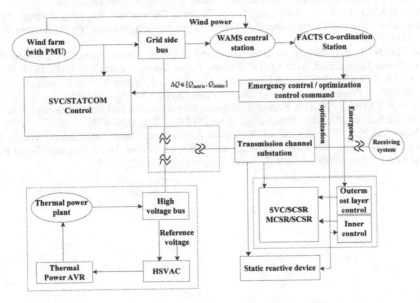

Figure 6. Voltage coordinated control based on multiple FACTS coordination.

When the voltage control is performed, the wind farm SVC is adjusted, so that the reactive exchange between the wind farm and the main network is in a smaller interval, and a reasonable reactive reserve is reserved. When the voltage fluctuates greatly, it is necessary to control the high-resistance of the bus line of the external transmission channel, the controllable high-resistance of the line, and the substation SVC. When the voltage fluctuates smoothly, it is necessary to operate the low-voltage compensation equipment, so that the reactive power distribution under the premise of voltage qualification is optimized and reasonable. For the overvoltage problem caused by the fan off-line and DC blocking, the dynamic reactive reserve of SVC, SVG and synchronous condensers are released through the partition to block the abnormal voltage reaction.

6   CONCLUSION

Aiming at the risk of the operation of the delivery system of the wind-fired energy base and the shortage of existing prevention and control measures, a four-layer closed-loop control system covering the multi-resource collaborative control objectives is proposed. Through the emergency communication private network, the network source control and voltage coordination control strategy are issued and coordinated, which effectively responds to the voltage and frequency fluctuations under the disturbance of the wind. The two types of coordination control policy configuration in the defense system are highlighted. The proposed strategy can fully exploit the regulatory potential of thermal power units, improve the level of source network coordination capability and voltage control capability in real time.

REFERENCES

XIAO Ch. & WANG N. 2010. Power Characteristics of Jiuquan Wind Power Base. *Automation of Electric Power System* 34(17): 64–67.

TANG Y. & ZHAO L. 2013. Impact of Wind Power Penetration on Angle Transient Stability of Wind-thermal Combined System. *Automation of Electric Power System* 37(20): 34–40.

TU J. & ZHANG J. 2015. Study on Wind Turbine Generators Tripping Caused by HVDC Con-tingencies of Wind-Thermal-Bundled HVDC Transmission Systems. *Power System Technology* 39(12): 3333–3338.

WU P. & CHEN H. 2016. Study on Interaction and Stability Characteristics of Bundled Wind-PV-Thermal Power Transmitted With AC/DC System. *Power System Technology* 40(7): 1934–1942.

GUO X. & MA S. 2012. HVDC Grid Connection Schemes and System Stability Control Strategies for Large-scale Wind Power. *Automation of Electric Power System* 36(15): 107–115.

ZHENG Ch. & TANG Y. 2014. A Survey on Typical Scenarios and Technology Needs for HVDC Participated into Stability Control. *Proceedings of the CSEE* 34(22): 3750–3759.

CHEN Zh. & CHEN Y. 2011. A Control Strategy of Active Power Intelligent Control System for Large Cluster of Wind Farms Part Two Coordination Control for Shared Transmission of Wind Power and Thermal Power. *Automation of Electric Power System* 2011, 35(21): 12–15.

Emerging Developments in the Power and Energy Industry – Dufo-López, Krzywanski & Singh (eds)
© 2020 Taylor & Francis Group, London, ISBN 978-0-367-27169-5

An improved phase current reconstruction strategy for brushless DC motors with single current sensor

Qian Zhang[1], Ming Feng & Yi sheng Fan
Beijing University of Science and Technology, Beijing, China

ABSTRACT: Phase current reconstruction using single current sensor (PCRSS) in conventional brushless DC motors can be used for the purposes of cost reduction and hardware imbalance elimination, etc. In this paper, the phase currents with the unipolar PWM modulation are reconstructed successfully. In particular, an algorithm of reconstructing the current of non-conducting phase is proposed for the first time. Moreover, the effect of freewheeling current is considered in detail. The accuracy of the proposed PCRSS with unipolar PWM is verified by the experiments.

Keywords: single current sensor, phase current reconstruction, BLDC, unipolar PWM

1 INTRODUCTION

The brushless direct current (BLDC) motors have been widely used for various industrial applications because of their high efficiency, low maintenance, greater longevity and compact construction. Most BLDC motor drives monitor the operation of electrical machines and maintain the current at the required level through a current control loop (Cui C & Liu G & Wang K 2015). Classical implementation of the current control loop requires current information provided by measurement of the winding currents, and the simplest method to obtain full information about three phase currents is to measure them directly by three current sensors. However, using three current sensors and the associated accessories increase the complexity, cost and size of a motor drive and will suffer from the problem of unbalanced sensor signals caused by hardware imbalance. To reduce the cost, in some drives, two of three phase currents are measured by two current sensors, and the third phase current is calculated based on Kirchhoff's current law and the detected currents. However, the imbalance caused by hardware still exists. To eliminate the imbalance in hardware and reduce the cost further, current reconstruction method using single current sensor was proposed based on the relationship between the DC-link current and phase currents, which requires only one current sensor or one shunt resistor in the DC-link. Because the balanced current signal and reduced hardware cost can be achieved by the PCRSS method, the PCRSS method is becoming more and more attractive for low-cost systems which is sensitive to the cost and high torque precision control systems that require high current detection accuracy.

Several PCRSS methods has been proposed for common used electrical machines, such as BLDC motors, induction motors and permanent magnet synchronous motors, etc. The basic concept of deducing motor winding currents from a single sensor has already been introduced since 1980s. Even though a lot of researches have been conducted on it, the PCRSS methods for reconstructing phase currents of BLDC motors are still immature. Song J H and Feyzi M R proposed two new methods to reduce the torque fluctuation by the current close-loop control, but they are

---

[1]Presently with JinShiDun Bearing Technology Co., Ltd(ZCJSD), Hebei, China

used to control the synthetic DC-link current and can't be used to reconstruct the phase current directly (Song J H & Feyzi M R 2004, Feyzi M R & Ebadpour M & Niapour S A K H M 2011). What's more, they can't take the influence of freewheeling current into account when commutation occurs. Nikolic M proposed a new hysteresis current control method to regulate the current at the required level, it can make the current waveform closer to square wave, but still can't obtain the freewheeling current circulating in the inverter (Nikolic M & Kovacevic J& Pjevalica N 2013). P. P. Acarnley presented one method of applying a series of observability criteria to estimating the phase currents (P. P. Acarnley 1993). The author's another paper presented a real-time current estimation method for brushless dc drives using a single DC-link current sensor (French C D & Acarnley P P & Jack A G 1993), but this method can be only used in a particular PWM mode and the current in non-conducting phase in unipolar PWM mode is neglected. Besides, there is a new topology to measure the freewheeling current during commutation, but it will increase the complex of hardware greatly, which makes it hard to be applied in industry, what's more, the current in non-conducting phase in unipolar PWM mode is neglected too. C. Xia proposed a reconstruction method for four-switch three phase BLDC motors (C. Xia & Z. Li & T. Shi 2009), but it can only be used in the drives with the unique four-switch topology and can't be used in the six-switch three phase BLDC motors drives.

In this paper, a reconstruction method is specifically developed for the unipolar PWM modulation and the current of non-conducting phase caused by unipolar PWM is taken into account for the first time. In addition, the effect of freewheeling on the reconstruction model current is also considered.

2 ANALYSIS

2.1 *Mathematical model of BLDC motors*

The control of BLDC motors is realized based on six-step mode and there is a commutation process between two adjacent steps. Therefore, the phase currents are divided into *conduction stage* and *commutation stage* in one current cycle. In the following analysis, three-phase and two-pole BLDC motor is used. The equivalent circuit and the back-EMF waveforms with 120° conduction mode are shown in Figure 1 (a) and (b), respectively. The unipolar PWM modulation with chopping at the inverter is selected as the voltage regulation method.

Ignoring the voltage drop caused by power switches and freewheeling diodes, the stator voltage equations of a typical BLDC motor can be represented as

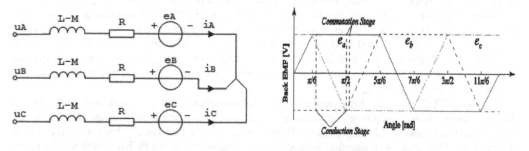

(a) BLDC motor   (b) Waveforms of back EMF

Figure 1.   Equivalent circuit of the BLDC motor and Waveforms of back EMF.

$$\begin{bmatrix} U_a \\ U_b \\ U_c \end{bmatrix} = \begin{bmatrix} r & 0 & 0 \\ 0 & r & 0 \\ 0 & 0 & r \end{bmatrix} \begin{bmatrix} i_a \\ i_b \\ i_c \end{bmatrix} + \begin{bmatrix} L_s & 0 & 0 \\ 0 & L_s & 0 \\ 0 & 0 & L_s \end{bmatrix} \frac{\mathrm{d}}{\mathrm{d}t} \begin{bmatrix} i_a \\ i_b \\ i_c \end{bmatrix} + \begin{bmatrix} e_a \\ e_b \\ e_c \end{bmatrix} + \begin{bmatrix} U_N \\ U_N \\ U_N \end{bmatrix}, \qquad (1)$$

where U_a, U_b and U_c represent the terminal voltage of three phases; i_a, i_b and i_c represent the phase currents; e_a, e_b and e_c represent the back electric motive force (EMF); U_N is the voltage of neutral point; L_s is the phase inductance and equal to L-M, L is the self-inductance and M is the mutual inductance.

2.2 Phase current reconstruction

There are several unipolar PWM modulation methods, such as PWM_on, on_PWM, Hon_LPWM and HPWM_Lon, however, there are only two possible conditions for any unipolar PWM method at any time, upper PWM or lower PWM. The following is based on upper PWM. Meanwhile, because six steps of BLDC motors are equivalent, the conduction stage of A-B (the current flows in from phase A and out from phase B) and commutation stage from A-B to A-C are taken as an example.

2.2.1 Conduction stage

The phase currents during PWM on state are shown in Figure 2 and reconstructed as

$$i_a = I_{dc}$$
$$i_b = -I_{dc}. \qquad (2)$$
$$i_c = 0$$

The currents during PWM off state are shown in Figure 3. It can be seen that the currents circulate in the inverter and can't be detected directly from DC-link. When the back EMF of non-conducting phase (phase C) is positive, i_c is equal to zero (Figure 3a); but when it is negative, current of non-conducting phase will occur (Figure 3b).

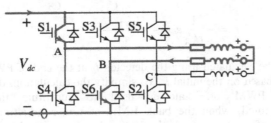

Figure 2.   Currents of conduction stage, PWM on state.

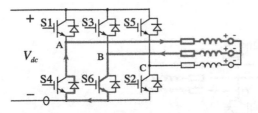

(a) Positive back EMF of non-conducting phase

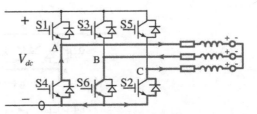

(b) Negative back EMF of non-conducting phase

Figure 3.   Currents of conduction stage, PWM off state.

As shown in Figure 3(a), when the back EMF of phase C is positive, the phase currents can be calculated as

$$i_a = -\frac{E}{r} + \frac{1}{rC}e^{-\frac{rt}{L_s}}$$
$$i_b = \frac{E}{r} - \frac{1}{rC}e^{-\frac{rt}{L_s}} \tag{3}$$
$$i_c = 0$$

The constant C can be calculated using the detected i_a or i_b at the end of PWM on state.

When the BEMF of phase C is negative, as shown in Figure 3(b). The voltage of neutral point can be gotten as

$$U_N = -\frac{1}{3}(e_a + e_b + e_c) \tag{4}$$

where e_a equals to E and e_b equals to $-E$, the value of e_c can be calculated based on the following equation

$$e_c = E - \frac{6\omega t_i E}{\pi} \tag{5}$$

where, the value of t_i begins to increase from zero when a new commutation instruction is given.

The value of phase currents can be calculated based on the voltage equations of phase A and phase C as

$$i_a = \frac{1}{rC_1}e^{-\frac{rt}{L_s}} - \frac{U_N + e_a}{r}$$
$$i_b = \frac{2U_N + e_a + e_c}{r} - \frac{1}{rC_1}e^{-\frac{rt}{L_s}} + \frac{1}{rC_2}e^{-\frac{rt}{L_s}} \tag{6}$$
$$i_c = -\frac{1}{rC_2}e^{-\frac{rt}{L_s}} - \frac{U_N + e_c}{r}$$

The constant C_1 can be calculated using the detected i_a at the end of PWM on state. Constant C_2 can be calculated based on the initial i_c. To simplify the calculation, the averaged value of e_c in the range of PWM off state is used to substitute the real time value.

What should be noted, when the back EMF of non-conducting phase (phase C) is negative and PWM off state is over, the induced current in phase C during PWM off state can't disappear immediately. Therefore, during PWM on state, the phase currents before current of phase C drops to zero are shown in Figure 4.

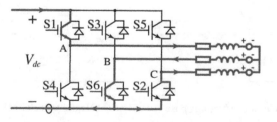

Figure 4. Currents of conduction stage, PWM on state, before non-conducting phase current dropping to zero.

The calculated phase currents as follows

$$i_a = I_{dc}$$
$$i_b = -I_{dc} - \frac{1}{rC}e^{-\frac{rt}{L_s}} + \frac{U_N + e_c}{r}$$
$$i_c = \frac{1}{rC}e^{-\frac{rt}{L_s}} - \frac{U_N + e_c}{r} \qquad (7)$$

where, the constant C can be calculated based on the i_c at the end of PWM off state. To simplify the calculation, the averaged value of e_c in the range of PWM on state is used to substitute the real time value.

2.2.2 Commutation stage

The currents of PWM on state and PWM off state are shown in Figure 5 and Figure 6 respectively. During PWM on state, the phase currents can be reconstructed as

$$i_a = \frac{U_{dc}}{r} - \frac{e_a + U_N}{r} + \frac{1}{rC}e^{-\frac{rt}{L_s}}$$
$$i_b = I_{dc} - \frac{U_{dc}}{r} + \frac{e_a + U_N}{r} - \frac{1}{rC}e^{-\frac{rt}{L_s}} \qquad (8)$$
$$i_c = -I_{dc}$$

where, the constant C can be calculated using the detected i_a at the end of PWM off state.

As shown in Figure 6, during PWM off, i_b can be detected directly, therefore the constant C can be calculated using the detected i_b at the end of PWM off. Therefore, the phase currents can be reconstructed as

$$i_a = \frac{1}{rC}e^{-\frac{rt}{L_s}} - \frac{U_N + E}{r}$$
$$i_b = I_{dc} \qquad (9)$$
$$i_c = -\frac{1}{rC}e^{-\frac{rt}{L_s}} + \frac{U_N + E}{r} - I_{dc}$$

where, the constant C can be calculated using the detected i_a at the end of PWM on state.

3 EXPERIMENTAL VERIFICATION

The experiments are implemented on an air compressor used in a fuel cell vehicle, which is shown in Figure 7(a). The microcontroller used in the drive is LPC2129, whose maximum frequency is 60 MHz. During reconstruction, the frequency of AD converter is 100 kHz and the accuracy is 10 bits.

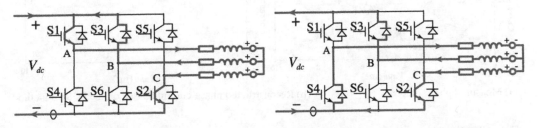

Figure 5. Currents of commutation stage, PWM on state.

Figure 6. Currents of commutation stage, PWM off state.

(a) Air compressor   (b) Motor drive

Figure 7.   The experimental devices.

The parameters are listed as follows:

Table 1.   Parameters of air compressor motors.

Motor parameters	Designed value	Unit
Rated power	12	kW
Rated speed	100 000	rpm
Pairs of pole	1	/
Line inductance	80	μH
Line resistor	40	mΩ
Rated voltage	250	V
Rated current	50	A

Figure 8 compares the measured currents of three phases with the reconstructed currents. It can be seen that the reconstructed current waveforms are in good agreement with the measured current waveforms. Table 2 further analyzes the numerical difference between the measured and reconstructed current of phase A, including the maximum and minimum values, the effective value and the root mean square error (RMSE). The ratio of the maximum and minimum errors to measured currents are 4.1% and 0.8%, respectively, and the ratio of RMSE to the effective value is only 5.7%. Furthermore, the effective value of the measured and reconstructed currents, which is the most important index to the motor, has almost no difference. Therefore, the three-phase current reconstruction algorithm proposed in this paper for the upper

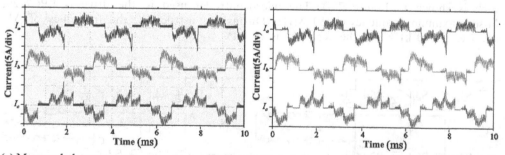

(a) Measured phase currents   (b) Reconstructed phase currents based on the PCRSS method

Figure 8.   Experimental comparison for upper PWM modulation.

Table 2. Current of phase A comparison for upper PWM modulation.

	Maximum value (A)	Minimum value (A)	Effective value (A)	RMSE (A)
Measured current	7.3	-12.1	2.8	0.16
Reconstructed current	7.0	-12.2	2.8	
Current error	0.3	0.1	0	RMSE/Effective value
Current error/Measured current	4.1%	0.8%	0	5.7%

PWM modulation not only can predict the three-phase current accurately, but also has good applicability.

4 CONCLUSIONS

With the unipolar PWM modulation, the current of the switch off phase which has been ignored in the traditional methods is calculated in detail. The freewheeling currents during the commutation stage are reconstructed based on an accurate calculation. Compared with the traditional methods, the proposed reconstruction algorithm expands the application range of PCRSS and can be applied to conventional BLDC motor drive systems without any extra hardware requirements.

ACKNOWLEDGMENT

This work was supported by the Ministry of Industry and Information Technology (MIIT) under Industrial Strong Foundation Engineering with the contract No. TC160A310.

REFERENCES

Acarnley PP. Observability criteria for winding currents in three-phase brushless DC drives[J]. IEEE transactions on power electronics, 1993, 8(3): 264–270.

C. Xia, Z. Li, and T. Shi, "A control strategy for four-switch three phase brushless DC motor using single current sensor," IEEE Trans. Ind. Electron., vol. 56, no. 6, pp. 2058–2066, Jun. 2009.

Cui C, Liu G, Wang K, et al. Sensorless drive for high-speed brushless DC motor based on the virtual neutral voltage[J]. IEEE Transactions on Power Electronics, 2015, 30(6): 3275-3285.

French CD, Acarnley PP, Jack AG. Real-time current estimation in brushless DC drives using a single dc link current sensor[C]//Power Electronics and Applications, 1993., Fifth European Conference on. IET, 1993: 445-450.

Feyzi MR, Ebadpour M, Niapour S A K H M, et al. A new single current strategy for high-performance brushless DC motor drives[C]//Electrical and Computer Engineering (CCECE), 2011 24th Canadian Conference on. IEEE, 2011: 000419-000424.

Nikolic M, Kovacevic J, Pjevalica N, et al. Real time FPGA implementation of brushless DC motor control using single current sensor[C]//Intelligent Systems and Informatics (SISY), 2013 IEEE 11th International Symposium on. IEEE, 2013: 23–27.

Song JH, Choy I. Commutation torque ripple reduction in brushless DC motor drives using a single DC current sensor[J]. IEEE Transactions on Power Electronics, 2004, 19(2): 312–319.

Operation optimization of an integrated energy system considering multiple redundant energy flow paths

Jianlin Yang, Mingxing Guo, Fei Fei & Aili Pang
State Grid Shanghai Municipal Electric Power Company economic and Technological Research Institute, Shanghai, China

Qifan Niu
Shanghai University of Engineering Science, Shanghai, China

ABSTRACT: To optimize the integrated energy management of industrial parks, this article presents an economic dispatch method considering the redundant energy flow paths of multiple devices. First, an energy center model is built based on the concept of an energy hub considering the real-time electricity price, hourly electricity, heat, and cooling load. The proposed model contains multiple energy conversion devices to provide a redundant energy flow path selection. Then, an optimal scheduling model is established. The new model takes the minimum cost of system operation as its objective and considers the constraints of equipment capacity and network power flow. An improved genetic algorithm is proposed to solve the new model. Finally, a coupling network containing five electrical nodes and four natural gas nodes is used as an example to validate the proposed method. Calculation results show that the model can optimize the selection of the energy flow path and reduce the operating cost.

1   INTRODUCTION

In order to solve increasingly serious energy and environmental problems, it is important to develop new energy sources and improve energy utilization efficiency. This introduces an urgent need for the integration and optimization of multiple energy sources (Hongbin 2016).

Multienergy complementarity can alleviate the disparity between supply and demand in some kinds of energy shortage through the complementarity of different energy sources. The concept of an energy hub is put forward in Geidl and Andersson (2007), which presents an approach for combined optimization of coupled power flows of different energy infrastructures such as electricity, gas, and district heating systems. Based on this theory, several energy center models with different structures are built in Carradore and Turri (2009). In Wei (2017), a multiobjective optimal power flow algorithm is proposed based on real-time electricity price, and the proposed model can minimize operation cost and pollution emission. In Zhaoyu (2017), the microgrid system is divided into three parts: supply side, conversion side, and load side. The aforementioned research results are based on the concept of the energy hub when analyzing and optimizing the dispatch of the electricity–gas system. However, the energy center model established considers too few devices. The cold and hot energy interactive equipment is not considered and the energy flow path is single.

Based on the above background, considering the demand response of real-time electricity price, this article first establishes a new energy center model. The new model contains multiple energy conversion devices based on energy hub theory and the redundant energy flow paths for collaborative optimization. Then, taking into account the correlation constraints, the energy optimal scheduling model is established. The genetic algorithm is used to solve the model. Finally, a coupling network containing five electrical nodes and four natural gas nodes is used as an example. The results show that the model can optimize the selection of energy flow paths and reduce operation costs.

2 INTEGRATED ENERGY SYSTEM MODEL

The electricity–gas coupling system model established in this article includes a power flow model, a natural gas pipeline network model, and an energy center model. The proposed model overcomes the shortcomings of the traditional model, such as neglecting the cooling load to simplify the calculation, containing limited energy flow path due to fewer conversion devices, etc. The new energy center model includes several energy conversion devices to provide flexible energy flow paths. At the same time, it is convenient to analyze the influence of equipment conversion efficiency and capacity on the energy distribution coefficient in an optimal dispatch.

2.1 Power flow model of power network

The power flow equations of the power network are established as follows:

$$P_{is,t} - U_{i,t} \sum_{j \in i} U_{j,t}(G_{ij} \cos \theta_{ij,t} + B_{ij} \sin \theta_{ij,t}) = 0 \tag{1}$$

$$Q_{is,t} - U_{i,t} \sum_{j \in i} U_{j,t}(G_{ij} \sin \theta_{ij,t} + B_{ij} \cos \theta_{ij,t}) = 0 \tag{2}$$

where $P_{is,t}$ $Q_{is,t}$ are respectively active and reactive power injected from outside at time t of node i; G_{ij} B_{ij} are respectively conductance and admittance of line section ij; $j \in i$ represents all nodes j directly connected to node i; and $\theta_{ij,t}$ is the voltage phase difference between node i and node j at time t.

2.2 Natural gas pipeline network model

Similar to Kirchhoff's current law in a circuit, the load of each node in a natural gas distribution network equals the total flow of inflow and outflow nodes.

$$L = AF \tag{3}$$

where L is the load vector of the load node, F is the flow vector of the branch, and A is the branch-node correlation matrix.

In a low-pressure pipeline network, the pipeline flow rate can be approximated as a function of the pipeline coefficient and pressure drop between pipelines:

$$f_n = S_{ij}\left(\frac{S_{ij}(p_i^2 - p_j^2)}{K_n}\right)^{1/2} \tag{4}$$

where f_n is the flow rate of the pipeline n; p_i, p_j are respectively pressure of the node i, j at both ends of the pipeline; S_{ij} is set equal to 1 when $p_i > p_j$, otherwise set equal to 0; K_n is the pipe constant of the pipeline.

2.3 Energy center model

Considering the actual demand for energy conversion equipment and multienergy load in industrial parks, this article constructs an energy center model as shown in Figure 1. In the model, electric energy and natural gas can meet users' demands for various energy loads through the conversion of coupling elements. The cold and hot energy can interact with each

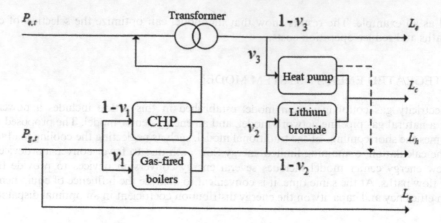

Figure 1. Energy center model.

other through a lithium bromide unit. The energy flow path can be characterized by an energy distribution coefficient. The electric-driven heat pump can change its working state according to the difference of heat and cold loads in winter and summer. So, it can improve the flexibility of system operation. The heat pump is applied to cooling in summer in this article.

The mathematical expression of the energy center model is established as follows:

$$\begin{bmatrix} L_{e,t} \\ L_{c,t} \\ L_{h,t} \end{bmatrix} = X \begin{bmatrix} P_{e,t} \\ P_{g,t} - L_{g,t} \end{bmatrix} \tag{5}$$

The matrix X is a coupling matrix, which consists of the energy distribution coefficient and the efficiency of energy conversion elements.

$$X = \begin{bmatrix} \eta_T(1-\nu_3) & \eta_A^e\eta_T(1-\nu_1)(1-\nu_3) \\ \eta_T\nu_3\eta_C & \eta_A^e\eta_T\eta_C\nu_3(1-\nu_1) + (\nu_1\eta_B + \eta_A^h(1-\nu_1))\eta_D\nu_2 \\ 0 & (\nu_1\eta_B + \eta_A^h(1-\nu_1))(1-\nu_2) \end{bmatrix} \tag{6}$$

From the coupling matrix, it can be seen that there is no direct coupling relationship between the heat load and the power supply, so the heat and cold loads can be combined into one.

The relationship between distribution coefficients ν_1 and ν_2 obtained from the equations is established as follows:

$$\nu_2 = 1 - \frac{L_h}{P_g(\nu_1\eta_B + \eta_A^h(1-\nu_1))} \tag{7}$$

Simplified mathematical expressions can be obtained from Equations (5), (6), and (7):

$$\begin{bmatrix} L_{e,t} \\ L_{c,t} + \eta_D L_{h,t} \end{bmatrix} = X^* \begin{bmatrix} P_{e,t} \\ P_{g,t} - L_{g,t} \end{bmatrix} \tag{8}$$

$$X^* = \begin{bmatrix} \eta_T(1-\nu_3) & \eta_A^e\eta_T(1-\nu_1)(1-\nu_3) \\ \eta_T\nu_3\eta_C & \eta_A^e\eta_T\eta_C\nu_3(1-\nu_1) + (\nu_1\eta_B + \eta_A^h(1-\nu_1))\eta_D \end{bmatrix} \tag{9}$$

where $L_{e,t}$, $L_{c,t}$, $L_{h,t}$ and $L_{g,t}$ are respectively electrical, cold, thermal, and natural gas load demand at time t; $P_{e,t}$, $P_{g,t}$ are electric, natural gas power purchased by energy center from

303

power grid and natural gas network at time t; η_T is the efficiency of the transformer; η^e_A and η^h_A arethe efficiency of generating and heating for the CHP unit, respectively; η_B is the thermal efficiency of gas-fired boilers; η_C and η_D are the energy efficiency coefficients of heat pumps and lithium bromide unit, respectively; v_1, v_2, and v_3 are energy distribution coefficients.

3 OPTIMAL DISPATCHING MODEL OF INTEGRATED ENERGY SYSTEM

3.1 Objective function

In this article, an optimal dispatching model of integrated energy system is constructed with the objective of minimizing the economic cost of system operation. The objective function is established as follows:

$$\min F = C_{e,t} P_{e,t}(v_{1,t}, v_{2,t}, v_{3,t}) + C_{g,t} P_{g,t}(v_{1,t}, v_{2,t}, v_{3,t}) \tag{10}$$

The objective function is divided into two parts: the cost of purchasing electricity and the cost of purchasing gas. $C_{e,t}$ and $C_{g,t}$ are the price of electricity and natural gas at time. $v_{1,t}$, $v_{2,t}$, and $v_{3,t}$ are respectively the energy allocation coefficients v_1, v_2, and v_3 of time t.

3.2 Constraint conditions

The model constraints mainly include the internal constraints of energy center, power system constraints, and natural gas pipeline constraints.

3.2.1 Internal constraints of the energy center

The output of energy conversion equipment is limited by the capacity of the equipment. All the output constraints of the equipment can be classified as the capacity constraints of the equipment. Therefore, according to the model structure, the internal constraints of the model can be obtained as follows:

$$P_{g,t}(1 - v_1) \leq S_A \tag{11}$$

$$P_{g,t} v_1 \leq S_B \tag{12}$$

$$\eta_T v_3 (P_{e,t} + P_{g,t} \eta^e_A (1 - v_1)) \leq S_C \tag{13}$$

$$v_2 (P_{g,t} \eta_B v_1 + P_{g,t} \eta^h_A (1 - v_1)) \leq S_D \tag{14}$$

$$0 \leq v_1 \leq 1$$
$$0 \leq v_2 \leq 1 \tag{15}$$
$$0 \leq v_3 \leq 1$$

$$0 \leq P_{e,t} \leq P_{e,\max}$$
$$0 \leq P_{g,t} \leq P_{g,\max} \tag{16}$$

where S_A is the capacity of the CHP unit, which does not consider the reactive power; S_B, S_C, and S_D are respectively the capacity of gas-fired boilers, heat pump, and lithium bromide units; and $P_{e,\max}$ and $P_{g,\max}$ are the upper limits for purchasing electricity and natural gas power, respectively, which are determined by the capacity of energy conversion equipment and the security constraints of the network.

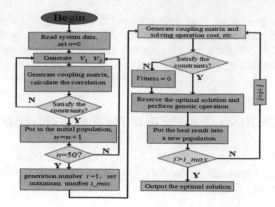

Figure 2.   Genetic algorithm flowchart.

3.2.2 *Power system constraints*

$$U_{i,\min} \leq U_{i,t} \leq U_{i,\max} \tag{17}$$

$$S_{ij,\min} \leq S_{ij,t} \leq S_{ij,\max} \tag{18}$$

where $U_{i,\max}$ and $U_{i,\min}$ are the upper and lower limits of node voltage amplitude respectively; $U_{i,t}$ is the voltage amplitude of the node i at time t; $S_{ij,\max}$ and $S_{ij,\min}$ are the upper and lower limits of the allowable power flow for line ij section respectively; and $S_{ij,t}$ is the power flow value of line ij section at time t.

3.2.3 *Natural gas pipeline network constraints*

$$P_{i,\min} \leq P_{i,t} \leq P_{i,\max} \tag{19}$$

$$f_{n,\min} \leq f_{n,t} \leq f_{n,\max} \tag{20}$$

where $p_{i,\max}$ and $p_{i,\min}$ are the upper and lower limits of node pressure respectively; $p_{i,t}$ is the voltage amplitude of the node i at time t; $f_{n,\max}$ and $f_{n,\min}$ are the upper and lower limits of the allowable pipeline flow for pipe n respectively; and $f_{n,t}$ is the pipeline flow value of pipe n at time t.

3.3 *Solving algorithm*

The genetic algorithm is one of the most popular modern optimization algorithms for solving power system optimization problems. The flow chart of the optimal scheduling model established by the genetic algorithm is shown in Figure 2.

4   APPLICATION

4.1 *Case study system*

In the coupling system in Figure 3, TS is the substation, AS is the gas source, and EH is the energy dispatching center. The voltage level of the distribution network is 10 kV. The natural gas pipeline network is the low-voltage distribution network. The gas source point pressure is 30 mbar and the power system and the natural gas system are coupled by EH. The real-time price of energy in a day is also shown in Figure 3.

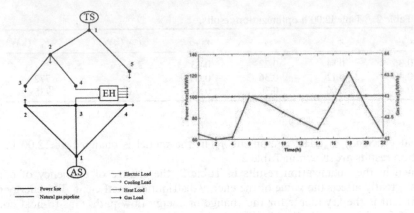

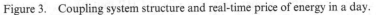

Figure 3.   Coupling system structure and real-time price of energy in a day.

4.2   Results and discussion

For convenience of analysis, the selected time step is 2 hours. After solving the proposed model, the energy allocation coefficients and real-time energy demand are obtained as shown in Table 1.

From the optimization results in Table 1, it can be seen that $v_{1,t}$ is higher between 0:00 and 04:00 h. This represents that most of the natural gas purchased by the system is used for heating of gas-fired boilers. The demand for heat load of the system is supplied mainly by gas-fired boilers, and the value of $v_{3,t}$ is higher throughout the day. This proves that electricity-driven heat pump is more economical for cooling at this time. The distribution coefficient v_1 at 06:00 h decreases significantly, $P_{e,t}$ decreases to 0, and $P_{g,t}$ increases from 400 kW to 800 kW. The reason is that the electricity price at 06:00 h increases from 62 $/MWh to 100 $/MWh. Inspired by the electricity price, the system chooses to buy more natural gas at a lower price. The gas turbine is used to generate electricity to meet the demand of electricity load. In addition, it can be seen that due to the increase of electricity price, the power purchased from 06:00 to 20:00 h is almost zero. Also, the natural gas power purchased increases greatly. The $v_{2,t}$ value rises while the $v_{3,t}$ value decreases. This indicates that the cooling of the electric-driven heat pump is no longer economical. At this time, the system operates mainly through lithium bromide units with lower operating costs. At 22:00 h, the electricity price is reduced from 90 $/MWh to 63 $/MWh. The cost of generating electricity is higher than that of purchasing electricity. The system chooses to purchase electricity again in order to meet most of the electricity load demand.

Table 1.   Optimization results.

Time (h)	$v_{1,t}$	$v_{2,t}$	$v_{3,t}$	$P_{e,t}$ (kW)	$P_{g,t}$ (kW)
0	0.83	0.21	0.14	240	342
2	0.85	0.26	0.12	234	407
4	0.86	0.32	0.11	228	472
6	0.20	0.37	0.06	0	849
8	0.13	0.35	0.06	0	784
10	0.08	0.34	0.07	0	747
12	0.06	0.30	0.07	0	735
14	0	0.29	0.08	15	654
16	0	0.33	0.04	0	664
18	0.02	0.29	0.08	0	705
20	0	0.31	0.06	0	763
22	0.85	0.22	0.12	232	419

Table 2. Time 12:00 h optimization results.

η_A^h	$v_{1,t}$	$v_{2,t}$	$v_{3,t}$	$P_{e,t}$ (kW)	$P_{g,t}$ (kW)
0.35	0.84	0.22	0.13	236	387
0.40	0.12	0.36	0.07	2	772
0.45	0.06	0.30	0.07	0	735

The heat efficiency of CHP equipment η_A^h in the model is changed at 12:00 h, and the optimization results are shown in Table 2.

As shown in the optimization results in Table 2, the change of efficiency of conversion equipment greatly affects the value of the energy distribution coefficient. The energy distribution coefficient is the key factor for the change of energy flow path. Gas-to-heat conversion efficiency (0.9) of gas-fired boilers is much higher than that of the CHP unit (0.45). Most of the natural gas purchased by the system is used in gas-fired boilers when the electricity price is low. However, there is the critical value for the distribution coefficient of energy limited by capacity. It means that the possible flow rate on the line cannot exceed the capacity of the terminal equipment and the capacity of the equipment can be selected according to the actual conditions.

From the analysis, it can be seen that the real-time electricity price system under the power market conditions is the main factor to drive the optimal dispatch of energy. The conversion efficiency of equipment is the key to the choice of energy flow path. Inspired by price, the system always chooses the most economical operation mode to save the cost for users.

5 CONCLUSIONS

In this article, the operation optimization of the integrated energy system considering redundant energy flow paths is studied. Based on this study, the following conclusions are obtained:

(1) An energy center model is established, which takes into account the independent cold and heat loads. The model also includes multiple energy conversion devices to provide flexible energy flow paths.

(2) In order to minimize the economic cost, an energy optimal dispatching model is established and solved by the improved genetic algorithm.

(3) Considering the energy optimal dispatching under real-time electricity price, the conversion efficiency of equipment is the key factor of energy flow path change. The capacity limitation of equipment provides a threshold for energy flow distribution.

REFERENCES

Carradore, L., & Turri, R. 2009. Modeling and simulation of multi-vector energy systems. *2009 IEEE Bucharest Power Tech Conference* 2009: 1–7.
Geidl, M., & Andersson, G. 2007. Optimal power flow of multiple energy carriers. *IEEE Transactions on Power Systems* 20(1): 145–155.
Hongbin, S. 2016. Energy management for multi-energy flow: challenges and prospects. *Automation of Electric Power Systems* 40(15): 1–8, 16.
Wei, L. 2017. Multi-objective optimal hybrid power flow algorithm for integrated local area energy system. *Proceedings of the CSEE* 37(20): 5829–5839.
Zhaoyu, C. 2017. Research on optimal day-ahead economic dispatching strategy for microgrid considering P2G and multi-source energy storage system. *Proceedings of the CSEE* 37(11): 3067–3077, 3362.

Emerging Developments in the Power and Energy Industry – Dufo-López, Krzywanski & Singh (eds)
© 2020 Taylor & Francis Group, London, ISBN 978-0-367-27169-5

Analysis of diesel engine pipeline vibration in a nuclear power plant

Jinchuan Bai, Weizhong Tang, Xiangyong Kong, Cong Jin, Zhangnan Xu, Liang Yuan & Shuzhou Li
Suzhou Nuclear Power Research Institute, Shenzhen, Guangdong, China

ABSTRACT: As important equipment of nuclear power plants, diesel engines perform the function of emergency power supply in case of power loss in the nuclear power plant. Taking the vibration problem of the fuel supply circuit of the diesel engine system of a nuclear power plant as an example, the cause of pipeline vibration is found by means of dynamic time-history analysis using PEPS 6.0, and the specific solution is given. At the same time, the stress of the pipeline under load conditions such as dead weight, pressure, thermal expansion, and earthquake is calculated. Finally, a general solution method is summarized to solve the problem of diesel pipeline vibration.

Keywords: diesel engine, PEPS, pipeline, vibrationPEPS

1 INTRODUCTION

In the process of conducting a vibration survey of the auxiliary pipeline of the diesel engine system in a nuclear power plant, it was found that the oil returning pipeline generally had excessive vibration conditions. Slight vibrations can be handled by the rigidity of the pipeline and the fixed brackets without plastic deformation to the pipeline. If the vibration of a diesel engine pipeline occurs with high frequency and large amplitude, the connection part between the pipeline and the components and the connection part between the pipeline and the support will generate a large alternating stress, which will cause fatigue damage of the pipeline, manifested mainly by cracking of the weld seam, deformation, damage to the instrument, etc. The function of the emergency diesel engine of the nuclear power plant is to provide spare equipment for the power of the medium- and low-voltage nuclear auxiliary equipment in case of power loss in the whole plant to ensure the safe shutdown of the nuclear power plant. According to the operation requirements, within 10 seconds after the emergency start signal is issued, the emergency diesel engine needs to be started up to the required speed and voltage. The fuel return line (hereinafter referred to as the return line) is an important device for maintaining the oil storage capacity of 8 m^3 of the daily fuel tank. If the oil return pipeline is in a state of excessive vibration, the fatigue damage of the pipeline material will be aggravated, and the probability of pipeline damage will increase, which will affect the emergency start of the diesel engine and threaten the safe and stable operation of the nuclear power plant.

2 ANALYSIS OF THE CAUSE OF PIPELINE VIBRATION

From the source of the pipeline vibration, the vibration of the pipeline can be divided into the forced vibration caused by the machine and the vibration caused by the fluid in the pipeline. The vibration of the return pipeline is caused mainly by the vibration of the diesel engine body. The analysis is as follows.

Figure 1.   Pipeline layout.

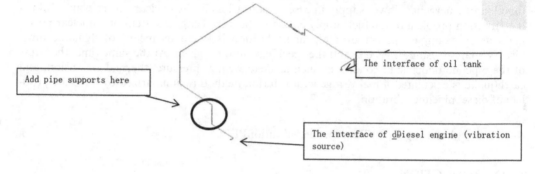

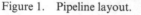

Figure 2.   Pipeline improvement diagram.

One end of the diesel oil return pipe is connected to the diesel engine body, and another end is connected to the hose. The diesel engine causes vibration of the equipment during the startup process, which drives the oil return line to vibrate. The relative positions of the diesel engine and the return line are shown in Figure 1. Due to the low overall stiffness of the return line, the diesel engine body is excited at the picture mark, causing the line to be forced to vibrate.

According to the on-site measurement, the vibration speed of the diesel return line is overprocessed for a long time, and the maximum vibration value is 47 mm/s, which exceeds the allowable limit of 42.8 mm/s. Therefore, the pipeline is modified to increase the pipeline support, and three pipeline supports are added to the pipeline. The specific position is shown in Figure 2.

After the increase of the line pipe support vibration measurements again, a significant improvement in vibration velocity is achieved, and the requirements in Table 1 are met. However, due to pipeline design changes, theoretical calculations are required.

3   APPLICATION OF PEPS IN PIPELINE VIBRATION CALCULATION

According to the relevant formula in the RCC-M (Markl 1955; Menon 2004; Lee et al. 2010), it is necessary to evaluate the stress of the diesel oil return pipeline under various

Table 1.   The speed value of pipeline vibration after adding pipe supports.

Direction-measurement(mm/s)	Measuring point 1	Measuring point 2	Measuring point 3	Measuring point 4	Measuring point 5
Horizontal	18.3	25.5	11.7	21.3	26.7
Vertical	24.3	15.4	10.9	25.7	\
Axial	14.6	19.4	9.1	\	24.3

criteria. The O, A, B, C, and D criteria respectively represent the design working conditions, normal working conditions, disturbance working conditions, emergency working conditions, and accident working conditions. The specific evaluation formulas are shown in Table 2.

Table 2. RCC-M grade pipeline assessment formula.

Conditions	Criterion	Loads	Evaluation equation
Design condition	O	P: Design pressure M_A: Torque generated by continuous load (self-weight, other continuous mechanical loads)	$S_{sl} = \frac{PD_o}{4t_n} + 0.75i\frac{M_A}{Z} \leq S_h \rightarrow (6)$
Normal condition	A	M_C: Torque generated by thermal expansion and fixed point displacement (caused by OBE; if not in Equation 10, consider it here)	$S_E = \frac{iM_c}{E} \leq S_h \rightarrow (7)$
		P: Operating pressure M_A: Torque generated by continuous load (self-weight, other mechanical loads) M_C: Torque generated by thermal expansion and fixed point displacement (caused by OBE; if not in Equation 10, consider it here)	$S_{sl} = \frac{PD_o}{4t_n} + 0.75i\frac{M_A}{Z} + i\frac{M_c}{Z} \leq S_h + S_Z \rightarrow (8)$
		M_D: Torque generated by fixed point nonrepetitive displacement (subsidence)	$\frac{iM_D}{E} \leq 3S_C \rightarrow (9)$
Disturbing condition	B	P: Greatest pressure M_A: Torque generated by continuous load (self-weight, other mechanical loads) M_B: Torque generated by accidental load (valve force, OBE)	$S_a = \frac{P_{max}D_o}{4t_n} + 0.75i\frac{M_A}{Z} + i\frac{M_B}{Z} \leq 1.2S_h \rightarrow (10)$
Emergency condition	C	P: Greatest pressure M_A: Torque generated by continuous load (self-weight, other mechanical loads) M_B: Torque generated by accidental load (valve force)	$S_a = \frac{P_{max}D_o}{4t_n} + 0.75i\frac{M_A}{Z} + i\frac{M_B}{Z} \leq 1.8S_h \rightarrow (10)$
Accident condition	D	P: Greatest pressure M_A: Torque generated by continuous load (self-weight, other mechanical loads) M_B: Torque generated by accidental load (valve force, SSE)	$S_a = \frac{P_{max}D_o}{4t_n} + 0.75i\frac{M_A}{Z} + i\frac{M_B}{Z} \leq 2.4S_h \rightarrow (10)$

i = stress index, $0.75\,i \geq 1.0$;
S_h = material allowable stress at design temperature (Sawa et al. 1991);
S_a = thermal expansion stress allows variation amplitude, $S_a = f(1.25S_c + 0.25S_n)$;
S_c = basic allowable stress of materials at room temperature;
S_n = basic allowable stress of the material at the highest temperature;
F = fatigue correction factor.

4 A GENERAL METHOD OF TIME-HISTORY ANALYSIS FOR VIBRATION OF OIL RETURN PIPELINE IN PEPS

4.1 *Static and seismic analysis*

The PEPS pipeline stress analysis program is based on the theory of structural mechanics, and some assumptions are made when establishing the mechanical equation of rod-like elements: the deformation of all rod-like elements in the structure is very small, and the displacement of each node does not change the shape of the elements. At this time, the force and displacement are linear and in accordance with Hooke's law, and the stress, strain, and displacement caused by each load in the elements can be superimposed (Lidiard 1980; Koenders & Sellmeijer 1992; Peng & Peng 2009).

Based on this assumption, when applying a PEPS pipeline stress analysis program, the analyzed pipeline system should meet or approach the above assumptions; otherwise it will cause large calculation errors. The principle of PEPS analysis of pipeline vibration is that, except for the first fixed end, all other constraints are assumed to be open. Based on the above assumption, the pipeline system will become a free statically determinate system. Then the displacement of each point in the pipe system caused by all external loads is calculated by structural mechanics theory, and the flexibility matrix is calculated. Finally, the loads required to restore the constraints in the pipeline to the original position or to a specified displacement (spring support position) are obtained (Matera & Shpargel 2006).

According to these loads, the forces, stresses, strains, and displacements of each element in the pipe system can be calculated. For different conditions, the corresponding results can be obtained by substituting different loads. When the point is additional displacement, the additional displacement will be superimposed with the displacement obtained by the above solution. For other special loads or special conditions, the superposition principle is also used (Rybicki & Stonesifer 1979).

According to the above calculation hypothesis theory and the calculation formula in Table 2, the stress analysis of pipeline under various working conditions can be carried out. The calculation results are shown in Table 3. It is concluded that the O/A/B/C/D evaluation criteria of pipelines under RCC-M specification meet the requirements.

4.2 *Fatigue analysis*

Due to the long-term vibration of the oil return pipeline exceeding the standard, the fatigue stress of oil return pipeline under vibration conditions needs to be analyzed. The fatigue failure of a pressure pipeline is manifested mainly in two types:

(1) Low cycle fatigue failure, in which stress changes slowly with time; generally in service life, the total number of stress cycles is less than 1×10^5 times. The objective of the design of this kind of pipeline is to simplify the calculation model by means of the concept of alternating stress amplitude and to determine the wall thickness of the structure by using the strength theory.

(2) The fatigue failure mode is high cyclic fatigue failure, in which stress varies rapidly with time, so that the strength theory can no longer be used to establish the calculation model and solve it. The typical case of this kind of fatigue failure is the mechanical vibration of a pipeline.

Table 3. Calculation results.

Pipeline No.	Stress ratio results				
	O	A	B	C	D
2LHP/1LHQ	0.136	0.297	0.264	0.293	0.238

311

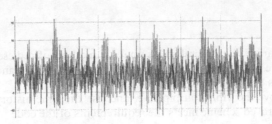

Figure 3. Vibration time-history spectrum.

The above pipeline system belongs to the second case, which needs to be solved by using the time-history analysis method in PEPS software. Time-history analysis is a relatively fine method. It can not only consider the redistribution of internal forces after the structure enters plasticity, but also record the whole process of the structure response. In the calculation process, the time step of vibration is divided into several sections. The integral method is used for each step and elastic analysis is carried out in each period.

According to the vibration control guideline of steam and water pipeline in a thermal power plant and ASME-OM3 (Smith & Thomas 1987), the allowable criterion of alternating stress intensity is evaluated. Vibration time-history spectra are collected in the field as shown in Figure 3.

The alternating stress intensity S_{alt} of the pipeline due to vibration must satisfy the following formula:

$$S_{alt} = \frac{C_2 K_2}{Z} M \leq \frac{S_{el}}{\alpha}$$

where
   C_2: = secondary stress index
   K_2 = local stress index
   M = maximum dynamic bending moment caused by vibration
   Z = pipeline section modulus
   $\alpha = 0.8$ according to ASME OM3
   $S_{el} = 0.8 S_a$
   S_{alt} = the alternating stress intensity of the material at 10^6 cycles at design temperature, equal to 103 MPa.

$$\sigma = 0.75i \frac{M}{Z} = \frac{3}{4} \times \frac{1}{2} C_2 K_2 \times \frac{M}{Z} \leq \frac{3}{8} \times \frac{S_{el}}{a}$$

According to the above formulas, it can be concluded that if the alternating stress of a pipeline under vibration is less than 38.63 MPa, it can meet the requirements. After analysis by PEPS software, the results are shown in Table 4.

According to the above conclusions, the fatigue stress of a pipeline under vibration meets the requirements of the code.

Table 4. Calculation results.

Serial number	Vibration stress of pipeline (MPa)	Allowable value (MPa)	Stress ratio
2	26.81	38.63	0.7

5 CONCLUSION

(1) This article summarizes the general stress analysis method of a diesel engine return pipe-line, and shows the calculation method and conclusion of stress analysis under different working conditions by using PEPS software.
(2) Modify and improve the vibration pipeline, and use the time-history analysis method to verify that the improved scheme meets the requirements of the code.

REFERENCES

Koenders, M.A., & Sellmeijer, J.B. 1992. Mathematical model for piping. Journal of *Geotechnical Engineering* 118(6): 943–946.

Lee, J.R., Jeong, H., Ciang, C.C., et al. 2010. Application of ultrasonic wave propagation imaging method to automatic damage visualization of nuclear power plant pipeline. *Nuclear Engineering and Design* 240(10): 3513–3520.

Lidiard, A.B. 1980. Applications of probabilistic fracture mechanics to light water reactor pressure vessels and piping. *Nuclear Engineering and Design* 60(1): 49–56.

Markl, A.R.C. 1955. Piping-flexibility analysis. *Transactions of ASME* 77(2): 12743.

Matera, A.G., & Shpargel, K.B. 2006. Pumping RNA: nuclear bodybuilding along the RNP pipeline. *Current Opinion in Cell Biology* 18(3): 317–324.

Menon, S. 2004. Piping Calculations Manual. New York: McGraw-Hill Professional.

Peng, L.C., & Peng, T.L. 2009. *Pipe Stress Engineering*. New York: ASME Press.

Rybicki, E.F., & Stonesifer, R.B. 1979. Computation of residual stresses due to multipass welds in piping systems. *Journal of Pressure Vessel Technology* 101(2): 149–154.

Sawa, T., Higurashi, N., & Akagawa, H. 1991. A stress analysis of pipe flange connections. *Journal of Pressure Vessel Technology* 113(4): 497–503. Smith, P.R., & Thomas, J.V.L. 1987. Piping and pipe support systems.

Integrated energy planning and technology application for an industrial park

Lu Wang
Guangdong Electric Power Design Institute Co., Ltd. of China Energy Engineering Group, Guangzhou, China

Shun Qi Zeng
Guangzhou Power Supply Co., Ltd., Guangzhou, China

Wenbo Xia, Pei Quan Liang & Yi Lu Zhong
Guangdong Electric Power Design Institute Co., Ltd. of China Energy Engineering Group, Guangzhou, China

Zi Wen Yu
Guangzhou Power Supply Co., Ltd., Guangzhou, China

He Li Peng
HaiNan Power Grid Co. Ltd., HaiKou, China

ABSTRACT: Integrated energy planning technology is widely used in power system planning and design. Based on a variety of energy requirements, a multilayer integrated energy planning method is discussed in this article, which includes a four-layer energy supply method, smart grid planning, and control system model application. Following the integration and application technology of the planning methods analysis, the application of integrated energy planning in the energy system for Mingzhu Industrial Park is introduced as a case study, with the conclusion that this verifies that an integrated energy planning method could effectively reduce the cost of distribution planning and improve load characteristics.

Keywords: industrial park energy system, integrated energy planning, smart grid

1 INTRODUCTION

As a large number of industrial parks have developed in China, higher requirements are put forward for green and reliable energy and power supply and and its intellectualization (Zhang 2016). Industrial parks generally require a complex energy system based on industrial load, covering a variety of energy systems that involve supply, transfer, and utilization of many kinds of energy.

In industrial parks, load characteristics are usually complex: higher requirements are placed on the operation and dispatch of a power distribution system, while energy waste, power shortage, and other problems are widespread, greatly affecting the operation efficiency and limiting economic and environmental benefits of the system. Therefore, to improve the construction planning of an industrial park energy system, it is necessary to configure an appropriate integrated energy system.

This article discusses the main methods and processes of integrated energy planning implemented in industrial parks, followed by the illustration and analysis of Mingzhu Industrial Park as a case study.

2 INTEGRATED ENERGY SYSTEM PLANNING

Integrated energy planning technology is widely used in power system planning and design, guiding the construction process of the energy network.

The integrated energy planning methods presented in this article include a power generating unit planning method, power grid and storage planning method, and control and scheduling system planning method. Algorithms and processes of the planning methods are analyzed. A method for application in an industrial park energy system is demonstrated.

3 PLANNING METHOD ANALYSIS

This project of integrated planning study is focused on a high ratio of renewable energy and high-efficiency energy consumption in the industrial park. The technology of energy storage capacity calculation and the joint reconstruction and expansion of the park electricity power grid and gas network will be studied to determine the optimum energy supply strategy of the industrial zone.

The construction of the integrated energy planning method is shown in Figure 1.

In the integrated technique study of multienergy allocation in the park, total energy demand, load characteristics, and alternative options of various users are analyzed. Coupling relationships of technical feasibility and engineering benefit for multienergy sources are studied using a multienergy collaborative planning and evaluation method (Zhang 2017).

3.1 Power supply planning method

In order to improve the energy efficiency of the park and renewable energy absorption, a multienergy collaborative planning evaluation method is applied to analyze the multitype energy optimization scheme. This planning method proposes a four-level energy hub model to realize multienergy coupling modeling for sources, network, load, storage, and other links (Wang 2016), and the relation between output and input is established by a transfer matrix, which reflects the complex multienergy coupling of an integrated energy system in the form of electricity, gas, heat and cold, and other factors.

Considering both equipment capacity planning and operation scheduling optimization, a model is established to optimize project strategy. A Source–Network–Load–Storage integrated planning framework would be carried out, in which the integrated energy planning framework is divided into an allocation layer, conversion layer, integration layer, and energy storage layer. In this method, energy inputs among different devices are analyzed. The energy conversion process is studied to calculate the output iteration and related parameters of devices, for configuring the energy storage system, respectively.

3.2 Power grid planning method

The objectives of power grid planning mainly include (1) increasing the reliability and power supply capacity of distribution network; (2) join various forms of energy storage devices; (3) improving the structure of control devices for users and distribution based on the demand of energy consumption; (4) effectively reducing the peak load value of the specific area; and (5) improving the consumption efficiency of power grid equipment.

Figure 1.   Construction of integrated energy planning.

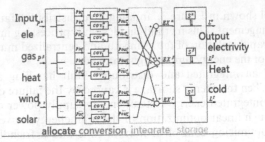

Figure 2.   Illustration of a four-level energy hub model.

According to the distribution of heat load and the forecast of heat consumption in the park collected during thermal data survey, the thermal pipeline architecture of the park could be planned and designed. The comprehensive benefits and cost of heating network construction are analyzed by using a multienergy coupling matrix, with a determined heating network planning scheme as a result (Yu & Qin 2014).

Combined with the power planning, the scale of the transmission network and distribution network in different periods should be analyzed, counting the intellectualization on the power smart grid structure. Moreover, the application of various technologies such as energy storage could improve the efficiency and controllability of the power grid at the same time. Based on the analysis of the load curve, the operation of energy storage could be simulated and the impact of the overall load of the park with load operation efficiency method could be analyzed. According to the results, the form and size of energy storage devices could be determined.

3.3   Control and scheduling system planning method

Based on the idea of multienergy collaboration and user interaction, the distributed resources should be effectively integrated in the industrial park.

Research on park control system shows that priority should be given to selecting industrial and commercial users who generally support a larger range of load regulation. In the terminal region, the user energy management substation is configured, with the monitoring terminal configured in the photovoltaic, energy storage, and active loads.

A self-optimizing real-time control model, which describes the operation and transformation of different energy systems, is constructed by analysis results of a multienergy scheme in industrial park. The model takes the operation cost and energy efficiency as the comprehensive optimization objectives.

The structure of the model is shown in Figure 3.

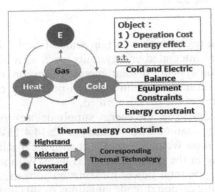

Figure 3.   Diagram of real-time control model.

Based on the model shown in Figure 3, an integrated distributed fault treatment method is applied in energy management. This method divides fault processing into two levels: local and substation. Through rapid isolation of faults in situ and centralized mutual supply of different energy, the reliability of the energy supply is guaranteed.

The framework of an integrated and distributed fault handling scheme in the park is designed as follows: When the energy shortage is caused by the failure of the cooling and heating equipment in the integrated energy system, energy supply power of each piece of equipment is obtained through linear optimization of the multienergy conversion. Thus, the whole system achieves self-optimizing control in normal time and a friendly interaction with the park, achieving a full energy control in a green industrial park.

3.4 *Energy system configuration*

In summary, on the basis of several kinds of energy consumption schemes, the optimization calculation is carried out. Meanwhile, the interaction between the adjustment and optimization of source network load storage schemes is analyzed. Thus, the integrated energy configuration of energy supply, integrated transmission, and user configuration are determined, achieving economic benefits and meeting relevant technical requirements for a control and dispatching system.

4 APPLICATION IN MINGZHU INDUSTRIAL PARK

4.1 *Fundamental situation of the park*

The National Key Project takes the factory enterprise group in Guangzhou's special industrial park as the research object. The planning area of Mingzhu Industrial Park is about 110 square kilometers, located in the north of Guangzhou with a superior geographical position. The demonstration area is located in the Mingzhu Industrial Park of Conghua, Guangzhou, covering an area of about 55 square kilometers. The users in the selected demonstration area are about 200 in number. In the discussed area, annual electricity consumption of park is about 250 million kWh, while annual heat consumption of the park is about 200,000 tons of steam. As for the energy source, energy input of the discussed area mainly includes solar energy, electricity, wind energy, and natural gas and so on.

The main task of an integrated energy system is to complete generation, transmission, and distribution of multiple energy sources and achieve multienergy coordination and green energy supply. To achieve these aims, the determined integrated energy project introduces a clean energy supply to reduce the emissions of the park, energy storage equipment to achieve the role of peak shifting and valley filling, and an energy management system to refine successfully the unified management of energy.

4.2 *Energy configuration of the park*

4.2.1 *Power supply*

Mingzhu industrial Park is a region with a certain scale of heat users, providing satisfactory ground for the application of integrated energy units, such as a high-proportion distributed photovoltaic power generation system for solar energy utilization, regional distributed energy for combined electricity and heat generation as multi-possibility coordination, etc. Overall, a green energy supply structure of the park is formed.

According to the optimized scheme, the energy supply devices installed in the area include an Aotou Distributed Energy Station with capacity of 2x14.4 MW, roof photovoltaic with capacity of 26 MW, and Tankou Waste Power Plant with a capacity of 2*12 MW. The total capacity of renewable energy in the discussed area is 44 MW, and the power generated by clean energy exceeds 80% of the total power in Mingzhu Industrial Park, thus reducing the standard coal consumption and carbon dioxide emissions by 40,000 tons and 130,000 tons.

4.2.2 *Power grid*

There are two 110 kV substations in the demonstration area. An intelligent substation configuration is considered in substation planning with introduction of a multifunction management system. In the meantime, a thermal pipe network is also equipped to supply steam and hot water. Combined with the results of iterative calculation of thermal energy substitution and comprehensive economic comparison, three thermal pipelines from the energy station to nearby DE (Distributed Energy) station are selected, together with a power grid and other types of network constituting the smart energy grid structure of the industrial park.

4.2.3 *Energy storage*

According to the energy storage planning method described in Section 3, the simulation operation optimization method is applied in the energy system for the purpose of equipment efficiency and investment profit in the park. The results are shown in Table 1 and Figure 4.

Based on the forecasting of energy consumption and the cost of energy storage, energy storage, 9.3 MW/44.75MWh in total, has been built after comprehensive analysis.

The site configuration of the energy storage is shown as follows:

4.3 *Energy management system of the park*

The industrial and commercial users who can support large-scale load regulation are selected as smart users. Substations of user energy management have been built in the region of the interactive zone. In addition, monitoring terminals in photovoltaic and energy storage equipment will be configured.

Table 1. Response results for different energy storages (MW, MWh).

Power	Capacity	Response	Electric
MW	MWH	Hour	MWh
4.1	20	13	20
5.1	25	6	9.75
6.1	30	3	4.54
7.2	35	2	2.25
9	44	1	0.37
9.3	47.2	1	0.07

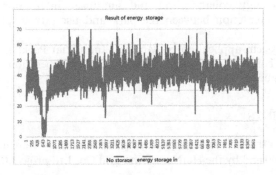

Figure 4.   Energy storage effect diagram.

Figure 5. Energy storage of Mingzhu Industrial Park.

The equipment provides information on energy demand and adjustable capacity for dispatching. Management system could respond to the control instructions of user and energy equipment of the park (Zheng 2017).

In research on the park dispatching system, the integrated energy management system is integrated in Mingzhu Industrial Park. The system monitors and collects energy operation data and coordinates and optimizes the distributed resources in the park. The system can dispatch orders from the strategy to the control system and the direct control resources of the park.

4.4 *Energy system of Mingzhu Industrial Park*

In summary, the energy system of the park achieves high energy efficiency by energy integration planning. Through an interactive mechanism to realize a flexible interaction with the users and energy system, energy can improve the observability and controllability of the park.

The industrial park will form an efficient and clean energy supply system based mainly on the distributed power and PV (Photovoltaic). In the demonstration area, the form of energy storage can flexibly adjust the energy use of the park and users. Construction of a smart distribution power grid and user equipment control framework will be improved.

5 CONCLUSION

Integrated energy would be widely used in cities in China. The planning method, which should be considered as the guide of the whole engineering project, plays a key role and reference for city construction.

This article proposes a new approach to integrate energy planning for an industrial park. According to the characteristics of the demonstration park, planning and control technology of multienergy collaboration and multiuser interaction is effectively integrated. An effective interaction between the park and users from planning to operation will be established.

From the energy system application of Mingzhu Industrial Park, the green development mode of the park is constructed through the optimization operation, energy storage, user interaction, and configuration of renewable energy in the park.

ACKNOWLEDGMENTS

The work described in this article is supported by National Key Research and Development grant 2016YFB0901300 and by the HaiNan power grid Co. Ltd project 070000KK52160001.

REFERENCES

Ju, P. Shen, Fu, & Wu, F. 2017. Research on distributed and interconnected online modeling of PS-IE Electric Power Automation Equipment 37(6) 2017 11–13.

Wang, L. 1970. Senior engineer with the rank of professor. His main research fields are integrated energy planning, smart grid technology application, power system planning, etc. E.mail:wlud%40163.com

Wang, Y., & Zeng, B. 2016. Multi-energy flow calculation method for integrated energy system containing electricity, heat and gas. *Power System Technology* 40(10) 2942–2950.

Yu, Y., & Qin, C. 2014. Expatiation on the basic ideas of smart grid. *Scientia Sinica Technologica* 2014 (6): 694–701.

Zhang, S., & Miao, A. 2016. Energy Internet as strong support for the development of smartcity. *Electric Power* 49(3); 13–17.

Zhang Dan Sha. 2017. Analysis and study on architecture of integrated smarter energy management system. *Sino-Global Energy* 22(4) 125–129.

Zheng Z., Han, B., Shan, X, et al. 2017. Analysis on key technologies for coordinated operation of advanced application software in transmission and distribution network. *Automation of Electric Power Systems* 41 (6): 122–128 (in Chinese).

Emerging Developments in the Power and Energy Industry – Dufo-López, Krzywanski & Singh (eds)
© 2020 Taylor & Francis Group, London, ISBN 978-0-367-27169-5

Study on treatment of hydro turbine torque characteristic curve based on load shedding test

Liang Fu
State Grid Hunan Electric Power Company Limited Research Institute, Changsha, China

Jiafu Wei & Bo Huang
Hunan Xiangdian Test Research Institute Co.Ltd, Changsha, China

ABSTRACT: At present, the torque characteristic of hydro turbine is extended and fitted by the experience and mathematical methods in transient process calculation of hydropower station. The calculation results are inconsistent with the testing results. In this article, the treatment of hydro turbine torque characteristic curve was studied based on load shedding test. The torque characteristic curve of the zero-opening was identified by the shedding test data. The whole torque characteristic curve was extended and fitted by the least square method. The torque characteristic curve obtained by this method was used to simulate the load rejection transition process, and the calculated results were compared with the test results. The calculation results of the extreme value and adjustment process are consistent with the test results, this torque characteristic curve can be used to accurately simulate the regulation process of the hydro turbine governor. The simulation accuracy of the transient process is greatly improved. The research of this article can effectively guide the safe operation of hydroelectric unit.

1 INTRODUCTION

In transient process calculation of hydropower station, the characteristic curve of hydraulic turbine had an important influence on the calculation results. The model hydro turbine comprehensive characteristic curve only includes the data of high efficiency and large guide vane opening of turbine, and it cannot meet the requirements of transient process calculation. So the characteristic curve needs to be extended to obtain the characteristics of the turbine in the small opening, low efficiency and braking conditions (Nguyen V.S 2004).

Hydro turbine torque characteristic curve mainly affects the unit speed in large fluctuation transient process. Because the turbine torque characteristics of the zero-opening are unknown, the torque characteristics of the turbine are often extended and fitted based on experience and mathematical methods (Nguyen V.S 2004, Zhang J.M et al. 2016, Zhang R.S et al. 2011, Cheng Y.C et al. 2003, Huang X.R & Liu D.Y. 2007), there is a big difference between the calculated results and the test results (Fu L & Wang Y.G. 2012)(as show in Figure 1). Although more accurate mathematical models can be obtained by internal characteristics method and CFD, there require detailed turbine structure and size parameters (Men C.S & Nan H.P. 2017, Luo X.Q et al. 2017, Li J.W et al. 2010, Li J et al. 2014), these parameters are often difficult to obtain actually.

This article gets the changes of unit speed under zero-opening of guide vane from load shedding test, and derives the unit torque characteristic equation. The torque characteristic curve of hydro turbine is expanded and fitted by the least square method. The test results are used to verify the torque characteristic curve of the hydro turbine finally.

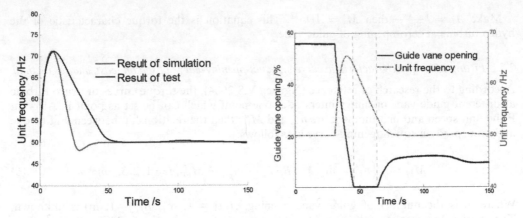

Figure 1.   Comparison of simulation and test results.   Figure 2.   Results of load shedding test.

2   THE TREATMENT OF TORQUE CHARACTERISTIC CURVE

2.1   *Acquisition of torque characteristic curve at zero-opening based on load shedding test*

At present, in the calculation of transition process, the relationship between the unit torque and unit speed at the zero-opening is assumed, and it is substituted into the whole characteristic curve and solved by least square method. The results obtained by this method have errors, and that cause the inconsistency between the simulation results and the test results (As shown in Figure 1).

Load shedding test results of guide vane opening and unit frequency are shown in Figure 2. It can be seen that the turbine speed decreases linearly with the time at zero-opening of guide vane, so that the change rule of the turbine speed with time can be obtained:

$$n = kt + b \tag{1}$$

Where n is the unit speed $r/\min$, t is the time, and k and b are coefficients, which can be obtained by test results (The test data in the dashed area in Figure 2).

After load shedding, the resistance torque of the unit is 0, and the equation of motion of the unit is obtained as (2). Equations (3) and (4) are obtained according to the similarity theorem of hydro turbine.

$$J\frac{\pi dn}{30dt} = M_t \tag{2}$$

$$n_1' = \frac{nD_1}{\sqrt{H}} \tag{3}$$

$$M_t = M_1'D_1^{3}H \tag{4}$$

Where J is the rotational inertia of the turbine generator set $t \cdot m^2$, M_t is the turbine dynamic torque kN·m, n_1' is the hydro turbine unit speed r/min, M_1' is the hydro turbine unit torque kN·m, D_1 is the hydro turbine runner diameter.

Substituting equation (1) into equations (2), (3), and (4) equation (5) can be yielded.

$$M_1' = J\frac{k\pi}{30D_1^{5}n^2}n_1'^{2} \tag{5}$$

Make $A_P = J \frac{k\pi}{30D_1{}^5 n^2}$, then $M_1' = A_P n_1'^2$. This equation is the torque characteristic of the hydro turbine at zero-opening of guide vane.

2.2 Fitting of torque characteristic curve on equal-opening-curve based on least squares method

According to the research of reference (Nguyen V.S 2004), the torque curves of hydro turbine at different guide vane openings intersect at one point which can be set as point P. At point P the unit speed and unit moment are $n_1'_p$ and $M_1'_p$, then the relationship between unit torque and unit speed on equal-opening-curve is as follows:

$$M_{1j}' = A_j \left(n_1'^2 - n_1'_p{}^2 \right) + B_j \left(n_1' - n_1'_p \right) + M_1'_p, j = 1, 2, 3....m \tag{6}$$

Where m is the number of guide vane opening, A_j (j = 1, m), B_j (j = 1, m) is unknown. The relationship between unit moment and unit speed of zero-opening is $M_1'_p = A_P n_1'_p{}^2$ which is substituted into the equation (6), and equation (7) can be obtained:

$$M_{1j}' = A_j \left(n_1'^2 - n_1'_p{}^2 \right) + B_j \left(n_1' - n_1'_p \right) + A_P n_1'_p{}^2, j = 1, 2, 3....m \tag{7}$$

$(M_{1i,j}', n_1'_{i,j}, a_j)$ is set as known point, among them $i = 1, 2, 3....L$, L is the number of points on each opening. The known points can be obtained by the model hydro turbine comprehensive characteristic curve. Then the quadratic error (w) between a single point $(M_{1i,j}', n_1'_{i,j}, a_j)$ and equation (7) is as follows:

$$w_{i,j} = \left[A_j \left(n_{1i,j}'^2 - n_1'_p{}^2 \right) + B_j \left(n_{1i,j}' - n_1'_p \right) + A_P n_1'_p{}^2 - M_{1i,j}' \right]^2 \tag{8}$$

Sum of quadratic errors (W) is as follows:

$$W = \sum_{j=1}^{m} \sum_{i=1}^{L} \left[A_j \left(n_{1i,j}'^2 - n_1'_p{}^2 \right) + B_j \left(n_{1i,j}' - n_1'_p \right) + A_P n_1'_p{}^2 - M_{1i,j}' \right]^2 \tag{9}$$

According to the least squares method, the sum of quadratic errors is the minimum and the equations are obtained as follows:

$$\begin{cases} \frac{\partial W}{\partial n_1'_p} = 0 \\[2mm] \frac{\partial W}{\partial A_j} = 0, j = 1, 2, 3...m \\[2mm] \frac{\partial W}{\partial B_j} = 0, j = 1, 2, 3...m \end{cases} \tag{10}$$

There are (2m+1) equations and (2m+1) unknowns of the equations (10), the unknowns are $n_1'_p$, $A_j(j = 1, 2, 3...m)$, $B_j(j = 1, 2, 3...m)$. Iterative method can be used to solve the equations (10) to get $n_1'_p$, $A_j(j = 1, 2, 3...m)$, $B_j(j = 1, 2, 3...m)$, then the data of torque characteristic curve can be obtained by adding equation (6).

2.3 Comparison of torque characteristic curves

Figure 3 is a three-dimensional schematic diagram of the hydraulic turbine torque characteristic curves obtained by the method proposed in reference (Nguyen V.S 2004) (original curve). Figure 4 is a three-dimensional schematic diagram of the torque characteristic curves obtained according to the method proposed by this article (new curve). Figure 5 is a two-dimensional

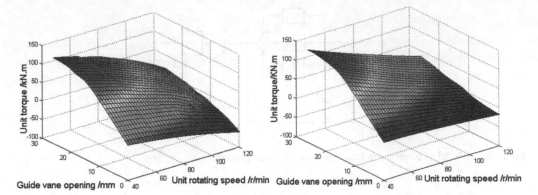

Figure 3. Torque characteristic curve obtained by original method.

Figure 4. Torque characteristic curve obtained by the method of this article.

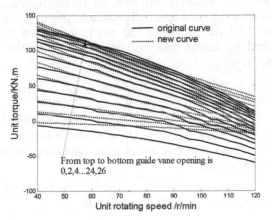

From top to bottom guide vane opening is 0,2,4...24,26

Figure 5. Comparison of characteristic curves.

comparison of the torque characteristic curves by the two methods. It can be seen that: 1) it is a big difference between the curves at zero-opening of guide vane. The unit torque of the new curve changes more smoothly with the unit speed, and the unit torque of each point is smaller than that of the point of original curve. 2) Compared with large guide vane opening, at small guide vane opening (a ≤ 4mm), the difference is greater in the two characteristic curves. 3) when the guide vane opening is greater than or equal to 6mm, in the range of unit rotation speed greater than or equal to 60r/min to less than or equal to 90r/min, the two characteristic curves are basically consistent. In the range of unit rotation speed less than 60r/min and greater than 90r/min, there are certain differences in the two characteristic curves.

3   COMPARISON OF SIMULATION RESULTS

3.1   *Simulation model*

The simulation method of large fluctuation transient process of water diversion power generation system has been mature. Characteristic line method is used to solve the continuity equation and motion equation of pressurized flow in water diversion system.

Load shedding test is usually carried out by directly disconnecting generator circuit breaker (GCB). In the operation of hydropower generating unit, the GCB suddenly disconnecting due to the electrical accident, which may also cause sudden load rejection of the unit.

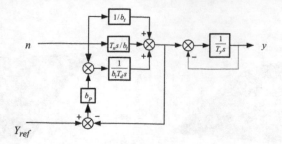

Figure 6. Model of governor in load rejection transient process.

After the GCB disconnecting, the hydro turbine governor is changed from load state to no load state, and the guide vane opening is closed to the no-load opening according to the closing law, and the governor automatically switches to the frequency mode. The unit is stabilized in no load operation at rated speed after the adjustment process. Governor model is shown in Figure 6, the PID parameters of the governor and the hydraulic servo system parameters were obtained by actual measurement according to the method described in reference (Chao H et al. 2013).

3.2 *Comparison of simulation results*

Different characteristic curves are used to simulate the load shedding transient process. The simulation results are compared with the test results as shown in Figure 7. Through comparison, it can be seen that the simulation results of the characteristic curve obtained of this article are more consistent with the test results.

3.3 *Influence of governor parameters on simulation results*

The new characteristic curve is used to calculate the load rejection transient process under different governor parameters. The calculation results are shown in Figure 8. Through the calculation, the governor parameters have no influence on the fast closing process of the guide vane and the fast rising and falling process of the unit speed in the load rejection process. The governor parameters mainly affect the adjustment process of the unit speed when the unit speed is falling close to rated speed. The smaller the governor parameters are, the more sensitive the governor is. The improper setting of the governor parameters will degrades the

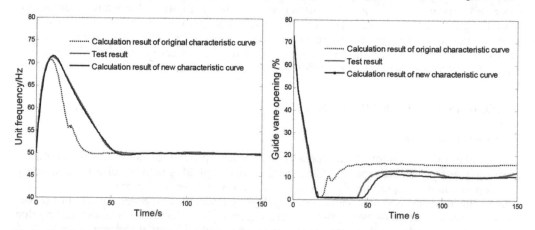

Figure 7. Comparison of calculate results and test results of the two characteristic curves.

325

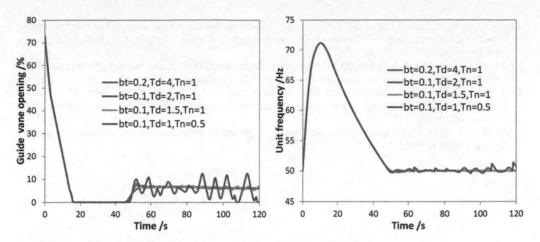

Figure 8.   Effect of governor parameters on load shedding transient process.

regulating quality of the governor, and even leads to the instability of the regulating system, causes the oscillation of the guide vane and the unit speed.

4   CONCLUSION

In this article, the treatment of hydro turbine torque characteristic curve is studied, and a fitting of torque characteristic curve based on load shedding test is proposed. Different characteristic curves are used to simulate the load shedding transient process of a hydropower unit, and the simulation results are compared with the test results. The characteristic curve obtained by this article can reflect the torque characteristic of the unit more accurately than the traditional method under the small guide vane opening. The calculation results using the characteristic curve obtained by this article are more consistent with the test results, and the simulation accuracy of the transient process is greatly improved. The characteristic curve obtained in this article was used to calculate the load rejection transient process under the different governor parameters. The calculation results show that: the governor parameters have no influence on the fast closing process of the guide vane and the fast rising and falling process of the unit speed in the load rejection process. The governor parameters mainly affect the adjustment process of the unit speed when the unit speed is falling close to rated speed.

REFERENCES

Nguyen V.S.2004. Study on boundary condition of transient process in hydropower station and Visualized Programming[D]. Wuhan: Wuhan University.
Zhang J.M et al. 2016. Code of electro-hydraulic regulating system and unit for hydraulic turbine [S], BeiJing: China Electric Power Press.
Zhang R.S, Liu Z, Wang L.C et al. 2011. Data interpolation by Delaunay triangulation for the combined characteristic curve of a turbine[J]. *Journal of Hydroelectric Engineering* 30(4): 197–201.
Cheng Y.C, Ye L.Q, Cai W.Y. 2003.Modeling of hyro-turbine hill chart by neural network[J]. *Journal of Huazhong University of Science and Technology Nature Science Edition* 31(6): 68–70.
Huang X.R, Liu D.Y. 2007.The treatment of hydroturbine synthetic characteristic curve by radial basis networks[J]. *Journal of Hydroelectric Engineering* 26(1): 114–118.
Fu L & Wang Y.G. 2012. Load Shedding Test and Simulation Analysis of Hydropower Station with Surge Shaft[J]. *Water Resources and Power* 30(6): 154–157.
Men C.S & Nan H.P. 2017.Improvement of Francis turbine internal characteristic model and its expanding application on outer characteristic[J]. *Transactions of the Chinese Society of Agricultural Engineering (Transactions of the CSAE)* 33(7): 58–66.

Luo X.Q, Li W.F, Feng J.J et al. 2017. Simulation of runaway transient characteristics of tubular turbine based on CFX secondary developmen[J]. *Transactions of the Chinese Society of Agricultural Engineering (Transactions of the CSAE)* 33(13): 97–103.

Li J.W, Liu S.H, Wu Y.L. 2010.3D unsteady turbulence simulation of the runaway transients of the Francis turbine[J]. *Large Electric Machine and Hydraulic Turbine*, 2010(6): 44–49.

Li J, Wang L, Liao W.L.2014.Internal flow characteristics analysis on S-shaped region of reversible pump-turbine[J]. *Transactions of the Chinese Society of Agricultural Engineering (Transactions of the CSAE)* 30(15): 106–113.

Chao H et al. 2013. Guide for modeling and testing of generator prime mover and governing system [S] Beijing: China Electric Power Press.

Emerging Developments in the Power and Energy Industry – Dufo-López, Krzywanski & Singh (eds)
© 2020 Taylor & Francis Group, London, ISBN 978-0-367-27169-5

Analysis of energy price distortion based on time-varying elastic production function

Peipei You, Chao Zhang & Xiao Gao
State Grid Energy Research Institute Co., Ltd., Beijing, China

Jiayin Wang
North China Electric Power University, Beijing, China

ABSTRACT: A reasonable energy price is an important endogenous force driving the sustainable development of energy production and consumption. However, affected by multiple factors such as government regulation, monopoly power, and market segmentation, the prices of several energy types in China have different degrees of distortion, which seriously hinder the optimal allocation of energy resources. Therefore, this article first analyzes the existing problems in China's energy price mechanism, then refers to the energy element to construct a time-varying elastic production function to measure the degree of distortion of energy price, and finally puts forward policy suggestions.

1 INTRODUCTION

The energy price mechanism is an important mechanism to guarantee realization of the energy development strategy of "Four Revolutions and One Cooperation." Through a scientific and reasonable energy price mechanism, it can effectively promote the reform of the energy system, ensure the safety of the energy supply, and promote energy development and innovation. At present, there are still many problems in China's energy price mechanism, which hinder building a safe, clean, low–carbon, and efficient modern energy system. Therefore, it urgently needs to be optimized and adjusted. This article focuses on the analysis of the existing problems in China's energy price mechanism, calculates the distortion of energy price, and finally puts forward policy recommendations.

2 PROBLEMS OF THE ENERGY PRICE MECHANISM IN CHINA

From 1985 until now, according to the unified arrangement of the country, price reform has made breakthrough progress. Natural monopoly industries such as coal, electricity, and oil have initially established a scientific and reasonable regulatory framework, laying a good foundation for optimizing the structure of the energy industry.

2.1 *The mechanism of the energy price system to ensure the safe supply of energy needs to be improved*

First, China has not yet established a pricing mechanism for energy that reflects the full factor cost of energy. At present, China's energy industry generally adopts a rate of return control method based on cost plus (Wu 2012), under which enterprises tend to overinvest (Zheng & Wu 2018). However, the government does not establish a scientific and standardized investment audit system when setting the regulated price. Therefore, the investment audit lacks an effective basis and methods, and there is a problem of not controlling the reasonable investment and scale of the project.

Second, the price mechanism of auxiliary services to ensure the safe and stable operation of the power system is missing. There is no compensation mechanism for flexible power sources such as pumped storage and gas and electrochemical energy storage. The cost allocation mechanism according to the degree of benefit is also unclear, which makes it difficult to ensure the auxiliary services play their role in maintaining the safety and stability of the power system.

Third, there is a lack of a modern professional supervision system that is compatible with the market-oriented reform and the modern energy system. China has not yet established a professional independent regulatory body and regulatory rules that separate the government and supervision. Besides, the supervisory personnel, professional standards, and supervisory ability are seriously insufficient.

2.2 *Lack of coordination in energy pricing mechanisms to promote clean alternatives*

First, China still needs to solve the problem of internalizing the external costs of traditional energy. Although China has imposed environmental taxes, the current energy price and tax system cannot reflect the ecological and environmental costs of various types of energy.

Second, the development of renewable energy faces multiple bottlenecks such as mechanism, market, and subsidy, and a top-level design of the price mechanism to promote the consumption of renewable energy is missing, resulting in no complementary force and poor coordination among the current renewable energy price policy, subsidy policy, and supplementary service compensation policy.

Third, the price mechanism to promote natural gas substitution has not yet been straightened out. On the one hand, the differential pricing system for natural gas has not yet been established and perfected, and seasonal gas prices and peak-shaving prices have not been fully promoted. So the changes in the relationship between natural gas prices and supply and demand cannot be reflected in a timely manner, which is not conducive to guiding the rational production and consumption of natural gas. On the other hand, the problem of cross-subsidy for domestic and nondomestic gas is prominent, inhibiting the large-scale use of natural gas in power generation, transportation, industrial fuel, and other fields.

Fourth, the electricity price mechanism to promote electricity substitution needs to be improved. The implementation scope of peak-valley (seasonal) time-of-use electricity price needs to be expanded. At the same time, the market bidding mechanism and auxiliary service compensation mechanism have yet to be established and perfected, which makes it difficult to meet the reasonable compensation income of electric energy substitution projects.

2.3 *The energy price mechanism and subsidy mechanism to promote the overall efficiency of the energy system urgently need to be improved*

First, there is a serious cross subsidy in electricity and gas prices (Lin & Liu 2016; Ye et al. 2017), which is not conducive to energy conservation in the whole society, resulting in a reduction in the overall energy efficiency of the energy system. In addition, cross-subsidies in electricity and gas prices deviate from the original intention of improving people's livelihood, and efficiency targeting urgently needs to be improved.

Second, there is a lack of a pricing mechanism and regulatory mechanism to promote the comprehensive energy efficiency of large-scale systems. On the one hand, there is a lack of unified planning and regulatory mechanism for electricity, gas, and heat networks and their supporting infrastructure, which makes it difficult to give full play to the complementary characteristics of electricity and heat networks. At the same time, the incentive mechanism to promote electricity, gas, and heat energy supply enterprises to save energy and reduce costs is not in place, and it is difficult to increase the enthusiasm of enterprises to save energy and reduce consumption.

Third, relevant policies to regulate the orderly development of self-owned power plants have not yet been put in place. At present, the social responsibilities of self-owned power plants such as cross subsidies, reserve capacity fees, renewable energy funds, and surcharges have not yet been put in place at the practical operation level. The coverage of self-owned power plants to fulfill social responsibilities nationwide is less than 50%. In addition, the support of a price mechanism to standardize the development of self-owned power plants needs to be improved (Lian 2017).

Fourth, the pricing mechanism for transmission channels needs to be straightened out. At present, it is difficult for China's transmission price to reflect the transmission distance and location signals, and the price transmission mechanism to promote long-distance large-scale transmission has not yet been straightened out, resulting in a low transmission economy of some lines and affecting the optimal allocation of transmission resources in the entire energy system.

Fifth, the price mechanism to promote demand-side management urgently needs to be improved. In recent years, the differences between China's peak load and peak-to-valley have increased rapidly. The electricity price mechanism to promote a demand-side response urgently needs to be established and perfected. Moreover, the electricity price level and structure are still unreasonable, which makes it difficult to play the guiding role of demand-side price signals effectively.

2.4 The price mechanism to promote energy innovation and development needs to be established and improved urgently

First, the flexible energy pricing mechanism to promote the development of new formats urgently needs to be improved. At present, with the deepening of energy marketization reform, new technologies and new modes such as electric vehicles and integration of charging and storage are continuously developing. However, it is difficult to reflect the difference in energy consumption costs in different periods effectively using the current pricing mechanism. It is crucial to speed up the improvement of a time-sharing pricing mechanism and accurately reflect the value of new formats in a more flexible way such as marketization.

Second, the price mechanism to promote the development of power grid-side energy storage needs to be improved. Internationally, power grid-side energy storage cost is usually included in transmission and distribution prices to support the development of this technology. Investment and cost of power grid-side energy storage have not been considered in the pricing of transmission and distribution prices in China, which is not conducive to the development of power grid-side energy storage.

Third, incentives such as subsidies to promote energy innovation and development need to be strengthened. The international community has given a large amount of incentive subsidies to the development of energy innovation. At present, China's subsidies in the field of energy innovation are not timely, flexible, and in place.

2.5 The energy price policy is inconsistent with the national energy planning, energy development, and management objectives

First, China's energy planning and utilization and pricing mechanisms lack overall planning. The price mechanism lacks effective guidance on the optimal distribution, rational utilization, and substitution of energy. For example, gas instead of coal and electricity instead of coal need a reasonable energy price comparison as support.

Second, the goal of energy price management is inconsistent with the goals of adequate energy supply and clean development. Establishing an efficient energy economy system and balancing the relationship between ecological environmental protection and economic development are important propositions in China's current macroeconomic development. At present, China's economy maintains medium- to high-speed development and clean energy utilization will push up the cost of the energy supply. It is difficult to form a Pareto improvement space with the current goal of reducing energy prices.

Third, the reform of energy prices is not in harmony with the goal of supervision of state-owned assets. Energy state-owned enterprises should not only actively participate in energy price reform, but also meet the requirements of supervision of state-owned assets. Energy price reform characterized by "strictly controlling costs" and "preserving and increasing the value of state-owned capital" is inconsistent in objectives.

3 MEASUREMENT OF ENERGY PRICE DISTORTION IN CHINA BASED ON A TIME-VARYING ELASTIC PRODUCTION FUNCTION METHOD

The degree distortion of energy prices is the deviation between the marginal output of energy and the actual price of energy in reality. At present, there is a large amount of research on the degree of distortion of factor markets. The main methods include a production function method, cutting-edge technology analysis method, shadow price calculation method, and general equilibrium method, while there is little research on the degree of distortion of energy price. When calculating the degree of distortion of energy price, we need to calculate the marginal output of energy first, so this article uses the production function method to calculate the degree of distortion of energy price in our country.

3.1 Construction of time-varying elastic production function considering energy input

Based on the Cobb–Douglas production function, this article uses energy as an element to construct a time-varying elastic production function to measure the degree of distortion of China's energy price. The specific steps are as follows:

Step 1. According to the Cobb-Douglas production function, a time-varying elastic production function including labor, capital, and energy can be constructed:

$$Y_t = A_t L_t^\alpha K_t^\beta E_t^\gamma \tag{1}$$

where Y_t = t-year economic output; A_t = the level of comprehensive technical; L_t = t-year labor input; K_t = t-year capital input; E_t = t-year energy input; = t-year elastic coefficient of labor output; = t-year elastic coefficient of capital output; and = t-year elastic coefficient of energy output.

Linearized equations can be obtained by taking natural logarithms for both sides of Equation (1):

$$\ln Y_t = \ln A_t + \alpha \ln L_t + \beta \ln K_t + \gamma \ln E_t + \mu_i \tag{2}$$

According to Equations (1) and (2), the time-varying elastic production function can be estimated by taking the data of each variable, and then the estimated values $\hat{\alpha}, \hat{\beta}, \hat{\gamma}$ of, can be obtained, so that the annual marginal output of energy input can be obtained.

$$MP_t^E = \hat{\gamma} \frac{Y_t}{E_t} \tag{3}$$

where MP_t^E = t-year marginal output of energy (i.e., shadow price).

Step 3. By comparing the actual energy price P_t^E with the shadow price MP_t^E, we can calculate the degree of distortion of energy prices in each year in China:

$$Dis_{E_t} = \frac{P_t^E - MP_t^E}{MP_t^E} \tag{4}$$

where Dis_{Et} = t-year distortion degree of energy prices.

3.2 *Empirical measurement and analysis*

This article estimates the distortion of energy prices in China from 2002 to 2016. Economic output (GDP) and labor force data come from the China Statistical Yearbook. Capital variable data are calculated according to the perpetual inventory method. Energy variable data come from the China Energy Statistical Yearbook. The development trend of these four variables from 2002 to 2016 is shown in Figure 1. It is difficult and complicated to obtain the energy price data for each year. This article uses the estimation method to determine the energy price for each year. The estimation steps are as follows: First, according to the energy consumption of coal, oil, and other energy (calculated in standard coal) and their prices (reduced by price according to the China Price Statistics Yearbook), the total cost of energy input for each year is calculated. Second, according to the total cost of energy input and total energy consumption in each year, the shadow price of energy in each year is estimated.

According to the data of GDP, labor force, capital and energy consumption, the elasticity of energy production is 0.5035 by using Eviews software in regression Equation (2). Furthermore, according to Equation (3), we can estimate the marginal energy output of China from 2002 to 2016, and the results are shown in Figure 2. Finally, according to Equation (4), we compare the actual energy price and shadow price, and the degree of distortion of China's energy price from 2002 to 2016 is calculated. The result is shown in Figure 2.

As shown in Figure 2, from 2002 to 2016, China's energy shadow price increased year by year, while the actual energy price showed a trend of increasing first and then decreasing slightly. The degree of distortion of energy price in China first decreases and then increases. In 2006, the degree of distortion of energy price was the lowest, at −0.57 (i.e., the actual energy price was 57% lower than the shadow price). In recent years, the degree of distortion of energy price increased further, reaching −0.79 in 2016. This shows that although reform of the dual-track energy price system and reform of the pricing mechanism have played a certain role, they are relatively limited. Especially in recent years, as China's energy resources are becoming

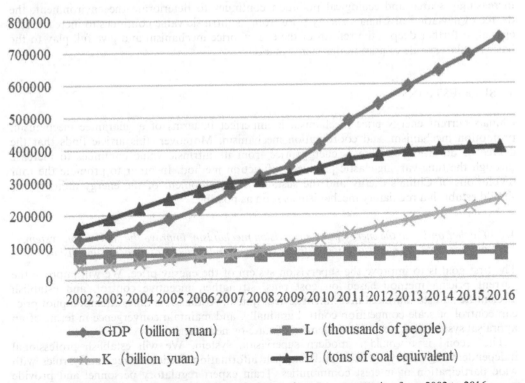

Figure 1. Development trends of GDP, labor, capital, and energy consumption from 2002 to 2016.

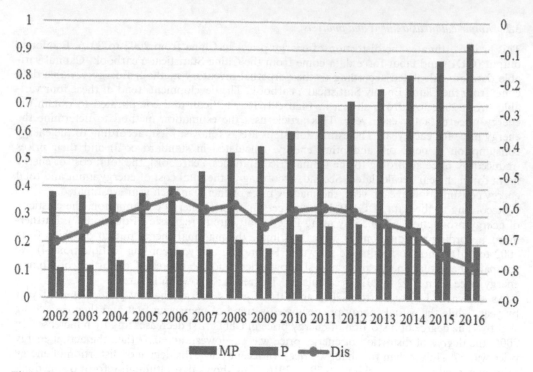

Figure 2. Degree of distortion of energy price in China.

increasingly scarce and ecological pollution continues to deteriorate the environment, the degree of deviation of China's energy price from its intrinsic value continues to increase. It is crucial to further deepen the reform of the energy price mechanism and give full play to the basic regulatory role of price in the energy market.

4 SUGGESTIONS

China's current energy price mechanism is imperfect in terms of a guarantee mechanism, promotion mechanism, and coordination mechanism. Moreover, this article finds that the degree of deviation of China's energy price from its intrinsic value continues to increase through the time-varying elastic production function method. In order to promote the four revolutions of China's energy and the sustainable development of the energy industry, we should establish a regulatory mechanism as soon as possible.

4.1 *Further optimize the energy price supervision mechanism, improve the supervision system, and promote the modernization of the energy system by means of modern supervision*

The first goal is to improve the supervision system of the energy price. We will improve the current pricing method based on cost plus, strengthen incentive control, and establish a scientific and standardized investment review system. At the same time, attempt a pilot price cap control or scale competition control gradually, and maintain convergence in terms of an appraisal system and income management of state-owned assets.

The second is to build a modern supervision system. We will establish professional independent regulatory agencies and establish information disclosure regulatory rules with wide participation of interest communities. Train expert regulatory personnel and provide necessary guarantees in staffing and funding.

The third is to link up with the price mechanism in national strategies, plans, and policies. On the one hand, the depth of China's energy strategy and energy planning needs to be linked with the price mechanism and market mechanism. On the other hand, top-level design should be further strengthened in system reform and market-oriented reform to ensure the coordination and unity of various reform objectives. In addition, it is suggested that government departments should objectively view the mutual restriction relationship among price management, sufficient energy supply, and clean development.

4.2 *Improve the energy price mechanism and promote market-oriented reforms in the energy sector so that prices fully reflect the relationship between market supply and demand and the scarcity of resources*

The first goal is to take appropriate measures to ease the cross-subsidy problem in electricity and gas prices gradually. For example, it is suggested that a national power universal service fund be established, which will be used by the National Development and Reform Commission. At the same time, we will improve the policy of tiered electricity prices for residents, appropriately raise electricity prices for residents based on "basic protection," and simultaneously lower the level of electricity prices for industry and commerce. To explore the "application for refund system," residents should first pay electricity charges according to reasonable electricity prices, and then apply to relevant departments for refunds according to the subsidy level announced by the government. In addition, a load rate electricity price is introduced to industrial and commercial users.

The second is to promote the application of residential time-of-use electricity price, gas price, and heat price. At the same time, we encourage the provincial price departments to formulate a residential peak-to-valley time-of-use electricity price suitable for electric vehicle charging. Dynamically adjust the price level of various types of time-sharing; dynamically adjust the peak-to-valley period and price difference according to peak-to-valley load changes. Encourage participation of electric vehicles in power peak regulation and actively obtain a reasonable income.

The third is to speed up the construction of energy market price rules and reduce government intervention. Based on China's national conditions and taking into full consideration the characteristics of China's economic development stage, energy endowment, and power grid structure, we should solve the problems of breaking interprovincial barriers and promoting large-scale consumption of clean energy from the market price mechanism. Gradually establish a fair and efficient market organization system and market rules.

The fourth is to improve the transmission and distribution pricing system and mechanism to meet the requirements of market-oriented reform. It is suggested that the government price department should perfect the price of special services and issue an access price policy on the user side to promote a fair burden. In addition, the government should adjust the price of auxiliary service and introduce a bidding mechanism in the auxiliary service market. In addition, in the low-voltage class with a large number of users and high operating cost, the "stamp method" is mainly adopted to share the cost.

The fifth is to improve the price mechanism of user demand-side management. We will improve the user-side time-sharing pricing mechanism, establish a high-reliability pricing mechanism and an interruptible pricing mechanism, innovate the user demand-side response mode by means of marketization, and increase the flexibility of user price response through the pricing mechanism and market mechanism.

4.3 *Establish and perfect supporting policies such as fiscal and taxation subsidies*

The first goal is to accelerate the internalization of environmental damage costs. We will further implement the environmental tax policy, assess and summarize problems in a timely manner, increase the intensity and standards of environmental tax collection, and gradually improve the carbon market construction to promote fair market competition through internalization of external costs.

The second is to speed up the implementation of policies for self-owned power plants to undertake cross-subsidy and other social responsibilities, specify the amount of cross-subsidy to be undertaken by self-owned power plants, and adopt a combination of collection and payment to ensure that funds are collected in place.

The third is to improve the financial subsidy mechanism for new energy technologies, effectively link up the existing subsidy mechanism and quota system for renewable energy, and innovate the subsidy mechanism through market-oriented means.

5 CONCLUSIONS

Energy is the focus of political, economic, military, and diplomatic attention of all countries in the world today. Deepening energy price reform will help to reflect the supply and demand of energy products and promote the healthy development of the energy industry. This article measures the degree of distortion of energy price in China through the time-varying elastic production function method and finds that the degree of deviation of China's energy price from its intrinsic value continues to increase. Therefore, the article proposes optimization of the supervision mechanism of energy price; further improvement of the energy price mechanism; and establishment and improvement of fiscal, tax, subsidy, and other supporting policies.

REFERENCES

Lian, W. 2017. Increase the intensity of advancement, accelerate the pace of reform, comprehensively deepen and conscientiously implement the market-oriented reform of the power system. *Macroeconomic Management* 5: 4–8.

Lin, B., & Liu, C. 2016. Chinese energy subsidy reform and effective energy subsidies. *Social Sciences in China* 10: 52–71, 202–203.

Wu, F. 2012. Thoughts on deepening China's energy price reform. *Theory & Practice* 8: 30–31.

Ye, Z., Wu, Y., Li, C., You, P., et al. 2017. The key problems and solutions of cross-subsidies in Chinese electricity tariffs. *Theory & Practice* 4: 20–24.

Zheng, X., & Wu, S. 2018. Regulation in electricity reform: necessity and initiative. *Theory & Practice* 1: 10–14.

Emerging Developments in the Power and Energy Industry – Dufo-López, Krzywanski & Singh (eds)
© 2020 Taylor & Francis Group, London, ISBN 978-0-367-27169-5

Calculation and experiment of short-circuit fusing characteristics of copper and copper-clad aluminum earthing cables

Ting Liu, Guangkai Yu, Bin Xiao & Kai Liu
China Electric Power Research Institute, Wuhan, China

ABSTRACT: In order to study the short-circuit resistant ability and fire protection measures of temporary earthing devices, the theoretical calculation and experiment of the copper and copper-clad aluminum(CCA) earthing cable in a short time fusing process has been conducted. It shows that the theoretical results of the fusing current of earthing cables in the adiabatic process are conservative, and the cross section has a certain safety margin through the calculation formula. Copper cable is prone to be fused than CCA cable under theoretical equivalent section. The addition of flame-retardant sheath can increase the fusing time of copper cable and has no obvious influence on the CCA cable. However, after the sheath is applied, the number of CCA splash beads is significantly reduced, which has a good protective effect against short-circuit fires.

Keywords: Safety engineering, earthing cable, copper, copper-clad aluminum, fusing characteristics

1 INTRODUCTION

Temporary earthing device is a kind of professional safety tool commonly used in the power industry. To prevent dangerous voltages and arcs generated by induced charge or unexpected resumption of power supply, it is necessary to ground after the power supply has been cut off (DAI, K. M. 2011, LI, J. X. 2008). In recent years, the casualties in power systems caused by the improper use of earthing devices accounted for a large proportion (LI, J. X. & LI, Y. C. 2013, WANG, C. Y. 2010).

The rated short-circuit current of the earthing cable under the rated short-circuit time should be less than the fusing current with an appropriate margin [ZHU, Z. M. 1994]. Regarding the calculation of the fusing current, (SSD 332 1989) and (ASTM F855 2015) provide the Onderdonk's equation for calculating the fusing current of copper material. (IEC 61230 2008) provides the relationship between current density, time and temperature rise of different materials such as copper, aluminum, silver and steel. Theoretical and experimental research on the electrical fusing problems in the case of rapid heating and slow heating of the wires has been carried out (Babrauskas, V. & Wichman, I. S. 2011, Choi, C. S et al. 2005, Wright, A. & Newbery, P. G. 2004). The variation laws of conductor resistance during the fuseing process has been studied in (Tanaka, T. & Yamasaki, M. 2004). Reference (LI, Y. 2018) studied the igniting ability and microstructural features of the melted beads after the $2.5mm^2$ CCA wire fused at 50–300A short-circuit current.

This paper derives the formula for calculating the fusing current of CCA wires based on Onderdonk's equation. Short-time fusing current test was carried out with $35mm^2$ copper wire and $60mm^2$ CCA wire which the copper layer volume ratio is 10% (CCA-10). The protective effect of flame-retardant sheath on short-circuit fire is studied, which provides a reference for the design and selection of copper and CCA earthding cable.

2 THEORETICAL CALCULATION OF INSTANTANEOUS FUSING CURRENT FOR CCA WIRES

2.1 Onderdonk's equation for copper wires

According to (SSD 332 1989) and (ASTM F855 2015), the fusing current of copper wires can be calculated using Onderdonk's equation:

$$I_m = A\sqrt{\frac{\log\left(\frac{T_m - T_A}{234 + T_A} + 1\right)}{33t}} \tag{1}$$

Where I_m is the RMS of fusing current, A; A is the wire's cross-sectional area, cmils; t is the time to withstand the current, s; T_m is the fusing temperature, °C; T_A is the ambient temperature, °C. For copper wires, by taking T_m=1083°C (melting point of copper) and $T_A = 40$°C and converting A (cmils) to S (mm^2), the equation above can be simplified as follows.

$$I_m = 283.7 S_{Cu}/\sqrt{t} \tag{2}$$

2.2 Promotion of Onderdonk's equation

For CCA wires, as the compound layer at the interface between copper and aluminum is usually no thicker than 20μm (Zhao, Z. S. & Dai, Y. K. 2011), the influence of this interface layer and the heat transfer between the copper and aluminum layers in the transient process can be neglected. The fusing characteristics of the copper layer can be expressed by Onderdonk's equation. Short-time fusing current of aluminum conductor is derived as follows:

As the duration of short-circuit current is very short and heat exchange between the wire and external environment is very limited, this process can be approximated as an adiabatic process. All electric energy generated by the short-circuit current is converted to the temperature rise. Then the equation for the temperature rise of the conductor per unit length in relation to time t is:

$$c\rho A \cdot d\theta(t) = \rho_A[1 + \alpha_A \cdot \theta(t)](I^2/A) \cdot dt \tag{3}$$

Where c is specific heat capacity, ρ is mass density, A is the cross-sectional area of earthing wire, ρ_A is the wire's resistivity at initial ambient temperature T_A, α_A is the temperature coefficient of resistance and I is the RMS of short-circuit current. Equation (3) can be simplified to:

$$\frac{d\theta(t)}{1 + \alpha_A \cdot \theta(t)} = \frac{\rho_A I^2}{c\rho A^2} dt \tag{4}$$

By solving equation (4) through integration on both sides at the same time, it can be obtained (when t = 0, $\theta(0) = 0$; and when t = t, $\theta(t) = \theta$):

$$\ln(1 + \alpha_A \cdot \theta) = \frac{\alpha_A \rho_A I^2}{c\rho A^2} \cdot t \tag{5}$$

Equation (5) can be transformed to:

$$\left(\frac{\alpha_A \rho_A}{c\rho}\right)\frac{I^2}{A^2} \cdot t = \ln\left(1 + \frac{\theta}{1/\alpha_A}\right) \tag{6}$$

It can be approximately assumed that the initial ambient temperature is within a certain range (for example 0–100°C) and $\alpha_A \rho_A$ is a fixed value.

$$\alpha_A \rho_A = \alpha_{20} \rho_{20} \tag{7}$$

Where α_{20} is the temperature coefficient of resistance, ρ_{20} is resistivity of the wire. For aluminum wires (LIU, S. X. 2011):

$$\alpha_{20} = 0.00429(1/^\circ C)$$
$$\rho_{20} = 2.87 \times 10^{-8}(\Omega \cdot m)$$
$$c = 889.5[J/(kg \cdot ^\circ C)] \tag{8}$$
$$\rho = 2703(kg/m^3)$$

Equation (6) can be expressed as:

$$5.12 \times 10^{-17} \frac{I^2}{A^2} \cdot t = \ln\left(1 + \frac{\theta}{1/\alpha_A}\right) \tag{9}$$

By converting the unit of cross-sectional area from square meter to cmils ($1m^2 = 1.98 \times 10^9$circ-miles) and converting the same to the base of common logarithm, which is:

$$87.2 \frac{I^2}{A^2} \cdot t = \log\left(1 + \frac{\theta}{1/\alpha_A}\right) \tag{10}$$

In addition, from equations (7) and (8), it can be obtained:

$$1/\alpha_A = \rho_A/(\alpha_{20}\rho_{20}) = [\rho_{20} + \rho_{20} \cdot \alpha_{20}(T_A - T_{20})]/(\alpha_{20}\rho_{20})$$
$$= \frac{1}{\alpha_{20}} + T_A - T_{20} = \frac{1}{0.00429} + T_A - 20 = 213 + T_A \tag{11}$$

By substituting equation (11) into equation (10), it can be obtained:

$$87.2 \frac{I^2}{A^2} \cdot t = \log\left(1 + \frac{\theta}{213 + T_A}\right) \tag{12}$$

When wire temperature reaches the fusing point T_m, temperature rise $\theta = T_m - T_A$ and current I at this point is the maximum short-circuit current I_m, which is substituted into equation (11) to obtain:

$$87.2 \frac{I_m^2}{A^2} \cdot t = \log(1 + \frac{T_m - T_A}{213 + T_A}) \tag{13}$$

After equation (13) is transformed, it can be expressed as:

$$I_m = A\sqrt{\frac{\log\left(\frac{T_m - T_A}{213 + T_A} + 1\right)}{87.2t}} \tag{14}$$

By taking $T_m = 660°C$ (fusing point of aluminum) and $T_A = 40°C$ and converting A (cmils) to S (mm²), the equation above can be simplified to:

338

$$I_m = 155.5 S_{Al}/\sqrt{t} \tag{15}$$

For CCA-10 wire, the ratio of copper layer cross-sectional area to aluminum layer cross-sectional area $\frac{S_{Cu}}{S_{Al}} = \frac{10}{90} = 0.111$. The impact of skin effect is neglected and it is assumed that the current flowing through the copper layer and aluminum layer at the same time is inversely proportional to resistance: $\frac{I_{Al}}{I_{Cu}} = \frac{R_{Cu}}{R_{Al}} = \frac{\rho'_{Cu}/S_{Cu}}{\rho'_{Al}/S_{Al}} = \frac{1.724}{2.87} \div 0.111 = 5.4$. If the total current through the wire is I, then:

$$I_{Cu} = 0.156I \tag{16}$$

$$I_{Al} = 0.844I \tag{17}$$

By substituting equation (17) into equation (15), the cross-sectional area of the aluminum layer is obtained:

$$S_{Al} = \frac{0.844I\sqrt{t}}{155.5} = \frac{I\sqrt{t}}{184.24} \tag{18}$$

Then the cross-sectional area of the copper layer is:

$$S_{Cu} = S_{Al} \cdot 0.111 = \frac{I\sqrt{t}}{1658} \tag{19}$$

In addition, by substituting equation (16) into equation (2), the minimum cross-sectional area required to ensure that the copper material does not fuse when current flows through the material is obtained:

$$S'_{Cu} = \frac{0.156I\sqrt{t}}{283.7} = \frac{I\sqrt{t}}{1819} \tag{20}$$

By comparing equation (19) with equation (20): $S'_{Cu} < S_{Cu}$. Therefore, aluminum material will fuse firstly when short-circuit current flows through the CCA wire. By substituting equation (17) and $S_{Al} = 0.9S$ into equation (15), it can be obtained that the relationship between the fusing current and cross-sectional area of CCA-10 wire:

$$I_m = 165.8 S_{CCA-10}/\sqrt{t} \tag{21}$$

2.3 Impact of skin effect

When AC current flows through the wire, skin depth δ can be expressed as (LIU, Q. P. et al. 2015):

$$\delta = \frac{1}{\sqrt{\pi f \mu \gamma}} \tag{22}$$

where f is current frequency, μ is the metal wire's magnetic permeability, γ is electric conductivity. Power frequency (50Hz) is the main component of short-circuit current. The skin depths of copper wire and CCA-10 wire are calculated respectively: $\delta_{Cu} = 9.4$mm, $\delta_{CCA-10} = 11.7$mm. Corresponding cross-sectional areas of cylindrical wires are 277.5mm^2 and 429.8mm^2 respectively, while the standard sizes of earthing cables are 25, 35, 50, 70, 95, 120mm^2 (DL/T 879

2004), which are much smaller than the critical value at which significant power frequency skin effect occurs. Therefore, the impact of skin effect can be neglected.

3 EXPERIMENT FOR FUSING CHARACTERISTICS OF EARTHING CABLES

3.1 Test arrangement

In order to verify the fusing characteristics of copper and CCA earthing cables at instantaneous current, fusing current test was carried out with the samples of copper and CCA-10 earthing cables. The test arrangement is shown in Figure 1 and Figure 2. 35mm^2 copper wire and 60mm^2 CCA-10 wire were selected for the test and bare wires were compared with cables of flame-retardant sheath.

3.2 Test results and analysis

Three fusing tests with different test current were carried out with bare and sheathed wires, the RMS and duration of the test currents were recorded and compared with the calculated values, as shown in Table 1 and Table 2.

According to the test results, the current - time relationship curves for the 35mm^2 copper cable and 60mm^2 CCA-10 cable were plotted in Figure 3. It can be seen that:

1) For copper cables, at the same short-circuit current, the tested fusing time is slightly longer than the calculated value. The reason is that a certain amount of energy dissipated into the surrounding environment during actual test.

2) At the same short-circuit current, CCA cable with a cross-sectional area equivalent to that of copper cable was less inclined to fusing. At short-circuit current levels of 12.9 kA, 16.3 kA and 21.7 kA, the fusing time of 60mm^2 CCA-10 bare wire was respectively 32.0%, 30.1% and 24.3% higher than that of 35mm^2 copper wire. Because the temperature usually did not reach the fusing point when wire parted. It was electro-dynamic force that pulled off the

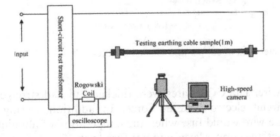

Figure 1.   Connection diagram of fusing characteristics test of earthing cable samples.

Figure 2.   On-site test arrangement and testing process.

Table 1. Copper wire test results.

No.	Wire type	RMS of test current /kA	Duration of test current/ms	Calculated fusing time/ms
1	35mm^2	12.94	616.49	588.82
2	Copper wire	16.33	392.31	369.73
3	(Bare wire)	21.72	226.67	208.99
4	35mm^2	12.91	671.03	591.56
5	Copper wire	16.31	411.90	370.64
6	(Sheathed wire)	21.51	237.56	213.10

Table 2. CCA-10 wire test results.

No.	Wire type	RMS of test current/kA	Duration of test current/ms	Calculated fusing time/ms
1	60mm^2	12.93	813.28	591.94
2	CCA-10 wire	16.31	510.05	372.02
3	(Bare wire)	21.60	281.31	212.11
4	60mm^2	12.98	826.03	587.38
5	CCA-10 wire	16.32	516.54	371.56
6	(Sheathed wire)	21.68	283.33	210.55

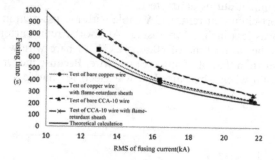

Figure 3. Fusing characteristics of copper and CCA-10 cables.

heated and softened wire. The CCA wires were thicker and had stronger capability in with-standing the short-circuit electro-dynamic force. In addition, during calculation, it was assumed that the entire wire would fuse when the temperature of aluminum wire reached the fusing point. But in fact, the copper layer had not fused yet.

3) At 12.9 kA short-circuit current, the provision of flame-retardant sheath increased the fusing time of copper cables by 8.9%. While the short-circuit current was increased to 21.5 kA, the provision of flame-retardant sheath increased the fusing time by only 4.9%. For CCA cables, flame-retardant sheath hardly affected the wire's fusing time. Because at the time of fusing, the splashing melted beads had already caused the sheath to fuse and burst. At 12.9 kA and 21.6 kA short-circuit currents, the flame-retardant sheath increased the fusing time by only 1.6% and 0.7%, but significantly reduced the number of splashing melted beads.

4 CONCLUSION

1) For CCA-10 cable, the relationship between fusing current I_m (kA), cross-sectional area S_{CCA-10} (mm^2) and fusing time t (s) is $I_m = 165.8 S_{CCA-10}/\sqrt{t}$. 35mm^2 copper cable and 60mm^2 CCA-10 cable had the same flow capacity.

2) At the same short-circuit current, the fusing time of CCA-10 cable was 24%~32% longer than that of copper cable with equivalent cross-sectional area. The result of theoretical calculation is more conservative. Therefore, a certain margin can be provided to meet the safety requirements.

3) The flame-retardant sheath can increase the fusing time of copper cables and provide better protection effect at lower short-circuit current. However, the flame-retardant sheath has no significant impact on the fusing time of CCA cables. But it can significantly reduce the number of splashing melted beads and provide effective protection against fires caused by short-circuit.

REFERENCES

ASTM F855 (ed.) 2015. *Standard specifications for temporary protective grouds to be used on de-energized electric power lines and equipment*. USA: ASTM International.

Babrauskas, V. & Wichman, I.S. 2011. Fusing of wires by electrical current. *Fire and Materials 12th International Conference and Exhibition* 769–778.

Choi, C.S et al. 2005. Analysis of dispersive characteristics and structures of copper wire melted by overcurrent. *IEEJ Trans* 125(12): 1327–1331.

DAI, K.M. 2011. Correct use of portable earthing and short-circuiting wires. *Electric Safety Technology* 13(11): 20–24.

DL/T 879 (ed.) 2004. *Portable equipment for earthing and short-circuiting for live working*. Beijing: China standard press.

IEC 61230 (ed.) 2008. *Live working - Portable equipment for earthing and short-circuiting*. International Electrotechnical Commission.

LI, J.X. (ed.) 2008. *Application of earthing and short-circuiting devices*. Beijing: China Electric Power Press.

LI, J.X. & LI, Y.C. (ed.) 2013. *Be safety, anti-shock: case detailed earthing wire use*. Beijing: China Electric Power Press.

LI, Y. 2018. Effects of the sparks spattered from the different energy short-circuit arcs on the ignitibility and microstructure of the copper- clad aluminum conductors. *Journal of Safety and Environment* 18(5): 1816–1822.

LIU, Q.P. et al. 2015. Introduction and simulation of skin effect of transmission line. *Electronic Measurement Technology* 38(6): 27–30.

LIU, S.X. (ed.) 2011. *Practical metal materials handbook*. Beijing: Mechanical Industry Press.

SSD 332 (ed.) 1989. *Technical standard of portable earthing and short-circuiting wires*. Beijing: China standard press.

Tanaka, T. & Yamasaki, M. 2004. Modelling of fuses for melting time and fusing current analysis. *Telecommunications Energy Conference* 671–675.

WANG, C.Y. 2010. Causes and preventive measures of "inductive electricity" in maintenance of transmission lines. *Electric Safety Technology* 12(9): 6–12.

Wright, A. & Newbery, P.G. (ed.) 2004. *Electric fuses*. UK: Institution of Electrical Engineers.

Zhao, Z.S. & Dai, Y.K. 2011. Study of copper clad aluminum wire annealing technology. *Electic Wire & Cable* (6): 6–8, 23.

ZHU, Z.M. 1994. The discussion of rated short-circuit current for portable earthing and short-circuiting wires. *East China Electric Power* (10): 23–25.

Emerging Developments in the Power and Energy Industry – Dufo-López, Krzywanski & Singh (eds)
© 2020 Taylor & Francis Group, London, ISBN 978-0-367-27169-5

Kinetic battery model application for capacity optimization of the wind/solar/battery stand-alone hybrid system

Zhonglei Shao* & Kwok Lun Lo
University of Strathclyde, Glasgow, UK

ABSTRACT: With the development of wind and solar energy, more local governments prefer to invest an environmental-friendly power system combining wind and solar energy especially for rural areas. Researchers use different softwares such as Homer and Matlab to achieve capacity optimization for a hybrid wind/solar/battery power system. The Kinetic Battery Model (KBM) is introduced in this article and is applied as a battery model in the combined system. Achieved results are compared and analyzed with a similar power system model which has no such a specific battery model part.

1 INTRODUCTION

Wind and Solar energy are attracting more and more attention during recent decades. For some rural areas, where major transmission lines are too costly to be installed, power systems combining wind turbines and photovoltaic (PV) panels prove to have good prospects (Z.L. Shao & K.L. Lo 2016).

It is not uncommon that in the daytime, solar irradiation intensity is high while at night wind energy resources are usually abundant. Due to the characteristics of wind and solar, energy storage devices are necessary to be included in a rural power system for reliability purposes (Z.L. Shao & K.L. Lo 2017). In this article, battery groups function as the energy storage device for a wind/solar power system.

Researchers have made considerable effort in modeling power production from wind turbines and PV panels. Few of published system include an independent theory model for the battery (Z.L. Shao & K.L. Lo 2016). Instead of matching a specific battery model in it, most researchers treat the difference between power generated by wind/solar and the demand directly as a charge or discharge power for the installed battery groups. As to this paper, the Kinetic Battery Model (Manwell JF & McGowan JG 1993) is applied in the optimization procedure. Results are compared with a model of wind/solar/battery (WSB) system without an independent battery model.

2 MATHEMATICAL MODEL OF DIFFERENT PARTS

2.1 *PV generation part*

The equation below form the power generated by the PV cells.

$$P_{pv} = P_{pv\_r} f_{pv} \left(\frac{\overline{G}_T}{\overline{G}_{T,STC}} \right) [1 + \alpha_p (T_c - T_{c,STC})] \tag{1}$$

Corresponding author: Email: zhonglei.shao@strath.ac.uk

Where P_{pv_r} is the rated capacity of PV cells and under the standard test condition where the radiation is 1 kW/m^2 and the temperature of a cell is 25°C F_{pv} is the PV derating factor due to some factors such as soiling of the panels, wiring losses, shading, material aging effect, and so on. G_T and G_{T_STC} are the solar incident radiation on the PV cells at each time step and the incident radiation at standard conditions, respectively. Tc and $T_{c,STC}$ are the PV cell temperature at each time step and the PV cell temperature under standard test conditions respectively. a_p is the temperature coefficient which also affects the power output of PV cells (Borowy B. S. & Salameh Z. M. 1996) (Borowy B. S. & Salameh Z. M. 1994).

Not only the factors mentioned above affect the performance of PV array output, but also some other parameters, such as the slope and the azimuth of a PV array, play a part in power production. In this article, these factors are taken into consideration when calculating effective incident solar radiation on the surface of the PV array.

2.2 *Wind power generation part*

Wind turbines have different power curves according to their types. Even from the same manufacturer, power output curve from a similar wind turbine type can be slightly different (Sunanda Sinha & S.S. Chandel 2015). Figure 1 shows a theoretical power curve and a power curve in real from the same wind turbine. And the real power output of a wind turbine at different time step (1 hour in this paper) can be calculated according to Figure 1.

The theoretical power can be calculated according to

$$P_w = \begin{cases} 0 & \text{if } V < V_{ci} \text{ and } V > V_{cout} \\ \frac{P_r}{V_r^3 - V_{ci}} V^3 - \frac{V_{ci}^3}{V_r^3 - V_{ci}^3} P_r & \text{if } V_{ci} < V < V_r \\ P_r & \text{if } V_r < V < V_{cout} \end{cases} \qquad (2)$$

where V_{ci}, V_{cout}, V_r are cut-in wind speed, cut-out wind speed, rated wind speed, respectively. And P_r is the rated power of a wind turbine.

2.3 *Battery model*

2.3.1 *Battery operation condition*
Battery groups function as a energy storage device in a WSB hybrid system. For a simplified battery operation process, charging power or discharging power from battery groups are decided by the difference between power served and power demanded at each time step (Xu L. et al. 2013).

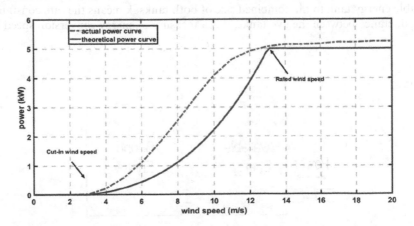

Figure 1.   Theoretical and practical wind turbine power curves.

If wind turbine and PV array serve more power than the demand, the surplus power in load will charge the battery

$$P_{charge} = P_{supplied} - P_{load} \qquad (3)$$

If wind turbine and PV array serve less power than demand, the battery discharge power to the load

$$P_{discharge} = P_{load} - P_{supplied} \qquad (4)$$

The state of charge (SOC), is used to measure of the amount of energy in a battery, 0% means no energy left and 100% means the battery is fully charged. During each time step, the state of battery charge should always satisfy

$$SOC_{min} < SOC(t) < SOC_{max} \qquad (5)$$

2.3.2 *Kinetic Battery Model application*

In this article, a typical lead acid battery kinetics model is used to represent the behavior of batteries. The Kinetic Battery Model is used to determine the amount of energy that can be absorbed or withdrawn from the battery bank at each time step (Manwell JF & McGowan JG 1993) (Manwell J. F. et al. 2005) it is shown in Figure 2.

There are total two storage tanks in this model, the available energy tank and the bound energy tank. The available energy is the amount of energy that can be instantly converted to power needed by the load. The bound energy tank stores the amount of energy that is chemically bound and can not be withdrawn immediately. All energy stored in bound energy tank should be transferred into the available energy tank first, then they are ready to be used.

At the end of each time step, energy stored in the two tanks are defined by

$$
\begin{aligned}
Q &= Q_1 + Q_2 \\
Q_{1,end} &= Q_1 e^{-k\Delta t} + \frac{(Qkc - P)(1 - e^{-k\Delta t})}{k} + \frac{Pc(k\Delta t - 1 + e^{-k\Delta t})}{k} \\
Q_{2,end} &= Q_2 e^{-k\Delta t} + Q(1 - c)(1 - e^{-k\Delta t}) + \frac{P(1 - c)(k\Delta t - 1 + e^{-k\Delta t})}{k}
\end{aligned}
\qquad (6)
$$

Where Q_1 is the available energy stored in the first tank at each time step; Q_2 is the bound energy stored in the second tank at each time step; c stands for the capacity ratio of the size of the available energy tank to the combined size of both tanks; k means the rate constant relates to the conductance between the two tanks, it is a measure of energy conversion speed between

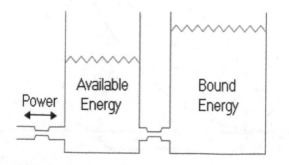

Figure 2.   Display of a Kinetic Battery Model.

345

the two tanks. Q_{max} is the maximum battery capacity, namely the total amount of energy the two tanks can contain.

Due to all these characteristic, battery charge and discharge power at each time step is limited by

$$P_{b,dmx} = \frac{-kcQ_{max} + kQ_1e^{-k\Delta t} + Qkc(1 - e^{-k\Delta t})}{1 - e^{-k\Delta t} + c(k\Delta t - 1 + e^{-k\Delta t})} \tag{7}$$

$$P_{batt,c\,max} = \frac{kQ_1e^{-k\Delta t} + Qkc(1 - e^{-k\Delta t})}{1 - e^{-k\Delta t} + c(k\Delta t - 1 + e^{-k\Delta t})} \tag{8}$$

Where Δt in all the above equation is the time step.

3 CASE STUDIES RESULTS AND ANALYSIS

A case study in which a village located at Qinghai Province, China is picked. This is a typical rural village with 10 houses and its demand is regarded as a load sample. Its longitude and latitude are 36°16.7'N, 101°15.0'E. Wind speed and solar irradiation data are all from NASA Surface meteorology and Solar Energy database (NASA 2018). The average hourly power of a total village is 625W, with a peak power of 3.2 kW, load factor is 0.2, peak load month happens in July. Daily energy output of a wind turbine, a PV array, and daily load consumption all through a year (365 days) are shown together in Figure 3,

Loss of load probability (LOLP) is a value which is used to measure the probability that a system demand will exceed supplied capacity during a given period (8760 hours in this article) (S. Dutta & R. Sharma 2012). LOLP 0% means a system configuration that can totally meet demand all through a year, while 100% means a system configuration that can not satisfy demand even for a single hour. In this article, the time step is set as 1 hour and the LOLP for a rural area is set as 5% at most, meaning that customers can only accept 438 hours power shortage in a year. When LOLP meets the requirement, a Wind turbine/PV array/Battery combination scheme with the least investment cost is chosen as the best one.

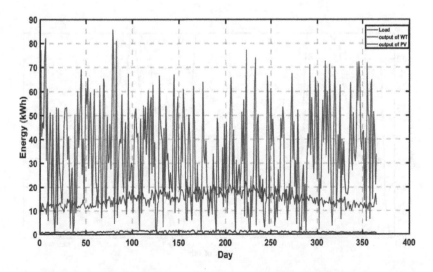

Figure 3.   Daily power output of a wind turbine and a PV array, daily load power.

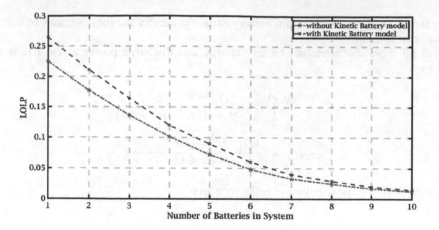

Figure 4.   LOLP vs number of batteries in two different optimization models.

3.1   *Scenario 1: system combied of 1 wind turbine and 10 PV arrays with different battery numbers with KBM or without KBM*

Figure 4 shows different LOLP value with a hybrid power system combining 1 wind turbine and 10 PV arrays, but the number of batteries ranges from 1 to 10. In Figure 4, it is obvious that in a WSB system, the LOLP will decrease correspondingly with number of batteries increasing in system.

When there is only one battery in this WSB system, the difference of LOLP is evident. LOLP from system optimization with KBM in is around 27% and LOLP from system optimization without KBM in is around 22.5%, differing almost 5%. When number of batteries reach ten, the difference of LOLP in these two conditions is not so apparent, because all batteries share the energy storage task together and the number of batteries in total is more than sufficient.

What is more, when Kinetic Battery Model is applied in optimization process, it can be concluded LOLP value always exceed that of system optimization without Kinetic Battery Model in. It makes sense since batteries model using KBM can not charge or discharge as much energy as that not using KBM. It means the combination of wind turbines, PV arrays and

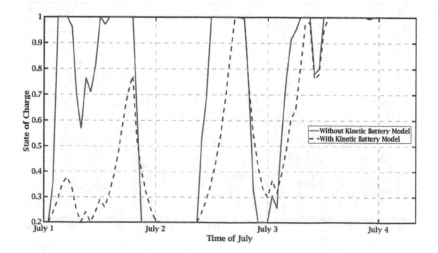

Figure 5.   Battery state of charge in 3 days in two different optimization models.

batteries calculated by a system optimization without KBM may not seem as reliable as it actually appears.

3.2 Scenario 2: system combination of 1 wind turbine, 10 PV arrays and 2 batteries with KBM or without KBM

In Figure 5, the WSB system combination is fixed as 1 wind turbine, 10 PV arrays and 2 batteries. The battery state of charge condition in the first 3 days of July is displayed. It shows that when battery is in charging condition, battery in system optimization without KBM have much more charging power at each time step compared that with KBM. Similarly, when battery is in discharging condition, battery in system optimization without KBM have higher discharge power than that with KBM.

What is more, when KBM is applied in system optimization, it is obvious the state of charge curve is smoother than that without KBM.

4 CONCLUSIONS

This article shows the basic principle of a specific Kinetic Battery model, which provides an ideal match with lead acid battery behavior. Results are compared between system optimization with KBM in or without KBM. On one hand, it is convincing that KBM model makes battery behavior close to what they are in reality. On the other hand, it offers help to promote development of hybrid renewable power systems such as wind/solar/battery power systems.

REFERENCES

Borowy, B.S. and Salameh, Z.M. (1994) Optimum photovoltaic Array Size for A Hybrid Wind/PV system. In IEEE Trans. On energy Conversion, 9, 482–488.

Borowy, B.S.; Salameh, Z. M. (1996) Methodology for optimally sizing the combination of a Battery Bank and PV Array in a Wind/PV Hybrid System. IEEE Trans.on Energy Conversion, 11, 367–375.

Manwell JF, McGowan JG (1993) Lead acid storage model for hybrid energy systems, Solar Energy, 50, 399–405.

Manwell, J.F., McGowan, J.G., Abdulwahid, U., & Wu, K. (2005, May), Improvements to the hybrid2 battery model, In Windpower 2005 Conference, American Wind Energy Association.

NASA Prediction Of Worldwide Energy Resources (2018). https://power.larc.nasa.gov/.

S. Dutta. R. Sharma. (2012) Optimal storage sizing for integrating wind and load forecast uncertainties. 2012 IEEE PES Innovative Smart Grid Technologies (ISGT).

Shao, Z.L. and Lo, K.L. (2016) An Overview of Research on Optimization of Integrated Solar/Wind Power Generation Systems. World Journal of Engineering and Technology, 4, 35–42.

Shao, Z.L. and Lo, K.L. (2017) A Case Study to Determine Optimal Capacity of a Stand-Alone Wind/PV/Battery System. Energy and Power Engineering, 9, 300–307.

Sunanda Sinha, S.S. Chandel. (2015) Review of recent trends in optimization techniques for solar photovoltaic-wind based hybrid energy systems. Renewable and Sustainable Energy Reviews, 50, 755–769.

Xu, L., Ruan, X.B., Mao, C.X., Zhang, B.H. and Luo, L. (2013) An Improved Optimal Sizing Method for Wind-Solar-Battery Hybrid Power System. IEEE Trans. Sustain.Energy, 4, 774–785.

Emerging Developments in the Power and Energy Industry – Dufo-López, Krzywanski & Singh (eds)
© 2020 Taylor & Francis Group, London, ISBN 978-0-367-27169-5

Optimal method for day-head dispatching of a multi-energy micro-grid

Peng Zhang*
School of Electrical and Information Engineering, Tianjin University, Tianjin, China

Yue Song*
School of Electrical and Information Engineering, Tianjin University, Tianjin, China

ABSTRACT: A multi-energy micro-grid is an important topic to explore the harmonious development of energy, economy, and society. The work reported in this article researched the current dispatching problem of a micro-grid that contains renewable energy. First, considering the economic and environmental aspects, multi-energy micro-grid dispatching models with the lowest operating cost and the lowest pollutant emission are established. Second, based on the typical equipment model of renewable energy technology, energy-saving technology, and electric energy alternative technology in a multi-energy micro-grid, a system operation constraint model, the balance of supply and demand of electricity, and cold/heat load are built in the heating and air-conditioning period. The model is a hybrid nonlinear optimization problem, and the particle swarm optimization algorithm with good global searchability is used in the model. Finally, through a specific example, the model and algorithm are validated, and the optimal energy micro-grid optimization dispatching scheme under the economic and environmental protection objectives is formed. Meanwhile, the reasons for the uniformity and contradiction of the scheduling scheme under different objectives are analyzed.

Keywords: multi-energy micro-grid, multi-objective, optimal dispatching, particle swarm optimization algorithm

1 INTRODUCTION

Energy is an essential basis for human survival and development (Jin 2005). How to make use of renewable energy and improve the efficiency of multi-energy utilization is a hot spot in multi-energy research (Stanislav 2009; Liu 2014). By rationally planning and operating optimal control of a multi-energy micro-grid, such as electricity, gas, and heat, a multi-energy micro-grid consisting of a distributed terminal integrated energy unit and a centralized energy supply network coupled with it are expected to become adaptive for humans. Multi-energy coupling and cascade utilization are effective ways to reduce the impact of distributed energy fluctuations on grids, foster the development of renewable energy, alleviate the shortage of fossil energy, and reduce environmental pollution (Wang 2010). From the perspective of energy utilization, multiple energy systems are correlated and complementary on different time scales, enabling energy storage and transfer over multiple time scales. Therefore, in the process of energy generation and utilization of integrated energy systems, the optimal scheduling of integrated energy systems has become a basic problem.

Existing research is mostly directed to multi-energy micro-grid systems. Sahin et al. (2012) deeply analyze the coupling relationship between natural gas and power systems in the

---

*Corresponding authors: zhangpeng1984@tju.edu.cn, 951696865@qq.com

United States, and propose an electric power risk assessment method that takes into account the operational constraints of natural gas pipelines. Alabdulwahab et al. (2015) propose a predistribution method for electric power units that takes into account the randomness of wind power. Xu et al. (2015) provide a dynamic model of a natural gas–electric hybrid system. Kamalinia et al. (2014) take into account wind power access, study the impact between the natural gas network and hydropower system, and put forward solutions for power system unit combination problems. Awad (2009) considers wind power output volatility to solve the problem, and propose an optimization model for cogeneration scheduling of a micro-grid. Zhou et al. (2012) establish a multi-objective energy-saving scheduling model with production cost, environmental cost, and cogeneration cost of cold and heat loads. The distributed energy system integrating cold, heat, and electricity is suitable for areas with a complete natural gas pipeline network, high power generation ratio, and high integration of grid gas network. At this stage, the domestic integrated energy system is more based on electricity, solar energy, wind energy, or geothermal energy, including renewable energy equipment (such as wind turbines, photovoltaics, etc.), energy-saving equipment (such as heat pumps, ice storage, etc.), and electrical energy substitute equipment (integrated energy supply systems such as electric boilers, electric floor heating, etc.). Little research has been carried out on optimization dispatching of an integrated energy system considering the aforementioned three types of equipment.

This article focuses on the optimal dispatching method of a multi-energy micro-grid system containing renewable energy technologies, energy-saving technologies, and energy-replacement technologies. Considering the equipment of the aforementioned three technologies, a mathematical model of multi-energy micro-grid optimal scheduling based on electricity is established. The model considers a variety of optimization objectives and the system constraints under different energy form requirements, and also uses heuristic algorithms to solve the problem. In the end, a concrete example is given. The simulation results show that the proposed model and algorithm can give an energy-optimized scheduling solution.

2 DISPATCHING MODEL OF A MULTI-ENERGY MICRO-GRID

The operation scheduling of the multi-energy micro-grid aims at meeting the requirements of the electrical load and the cooling/heating loads. Based on certain optimization criteria, the forecast values of various loads are used as a baseline. Using the optimal algorithm, the 24-hour dispatching schedule for the next day can be worked out with the optimal dispatching method for multi-energy micro-grid operation.

2.1 Objective function of optimal dispatching

The dispatching of multi-energy micro-grid operation scheduling contains multi-objectives. In this article, considering the economic and environmental conservation of the system operation, two mathematical models of optimal objectives are established.

The economic optimal criterion is based on the dispatching operation cost of the multi-energy micro-grid. The principal operating cost includes the electricity cost purchased by the community operation and the subsidy of renewable energy power generation. The economic optimal criterion objective function is shown in Equation (1):

$$F_1 = \sum_{t=1}^{T} C_t P_t^{TL} H - \sum_{n=1}^{N} \sum_{t=1}^{T} C^{RE_n} P_t^{RE_n} H \tag{1}$$

where T, H, C_t, P_t^{TL}, C^{RE_n}, and $P_t^{RE_n}$ respectively represent the total interval numbers of optimal dispatching of a multi-energy micro-grid, dispatching hours for each interval, hourly city electric power, powers of paralleled connection lines of the community, subsidy price of nth renewable energy, and generated power of nth renewable energy. H was set to 1 herein.

The environmental protection optimal criterion is based on the pollutant emissions of the integrated energy system during the dispatch period. For every 1 kW·h of mains saved, a corresponding saving of 0.4 kg of standard coals was saved, and meanwhile 0.272 kg of carbon dust, 0.997 kg of CO_2, 0.03 kg of SO_2, and 0.015 kg of NOx are reduced. Pollutant emission of new energy technology is converted into electricity consumption, and the ultimate environmentally friendly objective function is as shown in Equation (2):

$$F_2 = C \sum_{t=1}^{T} P_t^{TL} H \qquad (2)$$

where C represents total quantity of various pollutant discharges per kW·h electricity, which is about 1.3 kg/(kW·h); F_1 and F_2 represent both optimized objectives and evaluation indexes for the dispatching of a multi-energy micro-grid operation.

2.2 Constraints in multiple operation models

(1) In a multi-energy micro-grid based on different seasons and meteorological conditions, the demand for energy forms is different. The energy supply equipment and operating conditions of the multi-energy micro-grid will be quite different. This article discusses the system's constraint operational model of integrated energy-optimized scheduling in two cases. Heating period

In the case of the heating period, integrated energy-optimized scheduling is used mainly to meet the electrical load and thermal load demand in the multi-energy micro-grid. The main considerations for electrical heating technology and energy-saving technologies include electric boiler system and heat pump system (Zhu 2007; Liu 2011). A storage heating electric boiler system will combine with the heat pump system to provide a thermal load to the community to meet heating needs.

(i) Constraints of balance between supply and demand of electrical loads

$$P_t^{TL} + \sum_{n=1}^{N} P_t^{RE_n} = L_t^E + P_t^{HP-H} + P_t^B \qquad (3)$$

where L_t^E is the demand of electrical load except electric energy storage equipment in the community and P_t^B represents the quantity of electric energy consumption of the heat storage electric boiler, which can be worked out in Equation (4):

$$P_t^B = \sum_{i=1}^{N^B} \left[Q_{t,i}^B / \eta^{B + U_{t,i}^{B-H} P^{B-HWP-R}} \right] \qquad (4)$$

where P_t^{HP-H} represents electricity consumption for the heat supply of the heat pump, which can be worked out in Equation (5):

$$P_t^{HP-H} = \sum_{i=1}^{N^{HP}} \left(U_{t,i}^{HP} a_i^{HP-H} + b_i^{HP-H} Q_{t,i}^{HP-H} \right) \qquad (5)$$

where a_i^{HP-H} and b_i^{HP-H} represent the fitting parameters of the electric power-heating capacity curve of the i-th heat pump. N^{HP}, $Q_{t,i}^{HP-H}$, and $U_{t,i}^{HP}$ respectively represent the number of heat pumps that can be put into operation, the quantity of heat production of the i-th heat pump, and the operation status of the heat pump.

(ii) Constraints of balance between supply and demand of heating loads

$$\sum_{i=1}^{N^{HP}} Q_{t,i}^{HP-P} + \sum_{i=1}^{N^{B}} Q_{t,i}^{B-H} + Q_t^{HWT-H} = L_t^H \tag{6}$$

where $Q_{t,i}^{HP-P}$, $Q_{t,i}^{B-H}$, Q_t^{HWT-H}, and L_t^H respectively represent quantity of heat for heating of the ith heat pump, quantity of heat for direct heating of the ith heat storage electric boiler, quantity of heat for heating of the continuous hot water tank, and the requirement for heating of heat loads.

(iii) Constraints of the electric boiler for load distribution

$$Q_{t,i}^{B-H}\left(\sum_{i=1}^{N^{HP}} U_{t,i}^{HP} F_{t,i}^{HP} + \sum_{i=1}^{N^{B}} U_{t,i}^{B-H} F_{t,i}^{B-HWT} = U_{t,i}^{B-H} F_{t,i}^{B-HWT}\left(L_t^H - Q_t^{HWT-H}\right) \tag{7}$$

$$Q_{t,i}^{B-HWT} \sum_{i=1}^{N^{B}} U_{t,i}^{B-HWT} = U_{t,i}^{B-HWT} Q_t^{B-HWT} \tag{8}$$

$$Q_{t,i}^{B} = Q_{t,i}^{B-HW} + Q_{t,i}^{B-H} + Q_{t,i}^{B-HWT} \tag{9}$$

$$U_{t,i}^{B-HWT} + U_t^{HWT} \le 1 \tag{10}$$

In Equations (7)–(10), $U_{t,i}^{HP}$, N^{HP}, $F_{t,i}^{HP}$, $F_{t,i}^{B-HWT}$, $U_{t,i}^{B-H}$, $Q_{t,i}^{B-HWT}$, Q_t^{HWT-H}, $U_{t,i}^{B-HWT}$, U_t^{HWT}, and Q_t^{B-HWT} respectively represent the operational model of heat pump, number of heat pumps that can be put into operation, the rated flow of the heat exothermic circulating pump, the rated flow of thermal storage electric boiler system's exothermic circulation pumps, heating model setting of regenerative electric boiler for direct heat supply, heat storage quantity of the ith heat electric boiler, heat release of the water storage tank, heat storage model of the ith heat storage electric boiler, heat release model of the water storage tank, and heat storage quantity of the water storage tank.

(iv) Constraints of maximum and minimum operating capacity.

$$U_{t,i}^{B-HWT} \underline{Q}_i^{B-HWT} \le Q_{t,i}^{B-HWT} \le U_{t,i}^{B-HWT} \overline{Q}_i^{B-HWT} \tag{11}$$

$$U_{t,i}^{B-H} \underline{Q}_i^{B-H} \le Q_{t,i}^{B-H} \le U_{t,i}^{B-H} \overline{Q}_i^{B-H} \tag{12}$$

$$U_{t,i}^{B} \underline{Q}_i^{B} \le Q_{t,i}^{B} \le U_{t,i}^{B} \overline{Q}_i^{B} \tag{13}$$

In Equations (11)–(13), $\underline{Q}_i^{B-H}$ and、 $\overline{Q}_i^{B-H}$ respectively represent the minimum and maximum operating capacity limit for an electric boiler directly supplied to heating loads. $\underline{Q}_i^{B-HWT}$ and、 $\overline{Q}_i^{B-HWT}$ respectively represent the minimum and maximum operating capacity limit of heat storage of the electric boiler. $\underline{Q}_i^{B}$ and、 $\overline{Q}_i^{B}$ respectively represent the minimum and maximum operating capacity limit of the electrical boiler.

(Operational constraint of heat storage water tank

$$W_t^{HWT} = \left(1 - \varepsilon^{HWT}\right) W_{t-1}^{HWT} + \sum_{i=1}^{N^B} Q_{t,i}^{B-HWT} - Q_t^{HWT-HW} \tag{14}$$

$$\underline{W}^{HWT} \leq W_t^{HWT} \leq \overline{W}^{HWT} \tag{15}$$

$$\underline{Q}^{HWT} U^{HWT} \leq Q_t^{HWT} \leq U_t^{HWT} \overline{Q}^{HWT} \tag{16}$$

In Equations (14)–(16), ε^{HWT} represents the rate of self-heat release, $\underline{W}^{HWT}$ and $\overline{W}^{HWT}$ respectively represent the maximum and minimum heat operating capacity limit of the water storage tank and $\underline{Q}^{HWT}$, $\overline{Q}^{HWT}$ respectively represent the minimum and maximum operating capacity limit of the heat water storage tank.

Distribution restraints of loads of the heat pump

$$Q_{t,i}^{HP-H} \left(\sum_{i=1}^{NHP} U_{t,i}^{HP} F_{t,i}^{HP} + \sum_{i=1}^{N^B} U_{t,i}^{B-H} F_{t,i}^{B-HWT} \right) = U_{t,i}^{HP} F_{t,i}^{HP} \left(L_t^H - Q_t^{HWT-H} \right) \tag{17}$$

Restraints of maximum and minimum operating capacity limit of the heat pump

$$U_{t,i}^{HP} \underline{Q}_i^{HP-H} \leq Q_{t,i}^{HP-H} \leq U_{t,i}^{HP} \overline{Q}_i^{HP-H} \tag{18}$$

In Equation (18), $\underline{Q}_i^{HP-H}$, $\overline{Q}_i^{HP-H}$ respectively represent the maximum and minimum operating capacity limit of the heat pump unit.

During the heating period, the optimization variables are shown as Equation (19):

$$\left\{ U_{t,i}^{B-HW}, U_{t,i}^{B-HWT}, U_{t,i}^{B-H}, U_{t,i}^{HP-H}, Q_t^{HWT} \right\} \tag{19}$$

(2) Cooling period

In the case of the air conditioning period, the purpose of optimal dispatching of the multi-energy micro-grid is mainly to meet the electrical and cooling load demand in the community. The main considerations for cooling and energy saving technologies include dual-mode mainframe refrigeration system, onboard mainframe refrigeration system, heat pump system, and ice storage system (Swardt 2001; Zhang 2012).

(i) Constraints of balance between supply and demand of electrical loads

$$P_t^{TL} + \sum_{n=1}^{N} P_t^{RE_n} = P_t^E + P_t^{HP} + P_t^{CC} + P_t^{IC} + P_t^B \tag{20}$$

In Equation (20), P_t^{HP} represents the total consumption of electricity of cooling and cooling storage, P_t^{CC} represents the consumption of electricity of the airborne host, and P_t^{IC} represents the total consumption of electricity of cooling and ice making.

The expressions are shown as Equations (21), (22), and (23).

$$P_t^{HP} = \sum_{i=1}^{N^{HP}} \left(U_{t,i}^{HP-C} a_i^{HP-C} + b_i^{HP-C} Q_{t,i}^{HP-C} \right) \tag{21}$$

$$P_t^{CC} = \sum_{i=1}^{N^{CC}} \left(U_{t,i}^{CC} a_i^{CC} + b_i^{CC} Q_{t,i}^{CC} \right) \tag{22}$$

$$P_t^{IC} = \sum_{i=1}^{N^{HP}} \left[\left(U_{t,i}^{IC-C} a_i^{IC-C} + b_i^{IC-C} Q_{t,i}^{IC-C} \right) + \left(U_{t,i}^{IC-ICE} a_i^{IC-ICE} + b_i^{IC-ICE} Q_{t,i}^{IC-ICE} \right) \right] \qquad (23)$$

In Equations (21)–(23), a_i^{HP-C} and b_i^{HP-C} represent the fitting parameters of the electric power–refrigerating capacity curve of the ith heat pump. a_i^{CC} and b_i^{CC} represent the fitting parameters of the electric power–refrigerating capacity curve of the ith dual- mode mainframe refrigeration system, and a_i^{IC-ICE} and b_i^{IC-ICE} represent the fitting parameters of the electric power–ice making capacity curve of the ith dual-mode mainframe refrigeration system. $U_{t,i}^{HP-C}$, $U_{t,i}^{CC}$, and $U_{t,i}^{IC-C}$ respectively represent the cooling operational models of the heat pump, the cooling operational models of the dual-mode mainframe refrigeration system, and the ice-making operational model of the cooling operational models.

(ii) Constraints of balance between supply and demand of cooling loads

$$\sum_{i=1}^{N^{HP}} Q_{t,i}^{HP-C} + \sum_{i=1}^{N^{CC}} Q_{t,i}^{CC} + \sum_{i=1}^{N^{IC}} Q_{t,i}^{IC-C} + Q_t^{ICET-C} = L_t^C \qquad (24)$$

In Equation (24), $Q_{t,i}^{HP-C}$, $Q_{t,i}^{CC}$, $Q_{t,i}^{IC-C}$, Q_t^{ICET-C}, and L_t^C respectively represent the cooling capacity of the heat pump system, dual-mode mainframe refrigeration system, onboard mainframe refrigeration system, and the total need of cooling loads in the community.

(iii) Flow restraints of the cooling water pump

$$F^{HP-C} \sum_{i=1}^{N^{HP}} U_{t,i}^{HP-C} + F^{PCWP-HP} N_t^{PCWP-HP} + F^{CC} \sum_{i=1}^{N^{CC}} U_{t,i}^{CC} + F^{PCWP-ICE} N_t^{PCWP-ICE} = F_t^C$$

$$(25)$$

In Equation (25), F^{HP-C}, $F^{PCWP-HP}$, F^{CC}, and $F^{PCWP-ICE}$ respectively represent the rated flow of the chilled water pump of the ground source heat pump system, refrigeration cycle pump of the chilled water storage system, cooling pump of the host unit for the base load, and F_t^C represents the current flow of the system.

(iv) Distribution restraints of cooling loads of the heat pump

$$F_t^C Q_{t,i}^{HP-C} = U_{t,i}^{HP-C} F^{HP-C} \left(L_t^C - Q_t^{ICET-C} \right) \qquad (26)$$

(v) Restraints of the maximum and minimum operating capacity of the heat pump

$$U_{t,i}^{HP-C} \underline{Q}_t^{HP-C} \leq Q_{t,i}^{HP-C} \leq U_{t,i}^{HP-C} \overline{Q}_i^{HP-C} \qquad (27)$$

where $\underline{Q}_t^{HP-C}$ and $\overline{Q}_i^{HP-C}$ respectively represent the minimum and maximum operating capacity for the cooling operation of heat pump units.

(vi) Distribution restraints of cooling loads of the host unit for the base load

$$F_t^C Q_{t,i}^{CC} = U_{t,i}^{CC} F^{CC} \left(L_t^C - Q_t^{ICET-C} \right) \qquad (28)$$

(vii) The restraints of minimum and maximum operating capacity of the onboard mainframe refrigeration system for the base load

$$U_{t,i}^{CC} \underline{Q}_i^{CC} \leq Q_{t,i}^{CC} \leq U_{t,i}^{CC} \overline{Q}_t^{CC} \tag{29}$$

In Equation (29), $\underline{Q}_i^{CC}$ and $\overline{Q}_t^{CC}$ respectively represent the restraints of the maximum and minimum operating capacity of the onboard mainframe refrigeration system for the base load.

(viii) The distribution restraints of the dual-mode mainframe refrigeration system

$$F_t^C Q_{t,i}^{IC-C} = U_{t,i}^{IC-C} F^{PCWP-ICE} \left(L_t^C - Q_t^{ICET-C} \right) \tag{30}$$

(ix) The distribution restraints of ice-making of the onboard mainframe refrigeration system for the base load

$$Q_{t,i}^{IC-ICE} \sum_{i=1}^{N^{IC}} U_{t,i}^{IC-C} = U_{t,i}^{IC-C} Q_t^{IC-ICE} \tag{31}$$

(x) The maximum and minimum restraints of operating capacity of the dual-mode mainframe refrigeration system

$$U_{t,i}^{IC-C} \underline{Q}_i^{IC-C} \leq Q_{t,i}^{IC-C} \leq U_{t,i}^{IC-C} \overline{Q}_i^{IC-C} \tag{32}$$

$$U_{t,i}^{IC-ICE} \underline{Q}_i^{IC-ICE} \leq Q_{t,i}^{IC-ICE} \leq U_{t,i}^{IC-ICE} \overline{Q}_i^{IC-ICE} \tag{33}$$

In Equations (32) and (33), $\underline{Q}_i^{IC-C}$, $\overline{Q}_i^{IC-C}$ respectively represent the maximum and minimum restraints of the operating capacity of a dual-mode mainframe refrigeration system. $\underline{Q}_i^{IC-ICE}$ and $\overline{Q}_i^{IC-ICE}$ respectively represent the maximum and minimum restraints of the operating capacity of the dual-mode mainframe refrigeration system.

(xi) The operating model restraints of the dual-mode mainframe refrigeration system

$$U_{t,i}^{IC-ICE} + U_{t,i}^{IC-C} \leq 1 \tag{34}$$

(xii) The constraint of the maximum melting power restraint of an ice storage tank

$$W_t^{ICET} = \left(1 - \varepsilon^{ICET} \right) W_{t-1}^{ICET} + Q_t^{IC-ICE} - Q_t^{ICET-C} \tag{35}$$

$$\underline{W}^{ICET} \leq W_t^{ICET} \leq \overline{W}^{ICET} \tag{36}$$

$$U_t^{ICET-C} \underline{Q}^{ICET-C} \leq Q_t^{ICET-C} \leq U_t^{ICET-C} \overline{Q}^{ICET-C} \tag{37}$$

$$U_t^{IC-ICE} + U_t^{ICET-C} \leq 1 \tag{38}$$

where $\underline{W}^{ICET}$ and $\overline{W}^{ICET}$ respectively represent the minimum and maximum ice storage capacity of an ice storage tank. $\underline{Q}^{ICET-C}$ and $\overline{Q}^{ICET-C}$ respectively represent the maximum and minimum operating capacity when ice storage tanks are melting.

During the air-conditioning period, the optimization variables are shown as Equation (39):

$$\left\{ U_{t,i}^{B-HW}, U_{t,i}^{B-HWT}, U_{t,i}^{HP-C}, U_{t,i}^{CC}, U_{t,i}^{IC-C}, Q_t^{HWT}, Q_t^{ICET} \right\} \tag{39}$$

3 THE OPTIMAL ALGORITHM FOR DISPATCHING OF A MULTI-ENERGY MICRO-GRID

From the perspective of mathematical models, the dispatching of a multi-energy micro-grid is a hybrid optimization nonlinear problem. Common mathematical algorithms for solving such problems are mixed-integer stochastic programming, dynamic planning, heuristic algorithm, and so on (Liu 2012; Guo 2014). The optimal dispatching of a multi-energy micro-grid has characteristics of complex constraints and high solution dimension, thus making it difficult to obtain an analytical solution. Therefore, a heuristic algorithm is utilized to solve the problem in this article. Tabu Search has a good global searchability and hence is applied for the optimization problem.

In the optimal model of a multi-energy micro-grid, the single particle x of swarm optimization is defined as the matrix composed of the optimization variables. Taking the heating period as an example, the optimized particle matrix is shown in Equation (39). The fitness function $F(x)$ in particle swarm optimization is the objective function defined in this article, named Equation (1) or Equation (2).

3.1 *Simulation condition*

An energy-saving green community of a certain city in northern China was chosen as the research object. The community's integrated energy needs include winter heating and summer cooling. Power supplies include a city electrical supply and a distributed photovoltaic power system. The cold/heat source system adopts a composite energy system. The energy storage system includes an electric boiler, a heat pump system, an onboard host refrigeration system, water-ice chilling system, and an ice storage system. The number of basic energy storage equipment and basic performance parameters is shown in Table 1.

In addition to the integrated energy storage equipment at 24 points on a typical day in the community, the simulation data input contains the 24-hour electric loads, photovoltaic power output, cold/heat loads, peak-to-valley electricity price information and performance parameter of each storage device. The peak hours of the city are 08:30 to 11:30 h and 18:00 to 23:00 h and the peak hourly electricity price is 1.289 8 yuan/(kWh). The flat time periods are 07:00 to 08:30 h and 11:30 to 18:00 h and the usual electricity price is 0.8443 yuan/(kWh). The trough time is 23:00 to 07:00 h and the valley electricity price is 0.4188 yuan/(kWh). Through the analysis of historical electric load data and photovoltaic output data, the typical daily load curve and photovoltaic output curve are generated.

Load analysis of the park using DeST software yields a typical 24-point cold/heat load curve. The various loads are shown in Figure 1. The number of particle swarms is 100, the

Table 1. Quantity and specification of energy storage equipment.

Equipment	SpecificationV		Number
Electric boiler	2050 kW		6
Thermal storage	606 m^3		4
Water chilling system	3164 kW		2
Heat pump	Cooling	1195 kW	3
	Heat	1267 kW	
Water-ice chilling system	Cooling	3164 kW	2
	Heat	2339 kW	
Ice storage	231RTH		48

356

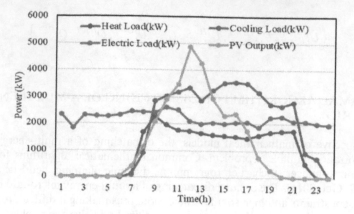

Figure 1.   Hourly load curves for a day.

maximum number of iterations is 1000, the individual acceleration factor is set to 2, and the global acceleration factor is set to 2.

3.2   *The results of simulation*

(1) Economic optimum criterion

Taking the economic optimality as the objective function, the PSO algorithm is used to calculate the equipment operation and dispatching plan for the heating and air-conditioning period under the typical daily load curve. The results are shown in Figure 2.

(2) Environmental optimum criterion

Taking the environmental optimization as the objective function, the equipment operation and dispatching plan for the heating and air-conditioning period under the typical daily load curve are calculated. The results are shown in Figure 3.

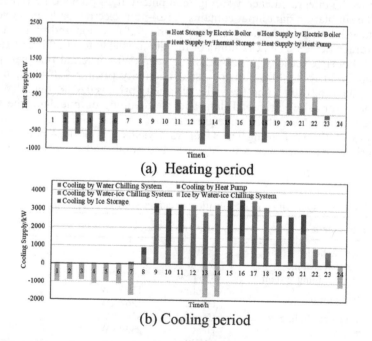

(a) Heating period

(b) Cooling period

Figure 2.   The dispatching plan under an economic optimum criterion.

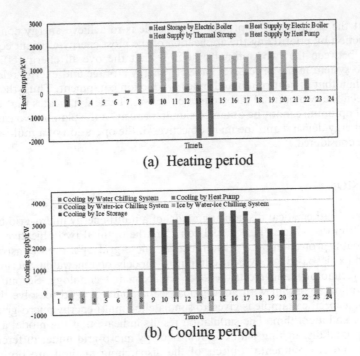

(a) Heating period

(b) Cooling period

Figure 3. The dispatching plan under an environmental optimum criterion.

The economic and environmental indicators of optimal dispatching plans under the two optimization criteria are shown in Table 2.

3.3 Analysis of results

The economic efficiency and environmental protection objectives of the dispatching of a multi-energy micro-grid are both unified and contradictory. One hand, the optimal dispatching under both targets tends to achieve maximum consumption of renewable energy. It can be seen from Figures 2a and 3a that in the period of large renewable energy generation, the cut light is avoided by converting the residual electricity into other energy forms, or the direct supply meets the cold/heat demand, or stores it for time-sharing use of energy. On the other hand, there are contradictions between the economic and environmental indicators of dispatching of multi-energy micro-grid. It can be observed in Figure 2a that the economic optimum criterion is to use the energy storage system at the low price, to realize the transfer of electricity to heat and time-sharing, thereby reducing operating costs. It can be observed in

Table 2. Evaluation indicators for dispatching plans.

	Heating period		Cooling period	
	Economic	Environment	Economic	Environment
Criterion	(yuan)	(kg)	(yuan)	(kg)
Economic	15,115	53,500	19,980	58,178
Environmental	19,020	49,080	22,137	57,472

Figure 3a that the environmental protection principle is to achieve energy conservation and emission reduction by prioritizing the application of the most energy-efficient equipment. The contradiction between the two optimal criteria is that the overall energy efficiency of the energy storage system in the process of capacity energy storage and energy release is significantly lower than that of heat pump and other production equipment. Thus, there are contradictions between the two optimal dispatching objectives. Table 2 indicates that economic and environmental optimum cannot be obtained simultaneously in the plan in two cases. The characteristics of energy demand and the main objectives of the operation of a multi-energy micro-grid should be considered.

4 CONCLUSION

First, an optimal mathematical dispatching model of a multi-energy micro-grid based on electricity was built in this article. This model considers the optimal two objectives of economic and environmental protection, and a multi-energy micro-grid operation constraint model is established, which is based on typical equipment models of renewable energy technology, energy-saving technology, and electric energy replacement technology. Second, the particle swarm optimization algorithm with good global searchability is used to solve the scheduling model. Finally, a specific example is given to simulate the multi-energy micro-grid dispatching schedule model and algorithm. The simulation results indicate that the model and algorithm can give the optimal dispatching plan of a multi-energy micro-grid under different objectives. The economic and environmental objects of the dispatching method are both unified and contradictory. The unification is that the two optimal dispatching methods both tend to realize the consumption and absorption of renewable energy and the improvement of the efficiency of comprehensive energy utilization. The paradox is that the economic criterion tends to make full use of the energy-storage system and the environmental criterion tends to use other energy-supply equipment that is energy efficiency. Economic and environmental optima cannot be met simultaneously in the dispatch plan. The characteristics of energy demand and main objectives of the operation of a multi-energy micro-grid should be considered in making an optimal dispatching plan.

ACKNOWLEDGMENTS

This work was financially supported by the National Key Research and Development Program of China (2017YFB0903300) and Natural Science Foundation of China (Grant No. 51807133).

REFERENCES

Alabdulwahab, A., Abusorrah, A., Zhang, X., et al. 2015. Coordination of interdependent natural gas and electricity infrastructures for firming the variability of wind energy in stochastic day-ahead scheduling. *IEEE Transactions on Sustainable Energy* 6(2): 606–615.

Awad, B., Chaudry, M., Wu, J, et al. 2009. Integrated optimal power flow for electric power and heat in a micro-grid. In Proceedings of the 20th International Conference and Exhibition on Electricity Distribution. IEEE, Prague, Czech Republic: 1–2.

Guo, L., Liu, W., Jiao, B., et al. 2014. Multi-objective optimal planning design method for stand-alone micro-grid system. *Proceedings of the CSEE* 34(4): 524–536.

Jin, H., Hong, H., Wang, B., et al. 2005. A new principle of synthetic cascade utilization of chemical energy and physical energy. *Science in China Series E: Technological Sciences* 48(2): 163–179.

Kamalinia, S., Wu, L., & Shahidehpour, M. 2014. Stochastic midterm coordination of hydro and natural gas flexibilities for wind energy integration. *IEEE Transactions on Power Systems* 5(4): 1070–1079.

Liu, B., Huang, X., & Li, J. 2014. Optimal sizing of distributed generation in a typical island micro-grid with time-shifting load. *Proceedings of the CSEE* 34(25): 4250–4258.

Liu, C. 2011. Study on Heimer small appliances Marketing Channel Management. Changchun: Jilin University.

Liu, M., Wang, C., Guo L., et al. 2012. An optimal design method of multi-objective based island micro-grid. *Automation of Electric Power Systems* 36(17): 34–39.

Sahin, C., Shahidehpour, M., & Erkmen, I. 2012. Generation risk assessment in volatile conditions with wind, hydro, and natural gas units. *Applied Energy* 96: 4–11.

Stanislav, P., Bryan, K., & Tihomir, M. 2009. Smart grids better with integrated energy system. In Electrical Power & Energy Conference (EPEC). IEEE: 1–8.

Swardt, D. 2001. A performance comparison between an air-source and a ground-source reversible heat pump. *International Journal of Energy Research* 25(10): 899–910.

Wang, C., & Li, P. 2010. Development and challenges of distributed generation, the micro-grid and smart distribution system. *Automation of Electric Power Systems* 34(2): 10–14.

Xu, X., Jia, H.J., Chiang, H.D., et al. 2015. Dynamic modeling and interaction of hybrid natural gas and electricity supply system in micro-grid. *IEEE Transactions on Power Systems* 30(3): 1212–1221.

Zhang, P., & Ma, Z.W. 2012. An overview of fundamental studies and applications of phase change material slurries to secondary loop refrigeration and air conditioning systems. *Renewable and Sustainable Energy Reviews* 16(7): 5021–5058.

Zhou, R., Ran, X., Mao, F., et al. 2012. Energy-saving coordinated optimal dispatch of distributed combined cool, heat and power supply. *Power System Technology* 36(6): 8–14.

Zhu, W. 2007. Application of solar water heating system in the existed residential building. Tianjin: Tianjin University.

Emerging Developments in the Power and Energy Industry – Dufo-López, Krzywanski & Singh (eds)
© 2020 Taylor & Francis Group, London, ISBN 978-0-367-27169-5

Research on integrated management system of power grid information of provincial and county based on the OMS system

Yujian Jia, Qiwei Lin, Bo Liu & Hongxia Liu
State Grid Jinan Power Supply Company, Jinan, Shandong Province, China

Lin Lin & Zheng Xu
State Grid Shandong Electric Power Company, Jinan, Shandong Province, China

ABSTRACT: For the data exchange mode between the scheduling service systems, this project deeply analyzes the data source, interaction mode and data transmission format of each switching mode. A set of system scheme is proposed, in which the adjustment of data connection mode and the configuration of data transmission format can be achieved using the configuration page. The basis of this solution is the outage management mode and outage management system (OMS system) of Shandong Province Power Grid. This solution realizes the goal of comprehensive management of power grid assets, intelligent analysis of power grid monitoring information, distribution network operation monitoring and fault assisted research as well as judgment.

Keywords: Data adjustment, OMS, Information analysis, Distribution network monitoring, Fault diagnosis

1 INTRODUCTION

To build a control data pool center, an integrated data interactive platform was established based on the province-city-county integrated scheduling management system (Operation Management System, abbreviated as OMS), which achieves the integration of the related system data of the municipal company's dispatch control center (Li, G. et al. 2011). Grid asset management is the fundamental guarantee for the regulation of production and management activities (Liu, W.X. et al. 2018). Accurate realizations of asset safety operation assessment and risk management are the key content of regulation and control work. Aiming at all data exchange methods between the scheduling service systems, this project deeply analyzes the data source, interaction mode and data transmission format of each switching mode. Finally, a set of system solutions capable of adjusting the data connection mode and configuring the data transmission format via the configuration page can be studied.

(1) When there is a need for the system to share data, the operation and maintenance personnel can achieve data sharing through the page configuration.
(2) By summarizing the current data relevant to regulation and control and drawing upon the advantages of its own production and management data integrity, the system fully introduces the fault and abnormal information, load level information and operating environment information of the equipment life cycle into the 220 kV substation of Jinan transformer assets. The system realizes multi-dimensional management and control of power grid assets by studying the comprehensive management method of assets.

(3) The in-depth mining analysis of data is utilized to establish a production management expert database, to achieve the fault processing aid decision-making of power grid, to conduct period forecasting of equipment safe operation and to research on the intervention evaluation of the overhaul engineering project.

(4) The massive historical telemetry data mining algorithm is studied. Implementation plan of the distribution network operating condition monitoring and fault diagnosis are proposed. The 10 kV line telemetry mutation early warning module is developed to provide technical support and decision-making basis for the refined network scheduling.

2 CONSTRUCTION OF UNIFIED DATA INTERACTION PLATFORM

The outage monitoring auxiliary analysis and expert decision-making system provide early warning for possible grid accidents and equipment anomalies. A more rational and scientific disposal strategy is offered in time by strategy calculation and model inference. The overall goal of this system is to realize the transformation of the monitoring business from extensive experience to intelligent analysis, and to achieve real-time online intelligent analysis of muti-dimensional, multilevel as well as muti-objective monitoring services.

The Shandong OMS project has been implemented in accordance with the requirements of the 'big operation' system of the State Grid Corporation, which has achieved the interconnection with EMS, coBase, TMR, PMS and integrated data platform, etc. Considering too many data sources and complex data structures, once the data sources are changed, the data structure will be complex with the poor flexibility of data updating. In the mean time, the data maintained by the OMS is tremendous. Because the data is scattered and the relevance is weak, it is a necessity to strengthen the management of the data sources, unify data access methods, plan data management models and establish blood relationships between the data, to effectively support the application construction of big data analysis. To the structure, unify and order data exchange between scheduling service systems, it is urgent to construct a unified data interactive platform technology based on OMS. The architecture design of the unified data interactive platform is shown in Figure 1.

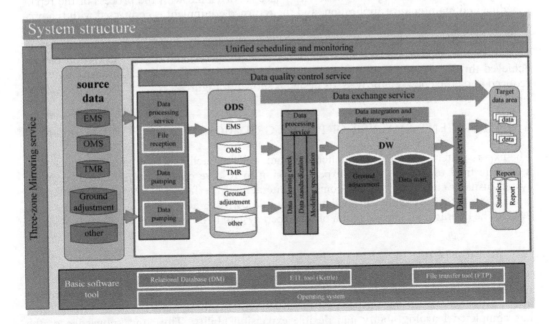

Figure 1. Unified data exchange platform architecture design.

Based on the IEC SG3 standard, the system establishes the panoramic data model with the grid model as the core and uses the application model of the primary and secondary basic equipment information parameters of the grid as the extended data model. The panoramic data model is divided into three parts: public model, application model and muti-dimensional analysis model. The data model is divided into three parts: public model, application model and multi-dimensional analysis model. The panoramic model achieves the connection between the core model of the power grid and the application model through the unified coding and normative semantic encapsulation. It finally makes the full-model layered modeling including the model and data, which serve as the basis for various applications and data services of the system.

3 RESEARCH ON INTEGRATED MANAGEMENT MECHANISM AND AUXILIARY FUNCTION OF GRID ASSETS

Though the grid has issued corresponding device naming conventions for the naming of grid equipment, it is difficult to ensure that the same device has the identical name in different systems in the actual device naming process, thereby making cross-system device correspondence highly different to achieve (Yin, Z. et al. 2016). The project plans to introduce a new fuzzy matching algorithm, which greatly improves the success rate of automatic matching of cross-system devices, and completes the full matching of devices in combination with the work of manual matching check.

At present, the success rate of the automatic matching function can be up to 75%. There are still numerous data that cannot be matched. Specific to the proportion of the equipment, the manual matching function of the cross-system equipment have been designed, which can achieve the manual selection of the device and complete the operation of the device-to-point, together with the already-matched examination and inspection for devices. Taking the equipment planning as the starting point, the management mode of the equipment life cycle is constructed after construction, service, and returning. The new equipment phase is associated with the new equipment management and new equipment commissioning processes, which projects the basic data entry and ensures the accurate maintenance of the newly-added basic data by the process. The equipment service phase is correlated with the process of the repair schedule and maintenance application, and the returned equipment is connected with the process of the equipment that is being returned. Any equipment operation is based on evidence, and the operation process is well documented. The information of each link of the process can be automatically written back into the basic database. Traceability and inspection can be achieved through leaving the trace, then the workload and error of manual intensive input are reduced and efficiency is improved. Its design is shown in Figure 2.

4 INTELLIGENT ANALYSIS OF POWER GRID MONITORING INFORMATION

The expert system is a smart computer program, and its intelligence is mainly manifested in a specific field to imitate human expert thinking to resolve complex problems. The expert system usually consists of human-computer interaction interface, knowledge base, inference engine, interpreter, comprehensive database and knowledge acquisition, etc (Tang, Y. et al. 2012). The operational logic diagram of the expert system is shown in Figure 3.

4.1 *Knowledge base*

Through the user-oriented rule maintenance method, the knowledge base creates a new way to transform the professional knowledge and operational experience in the field of monitoring attendants and maintenance experts into a structured reasoning model. The knowledge base has completed learning ability and flexible expansion ability. Thus, the knowledge of the

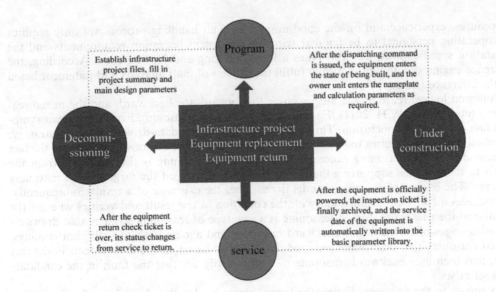

Figure 2. Schematic diagram of equipment replacement and returning for infrastructure projects.

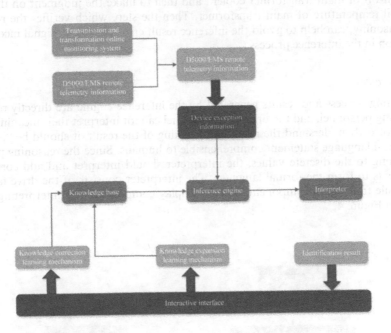

Figure 3. Expert system operation logic diagram.

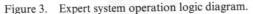

expert and the relevance of the structured storage are maintained (Gao, X. et al. 2005). The support for the optimization of the repeated signals are provided by the definition of the fault model, the signal grouping model, the logical reasoning model and the time window.

4.2 *Inference engine*

The conventional processing of equipment exception refers to the process of making the judgment, reasoning, and processing by the operating personnel following the operational procedures, experience and on-site conditions. This fault handling process not only requires the operating personnel to be familiar with the equipment operation requirements and the substation specifications, but also takes a long analyzing and judging time. According, the inference engine should automatically fulfill the process of fault analysis and judgment based on the inference strategy.

Common logical reasoning strategies cover the forward, the backward, and the mixed reasoning (et al. Chen, W.H. 2011). Forward reasoning is the reasoning from the known symptom fact to the fault conclusion. Thus it is also termed as the data-driven reasoning strategy. Forward reasoning requires to first input the facts related to the diagnostic object into the fact base, and then the reasoning proceeds. The backward reasoning is the reasoning from the target to the evidence supporting the target, thus it is also called the target-driven reasoning strategy. The backward reasoning model firs assumes the existence of a result. Subsequently, it concludes a rule that is consistent with the condition of the result and verifies whether the premise of the rule exists. Mixed reasoning is a new type of reasoning strategy that draws on the advantages of forward and backward reasoning and avoids both of their shortcomings. Mixed reasoning first determines the candidate fault set based on the known facts in the fact base, and then uses backward reasoning to further verify whether the fault in the candidate fault set exists.

As shown in the following figure, the logical reasoning for the alarm identification of the main transformer cooler calls the rules in the knowledge base to generate the analysis result of 'the full cut-off of main transformer cooler', and then to make the judgment on the variation trend of oil temperature of main transformer. Then the step, which verifies the result of the reverse reasoning, can help to avoid the inference result error caused by signal incompleteness or distortion in the inference process.

4.3 *Interpreter*

The reasoning process and results generated by the inference engine are directly presented to the operating personnel, and the operating personnel cannot interpret their meanings. To help the users correctly understand the result and meaning of the result, it should be "interpreted" into a natural language statement comprehensible to humans. Since the reasoning process and results belong to the discrete values, the interpreter should interpret and add corresponding modifier texts to form the formal language. The interpreter consists of the drive fact component, the rule translation component, and the display component. The interpreting process is as shown in Figure 4.

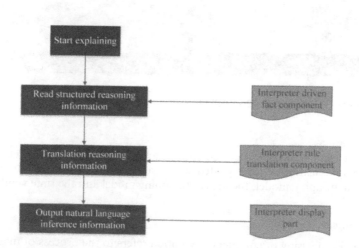

Figure 4.   Interpreter output interpretation result process.

5 CONCLUSION

Based on the OMS system, the province-city-county integrated scheduling management system, which integrates relatively independent and functionally dispersed operational data by the data exchange platform, has achieved the collection and standardization of power grid dispatching operation data in Jinan.

The project fully exploits the data and computing resources of the existing data resource pool and the main functions of the system cover: refined analysis modeling, real-time data analysis service, 10kV outline exception monitoring, 10kV branch line exception monitoring, 10kV tie line loop monitoring, 10kV busbar grounding monitoring, distribution network exception visualization monitoring and distribution network exception statistics analysis and display.

Based on the OMS system, the comprehensive information management system of provincial and county integrated power grids was studied. The operational rules of power grid dispatching operation and the operation experience of historical power grids were studied. The system overcomes the shortcomings of simple mathematical solution by using the rich knowledge of experts in the field of monitoring to solve the problem of comprehensive reasoning. Compared with other artificial intelligence methods, the expert system method is capable of avoiding the exception identification failure caused by the mutation fault model generated by the incomplete or distorted information. The fault diagnosis and judgment algorithm suitable for Jinan Power Supply Company was explored. This achievement is in a leading position in the Shandong province.

ACKNOWLEDGMENT

This study is financed by Science & Technology projects of State Grid Shandong Electrical Power Company.

REFERENCES

Li, G. & Ying, X.C. & Cheng, C.T. et al. 2011. Development and Implementation of Integrated Operation Management System for Provincial and Prefectural Power Plant. *Automation of Electric Power Systems* 35(16):79–81+93.

Liu, W.X. & Hao, Y.K. & Zhang, X.Y. et al. 2018. Digital Technology Based Key Technologies in Power Grid Asset Management and Its Application. *Power System Technology* 42(09):2742–2751.

Yin, Z. & Yan, B. & Zhou, B. et al. 2016. Whole Process Management and Application of Grid Equipment Model Based on Primary Equipment Sub-Responsibility Maintenance Method. *China Science and Technology Information* 2016(06):19–21.

Tang, Y. & Wang, Y.T. & Tian, F. et al. 2012. Research and Development of Stability Analysis Early-Warning and Control System for Huge Power Grids. *Power System Technology* 36(07): 1–11.

Gao, X. & Guo, C.X. & Zhang, J.J. et al. 2005. A Digital Communication Network Based Remote Maintenance System for Intelligent Electronic Device of Substation. *Power System Technology* 29 (23):62–67.

Chen, W.H. & Tsai, S.H. & Lin, H.I. 2011. Fault section estimation for power networks using logic cause-effect models. *IEEE Trans on Power Delivery* 26 (2):963–971.

Emerging Developments in the Power and Energy Industry – Dufo-López, Krzywanski & Singh (eds)
© 2020 Taylor & Francis Group, London, ISBN 978-0-367-27169-5

Piecewise analytical expressions for electromagnetic field distribution within an apertured shielding enclosure

W.X. Qian
State Grid Inner Mongolia East Electric Power Co., Ltd. Power Science Research Institute, Hohhot, China

H.B. Liu
State Grid Inner Mongolia East Electric Power Co. Ltd., Hohhot, China

B. Xiao, D. Wang & L.P. Bi
State Grid Inner Mongolia East Electric Power Co., Ltd. Power Science Research Institute, Hohhot, China

R. He, Y.F. Hu & C.Q. Jiao
State Key Laboratory of Alternate Electric Power System with Renewable Energy Sources, North China Electric Power University, Beijing, China

K.Q. Xu
Hengjingtong Electronic Technology Co., Ltd., Nanjing, China

ABSTRACT: An analytical model is presented for electromagnetic field distribution within an apertured shielding enclosure. The highlight of this model is that it contains very simple and close expressions of both the electrical field and magnetic field distributions. The inner space of an apertured shielding enclosure is divided into three regions: the region around the aperture is called the aperture region, the part that is close to the backboard can be defined as the wave-guide region, and the space between the aperture region and the waveguide region is described as the dipole region. The electromagnetic field distribution model of the aperture region can be replaced by an infinite perfect electric conducting plane with the same aperture. The electromagnetic field distribution model of the dipole region is described by using the dipole radiation model of Bethe's small aperture coupling theory. The electromagnetic field distribution model of the waveguide region is approximated by using the distribution of the waveguide mode field structure. This model is applied to an investigation of the shielding effectiveness of a rectangular shielding enclosure with a circle aperture under the illumination of a plane wave. The calculated results are in good agreement with those of a full-wave simulation.

1 INTRODUCTION

On the one hand, an apertured enclosure can be used as an interaction circuit or a high-frequency structure for the generation, coupling, transmission, conversion, and amplification of electromagnetic waves. On the other hand, it can be used as an electromagnetic shielding for electromagnetic interference suppression (Keqian & Dejie 2001). The study of this problem has continued since the previous century. The early research was based mainly on the analytical method, and later the numerical calculation method was developed using computer technology. At present, the numerical calculation method can be used for a variety of complex geometric models. The numerical calculation method, however, also has some disadvantages, such as the need for more storage space and computing time (Chen & Wang 2007). The analytical method has the advantages of uniqueness and clear physical meaning. At the same time, it can be used as a test case of the numerical calculation method. It is therefore still meaningful to find the analytical solution (Jianhong et al. 2016).

There are two main types of analytical methods: one is to use the equivalent dipole on the basis of Bethe's small aperture coupling theory (Bethe 1944; Collin 1990) to describe the small hole, and then combine it with the mode-expansion method (Chongqing & Yueyue 2014) to obtain the dipole excitation field distribution. The other is the equivalent circuit method, which uses the transmission line or surface impedance to describe the aperture and uses the equivalent of the waveguide mode transmission line to describe the field distribution in the enclosure at the same time (Robinson et al. 1998; Dehkhoda et al. 2008). However, neither of these analytical methods can be applied to the vicinity of the aperture because they cannot be used to characterize the field distribution near the aperture.

In this article, an approximate analytical model for calculating the internal field distribution is proposed for an apertured enclosure. The formula for calculating the electric field and the magnetic field in the enclosure is given directly. The main idea is to divide the inner space of the enclosure into three regions: the area around the aperture is called aperture region, the part that is close to the backboard can be defined as the waveguide region, and the space between the aperture region and the waveguide region is described as the dipole region. We can choose different analytical methods to calculate different field distributions. The electromagnetic field distribution model of the aperture region can be replaced by an infinite perfect electric conducting plane with the same aperture (Moser 1998). The electromagnetic field distribution model of the dipole region is described by using the dipole radiation model of Bethe's small aperture coupling theory. The electromagnetic field distribution model of the waveguide region is approximated by using the distribution of a waveguide mode field structure. In this article, the shielding effectiveness of a rectangular enclosure with an aperture is calculated. The effectiveness of the model is verified by comparing the calculated results with full-wave simulation results.

2 THE SURVIVABILITY ASSOCIATION MODEL OF A LARGE-SCALE NETWORK

2.1 *Model configuration*

Figure 1 shows a rectangular enclosure with an aperture. The origin of the coordinate system is o, and the three coordinate axes are respectively parallel to the three sides of the enclosure. The dimensions of the shield in the x, y, and z directions are a, b, and d. The aperture is located at the center of the face of the enclosure ($z = 0$), and the radius of the aperture is R. The plane wave has a wave vector k. The electric field intensity is E_0 (in the x direction), and the magnetic field strength is H_0 (in the y direction), which is perpendicular to the opening surface. The polarization direction of the plane wave is in the x-axis positive direction. $H_0 = E_0/\eta_0$, and $\eta_0 = \sqrt{\mu_0/\varepsilon_0}$. We assume that the enclosure material is a perfect electric conductor and its thickness is negligible.

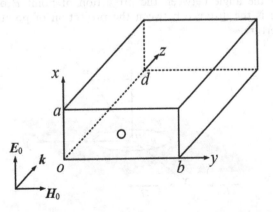

Figure 1. Schematic diagram of an aperture shielding enclosure under plane wave illumination.

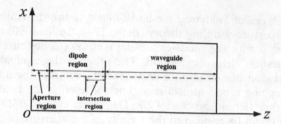

Figure 2. The distribution of the inner space of an aperture shielding enclosure.

2.2 The distribution of the inner space of the enclosure

Figure 2 shows the distribution of an aperture shielding enclosure. The inner space of the enclosure is divided into three regions: the region around the aperture is called the aperture region, the part that is close to the backboard can be defined as the waveguide region, and the space between the aperture region and the waveguide region is described as the dipole region.

2.3 The electromagnetic field distribution of the aperture region

As Figure 3 shows, the electromagnetic field distribution of the aperture region can be equivalent to an infinite perfect electric conducting plane with the same aperture. Jackson (1998) studied the case in which a tangential magnetic field penetrates from one side to the other through the aperture on an infinite perfect electric conducting plane. The results showed the other side of the magnetic field can be expressed as

$$H = -\nabla \Phi \tag{1}$$

$$\Phi = \frac{2H_1 R^2}{\pi} \int_0^\infty j_1(\beta R) e^{-\beta|z|} J_0(\beta \rho) \sin\varphi \, d\beta \tag{2}$$

$$j_1(\beta R) = \frac{\sin(\beta R) - \beta R \cos(\beta R)}{(\beta R)^2} \tag{3}$$

H_1 is the magnitude of the tangential component of the magnetic field on the surface of the aperture when the aperture is blocked, and it is numerically equal to twice the plane wave magnetic field H_0. J_0 is the first type of zero-order Bessel function. As shown in Figure 3, ρ is the distance from the projection of point P on the x–y plane to the center of the aperture. φ is the angle between the projection of point P on the x–y plane and the positive x-axis. z is the distance between the projection of point P on the z-axis and the center of the aperture.

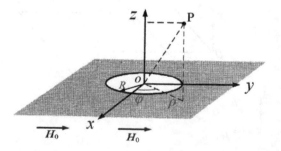

Figure 3. An infinite plane with an aperture.

The tangential component of H which is on the central axis of the aperture ($\rho = 0$, $z > 0$) can be expressed as

$$H = -e_\rho \frac{\partial \Phi}{\partial \rho} - e_\varphi \frac{\partial \Phi}{\rho \partial \varphi} = \frac{2H_0}{\pi} \left(\tan^{-1} \frac{R}{z} - \frac{Rz}{R^2 + z^2} \right) e_y \tag{4}$$

According to Equations (23a) and (23b) in Rahmat-Samii and Mittra (1977), the electric field at the center point on the aperture can be obtained as

$$E = -E_0 \frac{16j}{3} \frac{R}{\lambda} e_x \tag{5}$$

where E_0 is the electric field of the plane wave, and λ is the wavelength of the free space.

According to Bouwkamop (1954), the variation of the electric field along the central axis of the aperture satisfies the following formula:

$$E \propto \left[e^{jkz} - \frac{1}{2} \left(1 + \frac{z^2}{R^2 + z^2} + \frac{R^2}{jkR^2 + z^{2\frac{3}{2}}} \right) e^{jk\rho_1} \right] \tag{6}$$

where $k = 2\pi/\lambda$ and $\rho_1 = \sqrt{R^2 + z^2}$. From Equations (5) and (6), we can infer the expression of the electric field on the central axis of the aperture.

$$E = -j \frac{16RE_0}{3\lambda \left[1 - \frac{1}{2} \left(1 + \frac{1}{jkR} \right) e^{jkR} \right]} \times \left[e^{jkz} - \frac{1}{2} \left(1 + \frac{z^2}{R^2 + z^2} + \frac{R^2}{jkR^2 + z^{2\frac{3}{2}}} \right) e^{jk\rho_1} \right] e_x \tag{7}$$

2.4 The electromagnetic field distribution of the dipole region

According to Bethe's small aperture coupling theory, when the aperture size is much smaller than the wavelength, the field distribution of the region from the aperture to the dipole region can be described by the excitation field of the equivalent electric dipole moment p and the equivalent magnetic dipole moment m that is at the center of the aperture. p and m are determined by the plane wave. For the model shown in Figure 1, it is easy to find that when the plane wave has vertical incidence, the electric dipole moment is 0, and the magnetic dipole moment is

$$m = \frac{-16R^3 H_0}{3} e_y \tag{8}$$

It can be obtained that the magnetic field on the central axis of the aperture in the dipole region is

$$H = \frac{k^3 m}{4\pi} \left[-\frac{1}{kz} + \frac{j}{(kz)^2} + \frac{1}{(kz)^3} \right] e^{-jkz} e_y \tag{9}$$

The electric field is

$$E = -\frac{jk^3 m}{4\pi} \eta_0 \left[\frac{j}{kz} + \frac{1}{(kz)^2} \right] e^{-jkz} e_x \tag{10}$$

2.5 The electromagnetic field distribution of the waveguide region

According to the waveguide theory and Robinson et al. (1998), we can find that when the wave frequency is not significantly higher than the resonant frequency of the enclosure, the TE_{10n} modes exist mainly in the enclosure. The superposition effect of these models is equivalent to the TE_{10} waveguide mode which is shorter at the backboard of the enclosure ($z = d$). Therefore, the electric field and the magnetic field on the central axis of the aperture are

$$E = C_1 \sin k_g(d - z)e_x \tag{11}$$

$$H = C_2 \cos k_g(d - z)e_y \tag{12}$$

$$k_g = k\sqrt{1 - \left(\frac{\lambda}{2d}\right)^2} \tag{13}$$

where k_g is the propagation constant of the TE_{10} mode. In the enclosure, the dipole radiation model is applied to the area not far from the aperture, and the waveguide mode is applied for the area between the back of the enclosure and the part that is slightly farther from the aperture. From this, it can be inferred that there is an area in the enclosure whose dipole distribution is the same as the waveguide distribution; we call it overlapping area. From Equations (9)–(12), the following formulas can be obtained when z_1 is valued from the overlapping area:

$$C_1 \sin k_g(d - z_1) = -\frac{jk^3 m}{4\pi} \eta_0 e^{-jkz_1} \times \left[\frac{j}{kz_1} + \frac{1}{(kz_1)^2}\right] \tag{14}$$

$$C_2 \cos k_g(d - z_1) = \frac{k^3 m}{4\pi} e^{-jkz_1} \times \left[-\frac{1}{kz_1} + \frac{j}{(kz_1)^2} + \frac{1}{(kz_1)^3}\right] \tag{15}$$

Therefore,

$$C_1 = -\frac{jk^3 m\eta_0}{4\pi \sin\left[k_g(d - z_1)\right]} \times \left[\frac{j}{kz_1} + \frac{1}{(kz_1)^2}\right] e^{-jkz_1} \tag{16}$$

$$C_2 = \frac{k^3 m}{4\pi \cos\left[k_g(d - z_1)\right]} e^{-jkz_1} \times \left[-\frac{1}{kz_1} + \frac{j}{(kz_1)^2} + \frac{1}{(kz_1)^3}\right] \tag{17}$$

So, the electric field on the central axis of the aperture in the waveguide region is

$$E = -\frac{jk^3 m\eta_0}{4\pi \sin\left[k_g(d - z_1)\right]} \times \left[\frac{j}{kz_1} + \frac{1}{(kz_1)^2}\right] e^{-jkz_1} \sin k_g(d - z)e_x \tag{18}$$

The magnetic field is

$$H = \frac{k^3 m}{4\pi \cos\left[k_g(d - z_1)\right]} \left[-\frac{1}{kz_1} + \frac{j}{(kz_1)^2} + \frac{1}{(kz_1)^3}\right] \times \cos k_g(d - z)e^{-jkz_1}e_y \tag{19}$$

So far, a complete set of equations for calculating the internal field distribution of the enclosure has been given. The electric shielding effectiveness of a point in the enclosure can be calculated by $S_E = 20\log(E_0/E)$, and the magnetic shielding effectiveness can be expressed as $S_H = 20\log(H_0/H)$.

3 CALCULATION RESULTS AND DISCUSSION

Assume that the size of the shielding enclosure is 300 mm×300 mm×120 mm, and the diameter of the aperture is 20 mm. The shielding effectiveness observation point is located on the central axis of the aperture, as shown by the dashed line in Figure 2. The trend of the electromagnetic shielding effectiveness inside the enclosure with the change of observation point position is obtained by the method.

3.1 *Electric shielding effectiveness*

Figure 4 shows the trend of the electric shielding effectiveness with the change of the observation point at 0.5 GHz. We can see that the caculated results are in good agreement with those from full-wave simulation. In the range of 0–20 mm or one-time aperture diameter, the calculation result of the aperture region is in good agreement with the full-wave simulation result. In the range of 20–90 mm or 1–4.5 times the aperture diameter, the calculation result of the dipole region is in good agreement with the full-wave simulation. At 90–300 mm, that is, after 4.5 times the aperture diameter, we can see that agreement between the calculation result and full-wave simulation is generally very good.

3.2 *Magnetic shielding effectiveness*

Figure 5 shows the trend of the magnetic shielding effectiveness with the change of the observation point at 0.5 GHz. We can see that the agreement between the analytical method and full-wave simulation is generally very good. In the range of 0–20 mm or one-time aperture diameter, the calculation result of the aperture region is in good agreement with the full-wave simulation result. In the range of 20–90 mm or 1–4.5 times the aperture diameter, the calculation result of the dipole region is in good agreement with the full-wave simulation. At 90–300 mm, that is, after 4.5 times aperture diameter, the correspondence is slightly worse at 0.5 GHz, and the error is within 10 dB, which is acceptable. Figure 5 shows that when the frequency is 0.5 GHz, the magnetic shielding effectiveness of the waveguide region hardly changes with the position of the observation

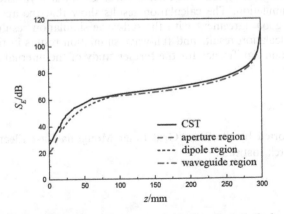

Figure 4. Trend of the electric shielding effectiveness with the change of observation point position (f = 0.5 GHz).

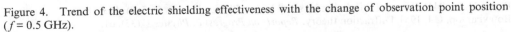

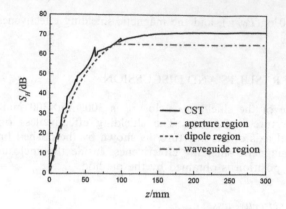

Figure 5.   Trend of the magnetic shielding effectiveness with the change of observation point position ($f = 0.5$ GHz).

point. The reason is that at this time $k_g = k\sqrt{1 - (\lambda/2d)^2}=0$, and the value of the cosine function in Equation (19) is 1, which is independent of the observation point position. In conclusion, the inner space of a shielding enclosure with an aperture is divided into three regions: the region within a one-time aperture diameter is defined as aperture region, 1–4.5 times aperture diameter is the dipole region, and the area after 4.5 times aperture diameter is called the waveguide region. It can be seen that the agreement between the analytical model and full-wave simulation is excellent, both qualitatively and quantitatively. Therefore, the analytical model is reliable.

4   CONCLUSION

An analytical model has been developed to calculate the inner space field distributions of an apertured shielding enclosure under plane wave excitation. The inner space of the enclosure is divided into three regions, according to the distance of the observation point to the aperture surface. The validity of the analytical model is verified by a comparison between analytical results and full-wave simulation results. The position of each region is clarified: the area within 1 times aperture diameter is called the aperture region, the space ranging from 1 to 4.5 times aperture diameter is defined as dipole region, and the part after 4.5 times the aperture diameter is described as the waveguide region. The calculation efficiency is obviously improved compared with the full-wave simulation. The calculation results show that the aperture region and the dipole region are in good agreement with the full-wave simulation results. But there is some error between the calculation results and full-wave simulation results in the waveguide region. The model has a certain significance for the further study of the internal shield distribution of the shielding enclosures.

ACKNOWLEDGMENTS

This project is supported by the State Grid Inner Mongolia East Electric Power Co., Ltd. Power Science Research Institute.

REFERENCES

Bethe, H.A. 1944. Theory of diffraction by small Holes. *Physics Reviews* 66(7–8): 163–182.
Bouwkamop, C.J. 1954. Diffraction theory. *Reports on Progress in Physics* 17(35): 61.

Chen, J., & Wang, J. 2007. A three-dimensional semi-implicit FDTD scheme for calculation of shielding effectiveness of enclosure with thin slots. *IEEE Transactions on Electromagnetic Compatibility* 49(2): 354–360.

Chongqing, J., & Yueyue, L. 2014. Analytical formulation for electromagnetic leakage from an apertured rectangular cavity. *Acta Physica Sinica* 63(21): 103–109.

Collin, R. 1990. *Field Theory of Guided Waves*. New York: Wiley-IEEE Press.

Dehkhoda, P., Tavakoli, A., & Moini, R. 2008. An efficient and reliable shielding effectiveness evaluation of a rectangular enclosure wwith numerous apertures. *IEEE Transactions on Electromagnetic Compatibility* 50(1): 208–212.

Jackson, J.D. 1998. *Classical Electrodynamics*. New York: Wiley-VCH Press.

Jianhong, H., Yanfei, G., Jieqing, F., & Luxing, J. 2016. An analytical model for shielding effectiveness of double layer rectangular enclosure with inner strip-shaped metallic plate. *Acta Physica Sinica* 65(4): 57–64.

Keqian, Z., & Dejie, L. 2001. *Electromagnetic Theory in Microwave and Photoelectronics*. Beijing: Electronic Industry Press.

Moser, J.R. 1988. Low-frequency low-impedance electromagnetic shielding. *IEEE Transactions on Electromagnetic Compatibility* 30(3): 202–210.

Rahmat-Samii, Y., & Mittra, R. 1977. Electromagnetic coupling through small apertures in a conducting screen. *IEEE Transactions on Antennas & Propagation* 25(2): 180–187.

Robinson, M.P., Benson, T.M., Christopoulos, C., et al. 1998. Analytical formulation for the shielding effectiveness of enclosures with apertures. *IEEE Transactions on Electromagnetic Compatibility* 40(3): 240–248.

Emerging Developments in the Power and Energy Industry – Dufo-López, Krzywanski & Singh (eds)
© 2020 Taylor & Francis Group, London, ISBN 978-0-367-27169-5

Analytical study of electromagnetic shielding characteristics of a wire mesh

W.X. Qian, H.B. Liu, B. Xiao, D. Wang & L.P. Bi
State Grid Inner Mongolia East Electric Power Co., Ltd. Power Science Research Institute, Hohhot, China

R. He, Y.F. Hu & C.Q. Jiao
State Key Laboratory of Alternate Electric Power System with Renewable Energy Sources, North China Electric Power University, Beijing, China

K.Q. Xu
Hengjingtong Electronic Technology Co., Ltd, Nanjing, China

ABSTRACT: The applicable range of the calculation formula for the shielding effectiveness of a wire mesh is discussed based on the surface impedance model. On this basis, when the shape of the mesh is parallelogram, the shielding effectiveness of a wire mesh is calculated by using full-wave simulation software, and then the mechanism of the wire mesh is analyzed. According to transmission line theory and the surface impedance of the wire mesh, a double-layer wire mesh is equivalent by the transmission line model. The calculated results are in good agreement with those of full-wave simulation. In addition, the shielding characteristics of a double-layer wire mesh are discussed.

1 INTRODUCTION

Metal plates and wire meshes are always used as electromagnetic shielding measures (Ott 1989, 2009; Clayton 1991). Although a metal plate can give relatively high shielding effectiveness, there are problems such as high cost and inconvenient use. Compared with a metal plate, a wire mesh is not only lightweight, but also can be applied in some situations that require high ventilation and lighting. Therefore, a wire mesh can sometimes be used instead of a metal plate to shield an electromagnetic disturbance. At present, a wire mesh is widely used in large-scale structures such as DC transmission valve halls (Weimin et al. 2008; Dingzhen et al. 2010) and small structures such as electronic equipment.

Research on shielding effectiveness of an infinite wire mesh has been conducted in China and abroad (Ramos 1994; Losito et al. 2010; Shuanglin et al. 2010; Lopez et al. 2013; Sarto et al. 2014; Hyun et al. 2016). Early research was based mainly on analytical theory, and later numerical calculation methods using computer technology were also widely used in research. Numerical calculation methods can be used to calculate a variety of complex models, but the calculation speed is slow. Moreover, there are high requirements on the performance and storage space of computers. The analytical method is not only fast and easy to implement, but also has clear physical meaning. Researching analytical methods is therefore still of great significance. The American scholar Casey proposed a wire mesh transfer function expression based on the surface impedance model of a wire mesh (Casey 1988). The expression is used mainly to solve the shielding effectiveness when the mesh hole of the wire mesh is square. The expression can only calculate the shielding effectiveness of a single-layer wire mesh. In practical applications, a wire mesh in the shape of a parallelogram or hexagon is often used, and when the shielding effectiveness of a single-layer wire mesh cannot meet the requirements and be replaced by a metal plate, a double-layer wire mesh is required instead.

This article discusses the applicable range of the analytical formula for shielding effectiveness of a wire mesh. Full-wave simulation software is used to study the shielding effectiveness of wire meshes with different shapes. On the basis of the simulation results, the shielding mechanism of the wire mesh is analyzed. Then based on the transmission line theory and the surface impedance calculation formula of the wire mesh, the double-layer infinite wire mesh is equivalent by the transmission line model (Robinson et al. 1998), and the approximate analytical model of a double-layer infinite wire mesh under plane wave illumination is given. By comparing the calculation results of the analytical model with the full-wave simulation, the accuracy of the analytical model is verified.

2 SURFACE IMPEDANCE MODEL AND ITS APPLICABILITY

2.1 *Surface impedance model of a wire mesh*

As Figure 1 shows, the surface impedance of an infinite wire mesh z_s can be calculated by the following formula according to Frederick et al. (1996):

$$z_s = R_s + j\omega L_s \tag{1}$$

$$R_s = \frac{4a}{\pi d^2 \sigma} \sqrt{j\omega\tau_w} \frac{I_0\left(\sqrt{j\omega\tau_w}\right)}{2I_1\left(\sqrt{j\omega\tau_w}\right)} \tag{2}$$

$$L_s = \frac{\mu_0 a}{2\pi} \ln \frac{1}{1 - e^{-\pi d/a}} \tag{3}$$

$$\tau_w = \frac{\mu_r \sigma d^2}{4} \tag{4}$$

The calculation formula of the shielding effectiveness of a single-layer wire mesh under the irradiation of a plane wave is

$$SE\Big|_{\text{单}} = 20\log_{10}\left|1 + \frac{z_0}{2z_s}\right| \tag{5}$$

where $z_0 \approx 377\Omega$, ω is electromagnetic wave angular frequency; σ, ε, and μ_r are electrical conductivity, dielectric constant, and magnetic permeability of the wire mesh respectively; a is the mesh diameter of the wire mesh; d is the wire diameter of the wire mesh; τ_w is the diffusion time of the material; and I_0 and I_1 are the first type modified Bessel function.

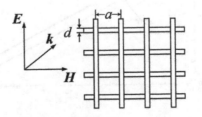

Figure 1.   Schematic diagram of a wire mesh.

2.2 Applicable frequency range of the wire mesh analytical formula

Ott (2009) gives the mutation frequency of the shielding effectiveness of a wire mesh as

$$f_{min} = \frac{2}{\tau_w} \left[\ln \left(1 - e^{-\pi d/a} \right)^{-1} \right]^{-1} \tag{6}$$

$$f_{max} = \frac{Z_0}{2\mu_0 a} \left[\ln \left(1 - e^{-\pi d/a} \right)^{-1} \right]^{-1} \tag{7}$$

The frequency of the electromagnetic disturbance that needs to be shielded in practical applications ranges from 100 kHz to 1 GHz. In order to verify the applicable frequency of an infinite wire mesh, we model it in full-wave simulation software. The mesh diameter of the model is 1 cm or 2 cm, the wire diameter is 1 mm, and the material is copper. The calculation results for infinite wire mesh mutation frequency calculated by Equations (6) and (7) are shown in Table 1. The shielding effectiveness of a wire mesh calculated by Equations (1)–(5) is shown in Figure 2.

Therefore, to meet the applicable range of the analytical formula, the mesh diameter of the wire mesh used in the following study is 1 cm or 2 cm, the wire diameter is 1 mm, and the calculation frequency is between 100 kHz and 1 GHz.

3 ANALYSIS OF THE SHIELDING MECHANISM OF A WIRE MESH

3.1 Analysis on shielding effectiveness of wire mesh with parallelogram mesh shape

At present, the research on wire mesh focuses mainly on a square mesh. However, in practical engineering, we often use other mesh types such as parallelograms and hexagons. Currently, there is no analytical formula for the shielding effectiveness of a wire mesh when the mesh shape is a parallelogram. Therefore, the numerical simulation model is established to study the shielding effectiveness of a wire mesh with a parallelogram shape.

Table 1. Mutation frequencies of a wire mesh when the wire diameter is 1 mm.

Mesh diameter	f_{min}	f_{max}
1 cm	84 kHz	11 GHz
2 cm	57 kHz	3.8 GHz

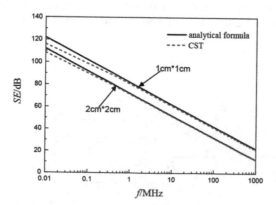

Figure 2. Shielding effectiveness of a single-layer wire mesh calculated by analytical formulas and CST.

377

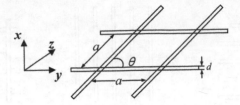

Figure 3.   Schematic diagram of wire mesh shape.

The numerical simulation model is modeled in full-wave simulation software CST. The boundary condition is selected to use period boundary condition and the results is calculated by the frequency domain solver. As shown in Figure 3, the mesh length of the wire mesh is a, and the angle between two wires is θ. Considering the symmetry of the angle, the variation of the wire mesh shielding effectiveness with θ is studied when θ is changing between 0 and 90°.

The mesh diameter of the wire mesh is 1 cm and the wire diameter is 1 mm. The material of the wire mesh is a perfect electric conductor. The irradiation of the plane wave is in the z direction, which is perpendicular to the surface of the wire mesh, and the electric field is along the positive direction of the x-axis. Figure 4 shows the variation of the shielding effectiveness of the wire mesh when θ varies from 0 to 90°. It can be seen that when θ is 30°, the shielding effectiveness is the smallest, and as θ increases, the shielding effectiveness also increases. When θ is increased to 90°, the electric field vector is parallel to the wire, and the shielding effectiveness reaches a maximum value.

3.2   *Shielding mechanism analysis* of a *wire mesh*

As noted in Section 3.1, the shielding effectiveness of a wire mesh increases with θ. The shielding effectiveness of an only horizontal wire mesh, an only vertical wire mesh, and wire mesh is compared in Figure 5, which shows the simulation model.

We can see from Figure 6 that the shielding effectiveness of the only vertical wire mesh is almost the same as that of the 1 cm × 1 cm wire mesh, and the shielding effectiveness of the only horizontal wire mesh is almost 0. Therefore, the wire parallel to the external electric field contributes the most to the shielding effectiveness, and the wire perpendicular to the external electric field contributes little to the shielding effectiveness. This may be explained by the fact that current is easily formed on the wire that is parallel to the external electric field, and on the other side of the wire mesh, the secondary field generated by the current on the wire may cancel out part of the incident field.

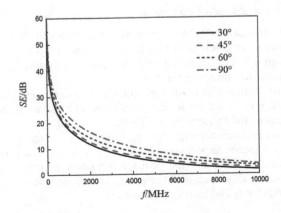

Figure 4.   Trend of wire mesh shielding effectiveness when θ is changing.

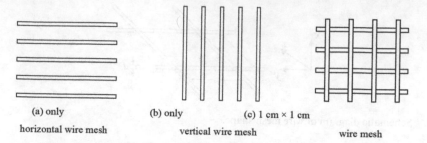

| (a) only | (b) only | (c) 1 cm × 1 cm |
| horizontal wire mesh | vertical wire mesh | wire mesh |

Figure 5. Comparison of the shielding effectiveness of different meshes.

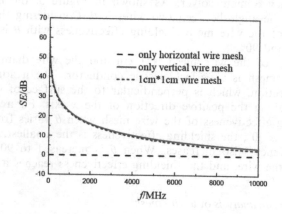

Figure 6. Comparison of the shielding effectiveness of an only horizontal wire mesh, an only vertical wire mesh, and a 1 cm × 1 cm wire mesh.

However, the surface of the wire that is perpendicular to the external electric field has almost no current flowing, and the secondary field generated on the other side of the wire mesh is almost 0.

4 ANALYSIS OF SHIELDING CHARACTERISTICS OF A DOUBLE-LAYER WIRE MESH

4.1 *Analysis model of a double-layer wire mesh*

As shown in Figure 7, the distance between two wire meshes when they are placed in parallel is p. A double-layer wire mesh is equivalent by the transmission line model, and the equivalent circuit is shown in Figure 8. The shielding effectiveness of a double-layer wire mesh can be obtained by transforming the voltage to point P.

The mesh diameter of the first wire mesh is a_1, and the wire diameter is d_1. The mesh diameter of the second wire mesh is a_2, and the wire diameter is d_2. The surface impedance z_{s1} and z_{s2} of the wire mesh can be calculated by Equations (1)–(4).

By Thevenin's theorem, combining z_0, v_0, and z_{ap} gives an equivalent voltage $v_1 = v_0 z_{s1}/(z_{s1} + z_0)$ and source impedance $z_1 = z_{s1} z_0/(z_{s1} + z_0)$. The propagation impedance of the transmission line is $z_c = z_0$, and the propagation constant is $\gamma = k_0 = 2\pi/\lambda$. Therefore, we transform v_1, z_1, and the short circuit at the end of the waveguide to P, giving an equivalent voltage v_2, source impedance z_2 and load impedance z_L:

$$v_2 = \frac{v_1}{\cos \gamma p + j\frac{z_1}{z_c}\sin\gamma p} \tag{8}$$

$$z_2 = \frac{z_1 + jz_c \tan \gamma p}{1 + j\frac{z_1}{z_c}\tan\gamma p} \tag{9}$$

$$z_L = z_{s2}z_0/(z_{s2} + z_0) \tag{10}$$

The voltage at P is $v = z_L v_2/(z_2 + z_L)$, and the electric shielding effectiveness of the double-layer wire mesh is

$$SE = -20\log_{10}\left|\frac{2v}{v_0}\right| \tag{11}$$

The difference between the shielding effectiveness of the double-layer wire mesh and the single-layer wire mesh is

$$SE|_{\text{diff}} = SE - SE|_{sigl} = -20\log_{10}\left|\frac{v(2z_s + z_0)}{v_0 z_s}\right| \tag{12}$$

4.2 Shielding characteristics of a double-layer wire mesh

The mesh diameter of a double-layer wire mesh is 2 cm, the wire diameter is 1 mm, and the material is copper. The shielding effectiveness of a double-layer wire mesh when p is 1.5 cm and 7 cm is calculated.

Figure 9 shows that when p is 1.5 cm and 7 cm, the calculation results of the analytical model are in good agreement with the full-wave simulation. Therefore, the feasibility of the analytical model is verified.

Figure 10 shows the shielding effectiveness of a single-layer wire mesh and double-layer wire mesh when p is 2 cm, 4 cm, 6 cm, and 8 cm. It can be seen that the shielding effectiveness

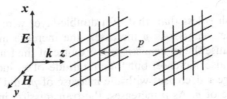

Figure 7.   Schematic diagram of double wire mesh.

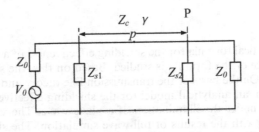

Figure 8.   The equivalent circuit of a double-wire mesh.

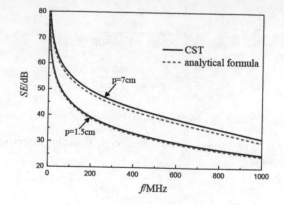

Figure 9. Shielding effectiveness calculated by analytical formulas and CST when $p = 7$ cm.

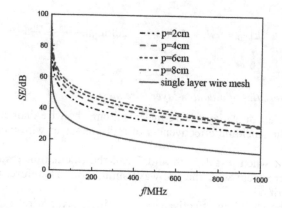

Figure 10. Trend of the shielding effectiveness of a single-wire mesh and double-wire mesh when the frequency is changing.

of a single-layer wire mesh is less than that of a double-layer wire mesh. The larger p is, the greater the shielding effectiveness. As p increases, the increase in shielding effectiveness is reduced. This may be because the transmission impedance of the transmission line increases as p increases, so v_2 and z_2 also increase, but the amplitude of v_2 increases much less than z_2. Therefore, the load voltage v decreases with the increase of p, and the shielding effectiveness increases with the increase of p. As p increases, the transmission impedance of the transmission line has a reduced effect on the wire mesh, and thus the increase in shielding effectiveness is reduced.

5   CONCLUSIONS

In this article, the analytical formula for the shielding effectiveness of a wire mesh is proposed, and the applicable range of the formula is studied. Based on this, the shielding mechanism of a wire mesh is studied. On the basis of the transmission line theory and the surface impedance formula of a wire mesh, an analytical model for the shielding effectiveness of a double-layer wire mesh is proposed under the illumination of a plane wave. The validity of the model is verified by comparison with the results of full-wave simulation. The shielding characteristics of a double-layer wire mesh is studied further. We have been able to establish some basic results:

(1) When the frequency of the incident wave is between the abrupt frequency of the wire mesh, the results of the analytical formula are consistent with those of full-wave simulation.
(2) When the electric field direction of the incident wave is parallel to the direction of the wire mesh, the shielding effectiveness of the wire mesh is the greatest.
(3) When the distance between two wire meshes is very close, the shielding effectiveness of the double-layer wire mesh is smaller than that of the single-layer wire mesh. When the distance is greater than a certain value, the shielding effectiveness of the double-layer wire mesh is higher than that of the single-layer wire mesh, and increases as the distance increases.

ACKNOWLEDGMENTS

This project is supported by the State Grid Inner Mongolia East Electric Power Co., Ltd. Power Science Research Institute.

REFERENCES

Casey, K.F. 1988. Electromagnetic shielding behavior of wire mesh screens. *IEEE Transactions on Electromagnetic Compatibility* 30(3): 298–306.

Clayton, R.P. 1991. *Introduction to Electromagnetic Compatibility*. New York: John Wiley & Sons.

Dingzhen, N., Weimin, M., Baoquan, W., et al. 2010. Shielding effectiveness and design requirement of UHVDC converter valve hall. *High Voltage Engineering* 36(2): 313–317.

Frederick, M.T., Ianoz, M.V., & Karlsson, T. 1996. *EMC Analysis Methods and Computational Models*. New York: John Wiley & Sons.

Hyun, S.Y., Jung, I.K., Hong, I.P., et al. 2016. Modified sheet inductance of wire mesh using effective wire spacing. *IEEE Transactions on Electromagnetic Compatibility* 58(3): 911–914.

Lopez, A., Voitech, L., & Neruda, M. 2013. Comparison among models to estimate the shielding effectiveness applied to conductive textiles. *Advances in Electrical & Electronic Engineering* 11(5): 387–391.

Losito, O., Barletta, D., & Dimiccoli, V. 2010. A wide-frequency model of metal form for shielding effectiveness applications. *IEEE Transactions on Electromagnetic Compatibility* 52(4): 75–81.

Ott, W.H. 1989. Noise reduction techniques in electronics systems. New York: John Wiley & Sons.

Ott, W.H. 2009. *Electromagnetic Compatibility Engineering*. New York: John Wiley & Sons.

Ramo, S., Whinnery, J.R., & Van Duzer, T. 1994. *Fields and Waves in Communication Electronics*. New York: John Wiley & Sons.

Robinson, M.P., Benson, T.M., Christopoulos, C., et al.1998. Analytical formulation for the shielding effectiveness of enclosures with apertures. *IEEE Transactions on Electromagnetic Compatibility* 40(3): 240–247.

Sarto, M.S., Greo, S., & Tamburrano, A. 2014. Shielding effectiveness of protective metallic wire meshes: EM modeling and validation. *IEEE Transactions on Electromagnetic Compatibility* 56(3): 615–621.

Shuanglin, W., Peiguo, L., & Jianguo H. 2010. Study on the shielding effectiveness of metal mesh to linear polarized electromagnetic wave. *Safety & EMC* 2: 66–68.

Weimin, M., Dingzhen, N., Baoquan, W., et al. 2008. Study on the shielding effectiveness of HVDC converter valve hall. *High Voltage Engineering* 34(11): 2400–2407.

Emerging Developments in the Power and Energy Industry – Dufo-López, Krzywanski & Singh (eds)
© 2020 Taylor & Francis Group, London, ISBN 978-0-367-27169-5

Protection scheme for flexible DC distribution grids based on a full-bridge modular multilevel conductor

Xiaodong Zhang
State Grid Corporation of China, Beijing, China

Hong Cao, Xinze Zhou, Xingguo Wang, Dingxiang Du & Wenrui Cai
State Key Laboratory for Security and Energy Saving, China Electric Power Research Institute, Beijing, China

Qiong Wu & Wenrui Cai
State Key Laboratory of New Energy Power System, North China Electric Power University, Beijing, China

ABSTRACT: At present, it is still difficult to realize arc-extinguishing of large DC currents for DC breakers. A converter with fault self-cleaning capability can cut off the fault current quickly without DC breakers, which provides a new direction for research on protection of flexible DC grids. This article proposes a protection scheme for flexible DC distribution grids based on a full-bridge modular multilevel converter (MMC). The discrete wavelet transform is used to extract the maximum modulus of the fault current traveling wave. The fault can be determined by comparing the polarities of the maximum modulus of the fault current waveform obtained at the two ends of the DC line. A startup criterion is designed to avoid the interference of an out-of-zone fault, which increases the reliability of protection. The simulation results in PSCAD/EMTDC verify the proposed protection scheme, which can be adopted in flexible DC distribution grids.

1 INTRODUCTION

At present, the construction and development of an urban DC distribution grid has become a research hotspot in the field of power engineering. The DC distribution grid has great advantages in the utilization efficiency of renewable energy compared with the AC grid. At the same time, the DC grid can avoid the problems of reactive power and frequency oscillation. Thus, it owns the future of the urban distribution grid. In China, the line commutated converter (LCC) has been widely used in long-distance and large-capacity DC transmission systems. But its reliability is not high owing to the existence of commutation failure. The voltage source converter (VSC) based on the fully controlled device does not have the risk of commutation failure, and the active and reactive power can be controlled independently so as to be more flexible. The modular multilevel converter (MMC) has been widely favored by scholars because of its stronger scalability, lower operating loss, and lower harmonics of output voltage waveform. It has good application prospects in the construction of an urban DC distribution grid. The half-bridge MMC has fewer internal power electronic devices and lower costs. But it still delivers power to the fault point through the antiparallel freewheeling diode after the AC system is locked (Xu et al. 2016; Wang et al. 2018). Therefore it must work synergistically with DC circuit breakers to remove the fault in the system. At this time, it is difficult to realize the arc-extinguishing of large DC currents safely and rapidly by employing DC breakers, so the protection scheme based on DC breakers needs to be improved further. MMCs based on a full-bridge submodule (FBSM) and clamping double submodule (CDSM) can cut off a fault current by converter blocking. They can eliminate a fault current rapidly without DC breakers. As there are many branches in a distribution grid, the cost to install DC circuit breakers in each branch is too high. Therefore, MMCs with fault isolation capability are more suitable for a flexible DC distribution grid. In this article, the full-bridge MMC is selected to research the protection scheme of the flexible DC distribution grid.

Among the existing research results, the literature (Jia et al. 2017) proposes a differential protection scheme that uses the energy difference of fault currents in different frequency segments to distinguish between internal and external faults. However, the protection is based on VSC converters and DC circuit breakers are required. The literature (De Kerf et al. 2011) proposes a scheme based on wavelet transform. But the protection scheme is applicable to HVDC transmission lines, not distribution grids. The protection still needs DC breakers.

In this article, the fault current characteristics of a flexible DC distribution grid are analyzed combined with the locking process of FBSM-MMC. The fault is located by polarities of the modulus maximums of the current traveling waves. The protection can distinguish between internal and external faults quickly and accurately.

2 BLOCKING PRINCIPLE OF FBSM-MMC

The topology of a flexible DC distribution grid is shown in Figure 1. There are photovoltaic power (PV) and wind turbine generators (WTGs) in the system. The system uses FBSM-MMC based stations to connect AC and DC lines. WTG and AC loads are also connected to the system through FBSM-MMC, while PV and DC loads are connected by a DAB converter. A mechanical switch is installed at the end of each DC line. The switch can be tripped to isolate the fault after the fault current is completely cleared.

DC faults are divided into three fault types: single-pole grounded faults, double-pole short-circuit faults, and disconnection faults. Only double-pole short-circuit faults can cause serious over-current, which has a greater impact on the distribution system (Hertem & Ghandhari 2010). Therefore, this article focuses on two-pole short faults. A FBSM-MMC structure is shown in Figure 2. The fault current of a double-pole short-circuit fault is mainly composed of two parts. One is the internal capacitor discharge, and the other is the energy feed from the AC system (Yang et al. 2012; Shi et al. 2015).

The FBSM-MMC will block rapidly due to the fast discharge of the internal capacitor. The fault current must flow through the submodule capacitor after the FBSM-MMC blocking. The capacitor provides a reverse electromotive force to cut off the direct electric connection between the power supply and the fault point. The flow path of fault current is shown in Figure 3. When a fault occurs, the FBSM-MMC will block in 3–5 ms, and the fault current will be cleared in around 10 ms. So we have to locate the fault before blocking, and to cut off the fault after the fault current cleared. This article uses initial wave head information of fault current traveling waves to locate the faults. The scheme meets the requirements of speed for protection.

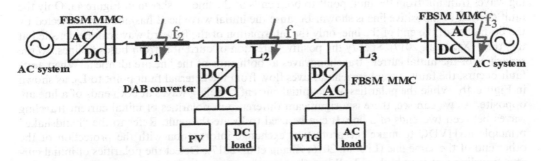

Figure 1. Topology of a flexible DC distribution grid.

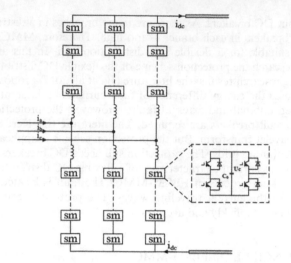

Figure 2. Structure of FBSM-MMC.

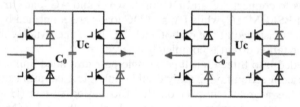

Figure 3. Flow paths of fault currents.

3 SCHEME OF DC DOUBLE-POLE FAULT LOCATION

3.1 Text and indenting analysis of polarities of current traveling waves

L_2 in Figure 1 is taken as the research object. In Figure 1, f_1 is an internal DC double-pole fault, f_2 is an external DC double-pole fault, and f_3 is an AC triple-phase short-circuit fault. The fault current traveling waves are caused by the fault, and transmit from the fault point to both sides of the lines, as shown as Figure 4. When a fault occurs, the voltage of the fault point can be regarded as the superposition of a normal voltage U_f and a fault voltage $-U_f$. The voltage traveling waves and current traveling waves on the lossless uniform transmission lines can be regarded as the superposition of a forward traveling wave and a reverse traveling wave (He 2017; Lin et al. 2017; Wenig et al. 2018). Therefore, when an f_1 fault occurs, fault current traveling waves transmit from the fault point to both ends of the line, as shown in Figure 4a. Only the fault current of a positive line is shown. Because the initial wave head has not been deflected by the boundary of the end of the line, only the information of the forward wave can be detected at this time (Xie et al. 2017). Specify the positive direction of current is from buses to lines, so the polarities of the initial current traveling waves at both ends of the line are identical. When an f_2 fault occurs, the fault current traveling waves flow from the external fault point to L_2, as shown in Figure 4b. While the polarities of the initial current traveling waves at both ends of a line are opposite. As we can see, there is a significant difference in polarities of initial current traveling waves between two ends of a line. It can be used to locate the fault. Refer to the "handshake" principle in HVDC to make the protection exchange information with the protection of the other end of the same line (Li et al. 2016; Zhang et al. 2017). Detect the polarities of initial current traveling waves at both ends. When they are the same, tripping switches at both ends after

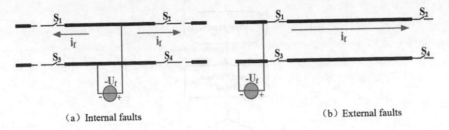

(a) Internal faults   (b) External faults

Figure 4. Flow paths of the current traveling waves.

fault current has been cleaned by MMC; when different, there is no tripping. After the faults have cleared, unlock the FBSM-MMC and put the system into operation again.

3.2 The extraction of a fault feature

Wavelet transform is a time-frequency analysis method, while discrete wavelet transform can identify local abrupt signals and capture singular points in high-frequency signals (Darwish et al. 2012). It is suitable for extracting steep wave heads of fault current. The selection of the mother wavelet has a great influence on the result. Commonly used mother wavelets include the Haar wavelet, Daubechies wavelet, B-spline wavelet, etc. The Daubechies wavelet has better smoothness and vanishing moment characteristics than the Haar wavelet, which can improve the integrity and accuracy of the extracted signals. While the high-order B-spline wavelet can guarantee the integrity of the extracted signals, its recursive operation process is relatively cumbersome and takes a longer time. Therefore, we select the Daubechies wavelet as the mother wavelet.

The Daubechies wavelet has no explicit analytical expression. In the case in which the coefficient $\{h_k\}$ is determined, the scaling function $\varphi(x)$ and the wavelet function $\psi(x)$ are obtained generally by the two-scale equation $\varphi(x) = \sqrt{2}\sum h_k\varphi(2x - k)$. Then, the Daubechies wavelet can be constructed. Considering that noise interference will decrease with the increase of the wavelet decomposition scale, this article selects the db4 wavelet as the mother wavelet and the decomposition scale is 4. The polarities of the fault current traveling waves are the same as those of the modulus maximums.

3.3 Protection criterion and its setting

For listing facts, use either the style tag List summary signs or the style tag List number signs. The protection criteria can be derived from the subsections 3.1 and 3.2 as Equation (1).

$$\begin{cases} I = 1, d_{m1} \cdot d_{m2} > 0 \\ I = 0, d_{m1} \cdot d_{m2} < 0 \end{cases} \tag{1}$$

In the equation, I is the characterization of action results of protection. When $I = 1$, protection acts; when $I = 0$, protection does not act. d_{m1} and d_{m2} are the modulus maximum of current traveling waves at both ends of the line. The startup criterion is proposed to avoid protection acting when non-DC side faults occur. The analysis process is as follows. The fault traveling waves can transmit from the fault point to both ends of the line in a lossless uniform transmission line. However, there are many inductive components in the system. These components exhibit a large impedance to high-frequency signals, forming a boundary to them (Tang et al. 2018). The traveling wave signals will be greatly weakened when they encounter the inductive components. When an AC fault occurs, the fault traveling wave will be affected by the arm inductance in MMC. So it is difficult to detect the traveling wave signals on the DC side. The the same reason is true for other non-DC faults. Therefore, the startup criterion can be shown as Equation (2):

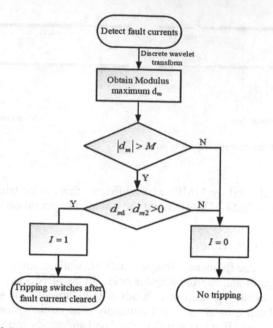

Figure 5. Flow chart of the protection process.

$$|d_m| > M \tag{2}$$

In Equation (2), M is the protection threshold. The modulus maximum is related to the decomposition scale. When the decomposition scale is known, the threshold value M can be determined by the system parameters. In this case, the M value is determined to be 0.04. In summary, the protection process is as follows. First the fault position can be located by Equations (1) and (2) when a fault occurs. Then the switches that meet the tripping conditions as objective switches are selected. The rapid increase of the fault current leads the FBSM-MMC to block. As a result, the fault current will be cleared. Finally, the objective switches will trip after the fault current cleared. The flow chart of the protection process is shown in Figure 5.

4 SIMULATION ANALYSIS

The DC distribution system model is built in PSCAD/EMTDC as shown in Figure 1. The FBSM-MMC on the left end takes control of DC voltage. The FBSM-MMC on the right end takes control of active power. When the current in a bridge arm exceeds 2 times its rated value, the MMC will block. The DC lines use overhead lines. The specific system parameters are shown in Table 1. The following is the simulation of an internal DC double-pole fault, an external DC double-pole fault, and an AC triple-phase short-circuit fault.

4.1 *An internal DC double-pole fault*

A DC double-pole fault occurs at f_1 at 0.2 s. Taking the positive line as the analysis object, the current measuring points are located at S_1 and S_2. The waveforms of fault currents at S1 and S_2 are shown in Figure 6.

A discrete wavelet transform is performed on the two fault currents, and the decomposition scale is 4. The modulus maximums extracted from the initial fault current traveling waves are d_{m1} and d_{m2}. The diagrams are shown in Figure 7. It can be seen from the figure that the

Table 1.   Parameters of a DC distribution system.

Parameters of the system	Numerical value
Voltage of DC lines	±10 kV
Voltage of AC systems	±110 kV
Voltage of photovoltaic power	±0.5 kV
Rated capacity of converter transformer	30 MVA
Rated capacity of DAB converter	10 MVA
Capacity of photovoltaic power	0.25 MW
Capacity of wind turbines	10 MW
DC load	2 MW
AC load	3 MW
Length of DC lines	10 km × 3
Resistance of DC lines	0.0318Ω/km
Inductance of DC lines	0.742 mH/km
Protection sampling frequency	10 kHz

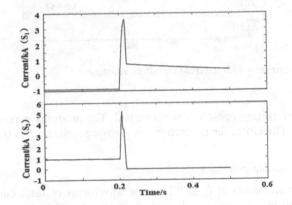

Figure 6.   Fault currents of an internal DC double-pole fault.

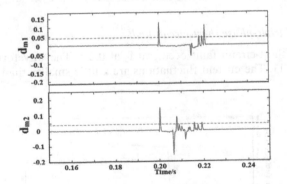

Figure 7.   Modulus maximums of an internal DC double-pole fault.

modulus maximums satisfy the starting criterion, and the polarities of the two maximums are the same. As a result, the fault can be judged to be internal.

The resistance of the protection scheme to the transition resistance has also been tested in this article. The modulus maximums of a metallic fault are compared with faults with transition resistance that range from 0.1Ω to 50Ω. The results are shown in Figure 8. As we can see, the modulus maximum does not change much with transition resistance below 1Ω. And it will

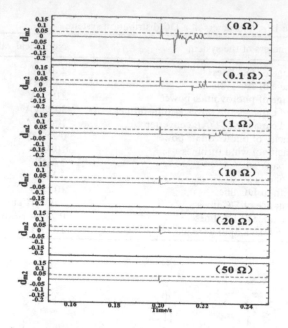

Figure 8.   Modulus maximums with different transition resistances.

change a little when transition resistance is above 10Ω. The modulus maximums always satisfy the startup criterion. Therefore, the protection has a strong resistance to transition resistance.

4.2   *An external DC double-pole fault*

A DC double-pole fault occurs at f_2 at 0.2 s. The waveforms of fault currents are shown in Figure 9. The modulus maximums of fault currents in an external fault are shown in Figure 10. The decomposition scale of the discrete wavelet transform is still 4. The two modulus maximums have opposite polarities. The conclusion is that the fault is external.

4.3   *An external AC triple-phase short-circuit fault*

An AC triple-phase short-circuit fault occurs at f_3 at 0.2 s. The waveforms of fault currents are shown in Figure 11. The current fluctuations are a little smaller than those in DC faults.

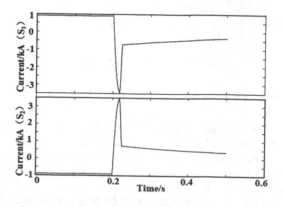

Figure 9.   Fault currents of an external DC double-pole fault.

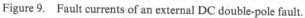

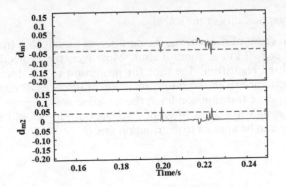

Figure 10.   Modulus maximums of an external DC double-pole fault.

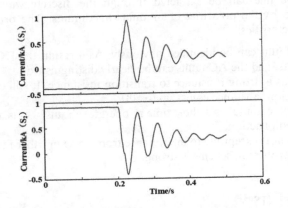

Figure 11.   Fault currents of an AC triple-phase short-circuit fault.

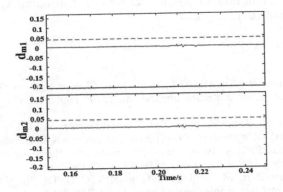

Figure 12.   Modulus maximums of an external DC double-pole fault.

The modulus maximums obtained in this fault are smaller, which does not satisfy the startup criterion. As a result, the mechanical switches at S_1 and S_2 will not trip. The reason for this phenomenon has been proposed in Section 3.3. The inductances in the bridge arm of FBSM-MMC block the fault current traveling waves to DC lines. For the same reason, neither the double-pole fault at PV nor the triple-phase short-circuit fault will cause the mechanical switches to trip. Because of space limitations, the simulation results of other non-DC faults are not presented in the text.

4.4 Analysis of the applicability of the scheme

In terms of the action speed of the scheme, the duration of signals that transmit in distribution lines that do not exceed 20 km is about 0.1 ms. Moreover, the protection scheme does not have a very complicated algorithm. The time for protection to act does not exceed 2–3 ms, which satisfies the requirements of speed for protection of flexible DC distribution grids.

In terms of the length of transmission lines, the traveling wave signals will not attenuate in the short-distance transmission lines. Most of distribution lines should be within 20 km. So the protection scheme can be applied to distribution grids.

5 CONCLUSION

A protection scheme for flexible DC distribution grids based on FBSM-MMC is proposed in this article. The modulus maximums of the initial fault current traveling waves at both ends of the line can be extracted through the discrete wavelet transform. The fault will be located by the polarities of modulus maximums. The protection scheme has the following characteristics:

(1) The features of faults can be identified obviously. As a result, the DC internal faults, the DC external faults, and the AC faults can be clearly distinguished.
(2) The protection has a strong resistance to transition resistance. It still can operate correctly even if the transition resistance comes up to 50Ω.
(3) The action of protection takes a short time to operate, about 2–3 ms, meeting the requirement for the speed of action.
(4) The protection is not susceptible to interference from noise or other factors. Therefore, the reliability of the protection scheme is strong.

ACKNOWLEDGMENTS

State Grid Corporation of China Science and Technology Project "Research of Fault Characteristics Analysis and Protection Strategy in Flexible DC Distribution Grids" provided funding (grant no. 521104180002).

REFERENCES

Darwish, H. A., Hesham, M., Taalab, A. I., et al. 2012. Close accord on DWT performance and real-time implementation for protection applications. *IEEE Transactions on Power Delivery* 25(4): 2174–2183.

De Kerf, K., Srivastava, K., Reza, M., et al. 2011. Wavelet-based protection strategy for DC faults in multi-terminal VSC-HVDC systems. *IET Generation, Transmission & Distribution* 5 (4): 496–503.

He, J., Li, B., Li, Y., et al. 2017. A fast directional pilot protection scheme for the MMC-based MTDC grid. *Proceedings of the CSEE*, 2017, 37 (23): 6878-6887+7078 (inChinese).

Hertem, D. V., & Ghandhari, M. 2010. Multi-terminal VSC HVDC for the European supergrid obstacles. *Renewable and Sustainable Energy Reviews* 14(9): 3156–3163.

Ke, J., Li, M., Bi, T., et al. 2017. Energy distribution-based differential protection for VSC-DC distribution lines. *Power System Technology* 41(9): 3058–3065 (in Chinese).

Li, M., Jia, K., Bi, T., et al. 2016. Fault distance estimation-based protection for DC distribution networks. *Power Systems Technology* 40 (3): 719–724 (in Chinese).

Lin, W., Jovcic, D., Nguefeu, S., & Saad, H. 2017. Protection of full bridge MMC DC grid employing mechanical DC circuit breakers. In IEEE Power & Energy Society General Meeting, Chicago, pp. 1–5.

Shi, B., Zhao, Y., & Sun, G. 2015. Research and implementation of protection scheme for MMC DC distribution network. *Southern Power System Technology* 39(9): 11–16 (in Chinese).

Tang, L., Dong, X., Shi S., Kong, M., & Qiu, Y. 2018. Principle and implementation of ultra-high-speed travelling wave based protection for transmission line of flexible HVDC grid. *Power System Technology* 42(10): 3176–3186 (in Chinese).

Wang Y., Wen W., Zhang, C., Chen, Z., & Wang, C. 2018. Reactor sizing criterion for the continuous operation of meshed HB-MMC-based MTDC system under DC faults. *IEEE Transactions on Industry Applications* 54(5): 5408–5416.

Wenig, S., Goertz, M., Hirsching, C., Suriyah, M., & Leibfried, T. 2018. On full-bridge bipolar MMC-HVDC control and protection for transient fault and interaction studies. *IEEE Transactions on Power Delivery* 33(6): 2864–2873.

Xie, Q., Li, H., Zhang, K., & Wu, J. 2017. Fault analysis and fast protection of DC bus in MMC-based HVDC grids. In 2nd International Conference on Power and Renewable Energy (ICPRE), Chengdu, 2017, pp. 225–228.

Xu, J., Zhao, C., Xiong, Y., Li, C., Ji, Y., & and An, T. 2016. Optimal design of MMC levels for electro-magnetic transient studies of MMC-HVDC. *IEEE Transactions on Power Delivery* 31(4); 1663–1672.

Yang, J., Fletcher, J. E., & O'Reilly. J. 2012. Short–circuit and ground fault analyses and location in VS-based DC network cables. *IEEE Transactions on Industrial Electronics* 59(10): 3827–3837.

Zhang, J., Zou, G., Xie, Z., Sui, H., & Sun, C. 2017. A fast non-unit line protection strategy for the MMC-based MTDC grid. In IEEE Conference on Energy Internet and Energy System Integration (EI2), Beijing, 2017, pp. 1–6.

Emerging Developments in the Power and Energy Industry – Dufo-López, Krzywanski & Singh (eds)
© 2020 Taylor & Francis Group, London, ISBN 978-0-367-27169-5

Measurement and analysis of the electromagnetic disturbance caused by the switching operation of distribution switchgear

Yan Wu & Lingyun Gu
China Electric Power Research Institute, Beijing, China

Yifan Hu, Ri He & Chongqing Jiao
State Key Laboratory of Alternate Electrical Power System with Renewable Energy Sources, North China Electric Power University, Beijing, China

ABSTRACT: In this article, we report the experimental results on the electromagnetic transients caused by the switching operation of a 6-kV switchgear. The transient disturbance voltage induced at the electronic current transformer (CT), the transient ground potential rise (TGPR), and the transient enclosure voltage (TEV) are measured when the circuit breaker of a 6-kV switchgear is open/closed. The results show that the disturbance waveform is similar to the damping oscillation waveform. The TGPR can reach 255.90 V, the TEV can reach 414.63 V, and the maximum disturbance voltage at the secondary side of CT does not exceed 100 V. The rise time of the pulses is between 8 ns and 16 ns, and the dominant frequencies are between 1 and 30 MHz. These parameters are useful for clarifying the immunity test requirement and improving the protection measures of the secondary equipment against the transit disturbance.

Keywords: current transformer, distribution switchgear, electromagnetic disturbance switching operation

1 INTRODUCTION

In a traditional power substation, the secondary equipment is concentrated in the relaying room. In recent years, with the development of the smart grid, a large amount of secondary equipment is arranged near the high-voltage primary equipment such as circuit breakers, disconnectors, and transformers, making these secondary devices in complex electromagnetic environments (Thomas et al. 1994; Lu et al. 2004; Tavakoli et al. 2013; Chen et al. 2015). At present, research at the Gas Insulated Substation shows that a switching operation can generate an obvious transient electromagnetic disturbance. The disturbance is coupled to the secondary circuit through conduction and can even interfere with the normal work of the secondary equipment (Zhang et al. 2013; Bai et al. 2016; Feng et al. 2016; Wang et al. 2017). Feng et al. (2016) show that a 9.9-kV transient disturbance voltage was observed on the port of the electronic transformer in a 500-kV substation.

In contrast, the operation voltage of distribution switch cabinet (below 10 kV) is much lower than that of the substation. However, the electromagnetic disturbance problem that resulted from the switching operation of distribution switchgear is also notable due to following reasons. In practice, several cases related to such a problem had been reported. For a 10-kV switch cabinet, the transient current at the secondary side of the current transformer is measured (Liu et al. 2008). It is shown that the peak of the transient current is nearly four times that of the normal operating current with the dominant frequencies of 15.625–62.5 MHz. However, no disturbance voltage data are provided.

The electronic transformer is the basic test equipment of a smart grid. It has the advantages of simple insulation structure, light weight, direct signal input into the protected device, and so on. Electronic transformer is a bridge between the high-voltage primary system and the low-voltage secondary system. As a result, the electromagnetic disturbance coming from the primary system can propagate along the transformer to the secondary system. Hence, there are great concerns about the electromagnetic disturbance characteristics induced at the secondary side of the electronic transformer, especially in the case when the primary side has a switch operation like the closing/opening of the circuit breaker. During the switching operations, a high-amplitude and high-frequency transient overvoltage is generated. It interferes with the secondary devices in complex coupling ways, making it impossible for them to work normally. The interference may go to the secondary devices directly without damping, causing them transmission data loss, waveform distortion, and other failures.

In this article, measurement of the transient electromagnetic disturbance caused by the switching operation of a 6-kV switch cabinet is carried out. The measured terms include not only the transient voltage of the secondary side of current transformer (CT), but also the voltage of the primary side, the ground potential rise (GPR), and the transient enclosure voltage (TEV). Then the waveform characteristics are analyzed in both the time domain and the frequency domain. This article is organized as follows. The measurement system and scheme of the experiment are introduced in Section 2. The typical waveforms are displayed in Section 3. The parameters extracted from the measurement are compared with those described in IEC 61000-4-18 in Section 4. Finally, the article is summarized in Section 5.

2 MEASUREMENT OVERVIEW

2.1 *Test circuit*

As shown in Figure 1, the switch cabinet is supplied by a single-phase power. The other phases are grounding. The measured CTs are arranged near the circuit breaker. The test is divided into two cases: open circuit and capacitive load. The resistance value of the capacitive load is 1800 pF.

2.2 *Measurement system*

In this test, the disturbance voltages are measured by an anti-interference measurement system that integrates the high-voltage probe, shield measurement device, and remote-control device. The measurement system adopted a PINTECH 6039A high-voltage probe as the sensor, which can measure signals from DC to 220 MHz and bear 40 kV at most. As shown in

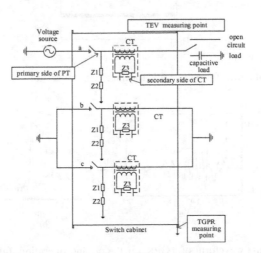

Figure 1.   Measurement circuit.

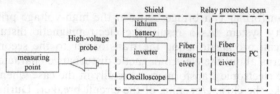

Figure 2. Sketch of the measuring system.

Figure 2, during the tests, the oscilloscope is powered by a lithium battery placed in a metal shielding box. With the application of a fiber, the signal can reach the remote computer. This system is highly reliable and practical (Zhang et al. 2016; Wu et al. 2017a,b,c).

2.3 *Measurement scheme*

In this test, we select the secondary sides of the CT and the top of the shell as measuring points. Each experiment was repeated at least three times, and the specific measuring conditions are as follows:

(1) Measure the transient ground potential rise (TGPR) at the grounding point and the transient enclosure voltage (TEV) with open circuit and closing/opening operations.

(2) Measure the secondary sides of the CT with an open circuit/load circuit and closing/opening operation.

3 CHARACTERISTICS OF AN ELECTROMAGNETIC DISTURBANCE

3.1 *Typical waveform of a TGPR*

A typical TGPR waveform under closing operation is shown in Figures 3 and Figure 4, and an opening operation is shown in Figures 5 and Figure 6. Figure 3 shows the time domain of a macro-pulse waveform collected by the oscilloscope. The waveform consists of two

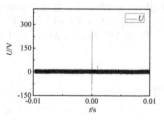

Figure 3. The macro-pulse waveform of TGPR for the closing operation.

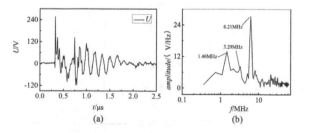

Figure 4. The micro-pulse waveform of TGPR for the closing operation. (a) The time domain waveform. (b) The frequency domain waveform.

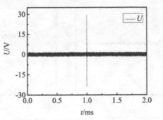

Figure 5.　The macro-pulse waveform of TGPR for the opening operation.

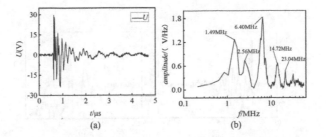

(a)　　　　　　　　　　　　(b)

Figure 6.　The micro-pulse waveform of TGPR for opening operation. (a) The time domain waveform. (b) The frequency domain waveform.

micro-pulses. The micro-pulse at the moment of switch operation ($t = 0$) is shown in Figure 4a. The frequency waveform shown in Figure 4b is obtained by Fourier transform of Figure 4a. The maximum peak of the disturbance voltage (u_{max}), the macro-pulse duration (t_{mac}), the number of micro-pulses (n), the micro-pulse duration (t_{mic}), the rise time (t_r), and the maximum frequencies (f_{max}) are extracted from Figures 3 and Figure 4, which were listed in Table 1. The following data processing is the same as previously.

By analyzing the data in Table 1 it can be concluded that u_{max} of TGPR is 255.9 V and f_{max} does not exceed 25 MHz.

3.2　Typical waveform of TEV

Typical micro-pulse waveform under the closing/opening operation are shown in Figures 7 and Figure 8. The relevant parameters are listed in Table 2.

By analyzing the data in Table 2, it can be concluded that u_{max} of TEV does not exceed 420 V and f_{max} does not exceed 30 MHz.

3.3　Effects of different loads and switch operations on the secondary side of CT

This experiment measures the differential mode voltage of the secondary side of the CT. The voltage parameters under various working conditions shown in Table 3 are extracted from measured waveforms. The f denotes the main frequency.

Table 1.　The waveform parameters of TGPR.

Type of operation	u_{max} (V)	t_{max} (µs)	n	t_{mic} (µs)	t_r (ns)	f_{max} (MHz)
Closing	255.90	21.24	2	1.76	8	6.21
Opening	29.50	2.27	1	2.27	6	23.04

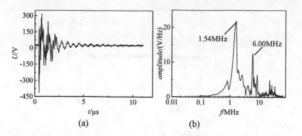

Figure 7. The micro-pulse waveform of TEV for closing operation. (a) The time domain waveform. (b) The frequency domain waveform.

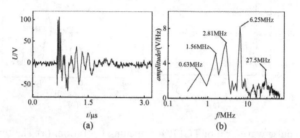

Figure 8. The micro-pulse waveform of TEV for the opening operation. (a) The time domain waveform. (b) The frequency domain waveform.

Table 2. The waveform parameters of TEV.

Type of operation	u_{max}/V	t_{max}/μs	n	t_{mic}/μs	t_r/ns	f_{max}/MHz
Closing	−414.63	280	2	8.13	16	6.00
Opening	106.10	1.63	1	1.63	14	27.50

From the data in Table 3, it can be concluded that u_{max} does not exceed 100 V, f does not exceed 25 MHz, and u_{max} under a closing operation is larger than under a opening operation. Furthermore, the t_{mac} is greatly shortened after the the capacitive load is connected.

4 COMPARISON BETWEEN THE MEASURED WAVEFORM AND EXISTING STANDARDS

The disturbance waveform generated by the switch operation is similar to the damped oscillation wave. In the past, standard damped oscillatory waveforms defined in IEC 61000-4-18 were used for assessment. According to IEC 61000-4-18, the damped oscillatory properties and parameters are as follows:

(1) The rise time of voltage (first peak value): 5 ns ± 30%
(2) Oscillation frequency: 3 MHz, 10 MHz and 30 MHz ± 10%
(3) Repeat rate: 5000/s ± 10%
(4) Macro-pulse duration: 3 MHz for 50 ms ± 20%; 10 MHz for 15 ms ± 20%; 30 MHz for 5 ms ± 20%
(5) Peak voltage of open circuit: 250 V–4 kV (± 10%)

In this test, it was found that the rise time of the disturbance waveform on the secondary side of the transformer was significantly higher than in the previous assessment standard, so

Table 3. Waveform parameters of disturbance voltage at the secondary side of VT.

Working conditions	Type of operation	Times	$|u_{max}|$ (V)	t_{max} (ms)	n	t_{mic} (μs)	t_r (ns)	f_{max} (MHz)
Open circuit	Closing	1	25.39	22.77	6	3.26	16	3.99
		2	20.85	30.78	7	2.08	8	3.6, 10.79, 11.69
		3	15.27	33.06	7	0.98	8	4.10, 20.49, 27.66
	Opening	1	10.73	47.46	10	1.33	16	3.81, 12.20, 20.58
		2	9.94	50.46	25	1.76	12	3.98, 11.93, 20.45
		3	8.04	46.99	9	1.1	12	3.68, 11.95, 20.22
capacity load	Closing	1	84.31	1.38	5	7.42	8	1.66, 3.73, 5.72
		2	43.89	1.07	3	4.52	14	1.77, 5.76, 15.07
		3	50.64	1.34	4	7.75	12	1.68, 14.33, 21.57
	open	1	5.95	5.95E-3	1	5.95	8	1.73, 5.85, 14.76, 22.34
		2	5.97	1.34E-3	1	1.34	10	2.26, 6.02, 15.06, 21.84

the previous test standard of anti-disturbance could not fully cover the distribution switch cabinet. According to the extraction and statistics of parameters in the experimental data:

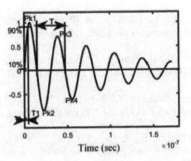

Figure 9. Single-pulse waveform of open circuit voltage under the switching operation.

Table 4. Parameters of the measured damped oscillation waveform.

Definition	t_r (ns)	t_{mic} (μs)	t_{mac} (ms)
Voltage of open circuit	12 × (1±50%)	3 ×(1±50%)	3 MHz: 20 ms ± 20% 10 MHz: 30 ms ± 20% 30 MHz: 40 ms ± 20%

(1) The differential mode voltage at the secondary side of the electronic transformer is no more than 30 V under an open circuit. The primary frequency is no more than 30 MHz. The measured rise time is between 8 and 16 ns, which exceeds the standard damped oscillations.

(2) The repetition rate is the ratio between the number of micro-pulses and the duration of the macro-pulse. The measured data are mainly between 191 and 495 times/s.

(3) The macro-pulse duration of standard waveform is longer at 3 MHz (50 ms + 20%) than measured and shorter at other frequencies.

Compared to the IEC61000-4-18, the standard is supplemented according to the actual measurement in this article. Specific waveform parameters are shown in Table 4.

5 CONCLUSION

(1) The peak voltage of TGPR caused by the switch operation described in this article is 255.90 V. Its main frequencies are between 1 and 25 MHz. The peak voltage of TEV caused by the switch operation describedin this article is 414.63 V. Its main frequency is between 1 and 30 MHz.

(2) For the secondary side of CT, the u_{max} are 225.39 V/84.31 V under an open circuit/capacitive load, the t_r are between 8 and 16 ns, and the f are no more than 30 MHz.

(3) For the secondary side of CT, the circuit connecting capacity load can reduce the macro-pulse duration of disturbance.

(4) According to the measured data, the anti-interference experiment of medium-voltage equipment is supplemented.

ACKNOWLEDGMENTS

This project is supported by the Key Laboratory Opening Fund Project of Energy Saving Technology for Distribution Transformer in Beijing (China Electric Power Research Institute).

REFERENCES

Bai, S.J., Zeng, L.C., Rong, M.Z., Li, Y., Chen, K., & Zhang, Y.R. 2016. Interference analysis and protection of GIS isolation switch operation on electronic current transformer. *High Voltage Electrical Appliances* 52(04): 54–62, 69.

Chen, W.J., Wang, H., Han, B., et al. 2015. Study on influence of the disconnector characteristics on very fast transient overvoltage in 1100kV gas insulated switchgear. *IEEE Transactions on Power Delivery* 30(4): 2037–2044.

Feng, L.M., Wang, X.B., Wu, L.Z., Wang, J.Y., & Liu, H. 2016, Study on electromagnetic disturbance of electronic transformer caused by VFTO in the 500 kV GIS substation. *Journal of Electrical Technology* 31(1): 85–90.

Li, P, Huang, D.C., Ruan, J.J., Niu, X.B., & Zhu, C.G. 2015. Electromagnetic interference of 10 kV switch cabinet on secondary intelligent equipment. *Power Grid Technology* 39(1): 110–117.

Liu, B, Ye, G.X., Tong, Y., & Huang, H. 2008. The effect of isolation switch switching operation on electromagnetic compatibility of electronic transformer. High voltage technology, 44(04):1204–1210.

Lu, B.X., Wang, Z.Z., Li, C.R., Ding, L.J., & Wang, W. 2004. Measurement and research of transient electric field caused by switching operation of the 500 kV substation. *Chinese Journal of Electrical Engineering* 24(4): 137–142.

Tavakoli, A., Gholami, H., Nouri, M., & Negnevitsky. 2013. Comparison between suppressing approaches of very fast transients in gas-insulated substations (GIS). *IEEE Transactions on Power Delivery* 28(1): 303–310.

Thomas, D.E., Wiggins, E.M., Salas, T.M., Nickle, F.S., & Wright, S.E. 1994. Induced transients in substation cables: Measurements and models. *IEEE Transactions on Power Delivery* 9(4): 1861–1868.

Wang, L.H., Qian, J.B., Ji, J.F., & Bu, Q.S. 2017. Modeling and characteristic analysis of switching transient interference in substation. *High Voltage Technology* 43(3): 960–965.

Wu, H.T., Jiao, C.Q., Cui, X., Liu, X.F., & Ji, J.F. 2017a. Development and application of electromagnetic disturbance measurement system for secondary equipment port of GIS substation. *High Voltage Electrical Appliances* 53(10): 68–75.

Wu, H.T., Jiao, C.Q., Cui, X., Liu, X.F., & Ji, J.F. 2017b. Simulation and analysis of electromagnetic disturbance caused by switching operation of GIS substation secondary equipment port. *High Voltage Technology* 43(10): 3387–3395.

Wu, H.T., Jiao, C.Q.,Cui, X., Liu, X. F., & Ji, J.F. 2017c. Transient electromagnetic disturbance induced on the ports of intelligent component of electronic instrument transformer due to switching operations in 500 kV GIS substations. *IEEE Access* 5(99): 5104–5112.

Zhang, J., Jiao, C.Q., & Li, S.J. 2016. Measurement and analysis of electromagnetic disturbance caused by circuit breaker operation in a 500kVGIS substation for the CT port of intelligent module. *High Voltage Electrical Appliances* 52(8): 29–36.

Zhang, Z.Y. 2013. Effect of GIS switching operation on electronic transformer. Harbin Institute of Technology.

Emerging Developments in the Power and Energy Industry – Dufo-López, Krzywanski & Singh (eds)
© 2020 Taylor & Francis Group, London, ISBN 978-0-367-27169-5

An evaluating method of cloud matter–element theory and its application for developmental form of an urban distribution network

Yanxun Qi
College of North China Electric Power University, Baoding, China

Aili Pang
State Grid Shanghai Electric Power Co., Ltd., Shanghai, China

Han Wu & Qing Xie
College of North China Electric Power University, Baoding, China

Lang Zhao
State Grid Economic and Technological Research Institute, Beijing, China

ABSTRACT: With the development of distribution network technology, the developmental level of a distribution network in different cities is obviously different and expressed in various forms influenced by the local economy and society. In this article, first indexes such as power supply quality, network structure, and equipment levels that characterized the operation condition and the future pattern of the urban distribution network were screened for evaluation. The current research hotspot of distribution network technology is summarized. In the analysis of the advantage of cloud matter–element theory applied in distribution network evaluation, the form was divided into five levels: demonstration, advanced, intermediate, low, and elementary based on the division standard. In the end, three representative cities, Suzhou, Beijing, and Hengshui, were selected as experimental cases for evaluation.

1 INTRODUCTION

At present, an ever growing quality of life and electric power are required by residents in Chinese international metropolitan areas. But in typical cities, it is difficult to possess the power distribution technology and equipment quality that has advanced with the times, which leads to various developmental degrees of the urban distribution network in different regions (Wang et al. 2018). Thus China plans to build world-class distribution networks in four years that are safe, reliable, high-quality, effective, green, low-carbon, intellectual, and interactive, taking the pattern city as the model for upgrading the distribution network of other cities (Li et al. 2018).

To realize the evolution of the form of a distribution network, its current development phase should be evaluated through the comprehensive evaluation method first to realize the form division, that is, to define the status of the current distribution network. Finally the development plan can be formulated by comparing the superior morphological characteristics. At present, the reliabilities in operation (Andrade et al. 2009) and the comprehensive benefits of planning schemes of urban distribution networks have been abundantly evaluated. The index system closely revolved around the operation cost and income (Liu et al. 2018). To assess development forms, the selected indicators should synchronously characterize the future pattern of distribution systems.

The system for evaluating a distribution network form is huge, where indicators have a complicated relationship. Thus the fuzzy data of some indicators increase the difficulties of the assessment process. Numerous methods have been successfully applied in distribution network evaluation. Qualitative evaluation methods include analysis hierarchy process (AHP) (Du et al. 2015), fuzzy (Wang & Sun 2010), Delphi method Chen et al. 2016, etc. Quantitative evaluation methods include the entropy weight method (Kadhem et al. 2017) and data envelopment analysis (DEA) (Pahwa et al. 2002). In the literature (Liang et al. 2017), cloud matter–element theory, combining a cloud model and extension method, was first applied in the comprehensive evaluation of life cycle distribution network planning. It can quantify the indicators described by natural language and has the advantage of randomizing the indicators' intervals. Thus the comprehensive benefit of the distribution network planning scheme can be evaluated scientifically. In summary, this article analyzes the connotation of morphological indicators and the corresponding index system is established. An evaluation model of cloud matter–element theory applied to the development form of an urban distribution network is proposed. To verify the rationality of our evaluation model, three representative urban distribution networks are selected in this article as research subjects.

2 CONSTRUCTION OF DEVELOPMENT FORM INDEX SYSTEM FOR AN URBAN DISTRIBUTION NETWORK

The selected indicators must be independent and complete and can be compared effectively under different social environments. Through the comparison between the relevant standard documents of the State Grid Corporation like the Technical Specification of Evaluation on the Distribution Network Development Plan and the Specification of Distribution Network Development Evaluation for new pattern urbanization and the reviewed literature, six first-class indexes were selected as follows: supply quality, network structure, equipment levels, power supply capability, green intelligent, and power grid efficiency.

The main features of future distribution systems' structures will be the harmonious and synchronous development of a large power grid and microgrid, AC–DC hybrid distribution system, and energy Internet that integrates multiple energy sources and realizes the interaction between supply and demand (Ma et al. 2015). These characteristics will eventually be reflected in the mature distribution network in that technological innovation was chosen as an additional first-level indicator to measure the adaptability of the current distribution network of the future development trend. Its third-class indexes were selected as follows: microgrid, AC–DC hybrid, and active distribution network. As a core important means to build world-class distribution networks, non-blackout operation technology was also involved where integration of multiple types of energy has been reflected in green intelligence. The final details are shown in Figure 1.

3 EVALUATION MODEL OF AN URBAN DISTRIBUTION NETWORK DEVELOPMENT FORM BASED ON CLOUD MATTER-ELEMENT THEORY

The cloud matter-element theory can be used to describe a qualitative concept by transforming matter–element information into three numerical features: expectation, entropy, and excess entropy. It can also randomize the index interval of membership grade. In this article, the following six steps are taken to evaluate the development form of an urban distribution network.

(1) Calculation of standard cloud for indexes

The form was divided into five levels: advanced, higher, intermediate, low, and primary based on division standard. Based on official reports and comparison of mass index data, the index intervals of membership grade in terms of different development form are shown in Table 1.

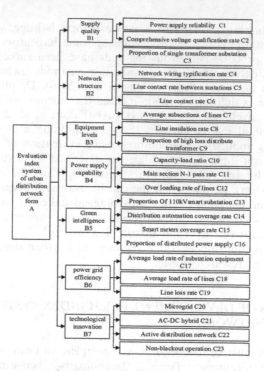

Figure1. Evaluation index system of an urban distribution network form.

Table 1. Index interval of membership grade.

| Index | Levels of development form | | | | |
	Advanced	Higher	Intermediate	Low	Primary
C1	[100,99.999]	[99.999,99.99]	[99.99,99.965]	[99.965,99.863]	[99.863,99.726]
C2	[100,99.99]	[99.99,99.97]	[99.97,99.95]	[99.95,98.79]	[98.79,97]
C3	0	[0–2]	[2–4]	[4–6]	[6–8]
C4	[100–90]	[90–75]	[75–60]	[60–45]	[45–0]
C5	[60–50]	[50–40]	[40–30]	[30–20]	[20–10]
C6	[100–80]	[80–60]	[60–40]	[40–20]	[20–0]
C7	[3.5–3]	[4–3.5]	[4.5–4]	[5–4.5]	[5.5–5]
C8	[100–80]	[80–60]	[60–40]	[40–20]	[20–0]
C9	0	[0–2]	[2–4]	[4–6]	[6–8]
C10	[2.2–2]	[2–1.8]	[1.8–1.6]	[1.6–1.4]	[1.4–1.2]
C11	[100–95]	[95–90]	[90–85]	[85–80]	[80–75]
C12	[2–0]	[4–2]	[6–4]	[8–6]	[10–8]
C13	[100–90]	[90–75]	[75–60]	[60–45]	[45–0]
C14	[100–90]	[90–80]	[80–70]	[70–60]	[60–50]
C15	[100–88]	[88–76]	[76–64]	[64–52]	[52–40]
C16	[100–80]	[80–60]	[60–40]	[40–20]	[20–0]
C17	[70–60]	[60–50]	[50–40]	[40–30]	[30–20]
C18	[70–60]	[60–50]	[50–40]	[40–30]	[30–20]
C19	[2–0]	[4–2]	[6–4]	[8–6]	[10–8]
C20	[100–90]	[90–75]	[75–60]	[60–45]	[45–0]
C21	[100–90]	[90–75]	[75–60]	[60–45]	[45–0]
C22	[100–90]	[90–75]	[75–60]	[60–45]	[45–0]
C23	[100–90]	[90–75]	[75–60]	[60–45]	[45–0]

Then the numerical features of the cloud are obtained according to the following equations:

$$E_x = (c_{\min} + c_{\max})/2 \tag{1}$$

$$E_n = (c_{\max} - c_{\min})/6 \tag{2}$$

where, extra-entropy can be adjusted according to the fuzzy degree of classification. In this article, it is taken as 0.5. Limitations of space force us to take the capacity-load ratio as an example to illustrate. Its standard cloud was: Advanced [2.1, 0.033, 0.5]; Higher [1.9, 0.033, 0.5]; Intermediate [1.7, 0.033, 0.5]; Low [1.5, 0.033, 0.5]; and Primary [1.3, 0033, 0.5].

(2) Date acquisition

Date are obtained from special Department of State Grid Corporation, website, and of power industry reports except for the parts of qualitative indicators scored by experts.

(3) Weight setting

In this article, the fast Delphi method and the entropy weight method are used to determine the weight of the index. The equations of the entropy weight method are

$$p_{ij} = b_{ij}/\sum_{i=1}^{N} b_{ij} \tag{3}$$

$$E_j = (\sum_{i=1}^{N} p_{ij} \ln p_{ij})/\ln N \tag{4}$$

$$q_j = (1 - E_j)/\sum_{j=1}^{M}(1 - E_j) \tag{5}$$

where N = the number of schemes, m = the number of indexes, b_{ij} = the original index data, p_{ij} = the standardized index data, q_j = index weight, E_j = entropy, and i and j represent the serial number of scheme and indicator, respectively.

The combination weight was calculated by Equation (6):

$$w_j = k_1 h_j + k_2 q_j \ (j = 1, 2, \ldots, 23) \tag{6}$$

where w_j = the combination weight of the jth indicator, k_1 and k_2 represent the coefficient of the Delphi and entropy method, respectively, h_j = index weight calculated by Delphi.

(4) The membership grade determination between every third-level index and each different form

For each development form, there is the matched E_n obtained by Equation (2). Then a random number y_i was generated according to the following normal distribution: $y = R_N$ (En, 0.5) (i =1, 2,..., m), and the probability of this random event was calculated as the process weight. Finally the relative-membership grade μ_j^k between the jth index and the given form was obtained by Equation (7).

$$\mu_j^k(b_{ij}) = \exp(-\frac{(b_{ij} - Ex)^2}{2y^2})(k = 1, 2, 3, 4, 5) \tag{7}$$

After repeating the above random process five times, five membership grades and five process weights were obtained. Then the final membership grades $\mu_j^k(b_{ij})$ were obtained by weighting them.

(5) Calculation of comprehensive membership grades

The comprehensive membership grades of the selected urban distribution network can be obtained by Equation (8).

$$Z_k = w_j * \mu_j^k(x) \ (k = 1, 2, 3, 4, 5) \tag{8}$$

where k is the serial number of the development form.

(6) Form level determination

It can be determined by the maximum membership grade within the sequence of $\mathbf{Z}$.

4 CASE ANALYSIS

4.1 *Assessment process*

To verify the rationality of the divide of the urban distribution network development form and the validity of cloud matter–element theory, the distribution network of three typical cities from south to north, Suzhou (SZ), Beijing (BJ), and Hengshui (HS), was selected in this study according to decreasing power consumption. The index data of 2017 was gathered and the weights were obtained by step (3) as Table 2.

Based on the 23 standard clouds calculated by Equation (7) and intervals in Table 1, the membership grade of all third-class indicators for each development form was obtained through

Table 2. Index data and weights of the urban distribution network development form.

Second-class indicators and weights	Third-class indicators and weights	Suzhou	Beijing	Hengshui
Supply quality 31%	Power supply reliability 9.7%	99.998	99.995	99.974
	Comprehensive voltage qualification rate 11%	100	99.99	99.988
Network structure 13%	Proportion of single transformer substation 3.9%	0	0	10
	Network wiring typification rate 3.6%	97	97	60
	Line contact rate between substations 3.2%	59.99	48.65	60
	Line contact rate 2.9%	98.2	100	85
	Average subsections of lines 2.9%	4.1	2.79	5.25
Equipment levels 9%	Line insulation rate 5.3%	81.82	72.12	90
	Proportion of high loss distribute transformer 3.3%	0	0	10
Power supply capability 13%	Capacity-load ratio 5%	1.95	2.03	1.7
	Main section $N - 1$ pass rate 4.1%	100	100	90
	Over loading rate of lines 4.1%	0	0	5
Green intelligence15%	Proportion of 110-kV smart substation 3.5%	31.24	25.46	5
	Distribution automation coverage rate 6.3%	100	75	70
	Smart meters coverage rate 3.4%	100	100	80
	Proportion of distributed power supply 3.9%	2.15	6.2	3.7
Power grid efficiency 9%	Average load rate of substation equipment 4%	39.97	51.2	29.9
	Average load rate of lines 4.2%	42.81	38.42	35.8
	Line loss rate 2.8%	3.79	6.05	3.39
Techno logical innovation 10%	Microgrid 3.3%	90	60	10
	AC–DC hybrid 3.2%	94	90	47
	Active distribution network 3.2%	95	95	57
	Non-blackout operation 3.2%	92	87	62

MATLAB® software as Table 3, which has been weighted by process factors. After weighted by the weights of all indicators, the comprehensive membership grades was shown as Table 4.

Afterwards, the membership grades of second-class indexes were also be calculated, the maximum membership degree among them was taken as the unit, and others scaled in proportion. The corresponding radar image is shown in Figure 2.

4.2 Discussion

Table 4 shows that according to the principle of maximum membership, the development level of Suzhou distribution network was obviously located in the advanced form. Though Beijing was located in the higher development form, the boundary between advanced and higher was not obvious, which indicated Beijing will soon step into the advanced form. Hengshui was located in the intermediate form.

Table 3. Membership grade of third-class indicators.

| Index* | Levels of development form | | | | |
	Adva-nced	Higher	Interm-ediate	Low	Primary
C3	0.113	0.06	0.003	0	0
C4	0.11	0.01	0	0	0
C5	0.056	0	0	0	0
C6	0.061	0.001	0	0	0
C7	0.005	0.056	0.059	0.024	0.001

* Here the part of indexes of Suzhou were given as the limitations of space.

Table 4. Comprehensive membership grades.

| Cities | Development form | | | | |
	Adva-nced	Higher	Interm-ediate	Low	Primary
SZ	2.161	1.324	1.024	0.620	0.391
BJ	1.550	1.638	1.149	0.671	0.508
HS	1.022	0.963	1.440	1.289	0.923

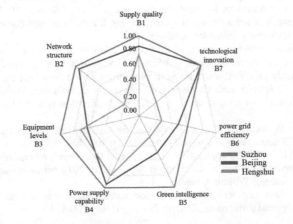

Figure 2. Radar image of the membership grades.

According to Figure 2, for the Beijing distribution network, its green intelligence was located in the advanced form. It can realize an evolution to the advanced form by strengthening the construction of a 110-kV smart substation and improving the proportion of distributed power supply. For the Hengshui distribution network, its network structure and green intelligence were located below the intermediate form, which should be improved principally. In addition, the power grids of the three cities' distribution networks all demonstrated limited efficiency. It can be greatly improved by calling off blind addition of line capacity.

5 CONCLUSION

This article summarized the research background of the development form of the urban distribution network. Based on the future structure and pattern characteristics, the new index of technological innovation was put forward. The evaluation index system of the urban distribution network development form was established that included six traditional benefit indexes of the distribution network. All the corresponding index intervals in terms of different forms were given.

An evaluation model applied for the development form of the urban distribution network based on cloud matter–element theory was proposed. It realized the randomization of the boundary of different development forms, which increased the accuracy of the assessment. Three typical cities were selected in the case analysis, whose development form of distribution network were respectively located. In addition, the advantages and disadvantages of the second-class index of the distribution network in three cities are analyzed. The trend of morphological evolution and relevant implementation measures is given.

ACKNOWLEDGMENTS

The authors acknowledge funding support from the State Grid Science and Technology Projects (Grant SGSHJY00GPJS1800060).

REFERENCES

Andrade, W.S., et al. 2009. Integrated reliability evaluation of distribution and sub-transmission systems incorporating distributed generation. In *Power Systems Conference and Exposition*, Seattle, WA.
Chen, J.M., et al. 2016. Study on quality fuzzy synthetic evaluation method for power transformer based on life cycle theory. In *2016 China International Conference on Electricity Distribution*, Xi'an, China, pp. 1–4.
Du, Z., et al, 2015. Diagnosis of smart distribution grid development. In *2014 IEEE PES Asia-Pacific Power and Energy Engineering Conference*, Hong Kong, China.
Kadhem, A.A. et al. 2017. Computational techniques for assessing the reliability and sustainability of electrical power systems: A review. *Renewable & Sustainable Energy Reviews* 80: 1175–1186.
Li, X., et al. 2018. Research of the assessment system for the life-cycle of first-class distribution network. *Power System Protection and Control* 46(9): 80–85.
Liang, T.J., et al. 2017. Comprehensive benefit evaluation of distribution network planning based on the life cycle cost theory and the cloud matter element theory. *Power System Protection and Control* 45(19): 1–6.
Liu, W., et al. 2018. Study on distribution network planning method based on reliability of cost-benefit analysis. In *2018 China International Conference on Electricity Distribution*, Tianjin, China.
Ma, Z., et al, 2015. Form and development trend of future distribution system. *Proceedings of the CSEE* 35(6): 1289–1298.
Pahwa, A., et al. 2002. Performance evaluation of electric distribution utilities based on data envelopment analysis. *IEEE Transactions on Power Systems* 18(1): 400–405.
Wang, J.M., & Sun, J.M. 2010. The application of multi-level fuzzy comprehensive evaluation method in technical and economic evaluation of distribution network. In *2010 International Conference of Management and Service Science*, Wuhan, China, pp. 1–4.
Wang, X.C., et al. 2018. A comprehensive evaluation model for power system with large-scale new energy considering regional differences. *Electric Power* 51(10): 178–184.

Emerging Developments in the Power and Energy Industry – Dufo-López, Krzywanski & Singh (eds)
© 2020 Taylor & Francis Group, London, ISBN 978-0-367-27169-5

Wind farm layout optimization for minimum levelized cost of energy considering land lease cost, noise emission, and wake loss

Shuwei Miao*, Han Li, Dan Li, Yuanhang Zhang & Baohua Yang
College of Electrical Engineering and New Energy, China Three Gorges University, Yichang, China

ABSTRACT: Wind farm installation on private land has to deal with three challenges: land lease cost, noise emission, and wake loss. In this article, three effective equations are used to evaluate the land lease cost, noise emission, and annual energy production due to wake loss. These equations are incorporated into an immune clone algorithm that determines the optimal wind farm layout. Then a novel wind farm layout optimization (WFLO) model that minimizes the land lease cost, noise emission, and wake loss is developed. The proposed WFLO model is used to lay out 28 wind turbine generators (WTGs) in a 1.5 km × 1.5 km wind farm consisting of 9 private lands and residences. Results show that the proposed model is a promising wind farm planning tool for handling engineering challenges from land lease cost, noise emission, and wake loss.

Keywords: levelized cost of energy, wake loss, wind farm, wind farm layout optimization

1 INTRODUCTION

With the accelerated depletion of mainstream energy resources such as oil and gas, countless efforts have been devoted to energy substitutes in the past decades (International Energy Agency 2014). Global wind capacity has sextupled during the past decade, expanding from 74 GW to 432.4 GW (Global Wind Energy Council 2014). In addition, it keeps growing rapidly and is expected to reach 2000 GW by 2030 (Global Wind Energy Council 2014).

As more public land is being exploited, private land with considerable wind resources would be an inevitable choice for future wind farm installation. In such a case, wind farm planners have to acquire permission from landlords for land leasing. Moreover, wind turbine generators (WTGs) may be erected in proximity to their residences. Then noise emissions from WTGs may create significant disturbance on nearby residences and result in complaints or protests against the wind farm project (Eltham et al. 2008). Furthermore, proximity to WTGs could lead to considerable wake losses that account for 10%–20% reduction on annual energy production (Barthelmie 2009). Owing to these facts, wind farm planners encounter three potential challenges:

- Alleviate noise emissions of WTGs
- Minimize wake losses related to WTGs
- Minimize land lease cost of WTGs

Wind Farm Layout Optimization (WFLO), as a powerful planning tool, strategically determines the placement of WTGs under various engineering requirements. A WFLO model that answers the above challenges would be of great benefit to the wind energy industry.

Generally, the existing WFLO models can be classified into two categories: the grid-based WFLO and the coordinate-based WFLO. The grid-based WFLO divides a wind farm into a set of square cells. The WTGs are placed at the center of each cell by a solution algorithm. Grady et al. (2005) placed 26–30 WTGs at 10 × 10 cells at a minimum cost of energy using a genetic algorithm. Notable improvements have been reported to solve the similar WFLO problem using

---

*Corresponding author: E-mail: jabker@163.com

Monte Carlo simulation (Marmidis et al. 2008), the improved genetic algorithm (Emami & Noghreh 2010), and the particle swarm optimization algorithm (Pookpunt & Ongsakul 2013). These works focused more on the effective solution algorithms rather than on the practical engineering requirements. On the other hand, Chen and MacDonald (2012) coded the landlords' participation and wind farm layout as a binary chromosome, and used a C++ library GAlib to search the optimal wind farm layout at a minimum cost of energy. Civil infrastructures and restrictions, such as road, forbidden, and low-load-bearing zones, have been considered in WFLO (González et al. 2010). In addition, WTGs of different hub heights could lead to a wind farm layout with higher power output (Chen et al. 2013).

In summary, as candidate positions for WTGs are fixed and limited, the grid-based WFLO method is a straightforward and intuitive technique. However, it suffers from a drawback: all WTGs must lie on the candidate positions, and possibly better positions lying between adjacent cells can never been selected.

Therefore, a WFLO that strategically determines WTGs' placement considering all three challenges is appealing and will be developed in this article. The main contributions of this article are as follows:

(1)   Based on a commonly accepted land lease treaty, an effective equation is proposed to evaluate land lease cost of any candidate wind farm layout.
(2)   Based on the international noise standard, ISO 9613-2, a scheme is designed to effectively evaluate the noise emission of any candidate wind farm layout.
(3)   A comprehensive WFLO model is proposed to search the optimal wind farm layout of minimum levelized cost of energy.

2   WIND FARM MODELING TECHNIQUES

2.1   *WTG power output modeling*

The power output of a WTG can be expressed as a function of wind speed (Billinton &Chowdhury 1992),

$$P(v) = \begin{cases} 0 & v \in [0, v_{ci}) \cup [v_{co}, +\infty) \\ (C_1 + C_2 v + C_3 v^2) P_r & v \in [v_{ci}, v_r) \\ P_r & v \in [v_r, v_{co}) \end{cases} \qquad (1)$$

where P_r, v_{ci}, v_r, and v_{co} are the rated capacity, the cut-in speed, the rated speed, and the cut-out wind speed of WTG, respectively. The values of coefficients C_1, C_2, and $C_3$are based on v_{ci} and v_r, and can be found in Billinton and Chowdhury (1992).

2.2   *Wind profile modeling*

The site-specific wind profile is modeled with a discrete joint probability distribution (DJPD) of the wind speed and the wind direction. This DJPD is constructed from collected wind profile samples. Let (V_1, θ_1), (V_2, θ_2), ..., (V_{NWP}, θ_{NWP}) be samples of wind profile that consist of N_{WP} observations, where V_i ($i = 1, ..., N_{WP}$) and θ_i ($i = 1, ..., N_{WP}$) are observed wind speeds and wind directions. Let the north, south, east, and west be 0°, 180°, 90°, and 270° in the wind rose, respectively. Then $\theta_i \in [0, 360]$ are angles that deviate from the north in a clockwise direction.

Divide this sample space into $Q \times S$ discrete states. Then the probability of the discrete state (V_q, θ_s) ($q = 1, ...,Q$, $s = 1, ..., S$) can be estimated by

where

$$p(V_q, \theta_s) = \sum\nolimits_{i=1}^{N_{wp}} I_{qs}(V_i, \theta_i)/N_{wp} \qquad (2)$$

$$\begin{cases} V_q = \frac{q\Delta V + (q-1)\Delta V}{2} \\ \theta_s = \frac{s\Delta\theta + (s-1)\Delta\theta}{2} \end{cases} \qquad (3)$$

and $\Delta V = V_{\max}/Q$ and $\Delta\theta = \theta_{\max}/S$ are the step sizes of wind speed and direction, where $V_{\max}$ and $\theta_{\max}$ are the maximum values of wind speed and wind direction calculated from the collected wind profile sample. $I_{qs}(\cdot)$ is the indicator function shown as follows.

$$I_{qs}(V_i, \theta_i) = \begin{cases} 1 & V_i \in ((q-1)\Delta V, q\Delta V] \text{ AND } \theta_i \in ((s-1)\Delta\theta, s\Delta\theta] \\ 0 & \text{Otherwise} \end{cases} \qquad (4)$$

where AND indicates the operator of logical conjunction. Table 1 illustrates the discrete joint probability distribution of wind profile in tabular form.

2.3 Wake modeling

A turbulent wake that reduces downstream wind speed is generated when a WTG transforms wind energy into electrical power. With known ambient wind speed and wind direction, many wake models have been proposed to determine the downstream wind speed. Among them, the Jensen model is widely used because of its simplicity and effectiveness.

According to Jensen model, the wind speed v_j is determined using (Kim et al. 2012)

$$v_j(V, \theta) = V \left[1 - \sum_{i=1}^{M_{up}} \left(1 - \sqrt{1 - C_T}\right) \left(\frac{A_{si}}{\pi r^2}\right) \left(\frac{r}{r + \alpha h_{ij}}\right)^2 \right] \qquad (5)$$

where M_{up} are the number of upstream WTGs for WTG$_j$. C_T is the thrust coefficient of WTG, which can be found in the WTG manual. A_{si} can be calculated by equations in Kim et al. (2012). The h_{ij} and d_{ij} of A_{si} will be determined with θ, which will be discussed in Section 3.

According to the analysis of Sections 2.1 and 2.2 and Equation (5), the annual energy production of a wind farm consisting of M WTGs can be expressed as follows.

$$AEP = 8760 \sum_{q=1}^{Q} \sum_{s=1}^{S} \sum_{j=1}^{M} p(V_q, \theta_s) P(v_j(V_q, \theta_s)) \qquad (6)$$

2.4 Noise modeling

In addition to turbulent wake, a WTG also produces noise emission during operation, which may create considerable disturbance to nearby residences. ISO 9613-2, an international noise standard, is used to quantify such noise emission that consists of eight frequencies between 63 Hz and 8 kHz (Standardization I O 1996).

According to the standard, the overall noise emission of eight frequencies that propagated from M WTGs to H_k can be evaluated as a function of (V, θ) shown in Equation (7) (Standardization I O 1996).

Table 1. A discrete joint probability distribution of wind speed and wind direction.

θ (°)	V (mph)			
	$0.5\Delta V$	$1.5\Delta V$	—	$(Q-0.5)\Delta V$
$0.5\Delta\theta$	p_{11}	p_{12}	—	p_{1S}
$1.5\Delta\theta$	p_{21}	p_{22}	—	p_{2S}
—	—	—		—
$(S-0.5)\Delta\theta$	p_{Q1}	p_{Q2}	—	p_{QS}

$$N_k(V, \theta) = 10 \log_{10}\left(\sum_{j=1}^{M} \sum_{z=1}^{8} 10^{0.1\left(N_{jk}\left(v_j, l_{jk}, f_z\right) + W(f_z)\right)}\right) \qquad (7)$$

where $W(\cdot)$ is the weighting coefficient, which is a function of f_z (Wikipedia. A-weighting).

Similarly, according to the analysis of Section 2.2 and Equation (7), the expected noise emission of H_k can be expressed as

$$EN_k = \sum_{q=1}^{Q} \sum_{s=1}^{S} p(V_q, \theta_s) N_k(V_q, \theta_s) \qquad (8)$$

3 WIND FARM LAYOUT OPTIMIZATION CONSIDERING LAND LEASE COST, NOISE EMISSION, AND WAKE LOSS

3.1 Assumptions

Three assumptions are considered for the development of new wind farm layout methods to evaluate the land lease cost, noise emission, and annual energy production due to wake loss. These assumptions can be easily modified for other applications.

Assumption 1. The numbers of WTGs M, residences K, and land divisions R are known and fixed as well as their parameters, such as v_{ci}, v_r, v_{co}, P_r, r, a and b, in which all WTGs and residences are homogeneous.

Assumption 2. According to Assumption 2, a set of geometrical simplifications are made for the wind farm area, private land, WTG, and residence as follows.

(1) Both the wind farm area and private land are assumed to be rectangles, which are designated as WF and L_t ($t = 1, \ldots, R$), respectively. Using the 2-D Cartesian coordinate, WF and L_t are characterized as $\{x_{\text{left}}, x_{\text{right}}\} \times \{y_{\text{low}}, y_{\text{up}}\}$ and $\{x_{Lt,\text{left}}, x_{Lt,\text{right}}\} \times \{y_{Lt,\text{low}}, y_{Lt,\text{up}}\}$, where x_{left}, x_{right}, y_{low}, y_{up}, and $x_{Lt,\text{left}}$, $x_{Lt,\text{right}}$, $y_{Lt,\text{low}}$, $y_{Lt,\text{up}}$ are the left, right, lower, and upper boundary coordinates of WF and L_t.

(2) Both the WTG and residence are assumed to be circles, which are designated as WTG_i ($i = 1, \ldots, M$) and $H_k(k = 1, \ldots, K)$, respectively. The r_f and r_h are the foundation radii of WTG_i and H_k.

Let the centers of WTG_i and H_k and boundary of L_t and WF be represented by $(x_{\text{WTG}i}, y_{\text{WTG}i})$, (x_{Hk}, y_{Hk}), $(x_{L1,\text{left}}, x_{L1,\text{right1}}, y_{L1,\text{low}}, y_{L1,\text{up}})$, and $(x_{\text{left}}, x_{\text{right}}, y_{\text{low}}, y_{\text{up}})$; then coordinates of M WTGs, K residences, R private lands, and a wind farm can form the $M \times 2$, $K \times 2$, $R \times 4$, and 1×4 dimensional matrixes as follows.

$$\begin{cases} \boldsymbol{xy}^{WTG} = \begin{bmatrix} x_1^{WTG} & y_1^{WTG} \\ \vdots & \vdots \\ x_M^{WTG} & y_M^{WTG} \end{bmatrix} \\[4ex] \boldsymbol{xy}^{H} = \begin{bmatrix} x_1^{H} & y_1^{H} \\ \vdots & \vdots \\ x_K^{H} & y_K^{H} \end{bmatrix} \\[4ex] \boldsymbol{xy}^{L} = \begin{bmatrix} x_{1,\text{left}}^{L} & x_{1,\text{right}}^{L} & y_{1,\text{low}}^{L} & y_{1,\text{up}}^{L} \\ \vdots & \vdots & \vdots & \vdots \\ x_{R,\text{left}}^{L} & x_{R,\text{right}}^{L} & y_{R,\text{low}}^{L} & y_{R,\text{up}}^{L} \end{bmatrix} \\[4ex] \boldsymbol{xy}^{WF} = \begin{bmatrix} x_{\text{left}} & x_{\text{right}} & y_{\text{low}} & y_{\text{up}} \end{bmatrix} \end{cases} \qquad (9)$$

These matrices $\boldsymbol{xy}^{WTG}$, $\boldsymbol{xy}^{H}$, $\boldsymbol{xy}^{L}$, and $\boldsymbol{xy}^{WF}$ represent the placements of WTGs, residences, and boundaries of private lands and wind farm. The objective of WFLO is to find the optimal placement of WTGs under various engineering requirements. Therefore,

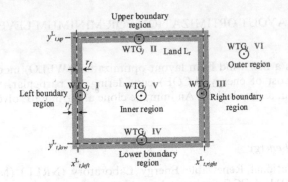

Figure 1. The relative location between WTG_j and L_t.

Table 2. Look-up table to determine the relative location between WTG_j and L_t.

Range of (x_{WTG_j}, y_{WTG_j})	Relative location
$x_{WTG_j} \in [x_{Lt,left} + r_f, x_{Lt,right} - r_f]$ AND $y_{WTG_j} \in [y_{Lt,low} + r_f, y_{Lt,up} - r_f]$	V
$x_{WTG_j} \in (-\infty, x_{Lt,left} - r_f) \cup (x_{Lt,right} + r_f, +\infty)$ AND $y_{WTG_j} \in (-\infty, y_{Lt,low} - r_f) \cup (y_{Lt,up} + r_f, +\infty)$	VI
Otherwise	I–IV

xy^{WTG} is considered as variable during the WFLO process, while xy^H, xy^L, and xy^{WF} are constants.

3.2 Evaluating land lease cost

The land lease treaty should be declared before evaluating the land lease cost for a particular xy^{WTG}. According to an engineering report (National Renewable Energy Laboratory 2011), an acceptable land lease treaty is given below. Other treaties can also be applied by modifying the following terms.

Land Lease Treaty. If a WTG is placed in L_t, then the land lease cost for this WTG will be charged as $P_r \times \rho_t$ dollars per year, where ρ_t is the unit land lease cost in \$/MW/yr.

It can be seen from this treaty that the relative location between WTG and private land is essential for evaluating land lease cost. Figure 1 illustrates the possible relative locations between WTG_j and L_t, where I–VI represent the left, upper, right, lower boundary, and the inner and outer regions. Based on geometrical analysis, a look-up table, Table 2, is constructed from Figure 1 to help identify the relative location between WTG_j and L_t.

The land lease cost of WTG_j at locations V–VI can be clearly evaluated. However, it could be difficult to evaluate land lease cost at locations I–IV, because WTG_j is located at both L_t and its nearby land. Such WTG placement should be avoided, which will be discussed with the WTG placement indicator vector defined in the following.

Definition 1. Define M dimensional vector $w = [w_1, w_2, \ldots, w_M]$, where, $\forall j \in M$, $w_j \in \{0, 1\}$. If $w_j = 0$, it indicates WTG_j is placed at the location V–VI. Otherwise, $w_j = 1$ indicates that WTG_j is at locations I–IV.

In Section 4, w will be used to eliminate the infeasible xy^{WTG}, in which some of WTGs are placed at locations I–IV.

4 WIND FARM LAYOUT OPTIMIZATION FOR MINIMUM LEVELIZED COST OF ENERGY

This section presents a new wind farm layout optimization (WFLO) model that enables the minimum levelized cost of energy (LCOE) considering, in particular, the land lease cost (LLC), noise emission, and wake loss. An immune clone algorithm to solve such model is also presented.

4.1 Levelized cost of energy

According to the National Renewable Energy Laboratory (NREL) (National Renewable Energy Laboratory 2011), LCOE can be evaluated by

$$LCOE = [ICC \times FCR + LLC + OM + LRC]/AEP \tag{10}$$

where LLC is the land lease cost (\$/yr) and AEP is the annual energy production (MWh/yr). ICC, OM, LRC, and FCR are the installed capital cost (\$/yr), operation and maintenance cost, levelized replacement cost (\$/yr), and the fixed charge rate (%), respectively.

4.2 Constraints

Constraints play an important role in the layout plan considering land lease cost, wake loss, and WTG noise emission. The following constraints are relevant to the proposed wind farm layout plan.

4.2.1 Boundary region constraint
According to the discussion in Section 3.2, WTG should not be placed at the boundary region. Then a feasible WTG location xy^{WTG} should ensure the following condition:

$$\forall j \in M, w_j = 0 \tag{11}$$

where w_j is the placement indicator of WTG_j according to Definition 1.

4.2.2 Expected noise emission constraint
According to a wind farm noise standard (NZS 68082010), the noise emission should not exceed the prespecified limit, which is usually set as 40 dB or 5 dB higher than background sound level. Then a feasible boundary xy^{WTG} should satisfy the constraint

$$\forall k \in K, EN_k \leq N_{k,\text{limit}} \tag{12}$$

where $N_{k,\text{limit}}$ is the prespecified limit, and EN_k can be evaluated with Equation (8).

4.2.3 Safety constraint
To avoid rotor collision, the safety distance between any two WTGs should be specified. Then the following constraint is required for a feasible xy^{WTG}:

$$\forall i,j \in M, i \neq j, \sqrt{\left(x_i^{WTG} - x_j^{WTG}\right)^2 + \left(y_i^{WTG} - y_j^{WTG}\right)^2} > 2r \tag{13}$$

Similarly, a WTG should not be placed in the residence area. Then a feasible boundary xy^{WTG} should ensure the following constraint will hold:

$$\forall j \in M, \forall k \in K, \sqrt{\left(x_k^{\mathrm{H}} - x_j^{\mathrm{WTG}}\right)^2 + \left(y_k^{\mathrm{H}} - y_j^{\mathrm{WTG}}\right)^2} > r_f + r_h \tag{14}$$

4.3 Objective function and penalty function

To eliminate the unfeasible WTG location xy^{WTG}, four penalty functions are proposed if constraints (13)–(16) are violated,

$$
\begin{cases}
g_1(w_j) = w_j r_f \\[2mm]
g_2(EN_k) = \max(0, EN_k - N_{k,\mathrm{limit}}) \\[2mm]
g_3\left(x_j^{\mathrm{WTG}}, y_j^{\mathrm{WTG}}\right) = \\[2mm]
\max\left(0, r_f + r_h - \sqrt{\left(x_k^{\mathrm{H}} - x_j^{\mathrm{WTG}}\right)^2 + \left(y_k^{\mathrm{H}} - y_j^{\mathrm{WTG}}\right)^2}\right) \\[3mm]
g_4\left(x_i^{\mathrm{WTG}}, y_i^{\mathrm{WTG}}, x_j^{\mathrm{WTG}}, y_j^{\mathrm{WTG}}\right) = \\[2mm]
\max\left(0, 2r - \sqrt{\left(x_i^{\mathrm{WTG}} - x_j^{\mathrm{WTG}}\right)^2 + \left(y_i^{\mathrm{WTG}} - y_j^{\mathrm{WTG}}\right)^2}\right)
\end{cases}
\tag{15}
$$

where $\max(\cdot)$ denotes the maximum value of a set. According to this function, four penalty vectors are defined as

$$
\begin{cases}
g_1 = [g_1(w_1), \ldots, g_1(w_M)] \\[2mm]
g_2 = [g_4(EN_1), \ldots, g_4(EN_K)] \\[2mm]
g_3 = \left[g_3\left(x_1^{\mathrm{WTG}}, y_1^{\mathrm{WTG}}\right), \ldots, g_3\left(x_M^{\mathrm{WTG}}, y_M^{\mathrm{WTG}}\right)\right] \\[2mm]
g_4 = \left[g_4\left(x_1^{\mathrm{WTG}}, y_1^{\mathrm{WTG}}, x_2^{\mathrm{WTG}}, y_2^{\mathrm{WTG}}\right), \ldots\right]
\end{cases}
\tag{16}
$$

Furthermore, an overall penalty function that evaluates the feasibility of xy^{WTG} can be expressed as

$$PF\left(xy^{\mathrm{WTG}}\right) = \left(\max(g_{C1}, g_{C2}, g_{C3}, g_{C4})\right)^2 \tag{17}$$

If the xy^{WTG} is feasible, then the function value of Equation (17) is zero. Otherwise, it would be the square of the maximum of all penalty functions.

With Equations (10) and (17), the proposed wind farm layout optimization for minimum LCOE can be expressed by the following unconstrained optimization problem:

$$\min_{xy^{\mathrm{WTG}}} F\left(xy^{\mathrm{WTG}}\right) = \mathrm{LCOE} + PF\left(xy^{\mathrm{WTG}}\right) \tag{18}$$

xy^{WTG} is the only variable for this objective function. Should xy^{WTG} not be feasible, $F(\cdot)$ would be extremely large. Then such xy^{WTG} probably will be eliminated by the optimization algorithm. Otherwise, such xy^{WTG} is saved for further consideration.

4.4 *Immune clone algorithm for wind farm layout optimization*

This section gives a detailed description of the immune clone algorithm (ICA) used in the proposed wind farm layout optimization. ICA simulates a learning technique inspired by the human immune system (Nicosia et al. 2004). It considers the optimization problem, solution, and solution quality as the antigen, antibody, and affinity, respectively. The clonal expansion and the affinity maturation are used to generate an optimal antibody (solution) that shows best affinity to the antigen (optimization problem). The unconstrained optimization problem formulated in Equation (18) refers to search the optimal WTG location xy^{WTG} in the continuous solution space. To facilitate the description of the immune clone algorithm for the proposed WFLO, the following definitions are given.

Definition 2. Define a $U \times M \times 2$ matrix $\mathbf{A} = [xy_1^{WTG}, ..., xy_U^{WTG}]$ as an antibody population of size U, where $\mathbf{A}(u)$ is the uth antibody containing the wind farm layout solution $xyWTGu$ for Equation (18).

Definition 3. Define the objective function $F(\cdot)$ in Equation (18) as the antigen and the U dimensional vector $\mathbf{AG} = [F(xy_1^{WTG}), ..., F(xy_{WTGu})]$ as the affinities for $\mathbf{A}$, where $\mathbf{AG}(u)$ is the uth antigen $F(xy_{WTGu})$. For the optimization problem of Equation (18), a smaller $\mathbf{AG}(u)$ indicates better affinity.

Definition 4. Define the U dimensional vector $\mathbf{TA} = [ta_1, ..., ta_U]$ as the ages of $\mathbf{A}$, where $\mathbf{TA}(u)$ is the uth age of the antibody.

Definition 5. Define DP as the number of clones for each antibody, AB as the maximum age of an antibody, GM as the maximum number of generations, SM as the maximum number of generations without improvement, and c as the scale parameter of inversely proportional hypermutation (IPH) operator (Nicosia et al. 2004).

The flow chart of the immune clone algorithm is given in Figure 2. $(xy^{WTG})_{op}$ and $(xy^{WTG})_{newop}$ are the best antibodies (solutions) selected from $\mathbf{A}_0$ and $(\mathbf{A}_0)_{new}$, respectively.

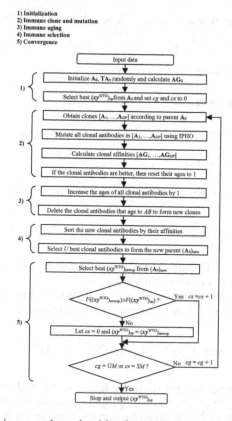

Figure 2. Flow chart of the immune clone algorithm for wind farm layout optimization.

cs denotes the counter that counts the number of generations without improvement and cg denotes the number of generations. The convergence condition for this optimization algorithm is described in the text that follows.

Convergence condition. The optimization algorithm stops when cs is greater than SM or cg is greater than GM, and outputs $(xy^{WTG})_{op}$ as the optimal solution. In other words, the optimization process converges if the maximum number of generations has been reached or the maximum number of generations without improvement is met.

As shown in Figure 2, the immune clone algorithm consists of five steps. Each step plays an important role in searching the optimal solution $(xy^{WTG})_{op}$.

(1) Initialization. An initial exploration of the solution space is conducted by randomly generating A_0 within xy^{WF}. AG_0 is evaluated and TA_0 is set accordingly.

(2) Immune clone and mutation. The exploration of the solution space is further enhanced by cloning and mutation. For example, A_0 is cloned into $[A_1, ..., A_{DP}]$, and brings more antibodies. Meanwhile, all antibodies in $[A_1, ..., A_{DP}]$ are mutated by IPH operator, which introduces variations for each candidate solution. The age of mutated antibodies that show improvement in affinity is reset.

(3) Immune aging. The diversity of the exploration is created by aging all antibodies. For example, antibodies that age to AB will be removed to avoid premature convergence.

(4) Immune selection. The selection of the elite antibodies directs the exploration toward the potential regions of solutions. For instance, U elite antibodies are selected from $[AG_1, ..., AG_{DP}]_{new}$ to form $(A_0)_{new}$ for further cloning and mutation.

(5) Convergence. This step checks the fulfillment of the convergence condition.

5 CASE STUDIES

5.1 *Wind profile data*

With the permission of the North Dakota Agricultural Weather Network, the hourly wind profile data collected between January 1, 2012 and December 31, 2014 at Baker were used for investigation.

5.2 *Baseline case*

A baseline case is established to facilitate the analysis.

Baseline case. The layouts of WTGs are planned on a 1.5 km × 1.5 km wind farm, which consists of nine private lands. Such a wind farm is subject to the Baker wind profile, where each land has a residence located at its center. The parameters of WTGs, house, land rent, and wind farm for this case are shown in Tables 3 to 6. The noise limits at all residences are set to 30 dB (New Zealand Standard 2010).

Table 3. WTG parameters.

v_{ci}	v_r	v_{co}	P_r	r	a	r_f
8.95 mph	24.61 mph	55.92 mph	1.5 MW	40 m	80 m	8 m

Table 4. House parameters.

b	r_h	$N_{k,\text{limit}}$ ($t = 1, ..., 9$)
6 m	10 m	30 dB

416

Table 5. Land rent at each private land in $/MW/yr.

ρ_1	ρ_2	ρ_3	ρ_4	ρ_5	ρ_6	ρ_7	ρ_8	ρ_9
7,000	6,000	6,500	7,500	7,250	5,500	6,750	6,250	6,300

Table 6. Parameters of wind farm.

M	K	R	T	Q	S	λ	τ	δ	β	ϕ	η
28	9	9	30 yr	39	36	38.9%	81.1%	5.7%	2,098,000 $/MW/yr	17,000 $/MW/yr	11,000 $/MW/yr

An optimal wind farm layout by the proposed WFLO model for the baseline case is shown in Figure 3. It can be seen that the WTGs are placed in a way such that WTG noise emissions on residence are minimized.

Table 7 shows the LCOE and other costs of the baseline case, and Table 8 presents the expected noise emission at each residence, respectively. The following observations can be made.

(1) The WTGs are placed in a way such that the land lease cost is reduced.

(2) The WTGs are placed in a way to alleviate wake loss.

The proposed WFLO model presents an appealing wind farm layout because it reduces wake loss, land lease cost, and noise emission.

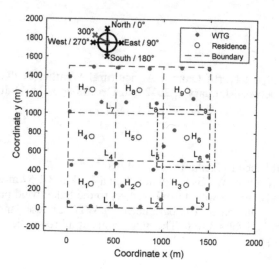

Figure 3. Optimal wind farm layout of the baseline case.

Table 7. *LCOE* and other costs of the baseline case.

LCOE ($/MW)	LLC ($/yr)	AEP (MWh/yr)	OM ($/yr)	LRC ($/yr)	ICC ($)	FCR (%)
60.6751	263,100	138,149	714,000	462,000	88,116,000	7.87953

Table 8. The expected noise emission at each residence.

H_1	H_2	H_3	H_4	H_5	H_6	H_7	H_8	H_9
28.365	27.933	26.863	25.261	26.089	27.915	27.258	27.238	28.473

6 CONCLUSION

In this article, a wind farm layout optimization (WFLO) model for the levelized cost of energy (LCOE) has been developed to deal with the challenges derived from land lease cost, noise emission, and wake loss. Derived look-up tables through geometrical analysis are used to determine the relative locations between WTG and private land. Then the immune clone algorithm is used to search the optimal WTG layout for minimum levelized cost of energy.

The proposed WFLO model has been evaluated considering the layout of 28 WTGs in a 1.5 km × 1.5 km wind farm of nine private lands and residences. Results have demonstrated that the proposed WTG layout plan reduces the impact of land lease cost, noise emission, and wake loss.

ACKNOWLEDGMENTS

This work was supported in part by the National Natural Science Foundation of China under Grant 51807109.

REFERENCES

Barthelmie, R.J., Hansen, K., Frandsen, S.T., et al. 2009. Modelling and measuring flow and wind turbine wakes in large wind farms offshore. *Wind Energy* 12(5): 431–444.

Billinton, R., & Chowdhury, A.A. 1992. Incorporation of wind energy conversion systems in conventional generating capacity adequacy assessment. Generation, transmission and distribution. *IEE Proceedings C. IET* 139(1): 47–56.

Chen, L., & MacDonald, E. 2012. Considering landowner participation in wind farm layout optimization. *Journal of Mechanical Design* 134(8): 084506.

Chen, Y., Li, H., Jin, K., et al. 2013. Wind farm layout optimization using genetic algorithm with different hub height wind turbines. *Energy Conversion and Management* 70: 56–65.

Eltham, D.C., Harrison, G.P., & Allen, S.J. 2008. Change in public attitudes towards a Cornish wind farm: Implications for planning. *Energy Policy* 36(1): 23–33.

Emami, A. & Noghreh, P. 2010. New approach on optimization in placement of wind turbines within wind farm by genetic algorithms. *Renewable Energy* 35(7): 1559–1564.

Global Wind Energy Council. Global Wind Report Annual Market Update 2014. Retrieved from: http://www.gwec.net/wp-content/uploads/2015/03/GWEC_Global_Wind_2014_Report_LR.pdf.

Global Wind Energy Council. Global Wind Energy Outlook 2014. Retrieved from: http://www.gwec.net/wp-content/uploads/2014/10/GWEO2014_WEB.pdf.

González, J.S., Rodriguez, A.G.G., Mora, J.C., et al. 2010. Optimization of wind farm turbines layout using an evolutive algorithm. *Renewable Energy* 35(8): 1671–1681.

Grady, S.A., Hussaini, M.Y., & Abdullah, M.M. 2005. Placement of wind turbines using genetic algorithms. *Renewable Energy* 30(2): 259–270.

International Energy Agency. Key world energy statistics 2014. Retrieved from: http://www.fossilfuelsreview.ed.ac.uk/resources/Evidence%20-%20Climate%20Science/IEA%20-%20Key%20World%20Energy%20Statistics.pdf.

Kim, H., Singh, C., & Sprintson, A. 2012. Simulation and estimation of reliability in a wind farm considering the wake effect. *IEEE Transactions on Sustainable Energy* 3(2): 274–282.

Marmidis, G., Lazarou, S., & Pyrgioti, E. 2008. Optimal placement of wind turbines in a wind park using Monte Carlo simulation. *Renewable Energy* 33(7): 1455–1460.

National Renewable Energy Laboratory. 2011 Cost of Wind Energy Review. Retrieved from: http://www.nrel.gov/docs/fy13osti/56266.pdf.

New Zealand Standard. NZS 6808:2010 Acoustics – Wind farm noise. Retrieved from: https://law.resource.org/pub/nz/ibr/nzs.6808.2010.pdf.

Nicosia, G., Cutello, V., Bentley, P.J., et al. 2004. Artificial immune systems. 10th International Conference on Artificial Immune Systems, 13–16.

Pookpunt, S. & Ongsakul, W. 2013. Optimal placement of wind turbines within wind farm using binary particle swarm optimization with time-varying acceleration coefficients. *Renewable Energy* 55: 266–276.

Standardization I O. 1996. Attenuation of sound during propagation outdoors–part 2: General method of calculation.

Wikipedia. A-weighting. Retrieved from: https://en.wikipedia.org/wiki/A-weighting.

Emerging Developments in the Power and Energy Industry – Dufo-López, Krzywanski & Singh (eds)
© 2020 Taylor & Francis Group, London, ISBN 978-0-367-27169-5

Investigation of the hydraulic forces on the radial vibration of hydro-turbine units

Yong Xu, Xiaoqiang Dang & Yunliang Chen
State Key Laboratory of Hydraulics and Mountain River Engineering, College of Water Resource & Hydropower, Sichuan University, Chengdu, China

Bin Wang
Sichuan Water Conservancy Vocational College, Chengdu, China

Yufang Huang
College of Mathematics, Southwest Jiaotong University, Chengdu, China

ABSTRACT: This paper presents a new analytical method to simplify the analysis of hydraulic forces on the radial vibration of hydro-turbine generator units. The torsional effects of hydraulic forces exerted on the runner were considered in the new analytical method. The hydraulic forces calculated by the new analytical method were exerted on the runner node of a 125 MW Kaplan turbine unit rotordynamic model to investigate the numerical response of the unit with uneven runner blade opening and uneven guide vane opening. Meanwhile, the field data of vibration for the unit with load are examined for comparing with the numerical response. The new analytical method on simplifying hydraulic forces might be suitable according to the comparison between the numerical response and the field data analysis.

1 INTRODUCTION

For hydro-turbine generator units, the water flow provides the original power by impinging on the turbine runner, and produces a complicated vibration problem induced by hydraulic forces. Excessive vibration of a unit in the hydropower plant represents that the unit is suffering from a drop in efficiency and always indicates that some malfunctions might exist in the unit. Investigation of the hydraulic forces on the radial vibration of hydro-turbine generator units may be helpful for the analysis of hydraulic-induced vibration problem. And this is useful for the diagnosis of hydraulic-induced malfunction and for the control of excessive vibration.

As the water flow in the runner is a complex spatial flow, calculation of the unbalanced hydraulic forces by an analytical method seems very difficult. In the past, estimation of hydraulic forces was generally experimental hypothesis-based approach. Schwirzer (T. Schwirzer 1977) suggested that hydraulic forces might be in a certain relationship with a power spectral density function, which could be used to analyze the random hydraulic forces exerted on the runner, according to the experimental measurement. Ma (Z. Ma & Y. Dong 1990) took advantages of the method to analyze the hydraulic forces of a hydro-turbine unit and discussed the dynamic response of the unit under the hydraulic forces. Kramer (E. Kramer 1981) suggested that the radial hydraulic forces could be considered as three components, a constant component, a component with the same operating frequency as the rotor and of constant magnitude and the remaining lateral force. Vibration measurements were used to estimate the remaining lateral force, taking advantages of the transfer function method. Chamieh (D. Chamieh, A. Acosta & C. Brennen 1985) designed a set of experimental device and measured the radial hydraulic forces exerted on a centrifugal pump. Actual measurements of the h ydraulic forces may be of high application value for dynamic analysis, but additional measuring equipment needs to be laid.

Finite element method (FEM) is an effective method for solving fluid-structure interaction problems. Taking the advantages of FEM, fluid-structure interaction problems in the runner chamber could be investigated in detail and more accurate results could be obtained than the estimation values. However, rarely researches have been reported for exploring the hydraulic forces on the hydraulic turbine, although the runner related fluid-structure interaction problems have been discussed by the FEM in many literatures (D. Liu, X. Liu & J. Li 2008) (C. Gu, X. Yao & Q. Chen 2001). This might be ascribed to a twofold reason: the tremendous calculation time for solving a fluid-structure interaction problem, and the requirement of frequently updating for the hydraulic forces in the dynamic analysis of a rotordynamic model for hydro-turbine units. Thus it might be impractical to update the hydraulic forces by a FEM method on the radial vibration analysis of hydro-turbine units. In contrast, an experimental hypothesis-based approach for the hydraulic forces might be more proper at present to meet the requirement of frequently updating.

Among the related literature for the dynamic analysis of hydro-turbine units, the unbalanced hydraulic forces are considered to be linear with the turbine blade tip clearance by Bettig and Han (B. Bettig & R. Han 1999), and computed using Alford's equation which is generally used for jet engine rotor modeling. The resulting force is a linear function of the turbine blade tip clearance and thus, the constant could be entered as a cross-coupling stiffness in the finite element equations of the rotor system. The power spectral density function method is used in references (R. Cardinali, 1993) (E. Kramer 1981) to determine the unbalanced hydraulic forces according to the experimental measurement. Feng (F. Feng & F. Chu 2001) takes the actual measured data of load and exerted on the turbine nodes of shaft system for dynamic response analysis. The analytical methods that already existed for the hydraulic forces on the radial vibration of hydro-turbine units might in general be qualitative or over-simplified, thus the actual characteristics of the hydraulic forces could not be accurately described.

In this paper, the authors suggest that the torsional effect of hydraulic forces exerted on the runner should not be ignored, when investigating the influence of hydraulic forces on the radial vibration of hydro-turbine units. a rotordynamic model for the shaft system of a Kaplan turbine unit is developed for the sake of investigating the hydraulic forces on the radial vibration. Simulations are proposed to check if the analytical method for hydraulic forces is suitable.

2 SIMPLIFYING THE HYDRAULIC FORCES

For convenience, we firstly suppose that there is axisymmetric water flow in the turbine runner. Consequently, the water flow in the turbine runner might be stable relative to the runner blades in a stable operating condition, i.e., the motion of flow would not change over time. In this situation, the movement of water flow in the runner could be expressed by a so called velocity triangle of a hydraulic turbine. Taking a point K in the runner blade for analysis, the velocity triangle of the hydraulic turbine could be drawn as shown in Figure 1. The runner is supposed to move with a uniform velocity u. The absolute velocity of water flow is v, and the water entering the runner with a relative velocity w. The absolute velocity v could be reduced into two mutually perpendicular components, the rotational v_u and the meridian velocity v_m components. The meridian velocity v_m could again be reduced into two mutually perpendicular components, the radial v_r and the axial velocity v_z components. For a Kaplan unit, the uniform velocity u at point K could be expressed as $u = R\Omega = 2\pi Rn/30$, where n represents the rotational speed of the unit in revolutions per minute, and R denotes the distance between the point K and the center of the shaft. And the meridian velocity v_m could be expressed as $v_m = Q/\pi(R^2 - r_h^2)$, where Q represents the discharge passing the turbine per second, ignoring the volumetric loss, and r_h denotes the radius of the turbine hub.

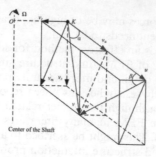

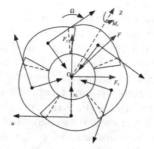

Figure 1. The velocity triangle of a hydraulic turbine runner.

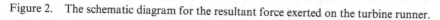

Figure 2. The schematic diagram for the resultant force exerted on the turbine runner.

A radial blade force acting upon the runner blade may arise from the radial velocity component v_r. In an ideal case, the resultant of all radial blade forces might be equal to or approximating zero, owing to the symmetry of the design. The resultant radial force F, however, shall not be equal to zero, as a consequence of the asymmetry of structure and the nonuniformity of flow in the practical operation of the unit. And it brings about the unbalanced hydraulic radial forces exerted on the runner. The schematic diagram for the resultant force exerted on the runner is shown in Figure 2. The resultant radial force F could be reduced into two mutually perpendicular components F_x and F_y, which will be exerted on the corresponding node of rotordynamic model respectively. Each component of the resultant will contain the three components of radial hydraulic force suggested by Kramer.

An axial blade force acting upon the runner blade may arise similarly from the axial velocity component v_z. The sum of all axial blade forces would constitute the axial thrust exerted on the runner. Each axial blade force on the runner blade may generate a torque to the center of the shaft in the axial plane. As it is similar with the unbalanced radial hydraulic force, it is not difficult to conclude that the resultant of all torques M_s is unbalanced due to the asymmetry. Since the unbalanced torques M_s, as is shown in Figure 2, exerted on the blades by the axial hydro-thrusts shall also generate a radial vibration, we suggest that the torsional effect of hydraulic forces exerted on the runner should not be ignored.

3   A ROTORDYNAMIC MODEL FOR THE SHAFT SYSTEM OF A TURBINE UNIT

A radial vibration model for a Kaplan type large hydro-turbine generator unit with vertical arrangement in a power plant is proposed for simulation. Figure 3 shows the

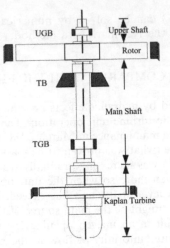

Figure 3.   The sketch of the shaft system of the hydraulic unit.

sketch of the shaft system of the hydraulic unit. The main shaft is hollow with inner diameter of 1290 mm and outer diameter of 1620 mm. The thrust bearing is arranged below the rotor with 18 pads and two tilting pad guide bearings are fixed along the shaft line above the rotor and the runner respectively, the distance of them is 12.9 m. The number of runner blades and guide vanes are 5 and 32 respectively. FE method is employed to obtain a satisfactory model of shaft system. The shaft system is divided into several elements, and each element equation is formed separately. Global equation of motion is then obtained by superposition of these equations.

The shaft is modeled as rotational Timoshenko beam elements. There are two nodes in one element. Each node has four degrees of freedoms (dof), two displacement coordinates, x_i, y_i, and two distortion angles, θ_{xi}, θ_{yi}. Generator rotor is modeled as a rigid element where the stiffness is neglected. Water turbine runner is modeled as rotational Timoshenko beam elements with different outer diameters. Dynamic characteristics of boundary conditions, such as force induced by guide bearings, are always of nonlinearity and depend on displacement, velocity, rotational speed etc. It is not easy to figure out a simple expression to describe the dynamic characteristics. Thus, the thrust bearing is modeled as a fictitious stiffness on the distortion angles. A database method is adopted for analyzing the guide bearings (Y. Xu, Z. Li & X. Lai 2013), where the damping and stiffness coefficients for bearings are considered to be changing with the rotor eccentricity and rotating speed based on a database of the damping and stiffness coefficients. Exciting forces could be of mechanical, hydraulic and electromagnetic origin. In our case, an unbalanced force of mechanical origin is exerted on the node of generator rotor. A periodic force with basic harmonic of hydraulic origin is exerted on the node of turbine runner. All the exciting forces mentioned above are exerted on the rotor system all over the simulation. Superposing all the element equations by nodes, the global equation of motion of the rotor system is constructed as follow:

$$M\ddot{q}(t) + C\dot{q}(t) + Kq(t) = F_{unbalance} + F_{hydraulic} + F_{boundary} \qquad (1)$$

where M is the mass matrix of system, C is the gyroscopic matrix added with damping matrix of system, K is the stiffness matrix of system, q is an arrangement of all generalized coordinates of the nodes. The force is assumed to be only exerted on the nodes. The equations of motion of the system construct a linear system with local nonlinearity in several nodes. Equation (1) can be solved by numerical integration methods such as the Wilson theta method and the Newmark beta method.

4 SIMULATION RESULTS COMPARED WITH THE FIELD DATA ANALYSIS

The vibration problems induced by hydraulic effects are one of the most common vibration phenomena during the hydraulic turbine unit operation. The unit investigated in this paper were implemented an exhaustive maintenance in March 2008. A runner blade was replaced by a new blade in the maintenance (what was proven to be inappropriate during the later operation). The vibration amplitude was increasing gradually when putting into operation after the exhaustive maintenance, even more so when the unit operating under a lower head in flood situations. The vibration level of the unit has exceeded the allowable value in October 2008. The new runner blade might be the main source of the failure, as the installation of inappropriate blade would result in an uneven opening for the runner blades. This would cause the asymmetry and the nonuniformity of flow in the runner chamber, leading to the hydraulic imbalance, and thus resulting in an excessive vibration for the unit gradually. Figure 4 shows the amplitude frequency spectrum of the field vibration data at the measuring point of turbine guide bearing in direction X (direction X and direction Y are two mutually perpendicular directions at the same measuring point) when the unit is undergoing excessive vibration with load. It can be seen in Figure 4 that the higher harmonic frequencies like 2 times, 3 times, and 4 times etc. are apparently involved in the frequency spectrum. It could be concluded that, under the failure of uneven opening for the runner blades, not only the harmonic frequency 5 times (the number of runner blade is 5) would be involved, but also for the other higher harmonic frequencies. In this section, the authors attempt to give an interpretation for this phenomenon based on the analytical method for the hydraulic forces and the rotordynamic model for the shaft system of the unit.

As is mentioned in section 2, the water flow passing the space between the blades will produce a blade force on the runner blade. The component of the blade force parallel to the direction of rotating movement will provide the power of the turbine, while the meridian component of the blade force acting upon each runner blade will bring about a radial force and a torque.

The force and torque on each runner blade should be different under a situation of uneven opening for the runner blades. As all the radial forces are passing through the axial center, what is clearly defined according to the design of the runner, the five radial forces on each blade are simplified as a resultant radial force F exerting on the displacement dofs at the node of runner of the rotordynamic model. The torques arising from axial blade force component are exerted on the distortion angle dofs at the node of runner of the rotordynamic model correspondingly with a phase interval of 72 degree. Since the runner blades are synchronously rotating with the unit, all the force and torque exerted will also synchronously rotate with the unit. Supposing that the torque in one blade become apparently increased than others due to the uneven opening for the runner blades, one torque with the value 120e4 Nm and others with 50e4 Nm is exerted on the rotordynamic model for simulation purpose. All the torques is

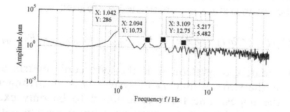

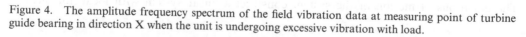

Figure 4. The amplitude frequency spectrum of the field vibration data at measuring point of turbine guide bearing in direction X when the unit is undergoing excessive vibration with load.

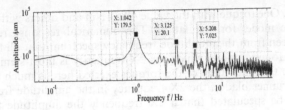

Figure 5. The amplitude frequency spectrum for the simulation results with a failure of uneven opening for runner blades.

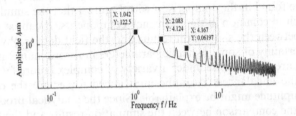

Figure 6. The amplitude frequency spectrum for the simulation results with a failure of uneven opening for runner blades.

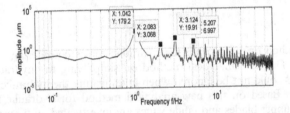

Figure 7. The amplitude frequency spectrum for the simulation results with a failure of uneven opening for both runner blades and guide vanes.

set to be rotating with unit synchronously. The numerical response results at the same measuring point and direction (the same with Figure 6 and Figure 7) are shown in Figure 5. It could be found in the amplitude frequency spectrum that only the odd multiple but even frequencies are appearance. This shall not appear in a simulation only exerted on the radial hydraulic force. It could be speculated that the rotating torques on the runner (to simulate an uneven opening for the runner blades) may contribute to the appearance of odd harmonic frequencies in the amplitude frequency spectrum.

Changes in the gate opening for guide vanes will change the discharge entering the runner chamber. When an uneven opening for guide vanes happened (the space between two guide vanes is jammed by a foreign object for example), the discharge entering the runner chamber is no longer uniform and symmetric. Supposing that the flow passing through each space of the guide vanes act upon the runner blade independently, to simulate a situation of uneven opening for guide vanes, 32 (the number of the guide vanes) radial forces and 32 torques should be set exerted on the runner blade, among which the radial forces could be simplified as a resultant radial force. The torques shall be exerted on the distortion angle dofs at the node of runner of the rotordynamic model correspondingly with a phase interval of 360/32 degree. The failure of uneven opening for guide vanes could be simulated as an increase or decrease in the radial force and torque compared with other ones. Noting that the guide vanes are fixed components and will not rotating, all the force and the torque exerted may neglect the effect of rotation. One torque with the value 4.5e4 Nm and others with 20e4 Nm is exerted on the rotordynamic model for simulation. The numerical response results are shown in Figure 6. It can be seen from the amplitude frequency spectrum that all the higher harmonic frequencies could be found, among which the 2X frequency is most significant in amplitude. Noting that although the magnitude of the torques exerted here is much smaller than that in uneven opening for runner blades, the 2X frequency in the amplitude frequency spectrum is significant. It could be speculated that a 2X frequency the amplitude frequency spectrum would most likely to be excited in a failure of uneven opening for guide vanes.

To simulate a situation with uneven opening for both runner blades and guide vanes, all the forces and torques in the above two cases are exerted on the distortion angle dofs at the node of runner of the rotordynamic model. The amplitude frequency spectrum is shown in Figure 7. It could be found in the comparison with Figure 4 that the simulated results and the field data have much in common in the frequency components. It can be seen that the difference in amplitude between the simulation results and the field data is obvious. This might be ascribed to the uncertainty of some dynamic parameters in rotordynamic model. And, what is more, the field data are the response under a variety of complex failure factors together, while in the simulation, the unbalanced hydraulic unbalanced forces are the only failure factors. The difference in amplitude might be explainable. Since the analytical process of the hydraulic force is logical and the comparison between the simulation results and the field data has much in common in the frequency components, the method for simulating the uneven opening of runner blades and guide vanes seems appropriate. It could also be concluded that the new analytical method on simplifying hydraulic forces might be suitable.

5 CONCLUSION

A new analytical method on simplifying the hydraulic forces considering torque effects is proposed in this paper. The new analytical method on simplifying the hydraulic forces suggested that the torque effects could not be ignored when investigating the hydraulic forces on the radial vibration of the unit. Based on this new analytical method for hydraulic force, simulation of uneven opening for runner blades and guide vanes are investigated, and many interesting results were obtained. The authors are aware of the fact that having much in common in the frequency components between the simulated results and the field data might be somewhat inadequate for supporting the conclusions, as the difference in amplitude is obvious and the flow in the runner chamber is not investigated in detail. But this does not mean the inference is baseless. The analytical process for this new analytical method is logically sound from a macroscopic point of view. And the difference in amplitude might be explainable as the field data are the response under a much more complexity failure factors than the simulation. The consistency in the frequency components between the simulated results and the field data indicates that the consideration of the torque effects would be necessary and the new analytical method is suitable.

ACKNOWLEDGMENTS

This work was supported by the National Natural Science Foundation of China, No. 51309172, was supported by the Open Research Fund of Key Laboratory of Fluid Machinery, Xihua University, No. szjj2013-008, and was supported by the Open Research Fund of Key Laboratory of Signal and Information Processing, Xihua University, No. szjj2014-017, which we would like to gratefully acknowledge.

REFERENCES

[1] T. Schwirzer, 1977, "Dynamic Stressing of Hydroelectric Units by Stochastic Hydraulic Forces on the Turbine Runner," Water Power & Dam Construction, pp. 39–44.

[2] Z. Ma, Y. Dong, 1990, "Dynamic Response of Hydroelectric Set by Hydraulic Lateral Force on Turbine Runner," Journal of Hydoelectric Engineering, No.2, pp. 31–40.

[3] E. Kramer, 1981, "Determining the Hydraulic Lateral Force of Pump-turbines," Water Power and Dam Construction, Vol. 33, pp. 50–54.

[4] D. Chamieh, A. Acosta, and C. Brennen, 1985, "Experimental Measurements of Hydrodynamic Radial Forces and Stiffness Matrices for a Centrifugal Pump-Impeller," Journal of Fluids Engineering, Vol. 107, No. 3, pp. 307–315.

[5] D. Liu, X. Liu, J. Li, 2008, "Vibration Analysis of Turbine Based on Fluid Structure Coupling," Fluid Power Transmission & Control, No. 1, pp. 21–25.

[6] C. Gu, X. Yao, Q. Chen, 2001, "Study on Fluid-Solid Coupling Dynamic Characteristics for the Component of Hydraulic Turbines," Large Electric Machine and Hydraulic Turbine, No. 6, pp. 47–52.

[7] B. Bettig, R. Han, 1999, "Modeling the Lateral Vibration of Hydraulic Turbine-generator Rotors," Journal of Vibration and Acoustics-Transactions of the ASME, Vol. 121, No. 3, pp. 322–327.

[8] R. Cardinali, 1993, "Dynamic Simulation of Non-linear Models of Hydroelectric Machinery," Mechanical Systems and Signal Processing, Vol. 7, No. 1, pp. 29–44.

[9] E. Kramer, 1981, "Determining the Hydraulic Lateral Force of Pump-turbines," Water Power and Dam Construction, No. 33, pp. 50–54.

[10] F. Fcng, F. Chu, 2001, "Dynamic Analysis of a Hydraulic Turbine Unit," Mechanics of Structures and Machines, Vol. 29, No. 4, pp. 505–531.

[11] Y.Xu, Z.Li, X.Lai, 2013, "Dynamic Model for Hydro-Turbine Generator Units Based on a Database Method for Guide Bearings," Shock and Vibration, Vol. 20, No. 3, pp. 411–421.

Emerging Developments in the Power and Energy Industry – Dufo-López, Krzywanski & Singh (eds)
© 2020 Taylor & Francis Group, London, ISBN 978-0-367-27169-5

A new protection principle of MMC-HVDC transimission lines with voltage polarity on both sides of the series inductor

Shuping Gao & Yue Gao
Xi'an University of Science and Technology, Xi'an, China

Zhenyu Hu
State Grid Tongchuan Electric Power Supply Company, Tongchuan, China

Guobing Song
Xi'an University of Science and Technology, Xi'an, China

ABSTRACT: Modular Multilevel Converters (MMC) has been applied in HVDC transmission system. The safe operation of the modular multilevel converter based HVDC system is inseparable from the reliable HVDC line protection. In this paper, through the series inductance on the HVDC line, the characteristics of the fault additional state of the different short-circuit faults in the DC line of the MMC-HVDC system are analyzed, and a protection strategy is obtained with voltage polarity on both sides of the series inductor. By analyzing the polarity of the voltage fault component of the positive and negative poles of the HVDC line, the fault pole can be judged. By analyzing the magnitude of the voltage fault component on both sides of the series inductance, internal or external fault can be distinguished. In the end, a simulation model of MMC-HVDC system is established in PSCAD. Based on the model, different faults type are simulated, and the protection algorithm is verified by Matlab. Simulation results showed that the proposed protection principle is correct and effective. This work is supported by Key support projects of the National Natural Science Foundation (U1766209) and National Natural Science Foundation of China (No. 51777166).

Keywords: Modular multilevel converters, fault analysis, relay protection, voltage fault component, HVDC transmission system

1 INTRODUCTION

Modular Multilevel Converters (MMC) is applied to voltage source High Voltage Direct Current (HVDC) transmission system as a new topological structure and adopts the sub-module (SM) series method. It has great market application value in grid interconnection and improvement of power quality, and has become an important development direction in the field of flexible HVDC transmission system (Zhang Xuena & Zhao Chengyong & Pang Hui & Lin Chang 2013).

The protection of the DC transmission line is crucial for the stable operation of the MMC-HVDC transmission system. The fast bypass switch is adopted to achieve the self-protection of the sub-module by throwing and cutting sub-module in (Guanjun Ding & Guangfu Tang & Zhiyuan He 2008 etc.). The fault characteristics of the bipolar short circuit are analyzed in detail in (Author, in press). The initial rise rate of the DC line current is estimated to quickly locate the fault, but only the bipolar short circuit fault is discussed in (Zhao Chengyong & Chen Xiaofang, & Cao Chungang 2011). The topological structure with severe DC fault ride-through capability is studied in (Ikhide M & Tennakoon S & Griffiths A 2015 etc.).

The possible short-circuit faults of the DC line of the double-ended MMC-HVDC system is analyzed in this article. According to results of fault analysis, a new protection principle of

MMC-HVDC transmission line is proposed, which can distinguish whether the fault occurs on the positive or negative side of the DC line and can identify that the fault is internal or external fault. Finally, a ±50 kV simulation model of the double-ended MMC-HVDC system is built in PSCAD, and protection algorithm is implemented in Matlab. A great deal of simulation results show that the proposed protection is correct and effective.

2 ANALYSIS OF FAULT CHARACTERISTICS OF MMC-HVDC TRANSMISSION SYSTEM

The schematic diagram of double-ended MMC-HVDC system is shown in Figure 1.

In Figure 1, MMC1 and MMC2 are two converter station, which adopt modular multilevel converter topology structure. MMC1 is on the rectifier side and MMC2 is on the inverter side. The transmission line uses cable line, which locates between MMC1 and MMC2. L is series small inductance in DC line. P_{1L} and P_{1R} are measuring devices, located on the two sides of series small inductance L on the positive line at MMC1 rectifier station; P_{2L} and P_{2R} are measuring devices, located on the two sides of series small inductance L on the positive line at MMC2 inverter station. N_{1L} and N_{1R} are measuring devices, located on the two sides of series small inductance L on the negative line at MMC1 rectifier station; N_{2L} and N_{2R} are measuring devices, located on the two sides of series small inductance L on the negative line at MMC2 inverter station. F1, F2, F3, and F4 are internal faults, and F5, F6, F7, F8, F9, and F10 are external faults.

At the moment when the fault occurs, according to the principle of superposition, the fault state can be equivalent to the superposition of the non-fault operation state and the fault additional state. The fault additional state will be analyzed as following.

2.1 *Polarity analysis of fault component of fault additional state network*

2.1.1 *Polarity analysis of fault component of monopole fault network*
An additional network diagram of the fault network of a single-pole fault is shown in Figure 2. In Figure 2, (a), (b), (c), (d), (e) and (f) are the equivalent circuit diagrams of fault states of F1,

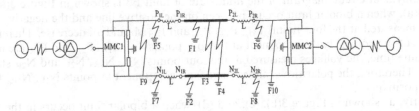

Figure 1.   Two terminal MMC-HVDC system.

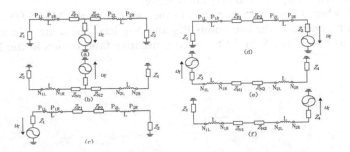

Figure 2.   Fault state network diagram of single electrode failure.

428

F2, F5, F6, F7 and F8, respectively, u_f indicates the voltage of the fault point to ground, Z_1, Z_2, Z_3 and Z_4 represent the equivalent impedance of the two sides of the AC side respectively, and Z_{P1}, Z_{P2}, Z_{N1} and Z_{N2} represent the equivalent impedance of the line to the fault point, respectively.

The equivalent circuit diagram of the fault state of fault F1 is shown in Figure 2(a). From Figure 2(a), when a fault occurs on the positive pole in the DC line of the MMC-HVDC system, the voltages measured at the four points P_{1L}, P_{1R}, P_{2L} and P_{2R} should all decrease, so the polarities of the voltage fault component at points P_{1L}, P_{1R} P_{2L}, and P_{2R} are all negative. In the same way, from Figure 2(c) and Figure 2(d), when a fault occurs beyond on the positive line of the MMC-HVDC system, the voltages measured at the four points P_{1L}, P_{1R}, P_{2L} and P_{2R} should all decrease, so the polarities of the voltage fault component at points P_{1L}, P_{1R}, P_{2L} and P_{2R} are also negative.

The equivalent circuit diagram of the fault state of fault F2 is shown in Figure 2(b). From Figure 2(b), when a fault occurs on the negative pole in the DC line of the MMC-HVDC system, the voltages measured at the four points N_{1L}, N_{1R}, N_{2L} and N_{2R} should all increase, so the polarities of the voltage fault component at points N_{1L}, N_{1R}, N_{2L} and N_{2R} are all positive. In the same way, from Figure 2(e) and Figure 2(f), when a fault occurs beyond on the negative line of the MMC-HVDC system, the voltages measured at the four points N_{1L}, N_{1R}, N_{2L} and N_{2R} should all increase, so the polarities of the voltage fault component at points N_{1L}, N_{1R}, N_{2L} and N_{2R} are also positive.

From the above analysis, when a fault occurs on the positive line or beyond the positive line of the MMC-HVDC system, the polarities of the voltage fault component at points P_{1L}, P_{1R}, P_{2L} and P_{2R} are negative. For the coupling between the positive line and the negative line, the polarities of the voltage fault component at points N_{1L}, N_{1R}, N_{2L}, and N_{2R} are also negative.

In the same way, when a fault occurs on the negative line or beyond the negative line of the MMC-HVDC system, the polarities of the voltage fault component at points N_{1L}, N_{1R}, N_{2L} and N_{2R} are positive. For the coupling between the positive line and the negative line, the polarities of the voltage fault component at points P_{1L}, P_{1R}, P_{2L} and P_{2R} are also positive.

2.1.2 *Polarity analysis of fault component of bipolar fault networktitle, author and affiliation frame*

An additional fault network diagram of a bipolar fault is shown in Figure 3 and Figure 3 (a), (b), (c) and (d) are equivalent circuit diagrams for faults F3, F 4, F9 and F10, respectively.

The equivalent circuit diagram of the fault state of fault F3 is shown in Figure 3(a). From Figure 3(a), when a bipolar fault occurs between on the positive line and the negative line, the voltages measured at the four points P_{1L}, P_{1R}, P_{2L} and P_{2R} should all decrease. Therefore, the polarities of the voltage fault component at the four points P_{1L}, P_{1R}, P_{2L} and P_{2R} are negative. At the same time, the voltages measured at the four points N_{1L}, N_{1R}, N_{2L} and N_{2R} should all increase. Therefore, the polarities of the voltage fault component at points N_{1L}, N_{1R}, N_{2L} and N_{2R} are positive.

Similarly, as shown in Figure 3(b), (c), and (d), when a bipolar fault occurs in the DC line or beyond the DC line of the MMC-HVDC system, the polarities of the voltage fault component at points P_{1L}, P_{1R}, P_{2L} and P_{2R} are negative, and the polarities of the voltage fault component at points N_{1L}, N_{1R}, N_{2L} and N_{2R} are positive.

In summary, when a positive fault occurs in the DC line of the MMC-HVDC system, the voltage fault components at the two points P_{1R} and N_{1R} have the same polarity and they are negative, such as fault F1, F5, and F6. When a negative fault occurs in the DC line of the

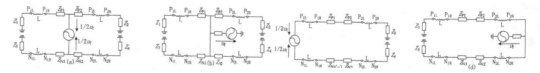

Figure 3.   Fault state network diagram of a bipolar fault.

MMC-HVDC system, the voltage fault components at the two points P_{1R} and N_{1R} have the same polarity and they are positive, such as fault F2, F7, and F8.

When a bipolar fault occurs on the DC line of the MMC-HVDC system, the voltage fault components at two points in P_{1R} and N_{1R} have different polarities.

Therefore, according to the voltage fault component polarities at the two points P_{1R} and N_{1R} are the same or not, fault pole can be found.

2.2 *Analysis on the variation of voltage fault component in network with fault state*

Take the positive fault of the DC line of the MMC-HVDC system as an example, as shown in Figure 2(a), (c) and (d), which are fault network of fault point F1, F5 and F6, respectively. From Figure 2(a), when a grounding fault occurs on the positive pole, the voltage at point P_{1R} rapidly drops, and the current flowing through the series of small inductors rapidly increases, so the inductor generates an induced voltage U_L.

$$U_L = L\frac{\mathrm{d}i_L}{\mathrm{d}t} \tag{1}$$

Therefore, the voltage at point P_{1L} is equal to the sum of the voltage at point P_{1R} and the inductor voltage U_L. Similarly, the voltage at point P_{2R} is equal to the sum of the voltage at point P_{2L} and the inductor voltage U_L.

When an external fault occurs on the fault point F5, the voltage at point P_{1R} is equal to the sum of the voltage at point P_{1L} and the inductor voltage U_L. And the currents at points P_{2R} and P_{2L} on the remote MMC2 side have not changed significantly, so the voltage fault components at the points P_{2R} and P_{2L} on the remote MMC2 side are almost equal. Similarly, when an external fault occurs on the fault point F6, the voltage fault components at the points P_{1R} and P_{1L} on the remote MMC1 side are almost equal.

Based on the above analysis, when an internal fault occurs in the positive line of the MMC-HVDC system, the voltage fault component at point P_{1L} is different from the voltage fault component at point P_{1R}. The voltage fault component at point P_{2L} and the voltage fault at point P_{2R} are also different. When an external fault occurs beyond the positive line of the MMC-HVDC system, on the remote side from the fault point F6 or F5 respectively, the voltage fault components at points P_{1L} and P_{1R} are the same or the voltage fault components at points P_{2L} and P_{2R} are the same. Due to the coupling between the positive and negative poles of the DC line of the MMC-HVDC system, when a positive fault occurs, the change of the negative pole is the same as that of the positive pole.

Similarly, when an internal fault occurs in the negative line of the MMC-HVDC system, the voltage fault component at point N_{1L} is different from the voltage fault component at N_{1R}, and the voltage fault component at point N_{2L} and the voltage fault component at point N_{2R} is also different. When an external fault occurs beyond the negative line of the MMC-HVDC system, on the remote side from the fault point F8 or F7 respectively, the voltage fault components at N_{1L} and N_{1R} are the same or the voltage fault components at N_{2L} and N_{2R} are the same. Due to the coupling between the positive and negative poles of the DC line of the MMC-HVDC system, when a negative fault occurs, the change of the positive pole is the same as that of the negative pole.

From Figure 3 (a) and (b), when short-circuit fault occurs on the positive and negative pole or the short-circuit ground fault of positive and negative poles occurs, the voltage at point P_{1R} rapidly drops, and the voltage at point P_{1L} is equal to the sum of the voltage at point P_{1R} and the inductor voltage U_L. Similarly, the voltage at point P_{2R} is equal to the sum of the voltage at point P_{2L} and the inductor voltage U_L. Similar to the situation of positive line, in the negative line, the voltage at N_{1R} is equal to the sum of the voltage at N_{1L} and the inductor voltage U_L, and the voltage at N_{2L} is equal to the sum of the voltage at N_{2R} and the inductor voltage U_L.

From Figure 3(c), when an external fault of point F9 occurs, the currents at points P_{2R} and P_{2L} on the remote MMC2 side have not changed significantly, so the voltage fault components at points P_{2R} and P_{2L} on the remote MMC2 side are almost equal. From Figure 3(d), the occurrence of F10 is similar to the situation of F9, the currents at points a P_{1L} nd P_{1R} on the remote MMC1 side have not changed significantly, so the voltage fault components at points P_{1L} and P_{1R} on the remote MMC1 side are almost equal.

To sum up the above analysis, when the bipolar fault occurs in the DC line of the MMC-HVDC system, the voltage fault component at point P_{1L} is different from the voltage fault component at point P_{1R}. The voltage fault component at point P_{2L} and the voltage at point P_{2R} are also different. When a bipolar fault occurs beyond the DC line of the MMC-HVDC system, on the remote side from the fault point F10 or F9 respectively, the voltage fault components at points P_{1L} and P_{1R} are almost equal or the voltage fault components at points P_{2L} and P_{2R} are almost equal.

Therefore, according to whether the voltage fault component of at the point P_{1L} and P_{1R} at the rectifier side is the same or not, and whether the voltage fault component at the point P_{2L} and P_{2R} at the inverter side is the same or not, an internal or external fault can be identified by using the end-to-end information.

3 IMPLEMENTION OF THE PROTECTION METHOD

At the initial stage when fault occurs on the MMC-HVDC system, the system can be approximately considered as linear. Therefore, according to the superposition principle, the fault state can be regarded as the superposition of the system's normal operation state and the additional fault state, there is

$$Ug = \Delta U + U \tag{2}$$

In (2), Ug is the voltage measured at the time of fault occurring, U is the voltage value of normal operation of system, ΔU is the voltage fault component.

The voltage fault component is equal to the measured voltage value after fault occurring minus the measured voltage value of N cycles before fault occurring. Due to the frequency deviation of the system, N is generally taken as 1 or 2.

From (2), the voltage fault components at the point of P_{1L}、 P_{1R}、 P_{2L}、 P_{2R} and N_{1R} can be obtained,which are UP_{1L}、 UP_{1R}、 UP_{2L}、 UP_{2R} and UN_{1R}.

From the above analysis, when an internal fault occurs on the DC line of MMC-HVDC transmission system, UP_{1L} is different from UP_{1R}, and UP_{2L} is different from UP_{2R}, and the criteria can be constructed as following. There is

$$\begin{cases} UP_{1L} - UP_{1R} > U_{set} \\ UP_{2L} - UP_{2R} > U_{set} \end{cases} \tag{3}$$

In (3), U_{set} is the threshold value. Its setting principle should meet the following two principles.

a) It can avoid the difference of the voltage fault component on both sides of series inductor when the most serious fault of external at the remote side occurs, that is ground fault at the remote side .

b) It is less than the difference of the voltage fault component on both sides of series inductor when the high impedance fault of internal at the far end of the DC line.

When the value of $UP_{1L} - UP_{1R}$ is larger than U_{set} and the value of $UP_{2L} - UP_{2R}$ is larger than U_{set}, the fault can be judged as internal fault. Otherwise, it can be judged as external fault.

From the analysis above, when a fault occurs on the positive DC line of the MMC-HVDC transmission system, the voltage fault components at two points P_{1R} and N_{1R} have the same polarity and the polarity of them are negative; when a fault occurs on the negative DC line of the MMC-HVDC transmission system, the voltage fault components at the two points P_{1R}

and N_{1R} have the same polarity and the polarity of them are positive; when a bipolar fault occurs on the DC line of the MMC-HVDC transmission system, the polarity of the voltage fault component at the two points P_{1R} an N_{1R} is different. Therefore, when a fault occurs on the DC line of the MMC-HVDC transmission system, whether the fault occurs on the positive pole, the negative pole, or the bipolar pole, the fault pole can be identified firstly according to the polarity of the voltage fault component at two points P_{1R} and N_{1R}.

Subsequently, according to (3), internal or external fault can be found. The internal or external fault discrimination method is as follows:

When a positive fault is determined, if the voltage fault component at point P_{1L} is not the same as the voltage fault component at point P_{1R}, and the voltage fault component at point P_{2L} and the voltage fault component at point P_{2R} are also different, then the fault is judged as an internal fault of the positive pole line, and the positive relay protection device operates; Otherwise, the fault is an external fault beyond the positive pole line, and the relay protection device can not operate.

When a negative fault is determined, if the voltage fault component at point P_{1L} is different from the voltage fault component at point P_{1R}, and the voltage fault component at point P_{2L} and the voltage fault component at point P_{2R} are also different, the fault is judged as an internal fault of the negative pole line, and the negative relay protection device operates; Otherwise, the fault is an external fault beyond the negative pole line, and the relay protection device can not operate .

When a bipolar fault is determined, if the voltage fault component at point P_{1L} is not the same as the voltage fault component at point P_{1R}, and the voltage fault component at point P_{2L} and the voltage fault component at point P_{2R} are also different, the fault is judged as an internal fault of a bipolar line, and the relay protection device in positive and negative line operates; Otherwise, it is an external fault and the relay protection device in positive and negative line can not operate.

4 SIMULATION VERIFICATION

A ±50 kV MMC-HVDC system is built in PSCAD, the number of model bridge arm submodules is 10, the AC side voltage is 23 kV, the DC side voltage is ±50 kV, the rated current of DC line is 0.3 kA, and the series inductance on both sides of the DC line is 0.1 H. The total line length is 20 kM.The time of fault occurring is set at 0.8s and the duration time of fault is 0.1 s. The sampling frequency of data is 100 kHz. And the protection algorithm is carried out in Matlab.

4.1 Unipolar fault simulation results

4.1.1 Simulation results of internal and external faults in the positive region
1) Simulation results when an internal fault occurs in the positive line
The simulation results are shown in Figure 4 when fault occurs at the fault point F1.

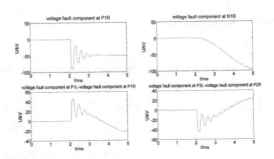

Figure 4. Simulation results of fault point F1.

From Figure 4, the voltage fault component at P_{1R} and the voltage fault component at N_{1R} have the same polarity and the polarity of them is negative, then the fault is determined to be a positive fault. In addition, from Figure 5, the difference between the voltage fault component at point P_{1L} and the voltage fault component at point P_{1R} is not zero, and the difference between the voltage fault component at point P_{2L} and the voltage fault component at point P_{2R} is not zero. Therefore, the fault is judged as an internal fault in the positive DC line.

2) Simulation results when an external fault occurs beyond the positive line at MMC1 rectifier station

The simulation results are shown in Figure 5 when fault occurs at the fault point F5 at MMC1 rectifier station.

From Figure 5, the voltage fault component at P_{1R} and the voltage fault component at N_{1R} have the same polarity and the polarity of them is negative, then the fault is determined as a positive fault. At the same time, the difference between the voltage fault component at point P_{1L} and the voltage fault component at point P_{1R} is not zero, and the difference between the voltage fault component at point P_{2L} and the voltage fault component at point P_{2R} is close to zero. Therefore, the fault is judged as an external fault at MMC1 rectifier station in the positive region.

3) Simulation results when an external fault occurs beyond the positive line at MMC2 inverter station

The simulation results are shown in Figure 6 when fault occurs at the fault point F6 at MMC2 inverter station.

From Figure 6, the voltage fault component at P_{1R} and the voltage fault component at N_{1R} have the same polarity and the polarity of them is negative, then the fault is determined as a positive fault. At the same time, the difference between the voltage fault component at point P_{1L} and the voltage fault component at point P_{1R} is close to zero and the difference between the voltage fault component at point P_{2L} and the voltage fault component at point P_{2R} is not zero. Therefore, the fault is judged as an external fault at MMC2 inverter station in the positive region.

A lot of simulation research have been carried out to verify the correctness of the proposed protection principle. Due to space limitations, the simulation results of internal and external faults in the negative pole region and bipolar fault are all omitted.

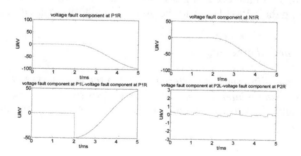

Figure 5.   Simulation results of fault point F5.

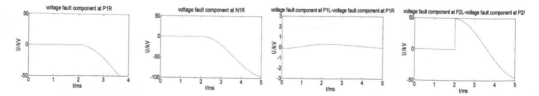

Figure 6.   Simulation results of fault point F6.

5 CONCLUSIONS

The fault network of MMC-HVDC system with DC series inductors is analyzed in this paper. The analysis results show that the fault pole can be distinguished by the polarity of the voltage fault components on both sides of the series inductance of the DC line. The polarities of voltage fault component of the positive region are the same and the polarities of them are negative, and the polarities of the voltage fault component of the negative region are the same and the polarities of them are positive. The polarities of the bipolar fault voltage component are different. In addition, according to the difference of the magnitude of the voltage fault component on both sides of the series inductor, internal or external fault can be distinguished. The magnitude of the voltage fault component on both sides of the inductor is different when internal fault occurs, and when external fault occurs, the magnitude of the voltage fault component on both sides of the inductor on one side is the same and the magnitude of the voltage fault component on both sides of the inductor on the other side is different.

According to the above characteristics, a new protection principle of MMC-HVDC transmission line is proposed. Based on the model, the simulation verification of protection algorithm is carried out in Matlab. A large number of simulation results show that the proposed method is correct and effective.

This work is supported by Key support projects of the National Natural Science Foundation (U1766209) and National Natural Science Foundation of China (No. 51777166).

REFERENCES

Zhang Xuena, Zhao Chengyong, Pang Hui, Lin Chang. A control and protection scheme of multi-terminal DC transmission system based on MMC for DC line fault[J]. Automation of Electric Power Systems, 2013, 37(15): 140–145.

LIU Siyuan, XU Dongxu, MEI Nian, FU Ying, LIU Chongru. Research on the effect of DC side fault on modular multilevel converter[J]. Power System Protection and Control, 2017, V45(10): 48–54.

Flourentzou N, Agelidis VG, Demetriades GD. VSC-based HVDC power transmission systems: an overview[J]. IEEE Transactions on Power Electronics, 2009, 24(3): 594–599.

Guanjun Ding, Guangfu Tang, Zhiyuan He. New technologies of voltage source converter (VSC) for HVDC transmission system based on VSC[C], Power and Energy Society General Meeting, 2008: 1-8.

Zhao Chengyong, Chen Xiaofang, Cao Chungang, et, al. Control and protection strategies for MMC-HVDC under DC faults[J]. Automation of Electric Power Systems, 2011, 35(23): 82–87.

YAO Zhiqing, ZHANG Qun, CHEN Peng, et al. Research on fault diagnosis for MMC-HVDC systems[J]. Protection and Control of Modern Power Systems, 2016, 1(1): 1–7.

Ikhide M, Tennakoon S, Griffiths A, et al. Fault detection in multi-terminal modular multilevel converter (MMC) based high voltage DC (HVDC) transmission system[C]//Proceedings of the 2015 50th International Universities Power Engineering Conference (UPEC). Stoke on Trent: IEEE, 2015: 1–6.

Tang Lianxiang, Boon-Teck Ooi. Locating and isolating DC faults in multi-terminal DC systems[J]. IEEE Transactions on Power Delivery, 2007, 22(3): 1877–1884.

BO Zhiqian, LIN Xiangning, WANG Qingping, etal. Development of power system protection and control[J]. Protection and Control of Modern Power Systems, 2016, 1(1): 1–8.

Zhao Chengyong, Xu Jianzhong, Li Tan. DC fault ride-through capability analysis of full-bridge MMC-MTDC systems[J]. SCIENCE Technological Sciences, 2013, 43(1): 106–114.

Jin Yang, Fletcher JE, O'Reilly J. Multiterminal DC wind farm collection grid internal fault analysis and protection design[J]. IEEE Transactions on Power Delivery, 2010, 25(4): 2308–2318.

Jin Yang, Fletcher JE, O'Reilly J. Short-circuit and ground fault analyses and location in vsc-based DC network cables[J]. IEEE Transactions on Industrial Electronics, 2012, 59(10): 3827–3837.

Xue Yinglin, Xu Zheng. DC fault ride-through mechanism and improved topology scheme of C-MMC[J]. Proceedings of the CSEE, 2013, 33(22): 63–70.

Antonopoulos A, Angquist L, Harnefors L, et al. Global asymptotic stability of modular multilevel converters[J]. IEEE Transactions on Industrial Electronics, 2014, 61(2): 603–612.

Emerging Developments in the Power and Energy Industry – Dufo-López, Krzywanski & Singh (eds)
© 2020 Taylor & Francis Group, London, ISBN 978-0-367-27169-5

Investigations of potential difference characteristics between two points on a grounding grid of a power substation

Y. Yu & Y.J. Li
*State Key Laboratory of Power Grid Safety and Energy Conservation, Relay Protection Research Department,
China Electric Power Research Institute, Beijing, China*

D.H. Wang
State Grid Zhejiang Electric Power Co., Ltd. Research Institute, China

A.Q. Guo & C.Q. Jiao
*State Key Laboratory of Alternate Electric Power System with Renewable Energy Sources, North China
Electric Power University, Beijing, China*

J.W. Sun
*State Key Laboratory of Power Grid Safety and Energy Conservation, Relay Protection Research Department,
China Electric Power Research Institute, Beijing, China*

ABSTRACT: With the intelligentization upgrade of power substations, a large amount of secondary equipment has been deployed dispersedly near the high-voltage primary equipment. The enclosures of these secondary devices are electrically connected to the grounding grid underground. When a fault current injection causes a potential rise in the grounding grid, there is a potential difference between any two points on the grounding grid conductors. For two pieces of secondary equipment located at different positions and connected by a secondary cable, the potential difference can induce common mode and differential mode disturbance voltages at the ports of the secondary cable, which may interfere with the normal and stable operation of the secondary system. In this article, the grounding grid of a 220-kV substation is modeled and the corresponding potential difference is calculated by means of CDEGS software. Also, the effects of the grid density and size, soil resistivity, positions of current injection, and potential observation on the potential difference are investigated. The results are helpful for guiding grounding grid design and evaluating the influence of ground potential difference on secondary equipment.

Keywords: CDEGS, substation, grounding grid, ground potential difference

1 INTRODUCTION

A grounding grid in a power substation is a grid grounding system consisting of a group of horizontal and vertical conductors buried in the soil to provide electrical safety and protection of both personnel and equipment. In recent years, with the intellectualization of substations, a large amount of secondary equipment has been deployed dispersedly near primary high-voltage equipment such as gas-insulated switchgear (GIS), transformers, and buses (Dong et al. 2016). Usually, the shell of each secondary device will be grounded to the grid beneath directly. In the meantime, any two pieces of secondary equipment located at different positions will be electrically connected to each other by a secondary cable. As a result, when there is a potential rise caused by the injection of a fault current in the grounding grid, a potential difference between two different points in the grounding grid will occur (Zhang et al. 2014). The potential difference can be applied to both ends of the secondary cable, and then common mode and differential mode disturbance voltage can be formed at the ports of the secondary cable. The common mode disturbance voltage may threaten the insulation of secondary cables and equipment, while the differential mode disturbance voltage can act as a noise superimposed into the normal signals, which may lead to signal distortion and even secondary equipment malfunction in some severe cases (Mao et al. 2008). Traditionally, grounding resistance, step voltage, and contact voltage are the main indicators for the evaluation of grounding grid performance (Reid 1988). Therefore, there are many studies on the characteristics of grounding resistance, step voltage, and contact voltage of grounding grids in the existing literature (Takahashi & Kawase 1991; Chang & Lee 2006; He et al. 2015; Li et al. 2017; Puttarach et al. 2017), but relatively few studies on the difference of ground potential between two points on grounding grids. In this article, the grounding grid of a 220-kV substation is modeled and the corresponding potential difference is calculated by using CDEGS software (Puttarach et al. 2007; Duanmu et al. 2016). Also, the effects of the grid density and size, soil resistivity, the positions of current injection, and potential observation on the potential difference are investigated. It is shown that the effect of the soil resistivity on potential difference is different from that on the step voltage and grounding resistance, in a case in which the fault current injected into the grid is fixed.

2 GROUNDING GRID MODEL

Figure 1 is a schematic diagram of the grounding grid structure of a 220-kV substation. The size of the grounding grid is 100.5 m × 84 m, and the depth is 0.6 m underground. The soil is assumed to be homogeneous and its resistivity is $100\Omega \cdot$ m. The grounding grid conductors are made of 30CCS185 copper-plated steel strands, the radius is 7.68 mm, the relative resistivity is 3.33, the relative permeability is 636, and the thickness of the copper layer is 0.254 mm.

The basic grounding grid model shown in Figure 1 is established in CDEGS. The upper-left coordinates of the grounding grid are (1, 1, 0.6). The x- and y-axis point to the length and width of the grounding grid respectively, and the z-axis points to the center of the earth vertically.

According to the layout of substation, each protection installation site is selected as the observation point, and the coordinates of the observation point are shown in Table 1. Short-circuit fault is equivalent to adding short-circuit current into the ground as the excitation source of the grounding grid. The short-circuit point is the current injection point, and the coordinates are shown in Table 2 (the selection of short-circuit points varies with different factors).

3 CDEGS SOFTWARE VERIFICATION

3.1 *Experimental verification*

The size and arrangement of the grounding grid are shown in Figure 2. The test grounding grid consists of three parts: a 30 × 30 mm^2 main grounding grid, a 4 × 4 mm^2 auxiliary grounding

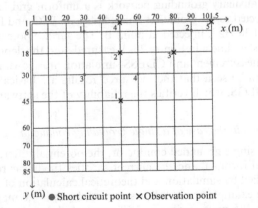

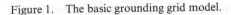

Figure 1. The basic grounding grid model.

Table 1. Coordinates of observation points (m).

Observation point number	x	y	z
1	50	45	0.6
2	50	18	0.6
3	80	18	0.6
4	50	1	0.6
5	101.5	1	0.6

Table 2. Coordinates of short-circuit points (m).

Short circuit point number	x	y	z
1	30	6	0.6
2	90	6	0.6
3	30	30	0.6
4	60	30	0.6

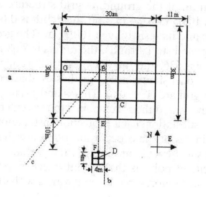

Figure 2. Schematic diagram of model validation.

grid, and a 30 m long horizontal suspended grounding electrode. The parameters are as follows: each grounding body is 55×5 mm^2 galvanized flat steel, its resistivity is 1.7e-7 Ω·m, relative permeability is 636; the buried depth of grounding body is 0.6 m; the grounding body of the main grounding network and auxiliary grounding network is a uniform grid layout. In Figure 2, flat steel 55×5 mm^2 is connected to the ground at four points—A, B, C, and D—as ground lead.

A current of 30 A and 30 Hz is injected at point B. The surface potential distribution at the position indicated by dashed line a in Figure 2 is measured, and the depth of polar penetration is 0.3 m. The results of measurement and CDESS simulation are shown in Figure 3.

From the figure, it can be seen that the measured results are basically consistent with the simulation results of CDEGS, which verifies the reliability of the software.

3.2 *Analytical verification in the case of a single grounding conductor*

For a simple model of a single grounded conductor, the potential difference can be calculated by an accurate analytical formula, and the current satisfies the linear reduction relationship. Therefore, it can be verified by simulation and theoretical calculation of a single conductor.

The parameters of the example are as follows: buried depth is 0.6 m; conductor length L is 50 m, cross section S is 185 mm^2, radius is 7.68 mm, relative resistivity is 3.33 (relative to copper, resistivity is 1.72e-8 Ω·m), relative permeability is 636; the coating is copper, with

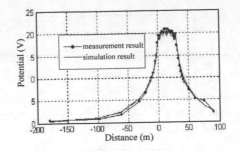

Figure 3.   Distribution of surface potential along line a.

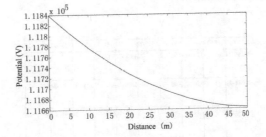

Figure 4.   Potential distribution of a single conductor.

a thickness of 0.254 mm and resistivity of 1.72e-8 Ω·m; relative permeability is 1; soil resistivity is 130 Ω·m, calculated by single-layer geodetic model. A direct current I of 23 kA is injected into the left side of the conductor.

The simulation results show that the potential at the left end of the conductor is the highest, reaching 11,839.20 V, and the potential at the right end is the lowest, 11,662.33 V. The maximum potential difference was 176.87 V. The potential distribution is shown in Figure 4. The abscissa represents the distance from the calculated point to the left end of the conductor.

In the case of DC, it is assumed that all 23-kA short-circuit currents flow in one direction. Considering the linear decrease of the current, the maximum potential difference on this conductor is

$$U = \frac{IL}{2S\sigma} = \frac{2.3 \times 10^4 \times 50}{2 \times 185 \times 10^{-6} \times 0.3 \times 5.8 \times 10^7} = 178.63\,\text{V} \qquad (1)$$

where σ is the conductivity of conductors.

The theoretical result of 178.63 V are basically consistent with the calculated result of 176.87 V, which is in line with expectations.

The current distribution of the conductor is shown in Figure 5. It can be seen from the figure that the current decreases linearly to 0, which is in line with the expectation.

4   INFLUENCING FACTORS OF GROUNDING GRID POTENTIAL DIFFERENCE

In order to analyze the influence of current injection point location, observation point location, grounding grid size, grid density, buried depth, soil resistivity, and vertical grounding electrode on the ground potential difference, the grounding grid model based on Figure 1 is simulated and calculated.

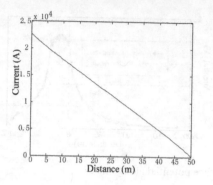

Figure 5. Conductor current distribution.

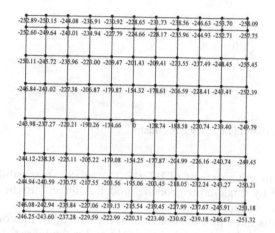

Figure 6. Distribution of ground potential difference.

4.1 Position of the observation point

The coordinates of the short-circuit point are (50, 45, 0.6), and the injection current of the short-circuit point is 23 kA. The ground potential of the injection point rises to 11,843 V. The difference between the grounding grid node and current injection point is calculated and marked in Figure 6.

As can be seen from the figure, the maximum difference of the ground potential in the grounding network is about 258 V, less than 10% of the increase of the ground potential. The ground potential at the corner of the grounding grid is the lowest (the gradient of the ground potential is the largest).

4.2 Short-circuit point position

The short-circuit current is 23 kA. The potential distribution and grounding resistance of the conductor at four short-circuit points in Table 2 are calculated as shown in Table 3.

It can be seen from the table that the ground potential at the short-circuit point is the highest, and the lower the ground potential near the center of the grounding network, the smaller the grounding resistance.

By calculating the potential difference between the points, it can be concluded that the closer the short-circuit point is to the center of the ground network, the smaller the maximum ground potential difference of each observation point is. The maximum potential difference occurs between the short circuit point and the corner of the ground network.

Table 3.   Potential distribution at different short-circuit points.

	Short-circuit point 1	Short-circuit point 2	Short-circuit point 3	Short-circuit point 4
1	11,609.40	11,588.87	11,637.94	11,661.86
2	11,678.23	11,631.87	11,655.07	11,681.71
3	11,593.09	11,769.02	11,577.89	11,665.39
4	11,714.17	11,651.57	11,641.01	11,653.10
5	11,579.43	11,912.05	11,560.68	11,638.03
circuit point	11,906.41	11,989.81	11,862.59	11,847.52
Grounding resistance (Ω)	0.5178	0.5214	0.5159	0.5152

4.3 Soil resistivity

When the short circuit point 1(30, 6, 0.6) fails, the injection current is 23 kA. Potential distributions of soil resistivity at 10 Ω·m, 100 Ω·m, and 1000 Ω·m are calculated respectively. The results are shown in Table 4.

Comparing the data in the table, it can be found that the greater the soil resistivity, the greater the grounding resistance, and the relationship between them is proportional. However, the potential difference between the calculated points is little affected by soil resistivity. The greater the soil resistivity, the smaller is the difference of potential under different soil resistivity.

4.4 Depth of the grounding grid

The coordinates of the short-circuit point are (30, 6, 0.6), and the injection current is 23 kA. The conductor potentials for burial depths of the local network of 0.2 m, 0.6 m, 1 m, and 2 m are shown in Table 5.

Table 4.   Potential distributions in different soil resistivities.

	10 Ω·m	100 Ω·m	1000 Ω·m
1	1155.41	11,609.40	116,149.19
2	1227.19	11,678.23	116,217.52
3	1138.99	11,593.09	116,132.94
4	1266.10	11,714.17	116,252.99
5	1125.46	11,579.43	116,119.27
Short-circuit point	1488.10	11,906.41	116,440.52
Grounding resistance (Ω)	0.0648	0.5178	5.0627

Table 5.   Potential distribution under different burial depths of the ground network.

	0.2 m	0.6 m	1 m	2 m
1	11,886.34	11,609.40	11,455.82	11,191.40
2	11,955.26	11,678.23	11,524.58	11,260.06
3	11,870.35	11,593.09	11,439.33	11,174.63
4	11,991.38	11,714.17	11,560.42	11,295.73
5	11,857.15	11,579.43	11,425.42	11,160.37
Short circuit point	12,173.17	11,906.41	11,752.72	11,488.12
Grounding resistance (Ω)	0.5295	0.5178	0.5111	0.4996

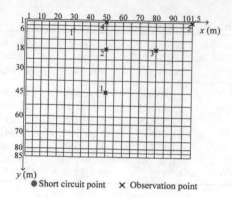

Figure 7. Model of encrypted grid ground network.

Comparing the data in the table, it can be found that the deeper the ground network is buried, the smaller the grounding resistance is, but the potential difference between the observation points is basically unchanged.

4.5 Grid density

Without changing the size of the grounding grid, the grid is refined and the structure of the grounding grid is shown in Figure 7.

When the short-circuit point 1 (30, 6, 0.6) fails, the injection current is 23 kA. The potential of each point is calculated as shown in Table 6.

The ground potential decreases and the grounding resistance decreases after the grid is densified. After sorting out the potential difference between the two cases, it is found that the potential difference decreases to about half of the original grid potential difference after the grid is densified.

4.6 The area of the grounding grid

The structure of the grounding grid after it is expanded and 20 m added around it is shown in Figure 8.

The structure of the grounding network after the grounding network is reduced is shown in Figure 9.

The potential of each point is calculated as shown in Table 7.

It can be seen from the table that the ground potential of each point decreases and the grounding resistance decreases after the grounding grid is enlarged. After the grounding grid is reduced, the ground potential and grounding resistance of each point increase. The short-circuit point potential and grounding resistance are inversely proportional to the arithmetic square root of the grid area.

Table 6. Potential distribution after finite grid (V).

	Before encryption	After encryption
1	11,609.40	11,082.25
2	11,678.23	11,118.15
3	11,593.09	11,071.89
4	11,714.17	11,135.01
5	11,579.43	11,063.38
Short circuit point	11,906.41	11,259.85
Grounding resistance (Ω)	0.5178	0.4897

441

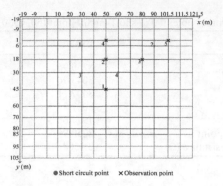

Figure 8.   Expanded grounding network model.

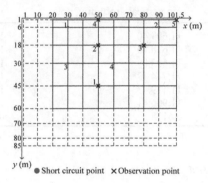

Figure 9.   Reduced grounding network model.

Table 7.   Potential at different sizes of the grounding grid (V).

	Unchanged	Expand	Reduce
1	11,609.40	8005.53	15,688.68
2	11,678.23	8061.64	15,756.25
3	11,593.09	7995.19	15,644.66
4	11,714.17	8085.45	15,792.86
5	11,579.43	7978.04	15,622.78
Short circuit point	11,906.41	8255.78	16,017.18
Grounding resistance (Ω)	0.5178	0.3591	0.6965

When the coordinates of the short-circuit point are (50, 45, 0.6), conductor potentials of different grounding grid sizes are calculated respectively. The calculation results are shown in Table 8.

From the distribution of potential difference under different short-circuit points, it can be seen that when the short-circuit point is close to the center of the grounding grid, the larger the area of the grounding grid is, the larger the potential difference is. When the short-circuit point is near the edge of the grounding grid, the larger the area of the grounding grid, the smaller the potential difference. The distribution of conductor potential differences under different grounding grid size and different short-circuit points is complex.

Table 8. Ground potential at short-circuit point (50, 45, 0.6).

	Expand	Reduce
1	8250.96	11,842.57
2	8043.96	11,641.14
3	8007.11	11,605.08
4	8007.34	11,613.92
5	7977.66	11,584.72
Short circuit point	8250.96	11,842.57
Grounding resistance (Ω)	0.3588	0.5150

Table 9. Potential distribution with vertical grounding pole.

	No vertical grounding pole	With vertical grounding pole
1	11,609.40	11,405.81
2	11,678.23	11,475.08
3	11,593.09	11,389.00
4	11,714.17	11,508.56
5	11,579.43	11,373.92
Short circuit point	11,906.41	11,701.40
Grounding resistance (Ω)	0.5178	0.5089

4.7 *Vertical grounding electrode*

The vertical grounding pole with a length of 1.5 m is added at each node of the grounding grid. The material of the vertical grounding pole is the same as that of the horizontal grounding grid. The calculated potential distribution is shown in Table 9.

From the data in the table, it can be seen that with the increase of vertical grounding poles, the conductor potential decreases and the grounding resistance decreases. The potential difference is basically unchanged.

5 CONCLUSIONS

(1) The maximum potential rise occurs at the current injection point and the minimum potential rise occurs at the corner of the grounding network. The maximum potential difference between two points in the grounding network generally does not exceed 10% of the potential rise at the current injection point. When the size of the grounding grid is constant, the closer the short-circuit point is to the center of grounding grid, the smaller the maximum potential difference is.

(2) On the premise of a constant injection current, the potential difference between two points in the grounding network is independent of soil resistivity. Relatively speaking, the grounding impedance varies linearly with soil resistivity.

(3) On the premise of a constant injection current, the potential difference between two points in the grounding network is independent of the depth of the grounding network.

(4) The ground potential difference changes inversely with grid density. The ground potential difference can be reduced by increasing the grid density.

(5) When the short-circuit point is close to the center of the grounding grid, the larger the area of the grounding grid, the greater the difference of the ground potential. When the short-circuit point is near the edge of the grounding grid, the larger the area of the grounding grid, the smaller the difference of the ground potential.

(6) On the premise of a constant injection current, the potential difference between two points in the grounding network is independent of the vertical grounding electrode.

ACKNOWLEDGMENTS

This project is supported by the Science and Technology Project of SGCC "Research on Outdoor Installation Protection Scheme and Key Technologies."

REFERENCES

Chang, C.N. & Lee, C.H. 2006. Computation of ground resistances and assessment of ground grid safety at 161/23.9-kV indoor-type substation. *IEEE Transactions on Power Delivery* 21(3): 1250–1260.

Dong, X., Wang, D., Zhao, M., et al. 2016. Smart power substation development in China. *CSEE Journal of Power and Energy Systems* 2(4): 1–5.

Duanmu, L., Zhao, X., Zhou, X., et al. 2016. Analysis of the experiment and simulation based on CDEGS of the shunt current coefficient of the grounding grid. *High Voltage Apparatus* 52(1): 154–161.

He, J., Zhang, B., & Zeng, R. 2015. Maximum limit of allowable ground potential rise of substation grounding system. *IEEE Transactions on Industry Applications* 51(6): 5010–5016.

Li, Z., Wang, S., Guo, F., et al. 2017. Study on characteristics of grounding impedance of large grounding grid. In IEEE International Conference on Environment & Electrical Engineering & IEEE Industrial & Commercial Power Systems Europe. IEEE.

Mao, B., He, J., Chen, S., et al. 2008. Electromagnetic interference in low voltage secondary system of substation caused by short circuit fault. In International Symposium on Electromagnetic Compatibility. IEEE.

Puttarach, A., Chakpitak, N., Kasirawat, T., et al. 2007. Substation grounding grid analysis with the variation of soil layer depth method. Power Tech, 2007 IEEE Lausanne. IEEE.

Puttarach, A., Chakpitak, N., Kasirawat, T., et al. 2017. Substation grounding grid analysis with the variation of soil layer depth method. Power Tech, 2007 IEEE Lausanne. IEEE.

Reid, I.A. 1988. Book review: IEEE Guide for Safety in AC Substation Grounding—ANSI/IEEE Std. 80-1986. *Power Engineer* 2(3): 126–128.

Takahashi, T., & Kawase, T. 1991. Calculation of earth resistance for a deep-driven rod in a multi-layer earth structure. *IEEE Power Engineering Review* 11(4): 63.

Zhang, B., Jiang, Y., Wu, J., et al. 2014. Influence of potential difference within large grounding grid on fault current division factor. *IEEE Transactions on Power Delivery* 29(4): 1752–1759.

Emerging Developments in the Power and Energy Industry – Dufo-López, Krzywanski & Singh (eds)
© 2020 Taylor & Francis Group, London, ISBN 978-0-367-27169-5

FEM analysis of iron core temperature distribution in a magnetically controlled shunt reactor with distributed magnetic valves

Li Tong
State Grid Zhejiang Electric Power Research Institute, Hangzhou, Zhejiang, China

Liu Huang & Yibo Tang
State Grid Zhejiang Electric Power Corporation, Lishui Power Supply Company, Lishui, Zhejiang, China

Caifei Hu, Xiaohui Li, Xueliang Fan & Liqun He
School of Rail Transportation, Soochow University, Suzhou, Jiangsu, China

ABSTRACT: In this article, the magnetic and thermal characteristics of a magnetically controlled shunt reactor (MCR) with distributed magnetic valves are studied via multiphysics coupling analysis. A full-size two-dimensional (2-D) simulation model of a 35-kV/10-MVA MCR with distributed magnetic valve is established in the finite element software ANSYS and COMSOL. The iron loss of MCR is obtained through the field-circuit coupling method. Then the steady-state thermal model of MCR is established in COMSOL and the 2-D temperature distribution of the iron core is simulated. Simulation results indicate that the temperatures of the iron core where magnetic valves are located are higher and more fluctuating than at other parts, to which more attention should be paid during the future design of MCRs.

Keywords: MCR, Distributed magnetic valves, Finite element, Iron loss, Temperature distribution

1   INTRODUCTION

With the development of the world's power engineering industry and power system, Flexible AC Transmission Systems (FACTS) technology is widely applied for power quality improvement. Recently, the magnetically controlled shunt reactor (MCR) has drawn much interest because of its advantages of continuously impedance adjusting, high operation reliability, and improved power system damping (Tümay et al. 2017). Finding wide application in HV/UHV transmission and reactive power compensation of distribution power grid, MCR has been actively implemented in Russia, the C.I.S countries, and also in China (Bryantsev et al. 2010; Wang et al. 2018).

The equivalent impedance of the MCR is regulated by changing the DC excitation of magnetic valves. A series of literatures have investigated the structure of the valves and the optimization methods (Chen et al. 2012; Wang et al. 2012; Chen & Wang 2017). The relationship with iron-core loss is also studied through a simulation model using finite element analysis software (Chen et al. 2012; Xiangzheng 2014). In existing studies, the magnetic valves are mostly a series of reduced sections on the core limb. However, in practical equipment the valves are stacked up by laminated silicon steel sheets and nonmagnetic materials such as epoxy resin sheets (i.e., distributed valves). Such a structure reduces leakage flux, but the joints inside the iron core become hotspots that have abnormal temperatures. This phenomenon and the thermal safety of MCRs deserve attention especially in high-voltage high-power applications.

In this article, the thermal characteristics of MCR with a distributed magnetic valve are studied by means of multiphysics coupling analysis. A full-size two-dimensional (2-D) simulation model of a 35-kV/10-MVA MCR with distributed magnetic valve is first established in the finite element software ANSYS MAXWELL 16.0. Its iron loss in practical operation through the field-circuit coupling method, the thermal model, and temperature distribution of MCR are acquired in COMSOL5.3a. Simulation results indicate that there are abnormal temperature spots where the magnetic valves are located, which needs to be considered in the future thermal design of an MCR.

2 PRINCIPLE OF THE MCR

The MCRs applied in a power system can be combined with a capacitor bank, as shown in Figure 1a. For each phase of the MCR, the configuration is depicted in Figure 1b. Theoretically, the magnetic valve is a small transverse area in the middle concave part of the limb. The two thyristors V_{T1} and V_{T2} are alternatively triggered at the two half-cycles of grid voltage and introduction of DC magnetizing current into the iron core. As the flux density at the valves increases, it gets into deeply saturated. Therefore, the inductance and capacity of the MCR can be adjusted smoothly.

3 STRUCTURE AND PARAMETERS OF AN MCR WITH DISTRIBUTED MAGNETIC VALVES

3.1 Distributed magnetic valve structure

The practical model of an MCR with distributed magnetic valve structure is illustrated in Figure 2, where the light blue parts represent the silicon steel, while yellow parts represent the epoxy resin sheets. Line 1 passes through the silicon steel and epoxy resin plates alternatively, while line 2 is located completely on the silicon steel. The magnetic valve can be regarded as parallel to the constricted silicon steel sheets that are discretely distributed in this area.

Based on a 35-kV/10-MVA MCR put into operation in a power substation in Zhejiang province, China, the FEM model of the MCR is established in ANSYS Maxwell 16.0. The size, material, and rated parameters are summarized in Table 1. The external circuit of the MCR is organized in MAXWELL Circuit Editor, so that the voltage and current waveforms are instantaneously applied in the FEM model of the MCR.

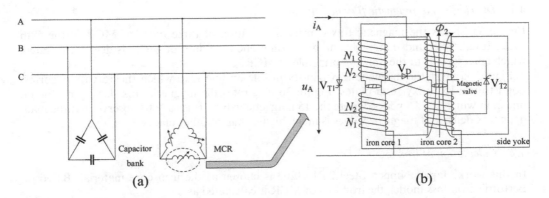

(a)   (b)

Figure 1.   Application of an MCR in a power system.

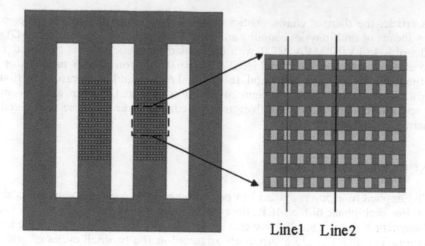

Line1 Line2

Figure 2.   Diagram of the 2-D structure of the MCR.

Table 1.   Specific parameters of the MCR.

Parameter	Value
Length (m)	2.534
Height (m)	2.832
Depth (m)	0.3
Iron core diameter (m)	0.416
Upper and lower yoke width (m)	0.416
Side yoke width (m)	0.416
Magnetic valve total height (m)	0.45
Number of magnetic valves	150
Number of turns ratio	1000/10
Rated capacity	10 MVA
Rated voltage	35 kV
Rated frequency	50 Hz
Rated current	165 A
Tapping ratio	0.01

4   RESULTS AND DISCUSSION

4.1   *Distribution of magnetic flux density*

Figure 3a Shows the magnetic flux density distribution of three types of MCR in the 59th cycle. It can be found that at this moment, only one core limb of MCR is deeply saturated, which corresponds to the working principle of MCR.

Figure 3a gives the magnetic flux density profile on line 1, in which the 15 valleys correspond to the 15 epoxy resin plates. Figure 3b illustrates the magnetic flux density profile on line 2, in which the 15 peaks denote the 15 magnetic valves. It is easy to observe that the magnetic flux density at magnetic valves is much higher than at other places.

4.2   *Calculation of iron loss*

In this work, Japan Nippon Steel 20HX1200 is chosen as the iron core material. Based on Bertotti's iron loss model, the iron loss of MCR is calculated as

$$P_{Fe} = P_h + P_c + P_e \tag{1}$$

447

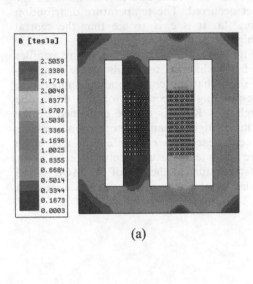

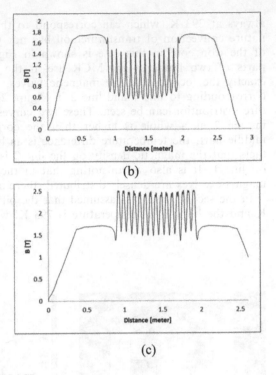

(a)

(b)

(c)

Figure 3.   Magnetic flux density distribution of the MCR.

where the hysteresis loss $P_h = K_h f B^2$, eddy current loss $P_c = K_c (fB)^2$, and additional losses $P_e = K_e (fB)^{1.5}$. Using the B–H curve of Nippon Steel Silica, and the embedded fitting method of ANSYS MAXWELL, the iron loss of MCR is obtained and the average loss intensity is 0.638 W/ kg. Because the iron core loss is proportional to the square of the magnetic flux intensity, the section where the magnetic valves are located has a higher loss intensity than other parts.

4.3 Steady-state temperature distribution of iron core of the MCR

Once the iron core loss of the MCR was obtained, the temperature distribution of iron can be obtained by solving the steady-state heat diffusion equations:

$$k\left(\frac{\partial^2 T}{\partial x^2} + \frac{\partial^2 T}{\partial y^2}\right) + q_v = 0 \tag{2}$$

where T [K] is the temperature, k is the thermal conductivity of the material, and q_v is thermal power density, i.e., the iron loss of MCR obtained in Section 4.2. To reduce the computational complexity, the interface between the solid core and the surrounding oil is modeled by Newton's law of cooling.

$$q = h(T_s - T_{oil}) \tag{3}$$

where q is the convective heat flux through the boundary, T_{oil} is the oil temperature, T_s is the surface temperature of iron core, and h is the convection heat transfer coefficient.

Similar to the method described in Lie et al. (2017), the steady-state temperature distributions of the iron core of the MCR are simulated using COMSOL. The simulation is carried out in two cases. In the first case, it is assumed that the oil outside the MCR is always at 293 K, which can correspond to the cold start case, and at this moment the nature convection of transformer oil has not yet occurred. The temperature distribution of the iron core in this case is shown in Figure 4a. It is easy to see that the central parts of two limbs of the MCR are at the highest temperature, and these parts are exactly the locations of the magnetic valves. Figure 4b shows the temperature profiles corresponding to line 1 and line 2 in Figure 2, from which more details of the temperature distribution can be seen. These two curves show that the corresponding temperature in line 2 as a whole is higher than the corresponding temperature of line 1. In the middle part, the temperature difference is as high as 4 K. This can be explained by the facts that the magnetic density on the line 2 is much higher than the corresponding area on line 1. It is also worth noting that in the central sections of two curves where the magnetic valves are densely distributed, there are wavy temperature fluctuations.

In the second case, it is assumed that the oil temperature at the top of the MCR is 363 K and the bottom oil temperature is 293 K, which corresponds to the thermal state, and

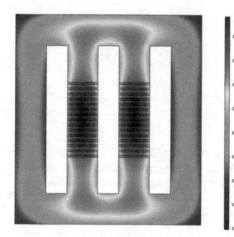

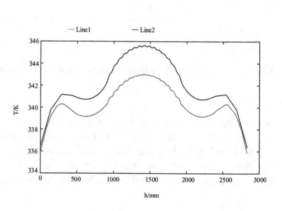

Figure 4a. Temperature distribution of MCR during cold start.

Figure 4b. Temperature profiles of MCR along line 1 and line 2 during cold start.

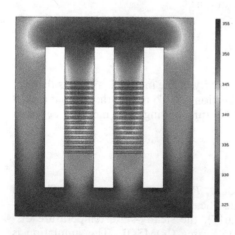

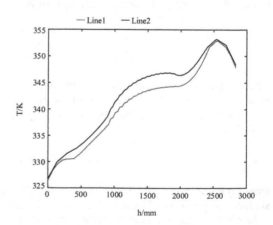

Figure 5a. Temperature distribution of MCR under a thermal stateb.

Figure 5b. Temperature profiles of MCR along line 1 and line 2 under a thermal state.

natural convection in the transformer oil has occurred. The 2-D temperature distribution of the iron core is shown in Figure 5a. Similar to the first case, the temperature on line 2 is higher than that on line 1 and there are also temperature fluctuations in the central sections of two limbs. The difference is that the highest temperature now appears at the top yoke of the MCR, but this is mainly due to the natural convection of oil.

In either case, it is easy to observe that in the central sections of two limbs of MCR, the heat source intensity is higher but distributed uniformly. In addition, with a big difference of thermal conductivity between epoxy resin and silicon steel, these sections can easily become local hotspot high-incidence areas. During long-term operation, it will cause aging and failure of insulation materials, to which more attention needs to be paid in equipment design and maintenance.

5   CONCLUSION

Based on the geometrical dimensions and parameters of an operating 10-MVA/35-kV MCR with distributed magnetic valves, this article establishes the magnetic thermal coupling simulation model of an MCR by using the finite element method. The distribution of magnetic flux density of the iron core is simulated, the iron core loss of MCR is calculated, and the steady-state two-dimensional temperature distribution of the iron core is analyzed. From the simulation, the following results can be obtained:

(1) The magnetic thermal coupling of the MCR as well as other power equipment can be simulated using the finite element method, and this can provide guidance during the design stages of the equipment.

(2) The temperature of the iron core where the magnetic valves are distributed is much higher and has more fluctuation than that of other parts. Considering that the heat dissipation in the central section of limb is difficult to determine, this deserves consideration in the future thermal design of an MCR.

ACKNOWLEDGMENTS

This work is sponsored by Technology Project of State Grid Zhejiang Electric Power Co., Ltd. (grant no. 5211DS170019) and National Natural Science Foundation of China (grant no. 51707127).

REFERENCES

Bryantsev, A., Bryantsev, M., Bazylev, B., et al. 2010. Power compensators based on magnetically controlled shunt reactors in electric networks with a voltage between 110 kV and 500 kV. In *Proceedings of the 2010 IEEE/PES Transmission and Distribution Conference and Exposition: Latin America (T & D-LA)*, November 8–10, 2010, São Paulo, pp. 239–244.

Chen, X., Chen, B., Tian, C., Yuan, J., & Liu, Y. 2012. Modeling and harmonic optimization of a two-stage saturable magnetically controlled reactor for an arc suppression coil. *IEEE Transactions on Industrial Electronics* 59: 2824–2831.

Chen, B., Gao, Y., Nagata, M., & Muramatsu, K. 2012. Investigation on harmonics suppression of saturable magnetically controlled reactor using nonlinear magnetic field analysis. In *Proceedings of the 6th International Conference on Electromagnetic Field Problems and Applications*, June 19–21, 2012, Dalian, pp. 1–4.

Chen, X., & Wang, B. 2017. Optimal design and modeling of the multi-stage saturable magnetically controlled reactor. In *Conference 2017 Progress in Electromagnetics Research Symposium – Spring*, May 22–25, 2017, St. Petersburg, pp. 76–81.

Liu, X., Yang, Y., Yang, F., & Jadoon, A. 2017. Numerical research on the losses characteristic and hot-spot temperature of laminated core joints in transformer. *Applied Thermal Engineering* 110: 49–61.

Tümay, M., Demirdelen, T., Bal, S., Kayaalp, R. I., Doğru, B., & Aksoy, M. 2017. A review of magnetically controlled shunt reactor for power quality improvement with renewable energy applications. *Renewable and Sustainable Energy Reviews* 77: 215–228.

Wang, Y., Shu, Z., & Chen, G. 2012. A novel continuously adjustable magnetic-valve controllable reactor and its modeling. In *Proceedings of the 7th International Power Electronics and Motion Control Conference*, June 2–5, 2012, Harbin, pp. 77–80.

Wang, P., Zou, Z., & Ma, X. 2018. Stability analysis of magnetically controlled reactor for reactive power compensation based on small-signal model. *IEEE Transactions on Industrial Electronics* 65: 8585–8594.

Xu, X. 2014. Research on magnetic valve structure optimization of magnetic controlled reactor. *Open Mechanical Engineering Journal* 8: 655–661.

Emerging Developments in the Power and Energy Industry – Dufo-López, Krzywanski & Singh (eds)
© 2020 Taylor & Francis Group, London, ISBN 978-0-367-27169-5

An effective health index calculation of 10 kV distribution transformer using fuzzy set theory and nonlinear fuzzy AHP

Lingjie Sun
Taizhou Vocational and Technical College, Zhejiang, Taizhou, China

ABSTRACT: This paper proposed an effective intelligent model to calculate a health index for distribution transformer using the nonlinear fuzzy analytic hierarchy process (NFAHP). The proposed method depends on the preventive test analysis of distribution transformer, and the key characteristics quantities of distribution transformer are chosen by the expert experiments and actual situation. The weights of characteristic quantities are obtained through the nonlinear fuzzy analytic hierarchy process (NFAHP). Then, the health index (HI) of distribution transformer is obtained through the fuzzy comprehensive evaluation. An intelligent maintenance model is built with filed example to demonstrate the effectiveness of the proposed model in assessing the health condition of distribution transformer. Example analysis shows that the method is correct, clear and close to the true health condition of the distribution transformer.

1   INTRODUCTION

10 kV Distribution transformer is one of the most important electrical equipment with complex structure and high cost in distribution system. It is the core part of distribution system. It is also the most important hub in the construction of grid defense measures. Its safe operation is directly related to the reliability and quality of the whole distribution system. Once a fault occurs, it will reduce the generating capacity and badly disturb people's normal life. If inconvenience comes, it will threaten the safety of life and property and hinder the steady and healthy development of the national economy. Therefore, mastering the health status of 10 kV distribution transformer can guide the condition-based maintenance of distribution transformer and ensure the safe and reliable operation of distribution system (Wang, M, A.J 2002, Ahmed E. B 2012).

Many 10 kV distribution transformers are distributed in different parts of the city, and its cost is relatively low. In operation and maintenance, general recoverable faults may be scrapped for some reason. According to the investigation, 10 kV distribution transformer generally has a running life of 10-15 years, and when there are defects, it will be scrapped directly. This phenomenon is more common in actual substations. The way of disposal has caused great economic losses invisibly.

At present, for the health assessment of 10 kV voltage distribution transformer, the expert scoring system is mainly used in the evaluation guideline of distribution equipment (Naderian A 2008, Ortiz F 2016) . This method is subjective and subject to the limitation of expert knowledge. The evaluation criteria and health grade cannot correspond one by one, and the related natural factors are not considered. The results need to be verified. At the same time, in the existing literature, there are few health assessments of 10 kV distribution transformer. Therefore, it is necessary to establish a complete theoretical system for the health status of 10 kV voltage distribution transformer to guide the actual condition maintenance.

In order to overcome the restriction of scoring system in the current guidelines, and considering that the actual 10 kV distribution transformer has few key features, the test index is a quantitative index, this paper adopts a fuzzy comprehensive evaluation model. The model is based on the combination of fuzzy set theory and non-linear fuzzy extended analytic hierarchy process. Nonlinear Fuzzy Analytic Hierarchy Process (NFAHP) has been developed in the

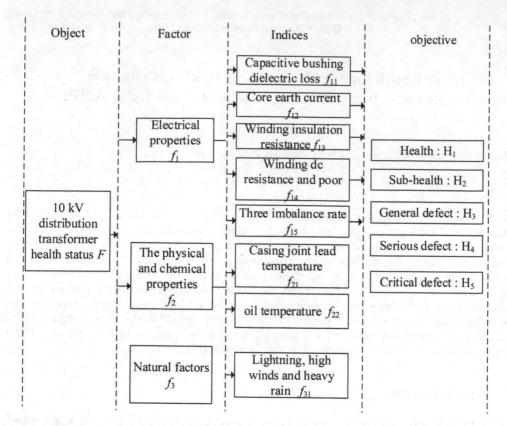

Figure 1. hierarchy structure of 10 kV transformer.

traditional hierarchical analysis method. It overcomes the characteristics of traditional AHP, such as 1-9 scale is not suitable for people's mind, and consistency test is difficult (Sun L 2016). This model establishes a theoretical system for guiding actual maintenance and is suitable for the health index calculation model of 10 kV distribution transformer.

2 HIERARCHICAL STRUCTURE OF THE 10 KV DISTRIBUTION TRANSFORMER

Considering the index system for health index calculation of 10 kV distribution transformer, as shown in Figure 1, when analyzing its structure, it mainly includes four levels, the first layer is the object layer, the second layer is the factor layer, the third layer is the index layer and the fourth layer is the object layer. The results of health index calculation for 10 kV distribution transformer can be reflected from the object layer. The factor layer can objectively present all aspects related to the health status of 10 kV distribution transformer. If it is expressed, there are F= $\{f_1, f_2, f_3\}$, which are electrical performance, physical and chemical performance and natural factors respectively. If we continue to divide the factor layers, then the corresponding index is produced. Tables 1-3 are the reference standard for evaluating external factors.

3 HEALTH INDEX CALCULATION MODEL

An intelligent health index model is presented in this paper to evaluate the status of 10 kV distribution transformer by using the fuzzy set theory, nonlinear fuzzy AHP. The main steps of this model are explained as below.

Table 1. Lightning, high winds and heavy rains level.

Code	Level	Description
0	Blue warning	Wind scale from 1-5, (or) level of lightning green, (or) rainfall blue warning
1	Yellow warning	Wind scale 6-7, (or) level of lightning yellow, (or) rainfall yellow warning
2	Orange warning	Wind scale magnitude 8, (or) level of lightning orange, (or) rains orange alert
3	Red warning	Takes more than 9, (or) level of lightning red, (or) rainfall red alert

Table 2. Transformer wind level monitoring results.

Code	Name	Description
0	calm	Wind speed 0.0- 0.2; the ground object: calm
1	soft air	Wind speed 0.3- 1.5; the ground object: soft air
2	light air	Wind speed 1.6- 3.3; the ground object: feel the wind
3	breeze	Wind speed 3.4- 5.4; the ground object: banners fluttered
4	gentle breeze	Wind speed 5.5- 7.9; the ground object: blowing dust
5	cool breeze	Wind speed 8.0 -10.7; the ground object: sapling sway
6	strong breeze	Wind speed 10.8 -13.8; the ground object: big tree sway
7	Strong wind	Wind speed 13.9- 17.1; the ground object: walking difficult
8	high wind	Wind speed 17.2- 20.7; the ground object: damaged branches
9	strong gale	Wind speed 20.8 -24.4; the ground object: a small damage of houses
10	whole gale	Wind speed 24.5- 28.4; the ground object: pull up a tree
11	storm wind	Wind speed 28.5- 32.6; the ground object: damage is significant
12	hurricane	Wind speed > 32.6; the ground object: destroy greatly

Table 3. Rainfall level monitoring results.

Code	Level	Description
0	Blue warning	12 hours: less than 5 mm of rainfall, 24 hours: less than 10 mm of rainfall
1	Yellow warning	12 hours: 5-14.9 mm of rainfall, 24 hours: 10-24.9 mm of rainfall
2	Orange warning	12 hours: 15-29.9 mm of rainfall, 24 hour: 25- 49.9 mm of rainfall
3	Red warning	12 hours: greater than 30 mm of rainfall, 24 hour: greater than 50 mm of rainfall

3.1 Step 1: Fuzzy theory

Based on the previous research and experts' experiences, the evaluation grades, relating to maintenance purpose, can be divided into 5 grades (health, subhealth, minor defect, major defect, and critical defect)

$$H = \{H_1, H_2, H_3, H_4, H_5\}$$
$$= \{health, sub\ health, general\ defect, serious\ defect,\ critical\ defect\}$$
(1)

Table 4. Assessment grades' relation to the maintenance.

Grade	Condition Description	Description
Health	Each property reaches the standard level, with sufficient margin for all the corresponding critical characteristic quantities and strong ability to resist risks and adapt to environment	One may properly delay the maintenance schedule
Sub health	All the properties can reach the standard level, but some of the critical characteristic quantities values are close to standard limit values. And ability to resist risks and adapt to environment declines	Conduct maintenance as originally planned. Pay attention to the parts which are close to standard limit values of characteristic quantities
General defect	Some of the critical characteristic quantities are out of limit, but comprehensive influence is small. There appear slight defects of ability to resist risks and adapt to environment	Arrange to carry out maintenance schedule in advance, intensify tour inspection, operation monitoring, on-ling inspection, etc
Serious defect	There appears serious degeneration of some properties, and corresponding critical characteristic quantities are out of limit. The comprehensive influence is large, and there exist obvious defects of ability to resist risks and adapt to environment	Timely arrange to carry out the maintenance schedule, intensify tour inspection, operation monitoring, on-ling inspection, etc. Defect elimination time is recommended to not exceed one week
Critical defect	The transformer cannot normally carry out the regulated functions, but its functions can be recovered after overhaul	Promptly arrange maintenance, and defect elimination time is recommended to not exceed 24 hours.

The relationship between the assessment grades and maintenance strategy is described in Table 4

3.2 Step 2: Nonlinear fuzzy AHP

The detailed steps of nonlinear fuzzy AHP are explained as follows.

On the basis of (Rezaei J 2013,Sun L 2016), the decision-maker supplies at most $n(n-1)/2$ pairwise comparisons c_{ij}, $i = 1,2, \ldots, n-1$, $j = 2,3,\ldots, n$, $j>i$. The goal is to find out the relative weight of the criteria w = $(w_1, w_2,\ldots,w_n)$ so that the ratios w_i/w_j can roughly fall into the scopes of the pairwise judgement c_{ij} which means $l_{ij} \leq w_i/w_j \leq u_{ij}$

As to every i and j, many w_i as well as w_j probably meet the aforementioned inequality. Nevertheless, various ratios w_i/w_j or wj/wi supply a diverse extent of contentment to the decision-maker. The contentment of the decision-maker could be examined by the membership function as

$$\text{Type I}: \mu_{ij}\left(\frac{w_i}{w_j}\right) = \begin{cases} \frac{\frac{w_i}{w_j}-l_{ij}}{m_{ij}-l_{ij}}, & \frac{w_i}{w_j} \leq m_{ij}; \\ \frac{u_{ij}-\frac{w_i}{w_j}}{u_{ij}-m_{ij}}, & \frac{w_i}{w_j} \geq m_{ij}; \end{cases} \tag{7}$$

$$\text{Type II}: \mu_{ji}\left(\frac{w_j}{w_i}\right) = \begin{cases} \frac{\frac{w_j}{w_i}-l_{ji}}{m_{ij}-l_{ij}}, & \frac{w_j}{w_i} \leq m_{ji}; \\ \frac{u_{ji}-\frac{w_j}{w_i}}{u_{ji}-m_{ji}}, & \frac{w_j}{w_i} \geq m_{ji}; \end{cases} \tag{8}$$

where the ratio w_i/w_j approximately meets the original judgment c_{ij} for the triangular fuzzy number of Type I, and the ratio w_j/w_i approximately meets the original judgment c_{ji} for the triangular fuzzy number of Type II.

The membership function (7) and (8) could adopt the following values:

$$\begin{cases} \mu_{ij}\left(\frac{w_i}{w_j}\right) \in (-\infty, 0), & \text{if } \frac{w_i}{w_j} < l_{ij} \text{ or } \frac{w_i}{w_j} > u_{ij} \\ \mu_{ij}\left(\frac{w_i}{w_j}\right) \in [0, 1], & \text{if } l_{ij} \le \frac{w_i}{w_j} \le u_{ij} \end{cases} \tag{9}$$

$$\begin{cases} \mu_{ji}\left(\frac{w_j}{w_i}\right) \in (-\infty, 0), & \text{if } \frac{w_j}{w_i} < l_{ji} \text{ or } \frac{w_j}{w_i} > u_{ji} \\ \mu_{ij}\left(\frac{w_j}{w_i}\right) \in [0, 1], & \text{if } l_{ji} \le \frac{w_j}{w_i} \le u_{ji} \end{cases} \tag{10}$$

It brings forth the peak of 1 when $w_i/w_j = m_{ij}$ or $w_j/m_i = m_{ij}$. The fuzzy preference programming is for the purpose of figuring out the optimal crisp priority vector w^* of the fuzzy feasible area P on the $(n-1)$- dimensional simplex Q^{n-1}

$$Q^{n-1} = \left\{ w_i \middle| \sum_{i=1}^n w_i = 1, w_i > 0 \right\} \tag{11}$$

with the following membership function

$$\mu_p(w) = \min_{ij} \left\{ \mu_{ij}(w) \middle| i = 1, 2, \dots, n-1, j = 2, 3, \dots, n, j > i \right\} \tag{12}$$

Usually, an optimal crisp priority vector with the maximum membership degree exists as below.

$$\lambda^* = \mu_P(w^*) = \max_{w \in Q^{n-1}} \min \left\{ \mu_{ij}(w) \right\} \tag{13}$$

Equation (13) can be changed in another way as

$$\begin{aligned} &\max \ \lambda \\ &s.t. \\ &\left. \begin{aligned} (m_{ij} - l_{ij})\lambda w_j - w_i + l_{ij}w_j \le 0, \\ (u_{ij} - m_{ij})\lambda w_j + w_i - u_{ij}w_j \le 0 \end{aligned} \right\} \text{ for fuzzy numbers of Type I,} \\ &\left. \begin{aligned} (m_{ji} - l_{ji})\lambda w_i - w_j + l_{ji}w_i \le 0, \\ (u_{ji} - m_{ji})\lambda w_i + w_j - u_{ji}w_i \le 0 \end{aligned} \right\} \text{ for fuzzy numbers of Type II,} \\ &\sum_{k=1}^n w_k = 1, \\ &w_k > 0, \\ &i = 1, 2, \dots, n-1, j = 2, \dots, n. \end{aligned} \tag{14}$$

The optimal priority vector w^* and λ^* can be obtained by solving the above mentioned non-linear programming issue. λ^* is demonstrated as consistency index, that is, the negative values of λ^* means that the pairwise comparisons are intensely contradictory, whereas the positive values λ^* shows that the pairwise comparisons are consistent. $\lambda^*=1$ refers to whole consistency.

4 CASE STUDY

Taking the 1# main transformer of a substation affiliated to Jiangsu Electric Power Company as an example, the health index is calculated and analyzed. The specifications of the

Table 5. Test data.

test parameters	Value		
	2014/4/8	factory default	waring value
Capacitive casing dielectric loss	0.02%	0.01%	1.5%
Core ground current	0.03 A	0.01A	0.1A
Winding insulation resistance	170 MΩ	200MΩ	10MΩ
Phase difference of winding dc resistance	0.4%	0	4%
Three unbalance rates	0.02%	0.01%	15%
Temperature of casing lead joint	18°C	6°C	70°C
oil temperature	19°C	6°C	85°C

Table 6. Weights of characteristic quantity.

Factor	Corresponding indices weights				
0.58	0.45	0.3	0.04	0.04	0.16
0.3	0.5	0.5			
0.12	0.33	0.33	0.33		

transformer are S11-2500 kVA/10 kV. The test report data are collected on April 8, 2014, shown in Table 5. The weather is sunny and the ambient temperature is 14 °C.

Based on the nonlinear fuzzy AHP method, the weights of characteristic quantity for power transformer is obtained and shown in Table 6.

The membership grade of the evaluation index can be obtained, shown in Table 7. From Table 7, we have

$$U(H)=\begin{bmatrix} 1 & 0 & 0 & 0 & 0 \\ 0.4 & 0.6 & 0 & 0 & 0 \\ 0.7 & 0.3 & 0 & 0 & 0 \\ 1 & 0 & 0 & 0 & 0 \\ 1 & 0 & 0 & 0 & 0 \\ 0.55 & 0.45 & 0 & 0 & 0 \\ 0.7 & 0.3 & 0 & 0 & 0 \end{bmatrix}$$

The result can be calculated as

$$\beta(H)=\{0.74, 0.26, 0, 0, 0\}$$

The final health index value of the equipment is obtained as

$$HI = 0.74 + 4 = 4.74$$

It can be seen that the transformer is in a healthy state, which is consistent with the actual health state of the transformer. The transformer runs well and can properly delay the maintenance plan.

5 CONCLUSIONS

According to the characteristics of 10 kV distribution transformer, a health index calculation model for 10 kV distribution transformer is proposed. The health index value of 10 kV

Table 7. The membership grade of index.

Membership grade	Evaluation grade				
	H_1	H_2	H_3	H_4	H_5
f_{11}	1	0	0	0	0
f_{12}	0.4	0.6	0	0	0
f_{13}	0.7	0.3	0	0	0
f_{14}	1	0	0	0	0
f_{15}	1	0	0	0	0
f_{21}	0.55	0.45	0	0	0
f_{22}	0.7	0.3	0	0	0

distribution transformer is obtained based on the fuzzy comprehensive evaluation. The weights are determined by the Nonlinear Fuzzy Analytic Hierarchy Process (NFAHP). Then, the health index of 10 kV distribution transformer is calculated by the Fuzzy Comprehensive Evaluation Method. Example shows that the model can effectively reflect the health status of distribution transformer and effectively guide the actual maintenance. Compared with the scoring method in the evaluation guidelines, this method can effectively improve distribution transformer assets management, and can accurately obtain the health index of distribution transformer under different operating conditions and service life.

ACKNOWLEDGMENT

The research was supported by the Scientific Research Projects of Zhejiang Education Department(Y201738354) and Taizhou Science and Technology Project (1802gy11)。

REFERENCES

Wang, M, A.J. Vandermaar and K.D. Srivastava, Review of condition assessment of power transformers in service, *IEEE Electrical Insulation Magazine*, 2002. 18(6): 12-25.

Ahmed E. B. Abu-Elanien, M. M. A. Salama, and M. Ibrahim, "Calculation of a health index for oil-immersed transformers rated under 69KV using fuzzy logic", *IEEE Transactions on Power Delivery*, 2012, 27(4): 2029-2036.

Naderian A, Cress S, Piercy R, et al. An Approach to Determine the Health Index of Power Transformers. *Conference Record of the 2008 IEEE International Symposium on Electrical Insulation. IEEE*, 2008:192-196.

Ortiz F, Fernandez I, Ortiz A, et al. Health indexes for power transformers: a case study. *IEEE Electrical Insulation Magazine*, 2016, 32(5):7-17.

Sun L, Liu Y, Zhang B, et al. An Integrated Decision-Making Model for Transformer Condition Assessment Using Game Theory and Modified Evidence Combination Extended by D Numbers. *Energies*, 2016, 9(9): 697-719.

Rezaei J, Ortt R, Scholten V. An improved fuzzy preference programming to evaluate entrepreneurship orientation. *Applied Soft Computing*, 2013, 13(5):2749-2758.

Sun L, Ma Z, Shang Y, et al. Research on multi-attribute decision-making in condition evaluation for power transformer using fuzzy AHP and modified weighted averaging combination. *IET Generation, Transmission & Distribution*, 2016, 10(15): 3855-3864.

Emerging Developments in the Power and Energy Industry – Dufo-López, Krzywanski & Singh (eds)
© 2020 Taylor & Francis Group, London, ISBN 978-0-367-27169-5

A two-settlement contract with risky power trading to reduce generation uncertainty and increase generation profit

Zhao Wang & K.L. Lo
Electronic & Electrical Engineering, University of Strathclyde, Glasgow, UK

ABSTRACT: With the increasing penetration level of renewable wind power participating in the electricity market, the uncertainty of the system supply is increased. The main uncertainty is the wind forecast error, hence increasing the uncertainty of wind generation. This uncertainty could reduce system stability, requiring an extra cost to maintain system stability. A better profit could be achieved by reducing the uncertainty. This article discusses different methods to reduce the uncertainty, such as feed-in-tariff, improved forecast accuracy, energy storage, and aggregating renewable wind generation. To aggregate renewable wind power generation, a two-settlement contract is applied to the day-ahead market. The two-settlement contract not only allows renewable wind power generators (RWPGs) to offer firm trade electricity power one day ahead, but also allows RWPGs to trade uncertain and risky next-day electricity power with each other. From our results, the two-settlement risky contract could significantly increase the profit of RWPGs. The average increases in profit is about 9%.

Keywords: renewable wind power generation, system uncertainty, two-settlement contract with risky trading

1 INTRODUCTION

Today renewable power generation has become more and more important in electricity power generation. In 2017, renewable power generation, including solar, wind, and nuclear, exceeded conventional coal and gas power generation for the first time in the UK. The consumption of renewable electricity is anticipated to be 30% of the total in 2020. Wind energy contributes to the majority of electricity generation from renewable energy sources. In the current electricity market, renewable power generators need to sign a one-day-ahead contract, to deliver a certain amount of electric power during certain time period on the next day, whereas wind generation could not be accurately forecasted a few hours before the committed period. This characteristic increases the uncertainty of the wind generation. To compensate for the uncertainty caused by forecast error, a few generators that burn conventional fuels are required to serve as backups. However, the amount of additional reserve capacity increases as the forecast error goes up, which could increase the generation cost. As a result, overall the system cost will increase.

To increase renewable wind generation penetration and reduce the electricity price to end users, different researchers proposed different market structures. Klessmann et al. (2008) use a feed-in-tariff method, which allows wind generation to access the system by setting the marginal cost of renewable generation at zero. At the beginning, the feed-in-tariff can be regarded as a reward mechanism to attract wind generation investment, although a feed-in-tariff is not a desirable market mechanism and is not suitable for the long term. Abbad (2010) introduces a penalty concept, which allows renewable wind power generator to participate in the electricity wholesale market requires a renewable wind power generator to sign a one-day-ahead contract to deliver a committed amount of electricity during a certain period on the next day. If the generator cannot not deliver the committed electricity on the next day, it will need to pay a penalty for the shortfall. Hence, all the RWPGs have the motivation to make sure its generation matches its committed target. Generally, there are two approaches to improve the generation: energy storage (National Renewable Energy Laboratory 2010) and forecast accuracy (Pinson 2013). The first approach is for energy storage to store overproduced energy in order to use it at a later time to compensate for the shortage. However, limited by the prohibitive cost of energy storage materials, an energy storage system is not economical for use in large systems. The second approach is to improve wind forecast accuracy. To accurately forecast wind one day ahead is still very difficult. Zhao et al. (2014) propose a method to reduce wind generation through aggregating wind generators at different locations. This method not only reduces wind forecast error but also reduces the cost of energy storage systems.

The authors of this article propose a two-settlement trading contract on a day-ahead power market. This two-settlement trading includes a firm electricity power trade and a risky electricity power trade. When signing the two-settlement contract, firm electricity power trading allows RWPGs to sell and to buy conventional firm electricity power on the day-ahead market, while risky electricity power trading allows RWPGs to trade next day wind power with each other. Using this two-settlement contract, an RWPG can sell and buy a fractional amount of uncertain wind power generation at a certain period on the next day, and the uncertainty of the system can be reduced under a market mechanism. All the RWPGs participating in the market can be regarded as a grand collation and there exists a competitive equilibrium. The equilibrium describes the marginal contribution of each RWPG to the coalition. It can be used to allocate profits. When RWPGs stay in the equilibrium, none of them have a motivation to deviate from the equilibrium because under the condition of equilibrium, every RWPG that stays in responds "optimally" to each other, and it could not achieve a better payoff by leaving.

2 TWO-SETTLEMENT TRADING CONTRACT

Define the expected profit of renewable wind power generator (RWPG) i, given $[\alpha_{ij}]$, $[p_{ij}]$:

$$\Pi_i = p_f s_i + \sum_{j \neq i} \mathbb{E}[p_{ij}\alpha_{ij}R_i] - \sum_{j \neq i} \mathbb{E}[p_{ji}\alpha_{ji}R_j] + \mathbb{E}\left[A(1^T R)\left(\tilde{R}_i - s_i\right)_+ - B(1^T R)\left(s_i - \tilde{R}_i\right)_+\right]$$

(1)

where $\tilde{R}_i \triangleq \sum_{j=1}^{N} \alpha_{ji}R_j$, $\tilde{R}_i$ is the sum of available risky power of other RWPGs to RWPG$_i$, p_f is the price of firm power RWPG$_i$ sold on day ahead market, s_i is the committed firm electricity power, R_i, R_j is the actual wind power generated by RWPG$_i$ and RWPG$_j$ on the committed delivery day respectively, α_{ij} is the fraction of the realized power R$_i$ sold to R$_j$. $p_{ij}\alpha_{ij}R_i$ means RWPG$_j$ bought α_{ij} fraction of R_i at price p_{ij}. R_i is the total random power available to RWPG$_i$. A is the reward coefficient of selling surplus power, B is the penalty coefficient of buying shortfall power.

In this expression, $p_f s_i$ is the payoff of selling firm electricity power.

$\sum_{j \neq i} \mathbb{E}[p_{ij}\alpha_{ij}R_i] - \sum_{j \neq i} \mathbb{E}[p_{ji}\alpha_{ji}R_j]$ is the expected payoff of selling fractions of R_i to the other RWPGs minus the expected payment of buying fractions of R_j from the other RWPGs.

$\mathbb{E}\left[A(1^T R)\left(\tilde{R}_i - s_i\right)_+ - B(1^T R)\left(s_i - \tilde{R}_i\right)_+\right]$ is the subtotal reward/ penalty from realized extra risky power or shortfall risky power. This part can be regarded as a market signal, at the committed time if the available risky wind power to RWPG$_i$ ($\tilde{R}_i$) is more than the committed firm power sold by RWPG$_i$, RWPG$_i$ will receive a reward; on the contrary, if the available risky power wind power to RWPG$_i$ is less than the committed firm power sold by RWPG$_i$, RWPG$_i$ will receive a penalty. In general, B>A, the penalty is very large while the reward is very low.

460

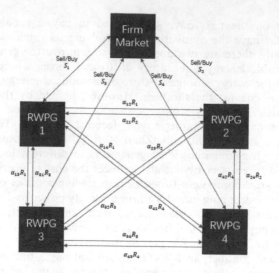

Figure 1.   The two-settlement contract trading of 4 RWPGs.

There exists an equilibrium, and at the equilibrium, $RWPG_i$ achieves its maximum profit. Max Π_i has the following constraints: (1) For $RWPG_i$, $\sum_{j=1}^{N} \alpha_{ij} = 1$, it means the total amount of risky power $RWPG_i$ sold to other RWPGs must equal the actual wind power generation on the committed time. (2) At equilibrium, every RWPG achieves a maximum profit from both a risky contract and a firm contract. Also at equilibrium, $RWPG_i$ sells its risky power at the same price as all the other $RWPG_j$. $\forall i$, $p_{i1} = p_{i2} = p_{i3} = \ldots = p_{iN} \overset{\Delta}{=} p_i$. If the price is different, $RWPG_i$ has an intention to sell all its risky power to whoever offers the highest price.

Figure 1 shows the two-settlement contract trading of 4 RWPGs.

3   CASE STUDY

A 4 RWPGs system was modeled to illustrate the two-settlement electricity trading contract with risky power trading can achieve a better payoff than the conventional firm electricity trading contract. The data sets of wind forecasts and actual renewable wind power generation data were obtained from Nord-pool Sweden. In our model $\tilde{R}_i(t)$ is the forecast renewable wind generation; $\varepsilon_i(t)$ is the forecast error which is a normal distribution, recorded as $\varepsilon_i(t) \sim N(0, \sigma^2)$; σ is obtained from the forecast error data of December 2016. The renewable wind power generation is

$$R_i(t) = \tilde{R}_i(t) + \varepsilon_i(t) \qquad (2)$$

In our model, the wind generation is averaged within an hour and we neglect intrahour wind generation fluctuation. The day-ahead price and real-time price are denoted as $p_{DA}(t)$ and $p_{RT}(t)$ respectively. We select the wind forecast, day-ahead price and real-time price data hourly for January 2017, so there are $24 \times 31 = 744$ hours of data.

On the day-ahead market, RWPGs sign a firm contract at $p_{DA}(t)$. At this time they do not know the $p_{RT}(t)$ on the committed hour; on the delivery day, if RWPGs cannot deliver the committed electricity power, they pay a penalty $B(t)$. On the contrary, if RWPGs can deliver more electricity power than the committed amount, they receive a reward $A(t)$. Also, $B(t) > A(t)$ to make sure RWPGs have the motivation to accurately deliver the committed electricity power. In our model, we set $B(t) = \max(1.3p_{DA}(t), 1.8p_{RT}(t))$, and $A(t) = \min(p_{DA}(t)/1.3, p_{RT}(t)/1.8)$. Figure 2 shows the 31–day-averaged (January 2017) profit achieved for two different scenarios.

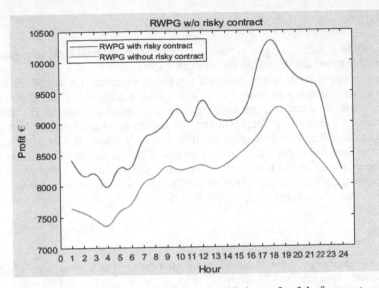

Figure 2. Comparison of the profit of a risky contract with the profit of the firm contract only.

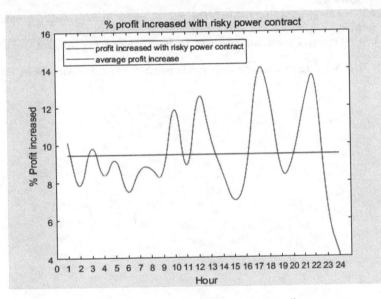

Figure 3. Percentage of profit increase with a two-settlement risky trading contract.

The red line represents the profit of an RWPG with a two-settlement contract that incorporates risky power trading, while the green line represents the profit of an RWPG without risky power trading. In the scenario without a risky trading contract, only conventional firm electricity trading is allowed. The graph shows the two-settlement with risky contract scenario can achieve a better profit than the one with only a conventional firm contract scenario.

Figure 3 shows the percentage of profit increase with the two-settlement risky trading contract. The red line shows the average profit increase for the same hour of a day in 31 days. The blue line shows the average profit increase over 744 hours. The two-settlement risky trading achieved 9% more profit than the conventional firm contract.

4 CONCLUSION

This article presents a methodology to enable an increase in renewable wind power penetration and yet be able to reduce the uncertainty of the system.

A conventional feed-in-tariff is not exactly a market mechanism, improved forecast accuracy suits only some systems, and energy storage requires a huge initial investment. A more efficient and economical method is aggregating renewable wind generation at different locations. Moreover, on a day-ahead market a two-settlement contract is applied to the aggregated wind generation. The two-settlement contract not only allows RWPGs to trade in firm power in the one-day-ahead market, but also allows RWPGs to trade uncertain and risky next-day electricity power with each other. From our analysis, the two-settlement risky contract could significantly increase an RWPG's profit. The two-settlement contract with risky trading could increase the average profit by about 9%.

REFERENCES

Abbad, J.R. 2010. Electricity market participation of wind farms: The success story of the Spanish pragmatism. *Energy Policy* 38(7): 3174–3179.

Klessmann, C., et al. 2008. Pros and cons of exposing renewables to electricity market risks—a compare son of the market integration approaches in Germany, Spain, and the UK. *Energy Policy* 36(10): 3646–3661.

National Renewable Energy Laboratory. 2010. The role of energy storage with renewable electricity generation. Technical Report.

Pinson, P. 2013. Wind energy: Forecasting challenges for its operational management. *Statistical Science* 28(4): 564–585.

Zhao, Y., et al. 2014. Wind aggregation via risky power markets. *IEEE Transactions on Power Systems* 30(3): 1571–1581.

Emerging Developments in the Power and Energy Industry – Dufo-López, Krzywanski & Singh (eds)
© *2020 Taylor & Francis Group, London, ISBN 978-0-367-27169-5*

Study on the characteristics of low voltage ride through for DFIG based on vector control and direct torque control technology

Yi Wan, Zhe Zhao, Zhongxiang Li, Liu Yang, Chao He & Qinyue Tan*
Department of Power and Electrical Engineering, College of Water Resources and Architectural Engineering, Northwest A&F University, Yang Ling, China

ABSTRACT: In the control methods of the Doubly-Fed Induction Generator (DFIG), the vector control technology has good dynamic performance and wide speed range, but the decoupling is complex and the response is slow. The direct torque control technology can control the torque and flux linkage of the DFIG directly, precisely and rapidly, but it is easy to generate large impulse current in the converter of the rotor side. In order to overcome the disadvantages of vector control and direct torque control technology, a new control model combining vector control and direct torque control is analyzed and set up. The new model reduces the dynamic response time by reducing the number of proportional integral controllers and makes the control process more stable. The effectiveness of the model is verified by the experiments of low voltage ride through (LVRT).

Keywords: doubly-fed induction generator, vector control, direct torque control, stator-voltage oriented, Low voltage ride through

1 INTRODUCTION

The doubly-fed wind generator is currently the mainstream machine type of the wind power turbines (Guo, X. M. 2008). Due to the small capacity of the excitation converter of the DFIG unit, the DC voltage between the grid-side converter and the rotor-side converter will fluctuate drastically when the wind fluctuates greatly, which will easily cause overload and damage of the converter. So a reliable and effective control strategy for the converter must be applied. At present, the main control methods of DFIG are vector control (VC) (Cheng, P. et al. 2017), direct torque control (DTC) (Gundavarapu A. et al. 2017), direct power control (DPC), and nonlinear control. Among them, the vector control appears earlier and it is more mature at home and abroad, which has solved the static and dynamic performance problems of the AC speed regulation system theoretically, and it has good dynamic characteristic and wide speed range. However, as the power grid becomes more and more complex and requirements of the power control become higher and higher, the vector control method shows its disadvantages of its dependence on parameters and the complexity of decoupling process (Mondal, S. & Kastha, D. 2015). By contrast, the direct torque control does not have complex coordinate transformation, and it directly controls the torque and flux linkage of the motor by hysteresis comparator, which has a simple process and a fast response speed (Wang, L. P. et al. 2007). Due to the above advantages, DTC is applied to wind power control gradually. However, in practical applications, DTC is not decoupled by voltage and current, so it is easy to generate large voltage and current fluctuations in the dynamic response process, which makes the motor produce large electromagnetic pulse or even cause the damage of the converter on

---

*Corresponding author: Email: qinyuetan@126.com

the rotor side. In order to overcome the shortcomings of vector control and direct torque control technology, a control model combining vector control and direct torque control was analyzed and set up in this paper. The new model reduced the dynamic response time by reducing the number of PI controllers and made the control process more stable.

2 MATHEMATICAL MODEL OF DFIG

The equations of stator and rotor voltage for the DFIG in the synchronous speed rotation d-q coordinate system can be obtained by Parker Transformation.

$$
\begin{cases}
u_{sd} = R_s i_{sd} + d\psi_{sd}/dt - \omega_s \psi_{sq} \\
u_{sq} = R_s i_{sq} + d\psi_{sq}/dt + \omega_s \psi_{sd} \\
u_{rd} = R_r i_{rd} + d\psi_{rd}/dt - \omega_{sl} \psi_{rq} \\
u_{rq} = R_r i_{rq} + d\psi_{rq}/dt + \omega_{sl} \psi_{rq}
\end{cases}
\tag{1}
$$

Where, ψ_{sd}, ψ_{sq}, ψ_{rd}, ψ_{rq} are the flux linkages of the stator and rotor on the d-q axis; i_{sd}, i_{sq}, i_{rd}, i_{rq} are the currents of the stator and rotor on the d-q axis; u_{sd}, u_{sq}, u_{rd}, u_{rq} are the voltages of the stator and rotor on the d-q axis; ω_r is the speed of rotor; $\omega_{sl} = \omega_s \text{-} \omega_r$ is the angular velocity of slip.

The equations of flux linkage are shown as follows.

$$
\begin{cases}
\psi_{sd} = L_s i_{sd} + L_m i_{rd} \\
\psi_{sq} = L_s i_{sq} + L_m i_{rq} \\
\psi_{rd} = L_m i_{sd} + L_r i_{rd} \\
\psi_{rq} = L_m i_{sq} + L_r i_{rq}
\end{cases}
\tag{2}
$$

Where, Ls, Lr are the self-inductance of the stator and rotor windings; Lm is the mutual inductance between the stator and rotor windings.

The electromagnetic torque equation is shown as follow.

$$
T_e = \frac{3P}{2}(i_{sq}\psi_{sd} - i_{sd}\psi_{sq})
\tag{3}
$$

Where, P is the pole-pair number of the motor.

The power equations of stator side are shown as follow.

$$
\begin{cases}
P_s = \frac{3}{2}(u_{sd}i_{sd} + u_{sq}i_{sq}) \\
Q_s = \frac{3}{2}(u_{sq}i_{sd} - u_{sd}i_{sq})
\end{cases}
\tag{4}
$$

3 CONTROL OF ROTOR SIDE CONVERTER OF DFIG

The dynamic characteristics of DFIG are controlled effectively by adjusting the output voltage of the back-to-back converter. The grid-side converter (GSC) adopts the control strategy of orientation of grid voltage generally, and it will not be described here. This section analyzes two control strategies of rotor side converters (RSC).

3.1 *The control of orientation of stator flux linkage*

The control strategy of rotor-side converter adopts the control strategy of orientation of stator flux linkage, that is, the d-axis of the d-q coordinate system is oriented on the stator flux linkage ψ_s, so $\psi_{sd} = \psi_s$, $\psi_{sq} = 0$, then the equations of stator flux linkage in (2) can be written as follow.

$$\begin{cases} \psi_s = L_s i_{sd} + L_m i_{rd} \\ 0 = L_s i_{sq} + L_m i_{rq} \end{cases} \tag{5}$$

So the stator current can be recorded as:

$$\begin{cases} i_{sd} = (\psi_s - L_m i_{rd}) L_s \\ i_{sq} = (-L_m/L_s) i_{rq} \end{cases} \tag{6}$$

Generally speaking, the resistance of stator winding is small and negligible. So the stator voltage in equation (1) can be written as follow.

$$\begin{cases} u_{sd} = d\psi_s/dt \\ u_{sq} = \omega_s \psi_s \end{cases} \tag{7}$$

When the DFIG is connected to the infinite power grid, the terminal voltage of the stator side can be kept unchanged. At this time, the flux linkage of the stator side remains unchanged, that is, $d\psi_{sd}/dt = 0$, and the equation of stator voltage can be obtained.

$$\begin{cases} u_{sd} = 0 \\ u_{sq} = \omega_s \psi_s = u_s \end{cases} \tag{8}$$

Substituting (7) and (8) into (4).

$$\begin{cases} P_s = \frac{3}{2} u_{sq} i_{sq} = -\frac{3 L_m}{2 L_s} \omega_s \psi_s i_{rq} \\ Q_s = \frac{3}{2} u_{sq} i_{sd} = \frac{3 \omega_s \psi_s (\psi_s - L_m i_{rd})}{2 L_s} \end{cases} \tag{9}$$

It can be known from (9) that after orientation of the stator flux linkage, the active power P_s on the stator side of the DFIG is controlled by the q-axis component i_{rq} of the rotor current, and the stator output reactive power Q_s is controlled by the d-axis component i_{rd} of the rotor current, thereby it will realize the decoupling control of stator-side power.

Substituting equation (6) into equation (2), the rotor side flux equation can be deduced.

$$\begin{cases} \psi_{rd} = \sigma L_r i_{rd} + (L_m/L_s) \psi_s \\ \psi_{rq} = \sigma L_r i_{rq} \end{cases} \tag{10}$$

In the formula (10) $\sigma = (1 - L_m^2)/L_r L_s$.

Substituting (10) into (1), the equation of rotor voltage can be obtained.

$$\begin{cases} u_{rd} = R_r i_{rd} + \sigma L_r \frac{di_{rd}}{dt} - \omega_{sl} \sigma i_{rq} \\ u_{rq} = R_r i_{rq} + \sigma L_r \frac{di_{rd}}{dt} + \omega_{sl} (\frac{L_m}{L_s} \psi_s + \sigma L_r i_{rd}) \end{cases} \tag{11}$$

According to equations (5) ~ (11), the control strategy of orientation of rotor side stator flux can be obtained.

3.2 The model of direct torque control

The equation of electromagnetic torque of asynchronous generator is shown as follow.

$$T_e = \frac{3P}{2} \frac{L_m}{\sigma L_s L_r} |\psi_r||\psi_s| \sin \delta \qquad (12)$$

Where, δ is the angle between the stator flux vector and the rotor flux vector, which is the torque angle.

The direct torque control does not have complicated coordinate transformation, it utilizes the tolerance control of the hysteresis comparator to determine the required voltage vector according to the characteristics of the voltage-type inverter switch selection. Therefore, it can adjust the inverter switching state to control ψ_r and δ, maintain the electromagnetic torque balance between the stator and rotor flux linkages, and then realize the control of the rotor side of the DFIG converter.

3.3 A new control strategy

The vector control strategy is accurate, but it has many coordinate transformations, complex decoupling and slow dynamic response. The direct torque control strategy has a fast dynamic response, but it is easy to generate large impulse current for the rotor side converter. According to the characteristics of the two control strategies, a new control strategy is designed.

Under the new control strategy, the stator flux is estimated by detecting the stator current, the stator voltage and the motor speed. Then, the active component of the rotor current can be obtained according to (13).

$$T_e = C_T \Phi I_2' \cos\varphi_2 \qquad (13)$$

4 LOW VOLTAGE RIDE THROUGH (LVRT) EXPERIMENT

Based on above analysis of control strategies, a DFIG model with a single unit capacity of 1.5MW is set up to simulate the dynamic response under different control strategies when the voltage dropped by 70% symmetrically. Simulation results of three control strategies are listed in Table 1, and dynamic characteristics are shown in Figure 1.

From the above figures and table, we can see that system response under DTC is faster than that under the VC, but the electromagnetic torque Tem and the DC bus voltage in the regulation process change greatly as shown in Figure (b) and (e), the system absorbs a large amount

Table 1. Simulation results of three different strategies.

Control Strategy		DTC	VC	The new control
Mechanical torque Tm	Degree of fluctuation(p. u.)	2.2	1.8	1.2
	Response time(s)	4.5	2.5	3
Electromagnetic torque Tem	Degree of fluctuation(p. u.)	2.5	2	1.5
	Response time(s)	1.2	1.5	1
Active power	Degree of fluctuation(MW)	6.5	2	1.5
	Response time(s)	2.5	1.5	1
Reactive power	Degree of fluctuation(MVar)	5	1.5	2
	Response time(s)	1.5	1.5	1.5
DC bus voltage	Degree of fluctuation(V)	450	150	100
	Response time(s)	1	1.5	0.8
Rotor speed	Degree of fluctuation(p. u.)	0.13	0.14	0.8
	Response time(s)	5	5	3

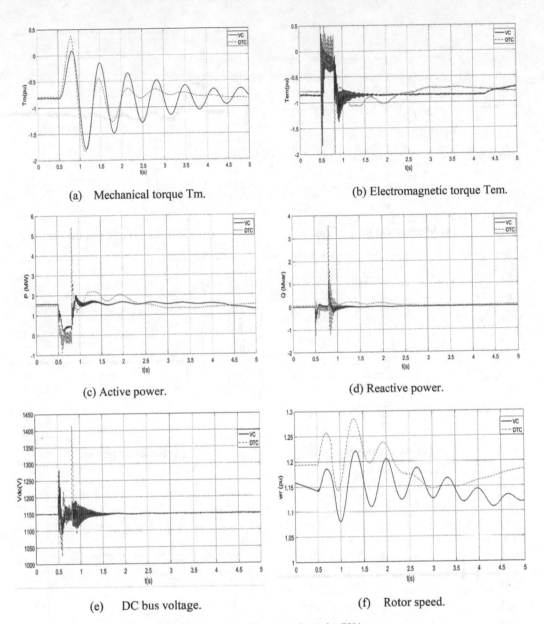

(a) Mechanical torque Tm.

(b) Electromagnetic torque Tem.

(c) Active power.

(d) Reactive power.

(e)   DC bus voltage.

(f) Rotor speed.

Figure 1. Dynamic characteristics when the grid voltage drops by 70%.

of reactive power when the fault occurs as shown in Figure (d). For VC, the fluctuation when the fault occurs is smaller than DTC as shown in Figure (e). The rotor flux linkage is stable rapidly near the grid drop at 0.1s, which indicates that the DTC can achieve the rotor flux linkage tracking much better, thus it can prevent the rotor side from getting overcurrent effectively; As can be seen from the analysis of Figure (e), VC has a good effect on preventing DC side from getting over voltage. When the grid voltage dropped severely, the low voltage rides through ability is strong, while the DTC controls directly and the response is faster, so the characteristic of dynamic response is better than VC.

The new control strategy controls the electromagnetic torque through the d-axis component of the rotor current, and realizes the torque control, which reflects the characteristics of direct torque control. At the same time, the rotor current is adjusted by PI controller through

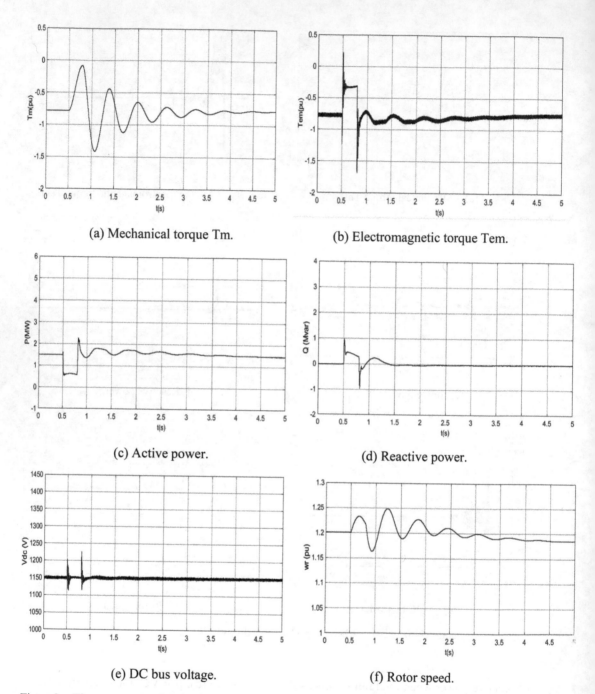

(a) Mechanical torque Tm.

(b) Electromagnetic torque Tem.

(c) Active power.

(d) Reactive power.

(e) DC bus voltage.

(f) Rotor speed.

Figure 2.   The response results of voltage symmetrical drops by 70% under the new strategy.

coordinate transformation, combined with the stator voltage and the flux linkage in the d-q coordinate system, the characteristics of vector control are reflected. The new control strategy combines the features of vector control and direct torque control, reduces the number of PI controllers and shortens the dynamic response time. The simulation results are shown in Figure 2.

Comparing Figure 2 and Figure 1, it can be seen that under the same simulation conditions, the dynamic response of the new control strategy is faster, the mechanical torque Tm and the electromagnetic torque Tem are less changed and the stabilization time is reduced. The fluctuations of active power and reactive power are small when the fault occurs as shown in Figure (c) and (d), which ensure that the grid is still connected when the voltage drops for a short period of time. The DC bus voltage changes less during the regulating process as shown in Figure (e) to ensure the safety of the rotor side converter; The rotor speed is quickly stabilized as shown in Figure (f), which ensure the stable operation of the generator. Analyzing Figure (a) to (f), it can be seen that the new strategy has a good effect on preventing DC side from getting over-voltage. When the grid voltage drops seriously, the ability of LVRT is strong. It combines the advantages of direct torque control and vector control, the dynamic response is faster and the fluctuation is smaller than VC and DTC.

5 CONCLUSION

The new control strategy proposed in this paper combines the characteristics of vector control and direct torque control. Compared with VC, the dynamic response of the new control strategy is faster, while the changes of the Tm and Tem are smaller than DTC about 1p.u. The fluctuation of active power and reactive power is smaller than that of DTC about 5MW and 3MVar respectively, which ensures that the grid connection is maintained during short-time voltage drop. The change of the DC bus voltage is smaller during the regulating process relative to DTC and VC about 350V and 50V respectively, the rotor-side converter is guaranteed to be safe. The rotor speed can be stabilized quickly, about 2 seconds faster than DTC. It can be seen that the new strategy has a good effect on preventing DC side from getting over-voltage. When the grid voltage is dropped severely, the ability of LVRT is strong, the dynamic response is faster and the fluctuation is smaller than VC and DTC.

ACKNOWLEDGEMENT

This work is financially supported by the National Natural Science Foundation of China (51577157).

REFERENCES

Guo, X.M. 2008. Direct power control of doubly-fed asynchronous wind turbines under abnormal grid conditions. Ph.D. dissertation, College of Electrical Engineering, Zhenjiang University, Hangzhou, China.

Cheng, P. & Nian, H.C, et al. 2017. Direct Stator Current Vector Control Strategy of DFIG Without Phase-Locked Loop During Network Unbalance. IEEE Transactions on Power Electronics 32 (1):284–297.

Gundavarapu, A. & Misra, H, et al. 2017. Direct Torque Control Scheme for DC Voltage Regulation of the Standalone DFIG-DC System. IEEE Transactions on Industrial Electronics 64(5):3502–3512.

Mondal, S. & Kastha, D. 2015. Improved Direct Torque and Reactive Power Control of a Matrix-Converter-Fed Grid-Connected Doubly Fed Induction Generator. IEEE Transactions on Industrial Electronics 62(12):7590–7598.

Wang, L.P. & Guo, S.Y, et al. 2007. Vector control of doubly-fed asynchronous generator. Converter Technology and Electric Traction 4: 31–35+43.

Emerging Developments in the Power and Energy Industry – Dufo-López, Krzywanski & Singh (eds)
© *2020 Taylor & Francis Group, London, ISBN 978-0-367-27169-5*

Research on electromagnetic disturbance and suppression measures at the secondary side of the voltage transformer in distribution switchgear

W.X. Qian, B. Xiao, D. Wang & L.P. Bi
State Grid Inner Mongolia East Electric Power co., LTD, Power Science Research Institute, Hohhot, China

H.B. Liu
State Grid Inner Mongolia East Electric Power co. LTD, Hohhot, China

Y.F. Hu, R. He & C.Q. Jiao
State Key Laboratory of Alternate Electric Power System with Renewable Energy Sources, North China Electric Power University, Beijing, China

K.Q. Xu
Hengjingtong Electronic Technology Co., Ltd, Nanjing, China

ABSTRACT: In this paper, the electromagnetic disturbance of the voltage transformer arranged inside the switchgear is measured at the moment of switching operation. Then the characteristics of the disturbance waveform are analyzed. The measurement results show that the peak voltage of disturbance can be as high as about 120V, and the frequency does not exceed 25MHz. At the same time, the voltage transmission characteristics of the voltage transformer are obtained based on the experimental data. It is found that the transfer function of voltage transformer remains stable in the low frequency (below 50 kHz), and rises in the high frequency (above 50 kHz). This makes high frequency disturbances easier to affect its secondary side through the transformer. Finally, a filter design method is proposed for electromagnetic disturbance of voltage transformers. It is used to inhibit the high frequency disturbance above a certain frequency through the transformer to the secondary equipment, and it does not affect the signal below this frequency.

1 INTRODUCTION

In modern power distribution systems, the electromagnetic transient phenomenon generated during the switching operation makes the power distribution equipment itself a powerful interference source. It will affect the electronic transformers distributed inside the switchgear (Wang Y 2015, Liu B et al. 2018). As basic test equipment of voltage/current, electronic transformer has the advantages of simple insulation structure, light weight, and direct signal input into the protected device and so on. It is a bridge between the high voltage primary system to the low voltage secondary system. As a result, the electromagnetic disturbance coming from the primary system can propagate along the transformer to the secondary system. Hence, there are great concerns about the electromagnetic disturbance characteristic induced at the secondary side of electronic transformer, especially in the case when primary side has a switch operation like the closing/opening of the circuit breaker (Tian B et al. 2017, Xin W F et al. 2017, Li Z H et al. 2017).

The transfer function of the voltage transformer (VT) can reflect its transmission characteristics (Wu M L 2004). If the voltage transformer has great transmission capability for disturbance in high frequency, the secondary equipment will undoubtedly suffer from various interferences from the primary side. In order to achieve the purpose of suppressing high frequency disturbance through the voltage transformer, it is necessary to install a low-pass filter at the secondary side of the transformer. The filter is a filter circuit composed of a capacitor, an inductor, and a resistor. It can effectively filter the frequency of a specific frequency or outside the frequency in the circuit to get a signal in a specific frequency range or eliminate a signal after a specific frequency range (Xu Y K et al. 2018, Liu Y T et al. 2010). At present, there are many different designs for filters, but the filtering effect of many software designed filters is not ideal (Wang Y et al. 2008, Lin K S et al. 2014). In this paper, based on the normalized low-pass filter, a filter design method is proposed with simple algorithm and good filtering effect. The so-called normalization is that the characteristic impedance is 1 Ω and the cutoff frequency (the frequency corresponding to 3 dB) is about 0.159 Hz.

In this paper, the electromagnetic disturbance measurement results of the voltage transformer are introduced. Then the transfer function of the voltage transformer is calculated and analyze its transmission characteristic is analyzed. In addition, a design scheme for suppressing the disturbance at the secondary side of the transformer is proposed. This filter can inhibit disturbance above any frequency. Finally, the effect of different load for attenuation performance of filter is calculated.

2 MEASUREMENT SCHEME

The measurement system shown in Figure 1 is used to collect the electromagnetic disturbance at the secondary side of the voltage transformer during the switch closing/opening operation. The high voltage probe acts as a sensor and the one end is connected to the port at the secondary side of the transformer, the other end is connected to the port of the oscilloscope. The oscilloscope will send the signal from the high voltage probe to the hard disk. In order to improve the accuracy of the measurement results, the oscilloscope is placed in the electromagnetic shielding box, which is independently powered by the DC power supply. Then the optical fiber is used for remote measurement operation. The reliability of this set of measurement systems has been verified in the ref. (WU H T et al. 2017a, 2017b, 2017c, 2017d).

3 VT OF ELECTROMAGNETIC DISTURBANCE

This experiment measures the differential mode voltage of the secondary side of VT. Figure 2 shows the macro pulses during switching operations. The voltage parameters under various working conditions shown in Table 1 are extracted for measured waveforms. The f means main frequency.

According to the data in table 3, it can find that the u_{max} of VT is 110.3V, and the f is no more than 25MHz. In addition, the t_{mac} and f in open circuit are much greater than capacity load. It means circuit connecting the capacity load can reduce the disturbance incoming VT.

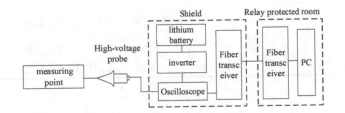

Figure 1.   Sketch of the measuring system.

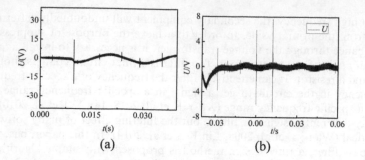

Figure 2. The macro pulse waveforms of VT: (a) closing operation; (b) opening operation.

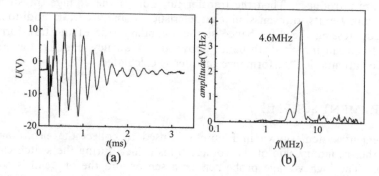

Figure 3. The micro pulse waveforms of VT: (a) closing operation; (b) opening operation.

Table 1. The waveform parameters of disturbance voltage at the secondary side of VT.

Working conditions	Type of operation	Times	$|u_{max}|$/V	t_{max}/ms	n	t_{mic}/μs	t_r/ns	f/MHz
Open circuit	close	1	19.94	32.15	6	2.74	12	4.6
		2	15.28	32.53	4	2.78	8	4.7
		3	16.68	26.08	4	2.5	12	4.81
	open	1	7.69	40.92	3	1.99	14	4.76, 13.64
		2	5.38	46.31	4	2.18	16	4.63, 13.43, 21.30
		3	4.57	78.84	8	2.02	9	4.61, 13.83, 21.52
capacity load	close	1	110.30	1.29	2	2.24	8	1.30, 2.70
		2	84.25	1.32	6	2.86	8	1.59, 2.87
		3	108.70	0.2	4	2.16	10	1.40, 2.80
	open	1	50.62	0.35	2	2.06	8	1.65, 3.02
		2	36.38	1.23E-3	1	1.23	12	2.69

4 ANALYSIS OF COUPLING MECHANISM

Figure 9 shows the transmission characteristics of VT. Transform the measured transient waveforms at the primary and secondary sides of VT into the frequency domain. Then use the peak voltage at secondary side divided by the peak voltage at primary side to obtain the voltage transfer function. Analyzing the characteristics of this transfer function, it can be obtained that the transmission function H remains basically unchanged between DC to 50kHz. With the increasing of frequency, H increases correspondingly and many resonance points appear. This is the reason why the high frequency disturbance is coupled to the secondary side through the transformer. At this time the voltage transformer is similar to a high-pass filter. The frequency of the disturbance waveform generated at the moment of switch operation is really high. The high-frequency component can be easily transferred to the secondary side through the voltage transformer, causing conducted interference to the secondary equipment. The ref. (S. W. Chen 2006) records several measurement result of the transfer function of the voltage transformer, which has the same rules as the results measured in this paper.

$$|H(j\omega)| = \frac{|U_2(j\omega)|}{|U_1(j\omega)|} \tag{1}$$

The transfer function H is obtained by formula (1). U_2 (j) is the frequency domain voltage of the secondary side, and U_1 (j) is the frequency domain voltage of the primary side.

5 MEASURES TO SUPPRESS THE ELECTROMAGNETIC DISTURBANCE

Figure 4 shows that high frequency disturbance can easily affect secondary equipment through the transformers. Therefore, it is necessary to take corresponding suppression measures at the secondary side of transformer. The following will introduce a filter design method, which can determine the parameters of each element according to the frequency range needed to be suppressed.

5.1 Design principle of filter

Figure 5 shows the circuit diagram of 2-5 order normalized low-pass filter.

Let's say the scaling factor A is cut-off frequency to be designed/reference cut-off frequency and B is characteristic impedance to be designed/reference characteristic impedance.

$$L' = \frac{L \times B}{A} \tag{2}$$

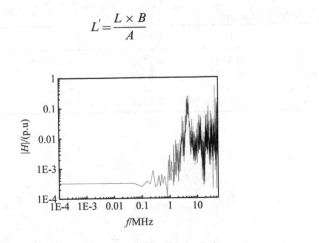

Figure 4.   The amplitude frequency characteristics of voltage transmission.

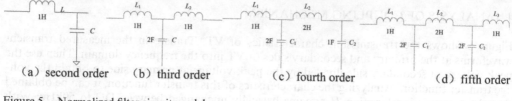

(a) second order　　(b) third order　　　(c) fourth order　　　(d) fifth order

Figure 5.　Normalized filter circuit model.

$$C' = \frac{C}{A * B} \tag{3}$$

It is assumed that the secondary equipment will be disturbed when the signal frequency exceeds 10kHz. If you want to design a low-pass filter whose cut-off frequency is 10kHz and characteristic impedance is 50 Ω, it is necessary to transform the cutoff frequency and characteristic impedance of the filter parameters. The values of the inductance and capacitance elements in the filter are calculated, as shown in Table 2.

The attenuation characteristic of the filter can be expressed by its transmission function, that is, the ratio of the frequency domain voltage of the filter's secondary side to that of the primary side. Pspice software is used for simulation and frequency sweep to facilitate analysis. The simulation circuit diagram is shown in Figure 6 and the result is shown in Figure 7. One end of the filter connects the ac power, the other end connects 50 Ω load. The attenuation characteristics of the filter can be obtained by calculating the ratio of the voltage at both ends of the load to the power supply voltage. As shown in Figure 5–Figure 8 (a), it is obvious that after 10kHz, the second-order filter (the red line part) has the slowest descending speed, and the 5th order filter (the blue line part) has the fastest descending speed. The higher the order of the filter is, the faster the descending speed is. In order to observe the attenuation characteristics of the filter more clearly, the formula $20\log(U_{out}/U_{in})$ is used to process the data of Figure 7 (a). The transformation to dB is shown in Figure 7 (b).

It can be seen clearly from the picture above that the simulation results of the filter with cut-off frequency of 10kHz show that the filter can block the frequencies above 10kHz. As the order of the filter increases, its attenuation becomes better. Therefore, this filter can be installed in the switch cabinet to suppress the high-frequency disturbance generated by the

Table 2.　Filter parameters.

	L1	C1	L2	C2	L3
Second order	0.8mH	0.32uF			
Third order	0.8mH	0.64uF	0.8mH		
Fourth order	0.8mH	0.64uF	1.6mH	0.32uF	
Fifth order	0.8mH	0.64uF	1.6mH	0.64uF	0.8mH

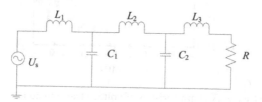

Figure 6.　Simulation circuit diagram of filter attenuation characteristics.

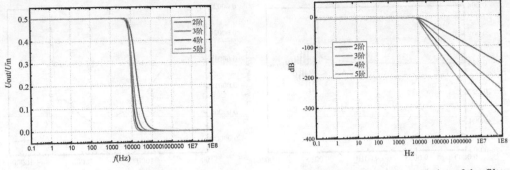

(a) the ratio of the load voltage to the supply voltage

(b) attenuation characteristics of the filter

Figure 7.　Comparison of attenuation characteristics of 2-5 order filters.

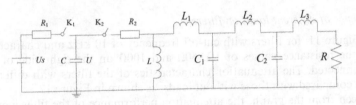

Figure 8.　Simulated circuit diagram of filter attenuation performance.

switching operation. A 3MHz damped oscillation wave is generated by the circuit and passed through a five-order filter. The simulation circuit is shown in Figure 8.

　　Observe the attenuation of damped oscillating waves which through the filters. Calculate the voltage at both ends of inductance element L and resistance element R. The parameters of the related elements are shown in Table 3. The calculation results are shown in Figure 9 and Figure 10. As can be seen from the figure, the inductor L produces a main frequency of 3.01MHz when $t = 2\mu s$, the peak value is 0.58 kV. After passing through the filter, the phase of the damped oscillating wave is delayed, and the peak value and frequency are greatly reduced. As can be seen from Figure 10, the waveform appears at $t=1.4ms$ with a peak voltage of 1.60mV and a frequency of 38.56kHz.

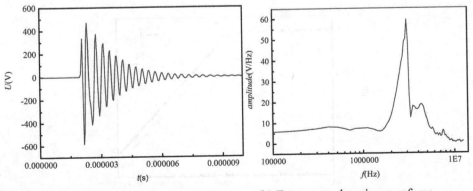

(a) Time domain waveform

(b) Frequency domain waveform

Figure 9.　Damped oscillating waves at 3MHz.

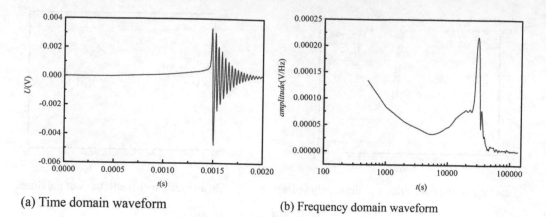

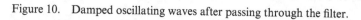

(a) Time domain waveform   (b) Frequency domain waveform

Figure 10.   Damped oscillating waves after passing through the filter.

5.2 *The influence of different loads on filters*

As shown in Figure 11, for filters with cut-off frequency of 10 kHz and characteristic imped-ance of 50Ω, the resistance values of 1Ω, 50Ω and 100Ω on the right side of the filter are respectively connected. The attenuation characteristics of the filters with different loads are calculated and compared with those without loads, as shown in Figure 12.

As can be seen from the graph, the attenuation performance of the filter increases slightly after loaded. The smaller the load resistance, the better the attenuation performance. When the load is above 50Ω, the attenuation performance is basically unchanged. It is concluded that the smaller the load resistance of the filter, the higher its attenuation ability.

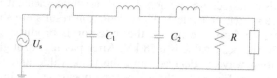

Figure 11.   Simulation circuit with different loads.

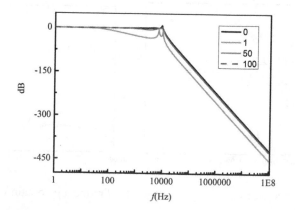

Figure 12.   Attenuation characteristics of filters with different loads.

6 CONCLUSION

(1) For the secondary side of PT, the u_{max} are 19.94V/110.3V under open circuit/capacitive load, the t_r are between 8 to 16 ns, and the f are no more than 25MHz. However, the peak voltage 110.3V will change with the change of capacitive load

(2) When the frequency of voltage transformer is lower than 50kHz, the transmission function is basically unchanged. With the increasing of frequency, the transmission function increases obviously, resulting in the easy passage of high-frequency harassment.

(3) The attenuation performance of the filter depends on the order and load of the filter. The higher the order or the smaller the load, the better the attenuation ability.

REFERENCES

Chen SW. 2006. Research on broadband transmission characteristics and equivalent circuit model of transformer in substation. North China electric power university (hebei).

Liu YT, Liu YB, Yin W. 2010. LC filter design method and comparison of simulation characteristics. Electronic measurement technology, 33(05):17-21.

Lin KS, Zhang L, Lin KW. 2014. Optimization design and simulation of butterworth low pass filter [J]. Journal of chongqing technology and business university (natural science edition), 31(06):58-62.

Li ZH, Yu J, Li ZX, Xu YC. 2017. Research status of electromagnetic compatibility of electronic transformer. High voltage electrical appliances, 53(04):220-226.

Liu B, Huang H, Tong Y, Deng XP. 2018. Electromagnetic compatibility testing system for electronic transformer based on on-off operation of isolation switch. High voltage technology, 44(03):1016-1022.

Tian B, Bao G, Li ZH, Li ZX. 2017. Research on electromagnetic interference of electronic transformer and its key technology [J]. Transformer, 54(06):30-35.

Wu ML. 2004. Study on broadband transmission characteristics of transformer in substation and transient characteristics of grounding grid [D]. North China electric power university (hebei).

Wang Y, Zheng M, Li ZH, et al. 2008. Normalized design of butterworth type low-pass filter. Ship electronic engineering, 38(01):61-64.

Wang Y. 2015. Research on electromagnetic compatibility of electronic transformer in intelligent substation. Huazhong university of science and technology.

WU HT, CUI X, LIU XF, JIAO CQ, et al. 2017a. "Characteristics of Electromagnetic Disturbance for Intelligent Component Due to Switching Operations via a 1100 kV AC GIS Test Circuit," in IEEE Transactions on Power Delivery, vol. 32, no. 5, pp. 2228-2237.

WU HT, JIAO CQ, CUI X, LIU XF AND JI JF. 2017b. Transient Electromagnetic Disturbance Induced on the Ports of Intelligent Component of Electronic Instrument Transformer Due to Switching Operations in 500 kV GIS Substations, in IEEE Access, vol. 5, no. 99, pp. 5104-5112.

WU HT, CUI X, JIAO CQ, LIU X, JI JF. 2017c. Development and application of electromagnetic disturbance measurement system for secondary equipment port of GIS substation. High voltage electrical appliances, V53(10): 68-75.

WU HT, JIAO CQ, CUI X, LIU X, JI JF. 2017d. Simulation and analysis of electromagnetic disturbance caused by switch operation of GIS substation on secondary equipment port. High voltage technology, V43(10):3387-3395.

Xin WF, Wang D, Kou XS, Fu HJ, Li P, et al. 2017. Study on the conduction coupling effect of transient electromagnetic interference through electronic mutual inductor. High voltage electrical appliances, 53 (04):164-170.

Xu YK, Gao F, Zhang Y, Li R. 2018. Research on a design method of LC low-pass filter. Electronic world, 158-159.

Emerging Developments in the Power and Energy Industry – Dufo-López, Krzywanski & Singh (eds)
© 2020 Taylor & Francis Group, London, ISBN 978-0-367-27169-5

Classification and identification of temperature status of gas insulated switch gear isolation switch contacts based on a learning vector quantization neural network

Yiping Ma & Xuntian Zhou
Ningbo Electric Power Design Institute Co., Ltd., Ningbo, Zhejiang Province, China

ABSTRACT: In the high-voltage power transmission and transformation system, the safety and reliability of gas-insulated switchgear (GIS) has become the key to maintaining safe operation of high-voltage power grids. The problem of local temperature rise of GIS is one of the important factors affecting the safe and stable operation of GIS. In this article, the temperature change of the GIS isolation switch contact is detected by changing the contact resistance and using fiber Bragg grating (FBG) sensors. Using experience to judge the thermal state of the isolation switch contacts, the traditional identification method involves significant subjectivity. For this reason, a learning vector quantization (LVQ) identification mode is proposed. The LVQ neural network has the advantages of simple mode, self-learning, and self-organization. It is very suitable for constructing the nonlinear mapping relationship between GIS isolation switch contacts temperature and working state to identify the thermal state of GIS isolation switch contacts.

Keywords: classification and identification, GIS isolation switch contacts, LVQ neural network

1 INTRODUCTION

With the vigorous development of China's power industry and the continuous increase of demand, gas-insulated switchgear (GIS), owing to its strong interrupting ability, low failure rate, easy installation and maintenance, and small footprint, has been widely used in power systems in China and abroad (Cong et al. 2014). However, while the GIS has the aforementioned advantages, problems of conductor loss and heat generation have also become increasingly prominent because of its tight seal, small size, strong current, and other structural and operating characteristics. Especially when GIS isolation switch contacts are in a bad connection, the contact resistance will increase, and overheating will occur when the load current flows (Li et al. 2009; Zhao t al. 2011). Based on the present artificial detection of its contact thermal state is too subjective. In order to monitor its contact state in a more convenient way, therefore, the learning vector quantization (LVQ) neural network identification mode is proposed to identify the contact status, which not only can prevent overheating and save time, but also can ensure safer and more reliable operation of GIS.

At present, there are mainly two kinds of measures used in the field to determine whether the GIS isolation switch contact is overheating: periodically measuring the circuit resistance and using infrared imagery to regularly monitor the temperature at fixed monitoring points (Xie et al. 2012; Chen et al. 2015; Sun et al. 2015). There are some shortcomings in these methods: the first method requires the GIS equipment to outage overhaul, while the second measure is difficult to achieve because it is difficult for the resolution and accuracy of infrared imaging technology to meet the requirements. Currently, infrared technology and grating fiber technology are mainly used in the industry to conduct online monitoring of the temperature of electrical equipment. In Li et al. (2012), Chen et al. (2015), and Wang et al. (2017), infrared temperature measurement method cannot destroy the temperature field and thermal balance inside GIS equipment, and can also solve the problems of high-voltage isolation and strong magnetic field interference. However, it is necessary to open a hole in the GIS shell to install an infrared temperature sensor, and its measurement accuracy is greatly affected by such factors as conductor metal surface emissivity and SF6 gas concentration. Fiber Bragg grating technology uses an optical wavelength as the monitoring quantity, which has the advantages of not being affected by electromagnetic interference, excellent insulation performance, small size, light weight, etc., and has been widely used in transformer, electrical machinery, switchgear, overhead transmission line, cable, and other power equipment, and the technology is relatively mature (Cong 2014). Fiber Bragg grating (FBG) temperature sensor is applied to GIS conductor contact temperature monitoring, which can accurately measure the temperature without damaging the electric field and temperature field inside the GIS. So its application prospect is very broad.

In this article, the installation location of sensors is determined, and the temperature rise experiment of GIS isolation switch contacts is described. The data obtained from the experiment are processed, and according to the normal operating range of the GIS isolation switch contact temperature, the contact state is classified and identified, and the contact state can be divided into normal, heated, and overheated. In this way, it is not only convenient for the staff to manage it, but also can prevent the overheating of the contact when the contact temperature is in a hot state, which plays an important role in the stable operation of the power system.

2 EXPERIMENTAL PRINCIPLE

2.1 *Fiber Bragg grating temperature sensor principle*

The fiber grating utilizes the photosensitivity of the optical fiber material to form a spatial phase grating by exposing the fiber core to ultraviolet light and its sensing principle is to obtain information by adjusting the central wavelength of FBG by external parameters. According to the grating theory, when the broadband continuous light passes through the FBG, the light wave matching the central wavelength of the FBG will be reflected, and the rest of the light waves will be transmitted directly (Wu 2013; Wang et al. 2016).

The relationship between the center wavelength of the fiber grating and the effective refractive index of the core and the grating modulation period is expressed as

$$\lambda_B = 2n_{\text{eff}}\Lambda \tag{1}$$

In the equation, λ_B is the center wavelength of FBG, n_{eff} is the effective refractive index of core, and Λ is the period of fiber gratings modulation.

$$\frac{\Delta\lambda_B}{\lambda_B} = \left(\frac{1}{\Lambda}\frac{d\Lambda}{dT} + \frac{1}{n_{\text{eff}}}\frac{dn_{\text{eff}}}{dT}\right)\Delta T = (\alpha + \xi)\Delta T \tag{2}$$

$$\alpha = \frac{1}{\Lambda}\frac{d\Lambda}{dT} \tag{3}$$

$$\xi = \frac{1}{n_{\text{eff}}}\frac{dn_{\text{eff}}}{dT} \tag{4}$$

In the equations, α is the coefficient of thermal expansion of the bare grating and ξ is the thermal-optical coefficient of the bare grating. Equation (2) shows that the impact of temperature change on the center wavelength of the fiber grating is caused by the thermal expansion effect and the thermo-optic effect, and wavelength-temperature demodulation can be realized by detecting the change of the incident wavelength and the reflection wavelength of the fiber

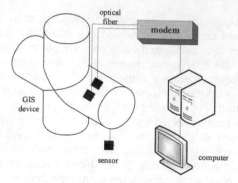

Figure 1.   Block diagram of the system.

grating, so the temperature value of the measured point can be calculated by measuring the variation in the center wavelength of the grating.

2.2   *Main components of the system*

The system is mainly composed of GIS, fiber Bragg grating sensor, demodulator, and host computer. A group of fiber Bragg grating temperature sensors are installed on the top of the outer wall of the disconnector, totaling eight sensors, and two sensors are installed 1 meter away from the bottom of the GIS to collect the environmental temperature. A high-precision FBG signal demodulation system is installed in the main control room to identify and classify the temperature and temperature rise rate information of the 10 temperature measuring points on the GIS.

3   LVQ NEURAL NETWORK

3.1   *Basic ideas of LVQ*

The LVQ neural network algorithm is a learning method for training the competition layer under supervision. Its structure is shown in Figure 2. The specific steps are as follows (Bao et al. 2011; Yuan et al. 2011; Wang et al. 2018).

Step 1: Set variables and parameters:

(1) Set input vector and target vector. $X(n) = [x_1(n), x_2(n), x_3(n), \cdots, x_N(n)]^T$ is the input vector (training sample).

(2) Set the weight vector and learning rate. $W_{i1}(n) = [w_{i1}(n), w_{i2}(n), w_{i3}(n), \cdots w_{iN}(n)]^T$, $(i = 1, 2, 3, \cdots, M)$ is the weight vector, $\eta(n)$ is a function of the learning rate, and n is the number of iterations.

Step 2: Initialize weight vector $W_i(0)$ and learning rate $\eta(0)$.

Step 3: Input the training sample, calculate the distance between the input vector and the weight training, find the weight vector with the smallest distance of the input vector, and find the winning neuron through the standard of the smallest Euclidean distance to realize the competition process of the neuron. The formula is as follows:

$$\|X - W_c\| = \min\|X - W_i\|, (i = 1, 2, \cdots, M) \tag{5}$$

Step 4: Select the expected error and determine the network structure; the training network has to go through several iterations to meet the error requirement. If it is necessary to further reduce the error, the competition layer node needs to be increased, but it will extend the training time.

Step 5: To determine whether the classification is correct, adjust the weight vector of the winning neuron according to the following rules (L_{w_c} denotes the class associated

481

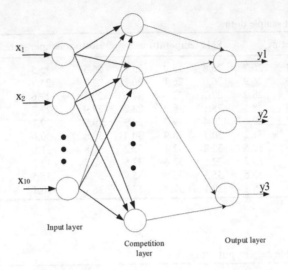

Figure 2. LVQ network structure diagram.

with the weight vector of the winning neuron, L_{w_i} denotes the class associated with the input vector).

If $Lw_c = Lw_i$, then

$$Wc(n + 1) = Wc(n) + \eta(n)[X - Wc(n)] \tag{6}$$

If $Lw_c \neq Lw_i$, then

$$Wc(n + 1) = Wc(n) - \eta(n)[X - Wc(n)] \tag{7}$$

Step 6: Adjust the learning rate $\eta(n)$, $\eta(n) = \eta(0)(1 - n/N)$; the learning rate affects the stability of the algorithm and the speed at which the weights converge. They are important parameters in the network training process (Li et al. 2012; Wang et al. 2012).

Step 7: Determine if the number of iterations exceeds N. If it is $n \leq N$, go to step 3; otherwise end the iterative process.

3.2 LVQ neural network learning algorithm

The LVQ network consists of three layers: input layer, competition layer, and output layer. As shown in Figure 2, the input layer and competition layer are fully connected, while the competition layer and linear output layer are partially connected (Lv et al. 2014; Zhao et al. 2017). As the LVQ network structure diagram shown in Figure 2, the input variables **X** are the top temperature of the case, the side temperature, the bottom temperature of the case, and the ambient temperature of the GIS device. Some data are shown in Table 1. The output vector y is the state value of the GIS isolation switch contact. There are three output elements in the output vector, which represent three categories: normal, hot, and overheat, as shown in Table 2.

The process of classification and recognition of the temperature state of GIS isolation switch contact by LVQ neural network is shown in Figure 3.

4 EXPERIMENTAL RESULTS AND ANALYSIS

This article mainly studies the classification effect of the LVQ neural network. The LVQ neural network of GIS conductor contact temperature mapping is composed of the input

Table 1. Partial input sample data.

Number	Top temperature[1]			Side temperature			Bottom temperature		Round temperature	
1	26.9	26.8	26.7	25.8	25.9	25.8	25.5	25.5	26.6	26.7
2	27.1	27.0	26.9	26.0	26.1	26.1	25.7	25.7	26.6	26.8
3	28.1	27.9	28.2	27.7	27.9	27.6	27.5	27.6	26.6	26.7
4	28.7	28.5	28.5	28.1	28.2	28.2	27.9	27.9	26.6	26.7
5	29.6	29.7	29.6	29.4	29.3	29.3	29.2	29.1	26.6	26.7
6	31.1	31.1	31.2	30.7	30.9	30.8	30.6	30.6	26.6	26.7
7	32.5	32.6	32.5	32.2	32.1	32.2	31.8	31.9	26.6	26.7
8	33.1	33.1	33.2	32.5	32.6	32.5	32.2	32.2	26.6	26.7
9	36.4	36.5	36.5	35.8	35.8	35.9	35.4	35.5	26.6	26.7
10	38.1	37.9	38.1	37.1	37.2	37.1	36.1	36.1	26.6	26.7

Table 2. Output category.

Serial number	1	2	3
Category	Normal	Heat	Overheat

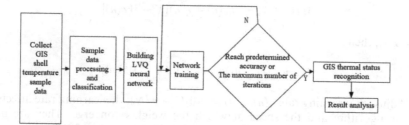

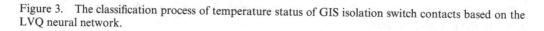

Figure 3. The classification process of temperature status of GIS isolation switch contacts based on the LVQ neural network.

layer, competition layer and output layer. Eighty sets of data obtained from the experiment were randomly selected from 95 groups as training samples, and the remaining 15 groups were used as test samples. Among them, too many or too few neurons in the competition layer will cause the performance of the network to decline, so the number of nodes in the competition layer should be determined comprehensively considering the time and error of network training. In order to determine the number of nodes, the experiment selects different numbers of nodes from competition layers for training.

In the process of eigenvalue selection and model generation, in order to maximize the network predictive performance, the expected error is set to 0.01, the iteration number is 100, the learning rate is 0.01, the training interval is 10, and the classification effect accuracy is high. For reasons of time and precision, the performance is best when the number of neurons in the competition layer is preferably 10, as shown in Figure 4.

---

1 All temperatures are in °C.

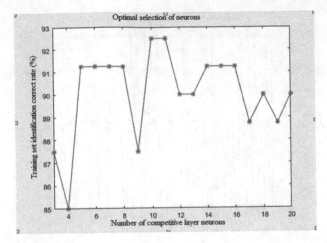

Figure 4. Relationship between number of neurons and LVQ network performance.

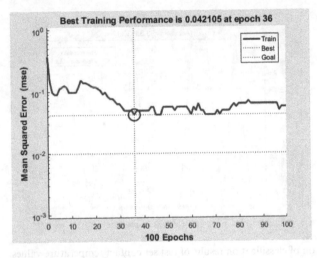

Figure 5. LVQ classification accuracy curve.

For classification using an LVQ neural network, the accuracy curve of the classification results is shown in Figure 5.

In the experiment, 10 neurons were used in the competition layer, the number of trainings was 100, and the weight vector of the training sample before and after training is different. Through experiments, it can be seen that the number of weights before and after training is different. There is only one weight before training, and there are three weights after training. This is the process of neural network learning. As can be seen from the classification accuracy curve, the optimal number of iterations is 36. The training samples are shown in Figure 6, and then the test samples are classified. The simulation results of the test samples are shown in Figure 7:

The aforementioned experiments show that the LVQ neural network has higher classification accuracy, and its training accuracy rate reaches 96.25%, which can classify the contact state of the GIS isolation switch, and the training rate is faster and the stability is better. The training test results are shown in Table 3.

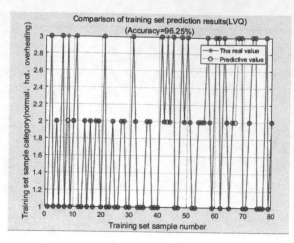

Figure 6.   Comparison of classification results of training set contact temperature values.

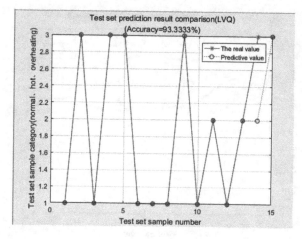

Figure 7.   Comparison of classification results of test set contact temperature values.

Table 3.   GIS isolation switch contact temperature state classification results.

	Normal (1)	Heat (2)	Overheat (3)	Correct rate (%)
Training set sample	35	24	21	96.25
Result	35	24	18	
Test set sample	7	2	6	93.33
Result	7	2	5	
Total set	42	26	27	

5 SUMMARY

The method of discriminating the thermal state of the GIS isolating switch contacts based on the manual method is subjective and has a large error. In this article, a GIS isolation switch contact temperature classification study based on a real-time temperature acquisition system is carried out. It studies the classification and identification of the temperature of a GIS isolation switch in normal, hot, and overheated states, which plays the role of equipment maintenance and accident prevention. It can be seen from the simulation results that the LVQ neural network has a fast convergence speed, strong approximation ability, and high classification accuracy, which can classify and recognize the temperature state of the GIS isolation switch contacts. An LVQ neural network is helpful for identifying and classifying the temperature state of the GIS isolation switch contacts, and thus plays a key role in the stable operation of the power grid.

REFERENCES

Bao, H., Kang, Q.S., & Zhou, M. 2011. A flame recognition algorithm based on LVQ neural network and image processing. *China Safety Science Journal* 21(6): 60–64.

Chen, Q., Li, Q.M., Cong, H.X., et al. 2015. On-line temperature monitoring for GIS disconnecting switch contacts based on multipoint-distributed fiber Bragg grating. *Transactions of China Electrotechnical Society* 30(12): 298–306.

Chen, R.G., Feng, X.Y., Gu, C., et al. 2015. Application of infrared thermometry technology for the fault diagnosis in GIS. *High Voltage Apparatus* 51(9): 190–194.

Cong, H.X., Li, Q.M., Qi B., et al. 2014. Online GIS switch contact temperature monitoring based on IR sensing. *Electric Power Automation Equipment* 34(3): 144–148.

Cong, H.X., Xing, J.Y., &Li, Q.M. 2014. Infrared online temperature monitoring technology for shielded GIS contacts. *Electric Power Automation Equipment* 34(9): 148–153.

Li, P.F., Yang, N., & Jing, J.F. 2012. Fabric defect classification based on radial basis function neural network. *Computer Measurement & Control* 20(10): 2751–2755.

Li, X., Wang, B.J., & Wang, X.Q. 2012. Research progress on temperature monitoring of GIS bus. *High Voltage Apparatus* 48(11): 130–138.

Li, Z.X. & Song, J.C. 2009. On-line temperature measurement system for contacts in HV switchgear. *High Voltage Apparatus* 45(2): 11–13.

Lv, F.C., & Zhang, B. 2014. Application of the LVQ neutral network to the recognition of GIS discharge types. *Electrical Measurement& Instrumentation* 51(18): 112–115.

Sun, G.X., Guan, X.Y., & Shu, N.Q. 2015. GIS conductor temperature inspection method and device based on fiber Bragg grating. *Diangong Jishu Xuebao/Transactions of China Electrotechnical Society* 30 (8): 316–321.

Wang, B., Du, X.X., & Jin, M. 2012. Application of breast tumor diagnosis based on learning vector quantization neural network. *Computer Simulation* 29(8): 171–174.

Wang, B.R., & Wu, Z.Q. 2018. Identification of radioactive nuclear elements using LVQ neural network. *Nuclear Electronics & Detection Technology* 38(4): 1965–1972.

Wang, E., Zhao, Z.G., Cao, M., et al. 2017. Multi point temperature monitoring of oil immersed transformer based on fiber Bragg grating. *High Voltage Engineering* 43(5): 1543–1549.

Wang, Y., Zhao, K., & Liu, J.P. 2016. Optical fiber Bragg grating temperature monitoring based on volume phase grating dispersion demodulation. *Laser & Optoelectronics Progress* 53(10): 113–118.

Wu, X.W., Shu, N.Q. Li, H.T., et al. 2013. Online temperature monitoring system based on FBG for GIS bus. *Electric Power Automation Equipment* 33(4): 155–160.

Xie, Z.Y., Jin, X.C., Cheng, D.P., et al. 2012. Application of fiber Bragg grating to temperature on-line monitoring of GIS busbar. *Engineering Journal of Wuhan University* 45(5): 658–661.

Yuan, H.X., Sun, Q., & Yu, X.Y. 2011. LVQ neural network apply in three dimensional object recognition. *Journal of Harbin University of Science & Technology* 16(5): 124–133.

Zhao, D.Z., Shu, N.Q., & Zhang, Z.L. 2011. On-line temperature monitoring system of contactor in GIS based on FBG sensor network. *Engineering Journal of Wuhan University* 44(4): 538–541.

Zhao, X.G., Wang, X. 2017. Tomato seed varieties recognition based on principal component analysis and LVQ neural network. *Acta Agriculturae Zhejiangensis* 29(8): 1375–1383.

Emerging Developments in the Power and Energy Industry – Dufo-López, Krzywanski & Singh (eds)
© 2020 Taylor & Francis Group, London, ISBN 978-0-367-27169-5

Reliability analysis of distribution network considering wind and photovoltaic power

Jian Liang
Shandong Luneng Intelligence Technology Co. LTD, Jinan, China

Xinrui Li
Hunan Changsha No.1 Middle School, Changsha, China

Song Zhang, Wensheng Li & Zhenhai Sun
State Grid Qingdao Electric Power Company, Qingdao, China

ABSTRACT: Distributed generation has the advantages of close to load center, flexible control and friendly environment. But its output needs attention for its random fluctuation and uncertainty. In this paper, the output characteristics of wind and photovoltaic power are mathematically modeled and the output probability density distribution function is calculated. The reliability of access to distributed generation is analyzed by an example. Results show that distributed generation can effectively reduce the outage time of distribution network and play a positive role in improving the reliability of power supply.

1 INTRODUCTION

Distributed generation(DG) generally has two operating states. It can run in parallel with the power distribution system, or in an island way. Compared with the traditional power generation method, DG is generally located close to the power users, which saves the cost of distribution network facilities and reduces the line loss. However, the DG power has a large upper and lower limit, which can reach between several kilowatts and dozens of megawatts (Thomas et al, 2001).

Wind power and photovoltaic power account for a large proportion of DG, so it is cleaner and more environmentally friendly than traditional generation. (Dan et al, 2006) took DG as backup in distribution network, and proposed the reliability analysis process of distribution network with DG based on analytic method. The calculation example shows that DG in grid-connected operation improves system reliability. (Yasser et al, 2009; 2010) calculated the probability distribution of renewable energy, and put forward a corresponding island division method considering probability characteristics.

Now more and more research on grid reliability has taken the energy storage into account. By studying the proportional distribution of renewable energy power and energy storage capacity allocation, the calculation method for the capacity ratio is proposed to meet the requirement of stabilizing the fluctuation of renewable energy output (Jun et al, 2014; Qionghui et al, 2017). Then the corresponding charging and discharging control strategy is developed (Shaobo et al, 2013; Ya et al, 2018).

Although wind and photovoltaic power are generally considered to have a certain impact on the stability of the power grid, changes in system reliability after their connection are rarely reported. In this paper, DG and distribution network were analyzed with mathematical model. The reliability problem of the distribution network with wind and solar energy was explored in detail.

2 PROBABILITY DISTRIBUTION CHARACTERISTICS OF DG OUTPUT AND LOAD IN DISTRIBUTION NETWORK

2.1 Model of wind power output

Complex wind speed changes are simulated by Weibull distribution (Shin et al, 2009), which are widely used and mature at present. The probability density function of wind speed is:

$$f(v) = \frac{k}{c} \cdot \left(\frac{v}{c}\right)^{k-1} \exp\left[-\left(\frac{v}{c}\right)^k\right] \tag{1}$$

Where, v represents the real-time wind speed, and k and c represent the shape and scale parameters respectively. The output of wind turbine has a certain function relation with the instantaneous wind speed (Hugo et al, 2008):

$$P_w = \begin{cases} 0 & v \leq v_{ci} \\ k_1 v + k_2 & v_{ci} \leq v \leq v_r \\ P_r & v_r \leq v \leq v_{co} \\ 0 & v_{co} \leq v \end{cases} \tag{2}$$

Where, P_r is rated output of wind turbine, v_r is rated wind speed, and v_{ci} is cut-in wind speed. For the wind speed is between v_{ci} and v_r most of the time, the formula is simplified as follows:

$$f(P_w) = \frac{k}{k_1 c} \cdot \left(\frac{P_w - k_2}{k_1 c}\right)^{k-1} \exp\left[-\left(\frac{P_w - k_2}{k_1 c}\right)^k\right] \tag{3}$$

2.2 Model of photovoltaic power output

According to the long-term statistical records of photovoltaic power plants, the probability density function of solar light intensity within several hours is the same as the Beta function, which can be expressed as follows:

$$f(r) = \frac{\Gamma(\alpha + \beta)}{\Gamma(\alpha)\Gamma(\beta)} \cdot \left(\frac{r}{r_{max}}\right)^{\alpha-1} \cdot \left(1 - \frac{r}{r_{max}}\right)^{\beta-1} \tag{4}$$

Where r and r_{max} are the actual illumination intensity and the maximal illumination intensity in this time period respectively, and α and β are the shape parameters of Beta distribution function. The approximate relationship between light intensity and photovoltaic power P_M is as follows:

$$P_M = r \cdot A \cdot \eta \tag{5}$$

Where M is the total number of battery packs in operation, A is the total area of all solar panels, and η is the photoelectric conversion efficiency of. According to the approximate relationship between photovoltaic power and light intensity, the probability density function of photovoltaic power is solved by the probability density function of light intensity, which is expressed as:

$$f(P_M) = \frac{\Gamma(\alpha + \beta)}{\Gamma(\alpha)\Gamma(\beta)} \cdot \left(\frac{P_M}{R_M}\right)^{\alpha-1} \cdot \left(1 - \frac{P_M}{R_M}\right)^{\beta-1} \tag{6}$$

Where, R_M is the maximum output power, $R_M = A \times \eta \times r_{max}$.

2.3 Load power demand model and its probability density function

In the distribution network system, there are various time-varying loads on system power side. In this paper, it is assumed that the changes on the load side can be represented by normal distribution, which is calculated as follows:

$$f(P) = \frac{1}{\sqrt{2\pi}\sigma_p} \cdot \exp\left(-\frac{(P - \mu_p)}{2\sigma_p^2}\right)^2 \tag{7}$$

Where, μ_p is the mathematical expectation value of distribution network load, and σ is the standard deviation. These two values can be calculated when the long-term load value of the demand side of distribution network is known.

According to the semi-invariant additivity, the probability density functions of wind power output, photoelectric power output and load power demand are transformed. The values of each order moment can be calculated respectively to obtain the semi-invariant variables, according to the power density function at the node where each power supply and power supply path are located. According to the functional relationship between the correlation coefficient of the Gram-Charlie series expansion and the semi-invariant value of the variable distribution, the normal power supply probability p can be calculated.

The probability distribution of wind power generation output is represented by the three-parameter Weibull distribution function as follows:

$$f(x) = \frac{k}{c}\left[\left(\frac{x-a}{c}\right)^{k-1} \exp\left[-\left(\frac{x-a}{c}\right)^k\right]\right] \tag{8}$$

$$\varphi(t) = \sum_{r=0}^{\infty}\left\{\sum_{i=0}^{r}\binom{r}{i}c^i a^{r-i}\Gamma(1+\frac{i}{k})\right\}\frac{(it)^r}{r!} \tag{9}$$

$$\alpha_r = \sum_{i=0}^{r}\binom{r}{i}c^i a^{r-i}\Gamma(1+\frac{i}{k}) \tag{10}$$

Where, equation (8) represents the specific Weibull distribution function calculated by the fitting method. Equation (9) is the characterization formula of Weibull distribution characteristic function. Equation (10) is the r-order moment of the probability density function of wind power output calculated according to the relationship between characteristic function and moment.

3 OUTAGE TIME CALCULATION OF DISTRIBUTION NETWORK WITH DG

Through the previous analysis, the reliability indexes of each evaluation area can be calculated. The outage time U_{sj} is considered in this paper, which is expressed as follows:

$$U_{sj} = \sum_{i}^{N_a}\alpha_i\lambda_i[(1-p_i)\gamma_i + p_i\gamma_{DG}] + \sum_{i}^{N_a}\lambda_{ai}^{''}[(1-p_i)\gamma_{ai} + p_i\gamma_{DG}] \tag{11}$$

In Equation (11), the first term indicates the outage time of class A fault in evaluation area. The second indicates the outage time in evaluation area when the system is still not operational before the completion of maintenance. Class B and class C faults have little effect on the outage time, so they are not considered.

α_i is fault probability of class A, λ_i is the outage rate of fault area i, λ''_{ai} is the planned maintenance rate. γ_{DG} is the DG switching time, γ''_{ai} represents the recover time from the scheduled maintenance in area i where class A fault occurs.

4 CASE STUDY

4.1 *Distribution network parameters with DG*

The entire distribution system is shown in Figure 1, where the PG&E69 node distribution system with DG is adopted. Where B is the circuit breaker, S is the isolation switch, F is the fuse, SK is the contact switch. The DG applied is shown in Table 1.

The parameters of the wind power and photovoltaic power are obtained according to the historical data, so as to obtain the values of k, c, and α, β. The whole distribution network is

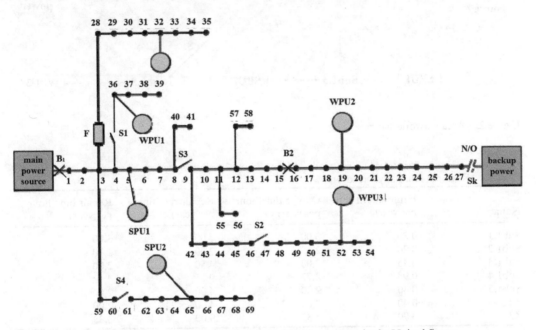

Figure 1.   Feeder wiring diagram of PG&E69 node distribution system in the United States.

Table 1.   Detailed parameter table of DG in distribution network.

Name	Power source	Maximum output(kW)	Total number of photovoltaic battery packs	Photoelectric conversion efficiency
WPU1	wind power	50		
SPU1	photovoltaic power	25	216	0.1344
WPU2	wind power	400		
WPU3	wind power	1300		
WPU4	wind power	40		
SPU2	photovoltaic power	10	216	0.1344

divided into regions by using the feeder partition method. The network diagram after the partition is shown in Figure 2. According to the formula, the reliability index parameters in each region are obtained, and the results are shown in Table 2.

4.2 *Reliability index calculation results*

By calculation, the outage time of the region changes with DG is shown in Table 3. After DG is configured, the average annual outage time of the evaluation area is significantly reduced. The outage time is about 3.68 hours per year without DG. When DG is added in the system,

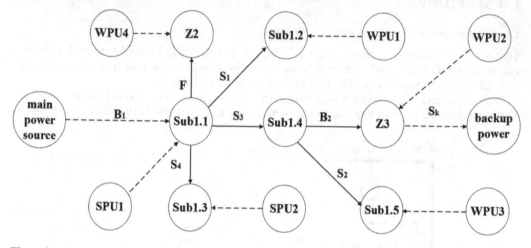

Figure 2.   Area network diagram.

Table 2.   Regional reliability indicators.

Name	Fault rate(time per year)	Outage time(hour per year)	Repair rate(time per year)	Repair time(hour per year)
Sub1.1	0.80	3.00	8.00	3.00
Sub1.2	0.20	2.00	2.00	2.00
Sub1.3	0.45	2.00	4.50	2.00
Sub1.4	0.98	2.72	9.75	2.72
Sub1.5	0.40	2.00	4.00	2.00
Z2	0.40	2.00	4.00	2.00
Z3	0.90	3.33	9.00	3.33

Table 3.   Regional reliability index changes with DG.

Name	Average outage time (hour per year)	
	Without DG	With DG
Sub1.1	3.4472	3.4412
Sub1.2	2.9381	2.9232
Sub1.3	3.9712	1.6745
Sub1.4	5.5301	4.1201
Sub1.5	2.9693	1.8391
Z2	3.8602	2.5657
Z3	3.0599	2.6310

the outage time declines by 25.54%, to 2.74 hours per year. Compared with the traditional distribution network system, the average outage time of the user side is significantly decreased, the normal power supply rate is improved, and the power quality of the system is improved significantly. Besides, Sub1.1, Sub1.4 and Z3 show significant hidden dangers in terms of fault rate and repair rate. However, after the addition of distributed power supply, the outage time has not been significantly reduced. Because the network in this area is relatively complex, the improvement effect of distributed power supply is not obvious.

5 CONCLUSIONS

In this paper, the output characteristics of wind and photovoltaic power are studied, and the mathematical model is carried out. Based on the historical data of wind and photovoltaic power, the output probability density distribution function is obtained. The distribution network is partitioned by an example, and the system reliability changes before and after adding distributed power supply are compared. The results are as follows:

(1) The addition of distributed power supply can effectively reduce the shutdown time of the system by 35.33%, and plays a positive role in improving the reliability of power supply.
(2) The distributed power supply is added to the area with independent power supply, and the improvement effect is more obvious.

ACKNOWLEDGMENTS

This work was supported by Science & Technology projects of State Grid Shandong Electric Power Company (2018A-037).

REFERENCES

Dan Z, Robert P.B., Kwa-Sur T, Rich S. et al. Impact of DG Placement on Reliability and Efficiency with Time-varying Loads. *IEEE Transactions on Power Systems*, 2006, 21(1): 419-427.

Hugo M, Peter K, Marilio C, et al. VPP Operating in the Isolated Grid. 2008, *IEEE.* "1-6.

Jun X, Zequn Z, Pan Z, et al. A capacity optimization method of hybrid energy storage system for optimizing tie-line power in micro grids. *Automation of Electric Power Systems*, 2014, 38(12): 19-26.

Qionghui L, Caixia W, Jing Z, et al. A roadmap for large scale energy storage for grid-level applications. *Energy Storage Science and Technology*, 2017, 6(1): 141-146.

Shaobo L, Minxiao H, Guopeng Z, et al. Capacity allocation of energy storage in distributed photovoltaic power system based on stochastic prediction error. *Proceedings of the CSEE*, 2013, 33(4): 25-33, 5.

Shin Y.O., Abeer G.E., Compound micro-grid installation operation planning of APEFC and photovoltaic with prediction of electricity Production using GA and numerical weather information. *International journal of hydrogen energy*34(2009): 8213-8222.

Thomas.A, Göran.A, Lennart.S. Distributed Generation: a Definition. *Electric Power System Research*, 2001, 57(3): 195-204.

Yasser M.A., Ehab F. Reliability Evaluation for Distribution System with Renewable Distributed Generation during Islanded Mode of Operation. *IEEE Transactions on Power Systems*, 2009, 24(2): 572-581.

Yasser M.A., Ehab F., Anne C. G. Supply Adequacy Assessment of Distribution System Including Wind-based DG During Different Modes of Operation. *IEEE Transactions on Power Systems*, 2010, 25(1): 78-86.

Ya M, Xinran L, Shujuan L, et al. Variable power control strategy of battery energy storage system participating in distribution network peak load shifting. *Electric Power Construction*, 2018, 39(4): 45-50.

Emerging Developments in the Power and Energy Industry – Dufo-López, Krzywanski & Singh (eds)
© *2020 Taylor & Francis Group, London, ISBN 978-0-367-27169-5*

Prediction of dissolved gas concentration in power transformer oil based on the multivariate ReLu deep belief networks model

Lei Su, Lu Chen & Peng Xu
State Grid Shanghai Electric Power Research Institute, Shanghai, China

Yingjie Yan & Gehao Sheng
Shanghai Jiaotong University, Shanghai, China

ABSTRACT: The development of the data analysis method of deep learning provides a new solution and technical means for the state prediction of power transformers. This article proposes a prediction method of dissolved gas concentration in transformer oil based on modified deep belief networks (DBNs). First, the multivariate ReLu–DBN model is established with the multivariate dissolved gas as the input of networks. Then, the networks are pretrained with a CD-k algorithm and the network structure parameters are fine tuned with an ReLu function to weaken redundant information and speed up network training. Finally, the example data are put into the prediction model to verify the validity of the algorithm. The results show that the model proposed in this article can excavate the inherent law of the data itself and overcome the drawbacks of low stability in the traditional methods.

Keywords: deep belief networks, dissolved gas concentration, power transformer, ReLu activation function

1 INTRODUCTION

The power transformer is one of the key pieces of equipment of the power system, and its operating state is related to the reliability of the power grid. During the operation of the transformer, a small amount of gas is dissolved in the insulating oil due to aging, electric and thermal faults, etc. The content of various components of the gas and the proportional relationship between the different components are closely related to the operating condition of the transformer. Through the dissolved gas analysis (DGA), some latent faults inside the transformer and the degree of their development can be discovered (Dong 2001; Wei & Liu 2015). This technology has been proved by a large number of fault diagnosis practices and is currently recognized internationally as an effective method for diagnosing early faults in transformers. Predicting the development trend of dissolved gas concentration in oil can provide an important basis for transformer state assessment, which has important practical significance (Wu 2015; Xu 2016).

Many scholars in China and abroad have performed a great deal of research on the prediction of dissolved gas concentration in oil. Generally, the traditional methods are gray model (GM), artificial neural network (ANN), support vector machine (SVM), and other methods (Luo 2001; Yang 2008; Bian 2012; Fu 2013). The traditional method often only considers the development trend of a certain gas in modeling, ignoring the correlation analysis between gases, making the prediction effect not scientific enough and the stability is poor. In order to solve this problem, Xiao (2006), Sima (2012), and Lin (2016) introduced the gray relational calculation, taking the correlation between the components into account, using the gray correlation analysis to predict the correlation of the input variables before the prediction, and eliminating the weaker correlation factors. Then the above prediction method was used to model the calculation. In addition, Tang (2013) improved the standardized mutual information variable selection algorithm for the selection of input variables, and introduced conditional mutual information to replace the original mutual information items in the feature selection evaluation standard, which improved the selection validity of input variables to some extent. Although the gray correlation and mutual information technology avoids the defects of single-component gas concentration prediction, the calculation amount is increased in the data preprocessing process, and the correlation threshold has a certain subjectivity. With the development of deep learning methods, the feature extraction ability of a deep belief network (DBN) shows certain characteristics in transformer fault classification modeling, pattern recognition, time series prediction, etc. Shi et al. (2016) provide new ideas for data processing in the power sector.

In conclusion, this article proposes a method for predicting dissolved gas concentration in oil based on the multivate ReLu–DBN model. Using DBN's powerful feature learning ability, a multihidden layer machine learning model is constructed to automatically extract the correlation characteristics between various gas components and environmental factors (temperature, oil temperature) for gas concentration prediction. It not only solves the problem of poor stability of single-variable prediction, but also avoids a large amount of preprocessing calculation and minimizes the process of manual intervention. The method uses a large amount of sample data on the engineering site to train and learn, and extracts the information characteristics and potential statistical laws under different changing trends. The experimental results show that the method has higher prediction accuracy.

2 FUNDAMENTALS OF THE DEEP BELIEF NETWORK

The deep belief network was proposed by G. Hinton in 2006. It is a generation model consisting of a multilayer restricted Boltzmann machine (RBM) for feature recognition, data dimensionality reduction, and classification prediction. The performance is outstanding, and it is widely used in image processing, speech recognition, and other fields (Bengio 2009; Yu 2014; Feng 2016). This article uses its automatic extraction of training data feature rules to achieve a prediction function.

The RBM is a probability map model that can be interpreted by a random neural network (Smolensky 1986). The visible layer v is composed of a spheroid v_i for inputting training data, and the hidden layer h is composed of a hidden element h_j for use as a feature detector. Figure 1 shows the RBM network structure. There is no connection inside one RBM layer but the layers are fully connected (Salakhutdinov & Hinton 2012).

RBM is an energy-based model. For a given set of states (v, h), the joint configuration energy is defined as follows:

$$E_\theta(v, h) = -\sum_{i=1}^{n_v} a_i v_i - \sum_{j=1}^{n_h} b_j h_j - \sum_{i=1}^{n_v} \sum_{j=1}^{n_h} h_j w_{j,i} v_i \tag{1}$$

where a_i and b_j are the offsets of the explicit element i and the hidden element j; $w_{j,i}$ is the connection weight between the explicit element and the hidden element; and $\theta = \{w_{j,i}, a_i, b_j\}$ is the model parameter.

Based on the energy function, the joint probability distribution of the state (v, h) is

$$P_\theta(v, h) = \frac{1}{Z_\theta} e^{-E_\theta(v,h)} \tag{2}$$

where $Z_\theta = \sum_{v,h} e^{-E_\theta(v,h)}$ is a normalization factor, also known as the partition function.

For the distribution $P_\theta(v)$ of a set of observations v, the edge distribution $P_\theta(v, h)$ of the joint probability distribution is also called the likelihood function and is defined as

$$P_\theta(v) = \sum_h P_\theta(v, h) = \frac{1}{Z_\theta} \sum_h e^{-E_\theta(v, h)} \tag{3}$$

Similarly, Equation (4) can be obtained:

$$P_\theta(h) = \sum_v P_\theta(v, h) = \frac{1}{Z_\theta} \sum_v e^{-E_\theta(v, h)} \tag{4}$$

According to the special network structure of RBM, the state and activation conditions of the explicit and hidden elements are independent. The activation probability of the jth hidden element is

$$P(h_j = 1|v) = \text{sigmoid}(b_j + \sum_{i=1}^{n_v} w_{j,i} v_i) \tag{5}$$

The same ith element activation probability is

$$P(v_i = 1|h) = \text{sigmoid}(a_i + \sum_{j=1}^{n_h} w_{i,j} h_j) \tag{6}$$

where $\text{sigmoid}(x) = 1/(1 + e^{(-x)})$ is the activation function of the neural network, mapping the variable x between 0 and 1.

After a training sample is given, the RBM is trained to adjust the parameter θ such that the probability distribution of the RBM representation under the control of the parameter is as close as possible to the distribution of the training data, and the likelihood of training RBM can be maximized as follows.

$$\ln L_{\theta, S} = \ln \prod_{i=1}^{n_s} P(v^i) = \sum_{i=1}^{n_s} \ln P(v^i) \tag{7}$$

where S is the training sample set and n_s is the number of training samples.

To find the maximum log-likelihood function represented by Equation (7), a stochastic gradient ascending method is usually used; see Equation (8).

$$\frac{\partial \ln L_{\theta, S}}{\partial \theta} = \frac{\partial \ln P(v)}{\partial \theta} = -\left\langle \frac{\partial E(v, h)}{\partial \theta} \right\rangle_{P(h|v)} + \left\langle \frac{\partial E(v, h)}{\partial \theta} \right\rangle_{P(v, h)} \tag{8}$$

where $\langle \rangle_P$ represents the mathematical expectation of the distribution P.

Because of the existence of the normalization factor Z_θ, the joint probability distribution P is more complicated. Hinton (2002) proposed a contrast divergence (CD) algorithm. In the training process, first map the explicit vector to the hidden element, then reconstruct the explicit vector with the hidden element, and then map the explicit vector. To the hidden element, repeat the above steps k times, and finally realize the rapid training and learning of RBM.

3 THE MULTIVARIATE RELU–DBN MODEL OF DISSOLVED GAS CONCENTRATION

3.1 *Dissolved gas concentration as model input*

The transformer under normal operation will decompose a very small amount of gas due to aging cracking of insulating oil and solid insulation, mainly hydrogen (H_2), methane (CH_4), ethane (C_2H_6), ethylene (C_2H_4), acetylene (C_2H_2), carbon monoxide (CO), and carbon dioxide (CO_2). When an abnormality occurs inside the transformer, the content of some gases will increase rapidly. For example, when the insulating oil is overheated, CH_4 and C_2H_4 are the main increased gas components, and show a strong correlation; H_2 and C_2H_2 levels increase at high energy discharge and show a strong correlation. Therefore, when predicting the concentration of dissolved gases in transformer oil, the interaction between gas components should be considered. According to the relevant data, the statistics of multiple faulty transformers indicate that the superheat fault accounts for 63%, and there is a certain correspondence between temperature and dissolved gas content. Therefore, the ambient temperature and oil temperature are taken as the correlation factors.

The single-variable sequence of dissolved gas concentration in transformer oil can be expressed as $X^m = \{x_{t-k}^m, x_{t-k+1}^m, \cdots, x_t^m, x_{t+1}^m \cdots\}$, where x_t^m represents the measured value of the mth gas at time t. Considering the correlation of each component and environmental factors, the sequence can be expressed as $X = \{X^1, X^2, \ldots\ldots, X^D\}$. D represents the number of related variables, so the gas concentration prediction can be regarded as a prediction problem with high multivariate complexity. If $x_{t+\tau}$ is related to its top K data, the prediction task can be described as $f(x_{t-K}, \ldots, x_{t-1}, \theta)$, where θ is the parameter vector in the f model. The predicted value is $f(x_{t-K}, \ldots, x_{t-1}, \theta)$, and the cost function is as shown in Equation (9).

$$J(\theta) = \frac{1}{n} \sum_{t=1}^{n} (x_{t+\tau} - \hat{x}_{t+\tau})^2 \tag{9}$$

Finding the optimal for the model f_θ minimizes the cost function J, and the predicted value is as close as possible to the actual value to achieve the prediction effect.

3.2 *Prediction model based on the multivariate ReLu–DBN*

The DBN consists of a multilayered RBM network and the last layer of logistic regression. The visible layer neurons are dissolved gas concentrations and oil temperature and ambient temperature vectors in seven kinds of oils such as H_2 and CH_4, wherein each state vector is a K-dimensional sequence.

The modeling and prediction steps of the dissolved gas concentration prediction method in transformer oil based on ReLu–DBN are as follows:

Step 1: Collect the chromatographic data of the transformer oil; select the characteristic parameters H_2, CH_4, C_2H_6, C_2H_4, C_2H_2, CO, CO_2 concentration, ambient temperature and oil temperature vector; and divide the training samples and test samples. The normalized method maps to [0–1], where the transfer function is, where x_{min} is the sample data minimum, and x_{max} is the sample data maximum.

Step 2: Pretraining of ReLu–DBN. First, the RBM is trained layer by layer without supervision, and the hidden feature information of the data is deeply excavated. In each training procedure, only one layer of RBM is trained. After the training, the lower layer of RBM is trained and stacked. Then, combined with the label of the sample, the backpropagation algorithm is used for supervised optimization, so as to achieve the effect of classification. The pretraining steps are shown in Table 1.

Step 3: Fine tuning. Sample data with labels are added to the top layer of the network, and supervised fine-tuning is carried out by back propagation algorithm. The commonly used

Table 1. Pretraining steps.

Pretraining steps	Context
Initialization	To a given training sample, set the training period *epoch* and learn the rate γ. Specify the number of visible layers n_v and hidden layer units n_h. Initialize the weight matrix W and offset vector a, b.
Training each layer of RBM using the CD algorithm	The training samples x are assigned to the display layer $v^{(0)}$, and the probability $P(h_j^{(0)} = 1|v^{(0)})$ that the corresponding hidden element is activated is calculated according to Equation (5). Extract a sample $h^{(0)} \sim (P(h^{(0)}|v^{(0)})$ from the calculated probability distribution and reconstruct the layer $P(v_i^{(1)} = 1|h^{(0)})$ with $h^{(0)}$. Extract a sample $v^{(1)} \sim (P(v^{(1)}|h^{(0)})$ from the reconstructed layer to calculate the probability $P(h_j^{(1)} = 1|v^{(1)})$ that the hidden element $v^{(1)}$ is activated. Weights are updated using the following equations: $W \leftarrow W + \gamma[(P(h^{(0)} = 1|v^{(0)})v^{(0)\mathrm{T}} - P(h^{(1)} = 1|v^{(1)})v^{(1)\mathrm{T}}]$, $a \leftarrow a + \gamma[v^{(0)} - v^{(1)}]$ and $b \leftarrow b + \gamma[P(h^{(0)} = 1|v^{(0)}) - P(h^{(1)} = 1|v^{(1)})]$.
Fine tuning	For the output layer data, according to the cost function equation (1), the gradient descent method is used to backpropagate to each layer of RBM, and the weighting parameters are fine-tuned.

activation functions in DBN networks are the sigmoid function and tanh function. Their derived function is shown in Equation (10) and Figure 1.

$$\mathrm{sigmoid}'(z) = \mathrm{sigmoid}(z) \cdot (1 - \mathrm{sigmoid}(z)) \tag{10}$$

$$\tan\mathrm{h}'(z) = 1 - \tanh^2(z) \tag{11}$$

In Figure 1, we can see that sigmoid function and tanh function have the characteristics of scaling in derivative and saturation value. That is, gradient error will be continuously attenuated when the error terms are propagated backward layer by layer, which will lead to the decrease of learning efficiency of the model. This article chooses the ReLU function as the activation function of the DBN network. The gradient function of the ReLU function is 1 on the positive half-axis side and 0 on the negative half-axis side. The activation function has relatively broad excitation space, and can have a unilateral suppression function. It has better convergence performance for a deep structure, so it can accelerate the training speed of the whole network.

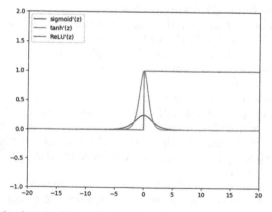

Figure 1. Derived function image of the activation function.

Step 4: The test samples are predicted using the network model obtained in Step 3. It is determined that the model input and output are both 27, and the number of hidden elements in the hidden layer is 100, that is, the network structure is 27-100-100-...-100-27. It is found experimentally that the number of network structure layers and training period have a certain relationship with the prediction accuracy, as shown in Figure 2. When the training period is fixed, the number of network layers is increased from 1 to 4 layers, and the prediction accuracy rate is greatly improved. From 4 layers to 7 layers, the accuracy improvement effect is weak. When the network structure is determined, as the training period increases, the prediction accuracy rate increases, and the trend is gradually slowing down. Based on the comprehensive model prediction effect and computational efficiency, this article determines that the network structure is 4 layers and the training period is 250.

4   CASE STUDY

This article takes the oil chromatographic data of one 220-kV transformer oil chromatography online monitoring device from October 7, 2010 to March 24, 2013 as an example. The monitoring period is one day. The 800 sets of monitoring data are used as training samples from October 7, 2010 to December 14, 2012, and 100 sets of monitoring data are used as prediction samples from December 15, 2012 to March 24, 2013.

The predictive evaluation criteria adopts the average relative percentage error, and its expression is

$$\delta = \frac{1}{N} \sum_{t=1}^{N} \left| \frac{\hat{x}_t - x_t}{x_t} \right| \times 100\% \tag{12}$$

Maximum relative percentage error, the expression is $\max \delta = \max \left| \frac{\hat{x}_t - x_t}{x_t} \right|, t = 1, 2, \ldots \ldots N$. In this equation, N is the number of test samples, $\hat{x}_t$ is the predicted value, and x_t is the measured value. The prediction results of ethylene C_2H_4 concentration are taken as an example for illustration.

It can be seen from Figure 3 that the ReLu–DBN network structure has better prediction effects and better fitting and predicting ability for data. Compared with the univariate DBN prediction results without considering the correlation factors, when the sample data appear frequently fluctuating, the multivariate ReLu–DBN model can better track the data change trend by extracting the mutual coupling relationship between the variables, and the prediction effect is more robust.

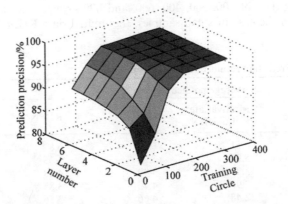

Figure 2.   Relationship of different layer numbers, training cycles, and prediction precision.

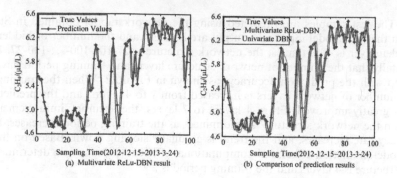

Figure 3. Prediction results of the multivariate ReLu–DBN model and comparison with univariate DBN.

Table 2. Prediction results of C_2H_4 by different methods.

Date	Future values (μL/L)	Multivariate ReLu–DBN (μL/L)	GM (μL/L)	RBFNN (μL/L)	SVM (μL/L)
12/15/2012	5.53	5.549	5.717	5.618	5.653
12/31/2012	4.66	4.705	4.805	4.759	4.772
1/15/2013	5.02	4.917	4.753	4.820	4.798
1/31/2013	6.01	5.639	5.499	5.530	5.511
2/15/2013	5.93	5.629	5.429	5.473	5.444
2/28/2013	6.49	6.291	6.233	6.267	6.244
3/15/2013	6.35	6.328	6.327	6.325	6.329
3/24/2013	6.00	5.942	5.916	5.935	5.921
Average prediction error %	—	2.480	4.660	3.560	3.890
Max prediction error %	—	8.590	11.98	10.01	10.04

The results of the prediction method proposed in this article with the results of the gray model (GM), the radial basis function neural network (RBFNN), and the support vector machine (SVM) prediction method are compared with results shown in Table 2. The average relative error of the multivariate ReLu–DBN model test is 2.48%, and the maximum test error is 8.59%. The multivariate prediction method based on the correlation factor is superior to other prediction methods in both the average prediction error and the maximum prediction error, and the accuracy is higher.

In order to analyze the impact of the training samples on the prediction results, the predictions of the training data of 100, 200, 400, 600, and 800 groups were tested. It can be seen from Table 3 that in the case of small samples, the multivariate ReLu–DBN model has the

Table 3. Prediction results of different training sets.

| Number of training samples | Average prediction error (%) | | | |
	Multivariate ReLu–DBN	GM	RBFNN	SVM
100	6.51	6.63	6.82	6.38
200	5.21	5.26	5.35	5.22
400	3.72	4.99	4.48	4.64
600	2.98	4.75	3.87	4.23
800	2.48	4.66	3.56	3.89

Table 4. Prediction results of other gases.

Dissolved gas	Average prediction error (%)			
	Multivariate ReLu–DBN	GM	RBFNN	SVM
H_2	2.32	4.83	3.17	3.76
CH_4	1.43	3.95	2.59	3.03
C_2H_6	1.57	4.11	2.98	2.93
C_2H_2	2.16	4.50	3.42	3.77
CO	2.78	5.48	4.74	5.15
CO_2	3.34	6.22	5.71	5.49

same prediction effect as GM, RBFNN, and SVM, but with the increase of the number of training samples, the trend characteristics and associated features extracted by the multivariate ReLu–DBN model are more comprehensive and detailed, and the prediction error is significantly reduced. It is better than the other three methods.

In the same way, the multivariate ReLu–DBN model is used to predict the other gas concentrations from the data from October 7, 2010 to March 24, 2013. The prediction results are shown in Table 4. It can be seen from Table 4 that the prediction average error of the multivariate ReLu–DBN method is lower than that of the GM, RBFNN, and SVM methods, and has high prediction stability and reliability.

5 CONCLUSION

In the prediction of dissolved gas concentration in transformer oil, there is a certain correlation between gas components and gas content and environmental factors. Based on this idea, this article proposes a multivariate prediction model based on the deep belief network. The model has the following characteristics:

(1) Based on the DBN theory, the model automatically extracts the feature information between the input data and statistically changes the state quantity, which has better generalization ability than the traditional prediction method and avoids introducing errors due to subjective selection of the threshold.

(2) Considering the correlation between gas content and gas composition and oil temperature and ambient temperature, it is more scientific than the single gas content prediction method, especially when the data fluctuate frequently; the fitting ability is stronger; and the prediction accuracy is higher.

(3) As the training set increases, the prediction accuracy of dissolved gas concentration in oil based on the DBN increases. However, the amount of dissolved gas concentration data in the oil is limited, and it is impossible to obtain prediction accuracy improvement under the big data platform. Therefore, the next phase will focus on the improvement and application of this article's prediction model under the power big data.

ACKNOWLEDGMENTS

This work was supported by the Science and Technology Program of Shanghai Electric Power Company, State Grid.

REFERENCES

Bengio, Y. 2009. Learning deep architectures for AI. *Foundations and Trends in Machine Learning*, 2(1): 1–127.

Bian, J., Liao, R., & Yang, L. 2012. Concentration prediction of gases dissolved in transformer oil based on weakening buffer operator and least square support vector machine. *Power System Technology*, 36 (2): 195–199 (in Chinese).

Dong, Q. 2001. *Fault Diagnosis of Power Transformer*. Beijing: China Electric Power Press, p. 37.

Feng, Y., Xiong, Q., & Shi, W. 2016. Speaker feature extraction algorithm based on restricted Boltzmann. *Chinese Journal of Scientific Instrument*, 37(2): 256–262 (in Chinese).

Fu, B. 2013. Transformer fault diagnosis of adaptive particle swarm optimization BP neural network. *Journal of Huaqiao University (Natural Science)* 34(3): 262–266.

Hinton, G. 2002. Training products of experts by minimizing contrastive divergence. *Neural Computation* 14(8): 1771–1800.

Hinton, G., & Salakhutdinov, R.R. 2006. Reducing the dimensionality of data with neural networks. *Science* 313: 504–507.

Hinton, G., Simon, O., & Yee, T. 2006. A fast learning algorithm for deep belief nets. *Neural Computation* 18 (7):1527–1554.

Lin, X., Huang, J., & Iong, W. 2016. Interval prediction of dissolved-gas concentration in transformer oil. *Electric Power Automation Equipment* 36 (4):73–77 (in Chinese).

Liu, K., Wang, P., Wang, W., et al. 2015. Electric field distribution in transformer oil under DC electric field. *Power System Technology* 39(6):1714–11718 (in Chinese).

Luo, Y., Yu, P., & Song, B. 2001. Prediction of the gas dissolved in power transformer oil by the grey model. *Proceedings of the CSEE* 21(3):65–69 (in Chinese).

Salakhutdinov, R.R., & Hinton, G. 2012. An efficient learning procedure for deep Boltzmann machines. *Neural Computation* 24(8): 1967–2006.

Shi, X., Zhu, Y., & Sa, C. 2016. Power transformer fault classifying model based on deep belief network. *Power System Protection and Control* 44(1): 71–76 (in Chinese).

Sima, L., Shu, N., Zuo, J., et al. 2012. Concentration prediction of dissolved gases in transformer oil based on grey relational analysis and fuzzy support vector machines. *Power System Protection and Control* 40 (19):41–46 (in Chinese).

Smolensky, P. 1986. *Information Processing in Dynamical Systems: Foundations of Harmony Theory*. Cambridge, MA: MIT Press, pp. 12–23.

Tang, Y., Gui, W., Peng, T., et al. 2013. Prediction method for dissolved gas concentration in transformer oil based on variable selection of mutual information. *Chinese Journal of Scientific Instrument* 34(7): 1492–1498 (in Chinese).

Wei, Z., Qi, B., Zuo, J., et al. 2015. A method to diagnose defects in oil-paper insulation of converter transformer based on image feature of partial discharge. *Power System Technology* 39(4): 1160–1166 (in Chinese).

Wu, G., Yao, M., Xin, D., et al. 2015. Experimental study on oil-impregnated paper with non-uniform thermal aging. *Power System Technology* 39(11): 3298–3304 (in Chinese).

Xiao, Y., Zhu, H., & Chen, X. 2006. Concentration prediction of dissolved gas-in-oil of a power transformer with the multivariable grey model. *Automation of Electric Power Systems* 30(13): 64–67 (in Chinese).

Xu, Z., Wang, K., & Sun, J. 2016. Research on characteristics during latent period of partial discharge developing process under direct voltage of oil-paper insulation. *Power System Technology* 40(2): 614–619 (in Chinese).

Yang, T., Liu, P., Li, Z., et al. 2008. A new combination forecasting model for concentration prediction of dissolved gases in transformer oil. *Proceedings of the CSEE* 28 (31):108–113 (in Chinese).

Yu, B., Li, S., Xu, S., et al, 2014. Deep learning: A key of stepping into the era of big data. *Journal of Engineering Studies* 6(3): 233–243 (in Chinese).

Zhang, X., Tang, J., Pan, C., et al. 2016. Research of partial discharge recognition based on deep belief nets. *Power System Technology* 40(10): 3272–3278 (in Chinese).

Zhou, X., Zhang, F., & Du, Z. 2016. A study on time series prediction model based on CRBM algorithm. *Journal of Zhejiang University (Science Edition)* 43(4): 442–451 (in Chinese).

Emerging Developments in the Power and Energy Industry – Dufo-López, Krzywanski & Singh (eds)
© 2020 Taylor & Francis Group, London, ISBN 978-0-367-27169-5

Feasibility study on reducing the grounding resistance of a transmission tower with conductive concrete

Xu Tian, Feng Pei, Xin Liu, Lulu Jia & Chenxing Deng
State Grid Jiangxi Electric Power Co., Ltd. Electric Power Research Institute, Nanchang, China

Xin Wang, Pinghao Yang & Hongbo Cheng
School of Electricity and Automation Engineering, East China Jiaotong University, Nanchang, China

ABSTRACT: Tower grounding is very important for the protection of power systems. Conductive concrete has great potential in reducing tower grounding resistance. The application schemes of conductive concrete are analyzed and simulated in this article, which concludes that the partially wrapping method weighs better both in resistance reduction and potential distribution improvement. Horizontal radial electrodes combined with conductive concrete is proposed to improve the grounding resistance under high soil resistivity. The influencing factors of the improvement effect are discussed. The results of the analysis can provide a theoretical basis and guidance for the application of conductive concrete in transmission tower grounding.

Keywords: conductive concrete, grounding resistance, transmission tower

1 INTRODUCTION

Reducing the grounding resistance of a transmission tower can effectively reduce device damage caused by lightning strikes (Zhou et al. 2015; Liang et al. 2017). With the development of a power grid, construction land for transmission lines is getting increasingly strained, and how to reduce the grounding resistance of transmission towers under high soil resistivity has become a problem (Rao et al. 2010; Lim et al. 2013; Gatta et al. 2017).

At present, methods such as adding a resistance reducing agent (Wang et al. 2017), laying an external grounding electrode (Tomaškovičová et al. 2016), laying a vertical grounding electrode (Zhou et al. 2017), and deep-hole blasting (Rong et al. 2017) are generally adopted to reduce the grounding resistance in areas with high soil resistivity. The effect of resistance reducing agents is good, but the metal corrosion is serious, which will shorten the working life of the grounding electrode. Installation of the external grounding electrode and the vertical grounding electrode is difficult, and there are special requirements of the topographic condition. Deep-hole blasting has high construction requirements, and it is easy to destroy the surroundings. It is necessary to find a new method to improve the tower grounding effect in steep areas with high resistivity.

As a new type of resistance-reducing material, conductive concrete can not only provide better resistance-reducing effects, but also protect the grounded metal (Min et al. 2017). There are many studies on the preparation of conductive concrete, but few of them are on the application of conductive concrete in the power industry (Xin 2012; Du et al. 2016). A small amount of research has concentrated on large-area grounding grids such as power substations (Zhang et al. 2014; Sassani et al. 2018). However, the application of conductive concrete in the grounding of transmission towers has not been reported yet. There is no specific analysis on the applicable conditions, applicable scope, and application methods of conductive concrete in the grounding of transmission towers.

In this article, conductive concrete is introduced to solve transmission tower grounding problems. Based on the simulation model built by CDEGS, two different schemes are put forward and studied. Influencing factors such as thickness and the resistivity of conductive concrete are also taken into account so as to provide guidance for the application of conductive concrete in the grounding of transmission towers.

2 SCHEME OF CONDUCTIVE CONCRETE TO REDUCE TOWER GROUNDING RESISTANCE

The grounding resistance of the tower is related to the diameter of the grounding conductor and the soil resistivity:

$$R = \frac{\rho}{2\pi L}\ln\frac{4L}{d} \tag{1}$$

where R is the grounding resistance, ρ represents the soil resistivity, L is the length of the grounding conductors, and d refers to the diameter of the grounding conductor.

After the grounding conductor is wrapped with conductive concrete, the resistivity ρ of the soil around the grounding conductor is reduced; on the other hand, the effective diameter d of the grounding conductor is increased, helping to reduce the grounding resistance, both of which can be seen from Equation (1).

In addition, it can be seen from Equation (1) that increasing L, the length of the grounded conductor, can reduce the grounding resistance too.

There are two schemes for covering conductive concrete, which are full laying and partial wrapping: the purpose of full laying is to fill the entire horizontal grounding grid with conductive concrete, as shown in Figure 1a; the partially wrapping is just to use a certain thickness of the concrete block to wrap each grounding conductor (including the horizontal emitter), without laying conductive concrete on any other areas of the grounding grid, as shown in Figure 1b.

The square box in Figure 1 is formed by horizontal grounding electrodes. The small boxes at the four corners refer to the tower base, which are connected to the grounding electrode by metal conductors. L_1 represents the length of the grounding electrodes.

In order to increase the equivalent length of the grounding electrodes, horizontal radial grounding electrodes are usually used. There are a variety of plans to set radial electrodes considering the local soil resistivity. Referring to engineering practice, when the soil resistivity ranges from 100 to 4000 Ω•m, five radial electrode setting schemes are shown in Figure 2, in which the radial dotted line represents the radial electrode and the symbol L2 indicates its length.

The higher the soil resistivity is, the more the electrodes will be set, which helps increase L.

3 THE SIMULATION MODEL OF APPLICATION OF CONDUCTIVE CONCRETE

CDEGS is used to simulate the schemes we proposed. There are two problems to be solved: how to build the transmission tower grounding model and how to build the conductive concrete model.

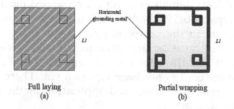

Full laying
(a)

Partial wrapping
(b)

Figure 1.   Scheme of conductive concrete application.

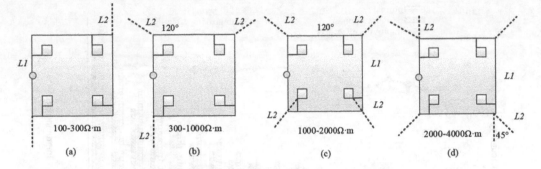

| 100-300Ω·m | 300-1000Ω·m | 1000-2000Ω·m | 2000-4000Ω·m |
| (a) | (b) | (c) | (d) |

Figure 2. Electrode setting schemes for different soil resistivities.

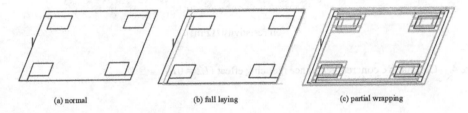

(a) normal   (b) full laying   (c) partial wrapping

Figure 3. Simulation models built by CDEGS.

Referring to Figure 2, we use the module MALZ in CDEGS to build the designed square boxes to simulate the tower base and the main grounding grid. In this model, common Φ12 mm galvanized steel is chosen as the grounding conductor material; its relative resistivity is 10 and the relative permeability is 300. The grounding grid is set to a depth of 0.8 m, and the red dots shown in Figure 2 represent the location where the 2000 A leakage current flows into the ground.

Accurate calculation of the transmission tower grounding resistance is very complicated when factors such as soil quality and moisture are taken into account. But essentially, soil resistivity is the the most obvious factor that directly influences the grounding resistance of an electrical grounding system. Similarly, although conductive concrete is a synthesis of different kinds of characteristics, when taken into the application simulation model, its resistivity is the one that we should put more focus on. We therefore choose the function "Arbitrary Heterogeneities Soil" in module MALZ to simulate conductive concrete. By modifying the size and the resistivity of the arbitrary heterogeneities of soil, any types of conductive concrete can be simulated. The electrical resistivity of the simulated conductive concrete is set to 25 Ω • m (Xin. 2012), and both the thickness and its horizontal size are set to 10 cm. The simulation models are shown in Figure 3.

4 SIMULATION RESULTS

4.1 *Application effect and scheme selection*

In order to compare the full laying scheme with the partial wrapping scheme, we measured both their resistance reduction and potential distribution improvement. To make the experiments simple, we set the length of the grounding electrode L2 as zero, and the results are shown as Figure 4.

It can be seen from Figure 4 that when the soil resistivity is low (about 100–600 Ω•m), the grounding condition is appropriate; even if conductive concrete is not used, the grounding standard can be met. But after conductive concrete is laid, the resistance of the grounding system can be reduced further. Under high soil resistivity (around 600–800 Ω•m), the laying of conductive concrete can reduce the grounding resistance to keep it from exceeding the

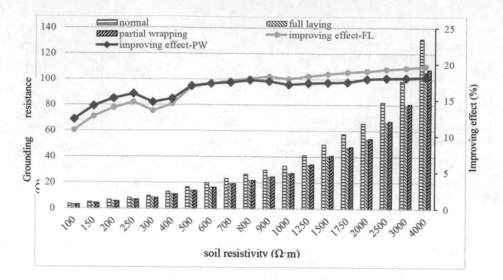

Figure 4. Conductive concrete resistance reducing effect ($L2 = 0$).

standard to meet the requirement. When the resistivity is further increased (>800 Ω•m), laying conductive concrete can reduce the grounding resistance, and the improvement can be more than 17.7%. However, the use of conductive concrete alone cannot make the grounding resistance drop down below the standard. So it is also necessary to coordinate with other measures, such as adding radial grounding electrodes.

Theoretically, when conductive concrete is applied, the potential distribution of the whole grounding grid can be improved to different degrees. The step and contact potential distribution of the grounding grid experimental results are shown in Figure 5 and Table 1.

It can be clearly seen from Table 1 that the maximum/minimum contact voltage and the maximum/minimum step voltage are significantly reduced after the application of conductive

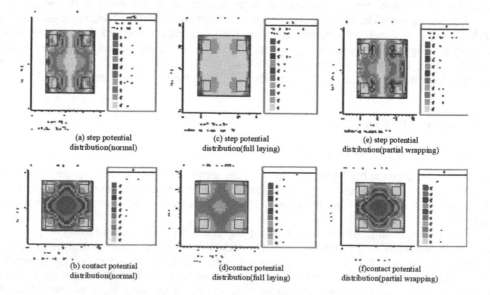

Figure 5. Step and contact potential distribution in different models.

Table 1. Voltage threshold in different models.

Voltage Model type	Contact voltage threshold (V)	Step voltage threshold (V)	Maximum contact voltage (V)	Minimum contact voltage (V)	Maximum step voltage (V)	Minimum step voltage (V)
Normal	373.7	1031.1	5700.16	377.20	1519.86	58.26
Fully laying	368.4	1009.9	1854.89	164.31	1024.28	15.19
Partially wrapping	373.7	1031.1	640.85	1.50	217.17	12.35

concrete, which means the safety of the surface can be effectively improved in grounding systems.

In Figure 5, taking (a), (c), and (e) for comparison, we can know that without conductive concrete, only the central area and four corner areas of the entire grounding grid can meet the safety threshold. But once conductive concrete is applied, whether fully or partially, almost every corner of the zone can meet the requirements of step voltage safety, except that some areas in fully laying model are still above the threshold. Figure 5b, d, and f shows that before applying conductive concrete, there is hardly any area in the whole zone under the contact voltage threshold. But after it is applied, both full laying and partial wrapping can effectively lower the contact voltage down below the threshold, in which the latter has a better performance.

Taking both resistance reduction and potential improvement into consideration, the partial wrapping method works better under the premise of consuming much less conductive concrete. So the partial wrapping method is more recommended in practical construction.

4.2 Feasibility of combined resistance reducing method

Conductive concrete does have an effect in reducing resistance, but the effect becomes limited when the soil resistivity grows over 500 Ω•m. We cannot neglect the fact that adding extra horizontal radial electrodes can also effectively improve the grounding resistance, but it takes up too much land. Hence, we suppose that when conductive concrete is combined with horizontal radial electrodes, the new combined resistance reducing method can work effectively in grounding systems without occupying too much land.

To verify our supposition, an experiment is designed as follows: (1) Perform the simulation (referring to Figure 2), keeping $L2$ still and changing the soil resistivity ρ from 500 to 4000 Ω•m; (2) take step (1) as the basis and adjust the value of $L2$ to obtain the variation of the grounding resistance under different soil resistivities and radial electrode lengths. The experiment results are shown in Figure 6.

Intuitively, the combined resistance reducing method can greatly shorten the minimum length of the radial electrode that meet the grounding requirement. And the improving effect becomes more obvious with the increase of soil resistivity. The minimum length of the radiation electrode can be shortened by 5 meters, and the minimum length of the radiation electrode can be reduced by 26.5%. That is to say, the method can effectively reduce the land occupied by grounding horizontal radial electrodes.

4.3 Influencing factors

In order to study how the thickness and the resistivity of the concrete influence the grounding resistance, first we take the 500 Ω•m tower grounding model and 25 Ω•m conductive concrete model to perform the simulation by changing the thickness of conductive concrete. The results are shown in Table 2. We thenkeep the thickness of conductive concrete at 10 cm, and change the resistivity of conductive concrete to continue the simulation. Table 3 shows the results.

506

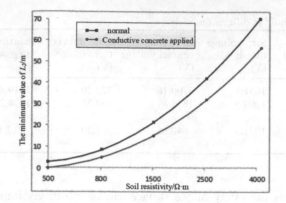

Figure 6. Comparison of the shortest electrode lengths under different soil resistivities.

Table 2. Effect of the thickness of conductive concrete.

Thickness (cm)	Grounding resistance (Ω)	Resistance reducing (%)
5	13.8135	16.53
10	13.7009	17.21
20	13.3495	19.33
30	13.3351	19.42
40	13.2622	19.86
50	13.2245	20.09
100	12.4476	24.78

Table 3. Effect of the resistivity of conductive concrete.

Resistivity (Ω·m)	Grounding resistance (Ω)	Resistance reducing (%)
0.5	13.5022	18.41
5	13.5532	18.10
10	13.5915	17.86
15	13.6286	17.64
20	13.6650	17.42
25	13.7009	17.21

From Table 2 we can see that when the thickness doubles from 10 cm to 20 cm, the resistance reducing effect increases by 2.12%. When it comes to the factor of resistivity, if we want to make it reach an effect like this, we have to drop the resistivity of conductive concrete from 25 Ω·m to less than 0.5 Ω·m. By contrast, increasing the thickness makes it easier for us to reduce the grounding resistance in practice.

5 CONCLUSIONS

In this work, we demonstrated that the conductive concrete is helpful in reducing the grounding resistance of a transmission tower. Conductive concrete can improve both the potential distribution and the grounding resistance. We put forward two conductive concrete application schemes, and demonstrated that the partial wrapping scheme is more appropriate. We found that as the soil resistivity grows, the resistance-reducing effect is better and when the soil resistivity is below 800 $\Omega \cdot m$, conductive concrete can effectively reduce the transmission tower grounding resistance. But when it is over 800 $\Omega \cdot m$, another resistance-reducing method should be added. Simulations proved the feasibility of the combined resistance-reducing method, which consists of horizontal radial electrodes and conductive concrete. With the convenience of CDEGS, we studied the influencing factors. The results provide a basic guidance that changing the thickness of conductive concrete is more recommended to reach a better resistance-reducing effect when conductive concrete is applied, instead of adjusting the resistivity of conductive concrete, because the grounding resistance is more sensitive to the former.

ACKNOWLEDGMENTS

This work was supported by the State Grid Jiangxi Electric Power Co., Ltd under grant no. 521820170024.

REFERENCES

Du, J., Tang, C., Jia, B., Zhang, D., & Miao, Q. 2016. Preparation and long-term stability study of steel fiber/graphite conductive concrete. *Key Engineering Materials* 680: 361–364.

Gatta, F.M., Geri, A., Maccioni, M., Lauria, S., Palone, F., & Pelliccione, G. 2017. Design approaches for EHV OHL "compact" tower grounding systems. In *2017 IEEE International Conference on Environment and Electrical Engineering and 2017 IEEE Industrial and Commercial Power Systems Europe (EEEIC/I&CPS Europe)*. IEEE, pp. 1–6.

Liang, X., & Wang, C. 2017. Factors affecting ground potential rise and fault currents along transmission lines with multigrounded shield wires. *IEEE Transactions on Industry Applications* 53(2): 888–900.

Lim, S.C., Gomes, C., & Ab Kadir, M.Z.A. 2013. Electrical grounding in troubled environment. *International Journal of Electrical Power & Energy Systems* 47: 117–128.

Min, C., Gao, P., Fei, G., Zhang, L., & Liu, H. 2017. Mechanical and smart properties of carbon fiber and graphite conductive concrete for internal damage monitoring of structure. *Construction & Building Materials* 142: 320–327.

Rao, N.P., Knight, G.S., Lakshmanan, N., & Iyer, N.R. 2010. Investigation of transmission line tower failures. *Engineering Failure Analysis* 17(5): 1127–1141.

Rongxun, Y.U., Huang, S., Zhu, Z., et al. 2017. Resistance reduction research of deep blasting method under complex soil conditions. *Insulators & Surge Arresters* 1: 91–95.

Sassani, A., Ceylan, H., Kim, S., Arabzadeh, A., Taylor, P.C., & Gopalakrishnan, K. 2018. Development of carbon fiber-modified electrically conductive concrete for implementation in Des Moines International Airport. *Case Studies in Construction Materials* 8:277–291.

Tomaškovičová, S., Ingemannielsen, T., Christiansen, A.V., Brandt, I., Dahlin, T., & Elberling, B. 2016. Effect of electrode shape on grounding resistances — part 2: Experimental results and cryospheric monitoring. *Geophysics* 81(1): WA159–WA172.

Wang, C., Tang, F., Li, Q., Zhang, Y., & Wang, X. 2017. Spray-coated superhydrophobic surfaces with wear-resistance, drag-reduction and anti-corrosion properties. *Colloids & Surfaces A: Physicochemical & Engineering Aspects*, 514: 236–242.

Xin, T. 2012. Test and study on electrical property of conductive concrete. *Procedia Earth & Planetary Science* 5(32): 83–87.

Zhang, D., Le, H., Yan, X., Yuan, T., & Li, J. 2014. Preparation of steel fiber/graphite conductive concrete for grounding in substation. In *2014 ICHVE International Conference on High Voltage Engineering and Application*. IEEE, pp. 1–4.

Zhou, L., He, J., Xu, H., Wang, P., Chen, Y., & Chen, S. 2017. Simulation of impact of vertical grounding electrode on impulse grounding resistance of substation grounding network. In *2017 2nd IEEE International Conference on Integrated Circuits and Microsystems (ICICM)*. IEEE, pp. 18–22.

Zhou, W., Liu, Y., Li, G., et al. 2015. Power frequency interference and suppression in measurement of power transmission tower grounding resistance. *IEEE Transactions on Power Delivery* 30(2): 1016–1023.

Emerging Developments in the Power and Energy Industry – Dufo-López, Krzywanski & Singh (eds)
© 2020 Taylor & Francis Group, London, ISBN 978-0-367-27169-5

Thermoeconomic analysis and multiobjective optimization of a closed isobaric–compressed air energy storagesystem

Kai Yan, Huanran Wang, Mingming Liu, Ruixiong Li & Chengchen Li
Xi'an Jiaotong University, Xi'an, Shaanxi, China

ABSTRACT: Compressed Air Energy Storage (CAES), a widely applied energy storage technology, is regarded as a promising approach to ameliorate the shortage of renewable energies. However, the efficiency of CAES is limited due to the throttling process in the system, which results in exergy losses. To avoid the influence of throttling effect, a closed isobaric compressed air energy storage (CI–CAES) system was proposed to improve the performance of CAES technology. In this new system, the low-pressure underground storage tank achieves higher inlet pressure of compressors compared with conventional CAES systems, and constant air pressure in the high-pressure underground storage tank leads to stable power output. In addition, industrial waste heat is utilized to increase the inlet temperature of turbines. The trade-off between thermodynamic performance and economic performance is investigated by a multiobjective genetic algorithm, where 65.98% and 1.12 k$/kW are selected as optimal solutions for the two objectives.

1 INTRODUCTION

The exploitation of renewable energies is increasing rapidly as a result of environmental and climatic issues. However, the utilization of renewable energy resources is hindered by their random and intermittent nature (Mahlia et al. 2014). Energy storage technologies are regarded as essential methods to address the common feature of renewable energies (Zhao et al. 2016).

At present, pumped hydro energy storage (PHES), compressed air energy storage (CAES), and a battery energy storage system (BESS) can be applied for bulk energy storage (de Boer et al., 2014). Among the aforementioned energy storage technologies, CAES is promising because of its relatively low cost, fast construction, and environmental friendliness (Zafirakis et al. 2013).

For conventional CAES, air is heated in a combustor by burning fossil fuel resources, which results in environmental pollution and a greenhouse effect (Raju & Khaitan 2012). Moreover, conventional CAES is limited by low system efficiency due to exergy losses in compression and expansion stage, and the system efficiency is around 40% to 50%. To increase the system efficiency of CAES, Yao et al. (2017) designed a novel system combined cooling, heating, and power based on CAES. More recently, a number of concepts about isobaric CAES systems have been proposed to avoid exergy losses by eliminating the throttling process. Nielsen et al. presented an isobaric adiabatic compressed air energy storage (IA–CAES) system.

Concepts relating to a closed isobaric compressed air energy storage (CI–CAES) system are presented in this article. The new system is characterized by a higher inlet pressure of the compression process compared with ambient pressure. In addition, to make the system more economical, the low-grade waste heat is utilized, mainly from industrial waste energy. This article focuses on the exergy efficiency and investment capital cost of the presented CI–CAES system in order to improve the exergy efficiency by taking the economic constraints into consideration.

Corresponding author: goodyankaigood@163.com

2 SYSTEM DESCRIPTION

A schematic diagram of the CI–CAES system is presented in Figure 1. The high-pressure underground storage tank is set deeper under the ground than the low-pressure underground storage tank. During the charging stage, compressors (Comp) are driven by renewable energy or excess electricity power to compress the air, the pressure of which is above ambient pressure. Air from the low-pressure underground storage tank (LP-UST) is compressed by a three-stage compression and finally stored in the high-pressure underground storage tank (HP-UST). In addition, after each compressor, compressed air enters an intercooler (IC) and is cooled to nearly ambient temperature. During discharging stage, compressed air in the HP-UST is released and flows into turbines (Tur) to generate electricity. Before each turbine, a reheater (RH) is set to reheat air while thermal energy to reheat air is from industrial waste heat, which can avoid consuming fossil fuels thoroughly.

Obviously, during both the charging and discharging stage, while air flows into and out of the LP-UST and HP-UST, the water level changes synchronously. Therefore, air pressure in the tanks remains constant due to the influence of the hydrostatic pressure of water. In addition, a throttling valve is not required to throttle the compressed air to a certain pressure in this system, which can avert exergy losses during the throttling process.

3 MODELING OF THE CI-CAES SYSTEM

The following assumptions are adopted to simplify detailed modeling of the proposed system:

(1) The air is regarded as an ideal gas.
(2) The air consists of 78.12% N_2, 20.96% O_2, and 0.92% Ar.
(3) Pressure fluctuation caused by water level changes is negligible.
(4) The heat loss and pressure drops in all pipes are negligible.
(5) The underground storage tanks (USTs) are adiabatic (Chen et al. 2018), the outlet temperature of which is assumed as ambient temperature.

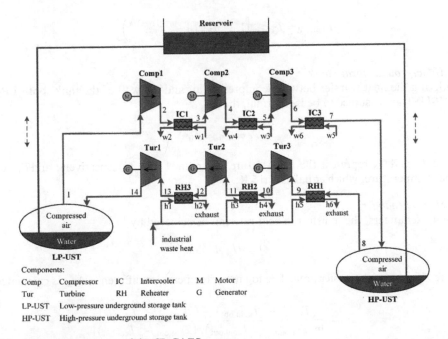

Figure 1.   Schematic diagram of the CI–CAES system.

3.1 Thermodynamic model

3.1.1 Compressor
The compression process is regarded as isentropic compression, and power consumption of the compressor can be formulated by considering energy and mass conservation:

$$\dot{W}_{\text{Comp}} = \frac{\kappa R_g T_{\text{Comp,in}} \dot{m}_{\text{Comp}} \left(\pi_{\text{Comp}}^{(\kappa-1)/\kappa} - 1 \right)}{(\kappa - 1)\eta_{\text{Comp}}} \tag{1}$$

where $\dot{m}_{\text{Comp}}$ stands for the mass flow rate of air, $T_{\text{Comp,in}}$ denotes the inlet air temperature of the compressor, κ denotes the adiabatic coefficient of air. π_{Comp} and η_{Comp} represent the pressure ratio and isentropic efficiency of the compressor, respectively.

The outlet temperature of the compressor can be calculated as follows:

$$T_{\text{Comp,out}} = T_{\text{Comp,in}} \left[\frac{\pi_{\text{Comp}}^{(\kappa-1)/\kappa} - 1}{\eta_{\text{Comp}}} + 1 \right] \tag{2}$$

3.1.2 Turbine
The expansion process is considered to be isentropic as well and the power produced by the turbine can be defined as

$$\dot{W}_{\text{Tur}} = \frac{\eta_{\text{Tur}} \kappa R_g T_{\text{Tur,in}} \dot{m}_{\text{Tur}} \left(1 - \pi_{\text{Tur}}^{(1-\kappa)/\kappa} \right)}{(\kappa - 1)} \tag{3}$$

where $\dot{m}_{\text{Tur}}$ denotes the air mass flow rate and $T_{\text{Tur,in}}$ is the inlet temperature of the turbine. π_{Tur} and η_{Tur} respectively stand for the pressure ratio and isentropic efficiency of turbine.

The outlet temperature and outlet pressure of turbine can be expressed as follows:

$$T_{\text{Tur,out}} = T_{\text{Tur,in}} \left[1 - \eta_{\text{Tur}} \left(1 - \pi_{\text{Tur}}^{(1-\kappa)/\kappa} \right) \right] \tag{4}$$

3.1.3 Underground storage tank
Considering the heat transfer between compressed air and the wall of the tank, both LP-UST and HP-UST are assumed to be isothermal, that is

$$T_1 = T_8 = T_0 \tag{5}$$

where T_1 and T_8 represent the temperature of stream 1 and 8, respectively, and T_0 denotes ambient temperature, which equals 273.15 K.

3.1.4 Heat exchanger
For heat exchangers, the transferred heat rate can be achieved by

$$\dot{Q} = U \cdot A \cdot \Delta t_{\text{m}} \tag{6}$$

where A is the heat transfer area. The log mean temperature difference Δt_{m} is calculated by

$$\Delta t_{\text{m}} = \frac{\left(T_{\text{hot,in}} - T_{\text{cold,out}} \right) - \left(T_{\text{hot,out}} - T_{\text{cold,in}} \right)}{\ln \left[\left(T_{\text{hot,in}} - T_{\text{cold,out}} \right) / \left(T_{\text{hot,out}} - T_{\text{cold,in}} \right) \right]} \tag{7}$$

3.2 Exergy model

Exergy analysis, which can identify the location and magnitude of energy losses during the system operation process (Lazzaretto & Tsatsaronis 2006; Zhang et al. 2012), provides a comprehensive and reasonable evaluation for the system (Yang et al. 2013). The exergy balance equation for steady flow working medium consists of physical exergy E_i^{ph} and chemical exergy E_i^{ch}:

$$E_i = E_i^{ph} + E_i^{ch} \tag{8}$$

and only physical exergy is considered in this article, which is defined as:

$$E_i^{ph} = m[(h_i - h_0) - T_0(s_i - s_0)] \tag{9}$$

3.3 Economic model

Economic analysis is significant for a CAES system, as it describes whether the system has potential for wide application. Capital investment cost is considered in this article, which is related to the size and performance of each component and capital investment models are illustrated in Table 1. In particular, it is suggested that a natural structure, e.g., lake is utilized as the reservoir in the proposed system to make the system more economical; thus it is unnecessary to construct a reservoir specially like that in a PHES plant and the cost of the reservoir is negligible.

To get the latest cost functions, the Marshall & Swift cost index is applied in the economic model. The total investment cost of the proposed system is calculated as

$$C_{tot} = \sum C_i \cdot \frac{I_p}{I_r} \tag{10}$$

where I_p and I_r are the Marshall & Swift cost index of the present year and reference year, respectively.

3.4 System evaluation criteria

3.4.1 Energy efficiency
Energy efficiency η_{en} in this article is defined as electricity-to-electricity efficiency and thus the waste heat is not considered. It can be given as

$$\eta_{en} = \frac{W_{Tur}}{W_{Comp}} \tag{11}$$

Table 1. Capital investment model of each component.

Component	Capital investment model (k\$) (Couper et al. 2009; Sadeghi et al. 2015)
Compressor	$C_{Comp} = 7.9(\dot{W}_{Comp})^{0.62}$ $\dot{W}_{Comp}$: power input, kW
Turbine	$C_{Tur} = 1.10(\dot{W}_{Tur})^{0.81}$ $\dot{W}_{Tur}$: power output, kW
Heat exchanger	$C_{He} = 130(A_{He}/0.093)^{0.78} \times 10^{-3}$ A_{He}: area of heat exchanger, m^2
Underground storage tank	$C_{UST} = 1218F_m \exp\left[11.662 - 0.6104(\ln V_{UST}) + 0.04536(\ln V_{UST})^2\right] \times 10^{-3}$ V_{UST}: volume of underground storage tank, m^3; F_m: material correction factor

where W_{Tur} is the electricity generated by turbines and W_{Comp} is the electricity consumed by compressors.

3.4.2 *Exergy efficiency*

Exergy efficiency η_{ex} evaluating the system performance is described as the ratio between the total exergy product $E_{P,tot}$ and total exergy fuel $E_{F,tot}$.

$$\eta_{ex} = \frac{E_{P,tot}}{E_{F,tot}} = \frac{E_{Tur}}{E_{Comp} + E_{wh}} \tag{12}$$

where E_{Tur} and E_{Comp} are respectively the total exergy of turbines and compressors; E_{wh} is the total exergy of waste heat.

3.4.3 *Total investment cost per total output power*

To achieve the appropriate investment cost of a certain system, an economic indicator, total investment cost per total output power (ICPP), is employed in this article:

$$ICPP = \frac{C_{tot}}{\dot{W}_{Tur}} \tag{13}$$

4 MULTIOBJECTIVE OPTIMIZATION

4.1 *Parameter setting for optimization*

Multiobjective optimization is performed to make the trade-off between thermodynamic performance (exergy efficiency) and economic performance (ICPP). The well-known nondominated sorting genetic algorithm II (NSGA-II) (Deb et al. 2002; Yusoff et al. 2011) is employed in this study. The lower bound and upper bound of decision variables are respectively listed in Table 2.

4.2 *Results*

The Pareto frontier for optimizing the proposed CI–CAES system is shown in Figure 2, which apparently reveals the conflict between exergy efficiency and ICPP. Point A achieves the lowest exergy efficiency with the lowest ICPP while point B achieves the highest exergy efficiency with the highest ICPP. Point C, a hypothetical ideal point that is out of the set of Pareto frontier solutions, is given in Figure 2. Therefore, point D is the closest point to the ideal point C on the Pareto frontier and makes the best trade-off between the two objectives. Values of objective and decision variables of points A, B, and D are illustrated in Table 3.

Table 2. Decision variables and their ranges.

Decision variables	Unit	Lower bound	Upper bound
Pressure of LP-UST	bar	15	75
Pressure of HP-UST	bar	90	135
Volume of HP-UST	m^3	600	1000
Pinch temperature difference	K	5	15
Mass flow rate of waste heat	kg/s	10	20

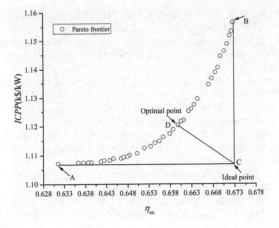

Figure 2.   Pareto frontier for the trade-off between exergy efficiency and ICPP.

Table 3.   Values of objective and decision variables of points A, B, and D.

Objective and decision variables	Unit	A	B	D
Exergy efficiency	—	0.6315	0.6727	0.6598
ICPP	k$/kW	1.1071	1.1566	1.1204
Pressure of LP-UST	bar	17.1015	17.4832	17.1130
Pressure of HP-UST	bar	129.0206	128.8035	128.9411
Volume of HP-UST	m^3	967.2323	960.8815	964.2645
Pinch temperature difference	K	14.9175	5.0000	8.9331
Mass flow rate of waste heat	kg/s	14.0365	10.0587	10.2485

5   CONCLUSIONS

In the proposed system, the hydrostatic pressure of water is utilized to make the internal pressure of underground storage tank keep constant and inlet pressure of compression process is higher than ambient pressure.

Based on multiobjective optimization, exergy efficiency is in conflict with overall capital investment cost, in other words, higher exergy efficiency leads to higher ICPP and the two objective functions cannot achieve optimum values at the same time. In addition, the optimal values of the pressure of LP-UST and the pressure and volume of HP-UST respectively are located in the approximate scope of 17.10–17.25 bar, 128.81–129.01 bar, and 961–967 m^3. The Pareto frontier, which is based on a genetic algorithm, presents a trade-off between exergy efficiency and ICPP. The final solution is selected by employing a decision-making procedure. The optimal exergy efficiency and ICPP are respectively 65.98% and 1.1204 k$/kW. Values of decision variables, i.e., pressure of LP-UST, pressure of HP-UST, volume of HP-UST, pinch temperature difference, and mass flow rate of waste heat are 17.1130 bar, 128.9411 bar, 964.2645 m^3, 8.9331 K, and 10.2485 kg/s, respectively. The results could provide a theoretical basis for the further engineering applications of this system.

REFERENCES

Chen, L.X., Xie, M.N., Zhao, P.P., Wang, F.X., Hu, P., & Wang, D.X. 2018. A novel isobaric adiabatic compressed air energy storage (IA-CAES) system on the base of volatile fluid. *Applied Energy* 210: 198–210.

Couper, J.R., Penney, W.R., Fair, J.R., & Walas, S.M. 2009. *Chemical Process Equipment* (revised 2nd edition). Oxford: Butterworth-Heinemann.

de Boer, H.S., Grond, L., Moll, H., & Benders, R. 2014. The application of power-to-gas, pumped hydro storage and compressed air energy storage in an electricity system at different wind power penetration levels. *Energy* 72(7): 360–370.

Deb, K., Pratap, A., Agarwal, S., & Meyarivan, T. 2002. A fast and elitist multiobjective genetic algorithm: NSGA-II. *IEEE Transactions on Evolutionary Computation* 6(2): 182–197.

Lazzaretto, A., & Tsatsaronis, G. 2006. SPECO: A systematic and general methodology for calculating efficiencies and costs in thermal systems. *Energy* 31(8): 1257–1289.

Mahlia, T.M.I., Saktisandan, T.J., Jannifar, A., Hasan, M.H., & Matseelar, H.S.C. 2014. A review of available methods and development on energy storage; technology update. *Renewable & Sustainable Energy Reviews* 33(33): 532–545.

Raju, M., & Khaitan, S.K. 2012. Modeling and simulation of compressed air storage in caverns: A case study of the Huntorf plant. *Applied Energy* 89(1): 474–481.

Sadeghi, M., Chitsaz, A., Mahmoudi, S.M.S., & Rosen, M.A. 2015. Thermoeconomic optimization using an evolutionary algorithm of a trigeneration system driven by a solid oxide fuel cell. *Energy* 89: 191–204.

Yang, Y., Wang, L., Dong, C., Xu, G., Morosuk, T., & Tsatsaronis, G. 2013. Comprehensive exergy-based evaluation and parametric study of a coal-fired ultra-supercritical power plant, *Applied Energy* 112(4): 1087–1099.

Yao, E., Wang, H., Wang, L., Xi, G., & Marechal, F. 2017. Multi-objective optimization and exergoeconomic analysis of a combined cooling, heating and power based compressed air energy storage system. *Energy Conversion and Management* 138: 199–209.

Yusoff, Y., Ngadiman, M.S., & Zain, A.M. 2011. Overview of NSGA-II for optimizing machining process parameters. *Proceedia Engineering* 15: 3978–3983.

Zafirakis, D., Chalvatzis, K.J., Baiocchi, G., & Daskalakis, G. 2013. Modeling of financial incentives for investments in energy storage systems that promote the large-scale integration of wind energy. *Applied Energy* 105((May): 138–154.

Zhang, X., He, M., & Zhang, Y. 2012. A review of research on the Kalina cycle. *Renewable & Sustainable Energy Reviews* 16 (7):,5309–5318.

Zhao, P., Gao, L., Wang, J., & Dai, Y. 2016. Energy efficiency analysis and off-design analysis of two different discharge modes for compressed air energy storage system using axial turbines. *Renewable Energy* 85: 1164–1177.

Emerging Developments in the Power and Energy Industry – Dufo-López, Krzywanski & Singh (eds)
© *2020 Taylor & Francis Group, London, ISBN 978-0-367-27169-5*

Experimental study and simulation of CO_2 membrane separation of coal-fired flue gas

Boyang Li, Juan Yu*, Zhongxiao Zhang & Yanchi Jiang
School of Mechanical Engineering, Shanghai Jiao Tong University, Shanghai, China

ABSTRACT: In the work reported in this article, an experimental single-stage membrane separation system for a carbon dioxide/nitrogen (CO_2/N_2) gas mixture was built using a hollow fiber membrane. The performance of the single-stage membrane separation system was investigated under different inlet gas pressures, total flow rates of inlet gas, and CO_2 volume fractions of inlet gas. The experimental results show that, when other conditions are constant, the volume fraction of CO_2 at the permeate end increases first and then decreases with the increase of inlet gas pressure, and the recovery rate of CO_2 in the system continues to rise. When the total flow rate of inlet gas increases, the volume fraction of CO_2 at the permeate end increases gradually, while the recovery rate of CO_2 in the system decreases. The influence of the CO_2 volume fraction of inlet gas on the experimental results is shown in the following aspects: when the CO_2 volume fraction of the inlet gas increases, the volume fraction of CO_2 at the permeate end will increase significantly, and the recovery rate of CO_2 in the system will rise in volatility. This article also models the single-stage membrane separation system and compares the simulation results with the experimental results. It was found that the model can simulate the experimental results well and can be used for further research.

Keywords: coal-fired flue gas, CO_2 membrane separation, experimental study, simulation

1 INTRODUCTION

In recent years, the environmental problems caused by global climate change have become increasingly severe, and the anthropogenic emissions of carbon dioxide (CO_2) have received wide attention from people in all walks of life. According to statistics, in 2017, China's total CO_2 emissions exceeded 9.2 billion tons, accounting for 27.6% of the total global emissions. It has become the world's largest carbon emitter for many years (British Petroleum Corporation 2018). Among them, CO_2 in flue gas, which is centrally emitted by coal-fired power generation, accounts for nearly one-third of the total emissions (Wu 2017). Therefore, the centralized capture and recovery of CO_2 in the flue gas of coal-fired power plants will greatly promote the realization of emission reduction targets.

At present, there are three main CO_2 capture technologies: precombustion capture, postcombustion capture, and oxycombustion removal. Among them, postcombustion capture technology is widely used, mainly including chemical absorption, membrane separation, adsorption separation, and cryogenic distillation (Luo et al. 2012). Chemical absorption is currently the most mature postcombustion carbon capture technology, and the method represented by amine absorption has been applied in some power plants. However, the disadvantages of chemical absorption, such as degradation of absorbent, high-energy consumption, and possible secondary pollution to the environment, limit its further development. In contrast, membrane separation has the advantages of modularization, nonpollution, and easy

---

*Corresponding author: E-mail: yujuan@sjtu.edu.cn

integration with existing coal-fired power plants. Some scholars have found that the energy consumption of membrane separation is comparable to or even lower than that of chemical absorption. For example, the results of Kundu et al. (2014) show that membrane separation provides minimal energy loss during postcombustion capture. Membrane separation saves 35%–55% of energy compared to the chemical absorption using a monoethanolamine solution alone. Brunetti et al. (2010) summarized that the energy consumption of membrane separation was 0.5–6 MJ/kg CO_2, while that of chemical absorption was 4–6 MJ/kg CO_2. Simulations by Hussain et al. (2015) also show that the total energy consumption of the membrane separation system is about 1.05 MJ/kg CO_2, which is much lower than that of the 3–6 MJ/kg CO_2 required by the chemical absorption. Therefore, membrane separation has strong competitiveness and is one of the options to replace chemical absorption for carbon capture from flue gas of coal-fired power plants in the future.

In the work reported in this article, a membrane separation system was built by using hollow fiber membrane module, and a single-stage membrane separation experiment was carried out to analyze the factors affecting the experimental results. On this basis, this article models the single-stage membrane separation system and compares the simulation results with the experimental results. The comparison found that the model can simulate the experimental results well and can be used for further research.

2 SINGLE-STAGE MEMBRANE EXPERIMENT AND SIMULATION

2.1 *Experimental system*

In the work reported in this article, a single-stage membrane experiment was carried out using a commercial hollow fiber membrane module. The membrane material of the module is polyimide.

The experimental system designed in this work is shown in Figure 1. The experimental system is composed mainly of the following parts: CO_2 cylinder gas, N_2 cylinder gas, mass flowmeter, filter, heating belt, thermocouple, hollow fiber membrane module, gas flowmeter, and gas analyzer. The schematic diagram of the hollow fiber membrane module in the experimental system is shown in Figure 2.

The gases used to simulate the flue gas of a coal-fired power plant in the experiment is CO_2 and N_2, which are supplied by cylinder gas. The volume fraction of CO_2 in the feed stream is

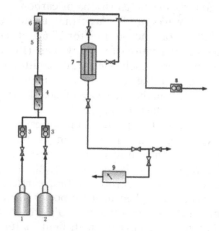

Figure 1.   Single-stage membrane separation. experiment system diagram.

1, CO_2 cylinder gas; 2, N_2 cylinder gas; 3, mass flowmeter; 4, filter; 5, heating belt; 6, thermocouple; 7, hollow fiber; membrane module; 8, gas flow meter; 9, gas analyzer.

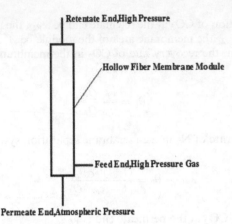

Retentate End,High Pressure

Hollow Fiber Membrane Module

Feed End,High Pressure Gas

Permeate End,Atmospheric Pressure

Figure 2.   Schematic diagram of hollow fiber membrane module.

Table 1.   Single-stage membrane separation experimental conditions.

Serial number	Feed pressure (MPa)	Total flow rate of inlet gas (L/min)	CO_2 volume fraction of inlet gas (%)
1	0.42		
2	0.62		
3	0.75	33.5	16.7
4	0.87		
5	1.05		
6			11.0
7			13.5
8	1.05	46.9	15.3
9			17.9
10			20.4
11		29.0	
12		33.2	
13		39.7	
14	1.05	43.8	17.7
15		48.1	
16		57.2	

achieved by means of a mass flowmeter. The experiment was carried out at room temperature (25°C). The feed pressure required to separate the gas is supplied by the cylinder gas, and the pressure regulating valves are used to achieve the desired feed pressure.

In this work, single-stage membrane separation experiments were carried out for different inlet gas pressures, total flow rate of inlet gas, and CO_2 volume fraction of inlet gas. The specific experimental conditions are shown in Table 1.

2.2   *Evaluation indicators of the experimental results*

In this work, two parameters, the volume fraction of CO_2 at the permeate end and the recovery rate of CO_2 in the membrane separation system, were selected as the evaluation indicators of the experimental results.

Let the CO_2 volume fraction(molar fraction) of the inlet gas be x_{FC}, the feed pressure be p_F, the total flow rate of inlet gas be Q_F(STP), the gas pressure at the permeate end be p_P, the gas flow rate at the permeate end be Q_P(STP), the volume fraction of CO_2 at the permeate end be y_{PC}, the gas pressure at the retentate end be p_R, the gas flow rate at the retentate end be

Q_R(STP), the volume fraction of CO_2 at the retentate end be x_{RC}, the permeance of CO_2 be J_{CO2}, the permeance of N_2 be J_{N2}, the membrane area of the module be A, and the height of filaments in the module be H. Then, the recovery rate of CO_2 in the membrane separation system (θ_1) is defined as

$$\theta_1 = \frac{y_{PC}Q_P}{x_{FC}Q_F} \tag{1}$$

Similarly, the recovery rate of N_2 in the membrane separation system (θ_2) is defined as

$$\theta_2 = \frac{(1 - y_{PC})Q_P}{(1 - x_{FC})Q_F} \tag{2}$$

The volume fraction of CO_2 at the permeate end is y_{PC}.

2.3 Single-stage membrane simulation

In this work, the single-stage hollow fiber membrane is simulated. In the simulation, it is assumed that the flow inside the hollow fiber membrane module is approximately countercurrent and the pressure drop on the permeate side is neglected. Under the conditions of this experiment, the performance of membrane module is approximately considered to be unchanged.

First, the initial values of θ_1 and θ_2 are assumed. Then, the parameters of feed, permeate, and retentate ends are obtained by using conservation Equations (3)–(5) and Equations (1) and (2) and known quantities Q_F and x_{FC}.

$$Q_F = Q_P + Q_R \tag{3}$$

$$Q_F x_{FC} = Q_P y_{PC} + Q_R x_{RC} \tag{4}$$

$$Q_F(1 - x_{FC}) = Q_P(1 - y_{PC}) + Q_R(1 - x_{RC}) \tag{5}$$

Next, set the volume fraction of CO_2 on the permeate side at the height H of the filaments (retentate end) to be y_{HC}; then according to the definition of the volume fraction, the following equation can be obtained:

$$y_{HC} = \frac{J_{CO2}(p_R x_{RC} - p_P y_{HC})}{J_{CO2}(p_R x_{RC} - p_P y_{HC}) + J_{N2}[p_R(1 - x_{RC}) - p_P(1 - y_{HC})]} \tag{6}$$

Combining the relevant parameters solved by the above steps and the known quantities J_{CO2}, J_{N2}, p_R, p_P, Equation (6) can be solved to obtain y_{HC}. Therefore, using the logarithmic averaging method, the average partial pressure difference of CO_2 on both sides of the module membrane is obtained:

$$\overline{p_R x - p_P y}_{CO2} = \frac{(p_F x_{FC} - p_P y_{PC}) - (p_R x_{RC} - p_P y_{HC})}{\ln\left(\frac{p_F x_{FC} - p_P y_{PC}}{p_R x_{RC} - p_P y_{HC}}\right)} \tag{7}$$

Thus, the flux of CO_2 through the membrane (Q_{CO2}) is

$$Q_{CO2} = J_{CO2}A\overline{p_R x - p_P y}_{CO2} \tag{8}$$

So the calculated CO_2 recovery rate of the system (θ_1^C) is

$$\theta_1^C = \frac{Q_{CO2}}{Q_F x_{FC}} = \frac{J_{CO2}A\overline{p_R x - p_P y}_{CO2}}{Q_F x_{FC}} \tag{9}$$

Similarly, the calculated N_2 recovery rate of the system (θ_2^C) can be obtained. Iterative operations are continued until $(\theta_1^C - \theta_1)^2$ and $(\theta_2^C - \theta_2)^2$ reach the range of allowable error. At this point, the iterative calculation is considered to converge, and the relevant parameters obtained are simulated values.

3 RESULTS AND ANALYSIS

In this article, the above conditions are experimental and the experimental results are compared with the results of simulation. It is found that the inlet gas pressure, total flow rate of inlet gas, and CO_2 volume fraction of inlet gas have significant effects on the experimental results of a single-stage membrane separation system. The analysis is performed separately below.

3.1 *Inlet gas pressure*

It can be seen from Figure 3a that the volume fraction of CO_2 at the permeate end increases first and then decreases with the increase of the inlet gas pressure, and the maximum value is obtained when the inlet gas pressure is about 0.62 MPa He & Hägg (2011) also obtained similar changes in the study. This may be because when the inlet gas pressure is low, the low CO_2 partial pressure difference on both sides of the membrane limits the penetration of CO_2. After the inlet gas pressure is increased, the partial pressure difference of CO_2 on both sides of the membrane increases, and the volume fraction of CO_2 on the permeate side increases. When the inlet gas pressure exceeds 0.62 MPa, a further increase in the inlet gas pressure causes more N_2 to pass through the membrane, thereby reducing the volume fraction of CO_2 on the permeate side.

Figure 3b shows that as the inlet gas pressure increases, the recovery rate of CO_2 gradually increases. This is because the increased inlet gas pressure causes more CO_2 to pass through the membrane, resulting in an increase in CO_2 recovery.

3.2 *Total flow rate of inlet gas*

It can be seen from Figure 4a and b that as the total flow rate of the inlet gas increases, the volume fraction of CO_2 at the permeate end gradually increases, and the recovery rate of CO_2 decreases. Kim et al. (2017) reached the same conclusion in their article.

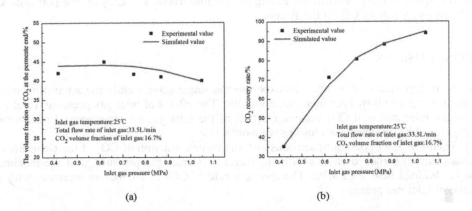

(a)   (b)

Figure 3.   Effect of inlet gas pressure on experimental results.

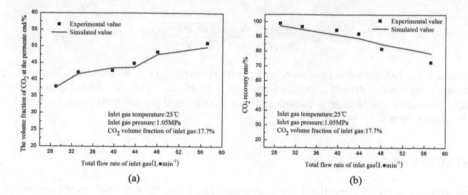

Figure 4. Effect of total flow rate of inlet gas on experimental results.

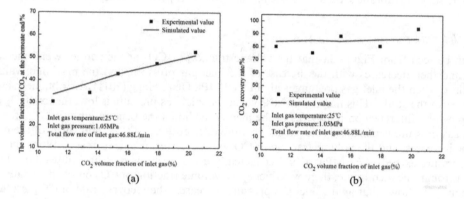

Figure 5. Effect of CO_2 volume fraction of inlet gas on experimental results.

3.3 CO_2 volume fraction of inlet gas

From Figure 5a and b, it can be seen that as the CO_2 volume fraction of inlet gas increases, the volume fraction of CO_2 at the permeate end will increase significantly, and volatility of the recovery rate of CO_2 in the system will rise. This is because, when the inlet gas pressure is constant, increasing the CO_2 volume fraction of inlet gas will significantly increase the partial pressure difference of CO_2 on both sides of the membrane, thereby promoting the transmembrane transport of CO_2 and thus increasing the volume fraction of CO_2 on the permeate side and the recovery rate of CO_2 in the system.

4 CONCLUSIONS

In the work reported in this paper, an experimental single-stage membrane separation system was built using a hollow fiber membrane module. The effects of inlet gas pressure, total flow rate of the inlet gas, and CO_2 volume fraction of the inlet gas on experimental results were investigated. The following conclusions were obtained:

(1) When the other conditions are constant, the volume fraction of CO_2 at the permeate end increases first and then decreases with the increase of inlet gas pressure, and the maximum value is obtained near 0.62 MPa. The recovery rate of CO_2 in the system increases with the increase of inlet gas pressure.

(2) When the other conditions are constant, after increasing the total flow rate of inlet gas, the volume fraction of CO_2 at the permeate end increases gradually, while the recovery rate of CO_2 in the system decreases.

(3) When the other conditions are constant, the volume fraction of CO_2 at the permeate end will increase significantly with the increase of the CO_2 volume fraction of inlet gas, and the volatility of the recovery rate of CO_2 in the system will rise.

In addition, this article models the single-stage membrane separation system and compares the simulation results with the experimental results. The comparison found that the model can simulate the experimental results well and can be used for further research.

ACKNOWLEDGMENTS

The authors of this article gratefully acknowledge financial support from National Key R&D Program of China, grant no. 2017YFB0603403.

REFERENCES

British Petroleum Corporation. 2018. *2018 BP World Energy Statistics Yearbook*. Beijing: Beijing Grammy Digital Graphic Production.

Brunetti, A., Scura, F., Barbieri, G., & Drioli, E. 2010. Membrane technologies for CO_2 separation. *Journal of Membrane Science* 359: 115–125.

He, X., & Hägg, M._B. 2011.Hollow fiber carbon membranes: Investigations for CO_2 capture. *Journal of Membrane Science* 378: 1–9.

Hussain, A., Farrukh, S., & Minhas, F. T. 2015. Two-stage membrane system for post-combustion CO_2 capture application. *Energy Fuels* 29: 6664–6669.

Kim, S. H., Kim, J.-K., Yeo, J.-G., & Yeo, Y.-K. 2017. Comparative feasibility study of CO_2 capture in hollow fiber membrane processes based on process models and heat exchanger analysis. *Chemical Engineering Research and Design* 117: 659–669.

Kundu, P. K., Chakma, A., & Feng, X. 2014. Effectiveness of membranes and hybrid membrane processes in comparison with absorption using amines for post-combustion CO_2 capture. *International Journal of Greenhouse Gas Control* 28: 248–256.

Luo, Z., Fang, M., Li, M., & Lin, G. 2012. *Carbon Dioxide Capture, Storage and Utilization Technology*. Beijing: China Electric Power Press.

Wu, L. 2017. Simulation for CO_2 capture by using membrane gas separation from flue gas in power plant. Qingdao: Qingdao University of Science and Technology.

Emerging Developments in the Power and Energy Industry – Dufo-López, Krzywanski & Singh (eds)
© 2020 Taylor & Francis Group, London, ISBN 978-0-367-27169-5

Design of the electric power emergency repair system and its software architecture

J.S. Cui, H. Gang & P. Qiu
State Grid Liaoning Electric Power Co. Ltd Jinzhou Power Supply Company, Jinzhou, China

J.X. Zhang
Northeastern University, Shenyang, China

ABSTRACT: With the rapid development of China's economy, electricity, as the basis of the national economy, has increasingly drawn attention because of its highlighted importance. It is necessary to maintain the normal power supply to users. In particular, it is important to quickly and scientifically solve the problem of slow decision-making and insufficient scheduling in the existing electric emergency repair system. Therefore, design and implementation of efficient power emergency repair system software architecture and networking construction are of important practical significance. Based on the Global Positioning System (GPS), General Packet Radio Service (GPRS), Geographic Information System (GIS), and other technologies, this article first analyzes the structure of the general dispatch of the electric emergency repair system and the structural functions of each component. Second, it introduces the geographical location of each mobile terminal and the communication between various subsystems. Finally, it delivers the software architecture of the system operation. The real-time monitoring of the system repair process and the efficiency of power emergency repair scheduling are both realized.

1 INTRODUCTION

With the continuous development of China's power system, the scale of the power grid continues to expand, and the number of power users has increased year by year. Maintaining users' normal power supply is the basis of ensuring high-quality services. However, various pieces of power equipment installed in an external complex circumstance may malfunction at any time, resulting in large-scale power outages in local areas, which leads to huge economic losses (Zhou 2018).

At present, China has put forward the theory of emergency management of a power system, provided a theoretical basis for the development of power emergency management, and also clearly put forward technical countermeasures for power emergency management (Tian et al. 2007). Tian et al. (2008) proposed the concept and framework of the power emergency management platform for the research of a power emergency management platform. Xu (2014) established a user complaint, consultation, and fault repair system with 95598 as the acceptance phone number. In the above study, there are application software platform, telephone, fax, short message service (SMS), and other communication tools between the district- and county-level power supply companies and the on-site repair units. The communication efficiency is low, and the real-time and high-efficiency transfer of the repair system cannot be realized.

In view of the shortcomings in the current repair process, the Global Positioning System (GPS) is used for real-time monitoring and positioning of the repair vehicles. The General Packet Radio Service (GPRS) is used for the transmission and acceptance of power data. The Geographic Information System (GIS) is used for map display during the repair processes. By adopting GPS, GPRS, and GIS systems, real-time monitoring of the repair process can be realized.

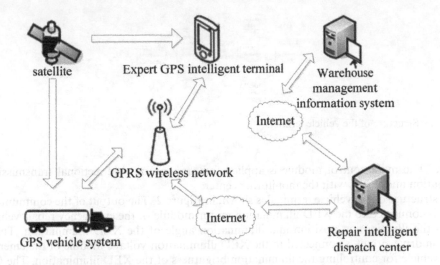

Figure 1. Schematic diagram of the electric emergency repair system.

This article focuses on the electric power emergency repair system and its software architecture. Section 2 presents the system configuration, followed in Section 3 by a detailed description of the functions of each component in the system. Next, Section 4 presents the software architecture of the system. Finally, Section 5 concludes this article.

2 THE STRUCTURE OF AN ELECTRIC POWER EMERGENCY REPAIR SYSTEM

The electric power emergency repair system consists of four parts: (1) the electric power emergency repair dispatch center, (2) the electric power repair vehicle system, (3) the electric power repair expert GPS intelligent terminal system, and (4) the electric power repair monitoring management information system. The intelligent dispatching center mainly includes an information server, intelligent fault location, and repair scheduling system; the vehicle-mounted system includes XED energy-saving lighting module, GPS satellite receiving module, GPRS wireless communication module, and extended function module; an expert GPS terminal is an intelligent terminal equipped for expert emergency repair. The terminal includes a GPS satellite receiving module, a GPRS wireless communication module, and an extended function module. The monitoring management information system includes a GIS geographic information module, a positioning monitoring system module, a state monitoring system module, and a remote image capturing module (Zhong 2008). Figure 1 is a schematic diagram of the overall structure of the system.

The system uses the GPS system for real-time monitoring and positioning of the repair vehicle and the required repair experts, and uses the GPRS system for power data communication, and the GIS system for the map display of the repair process to realize the system repair. The real-time monitoring of the process, the instant dispatch of the repaired vehicle and the efficient management of the repair work can greatly shorten the repair process and reduce the difficulty of repairing and facilitate the power users, while avoiding huge economic losses.

3 CONFIGURATION OF THE ELECTRIC EMERGENCY REPAIR SYSTEM

3.1 *Electric power repair vehicle terminal*

The electric power repair vehicle terminal is mainly composed of an XED energy-saving lighting module, GPS receiving module, GPRS wireless communication module, and main control

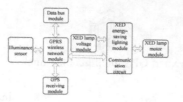

Figure 2. Structure of the vehicle terminal.

module. The main control module is applied for carrying out bidirectional transmission of information and voices with the monitoring center.

The structure of the vehicle terminal is shown in Figure 2. The output of the communication circuit is connected to the XED illumination motor module of the emergency repair vehicle to control the illumination position and illumination angle of the XED illumination. The illumination drive circuit is connected to the XED illumination voltage module of the emergency repair vehicle for controlling the illumination brightness of the XED illumination. The GPRS wireless communication module is respectively connected with the network interface of the communication circuit, the data sending end of the GPS receiving module, and the data bus module of the data entry and report analysis unit and the illuminance sensor data of the emergency repair vehicle. The GPRS wireless communication module is used for realizing bidirectional transmission of data between the vehicle navigation terminal unit and the monitoring information management unit, the intelligent scheduling unit, and the data entry and report analysis unit. The GPS receiving module is used to achieve precise positioning of the electric power repair vehicle.

The flow chart of the operation of the vehicle mobile terminal is shown in Figure 3. When the monitoring center issues a positioning information calling command, the GPS receiving module is started to receive the satellite signal, and the positioning information is calculated and transmitted to the main control module via the serial line. The main control module is the core of the vehicle terminal, which processes the data sent by the GPS receiving module, sending the processed information to the monitoring center through the GPRS transceiver module and providing the current status, working conditions, and current position information of the vehicle device to the vehicle. There is a real-time monitoring display module, receiving control, scheduling, and configuration command information from the processing and monitoring center from the GPRS wireless communication module (Li & Piao 2006). The GPRS wireless communication module is configured to receive the response to the master station command, sending the command to the main control module and uploading the vehicle information to the monitoring center in real-time. After the GPRS wireless communication module logs into the GPRS network, it automatically connects to the monitoring center to report its IP address, and maintains the connection of the link, monitoring the connection status of the link at any time, and automatically reestablishing the link once an abnormality occurs.

3.2 *Electric power repair intelligent terminal system*

The system is equipped with a GPS intelligent terminal for each repair expert, so that in the event of a sudden major power failure, the required repair experts can be quickly found. Thus the repair work can be supported by experts and an effective repair plan can be obtained. The GPS intelligent terminal system is mainly composed of a GPRS communication module, a GPS positioning module, and a control module. The basic structure is shown in Figure 4.

The GPRS communication module realizes bidirectional communication (including data and voice communication) between the expert and the dispatch center and the vehicle-mounted system, transmitting its own positioning information to the dispatch center and receiving various commands sent by the dispatch center, such as the location to wait for pick-up, the vehicle license, and the real-time position of the vehicle.

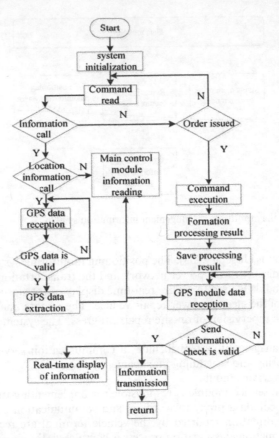

Figure 3. Vehicle mobile terminal flow chart.

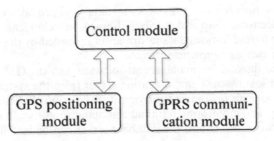

Figure 4. Power repair expert GSP intelligent terminal structure.

3.3 *Electric power repair monitoring management information system*

The electric power repair monitoring management information system includes a GIS geographic information module, a positioning monitoring system module, a state monitoring system module, and a remote image capturing module. The data interface of the GIS geographic information module is respectively connected with the data of the positioning monitoring system module, the state monitoring system module, and the remote image capturing module. The data interfaces of the positioning monitoring system module, the status monitoring system module, and the remote image capturing module are respectively connected with the data bus module of the data entry and report analysis unit. Among them, the GIS geographic information module is used to display dynamic map data and directly reflect the

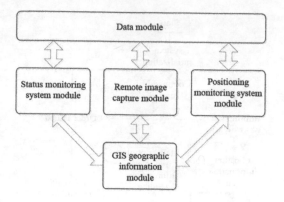

Figure 5. Structure of the monitoring management information system.

overall operational status (Joseph 2003). The positioning monitoring system module is used to realize the electronic display of the power network and the transportation network. The status monitoring system module is used to realize real-time display of the status of the power equipment and the status of the electric power repair vehicle. The remote image capture module is used to implement the supervision of on-site repair progress. The system structure is shown in Figure 5.

The monitoring management system includes a communication service module, a central database, and monitoring and scheduling management software.

(1) Communication service module

The communication service module is responsible for implementing functions such as data communication protocol, data proxy forwarding, and communication device status monitoring. The data and information reported by the vehicle terminal are received and extracted. After the vehicle information classification process is completed, the classification information is transmitted to the database server.

(2) Central database

The central database mainly stores and backs up data, and provides various data for system operation, such as electronic map and attribute library, vehicle location database, and user information database to save various data information uploaded by the vehicle terminal.

(3) Monitoring dispatch management software

The monitoring and dispatching management software, i.e., the GIS platform, extracts the classified information sent by the communication server from the system database, and completes functions such as map positioning, driving display, control command publishing, map data search and query, and monitoring terminal command response. In the specific implementation, the communication server and the database server can be installed in one server or in distributed servers.

3.4 *Data transmission*

GPRS is a highly efficient way to transmit data based on packet switching, including SGSN service support node, GGSN gateway support node, etc. (Zhong 2001).

The information between the remediation dispatch center and the vehicle terminal can be directly or indirectly transmitted to the dispatch center data server through various communication channels. In the past, there were mainly several ways to achieve data transmission: (1) radio transmission; (2) wired lines, such as data lines, dial-up Internet access, etc.; and (3) data transmission method of GSM short message function. When GPRS transmits data, its transmission rate is high, it is always online, and the access speed is fast. It is precisely due to many advantages and its application is more and more widely accepted by more and more people.

Table 1. Comparison of several data transmission communication methods.

Transfer method	Radio	Wired line	GSM short message	GPRS
Cost of investment	High	General	Low	Low
Operating costs	High	Higher	General	Low
Data processing	Complex	General	Complex	Simple
Coverage	Difference	Difference	Better	Better
Data reliability	General	Good	General	Good
Real-time	Good	Good	Difference	Good
Construction and maintenance	Difference	Difference	General	Good
Equipment monitoring	Good	Good	Difference	Good
Comprehensive theory	Not suitable	Not suitable for field use	Not suitable	Recommended

According to the characteristics of electric power emergency repair data transmission and application requirements, several data transmission methods are compared, as shown in Table 1.

It can be seen from Table 1 that the use of the radio transmission mode requires the construction of a large number of wireless transmitting and receiving base stations, occupying dedicated frequency points, high investment costs, difficult maintenance, poor data reliability, and unsuitability for multipoint data transmission. A wired line needs to build or lease special lines, which carries a high investment cost, high cost of a wired dialing method, poor coverage, and unsuitability for outdoor multipoint data transmission. GSM short message mode has disadvantages such as low real-time performance and low equipment monitoring capability. The data transmission mode based on GPRS has the characteristics of low investment, low operating cost, good data transmission reliability, strong real-time data transmission, remote control of equipment, wide monitoring range, and large transmission capacity of the system.

4 POWER EMERGENCY REPAIR DISPATCHING SYSTEM SOFTWARE

4.1 System network structure selection

The B/S structure, namely, the browser/server mode, is currently the most popular and widely used network structure mode. This structure mode is divided into server and browser, and the server is the web system to achieve its main functions. On the platform, almost all the calculations, queries, and other functions are completed on the server side. The most important application software on the browser side is the browser. The browser is used to display the results of the server-side operation and issue corresponding instructions (Ding 2010). This mode concentrates the functionality of the client in client\server mode on the server, leaving only the browser's related functionality. This greatly reduces the performance requirements for the client and simplifies system development and maintenance. Based on the many advantages of the B/S structure described earlier, the work described in this paper uses a multilayer B/S architecture to build the system to accommodate large-scale data collection and data sharing (Lu 2010).

4.2 System software structure

Because the system uses a multilayer B/S network architecture, the application logic, user interface, and database operation module are separated, which improves the simplicity of system maintenance and use. At the same time, in order to prevent invasion by a virus, all layers are exchanged data on the system support platform to achieve the confidentiality of the files. For the user interface, the WWW browser is used to implement data operation and

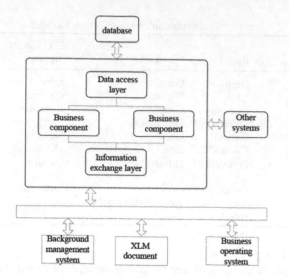

Figure 6.   System software structure diagram.

management with the browser as the platform, which reduces the cost and workload of system maintenance and upgrade, greatly simplifies the load on the client computer, and reduces the user's overall cost (Zheng 2014). The software structure diagram of the system is shown in Figure 6 and the software is implemented by Java.

5   CONCLUSION

Emergency repair of electric power accidents requires maintenance services such as accident handling, damaged equipment repair, power grid recovery, and emergency rescue at the fastest speed to avoid or recover from accidents. The electric power emergency repair system and its software architecture are designed. The system uses GPS, GPRS, and GIS technologies to provide full play to their respective advantages, accurately and quickly completing the positioning and implementation monitoring of the repair vehicle. On the other hand, the system can accurately and safely send, receive, and transmit information, which greatly improves the efficiency of electric power emergency repair and provides a friendlier and more efficient platform for power companies to better serve society.

ACKNOWLEDGMENTS

This work was supported by funding from the State Grid Liaoning Electric Power Corporation under the project "The Electric Power Emergency Repair System Based on a Novel Energy-Saving Light Sources."

REFERENCES

Ding, L. 2010. Design and implementation of ERP system sales management subsystem. Beijing University of Posts and Telecommunications.

Joseph, E. 2003. Responses to the design, application, and management of Geographic Information System (GIS) datasets (generated models) for managing extension programs. University of Wisconsin–Madison.

Li, Z., & Piao, Z. 2006. Design and research of electric power intelligent inspection system based on GPS/GPRS. *Agricultural Network Information* 8: 15–17.

Lu, H. 2010. Design and implementation of oilfield development information system. University of Electronic Science and Technology.

Tian, S., Chen, X., & Zhu, C., et al. 2008. Study on electric power emergency management platform. *Power System Technology* 32(1): 26–30.

Tian, S., Zhu, C., Chen, X., et al. 2007. Theory of electric power emergency management and its technological countermeasures. *Power System Technology* 31(24): 22–27.

Xu, J. 2014. Design and implementation of power failure repair information system based on mobile terminal. University of Electronic Science and Technology.

Xu, Z. 2004. Research on original image database system based on WEB. People's Liberation Army Information Engineering University.

Zheng, Q. 2014. Design and implementation of electric emergency command system. Jilin University.

Zhong, A. 2008. Research on intelligent dispatching system of electric power repair based on GIS. Chongqing University.

Zhong, Z. 2001. *GPRS General Packet Radio Service*. Beijing: Posts & Telecom Press.

Zhou, X. 2018. Application of satellite communication technology in emergency repair of electric power emergency communication. *Communication World* 8: 170–171.

Emerging Developments in the Power and Energy Industry – Dufo-López, Krzywanski & Singh (eds)
© 2020 Taylor & Francis Group, London, ISBN 978-0-367-27169-5

Experimental investigation and mechanism analysis of NO_x reduction by CH_4 in the absence of oxygen

Zhixiang Zhu, Zhongxiao Zhang, Juan Yu & Degui Bi
School of Mechanical Engineering, Shanghai Jiao Tong University, Shanghai, PR China

ABSTRACT: Experimental and theoretical studies have been performed to explore NO_x reduction by CH_4 without the presence of oxygen. GRI-Mech 3.0 is proved to be valid for modeling NO_x abatement by injection of methane and the agreement is within 20%. The NO can be reduced to nitrogen intermediates only with CH_4 converted into CH_i, which initially results from thermal decomposition of CH_4. The nitrogen intermediates are subsequently converted into N_2 or NO depending on the reaction conditions. The impact of temperature and CH_4 to NO ratio on NO_x reduction is also investigated. The increase of temperature and CH_4 to NO ratio induces a significant increase of NO_x reduction efficiency. But extremely high temperature, i.e., above 1400°C, has an adverse effect on NO_x reduction. The oxygen concentration has a remarkable effect on NO_x reduction and injection of CH_4 into regions characterized by reducing atmosphere and high temperature stimulates the NO reduction process.

Keywords: CH_4, GRI-Mech 3.0, NO_x reduction, oxygen concentration, temperature

1 INTRODUCTION

Reburning, a term first introduced by Wendt et al. (1973), is a process whereby a hydrocarbon fuel is injected immediately downstream of the combustion zone. The concept of reburning was first applied to a full-scale boiler by Mitsubishi in the early 1980s, more than 50% reduction of NO being achieved. To date, numerous studies have been conducted with reburning in laboratory, pilot-scale, and full-scale burners (Nazeer et al. 1999). Reburning tests were performed in a 300-kW boiler simulator facility to investigate the impact of reburn fuels on NO_x reductizon. Basic reburning performance approaches that of natural gas (70%) and NO_x reduction achieves 85%–95% via advanced reburning at a reburn heat input of 10% (Maly et al. 1999). Bilbao et al. (1997) also evaluated the use of different hydrocarbon fuels for gas reburning in a quartz tube reactor at relatively low temperature, i.e., at and under 1100°C. The effects of stoichiometry, residence time, and temperature were evaluated for different gaseous reburn fuels and results showed a strong influence of residence time and stoichiometry (Spliethoffet al. 1996). Zhong et al. (2002) studied the influence of different kinds of pulverized coals and their chars on NO_x reduction. The NO_x reduction capacities of the lignites and lignite chars were appreciably stronger than those of the bituminous coals and bituminous coal chars. Chen et al. (2010) used lignite as the reburning fuel and found that a heterogeneous mechanism was much more important than a homogeneous mechanism. Liu et al. (1997) used 13 coals and 1 char as reburning fuels and found that the NO_x reduction efficiency by coal reburning decreased with increasing fuel stoichiometry and increased with decreasing coal rank. Meanwhile, the influence of reburning fuel fraction, stoichiometric ratio, and so on has been widely explored (Ballester et al. 2008; Kim et al. 2012). Prada et al. (1998) studied the kinetic model for different gaseous reburning fuels at a temperature range between 973 and 1371 K and the major NO removal pathway is found to be the HCCO+NO reaction in all cases.

To summarize, the reburning process has been well understood both experimentally and theoretically. But to the authors' knowledge, a reburning process using CH_4 as reburning fuel at high temperature, i.e., above 1200°C, is seldom investigated. Furthermore, oxygen was indispensable in previous works. Therefore, the objective of the present work is to investigate NO_x reduction by methane at temperatures between 1200 and 1450°C in the absence of oxygen.

2 EXPERIMENTAL INTRODUCTION

A bench scale platform, consisting of gas feeding unit, a high-temperature reaction system, and gas analytical instruments, has been established to investigate the NO_x reduction by CH_4 in the absence of oxygen, as shown in Figure 1. A quartz tube with a diameter of 36 mm and a length of 2373 mm is adopted as the reactor. There are two gas inlets of the reactor. One gas inlet is used to introduce a main gas stream containing 16% CO_2, 490–520 ppm NO, and a balance gas (argon) and the other is utilized to introduce a reducing agent (CH_4). The mole ratio of CH_4/NO ranges from 0.5 to 1.5. The quartz tube is electrically heated and the tube wall is insulated to prevent heat losses to environment. Prior to the reburning process, a movable thermocouple is used to characterize temperature distribution and temperature profile the along axial direction of the reactor as shown in Figure 2. A 1000-mm length of uniform temperature profile is located in the middle of the reactor. A water-cooled probe made of stainless steel is employed to sample

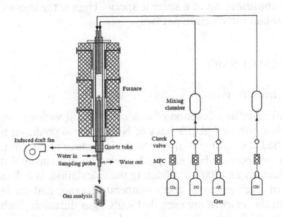

Figure 1. Schematic diagram of the experimental system.

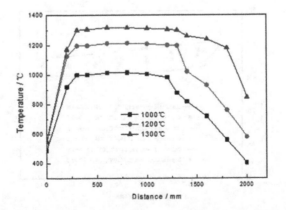

Figure 2. Temperature distribution along the axial direction of the quartz tube.

and quench the flue gas and the concentrations of O_2, CO, NO, and CO_2 are monitored by means of an Optima 7 gas analyzer.

The experiments are performed at atmospheric pressure and the reaction temperature range is within 1200–1450°C. The whole mass flowrate is adjusted to maintain a constant velocity of 0.7 m/s.

3 KINETIC MODELING

GRI-Mech 3.0 chemistry mechanism, consisting of 53 species and 325 reactions, is employed in this article. It is an optimized mechanism designed to model natural gas combustion, including NO formation and reburn chemistry (Shen et al. 2004). The mechanism is valid in a wide range of temperature, from 1000 K to 2500 K, and predictions agree well with the experimental data for lower O_2/CH_4 (Han et al. 2003).

A plug flow reactor (PFR) is adopted in calculations. Assumptions are proposed in PFR that physical parameters, like temperature and concentration, vary only along the axial direction of the reactor and remain constant in the radial direction. Therefore, all particles in the reactor have the same residence time. The calculations assume that simulated flue gas and reducing agent are perfectly premixed. Therefore, the influence of finite mixing rate is neglected, which usually leads to overestimated NO_x reduction efficiency.

Rate of production analysis is an effective method to analyze chemical reaction mechanisms. The absolute value of production rate indicates the influence of the elementary reaction on the production or consumption of a specific specie. Hence, the evolution of nitrogen specie is traced in order to obtain the reaction pathway.

4 RESULTS AND DISCUSSION

4.1 *The influence of temperature*

The NO_x reduction efficiency as a function of residence time at various temperatures is presented in Figures 3 and 4. It is apparent that efficiency of NO_x removal predicted by the model is slightly higher than that obtained in experiments owing to the ignorance of finite mixing rate in the mixing process. However, the comparison between predictions and experimental data points out a very good agreement, particularly at 1200°C, validating the calculation. It is because that temperature is a dominant factor for reaction rate at a low temperature range and the influence of mixing can be neglected. But when the temperature exceeds 1300°C, gas diffusion gradually plays an important role and the impact of the mixing process can no longer be ignored.

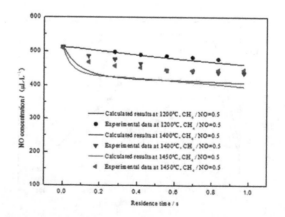

Figure 3.   Impact of temperature on NO_x reduction with $CH_4/NO = 0.5$.

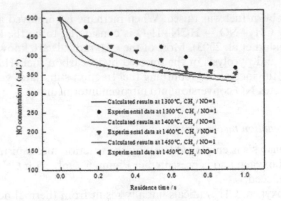

Figure 4. Impact of temperature on NO_x reduction with $CH_4/NO = 1$.

NO_x abatement by CH_4 depends in a significant way on temperature. The increase of temperature facilitates the NO_x removal process. The maximum NO_x reduction efficiency increases from 6.95% to 12.69% with a CH_4 to NO ratio of 0.5 as temperature increases from 1200°C to 1450°C and increases from 26.41% to 28.95% with a CH_4 to NO ratio of 1 as the temperature rises from 1300°C to 1450°C . A peak of NO reduction is clearly observed at both 1400°C and 1450°C and the peak values are almost the same. Therefore, the optimal temperature for NO_x reduction in the absence of oxygen is within 1400–1450°C, which differs from conclusions obtained in previous studies. Wang et al. investigated the impact of temperature in a reburning zone on NO_x reduction and found that optimal reaction temperature is between 1200 and 1350°C with excess air ratio of 0.85/0.9 and residence time of 0.6 s (Wang et al. 2009). Chen et al. found that the NO_x reduction efficiency at a stoichiometry of 0.9 decreased with temperature and the effect of temperature diminished when the temperature was above 1200°C (Chen et al. 1989). The difference between this research and previous work is largely attributed to the absence of oxygen, and a detailed analysis of the mechanism will be presented in Section 4.3.

4.2 The influence of CH_4 to NO ratio

CH_4 to NO ratio has a remarkable effect on NO_x reduction, as illustrated in Figure 5. It is of interest to note that the increase of CH_4 to NO ratio from 0.5 to 1.0 induces an increase of NO removal efficiency from 15.00% to 29.50%. The results are consistent with previous researches. Hampartsoumian et al. (2003) observed an increase in NO abatement when the

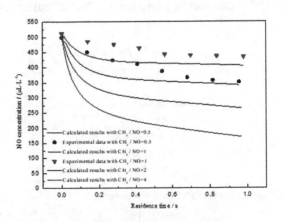

Figure 5. Impact of CH_4 to NO ratio on NO_x reduction at 1400°C.

amount of injected reburn fuel was raised. When methane is employed as the reburning fuel, the reaction pathway $CH_3 + NO \rightarrow HCN + H_2O$ is considered to be the key step for reducing NO emission (Frassoldati et al. 2003). Most of the CH_i radicals are known to be generated by the partial oxidation and pyrolysis of the injected hydrocarbon fuel (Hampartsoumian et al. 2003). Consequently, the increase of reburning fuel fraction stimulates generation of CH_i radicals and hence promotes NO conversion into nitrogen intermediates.

4.3 *Analysis of the reaction mechanism*

Rate of production analysis is employed to study the reaction mechanisms and reaction pathway in the absence of oxygen and is presented in Figure 6. Only with CH_4 converted into CH_3 or CH_2 can NO be reduced.

In the absence of oxygen, CH_3 radicals initially result from thermal decomposition of CH_4 via reaction (1) and then mainly from regeneration from reactions (2) and (3):

$$CH_4(+M) \rightleftharpoons CH_3 + H(+M) \tag{1}$$

$$H + CO_2 \rightleftharpoons CO + OH \tag{2}$$

$$CH_4 + OH \rightleftharpoons CH_3 + H_2O \tag{3}$$

Since reaction (1) is strongly temperature dependent, the decomposition of CH_4 is hindered when the temperature is below 1200°C and NO reduction is trivial.

NO is mainly converted into HCN with the presence of CH_3 radicals via the following reaction:

$$CH_3 + NO \rightleftharpoons HCN + H_2O \tag{4}$$

HCN is then directly or indirectly converted into HNCO, which is an important nitrogen intermediate. Subsequently, HNCO can be converted into N_2 via the following reactions:

$$HNCO + OH \rightleftharpoons NCO + H_2O \tag{5}$$

$$NCO + NO \rightleftharpoons N_2O + CO \tag{6}$$

$$N_2O(+M) \rightleftharpoons N_2 + O (+M) \tag{7}$$

$$NCO + NO \rightleftharpoons N_2 + CO_2 \tag{8}$$

On the other hand, HNCO can also be converted into NO through the following reactions:

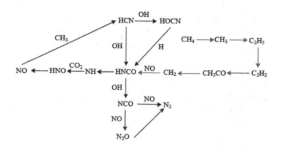

Figure 6.   Reaction mechanisms for CH_4/NO without oxygen.

$$HNCO + M \rightleftarrows NH + CO + M \qquad (9)$$

$$NH + CO_2 \rightleftarrows HNO + CO \qquad (10)$$

$$HNO + M \rightleftarrows H + NO + M \qquad (11)$$

Increasing the temperature favors the generation of NO via reaction (11). As mentioned earlier, the increase of temperature stimulates the thermal decomposition of CH_4 and accelerates NO conversion into HCN. Thus, the effect of temperature, especially high temperature, is determined by competition of thermal decomposition of CH_4 and HNO.

In the presence of oxygen, CH_i radicals are generated via the interaction of CH_4 and O_2. As the temperature requirement for CH_4/O_2 reactions is lower than that for thermal decomposition of CH_4, the optimal temperature for the reburning process is relatively low. According to the study of Smoot, under fuel-rich conditions, the formation of HCN relies strongly on the concentration of hydrocarbon species via the following reactions (Smoot et al. 1998):

$$CH_i + NO \rightarrow HCN + \cdots \qquad (12)$$

HCN is ultimately converted into N_2 via the following reactions:

$$HCN + O \rightarrow NCO + H \qquad (13)$$

$$NCO + H \rightarrow NH + CO \qquad (14)$$

$$NH + H \rightarrow N + H_2 \qquad (15)$$

$$N + NO \rightarrow N_2 + O \qquad (16)$$

However, under fuel-lean conditions, hydrocarbons react with oxygen and/or hydroxyl radicals via the following reaction to form CO:

$$CH_i + O \rightarrow CO + H + \cdots \qquad (17)$$

The competition of reactions (12) and (17) determines the final NO_x concentration.

Thus, an interesting NO reduction efficiency can be achieved by maximizing the exposure of NO to CH_i and minimizing CH_i interaction with oxygen. It also indicates that injection of CH_4 into a highly reducing atmosphere may facilitate NO reduction process. To validate the

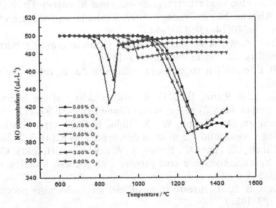

Figure 7. NO_x reduction under different oxygen concentrations with a mole fraction of 0.5.

idea and the scheme proposed, further calculations are performed in Chemkin Pro 17.0 and NO_x reduction efficiency as a function of temperature with various oxygen concentrations from 0% to 5% is illustrated in Figure 7. Obviously, a peak of NO_x reduction efficiency is observed with oxygen concentration of 0.05% at a temperature of about 1300°C. The optimal temperature for NO_x reduction shifts to a lower temperature with increasing oxygen concentration because the presence of oxygen has a remarkable effect on the production of CH_i via interaction between CH_4 and O_2 at relatively low temperature, i.e., below 1200°C . But when the oxygen concentration is high enough, hydrocarbon species are prone to be oxidized at extremely high temperature, which has an adverse effect on NO_x reduction.

5 CONCLUSION

This article describes experimental and theoretical research on NO_x reduction by CH_4 in the absence of oxygen. The CH_3 radicals are essential for NO reduction and initially result from thermal decomposition of CH_4, which is strongly temperature dependent. At extremely high temperatures, thermal decomposition of HNO inhibits NO removal. The final NO concentration is determined by the competition of thermal decomposition of CH_4 and HNO.

The influence of temperature and CH_4 to NO ratio on NO_x reduction has been investigated. With the increase of temperature, NO_x reduction is promoted because increasing temperature favors generation of CH_i radicals. But when the temperature exceeds 1400°C, the increase of temperature suppresses NO reduction due to thermal decomposition of HNO. Furthermore, increasing the CH_4 to NO ratio induces a significant NO reduction.

At a temperature range from 600 to 1450°C, a peak of NO_x reduction efficiency is observed with oxygen of 0.05% at temperatures of about 1300°C because trace oxygen is prone to convert NO into HCN rather than oxidize hydrocarbon species. However, the optimal temperature for NO reduction increases as oxygen concentration declines. Therefore, injection of methane into regions characterized by a highly reducing atmosphere and high temperature is crucial to increase NO removal efficiency.

ACKNOWLEDGMENTS

The authors gratefully acknowledge the financial support from National Key R&D Program of China, grant no. 2018YFB0604202.

REFERENCES

Ballester, J., Ichaso, R., Pina, A., González, M., & Jiménez, S. 2008. Experimental evaluation and detailed characterization of biomass reburning. *Biomass and Bioenergy* 32(10): 959–970.
Bilbao, R., Millera, A., Alzueta, M., & Prada, L 1997. Evaluation of the use of different hydrocarbon fuels for gas reburning. *Fuel* 76(14–15): 1401–1407.
Chen, S., Kramlich, J., Seeker, W., & Pershing, D. 1989. Optimization of reburning for advanced NOx control on coal-fired boilers. *Air Repair* 39(10): 1375–1379.
Chen, W., & Ma, L. 2010. Effect of heterogeneous mechanisms during reburning of nitrogen oxide. *Aiche Journal* 42(7): 1968–1976.
Frassoldati, A., Faravelli, T., & Ranzi, E. 2003. Kinetic modeling of the interaction between NO and hydrocarbons at high temperature. *Combustion and Flame* 135(1–2): 97–112.
Hampartsoumian, E., Folayan, O., Nimmo, W., & Gibbs, B. 2003. Optimisation of NOx reduction in advanced coal reburning systems and the effect of coal type. *Fuel* 82(4): 373–384.
Han, X., Rückert, F., Schnell, U., Hein, K., Koger, S., & Bockhorn, H. 2003. Computational modeling of NOx reburning by hydrocarbons in a coal furnace with reduced kinetics. *Combustion Science and Technology* 175(3): 523–544.
Kim, H., Baek, S., & Kim, S. 2012. Investigation of fuel lean reburning process in a 1.5 mW boiler. *Applied Energy* 89(1): 183–192.

Liu, H., Hampartsoumian, E., & Gibbs, B. 1997. Evaluation of the optimal fuel characteristics for efficient NO reduction by coal reburning. *Fuel and Energy Abstracts* 76(11): 985–993.

Maly, P., Zamansky, V., Ho, L., & Payne, R 1999. Alternative fuel reburning. *Fuel* 78(3): 327–334.

Nazeer, W., Jackson, R., Peart, J., & Tree, D. 1999. Detailed measurements in a pulverized coal flame with natural gas reburning. *Fuel* 78(6): 689–699.

Prada, L., & Miller, J. 1998. Reburning using several hydrocarbon fuels: A kinetic modeling study. *Combustion Science and Technology* 132(1–6): 225–250.

Shen, B., Yao, Q., & Xu, X. 2004. Kinetic model for natural gas reburning. *Fuel Processing Technology* 85(11): 1301–1315.

Smoot, L., Hill, S., &d Xu, H. 1998. NOx control through reburning. *Progress in Energy & Combustion Science* 24(5): 385–408.

Spliethoff, H., Greul, U., Helmut, R., & Hein, K 1996. Basic effects on NOx emissions in air staging and reburning at a bench-scale test facility. *Fuel* 75(5): 560–564.

Wang Z, Dong Y, Chun, Y., & Sun, S. 2009. Influence of temperature change on conversion characteristics of NOx in the reburning zone. In *International Conference on Energy & Environment Technology*. IEEE.

Wendt, J., Sternling, C., & Matovich, M. 1973. Reduction of sulfur trioxide and nitrogen oxides by secondary fuel injection. *Symposium (International) on Combustion* 14(1): 897–904.

Zhong, B., Shi, W., & Fu, W. 2002. Effects of fuel characteristics on the NO reduction during the reburning with coals. *Fuel Processing Technology* 79(2): 93–106.

Emerging Developments in the Power and Energy Industry – Dufo-López, Krzywanski & Singh (eds)
© 2020 Taylor & Francis Group, London, ISBN 978-0-367-27169-5

An identification method of weak links of overhead lines in distribution network based on typhoon scenario simulation

Gonglin Zhang
State Grid Fujian Electric Power Co., Ltd. Electric Power Research Institute, Fuzhou, China

Qingliang Wang & Zili Yin
State Grid Fujian Electric Power Co., Ltd. Fuzhou, China

Yunfei Mu & Lin Li
Key Laboratory of Smart Grid of Ministry of Education, Tianjin University, Tianjin, China

ABSTRACT: In order to reduce load loss and economic loss of distribution network when struck by typhoon, this paper proposed an identification method of weak links of overhead lines in distribution network based on typhoon scenario simulation. Firstly, the Batts model is used to describe the wind speed and attenuation process after typhoon landing. Then, the failure rates of pole and line are established based on the relationship between wind load and anti-bending strength of the pole. Further, the line vulnerability index is defined considering the line failure frequency and the importance of load. The weight of each factor in the index is determined by analytic hierarchy process. Finally, the identification method of weak links of overhead lines based on Monte Carlo method in distribution network is given. The proposed model is implemented on an IEEE RBTS BUS6 and the results are presented to show that the weak lines determined by the method can be used as the rule for reinforcing concrete poles and distributing emergency generators, which can effectively reduce the load loss and economic loss of the distribution network.

1 INTRODUCTION

In recent years, the frequency of typhoon disaster has increased significantly because of global warming. According to statistics, the economic loss caused by typhoons in China reaches 30 billion yuan (Duan Yihong & Chen Lianshou et al. 2012). For power distribution systems, typhoon often causes large-scale blackouts and may even escalate into catastrophic breakdown of the power grid. For example, the severe typhoon "Davei" caused the collapse of Hainan Power Grid in 2005 and a large number of users were affected by the disaster (Wu Yongjun & Xue Yusheng et al. 2016). In 2006, the typhoon "Sangmei" landed in Wenzhou with a maximum wind speed of 68m/s, leading to the breakdown of 510 distribution lines (Wu Mingxiang & Bao Jianqiang et al. 2007). The economic loss was nearly 200 million yuan. Especially in the underdeveloped areas such as rural areas, the poles in distribution network are mostly concrete poles. The typhoon wind speed is usually far higher than designed withstanding wind speed of pole (Peng Xiangyang & Huang Zhiwei et al. 2010), and thus causing high failure rate of concrete pole, which has become the focus of emergency rescue. However, it is unpractical to conduct large-scale reinforcement of the overhead lines. Also only limited emergency supplies can be allocated to prevent typhoon ahead of time. Although the meteorological department can make accurate forecast before typhoon, the problems of thoughtless material allocation and bad decisions still exist in power company when facing emergency situation.

Some existing researches in this field are as follow: a method to differentiate the output warning results based on typhoon forecast information was proposed in (Huang Yong & Wei Ruizeng 2018); a quantitative assessment framework for grids operating under typhoon weather was prosed in (Y. Yang & W. Tang 2018), using the typhoon wind field model; a N-k weak link identification strategy based on component vulnerability index was proposed in (Ming Wang & Yingmeng Xiang 2016); a reliability tracking method to identify the weak links of the power system was proposed in (Ji Jing & Xie Kaigui 2011), through tracking analysis, the contribution of each component to the system reliability index can be obtained.

This paper proposes a weak link identification method for overhead lines of distribution network based on typhoon scenario simulation. The example shows that the method can make the emergency supplies to be more effectively allocated and further reduce the load loss and economic loss. It has great guiding significance and theoretical reference value for wind-proof work in coastal areas.

2 BATTS TYPHOON MODEL

In the Batts model, the wind speed calculation steps at each point are as follows. R_{max} is the distance from the center of the cyclone to the strongest wind belt; V_{gx} is air flow velocity caused by pressure gradient force.

$$R_{max} = \exp(-0.1239\Delta P^{0.6003} + 5.1043) \tag{1}$$

$$V_{gx} = K\sqrt{\Delta P} - 0.5 \cdot R_{max}f \tag{2}$$

where ΔP is the pressure difference between the edge of tropical cyclone and meteorological center (hPa); f is the earth's rotation Coriolis force coefficient; K is a constant coefficient.

Usually the average maximum wind speed V_{Rmax} occurs at R_{max} in the wind field.

$$V_{R_{max}} = 0.865V_{gx} + 0.5V_T \tag{3}$$

where V_T is the moving speed of typhoon (m/s), which can be obtained from the typhoon forecasting information.

The calculation formula of each point in the wind field is as follows.

$$\begin{cases} V_{rin} = V_{R_{max}} \frac{r}{R_{max}}, & r \leq R_{max} \\ V_{rout} = V_{R_{max}} \left(\frac{R_{max}}{r}\right)^x, & r > R_{max} \end{cases} \tag{4}$$

where r is the distance from the overhead line to the center of typhoon field; V_{rin} and V_{rout} are the wind speeds in the wind field; $x \in [0.5, 0.7]$.

After typhoon landing, it can be seen from (5) that the air pressure difference at the center of the cyclone is continuously decreasing. From (1) to (4), the wind speed at each point in wind field is continuously reduced.

$$\Delta P(t) = \Delta P_0 - 0.675(1 + \sin\beta)t \tag{5}$$

where ΔP_0 is the central air pressure difference when typhoon lands (hPa); $\Delta P(t)$ is the central air pressure difference at time t after typhoon landing (hPa); β is the angle between the typhoon and the typhoon moving direction when the typhoon landed on the coastline.

3 MODEL OF POLE FAILURE RATE

According to statistics of historical fault data, most faults caused by typhoon are concrete pole failures, and the probability of disconnection of overhead lines is extremely low (Xu Dan & Tang Wei 2011). This paper only considers failure of poles.

3.1 Wind load

In the event of a typhoon disaster, the pole is mainly subjected to three external forces: wind load on the conductor, wind load on the tower and wind load on the insulator.

Considering the three kinds of loads, the bending moment M_x is calculated according to formula (6)–(9).

$$M_x = (w_{xz}h_1 + 2w_{xz}h_2 + w_{sv}\overline{h})(1 + m_x) \tag{6}$$

$$w_{sv} = \beta\mu_z\mu_s F \frac{v^2}{1600} \frac{D_0 + D_x}{2} h_1 \tag{7}$$

$$w_{xz} = w_x + w_z \tag{8}$$

$$\overline{h} = \frac{h_1}{3} \frac{2D_0 + D_x}{D_0 + D_x} \tag{9}$$

where h_1 is the distance from the section to top of the pole (m). Because collapse usually occurs at the root of the pole, h_1 is set to the whole height. h_2 is the distance from the section to crossbar (m); w_{xz} is the total wind load on lines and insulators (kN); $\overline{h}$ is the height between the section and the action point of pressure joint force on the pole(m); F is the projected area of the pole (m²); D_0 is the top diameter of the pole (m); D_x is the diameter of the section (m); m_x is the additional bending moment coefficient generated by disturbance.

3.2 Pole failure rate model

Considering the difference in the manufacturing process of concrete poles, there is a certain dispersion of the anti-bending strength. It generally conforms to a normal distribution, and its probability density function is shown in (10).

$$f_R(M_p) = \frac{1}{\sqrt{2\pi}\delta_p} e^{-\frac{1}{2}\left(\frac{M_p - \mu_p}{\delta_p}\right)^2} \tag{10}$$

where μ_p is the mean value of the bending strength of the pole; δ_p is the standard deviation of the bending strength of the pole.

The strength variables and load effects of the pole can be obtained from (6)–(10). Thus, the criterion Z is established as follows.

$$Z = M_p - M_x \tag{11}$$

Since M_p obeys to a normal distribution, Z obeys to a normal distribution. The following criteria is obtained.

$$\begin{cases} Z > 0 & \text{normal} \\ Z < 0 & \text{faulty} \end{cases} \tag{12}$$

For a specific bending moment M_x, the failure rate of the pole is:

$$P = P\{(M_p - M_x < 0\} = \int_0^{M_x} \frac{1}{\sqrt{2\pi}\delta_p} e^{-\frac{1}{2}\left(\frac{M_p - \mu_p}{\delta_p}\right)^2} \, dM_p \qquad (13)$$

3.3 *Correction model of pole failure rate*

During the entire operation of the pole, the number of failure times and the running years follows the 'bathtub curve' as shown in Figure 1.

Based on the data recorded by the power department, the parameters of the bathtub curve are further obtained by least squares fitting. The correction factor $\lambda(T)$ of pole failure rate is expressed by Weibull distribution, as shown in (14).

$$\lambda(T) = T^{(\beta-1)} \qquad (14)$$

where T is the operating period of the pole; in the 'Infant Mortality', $\beta<1$; in the 'Steady-state', $\beta=1$; in the 'Wearout Failures', $\beta>1$. The corrected pole failure rate is given in (15).

$$P_f = \lambda(T)P \qquad (15)$$

For a overhead line of the distribution network, the poles are connected in series. The overhead line failure rate can be represented by a pole series model.

$$P_2 l = 1 - \prod_{i=1}^{m}(1 - P_2 f(i)) \qquad (16)$$

where $P_f(i)$ is the failure rate of the i-th pole; m is the number of poles in one line.

4 IDENTIFICATION METHOD OF WEAK LINK IN TYPHOON SCENARIO

The main method to identify weak links is divided into four parts: selecting vulnerability index, calculating component vulnerability index, sorting vulnerability index and selecting the first k components to get N-k weak links.

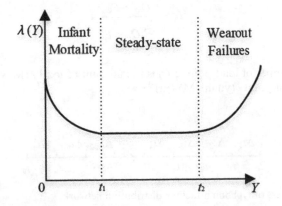

Figure 1. Tub curve of pole failure rate.

4.1 *Vulnerability index of lines*

Considering the line failure frequency F_k and the load importance I_k, the vulnerability index of overhead line V_k is defined in (17).

$$V_k = z_1 \times F_k + z_2 \times I_k \tag{17}$$

$$z_1 + z_2 = 1 \tag{18}$$

where z_1 and z_2 are weight coefficients.

The Monte Carlo Sampling method is used to simulate the typhoon landing process. The invading process of the typhoon is divided into n time intervals Δt, as shown in Figure 2. The wind speed of Δt_i is expressed by the average wind speed.

$$V(\Delta t_2 i) = \frac{\int_{t_{2i-1}}^{t_{2i}} V(t)}{\Delta t} \tag{19}$$

For Δt_1, the failure rate $P_f(k)$ of link k is obtained from (13)–(16). If $S_k(1)$ represents the state of line k, then a random number R in the interval [0,1] is generated for line k, which obeys to the uniformity distribution.

$$S_k(1) = \begin{cases} 0, R > P_l(k) & \text{normal} \\ 1, 0 \le R \le P_l(k) & \text{faulty} \end{cases} \tag{20}$$

Complete the sampling of the K lines of the distribution network. On the base of the state of each line after Δt_1, continue to sample the line state after Δt_2 (if line k is faulty at Δt_1, the fault state is maintained in the sample after Δt_1) until the completion of typhoon attacking process. Finally, the final state S with K line system is obtained:

$$\mathbf{S} = (s_1, \cdots, s_i \cdots, s_K) \tag{21}$$

When the number of typhoon simulation times is sufficient (N times), F_k is shown in (22).

$$F_k = \frac{1}{N} \sum_{i=1}^{N} S_k(i) \tag{22}$$

The load importance connected with line k is as shown in (23).

$$I_k = \frac{Q_k \times E_k}{\sum_{i=1}^{M} Q_i \times E_i} \tag{23}$$

where M is the number of load points; Q_i is the amount of load i (MW); E_i is the economic value per megawatt of point i (yuan•(MW•h)$^{-1}$).

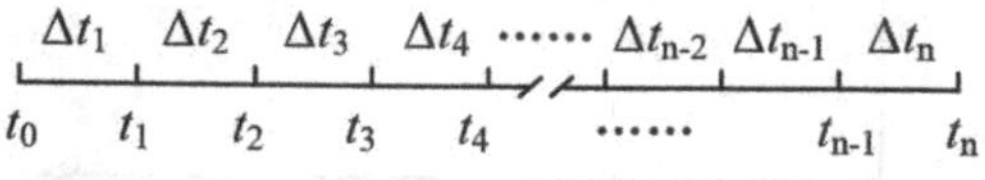

Figure 2.   Timing process of typhoon attacking distribution network.

4.2   Weight determination method of vulnerability index

Analytic Hierarchy Process (AHP) is a systematic and hierarchical analysis method combining qualitative and quantitative methods. It uses empirical judgment to quantify the importance of various standards. To determine the weight coefficient of each factor in V_k, the basic steps are as follows:

1) Building a hierarchical model. Taking the identification weak links as the target layer, considering the two factors of line failure frequency and load importance in the criterion layer, each line is used as an alternative.

2) Construct a pairwise comparison matrix. When comparing the importance of F_k and I_k factors in the comparison criteria layer to the target layer, the comparison matrix C is constructed using the 1-9 comparison scale. C is shown in (24).

$$C = \begin{bmatrix} c_{11} & c_{12} \\ c_{21} & c_{22} \end{bmatrix} \tag{24}$$

where c_{11} and c_{22} are the relative importance of F_k and I_k compared with themselves, and the value is 1; c_{12} is the importance of F_k relative to I_k, and c_{21} and c_{12} are reciprocal to each other.

3) Calculate the weight vector and check consistency. Calculating the maximum eigenvalue of C and the corresponding eigenvectors, and the consistency index is used for consistency checking. If the test passes, the feature vector (after normalization) is the weight vector; if it does, C needs to be reconstructed.

5   REFERENCES WEAK LINK IDENTIFICATION METHOD

Define the loss of load index (LLR) under typhoon scenario to compare the changes before and after the improvement of the selected weak links, as shown in (25). Convergence of the LLR results is used as a criterion for stopping the simulation of the typhoon process, as shown in (26).

$$LLR = \frac{1}{N} \sum_{i=1}^{N} llr(i) \tag{25}$$

$$U = \frac{\sqrt{\frac{1}{N} \sum_{i=1}^{N} (llr(i) - LLR)^2}}{LLR} \tag{26}$$

Where N is the simulation times of the typhoon process; $llr(i)$ is the load loss determined by the final state S of the system after the ith simulation; U is the variance coefficient, and the sampling is stopped when the value is less than the given value.

Before the arrival of typhoon, according to the weather forecast information, the power department usually takes precautionary measures such as reinforcing poles and distributing emergency generators to reduce the load loss after typhoon disaster. The traditional material distribution method regards the line where the load is larger and the economic value is higher as the weak links. Table A2 of Appendix A gives the average load level of each load point and economic losses of unit power. The example gives a comparison of the load loss after taking the traditional method and the method in this paper.

6   CASE STUDY

In this paper, the IEEE RBTS BUS6 feeder system shown in Figure 3 is used as a test example. The lengths of the feeder lines and the load node data of the system are shown in Table A1 and

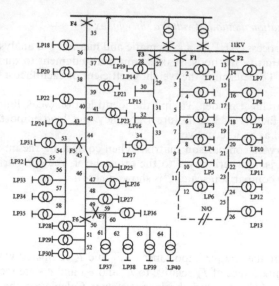

Figure 3.   Structure diagram of IEEE RBTS BUS6 process.

Table A2 of Appendix A. Figure 4 is a schematic diagram of a typhoon attacking the distribution network. When the typhoon lands, the center air pressure difference $\Delta P_0 = 25$hPa. Assume that the typhoon lands at O_1, and the path is O_1-O_4 after landing with an angle of 45°. The red five-pointed star in the figure indicates the geographical position of the distribution network, and the coordinates are (100km, 100km). The radius of the circle O_1 and circle O_4 gradually decreases, indicating the attenuation process after the typhoon landed.

According to the "Code of Operation for Distribution Network", the average span of the poles of the 35kV distribution network is 300m, and the average pitch of the poles of the 10kV overhead line is 100m. The design wind speed of the pole is about 30 m/s.

The vulnerability index of each overhead line is calculated. After the LLR calculation results converge, under the typhoon intensity, C is constructed. After the sampling of the typhoon process is completed, F_k of each line is statistically obtained. Figure 5 shows the calculation results of the vulnerability index of some overhead lines.

According to the materials and staffing allocation ability, the top 9 lines with the highest vulnerability index are selected to install wind-proof cables and emergency generators to reduce the load loss after the typhoon. According to the calculation results of the line vulnerability index in Figure 5, the weak links in this scenario are 30, 32, 64, 58, 28, 34, 62, 52, 63.

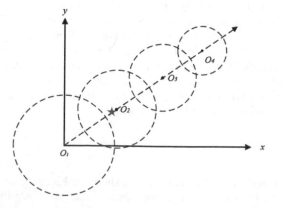

Figure 4.   Schematic diagram of typhoon landing F4 system.

545

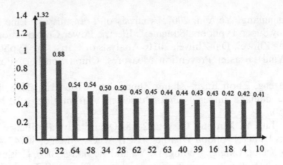

Figure 5.   Results of vulnerability index of overhead lines.

Table 1.   Load and economic loss of the system after eliminating weak links.

	Line numbers of weak links	LLR(MW)	Economic loss (ten thousand yuan)
Tradition method	30,32,43,64,47,62,39,58,38	2.6581	3.7699
Method in this paper	30,32,64,58,34,28,62,52,63	2.2273	3.2964

Similarly, according to the traditional method, 30, 32, 43, 64, 47, 62, 39, 58, 38 are selected as the weak link of the lines.

Table 1 compares the result of system load loss and economic loss using the proposed method and the traditional method. When calculating the economic loss, set the line rescue time to 24 hours. It can be seen from Table 1 that compared with the traditional method, the load loss and economic loss of the system in the typhoon scenario are significantly reduced, and the LLR is reduced by 4.56% after weak link improvement through considering line fault frequency and load importance factors.

7   CONCLUSION

This paper proposed an identification method of weak link of overhead line in distribution network based on typhoon scenario simulation. When setting up a disaster scenario, using the typhoon scenario to simulate the whole process of typhoon; when defining the line vulnerability index, consider the two factors of the line failure frequency and the importance of load to determine the weak link of the distribution network. The method of this paper can provide some guidance for the power company before typhoon landing.

ACKNOWLEDGEMENTS

This work is financially supported by the Research and demonstration of management and control verification for distribution network dispatching based on battalion coordination information(State Grid Corporation Science and Technology Project).

REFERENCES

Duan Yihong, Chen Lianshou, Xu Yinglong, Qian Chuanhai, 2012, Current Status and Suggestions of Typhoon Monitoring and Forecasting System in China. China Engineering Science, Mag, 14(09): 4–9.
Wu Yongjun, Xue Yusheng, Xie Yunyun, Wang Wei, Duan Ronghua, Huang Wei, 2016, The Temporal and Spatial Effects of Typhoon and Rainstorm on Power Network Failure Rate. Automation of Electric Power Systems, Mag, 40(02): 20–29.

Wu Mingxiang, Bao Jianqiang, Ye Yin, 2007, Analysis of Transmission Line Accidents in Wenzhou Power Grid Caused by Super Typhoon "Sangmei". Electric Power Construction, Mag, 28(9): 39-41.

Peng Xiangyang, Huang Zhiwei, Dai Zhiwei, 2010, Analysis of Causes of Typhoon Damage on Distribution Lines and Wind Disaster Prevention Measures. China Southern Power Grid Technology, Mag, 4(1):99–102.

Huang Yong, Wei Ruizeng, Zhou Enze, Zhang Zhuangling, Hou Hui, 2018, Early warning method of transmission line damage under typhoon disaster. Automation of Electric Power Systems, Mag, 42(23): 142–150.

Y. Yang, W. Tang, Y. Liu, Y. Xin and Q. Wu, 2018, Quantitative Resilience Assessment for Power Transmission Systems Under Typhoon Weather. IEEE Access, Mag, 6, 40747–40756.

Ming Wang, Yingmeng Xiang, Lingfeng Wang, D. Yu and Jie Jiang, 2016, Critical line identification for hypothesized multiple line attacks against power systems. IEEE/PES Transmission and Distribution Conference and Exposition (T&D), Con, 1–5.

Ji Jing, Xie Kaigui, Cao Wei, Hu Bo, Wu Weijie, 2011, Analysis of Weak Link Identification and Reliability Improvement of Guangdong Power Grid. Automation of Electric Power Systems, Mag, 35(13): 98–102.

Xu Dan, Tang Wei, 2011, Reliability Evaluation of Complex Distribution Network Based on Regional Accessibility Analysis. Transactions of China Electrotechnical Society, Mag, 26(6): 172–178.

APPENDIX A

Table A1. Feeder Type and Lengths BUS6.

Length (km)	Feeder Number
0.6	2,3,8,9,12,13,17,19,20,24,25,28,31,34,41,47
0.75	1,5,6,7,10,14,15,22,23,26,27,30,33,43,61
0.8	4,11,16,18,21,29,32,35,55,
0.9	38,44,
1.6	37,39,42,49,54,62
2.5	36,40,52,57,60,
2.8	34,46,50,56,59,64,
3.2	45,51,53,58,63,
3.5	48

Table A2. Customer Data of IEEE RBTS.

Load Point	Load Level (MW)	Economic Value of Unit Load (yuan·(MW·h)$^{-1}$)
15	1.6391	790
16	0.9025	790
32,37	0.1929	790
20,30,34	0.2501	790
21,35	0.2633	790
24,40	0.3057	790
26,38	0.2831	790
14,17	0.4697	570
5,6	0.2163	570
2,4,11,19	0.1808	570
12,13,22	0.2070	570
1,3,9	0.1775	570
7,8,10,18,23	0.1659	570
27,29,33,39	0.1585	570
25,28,31,36	0.1554	570

Emerging Developments in the Power and Energy Industry – Dufo-López, Krzywanski & Singh (eds)
© 2020 Taylor & Francis Group, London, ISBN 978-0-367-27169-5

Research on the VIENNA rectifier based on backstepping control

Junrui Wang, Sining Jia, Chuang Wang, Shang Xiang, Mengyue Zhang & Xiang Shan
School of Electrical and Information Engineering, North Minzu University, Yinchuan, China

ABSTRACT: Through the analysis of the basic principle of the three-phase VIENNA rectifier, the mathematical model under the d–q coordinate axis is established. This article then
adopts a backstepping control strategy to control the DC voltage and the d–q axis voltage. On
the one hand, it enhances the stability of the DC side output voltage; on the other hand, it
realizes the rapidity of the grid side current and stability of control. In addition, a simplified
algorithm based on two-level SVPWM is transplanted into the VIENNA rectifier. Finally, the
simulation platform is built by MATLAB®/Simulink software to verify the rationality and
effectiveness of the algorithm, so that the VIENNA rectifier control system has good stability
and dynamic performance, and maintains a good control effect on the basis of adding
disturbance.

1   INTRODUCTION

In the past few decades, the progress of power semiconductor switching device technology has
promoted the rapid development of power electronic converter device technology. Also, the
various converters based on a pulse width modulation device have emerged, such as variable-
frequency drive, inverter power supply, high-frequency switching power supply, and these
devices have been applied to all areas of the national economy. However, these converter
devices generally need a rectifier link to obtain DC voltage. Due to the conventional diode
uncontrolled rectifier circuit or thyristor phase-controlled circuit, it causes serious "grid pollution." The fundamental measure to control this "contamination" is to require the network to
measure the current as sinusoidal and run at unit power factor. According to whether the
energy can flow bidirectionally, the pulse-width modulated (PWM) rectifier is divided into
two different topologies, namely reversible PWM rectifier and irreversible PWM rectifier
(Zhang & Zhang 2012). A reversible PWM rectifier represented by a two-level PWM rectifier
overcomes the disadvantages of large current harmonics and low power factor of the traditional rectifier circuit. However, with the multilevel PWM rectifier gradually becoming
a research hotspot, it requires more and more switching devices, and the control is more and
more difficult. On this basis, the VIENNA rectifier is an irreversible PWM rectifier, which
solves the problems encountered with multilevel PWM rectifiers. It has few power switching
devices, small switching tube stress, no bridge arm straight-through problem during operation,
and no need to set dead band. Therefore, it has been more and more widely considered
(Zhang et al. 2018).

Currently, there is more and more research on VIENNA rectifier control strategy. Song
et al. (2013) adopt hysteresis current control technology, which has the advantages of simple
hardware equipment and small current deviation, but there are problems such as unsound
switching frequency, large current harmonics on the grid side, and poor immunity of grid
voltage. Wei et al. (2013) use single-cycle control technology, using analog devices with reset
integrators, pulse generators, etc. It has simple control, fast dynamic response, and strong
resistance to power disturbance, but it has problems of poor load disturbance rejection and
system stability error.

Backstepping control is very flexible, and the key is the special handling of nonlinear links. For some nonlinear systems, feedback linearization cannot be used to achieve the target, and backstepping control can solve this problem well. Since the introduction of backstepping control, the uniqueness of its controller design process, and its superior processing ability for uncertain problems, have allowed it to become more and more popular among relevant scholars in recent years. Backstepping control realizes the design of the controller through step-by-step backstepping according to the control target, which can overcome the general change of system parameters during the running of the system. However, the design of the controller relies on accurate mathematical models. In order to achieve global stability, the physical quantities of the system can reach the ideal equilibrium point. In practical applications and experiments, it is usually combined with the Lyapunov adaptive law to achieve control objectives.

First, the topology of the VIENNA rectifier is introduced, and the mathematical model of the rectifier in the d–q coordinate system is established. Second, the modulation technology improved based on two-level SVPWM used in this model is briefly introduced. Finally, MATLAB®/Simulink software is used to build a simulation model to verify the correctness and feasibility of the control strategy and parameter design.

2   VIENNA RECTIFIER TOPOLOGY MODEL AND MODELING

2.1   *Modeling of the VIENNA rectifier*

In 1994, J. W. Kolar proposed a two-quadrant midpoint clamped three-level PWM rectifier topology (VIENNA) rectifier. The topology evolved from a diode rectifier. A bidirectional switch is connected to each of the three-phase input terminals of a, b, and c and the midpoint of the DC bus capacitor to form a three-level structure. The bidirectional switch is commonly emitted by two insulated gate bipolar transistors with antiparallel diodes. The poles are reversely connected in series. In this article, the main circuit topology bidirectional switch is selected in this form, as shown in Figure 1.

The circuit is a current-driven power factor correction rectifier. The voltage at both ends of the power switch tube is determined by the state of the switch tube itself and the direction of the input current. In Figure 2, U_a, U_b, and U_c are symmetric three-phase AC sources; L_s is the filter inductor; and R_s is the input resistance. D_{ap}, D_{bp}, D_{cp}, D_{an}, D_{bn}, and D_{cn} are common diodes; S_{a1}, S_{a2}, S_{b1}, S_{b2}, S_{c1}, and S_{c2} are three-phase switches; C_1, C_2 are the load side upper and lower capacitors; and R_l is the load side resistance. Taking the A phase as an example, the input current is positive, the switch is on, and the switch tube is clamped in the positive pole

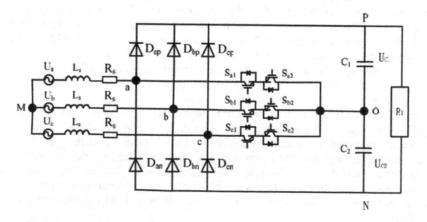

Figure 1.   Main circuit topology of the VIENNA rectifier.

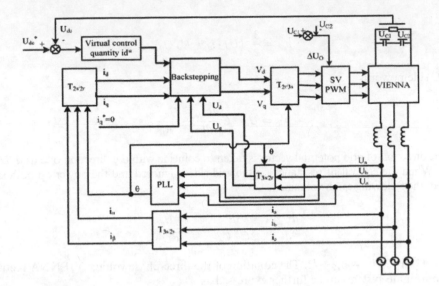

Figure 2. VIENNA rectifier backstepping control structure frame.

of the DC side. At this time, the voltage at both ends of the switch is $U_{ao} = U_{c1} = U_{dc}/2$; If the input current is negative, the switch is switched on, and the switch tube is clamped on the negative pole on the DC side, and $U_{ao} = U_{c2} = -U_{dc}/2$. When the switch tube is turned off, regardless of the current is positive or negative, the switch tube is clamped at the neutral point O on the DC side (Li et al. 2015).

The power grid is in an ideal state of equilibrium and VIENNA rectifier works in continuous current mode. Based on the above working process, the derivation process is omitted. The mathematical model under the a, b, c coordinate system is transformed into the d–q coordinate system by using the rotating coordinate transformation, where i_d, i_q is the current on the rotating coordinate axis; U_d, U_q is the voltage on the rotating coordinate axis; and S_{dp}, S_{dn}, S_{qp}, and S_{qn} are the single-pole switch values of the a, b, c coordinate system transformed to the d–q axis.

$$
\begin{cases}
L_s \frac{di_d}{dt} = -R_s \cdot i_d + \omega L_s \cdot i_q - S_{dp} \cdot U_{c1} + S_{dn} \cdot U_{c2} + U_d \\
L_s \frac{di_q}{dt} = -R_s \cdot i_q - \omega L_s \cdot i_d - S_{qp} \cdot U_{c1} + S_{qn} \cdot U_{c2} + U_q \\
C_1 \frac{dU_{c1}}{dt} = S_{dp} \cdot i_d + S_{qp} \cdot i_q - \frac{1}{R_l} \cdot U_{c1} - \frac{1}{R_l} \cdot U_{c2} \\
C_2 \frac{dU_{c2}}{dt} = -S_{dn} \cdot i_d - S_{qn} \cdot i_q - \frac{1}{R_l} \cdot U_{c1} - \frac{1}{R_l} \cdot U_{c2}
\end{cases}
\tag{1}
$$

In the ideal state, it is assumed that the capacitance voltage on the DC side is equal, i.e., $U_{c1} = U_{c2} = U_{dc}/2$, $C_1 = C_2 = C$, make $v_d = (S_{dp} - S_{dn})$, $v_q = (S_{qp} - S_{qn})$, reduction is given by

$$
\begin{cases}
L_s \frac{di_d}{dt} = -R_s \cdot i_d + \omega \cdot L_s \cdot i_q - \frac{1}{2} \cdot U_{dc} \cdot v_d + U_d \\
L_s \frac{di_q}{dt} = -R_s \cdot i_q - \omega \cdot L_s \cdot i_d - \frac{1}{2} \cdot U_{dc} \cdot v_q + U_q \\
C \frac{dU_{dc}}{dt} = v_d \cdot i_d + v_q \cdot i_q - \frac{2 \cdot U_{dc}}{R_0}
\end{cases}
\tag{2}
$$

According to the power balance equation (Zhang & Cheng 2009), the active power absorbed from the AC source (P_{ac}) and the active power delivered to the converter DC side (P_{dc}) are expressed by

$$p_{ac} = \frac{3}{2} \cdot (U_d \cdot i_d + U_q \cdot i_q) \tag{3}$$

The DC side power is

$$p_{dc} = U_{dc} \cdot C \frac{dU_{dc}}{dt} + \frac{U_{dc}^2}{R_l} \tag{4}$$

Since the direction of grid potential vector is taken to coincide with the direction of axis d, $U_d = E$, $U_q = 0$. When the equivalent resistance on the grid side is ignored and the power on both sides is balanced, the following can be obtained:

$$\frac{dU_{dc}}{dt} = \frac{3E}{2C \cdot U_{dc}} \cdot i_d - \frac{U_{dc}}{R_l \cdot C} \tag{5}$$

Set $V_d = U_{dc} \cdot v_d/2$, $V_q = U_{dc} \cdot v_q/2$. The equation of the three-phase voltage VIENNA rectifier in a d–q coordinate system can be further expressed as

$$\begin{cases} L_s \frac{di_d}{dt} = -R_s \cdot i_d + \omega \cdot L_s \cdot i_q - V_d + U_d \\ L_s \frac{di_q}{dt} = -R_s \cdot i_q - \omega \cdot L_s \cdot i_d - V_q + U_q \\ \frac{dU_{dc}}{dt} = \frac{3E}{2C \cdot U_{dc}} \cdot i_d - \frac{U_{dc}}{R_l \cdot C} \end{cases} \tag{6}$$

2.2 Design of the backstepping controller

The design process of the backstepping control is based on the expected value that the output needs to meet, and the complex nonlinear system is decomposed into subsystems that do not exceed the order of the system. Then, a partial Lyapunov function and an intermediate virtual control quantity are designed for each subsystem, and finally the controller of the entire system is obtained. For the three-phase VIENNA rectifier, the main goal of its control is to achieve unit power factor and DC side voltage stability. Starting from the voltage outer loop, first define the voltage error on the DC side as $e_1 = U_{dc} - U_{dc}^*$, where U_{dc}^* is the expected output voltage on the DC side (Cheng 2014).

Select e_1 as the virtual state variable, i_d as the virtual control function, and select $V_1 = e_1^2/2$ as the Lyapunov function, and take the derivative of the function.

$$\frac{dV_1}{dt} = e_1 \cdot \frac{de_1}{dt} = e_1 \cdot \left(\frac{3E}{2C \cdot U_{dc}} \cdot i_d - \frac{U_{dc}}{R_l \cdot C} \right) \tag{7}$$

In order to obtain $dV_1/dt \leq 0$, make

$$\frac{3E}{2C \cdot U_{dc}} \cdot i_d - \frac{U_{dc}}{R_l \cdot C} = -k_1 e_1 \tag{8}$$

When $k_1 > 0$, by means of Lyapunov stability, it can be known that the actual value of DC side voltage can be stable to track the DC side voltage, and the voltage tracking is set to zero.

$$i_d = \frac{2C \cdot U_{dc}}{3E} \cdot \left(\frac{U_{dc}}{R_l \cdot C} - k_1 e_1 \right) \tag{9}$$

In order to achieve complete decoupling and output voltage stability of the three-phase voltage VIENNA rectifier, the following assumed current function is selected, where i_d and i_d^* are current expectation values for the d-axis and q-axis, respectively.

$$\begin{cases} i_d^* = \frac{2C \cdot U_{dc}}{3E} \cdot \left(\frac{U_{dc}}{R_l \cdot C} - k_1 e_1 \right) \\ i_q^* = 0 \end{cases} \tag{10}$$

The next step is to realize the d-axis current tracking, take the error $e_2 = i_d - i_d^*$ of the d-axis current, select e_1, e_2 as a new state variable, constitute a subsystem, and set up a new Lyapunov function $V_2 = V_1 + e_2^2/2$, and take the derivative of this function.

$$\frac{dV_2}{dt} = -k_1 \cdot e_1^2 + e_2 \cdot \left(\frac{1}{L} \cdot (U_d - R_s \cdot i_d + L \cdot \omega \cdot i_q - V_d) - M \right) \tag{11}$$

Among them,

$$M = \frac{2C}{3E} \left(\frac{3E}{2C \cdot U_{dc}} \cdot i_d - \frac{U_{dc}}{R_l \cdot C} \right) \left(\frac{2U_{dc}}{R_l \cdot C} - k_1 \cdot e_1 - k_1 \cdot U_{dc} \right)$$

In the above equation, i_d is actually controlled. In order to satisfy the dV_2/dt for the above equation, by setting $[(e_d - R \cdot i_d + l \cdot \omega \cdot i_q - u_d) - M]/L = -k_2.e_2$, where $k_2 > 0$, the actual control can be obtained.

$$V_d = U_d - R_s \cdot i_d + L \cdot \omega \cdot i_q + L \cdot k_2 \cdot e_2 - L \cdot M \tag{12}$$

Similarly, the actual control of the q-axis also can be obtained.

$$V_q = U_q - R_s \cdot i_q + L \cdot \omega \cdot i_d + L \cdot k_3 \cdot e_3 \tag{13}$$

In theory, the backstepping adjustment coefficient can be combined with the optimal control theory and specific system control performance requirements for parameter optimization design, but the calculation is more complex. Generally, engineering requirements can be satisfied by adjusting values according to experience (Wang et al. 2017). Given the above virtual control and feedback control, the VIENNA rectifier is exponentially asymptotically stable, and the virtual control introduced in backstepping control is essentially a static compensation. The stability of the former subsystem depends on the virtual control of the later subsystem. Therefore, tracking of the d–q axis current and stabilization of the DC side voltage can be achieved by backstepping control, and its control structure is shown in Figure 2.

3 VIENNA RECTIFIER MODULATION STRATEGY BASED ON TWO-LEVEL SVPWM

The main reason for the complexity of the three-level rectifier SVPWM modulation algorithm is that it has more control objectives, in addition to ensuring that the output voltage is consistent with the reference voltage, and also controls the operating state of the converter itself. In this article, a three-level space vector simplification algorithm based on two-level SVPWM is adopted (Jae et al. 2001). The algorithm decomposes the three-level space vector into six small two-level regions, and solves the three-level vector action time by using the two-stage SVPWM vector action time calculation method, which simplifies the calculation.

4 SYSTEM SIMULATION AND RESULT ANALYSIS

4.1 *Comparison of voltage and current stability between proportional integral and backstepping control*

In order to verify the correctness of the analysis, a simulation platform based on backstepping control was built in MATLAB®/Simulink. The VIENNA parameters are: three-phase AC voltage $e = 100$ V, grid frequency $f = 50$ Hz, input inductance $L_s = 1$ mH, input resistance $R_s = 0.1$ Ω, DC side reference voltage $U_{dc} = 200$ V, DC side upper and lower capacitance = 6000 μF, and load resistance = 50 Ω.

By comparing the waveform of the DC side voltage in Figure 3 with a proportional integral (PI) double closed-loop control strategy, it can be clearly seen that the stability of the back-stepping control strategy is better. In the case of adding step disturbance, the output voltage of the DC side under backstepping control is basically free from system interference and maintained at 200 V. On the contrary, when the DC side under PI control is disturbed, the waveform obviously changes, and the amplitude of change is relatively large. Although it tends to be stable, it may cause great damage to the rectifier. It can be seen from Figure 4 that both the A phase current and the voltage waveform can achieve the same phase and maintain the stability of the A phase current and voltage after being disturbed, which proves the effectiveness of backstepping control.

As mentioned earlier, the current of the d–q axis can be stabilized within a certain range after being disturbed, and the current of the d–q axis can be stabilized more quickly after being disturbed under backstepping control (Figure 5), which shows the advantages of bac-stepping control. However, the dynamic response of the d–q axis current is fast, but there is much clutter, and the phase A current harmonics are not significantly improved compared with the PI control, requiring further improvement later. On the whole, the research on

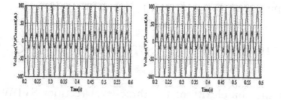

Figure 3.   Waveform of DC side voltage under PI control strategy and backstepping control.

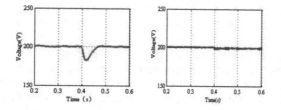

Figure 4.   A phase current and voltage waveform under PI control and backstepping control.

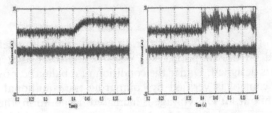

Figure 5.   Current waveform on the d–q axis under PI control and backstepping control.

VIENNA rectifiers based on backstepping control achieves the desired effect and lays a theoretical foundation for practical application.

4.2   *Effect of parameter changes on the system*

In addition, the backstepping controller parameters k_1, k_2, and k_3 also have a certain impact on the stability of the system. The first step to observe the influence of k_1 on the system, is to keep k_2, k_3 constant and set different k_1 values; the results are as shown in Figure 6.

When $k_1 = 200$, the output voltage of the system cannot be stable at the expected value, but when k_1 increases to the range of 2000–4000, there is almost no difference in the output voltage stability value. As shown in Figure 6, there is no significant difference in output voltage when k_2 is between 1000 and 2500, but the output result is closest to the expected value when $k_2 = 2500$. As k_2 increases, at $T = 0.4$ s, the system is more and more affected by disturbance.

Keep k_1 and k_2 unchanged and observe the change of k_3, as shown in Table 1.

When k_1 and k_2 remain unchanged, the output DC voltage is almost unaffected by k_3. However, k_1, k_2, and k_3 interact with and influence each other. Only by taking into account the three factors can influencing parameters suitable for the system be found.

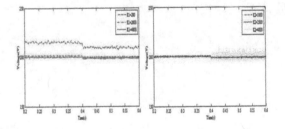

Figure 6.   Influence of k_2 and k_1 on the system.

Table 1.   Output voltage of DC side when k_3 changes.

k_1	k_2	k_3	Before adding disturbance	After adding disturbance
4000	2500	8000	199.4–200.4	198–200
4000	2500	10,000	199.2–200.5	198–200
4000	2500	12,000	199.2–200.5	198–200
4000	2500	15,000	199.2–200.5	198–200
4000	2500	18,000	199.5–200.5	198–200
4000	2500	20,000	199.4–200.4	198–200

5 CONCLUSION

Based on the analysis of the VIENNA rectifier, the mathematical model of d–q coordinates is established, and based on the simplified SVPWM model, the backstepping controller conforming to the system is designed. Under the condition of a balanced power grid, the effectiveness of backstepping control is verified by comparing the traditional PI and backstepping control. The stability of output DC voltage is increased. In addition, the most suitable correlation coefficient of the system is found by debugging the backstepping controller. In conclusion, VIENNA rectifier backstepping control has a certain application prospect.

ACKNOWLEDGMENTS

This work was supported by the National Ningxia Key Research and Development Program (2018BEE03016).

REFERENCES

Cheng, J. 2014. Research on nonlinear control of three-phase voltage type PWM rectifier. China University of Mining and Technology.

Jae, H.S., Chang, H.C., & Dong, S.H. 2001. A new simplified space-vector PWM method for three-level inverter. *IEEE Transactions on Power Electronics* 16(4): 545–550.

Li, X., Zeng, Q., & Wang, J. 2015. Study on direct power control strategy of three-phase Vienna rectifier. *Power Electronics Technology* 49(11): 90–92.

Song, W., Huang, J., Zhong, Y., & Wang, L. 2013. Hysteresis current control method of Vienna rectifier with midpoint potential balance control. *Power Grid Technology* 37 (7): 1909–1914.

Wang, J., Zhang, M., Zhang, J., Cao, X., & Shi, H. 2017. Backstepping direct power control of PWM rectifier under unbalanced power grid. *Acta Solarica Sinica* 38(11): 2998–3004.

Wei, Z., Chen, X., Fan, Z., & Gong, C. 2013. Single-period control of three-phase three-level VIENNA rectifier output midpoint potential analysis and control method. *Chinese Journal of Electrical Engineering* 33(15): 29–37.

Zhang, C., & Zhang, X. 2012. *PWM Rectifier and Its Control*. Beijing: Mechanical Industry Press.

Zhang, J., & Cheng, M. 2009. Modeling and control of three-phase PWM converter for wind power generation. *Power Grid Technology* 33(20): 43–48.

Zhang, J., Xie, Y., & Shi, Z. 2018. Parameter design method of VI controller for VIENNA rectifier. *Electric Drive* 48(3): 55–61.

Emerging Developments in the Power and Energy Industry – Dufo-López, Krzywanski & Singh (eds)
© *2020 Taylor & Francis Group, London, ISBN 978-0-367-27169-5*

Research on subsynchronous oscillation modal identification based on the FastICA–ARMA algorithm

Jiaqi Wang & Juan Li
Beijing Information Science and Technology University, Beijing, China

ABSTRACT: For the subsynchronous oscillation phenomenon of the power system, this article proposes a modal parameter identification method with strong anti-noise and high recognition accuracy. The method uses the Fast independent component analysis algorithm to improve the autoregressive moving average (ARMA) algorithm. The FastICA algorithm can effectively separate the original signal from the noise. The FastICA algorithm has better anti-noise ability. First, the original signal is analyzed and preprocessed by the FastICA algorithm, and then the modal parameters of the original signal are identified according to the established ARMA model. The research results show that compared with the traditional ARMA algorithm identification results, this method can effectively improve the signal-to-noise ratio and provide more accurate identification results.

Keywords: ARMA, independent component analysis, modal identification, subsynchronous oscillation

1 INTRODUCTION

In recent years, with the expansion of the scale of modern power systems and the complexity of the interconnection structure of the power grid, the problem of subsynchronous oscillation (SSO) caused by this has become increasingly prominent. It has caused a great threat to the safety and stable operation of the power system (Xiao et al. 2017). In severe cases, it can directly cause the fracture of the rotor shaft of a large steam turbine generator, causing a major accident. In view of the current situation of power network structure interconnection, it is very important to identify the modal parameters of SSO in the process of system operation (Zhu et al. 2017).

Commonly used methods for modal parameter identification are the Prony analysis method, random subspace method, and autoregressive moving average (ARMA) analysis method (Liu et al. 2014). The Prony analysis method is very sensitive to noise. When the noise is large, it is difficult to extract the required signal matrix, and the modal parameters of the system cannot be accurately identified (Zhang & Jin 2016). The random subspace method is suitable for parameter identification of linear structural responses under stationary random excitation, but this method is prone to false modal or modal loss, and the calculation amount is relatively large, which is generally not suitable for large and complex structures (Liu 2008). The ARMA model analysis method is based on measured data to process dynamic random data through a parametric model. A linear dynamic model is widely used in system identification, prediction, and control, but the method has limited anti-noise ability (Vu et al. 2011). In He (2012), the ARMA model is applied to the modal parameter identification in time-varying systems. The effectiveness of the method is demonstrated by analyzing this example. The FastICA algorithm is a commonly used algorithm in the independent component analysis method, which can quickly extract independent components from the mixed signal. In (Zhao et al. 2018), the combination of FastICA and matrix beam algorithm is used to identify the modal parameters, which effectively improves the accuracy of the identification results of the matrix beam algorithm. The effectiveness of the algorithm is verified by an example simulation model.

In conclusion, in order to improve the accuracy of modal recognition results, noise interference is greatly reduced. This article chooses the method of combining FastICA algorithm with ARMA model analysis. First, the original signal is pretreated by the FastICA algorithm, and then the modal parameters are identified by ARMA model analysis. Choosing the appropriate ideal signal and the oscillation signal of the first IEEE benchmark model for analysis, and comparing with the traditional ARMA algorithm, the effectiveness of this method is proved.

2 ARMA MODEL ANALYSIS

The ARMA model analysis method is to establish the ARMA model to directly analyze the modal parameters of ordered observation data. It is the most widely used identification method for studying time series. This method was first proposed by Akaike in 1969. The ARMA model was used in the modal identification of white noise environment excitation (Akaike 1969). Since then, the modal identification study of the ARMA model method has been started.

2.1 Introduction to the ARMA algorithm

Establish an ARMA model for a smooth, zero-mean timing $\{x_t\}$, as follows:

$$x_t - \sum_{i=1}^{n} \varphi_i x_{t-i} = a_t - \sum_{j=1}^{m} \theta_j a_{t-j} \tag{1}$$

where x_t $(t = 1,2,\ldots,L)$ is the element of the time series$\{x_t\}$ at time t. φ_i $(I = 1,2,\ldots,n)$ is the parameter of the autoregressive model, θ_j $(j = 1,2,\ldots,m)$ is the parameter of the sliding regression model; a_t is the Gaussian white noise $(a_t \sim NID(0,\sigma_a^2)$, where σ_a is the standard deviation of a_t, and NID indicates a normal distribution. Equation (1) can be called the nth order autoregressive mth order moving average model, which is written as the ARMA(n,m) model.

2.2 ARMA model parameter estimation

First, introduce a time-shift operator B, where $B_{jxt} = xt - j$. Substituting into Equation (1),

$$\left(1 - \varphi_1 B - \varphi_2 B^2 - \cdots - \varphi_i B^n\right)x_t = \left(1 - \theta_1 B - \theta_2 B^2 - \cdots - \theta_j B^m\right)a_t \tag{2}$$

Simplifying yields

$$\phi(B)x_t = \theta(B)a_t \tag{3}$$

Then,

$$x_t = \frac{\theta(B)}{\phi(B)} a_t \tag{4}$$

where $\theta(B)/\phi(B)$ is the impulse response function of the system, and its transfer function can be obtained by Laplace transform. The transfer function is

$$H(z) = \frac{\sum\limits_{k=0}^{m} \theta_k s^{-k}}{\sum\limits_{k=0}^{n} \varphi_k s^{-k}} \tag{5}$$

Since the frequency and damping ratio of the system are related to the denominator of the transfer function, the characteristic equation of the denominator polynomial is

$$\sum_{k=0}^{p} \varphi_k s^{-k} = 1 - \varphi_1 s^{-1} - \varphi_2 s^{-2} - \cdots - \varphi_n s^{-n} = 0 \qquad (6)$$

Solving the above equations can obtain the conjugate eigenvalues λ_i and λ_i^* of the discrete system. Since the power system is a continuous system, it is necessary to transform the model parameters obtained from the eigenvalue analysis to obtain

$$\lambda_i = e^{\left(-\xi_i \omega_i + j\omega_i \sqrt{1-\xi_i^2}\right) \bullet \Delta t} \qquad (7)$$

$$\lambda_i^* = e^{\left(-\xi_i \omega_i - j\omega_i \sqrt{1-\xi_i^2}\right) \bullet \Delta t} \qquad (8)$$

Then, the damped natural frequency f_i and the damping ratio ζ_i of the subsynchronous oscillation mode parameters are obtained, respectively, and the calculation formula is as follows:

$$\begin{cases} f_i = \frac{\sqrt{\ln \lambda_i \ln \lambda_i^*} \sqrt{1-\xi^2}}{2\pi\sigma} \\ \xi_i = -\frac{\ln \lambda_i}{\sqrt{\ln \lambda_i \ln \lambda_i^*}} \end{cases} \qquad (9)$$

In this article, the ARMA parameters of the Prony method time series response fitting are selected for modeling. This method can be mainly used for time domain IIR filter design, exponential signal modeling, and system identification (Wang & Hu 2006). The called Prony function is:

$$[A \ B] = \text{prony}(H, NB, NA) \qquad (10)$$

where the parameters A and B are the coefficient vectors of the numerator and denominator of the transfer function of ARMA model, namely the moving average coefficient θ_j and the autoregressive coefficient φ_i; H is the time domain impulse response vector; NB and NA are the denominators of the transfer function and the order of the molecules,that is the is the sliding average order m and the autoregressive model order n.

3 INDEPENDENT COMPONENT ANALYSIS

ICA is an independent component analysis, which is a blind source separation algorithm method. This method is used to decompose the observed signal data into statistically independent signal components, which are generally used to solve the problem of extracting independent signals from mixed signals (Wang 2016).

Express the ICA model in the form of a vector:

$$X = AS \qquad (11)$$

where $X = [x_1, x_2, \ldots x_m]$, and all components are independent of each other, A is an unknown mixing matrix, and S is an independent signal to be solved, $S = [s_1, s_2, \ldots s_n]$. The purpose of the ICA algorithm is to find the separation matrix W under the condition that the observed signal X is known, but the independent source signal S and the mixture coefficient matrix A are unknown, so that it and the reconstructed signal Y satisfy the following relationship:

558

$$Y = WX = WAS = \hat{S} \tag{12}$$

$\hat{S} = [\hat{s}_1, \hat{s}_2, \ldots \hat{s}_n]^T$ is an estimated value of S of the independent element, and ω_j ($j = 1,2,\ldots,n$) is a decomposition vector. It can be seen from the separation model that when $W = A^{-1}$, the separation effect is most ideal. In this article, the FastICA method proposed by the Finnish scholar Hyvarinen is used to obtain independent elements. This method is simple in concept, fast in operation, and stable. The steps of the algorithm are summarized as follows:

(1) First, the observed data X are centralized to make its mean zero.

$$E\{x - \bar{x}\} = E\{x\} - \bar{x} = 0 \tag{13}$$

(2) After whitening the zero-mean data, the correlation between the signals is removed by calculating the covariance matrix of the data.

(3) Set the maximum number of iterations (to prevent the bad convergence threshold from causing the algorithm's infinite loop) and the convergence threshold, which are set to 10000 and 0.00001 respectively.

(4) Randomly select an initial weight vector Wp and define its mathematical expression as

$$W_p = E\{Zg(W^T Z)\} - E\{g'(W^T Z)\} W \tag{14}$$

$E(\cdot)$ is the mean operation, $g(\cdot)$ is a nonlinear function, where $g(\cdot)$ is usually available: $g_1(y) = \tanh(y)$, $g_2(y) = y\exp(y_2/2)$, $g_3 = y_3$.

(5) Orthogonalization:

$$W_p = W_p - \sum_{j=1}^{p-1} \left(W_p^T W_j\right) W_j \tag{15}$$

(6) Standardization treatment:

$$W_p = W_p / \|W_p\| \tag{16}$$

(7) Determine whether W_p converges, and does not converge to return to step (5)

(8) Let $p = p+1$. If $p \leq m$ (m is the number of signals), return to step (4); otherwise end the iteration.

4   ARMA MODAL IDENTIFICATION BASED ON FASTICA ALGORITHM

The specific implementation method of ARMA modal parameter identification based on the FastICA algorithm is as follows:

(1) Denoising preprocessing of the original signal using the FastICA algorithm

(2) Establishing an ARMA model and call the Prony function to estimate the modal parameters

(3) Using the traditional ARMA algorithm and the improved ARMA algorithm to identify the modal parameters of the original signal and compare the identification results

According to the above steps, two test signals are selected to verify the FastICA algorithm and prove its validity.

4.1 Ideal signal analysis

The formula for defining the vibration system free response signal is:

$$x(t) = Ae^{-\xi\omega_n t}\cos(\omega_d t + \varphi) \tag{17}$$

where A is the amplitude, ω_n is the undamped natural frequency, ω_d is the damped natural frequency, and ζ is the damping ratio.

The damping ratio is calculated according to the following formula:

$$\omega_d = \omega_n\sqrt{1 - \zeta^2} \tag{18}$$

Construct the ideal signal, as follows:

$$\begin{aligned} X(t) &= 1.5e^{-0.15t}\cos(2\pi \times 12t \times \pi/4) \\ &+ 2.5e^{-0.21t}\cos(2\pi \times 23t \times \pi/5) \\ &+ 3e^{-0.5t}\cos(2\pi \times 30t \times \pi/3) \end{aligned}$$

The frequency fi of the ideal signal $X(t)$ (corresponding to the damping natural frequency ω_d in the formula) is 12Hz, 23Hz, 30Hz. The theoretical results of the damping ratio ζ are 0.1989%, 0.1453%, and 0.2653%, respectively. In order to verify the accuracy of the ARMA algorithm identification, ARMA identification is performed on the original signal without noise, as shown in Table 1.

It can be seen from the results in Table 1, the ARMA algorithm accurately recognizes the modal parameters of the signal, which is completely consistent with the theoretical value and proves the effectiveness of the method.

A Gauss white noise with a mean value of 0 and a variance of 0.4 is added to the ideal signal. The sampling frequency is 100Hz, and the sampling time is 3s. The spectrum analysis diagrams of the modal separation diagram of the noisy signal processed by FastICA are shown in Figures 1 and 2. It can be seen from the figures that FastICA can effectively separate the frequency signals and noise.

After the FastICA processed signal, the ARMA method is used to identify the modal parameters, and the results are compared with the traditional ARMA identification results, as shown in Table 2. According to the results of Table 2, the ARMA identification result processed by FastICA is closer to the modal parameters of the original signal, which effectively improves the signal-to-noise ratio of the identification result and achieves greater accuracy.

4.2 IEEE first benchmark model identification analysis

The first benchmark model of IEEE is selected as the research object, and then the simulation analysis is carried out. As shown in Figure 3, the model shafting of IEEE includes six mass blocks: exciter (EXC), generator (GEN), high pressure cylinder (HP), medium pressure cylinder (IP), and two low-pressure cylinders (LPA, LPB). The natural oscillation frequency of the

Table 1. Ideal signal modal parameter identification results.

Model	Frequency (Hz)	Damping ratio (%)	Attenuation factor
1	12	0.1989	−0.15
2	23	0.1453	−0.21
3	30	0.2653	−0.50

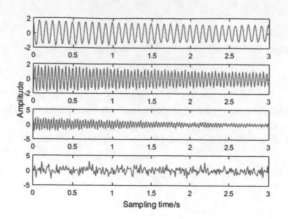

Figure 1.　Noisy signal FastICA separation signal diagram.

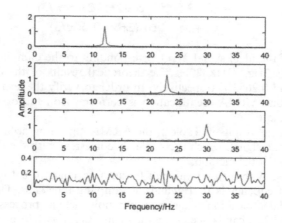

Figure 2.　Spectral diagram of the signal separated by FastICA.

Table 2.　Comparison of modal parameter identification results of noisy ideal signals.

Model	ARMA		FastICA-ARMA	
	Frequency (Hz)	Damping ratio (%)	Frequency (Hz)	Damping ratio (%)
1	12.85	18.83	12.00	0.32
2	26.43	7.87	23.00	0.17
3	36.29	19.73	29.96	0.58

generator shafting is 15.71Hz, 20.21Hz, 25.55Hz, 32.28Hz, and 47.75Hz. Owing to the effect of damping, the 47.75Hz high-frequency modes generally do not occur, so this article identifies the parameters of the first four modes.

Taking the generator's speed difference signal as the sampling signal, at 1.5s, a three-phase short circuit is set at node B in the model, and the short-circuit fault duration is 0.075s. For the convenience of analysis, select the 3 s time after the fault occurs. The oscillating signal is used as an analysis signal. In order to verify the noise immunity of FastICA, a Gaussian white noise with a mean of 0 and a variance of 0.4 was added to the original speed signal. After FastICA processing, the results are as follows.

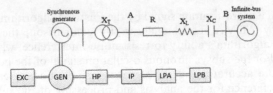

Figure 3.　IEEE first benchmark model system connection diagram.

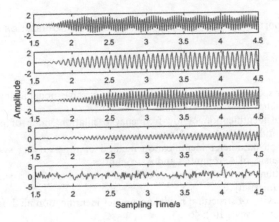

Figure 4.　Noisy signal FastICA separation signal diagram.

Table 3.　Comparison of IEEE signal modal parameter identification results.

Model	ARMA		FastICA-ARMA	
	Frequency (Hz)	Damping ratio (%)	Frequency (Hz)	Damping ratio/(%)
1	15.8344	1.44	15.8830	−0.24
2	20.3627	0.03	20.3178	−0.41
3	–	–	25.5414	−0.10
4	29.8020	9.81	32.2814	−0.17

It can be seen from the results of Figure 4 that FastICA also effectively separates the frequency signals of the IEEE rotational speed difference signal and removes the noise. Similarly, the ARMA identification results after FastICA processing are compared with the identification results before processing, as shown in Table 3.

From Table 3, it can be seen that the phenomenon of leakage identification occurs with traditional ARMA identification, and the ARMA algorithm processed by FastICA can effectively identify the frequency of the signal and the result is close to the actual value. Moreover, it can be seen from the results of the damping ratio that the four modes are divergent states, indicating that the system has a subsynchronous oscillation phenomenon.

5　CONCLUSION

Aiming at the subsynchronous oscillation of the power system, this article proposes a more modal parameter identification method with stronger anti-noise. First, the original signal is preprocessed to obtain a more accurate signal by the FastICA algorithm, and then the parameters are identified by ARMA model analysis. By comparison, the improved algorithm improves the ARMA algorithm's ability to resist noise interference, which proves the superiority of the method. For the subsynchronous oscillation signal of the power system, the identification result is more accurate, and the signal-to-noise ratio is greatly improved, which provides a powerful reference for the analysis and processing of the subsynchronous oscillation signal of the power system in practical engineering.

ACKNOWLEDGMENTS

This project was supported by the National Natural Science Foundation of China (51477010).

REFERENCES

Akaike, H. 1969. Power spectrum estimation through autoregressive model fitting. *Annals of the Institute of Statistical Mathematics* 21(1): 407–419.

Gong, X. 2011. On-line identification of low-frequency oscillation mode of power system based on ARMA recursive algorithm. Chongqing University.

He, J. 2012. Research on parameter identification method of time-varying system based on ARMA model. Nanjing University of Aeronautics and Astronautics.

Liu, D. 2008. Identification of modal parameters of beam bridges based on stochastic subspace method. Southwest Jiaotong University.

Liu, Y., et al. 2014. Overview of structural modal parameter identification methods under environmental excitation. *Engineering Science* 31(4): 46–53.

Vu, V. H., et al. 2011. Operation modal analysis by updating autoregressive model. *Mechanical Systems and Signal Processing* 25(3): 1028–1044.

Wang, J. 2016. Research on the influence of wind power grid-connected on low frequency oscillation of power grid based on FastICA and Prony algorithm. Lanzhou Jiaotong University.

Wang, J., & Hu, X. 2006. *Application of MATLAB in Vibration Signal Processing*. Beijing: China Water Resources and Hydropower Press.

Xiao, X.,et al. 2017. Summary of research on subsynchronous oscillation of new energy power system. *Transactions of China Electrotechnical Society* 32(6): 85–97.

Zhang, C., & Jin, T. 2016. Identification of low frequency oscillation mode of power system based on ISPM and SDM-prony algorithm. *Power System Technology* 40(4): 1209–1216.

Zhao, L., et al. 2018. Sub-synchronous oscillation modal parameter identification based on FastICA-MP algorithm. *Power System Protection and Control* 46(8): 37–42.

Zhu, G., et al. 2017. Summary of research on subsynchronous oscillation of power system[J]. *Journal of Shanghai Dianji University* 20(3): 155–162.

Emerging Developments in the Power and Energy Industry – Dufo-López, Krzywanski & Singh (eds)
© 2020 Taylor & Francis Group, London, ISBN 978-0-367-27169-5

Experimental and modeling investigations on impacts of deeply air staging in a 130 t/h tangentially firing furnace of pulverized coal

Wei Liu, Jian Zhang, Rui-Jie Cao, Xin-Wei Guo & Zhong-Xiao Zhang
University of Shanghai for Science and Technology, Shanghai, China

ABSTRACT: Through a series of experiments and CFD numerical simulations of 130t/h pulverized-coal-fired furnace, the influences of deeply air staging firing on combustion pattern and NOx emission control were discussed in the present paper. The modeling results show the NOx emissions were effectively decreased by 36.4% with more separated-over-fire-air (SOFA) ratio feeding of 25%. The CFD modeling results were verified through the on-site operation data. Furthermore, the adverse issues were discussed regarding the highly air-staging technology. The predicted average combustion temperature in the burner zone of furnace was reduced by 120°C with comparison of different SOFA ratios increased, while the flue-gas temperature increased by 80°C above the SOFA nozzle. More CO emission occurs on the top of furnace. Ash slagging phenomena were observed during ash sampling near SOFA, as is due to unburned char with higher particle temperature and bigger tangential circle diameter of flame pattern on basis of modeling results.

Keywords: Air staging, CFD modeling, NOx emission, Slagging, Pulverized coal

1 INTRODUCTION

Nitric oxides (NO_x) released from coal-fired power station boilers are important atmospheric pollutants. There are many ways to reduce the NO_x emissions of coal-fired boilers, e.g. air staging combustion, low-NOx burner, flue gas recirculation, SNCR (Wang & Zou (2008); Wang et al. (2009); Sun et al. (2008)). Air staging combustion is one of the most popular methods widely used in coal-fired power plants. The air staging technology is to arrange air-feeding distribution ratio of SOFA (separated over fire air) nozzle, which can form the reducing zone with lower oxygen level within the flame and then cause the overall NOx emission decreased effectively. Generally, NOx emission is reduced by 20-30% (Kuang et al. (2010); Qi et al. (2013); Zhou et al. (2010)). However, the utilization of air staging technology also brings about some problems, such as the overheating of the pipelines at the top of the furnace (superheater and reheater) (Yan et al. (2000); Zhang (2008)); the delayed flame along the furnace height, the unburned carbon particle and CO emission increased at the exit of the furnace chamber, even slagging problems (Zhou (2008); Wang et al. (2004); Hao (2007); Li et al. (2014); Zhang et al (2007)).

In this paper, a de-NOx combustion retrofit was introduced, as designed for a 130t/h tangentially firing furnace, involving updated low-NOx burner and deep air-staging allocation. Through a serious of experimental and modeling investigation, the main factors of deep air-staging technology were discussed, e.g. NOx emission, temperature distribution of flue gas within the furnace, CO emission, slagging problem. Those researches look useful to direct other retrofit design and similar operation in the tangentially firing furnaces of pulverized coal.

2 BRIEF INTRODUCTION OF BOILER RETROFIT

The retrofit boiler is a tangentially fired pulverized-coal boiler with 130t/h steam generation. The pressure and temperature of produced steam is 6.8MPa and 480°C. The furnace height is 21.7m with the sectional area of 6400 x 6400 mm (depth and width).
Prior to retrofit, the boiler had no over-fire air (OFA). And the NO_X emission was
900mg/Nm³. Figure 1 (a) expresses the three-dimensional structure of the boiler. And the
location difference of nozzles along the height before and after the retrofit, as shown in
Figure(b) and (c). The pulverized coal burner is arranged in two layers. The burner was
updated to a horizontal bias burner and a vertical bias burner. The original secondary
air nozzles include the upper, middle and lower layers. After the retrofit, the middle and
lower secondary-air nozzles were only retained. Two layers of SOFA wind were designed
and added in the near upper part of the burner area, and the injection angle is 15°C of
SOFA. Two Layers of SOFA nozzles could be controlled separately.

3 APPROACH OF CFD MODELING

The CFD simulation of the boilers was carried out, and the mesh generation is shown
in Figure 2. Firstly, we conducted an independence test on the grids. With comparison
of different overall grid numbers, including 0.45, 0.84 and 1.34 million, the aerodynamic
predictions show the predictions of 0.84 and 1.34 million grids are almost the same,
which looked much different from 0.45 million. Therefore, 840,000 grids were chosen for
further studies, taking into account both calculation cost of time spent and modeling
accuracy.

（a）the structure of boiler, （b）the location of the burner nozzles before the retrofit,（c）the location of the burner nozzles after the retrofit

Figure 1. Introduction of boiler structure.

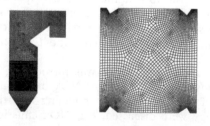

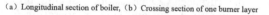

（a）Longitudinal section of boiler, （b）Crossing section of one burner layer

Figure 2. Grid generation of the furnace.

Table 1. Reaction rate used in numerical simulations (Andersen et al. (2009)).

No.	Process	Reaction kinetics	A	b	$E, J/kmol$
1	Devolatilization		5.910×10^7	0	180.00
2	Coal char	$C(s)+0.5O_2=CO$ $d[C(s)]/dt=A\exp(-E/RT)$	0.107	0	117.00
3	Volatile combustion	$CH_4+1.5O_2 \rightarrow CO+2H_2O$ $d[CH_4]/dt=AT^b$ $\exp(-E/RT)[CH_4]^{0.7}[O_2]^{0.8}$	5.012×10^{11}	0	200.0
		$CO+0.5O_2 \rightarrow CO_2$ $d[CO]/dt=AT^b\exp(-E/RT)[CO]$ $[O_2]^{0.25}[H_2O]^{0.5}$	2.239×10^{12}	0	170.0

The following models were used for the simulation based on the software, Ansys-FLUENT-15.0. To calculate gas phase turbulent flow, Realizable k-εmodel was used. The discrete phase (DP) model was applied to analyze the motion of pulverized coal. In the coupling process between Euler field and Lagrangian field, 5120 coal particle trajectories are tracked in each iteration. The finite rate-eddy dissipation model is chosen for influence of the gas-phase turbulence on the combustion process. We use the discrete-coordinate (DO) model for the radiation heat transfer. Based on the post-processing method, the proportion of the generated NOx, including thermal NOx and fuel-type NOx considered. The kinetics of volatile releasing rate and char combustion rate was shown in Table 1 (Tang et al. (2003); Zhang et al. (2015); Zhang et al. (2013)). Such reaction rate of devolatilization and char -O_2 used were obtained in the thermal gravimetric test of pulverized coal samples. In addition, for volatile combustion, the composition of CHxOy in volatile is obtained according to the composition of coal, assuming that the combustion rate of volatile is the same as that of CH_4.

4 EXPERIMENTAL TESTS OF FURNACE OPERATION AFTER RETROFIT

The characteristics of coal used during boiler operation are shown in Table 2. Proximate analysis and elemental analysis of coal were tested, and nct calorific value (Low heating value) of coal was tested.

After the low-NOx combustion retrofit, the boiler cold-case test was carried out. including primary air balance in the chamber, secondary air (SA) percentage test, and tangential circle diameter test. Figure 3 is the primary-air velocity measured in the cross position in the furnace, i.e. the tangential velocity in the furnace. The positions of the four different corners are marked in the Figure 1. The two figures (a) and (b) discussed the burner of 1# and 3#, 2# and 4#, respectively. The diameter of the primary air velocities were about 3.6-4.0m/s, and the flow fields are well organized. Figure 4 shows the effect of the opening rate of secondary air on NOx emission under the steam load of 130 t/h. When the opening ratio of secondary air nozzles decreases, the amount of NOx decreases gradually. In addition, considering the proper ignition of pulverized-coal flame requires enough amount of secondary air, the middle secondary air ratio was set as 10% in the test. With the proportion of SOFA air increases, the concentration of NO_x decreases significantly, when the concentration of oxygen remains

Table 2. Proximate and ultimate analyses of coal used.

proximate analysis, wt%						elemental analysis, wt%				
M_{ar}	M_{ad}	A_{ar}	V_{ar}	FC_{ar}	$Q_{net.ar}$, (MJ/Kg)	C_{ar}	H_{ar}	O_{ar}	N_{ar}	S_{ar}
7.1	0.72	33.14	14.79	44.79	19.2	50.8	2.86	4.32	0.98	0.8

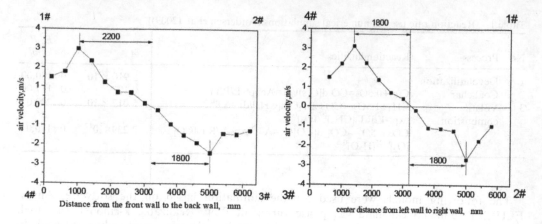

Figure 3.   Tangential velocities tested in furnace.

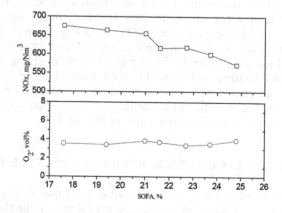

Figure 4.   Effects of SOFA ratio on NOx emission at 130t/h case.

stable at 3.5%. This is similar to the other large boilers, where the proportion of SOFA air could be over 30% in order to minimize the emission of NOx.

5   INFLUENCE OF DEEP AIR-STAGING TECHNOLOGY

5.1   *On NOx emission*

Table 3 provides the comparison between numerical results and the experimental data. The error were within 5%, verifying the accuracy of CFD modeling.

5.2   *On flue gas temperature distribution*

The flame temperature distribution in the furnace was predicted, as shown in Figure 5. With the proportion of SOFA increases, the flame is postponed along the furnace height. On other hand, the stochiometric coefficient of oxygen in the combustion zone reduces, and therefore the pulverized coal was not burnt, leading to the delay of combustion. The temperature decreases in the main combustion zone.

Table 3. Comparison between numerical simulation and experimental data.

SOFA ratio%	NO_2,mg/nm^3		O_2,%	
	Prediction	Test Data	Prediction	Test Data
Before retrofit		900		
17.6	659	675	3.795	3.565
21.5	614	616	3.600	3.705
23.7	606	599	3.382	3.515
24.8	594	572	3.450	3.915

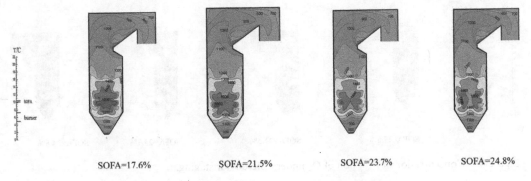

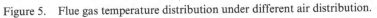

SOFA=17.6%   SOFA=21.5%   SOFA=23.7%   SOFA=24.8%

Figure 5.   Flue gas temperature distribution under different air distribution.

5.3   On the concentration profiles of NO, CO and O_2

Figure 6 shows the concentration distribution of NO^2 with different air distributions. After increasing the proportion of SOFA, the concentration of NOx reduced significantly. Before the low-NOx retrofit of the furnace, the concentration of NOx is 900 mg/Nm^3. With allocation of SOFA ratio 24.8%, NOx emission was reduced by 36.4% .

Figure 7 shows the concentration distribution of CO with different SOFA distributions. After increasing the proportion of SOFA, the concentration of CO increases in the main combustion zone of the furnace and along its height direction. After the retrofit, the emission of NOx was controlled, however incomplete combustion occurs in the furnace.

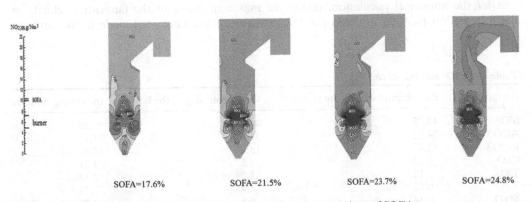

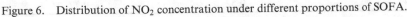

SOFA=17.6%   SOFA=21.5%   SOFA=23.7%   SOFA=24.8%

Figure 6.   Distribution of NO$_2$ concentration under different proportions of SOFA.

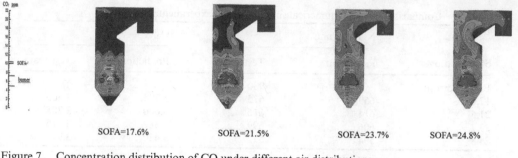

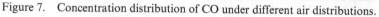

Figure 7.   Concentration distribution of CO under different air distributions.

Figure 8.   Concentration distribution of O_2 under different air staging.

The concentration distribution of O_2 under different air distributions is shown in Figure 8. With the increase of the proportion of SOFA, the concentration of O_2 decreases gradually in the combustion zone of the furnace and increases in higher area, which demonstrates effectiveness of the air-staging technology.

5.4   On slagging phenomena above SOFA nozzle

Compared with no air staging, slagging occurs nearby the SOFA nozzles after the retrofit. Sample XRF analysis was carried out on the ash slagging samples, shown in Table 4. The fusion temperature of ash samples are above 1500°C. The slagging is probably due to the higher temperature of burning particle surface. Beyond the ignition of char, the temperature is generally enhanced to higher temperature. The particle temperature is assumed beyond the fusion temperature nearby the SOFA nozzle. More CO concentration was predicted at the top of the furnace chamber, as verify the unburned carbon particle.

After the numerical calculation, taking the maximum value of the tangential velocity at the intersection position to determine the diameter of the tangential circle in the furnace,

Table 4.   XRF analysis of coal ash.

composition	the ash produced by the raw coal	the bottom slag of the boiler	the coking samples
SiO2	44.75	46.64	45.28
Al2O3	34.52	31.49	31.77
Fe2O3	12.47	13.22	13.67
CaO	3.59	3.57	3.91
TiO2	2.12	1.76	2.14
K2O	0.92	1.48	1.11
MgO	0.44	0.52	0.77

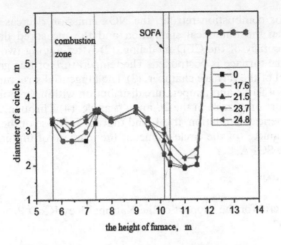

Figure 9.   The tangential diameter along the direction of the furnace height.

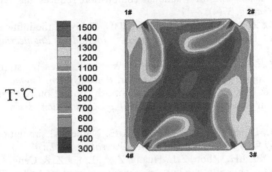

Figure 10.   Temperature distribution at the burner layer.

and the diameter of the tangential circle along the height of the furnace is shown in Figure 9. As the SOFA ratio increases, the diameter of the tangential circle increases along the height of the furnace, especially the flow fields above SOFA.

6   IMPACTS OF FURNACE STRUCTURE DESIGNED

Generally, the burner nozzles of large boilers are located at four corners, which ensures a symmetrical distribution of the flame during combustion. If the burner nozzles are not symmetrical in a level, flame temperature would be influenced. It can be seen from Figure 10 modeling result, that the local temperature near Corner 2# and Corner 4# was about 80 °C higher than that of Corner 1# and Corner 3#. Therefore, Corner 2# and Corner 4# risk to be damage, as observed in the operation experiments

7   CONCLUSION

The industrial tries were conducted to show the influence of deep air-staging technology in a 130t/h power plant furnace with pulverized coal as fuels. The main conclusion can be drawn as follows through experimental test and CFD modeling.

(1) After the low-nox combustion retrofit, the NOx emission decreased by 36.4%, and the relative error between the numerical simulation and the measured data was within ±5%, which proved the accuracy of the CFD modeling. (2) After adding two layers of SOFA, the flame at the top of the furnace is postponed. The temperature of flue gas increased by about 80°C at the nose level of the furnace chamber. (3) The tangential structure of the burner make a role on the tability of flame and temperature distribution within the boiler, which leads to an increase of temperature of nozzles in the 2# and 4# angle. (4) Through the analysis of experimental data and numerical simulation, it is found that the increase of the proportion of SOFA leads to the larger diameter of the circle diameter, the increase of unburned carbon and slagging trend nearby the SOFA.

ACKNOWLEDGMENTS

We are sincerely grateful for the support of the National Key R&D Program of China, Grant No. 2018YFB0604202.

REFERENCES

Wang, H & Zou, L.J. 2008. The air fractional combustion reduces the NOx emissions technology research. *Industrial Heating*, 37(4): 8–11.

Wang, H.J., Fang, Q.Y., Zhou, H.C. & Zeng, H.C. 2009. Study of the influence of the deep air-staged burning on a low-volatile coal combustion process. *Journal of Engineering for Thermal Energy & Power*, 24(6): 777–781.

Sun, G.C., Yan, X.Z., Chen, D.L & Wang, X.W. 2008. Present status and development of de-nitrification technology on coal-fired utility boiler. *Power System Engineering*, 24(2): 1–4.

Kuang, J.G., Lin, Z.C. & Fan, W.D. 2010. Influence of ash content on NOx emission characteristics of bituminous coal in air stage combustion. *Journal of Combustion Science and Technology*, 16(6): 553–559.

Qi, X.J., Li, F.R., Li, J., Ding, L.W. & Mao, L.H. 2013. Numerical simulation of NOx emission from a 300MW tangentially fired boiler. *Thermal Power Generation*, 42(2): 49–53.

Zhou, J.H., Zhao, C.J., Xu, J.H., Zhou, Z.J., Huang, Z.Y., Liu, J.Z. & Ceng, K.F. 2010. Application of air-staged and low NOx emission combustion technology in plant boiler. *Proceedings of the CSEE*, 30(23): 19–23.

Yan, Z.Y., Zhang, H.J., Qiu, G.M., Liu, Q.W. & Yuan, Y.L. 2000. Technical reformation and experimental analysis on NOx exhaust reduction by classifired combustion at boiler. *Journal of Power Engineering*, 20(4): 764–769.

Zhang, W.G. 2008. Numerical simulation of air-staged combustion in utility boiler. *Power System Engineering*. 24(5): 9–12.

Zhou, L.X. 2008. Application of low oxygen combustion technology in 410 t/h boiler. *Electrical Equipment*, 9(3): 72–75.

Wang, E.L., Zhang, H.Y., Luo, Y.H. & Peng, L. 2004. Analysis of low NOx combustion techniques and its application for the coal fired power plants in China. *Power Engineering*, 24(1): 23–28.

Hao, X.M. 2007. Study on two-stage air over-fire technology of staged-air combustion technology. *Clean Coal Technology*, 13(2): 70–73.

Li, Y., Yang, Z.C & Chen, H.W. 2014. Experimental study of air-staged combustion on a burning coking coal boiler. *Power System Engineering*, 30(1): 19–22.

Zhang, X.H., Sun, R., Sun, S.Z & Qin, M. 2007. Effects of combined application of OFA and horizontal bias burner on NOx emission characteristics. *Proceedings of the CSEE*, 27(29): 56–61.

Andersen, J., Rasmussen, C.L., Giselsson, T. & Glarborg, P. 2009. Global combustion mechanisms for use in CFD modeling under Oxy-fuel conditions. *Energy & Fuels*, 23(3): 1379–1389.

Tang, Z.G., Zhu, Q.L. & Tang, B.G, 2003, NOx emissions researched of air staging combustion burning pulverized coal. *Power System Engineering*, 19(3): 7–9.

Zhang, J., Wang, Q., Wei, Y. & Zhang, L. 2015. Numerical modeling and experimental investigation on the use of brown coal and its beneficiated semicoke for coal blending combustion in a 600 MW$_e$ utility furnace. *Energy & Fuels*, 29(2): 1196–1209.

Zhang, J., Prationo, W., Zhang, L. & Zhang, Z.X. 2013. Computational fluid dynamics modeling on the air-firing and Oxy-fuel combustion of dried victorian brown coal. *Energy & Fuels*, 27(8): 4258–4269.

Emerging Developments in the Power and Energy Industry – Dufo-López, Krzywanski & Singh (eds)
© 2020 Taylor & Francis Group, London, ISBN 978-0-367-27169-5

Wind power ramp forecasting based on deep metric learning

Zixin Chen
Chifeng Branch of China Datang Corporation, Chifeng City, Inner Mongolia Autonomous Region, China

Yongqian Liu, Aimei Lin, Shuang Han, Li Li & Jie Yan*
State Key Laboratory of Alternate Electrical Power System with Renewable Energy Sources, North China Electric Power University, Beijing, Changping District, China

ABSTRACT: The wind power ramp is a large and fast variation of wind power in a short period of time and poses a serious threat to the safe, stable, and economic operation of the power system. Forecasting the wind power ramp has great significance for mitigating these challenges. Using similar data as training samples is beneficial to the model accuracy. This article presents a method for forecasting a wind power ramp based on deep metric learning. Different from previous work, the proposed model takes numerical weather prediction sequences as model input and output to evaluate the similarity between them. A fully connected neural network with metric learning function is established in order to obtain a training sequence similar to the actual situations. A case study is conducted based on operational data of a Chinese wind farm. Results show that proposed model can extract the actual power ramp characteristics and thus improve the forecasting accuracy.

Keywords: deep learning, forecast, wind power ramp

1 INTRODUCTION

Due to the characteristics of random fluctuations of wind, especially the occurrence of gusts or extreme weather, wind power fluctuates greatly in a short period of time (the so-called ramps) (Qi & Liu 2013). The wind power ramp can lead to an imbalance of active power and instability of frequency in the power system, and even cause large-scale load loss, which seriously threatens the safe, stable, and economic operation of the power system (Zhang et al. 2018). Therefore, a wind power ramp forecasting with high accuracy has great significance for the power system to formulate countermeasures in advance, to reduce the adverse impact of the wind power ramp.

Many researchers, in China and abroad, have carried out studies on wind power ramp forecasting. The forecast method can be divided into two approaches: direct forecasting and indirect forecasting, according to whether it is forecasting the wind power.

(1) Direct forecasting: Forecasting the ramp characteristics or the occurrence of wind power ramp event directly, according to the definition of wind power ramp. Zheng and Kusiak (2009) established multivariate time series based on five different data mining algorithms, to forecast the ramp rate, and support vector machines (SVMs) is proved to have the highest forecasting accuracy. A hybrid forecasting wind power ramp event model based on a orthogonal test and support vector machine (OT-SVM) was proposed by Liu et al (2017);

---

*Corresponding author. yanjie_freda@163.com

this model determined the input parameter through an orthogonal test, improving the forecasting accuracy effectively. Zareipour et al. (2011) proposed a forecasting and classification wind power ramp event method based on SVMs.

(2) Indirect forecasting: Forecasting the ramp characteristics or the occurrence of wind power ramp event based on the wind power forecasting result. A statistical model and physical model are widely used for forecasting wind power. The objective of the statistical model is to build mapping relations between the inputs and outputs of model, many algorithms are widely used to build it, such as time series analysis method (Yu et al. 2019), neural network (Li et al. 2015), Kalman filtering (Federico & Massimiliano 2012; Yang et al. 2017), an ensemble method (Shi et al. 2011; Ye & Liu 2011), and so on. The objective of the physical model is to get the wind speed and direction at the hub height based on the numerical weather prediction and the terrain around the wind farm, and then forecast the wind power by a power curve. The numerical analysis methods commonly used in physical models are analytical (Feng et al. 2010) and computational fluid dynamics methods (Liu et al. 2014). The deep learning also widely used in the field of wind power ramp forecasting (Wang et al. 2017; Yan et al. 2018). Ouyang corrected the wind power forecasting residual by Markov switching auto-regression (MSAR) model, then improving the wind power ramp event forecasting accuracy (Ouyang et al. 2019). Qiu proposed an ensemble method composed of empirical mode decomposition (EMD), kernel ridge regression (KRR), and random vector functional link (RVFL) network to forecast the wind power ramp and ramp rate (Qiu et al. 2017). Probabilistic forecasting of a wind power ramp method using autoregressive logit models was proposed by Taylor (2017). Dorado-Moreno et al. (2017) used reservoir computing to establishe a wind power ramp event forecasting model. In addition, the improved swinging door algorithm is widely used in forecasting and detecting ramp events (Cui et al. 2015; Zhang et al. 2017; Ouyang et al. 2019).

Most of the ramp research aims to achieve the purpose of forecasting wind power ramp events or ramp characteristics by establishing the mapping relationship of the data. However, sometimes the complication or inconspicuous mapping relationship of data may cause poor learning effects of the model, which leads to low forecasting accuracy.

The wind power ramp forecasting model based on deep metric learning is proposed in this article. Different from previous works, the proposed model from the perspective of data similarity, describes the characteristics of actual situations by the volatility features of historical sequences as much as possible.

2 DEEP METRIC LEARNING DEPENDS ON A FULLY CONNECTED NEURAL NETWORK

2.1 Deep metric learning

The objective of metric learning, also called similarity learning, is to obtain a metric matrix or similarity function that effectively reflects the similarity (or distance) between samples by a given set of training samples and to measure the similarity of samples. A distance function is necessary when identifying the similarity of sample features. The Mahalanobis distance is a widely used metric learning distance function. It is used to obtain a metric matrix $M \in R^{d \times d}$ from the training set X, to measure the distance between sample x_i and x_j. The calculation method as shown in Equation (1).

$$d_M\left(x_i, x_j\right) = \sqrt{\left(x_i - x_j\right)^T M\left(x_i - x_j\right)} \tag{1}$$

The metric matrix M is a symmetric positive semidefinite matrix to ensure square validity, so M can be decomposed into Equation (2):

$$M = W^T W \tag{2}$$

where $W \in R^{p \times d}, p < d$.

Equation (1) can be decomposed into Equation (3):

$$d_M(x_i, x_j) = \sqrt{(x_i - x_j)^T W^T W (x_i - x_j)} = \|Wx_i - Wx_j\|_2 \tag{3}$$

According to Equation (3) we know the Mahalanobis distance metric projects x_i into a lower dimensional space by finding a linear transformation. So, traditional metric learning cannot describe the nonlinearity of samples effectively.

In order to solve the limitations of traditional metric learning methods, the method of nonlinear distance learning or similarly deep learning has appeared in recent years (Cheng et al. 2018). This method can significantly improve the learning ability of metric learning for complex distribution data.

The method of measuring similarity between samples directly by fully connected neural network is proposed in this article. It is used to find similar wind sequences in historical sequence and pick them out. This is the first and important step to establishing a wind power ramp forecasting model.

2.2 *Fully connected neural network*

As mentioned in Section 1.2, the fully connected neural network (FCNN) with deep metric learning ability is established to forecast wind power ramp directly. The fully connected neural network, extracting the data feature by the weight value, has a long history in the development process of a deep neural network (Zhou 2018). The structure of a fully connected neural network includes input layer, hidden layer, and output layer. Each neuron accepts its inputs, does some calculation on the data based on its activation function and weights, and then passes on the output as an input to the next neuron.

2.2.1 *Optimization*

The stochastic gradient descent method (SGD) is used to minimize the FCNN loss function in this article, which can improve the computational efficiency of the model. The basic principle of SGD is that along the gradient direction of θ is the fastest rising direction of the loss function $J(\theta)$, so the minimum value of the loss function can be found in the opposite direction of the gradient. The main steps of the SGD are:

(1) The loss function of model is

$$J(\theta) = \frac{1}{2m} \left[h_\theta(x^j) - y^j \right]^2 \tag{4}$$

(2) Determine the gradient direction, that is, the partial derivative of the loss function.

$$\frac{\partial}{\partial \theta} J(\theta) = \frac{1}{m} \sum_{j=0}^{m} \left[h_\theta(x^j) - y^j \right] x^j \tag{5}$$

(3) Update all θ until all gradient's descent distance less than the set error ε. The updating expression of θ is:

$$\theta = \theta - \alpha \frac{1}{m} \sum_{j=0}^{m} \left[h_\theta(x^j) - y^j \right] x^j \tag{6}$$

2.2.2 *Initialization*

A stacked auto-encoder (SAE) is adopted in this article to initialize the weight and bias of FCNN. The proposed use of SAE, that is Bengio. inspired by the auto-encoder (AE) algorithm (Hinton et al. 2006), is to migrate the greedy layer-by-layer pretraining algorithm in deep belief nets (DBNs) to the automatic encoder (Schölkopf et al. 2007).

The SAE consists of multiple AEs. The training mechanism can be divided into layer-by-layer pretraining and fine-tuning. These two parts can prevent local optimal solutions, improve the convergence of the model, and extract data features better. Figure 1 is the training mechanism of SAE.

(1) Layer-by-layer pretraining

The input vector X is trained by the auto-encoder to obtain the parameters $\{W_1, b_1\}$. The output of the first hidden layer is used as the input of the next auto-encoder, and the training continues to obtain the parameters $\{W_2, b_2\}$. Repeat the steps until the last auto-encoder is trained.

(2) Fine-tuning

Each hidden layer is initialized by a sequence of weights and bias obtained from the first step, and then minimize the loss function of the entire network by fine-tuning. The loss function is

$$J_0 = \frac{1}{N} \sum_{i-1}^{N} \left(y_i - y_i' \right)^2 \tag{7}$$

where y_i is the forecasting value and y_i' is the real value.

3 WIND POWER RAMP FORECASTING MODEL BASED ON DEEP METRIC LEARNING

3.1 *Model structure*

The wind power ramp forecasting model structure proposed in this article is shown in Figure 2. The forecasting model is based on a fully connected neural network, and takes numerical

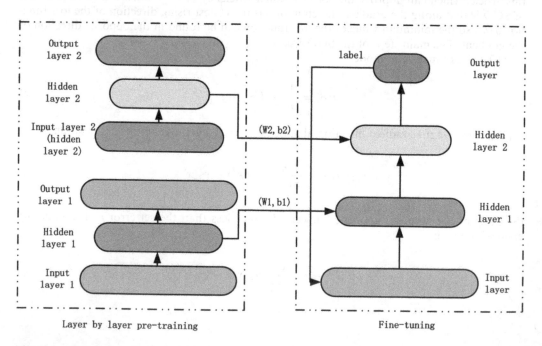

Figure 1. Training mechanism of SAE.

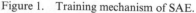

575

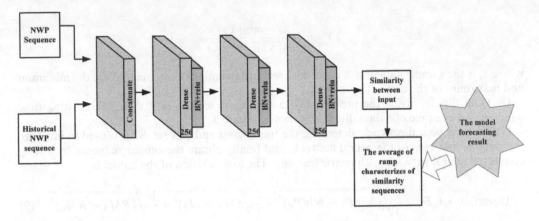

Figure 2. The model structure of wind power ramp forecasting based on deep metric learning.

weather prediction sequences as the model input and output to evaluate the similarity between them. The neural network includes a connection layer and three hidden layers. The connection layer connects the two inputs of the one-dimensional numerical weather prediction feature vectors into a two-dimensional vector and passes it to the hidden layer. The ReLu function is selected as the activation function of the hidden layer. It has the characteristics of unilateral suppression that keeps the positive values and turns all negative values into 0. Unilateral suppression makes the neurons in the neural network have sparseness, so that the neural network can mine the sample features better and fit the training samples. In addition, compared with other activation functions, the ReLU function has outstanding expressive ability, especially in deep neural networks, and it does not have the problem of gradient disappearance, which can maintain the convergence rate of model in a stable state.

The output of the model is the similarity between the predicted numerical weather prediction sequence and the historical one. According to the result, the average value of the ramp characteristics of the most similar historical sequences is selected as the forecasting result of the model, which achieves the purpose of forecasting the wind power ramp characteristics directly.

A power ramp can be defined by the following characteristics (Gallego-Castillo et al, 2015):
(1) Magnitude (ΔP_r): the variation of wind power
(2) Duration time (Δt_r): the time period of large variation takes place
(3) Ramp rate ($\Delta P_r/\Delta t_r$): the change rate of wind power
(4) Timing (t_0): can be defined as the starting time or the central time of the ramp
(5) Direction: the change trend of wind power; up-ramp means power increase and down-ramp means power decrease.

Magnitude, the variation of wind power, is more comprehensive in describing the severity of the ramp in engineering application. Therefore, this article selects the magnitude as the forecasting result to express the forecasting effect of the model.

3.2 Modeling process

The modeling process of the model includes:
(1) Data cleaning and normalization: the mean imputation method is used to correct the unreasonable data. The purpose of normalization is to eliminate the influence of dimension, and the min-max normalization method is adopted in this paper. The min-max normalization calculation is as follows:

$$X' = \frac{x_i - \min(X)}{\max(X) - \min(X)} \tag{8}$$

where x_i is the sample data and X is the data set, and the $\min(X)$ and $\max(X)$ are the minimum and maximum of the data set.

(2) Data divide: divide the processed data into short sequences within a forecasting time window. An example of a data divide is shown in Figure 3.

(3) Training neural network: determine the training set and text set. SAE is used to initialize the weights and bias of the neural network, and finally obtain the similar sequence by a fully connected neural network with metric learning. The loss function of the model is

$$\text{Deviation } (A, B) = \frac{1}{3}\sqrt{(NWP_A - NWP_B)^2} + \frac{1}{3}\sqrt{(M_A - M_B)^2} + \frac{1}{3}\sqrt{(RM_A - RM_B)^2} \tag{9}$$

where NWP_A and NWP_B are the wind speeds of input NWP sequence, M_A and M_B are the wind speeds of the wind farm operation data, RM_A and RM_B are the wind power magnitude of the wind farm operation data.

(4) Forecasting wind power ramp range: based on the output of trained model, select the historical sequences with higher similarity and take the average value of their ramp characteristics as the forecasting result.

4 CASE STUDY

4.1 Data

Take a wind farm in China as a case to verify the model forecasting result. The operating data include the wind speed, wind direction sine and cosine, and the output power of the wind farm for one year, with a time resolution of 5 min. The data of NWP include the wind speed and wind direction; the time period and resolution are the same as in the operation data. The data are divided into training samples and test samples in a ratio of 9:1. The accuracy of the forecasting result is compared with the model based on SVM, and in order to ensure the fairness of the comparison, each model in the case used the same training samples and test samples.

4.2 Evaluation indices

Root mean square error (RMSE) and mean absolute error (MAE) are used to evaluate the forecasting accuracy of the model. RMSE reflects the deviation between the forecasting value

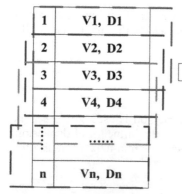
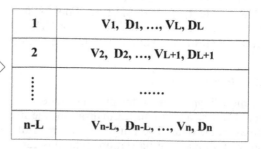

* L is the number of samples in the forecasting time windows and n is the number of samples

Figure 3.   Example of a data divide.

Table 1. RMSE and MAE for different model parameters.

Forecasting time window	K*	RMSE (%)	MAE (%)
1 hour	5	11.16	7.89
	10	10.89	7.73
	20	10.73	7.65
4 hours	5	21.63	15.70
	10	21.33	15.43
	20	21.13	15.24

*The number of similar historical sequences for an average value.

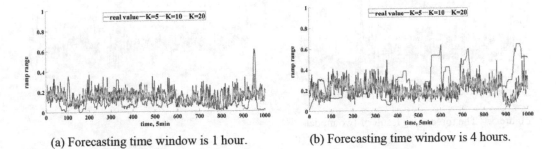

(a) Forecasting time window is 1 hour. (b) Forecasting time window is 4 hours.

Figure 4. The forecasting result for different model parameters.

and real value, and MAE reflects the actual situation of the error between the forecasting value and the real value. The calculation formula can be found in Blonbou (2011).

4.3 Forecasting result study

4.3.1 Influence of model parameters on the forecasting result study

The basic method of this forecasting model is to find the historical sequences with higher similarity, and then take their average power ramp range as the result. However, whether the number of similar sequences will affect the forecasting result is unknown, so this section compares the forecasting result by selecting different numbers of similar sequences and forecasting time windows. Table 1 shows the RMSE and MAE for different model parameters.

Figure 4 reflects the forecasting result of different numbers of similar historical sequences.

From Table 1 and Figure 4 we can find that with the number of similar historical sequences increasing, the forecasting accuracy has improved in the same forecasting time window. When K is 20, the forecasting result has the smallest RMSE and MAE, but the change trend of the forecasting result could not fit well with the real value. That is because the average value reduces the difference that exists between the similar historical sequences. When K is 5, the change trend of forecasting result is fit better, but the accuracy is reduced. So, this article selects the average of 10 similar historical sequences as results.

4.3.2 Forecasting result of power range study

Based on the result obtained in Section 4.3.1, this section selected a wind power ramp forecasting model based on a support vector machine as the reference model and compared the forecasting results of two models. Table 2 is the RMSE and MAE of the two models to forecast the ramp range.

Figure 5 reflects the forecasting results of two models in different forecasting time windows.

From Table 2 and Figure 5 we can find that the forecasting model proposed in this article improves the forecasting accuracy for the wind power ramp. With the increase of the

Table 2. RMSE and MAE for different forecasting models.

Forecasting time window	Forecasting model	RMSE (%)	MAE (%)
1 hour	Deep metric learning	10.89	7.73
	SVM	12.71	8.53
4 hours	Deep metric learning	21.33	15.43
	SVM	22.35	17.31

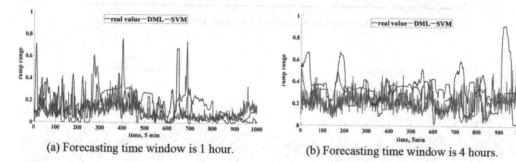

(a) Forecasting time window is 1 hour.　　　(b) Forecasting time window is 4 hours.

Figure 5. The forecasting result of two models.

forecasting time window, the accuracy decreased because in long forecasting time window, the accuracy influenced by the data feature was weakened.

5 CONCLUSION

This article proposes a method for forecasting a wind power ramp based on deep metric learning, which can forecast wind power ramp directly. This method, from the perspective of data similarity, aims establish a fully connected neural network to obtain the similarity between a historical sequence and the actual situations. Then, the average value of ramp characteristics of these sequences is taken as the forecasting result.

Based on operation data of a Chinese wind farm and taking a wind power ramp forecasting model based on a support vector machine as reference, the results show that the forecasting model proposed in this article can extract the actual power ramp characteristics and thus improve the forecasting accuracy.

ACKNOWLEDGMENTS

This work was supported in part by the project of the National Natural Science Foundation of China funded project (51707063), the project of China Datang Corporation Ltd.

REFERENCES

Blonbou, R. 2011. Very short-term wind power forecasting with neural networks and adaptive Bayesian learning. *Renewable Energy* 36(3): 1118–1124.

Cheng, G., et al. 2018. When deep learning meets metric learning: Remote sensing image scene classification via learning discriminative CNNs. *IEEE Transactions on Geoscience and Remote Sensing* 56(5): 2811–2821.

Cui, M., et al. 2015. An Optimized swinging door algorithm for identifying wind ramping events. *IEEE Transactions on Sustainable Energy* 7(1): 150–162.

Dorado-Moreno, M. 2017. Robust estimation of wind power ramp events with reservoir computing. *Renewable Energy* 111(2017): 428–437.

Federico, C., & Massimiliano, B. 2012. Wind speed and wind energy forecast through Kalman filtering of numerical weather prediction model output. *Applied Energy* 99(2012): 154–166.

Feng, S., et al. 2010. Study on the physical approach to wind power prediction. *Proceedings of the CSEE* 30(2): 1–6.

Gallego-Castillo, C., et al. 2015. A review on the recent history of wind power ramp forecasting. *Renewable & Sustainable Energy Reviews* 52: 1148–1157.

Hinton, G. E., et al 2006. A Fast learning algorithm for deep belief nets. *Neural Computation* 18(7): 1527–1554.

Li, S., et al. 2015. Wind power forecasting using neural network ensembles with feature selection. *IEEE Transactions on Sustainable Energy* 6(4): 1447–1456.

Liu, Y., et al. 2014. Research on error distributions of wind power prediction based on CFD pre-calculated flow fields. *Modern Electric Power* 31(5): 64–69.

Liu, Y., et al. 2017. A hybrid forecasting method for wind power ramp based on Orthogonal Test and Support Vector Machine (OT-SVM). *IEEE Transactions on Sustainable Energy* 8(2): 451–457.

Ouyang, T., et al. 2019. Prediction of wind power ramp events based on residual correction. *Renewable Energy* 136(2019): 781–792.

Qi, Y., & Liu, Y. 2013. Finite control of high risk wind power ramping. *Proceedings of the CSEE* 33(13): 69–75.

Qiu, X., et al. 2017. Short-term wind power ramp forecasting with empirical mode decomposition based ensemble learning techniques. In *2017 IEEE Symposium Series on Computational Intelligence (SSCI)*.

Schölkopf, B., et al 2007. Greedy layer-wise training of deep networks. *Advances in Neural Information Processing Systems* 19:153–160.

Shi, H., et al. 2011. A Short-term wind power prediction method based on wavelet decomposition and BP neural network. *Automation of Electric Power Systems* 35(16): 44–48.

Taylor, J. W. 2017. Probabilistic forecasting of wind power ramp events using autoregressive logit models. *European Journal of Operational Research* 259(2): 703–712.

Wang, H., et al. 2017. Deep learning based ensemble approach for probabilistic wind power forecasting. *Applied Energy* 188(2017): 56–70.

Yan, J., et al. 2018. Forecasting the high penetration of wind power on multiple scales using multi-to-multi mapping. *IEEE Transactions on Power Systems* 33(3): 3276–3284.

Yang, M., et al. 2017. Real-time prediction for wind power based on Kalman filter and support vector machines. *Journal of Northeast Electric Power University* 37(2): 45–51.

Ye, L., & Liu, P. 2011. Combined model based on EMD-SVM for short-term wind power prediction. *Proceedings of the CSEE* 31(31): 102–108.

Yu, R., et al. 2019. LSTM-EFG for wind power forecasting based on sequential correlation features. *Future Generation Computer Systems* 93(2019): 33–42.

Zareipour, H., et al. 2011. Wind power ramp events classification and forecasting: A data mining approach. In *2011 IEEE Power and Energy Society General Meeting*.

Zhang, D., et al. 2018. Review and prospect of research on wind power ramp events. *Power System Technology* 42(6): 1783–1792.

Zhang, J., et al 2017. Ramp forecasting performance from improved short-term wind power forecasting. *Energy* 122(2017): 528–541.

Zheng, H., & Kusiak, A. 2009. Prediction of wind farm power ramp rates: A data-mining approach. *Journal of Solar Energy Engineering* 131(3): 376–385.

Zhou, X. 2018. Implementation and optimization of fully connected neural network on FPGA. University of Science and Technology of China.

Emerging Developments in the Power and Energy Industry – Dufo-López, Krzywanski & Singh (eds)
© 2020 Taylor & Francis Group, London, ISBN 978-0-367-27169-5

Research on the elements and model of the competitive electricity retail market

Xuanyuan Wang
State Grid Jibei Electric Power Company, Beijing, China

Dunnan Liu
North China Electric Power University, Beijing, China

Zhen Liu
North China Electric Power Research Institute, Co. Ltd., Beijing, China

Xiaotong Li
North China Electric Power University, Beijing, China

ABSTRACT: With the liberalization of the electricity market, the number of electricity companies and the types of transactions are increasing. There are more and more uncertainties in the market, and the operating conditions and disturbance scenarios are more complex and unpredictable. Traditional electricity market model has been difficult to cope with the complex and changeable market environment and new subjects. Therefore, it is urgent to study the operation mode and trading mechanism of competitive retail market. The operation mode of competitive retail market is studied, a competitive retail market model is constructed based on the relationship among the main players in the market, and the trading mechanism of competitive retail market are put forward from the aspects of electricity tariff composition and settlement.

Keywords: competitiveness, electricity retail market, market model, trading mechanism

1 INTRODUCTION

Chinese power industry has been changing to a mature power market. The goal of power industry reform is supposed to be reducing electricity price and improving efficiency. Combining the two kinds of contradictory goals, the power enterprises can get more capital and market by reducing electricity price, improving efficiency and increasing customer satisfaction. It also makes the funds play a more efficient role, so that the power facilities can be continuously expanded to meet the growing needs of national economic and social development (Liu Junzhe 2016).

Current reform mode only introduces competition mechanism in Generation Side. In order to make power industry go to market, it is necessary to reform the system of power selling side (Hu Zhaohui & Wang Haisheng 2018). After introducing the market mechanism into the power selling side, through the price competition mechanism in the power retail market, the power generation enterprises are forced to operate according to the feedback price information (Wu Qiming 2016). Finally, an inverted mechanism to reduce the cost is formed on the power generating side, thus realizing the goal of power market reform. From the practice of electricity market reform in western developed countries, competitive electricity market has been established very early (Chen Lei 2012). Foreign scholars have made an economic analysis of monopoly, regulation, competition and liberalization based on the experience of regulation reform of British telecommunications, electricity and tap water. They believe that competition has a stronger driving force for the development of public utilities. Based on the actual situation of the electricity market in the United States, this paper demonstrates the trend of deregulation and gradual liberalization of the electricity market, which provides a basis for deregulation of the electricity market. Some studies have found that the first driving force of regulation reform in the power industry is the existence of "power price differentials" (Luo Qin & Song Yiqun 2015). The greater the power price differentials, the stronger the driving force of regulation reform in countries with larger power price differentials. The benefits of deregulation in the power industry have also been studied by a large number of scholars, mainly affected by risks and rewards (Pan Shuangshuang & Tong Jun 2017). Many other scholars have studied the practice of deregulation and deregulation of power market in various countries (Mao Hua, Liu Zhongyi & Yan Dawei 2017).

Although there were many domestic literatures on the operation mode of electricity market, most of these studies were concentrated on traditional electricity market, and the trading rules and modes of electricity market in the initial stage of development were not perfect. Based on which this paper studies the structure model of competitive electricity market and puts forward the operation mode of competitive electricity market.

2 COMPETITIVE ELECTRICITY RETAIL MARKET MODEL

2.1 Overview

The model of competitive electricity retail market in China is constructed by establishing a free competitive power commodity sales channel. Establishing a competitive electricity retail market is the ultimate goal of China's electricity market reform, and the competitiveness of the electricity retail market model is the key factor to the effectiveness of the model. In this model, the competitiveness is mainly related to the autonomy of market participants' trading behavior, the selectivity of trading channels and the controllability of risks in trading.

2.2 Characteristics of model

Competitive electricity retail market model has different characteristics from traditional electricity retail model. Firstly, the service attributes of electricity commodities have been fully reflected. Under the premise of monopoly operation on the power selling side, consumers' understanding of power commodities is mostly based on the physical characteristics of "electricity," and they have no right to consider their service attributes. With the introduction of competition mechanism in the electricity retail market and consumers' right of choice, the requirement of this service attribute has become an important part of the competition in the electricity retail market. As a competitive commodity, power supply should include reliability, safety, timeliness and economy of power supply besides

Table 1. Characteristics of competitive electricity retail market model.

Market type	Traditional electricity retail market	Competitive electricity retail market
Point 1	Only physical characteristics such as voltage and frequency were considered.	The service attributes as power supply reliability, security and economy are fully considered.
Point 2	Power grid companies have monopoly power over consumers	Users have the right to consume
Point 3	Traditional business links: generation, distribution, transmission and sale	Business links are completely independent
Point 4	Centralized technical support system	Independent technical support system can be owned

traditional physical indexes such as voltage and frequency. With the clarification of the service attributes of electric power commodities, more and more attention has been paid to the product heterogeneity of electric power commodities. This also provides a theoretical basis for power enterprises to create product differences and extend the content of power supply services. From the perspective of social welfare, the above model is also of positive significance to guarantee the interests of consumers and improve the efficiency of resource allocation.

Users have the right of consumption choice. In this model, all users can either purchase electricity directly from an independent power plant or choose a specific power retail company to supply electricity. Electric power retailers can be newly established companies providing exclusive electricity sales services, or they can be formed from the traditional power supply companies which are separated from the electricity sales business. Distribution companies can also conduct retail competition as a member of retailers, but because users have the option, if the retail price of distribution companies is not competitive, users will choose other retailers. Therefore, distribution companies no longer have monopoly rights to consumers, which is the biggest feature of competitive electricity retail market.

The nature of business links has changed. In the competitive electricity retail model, the traditional business links have changed. Generation, distribution, transmission and sale of electricity are completely independent: generation side is the energy producer; transmission is equivalent to logistics function, only responsible for the transmission of electricity without participating in transactions, because the transmission network is related to the security of the country, so it is managed by the government monopoly; distribution, like transmission, is managed by consumers through distribution. Provide transmission and transformation services and charge service fees. At the same time, in this mode, the nature of independent operating system or power pool has also changed, from the original buying mechanism to the trading intermediary, they do not directly participate in the transaction, no longer determine the price nor bear market risk, but through the transaction matching business between buyers and sellers to obtain service costs.

The technical support system is simple and feasible. The realization of competitive electricity retail market does not need centralized technical support system as a guarantee. In order to supply power to users, as a power supplier, it is necessary to have its own management software, which includes user management system, information release system, measurement system, contract management system and settlement system. It can be seen that the complexity of the software system is greatly reduced because it does not involve the management and safe operation of the power system. Many power suppliers can even cooperate with companies engaged in metering services to hand over complex automatic metering of electricity charges to metering companies, thereby further reducing the complexity of their own automation systems. Each end-retail user can get information about the service items and prices of power suppliers in the market through various open media, freely choose the most advantageous service providers for himself, and also seek the help of power brokers, through which power intermediaries can arrange the most advantageous power services for themselves.

3 ELEMENT RELATIONS IN COMPETITIVE ELECTRICITY RETAIL MARKET MODEL

The complex market relationship formed by various market players through various market channels is the competitive retail market model. The specific relationships among the elements are as follows:

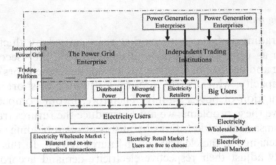

Figure 1.   Frame of competitive electricity retail market.

3.1 Power generation enterprises

The power generation enterprises are responsible for the production of power commodities. They can transfer electricity directly to the user through the power grid, or to the spot electricity market, such as power pool or power futures market, and then to the user through the power trading intermediaries.

3.2 Electricity retailers

The emergence of electricity retailers is a necessary condition for competition in the electricity retail market. Only when many power retailers participate in the competition in the electricity retail market can they truly reflect the original intention of the market reform of "openness, fairness and justice," "improving efficiency" and "protecting consumers' rights and interests." Electric power retailers undertake the services of end-users who purchase electricity from power trading markets or generators and deliver it to users through distribution networks in accordance with the company's electricity sales plan.

The power retailer can be transformed from the original distribution company or a newly formed company. In the electricity retail market, the competition between power retailers is not only reflected in the price of the products provided, but also in the service of power supply. In the face of fierce competition, retailers will inevitably work hard on service and management, by improving service content and quality, reducing operating costs and other measures to protect the original contract users, and develop new contract users. In the model, because of the large number of power retailers, it is impossible for a single power supplier to monopolize the power supply or for a few retailers to manipulate the retail price. Retail electricity prices will gradually stabilize at a lower level as the complete competition among many power suppliers becomes lower and lower.

3.3 Electricity users

Electric power users refer to all consumers who need to use electricity, including individuals, families, enterprises and institutions. In Zhejiang power grid, power users are classified into six categories: residential users; large industrial users; ordinary industrial or non-industrial users; commercial users; military, prison, nursing home users; agricultural users. The majority of small users, especially resident users, do not have the right to choose power suppliers independently, so competitiveness cannot be reflected in the electricity retail market. In the competitive electricity retail market model, because of the design of the power retailer, the majority of resident users also get the right of consumption choice by choosing the power retailer, which has epoch-making significance for the competition of the electricity retail market. In the competitive electricity retail market, power suppliers and end retail users need to find each other in the retail market and sign various agreements for electricity transactions to meet their needs.

The spot electricity trading is faced with market risks caused by fluctuations in electricity market demand, such as price risk, supply and demand risk, etc. Secondly, spot electricity trading lacks planning and control of electricity demand, and generators cannot determine the effective generation capacity, which is easy to form a situation of short supply or oversupply, resulting in the interruption of power supply. Therefore, in our competitive electricity retail model, in order to solve these problems, we must introduce the futures trading model.

Both sides of the transaction trade futures by means of open bidding or computer matching through the futures exchange in the electric power futures market. In this trading mode, power consumers can actively adjust their respective electricity demand according to the change of electricity price, thus introducing price elasticity coefficient into the load side, avoiding the phenomenon that the load side passively accepts the results of electricity trading on the generation side in the past electricity trading, and weakening the monopoly control position of the generation side in the past electricity trading. Therefore, the price of electric power futures can better reflect the actual value of electric power commodities. In addition, in the competitive electricity retail market model, the market power of the system dispatching center will also be limited.

The system dispatching center is responsible for power system security while participating in the formulation of power trading plan in the existing mode. Therefore, the system dispatching center has the absolute right to intervene in the trading plan, while the power traders lack the means to actively adjust the trading plan. But in the futures trading of the new model, the system dispatching center does not participate in the futures trading. Its impact on futures trading is only to determine the delivery plan of futures in the futures and spot market, but it will not affect the behavior of both sides of futures trading in hedging. Therefore, the system dispatching center no longer has the decisive ability to intervene in the power trading behavior, and no longer has the status above both sides of the power trading. At the same time, the system dispatching center has the necessary means to adjust the operation status of the power system, and has a certain ability to control the operation of the system, so as to maintain the safe operation of the power system. Therefore, on the premise of ensuring the safe operation of the system, the market position between the system dispatching center and the trader in the model is equal. Futures trading, supplemented by spot trading, is the inevitable trend of the development of China's electricity retail market. It not only makes the market more commercial, but also makes China's electricity retail market more secure in risk prevention.

3.5 *Trading channels*

In order to compete in the electricity retail market, one of the most important issues is to break the monopoly of distribution companies on their own distribution exclusive areas. Only by opening up the distribution network can it be possible to provide retail power transshipment business. In the traditional monopoly mode of vertical integration, electricity transaction is along the channel of "generation-transmission-distribution-power supply," and finally reaches the users. This kind of trade channel is the basic trade channel that has existed since the emergence of the electric power industry. In this kind of channel, users are actually in an undeniably weak position. In the competitive electricity retail market model, because of the liberalization of the prohibition of electricity retail, there are many kinds of auspicious channels for electricity commodity trading.

There are three trading channels in the competitive electricity retail market:

- The traditional channel, namely "generation-distribution-retail-user" mode. This channel evolves from the traditional "generation-transmission-distribution-sale" channel. In this channel, the traditional distribution function is divided into two kinds of business: power grid and non-power grid. Power grid business and power transmission together are only used as a tool of power commodity transmission, but no longer as a channel of market participants. In this channel, the electricity generated by power generation enterprises enters the power pool through the transmission network, while the power retailers choose the appropriate distributor through the network or purchase electricity directly from the power pool, and then sell it to the end users. Because of the large number of retailers, users have the right to choose retailers.

- The distribution channel, namely "generation-retail-user" mode. In this channel, the function of the power pool is gradually degraded, and the power retailer can purchase the required power directly from the power producer according to the electricity sale plan, and then sell it to the end user. Compared with the traditional channel, because the distribution channel is omitted, the transaction is more efficient. In this channel, the commercial function of the distributor is weaker, and it has completely withdrawn from the sales competition and only assumed the responsibility of power grid maintenance.
- The direct marketing channels, namely "generation-user" mode. This kind of channel is more efficient because of fewer transaction links, and it is also conducive to reducing transaction costs. From the current development of China's electricity market, direct sales can only be targeted at large consumers, while ordinary residents do not have the conditions to purchase electricity directly from power plants. However, with the establishment and development of China's competitive retail electricity market, it is inevitable that all users can purchase electricity directly from power producers.

In conclusion, power users can either obtain electricity supply through the power retailer or purchase it directly from the power generation company in the competitive electricity retail market. Various sales channels fully guarantee consumers' right of consumption choice, and also fully reflect the competitiveness of the market.

4 COMPOSITION AND MEASUREMENT OF ELECTRICITY PRICE IN COMPETITIVE ELECTRICITY RETAIL MARKET MODEL

The user tariff calculation is different from that of the traditional power supply company because of the increase of the power retailer channel in the competitive electricity retail market model.

4.1 *Electricity price composition of traditional power companies*

Because of its monopoly position, the traditional vertical integration power companies do not have a detailed description of the composition of electricity charges. In the list of electricity charges with two-part or three-part tariffs, they generally include monthly fixed electricity charges, capacity charges, energy charges and other charges. The monthly fixed electricity cost is used to recover the investment of power infrastructure; the capacity cost reflects the fixed cost of the power company, which is used to recover the investment to meet the maximum load of the users; the energy cost reflects the actual electricity cost of the users, which should fully compensate the cost of power generation; the other costs are used for the electricity subsidy plan or demand side of low-income families. Some of the management plans also include electricity retail tax. Table 6–1 is a two-part tariff sheet, which is set by a traditional power company at a fixed rate.

4.2 *Electricity tariff composition in competitive electricity retail market*

In the competitive electricity retail market, the user's electricity fee is mainly composed of the following four kinds of fees:

4.2.1 *Electricity purchasing in wholesale electricity market*
The wholesale price of electricity is the price that power retailers buy electricity from the wholesale market, which reflects the fuel cost and management cost of power production.

4.2.2 *Agency*

This part of the fee is paid to the power broker. Users usually choose power producers, distributors or brokers to arrange power supply services for themselves. The agent must select the appropriate power transshipment supplier and the appropriate transmission network to transfer the contracted electricity from the generator to the meter metering Office of the user's power supply service provider, negotiate the lowest tariff and transshipment fee on behalf of the user, and help the user sign the power supply contract. There are at least three different types of contracts: contracts between power producers and retail users, contracts between transmission networks and retail users, contracts between power service providers and retail users. The agency fee is generally 0.5%~2.0% of the agency price.

4.2.3 *Transmission network charges*

This part of the cost is to transfer electricity from the power producer through the high-voltage transmission network to the high-voltage grid usage fee of the meter metering Office of the power supply service provider belonging to the retail user. The fees for transmission network crossing are set by independent regulators, usually including line loss fees, congestion fees, grid equipment usage fees, grid expansion fees, management fees of transmission companies, ancillary service fees of transmission companies to customers, etc.

4.2.4 *Distribution network charges*

This part of the cost refers to the low-voltage power grid usage fee that transfers electricity from the meter metering Office of the power supply service provider to the meter installation Office of the retail user through the low-voltage distribution network. Distribution network fees are set by independent regulators, usually including distribution line loss fees, transformer loss fees, user meter fees and grounding investment recovery fees. This not only enables users to clearly understand the real process of electricity consumption, but also facilitates the distribution of electricity charges in various retail transport links.

4.3 *Electricity price measurement of traditional power companies*

In order to compete for power supply services in the retail electricity market, the key problem to be solved is the measurement and settlement of user's electricity fees. Because there are different types of meters installed by different regions and users in China at present, there will be replacement costs when users re-select retailers, which will create barriers to entry for new retailers. Therefore, it is necessary to adopt a unified measurement method and measurement system in the competitive electricity retail market model.

The main reason for the lack of demand response in the current electricity retail market is the lack of "demand response" motivation of power users. In China's electricity sales, wholesale electricity prices fluctuate sharply, and the price difference between peak and valley is very large, while retail electricity prices accepted by retail users are fixed prices spread according to time. Therefore, there is a lack of attention to demand side in the power industry. Secondly, the current way of electricity tariff measurement does not support user demand response. For a long time, our electricity tariffs are calculated on a monthly or annual basis. Only a fixed price cannot reflect the price changes in the real-time electricity market, and users cannot respond accordingly to the price changes. Therefore, the key to forming an effective market demand in the electricity retail market is time-sharing measurement. "Time-sharing metering" means that on the basis that users can understand the information of price changes through certain channels, the electricity meter must record the electricity consumption by hour; the electricity bill should show the electricity price at different times, and then multiply the electricity consumption as a time-sharing electricity charge in a month. Time-sharing measurement is not only a prerequisite for retail access, but also a prerequisite for generating competition.

5 CONCLUSIONS

The competitive electricity retail market model has different characteristics from a traditional electricity retail model. The service attribute of electric power commodities has been fully reflected. Distribution companies no longer have monopoly right to consumers and users have the choice, which is the biggest feature of competitive electricity retail market. The nature of the operation has changed. The power generation, the distribution, the transmission and the sale are completely independent. From the perspective of social welfare, the competitive electricity market model is of positive significance to guarantee the interests of consumers and improve the efficiency of resource allocation.

ACKNOWLEDGMENT

We are very grateful to the technology project, Research on Power Market Operation Mechanism of the State Grid Jibei Electric Power Company, for sponsoring this paper.

REFERENCES

Chen Lei, 2012. The international comparative study on regulation reform of power industry, Fujian Normal University.

Hu Zhaohui, Wang Haisheng, 2018. Consideration on Competitive Electricity Sale Market under the Situation of Electricity System Reform. *Economic & Trade* (11):222–224.

Luo Qin, Song Yiqun, 2015. Marketing Strategy in Competitive Retail Market Considering Interruptible Load. *Automation of Electric Power Systems* 39(17):134–139.

Liu Junzhe, 2016. A Study on the Social Capital Entering the Power Sales Market of China, Yunnan University.

Mao Hua, Liu Zhongyi, Yan Dawei, 2017. User-oriented Competitive Power Sale Strategy for Power Grid Enterprises. *China Power Enterprise Management* (28):60–61.

Pan Shuangshuang, Tong Jun, 2017. Competitive Strategy of Big Customer Electricity Market under the Reform of Electricity Sale Side. *China Power Enterprise Management* (01):42–45.

Wu Qiming, 2016. Consideration on Competitive Electricity Sale Market under the Situation of Electricity System Reform. *Modern SOE Research* (08):31–33.

Emerging Developments in the Power and Energy Industry – Dufo-López, Krzywanski & Singh (eds)
© 2020 Taylor & Francis Group, London, ISBN 978-0-367-27169-5

A tie-line power smoothing method for an office building Microgrid by scheduling thermal mass of the building and plug-in electric vehicles

Bo Zhao, Xuesong Zhang & Chouwei Ni
State Grid Zhejiang Electric Power Research Institute, Hangzhou, China

Xiaolong Jin, Lin Li & Lijia Du
Key Laboratory of Smart Grid of Ministry of Education, Tianjin University, Tianjin, China

ABSTRACT: Aiming to facilitate a high penetration of renewable generation and the low energy consumption technologies at the demand side, there is significant development of low-carbon building Microgrids integrated with renewable generation. However, renewable generation is usually intermittent, uncertain and uncontrollable, which induces power mismatches between power demand and supply for the low-carbon building Microgrids. Therefore, a two-layer method for an office building Microgrid is proposed to smooth the tie-line power by scheduling the thermal mass of the building and the electric vehicles at two different time scales. A Vehicle-to-Building control strategy was developed to dispatch the electric vehicles as a flexible resource. Numerical studies demonstrated that the proposed method can smooth the fluctuations of the tie-line power for the office building Microgrid.

1 INTRODUCTION

Increasing attention is being paid to technologies in renewable energy and energy efficiency improvement due to the rapid growth of global energy use and environmental deterioration (J. Keirstead, M. Jennings and A. Sivakumar, 2012). Aiming to facilitate a high penetration of renewable generation and the low energy consumption technologies at the demand side, there is significant development of low-carbon buildings integrated with renewable generation (D. H. W. Li, S. K. H. Chow, E. W. M. Lee, 2013).

The existing research work has made good contributions to the energy management for a building Microgrid. A multi-objective dispatch model was proposed in (D. Zhang, S. Evangelisti, P. Lettieri, et al, 2016) to minimize the daily operating cost and the pollutants emission. An electric chiller (EC) was used as the cooling system of a building in (Y. Zhao, Y. Lu, C. Yan, et al, 2014) and the electric power consumption of the EC was dispatched using a nonlinear programming method for cost saving.

However, the existing research work mainly focuses on reducing operating cost of the building Microgrid. The renewable generation is usually intermittent, uncertain and uncontrollable, which induces power mismatches between power demand and supply for the building Microgrid. The fluctuations of tie-line power of the building Microgrid are also induced by the mismatches. Therefore, both operating cost reduction and smoothing fluctuations of tie-line power should be considered in the building Microgrid energy management. Moreover, the flexibility of the building with heat inertia and the flexibility of the plug-in EVs in the building Microgrid haven't been fully explored in the tie-line power smoothing for the building Microgrid.

Therefore, a two-layer method for an office building Microgrid is proposed to smooth the tie-line power by scheduling the thermal mass of the building and the electric vehicles at two different time scales. An office building is simplified to a lumped thermal mass and modelled as a simplified thermal storage system, namely the virtual energy storage system (VESS). The

VESS is dispatched as a flexible resource to smooth the fluctuations of tie-line power of the building Microgrid. A Vehicle-to-Building control strategy is developed to dispatch the electric vehicles as another flexible resource. The VESS and the EVs under V2B control are dispatched at two different time scales considering their different response characteristics, which makes the method more efficient. Numerical studies demonstrated that the proposed method can smooth the fluctuations of the tie-line power for the office building Microgrid.

2 MATHEMATICAL MODEL OF THE OFFICE BUILDING MICROGRID

2.1 *Configuration of the office building Microgrid*

An office building Microgrid is adopted as the test system in this paper, as shown in Figure 1. The building Microgrid includes a photovoltaic system, an EC and a V2B system (The EVs under V2B control is abbreviated as V2B system in this paper).

2.2 *Model of virtual energy storage system*

Considering a summer cooling scenario, a building is modeled as a single isothermal air volume (JA. Duffie and WA. Beckman, 1991). Then, the model of VESS is developed considering the thermal performance of the building. The cooling energy generated by the EC is stored in the building when the electricity price is low. In that case, the VESS is charged in the view point of the Microgrid. In the same way, the cooling energy generated by the EC is discharged in the building when the electricity price is high. In that case, the VESS is discharged in the view point of the Microgrid. The indoor temperature comfort zone and temperature set-point are considered in the model of VESS to maintain the customer comfort level. The specific model of VESS can be found in our previous study(X. Jin, X. Wang, Y. Mu, et al, 2016).

2.3 *The Vehicle-to-Building (V2B) model*

The V2B model consists of three models: 1) the EV battery model; 2) the model of EV mobility behaviour; and 3) the operation constraints of the EV. The generic EV battery model developed in (Y. Mu, J. Wu, J. Ekanayake, et al, 2013) is used to obtain the relationship between the SOC and the charging/discharging behaviours of the EVs.

- EV battery model

The type of the EVs is determined as M1 with Home Based Work (HBW) transportation pattern (EU Merge Project, 2010). And the probability density function (pdf) of EV battery

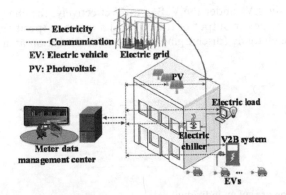

Figure 1.   Configuration of the office building Microgrid.

capacity (Cap) of M1 EV is defined as the truncated Gamma distribution, as shown in Eq. (1).

$$\begin{cases} f(Cap, \alpha, \beta) = \begin{cases} 0, & -\infty < Cap < Cap_{min} \text{ or } Cap_{max} < Cap < +\infty \\ f_0(Cap, \alpha, \beta) \Big/ \left(\int_{Cap_{min}}^{Cap_{max}} f_0(Cap, \alpha, \beta) dCap \right), & Cap_{min} \leq Cap \leq Cap_{max} \end{cases} \\ f_0(Cap, \alpha, \beta) = Cap^{\alpha-1} e^{-\frac{Cap}{\beta}} \Big/ (\beta^{\alpha} \Gamma(\alpha)) \end{cases} \quad (1)$$

Where α and β are the shape parameter and the scale parameter of the *pdf* of Gamma distribution; Cap_{min} and Cap_{max} are the minimum and maximum battery capacities of M1 EV. For M1 EV, the α is 4.5, the β is 6.3, the Cap_{min} is 10 kW and the Cap_{max} is 72 kW.

- Model of EV mobility behaviour

It is assumed that all the EVs investigated in this paper are charged at home. They are charged to their expected SOCs ($SOC_{d,\,home}$) before travelling to satisfy their energy demands travelling from home to work. It is assumed that SOC drops linearly with the travel distance. The SOC of an EV when it arrives at the office (SOC_{in}) is calculated by Eq. (2). Since the EVs are charged to $SOC_{d,\,home}$ at home, the minimum expected SOC at the plug-out time when they leave the office ($SOC_{d,\,office}$) is set to be the same as the SOC_{in}.

$$SOC_{in} = SOC_{d,home} - (D_{h-w} \times C_e)/Cap \quad (2)$$

The distribution of M1 EV of the C_e can be found in (EU Merge Project, 2010); D_{h-w} is assumed to be half of the EV's daily travelling distance (D) according to HBW EVs' mobility behavior.

D basically follows a Normal distribution, as shown in Eq. (3) (M. Wang, Y. Mu, H. Jia, et al, 2017).

$$f(D, \mu_d, \sigma_d) = e^{-(D, -\mu_d)^2/2\sigma_d^2} \Big/ \left(\sigma_d \sqrt{2\pi} \right) \quad (3)$$

Where μ_d is the mean value of daily travelling distance; σ_d is the standard deviation. For HBW vehicles, the μ_d is 35.9 km and the σ_d is 19.6 km (X. Dong, Y. Mu, H. Jia, et al, 2016).

The plug-in time (t_{in}) when the EVs arrive at the office and the plug-out time (t_{out}) when the EVs leave the office are determined based on the office time, which are assumed to follow the Normal distribution (Z.Wang, L.Wang, A.I.Dounis, et al, 2012). Based on the information supplied by the battery characteristics and mobility behaviors, the Monte Carlo simulation method is used to generate the Cap, SOC_{in}, t_{in} and t_{out} for all the EVs.

- Operation constraints of the EV

In order to dispatch the EVs under the V2B control effectively, the charging and discharging power is defined as the power output of the EV, and a negative (positive) value of the power output represents the charging (discharging) process. The operation area of an individual EV is shown in Figure 2.

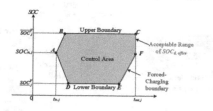

Figure 2. The operation area of an individual EV.

For an individual EV, the operation area (the shaded area shown in Figure 2) is limited by the boundaries of power output and the boundaries of state of charge (SOC). The charging boundaries of 'A-B' and 'A-D' are natural boundaries constrained by the rated charging and discharging power of an individual EV; the charging boundaries of 'D-E' and 'B-C' are natural boundaries constrained by the EV battery maintenance constraints, as shown in Eq. (4); the charging boundary of 'E-F' is a forced boundary to guarantee the minimum expected SOC ($SOC_{j,d,office}$) at t_{out}; and the charging boundary of 'C-F' is a flexible boundary with acceptable range of $SOC_{j,d,office}$, as shown in Eq. (5).

$$\underline{SOC}_j^V \leq SOC_{j,t} \leq \overline{SOC}_j^V \tag{4}$$

$$SOC_{j,t_{out}} \geq SOC_{j,d,office} \tag{5}$$

Then, the real-time power output of the V2B system ($P_{V2B,t}$), the upper boundary ($P_{V2B,t}^{upper}$) and the lower boundary ($P_{V2B,t}^{lower}$) of the $P_{V2B,t}$ under the V2B control are shown in Eq. (6).

$$\begin{cases} P_{V2B,t} = \sum_{j=1}^{N_t} P_{j,t}^V \\ P_{V2B,t}^{upper} = \sum_{j=1}^{m_t} \left(-P_{j,t}^{V,c}\right) + \sum_{j=1}^{n_t} \overline{P}_{j,t}^V \\ P_{V2B,t}^{lower} = \sum_{j=1}^{m_t} \left(-P_{j,t}^{V,c}\right) + \sum_{j=1}^{n_t} \underline{P}_{j,t}^V \end{cases} \tag{6}$$

Where N_t is the total number of EVs at time t; m_t is the number of EVs operating on the forced-charging boundary ('E-F') at time t; n_t is the number of the other EVs ($n_t = N_t - m_t$).

3 FORMULATIONS OF THE TIE-LINE POWER SMOOTHING METHOD

3.1 Framework of the tie-line power smoothing method

Due to the forecasting errors of the renewable generation, electric load demand, outdoor temperature and solar radiation, mismatches exist in both day-ahead and intra-hour schedules of the tie-line power. Therefore, a two-layer tie-line power smoothing method is conducted to balance the mismatches in both short term in VESS (15 minutes) and ultra-short term in V2B system (1 minute). At the master layer, the optimal dispatch program is used to generate operating schedules for the VESS (charging/discharging power of the VESS) and the V2B system (charging/discharging power of all EVs). At the client layer, the operation information of the VESS (occupied hours of the building and the indoor temperature comfort zone) as well as the upper and lower boundaries of power output of the V2B system are uploaded to the master.

3.2 Formulations of the tie-line power smoothing method

An optimal dispatch program is used to dispatch the VESS in 15 minutes to make the actual tie-line power schedules follow the day-ahead tie-line power schedules. The optimal dispatch program can be found in our previous work (X. Jin, X. Wang, Y. Mu, et al, 2016). The V2B system is dispatched in 1 minute to further smooth the fluctuations of the tie-line power. Call the optimal dispatch program in 1 minute to obtain the V2B control signals (i.e., the target power output of the V2B system, $P_{V2B,t''}^{tar}$). The dispatch objective in 15 minutes and in 1 minute are formulated in Eq. (7-a) and Eq. (7-b) respectively.

$$\min \left\{ \left[P_{ex,t'} - P_{ex,t'}^{set} \right]^2 \right\}, \ \forall t' \in T' \ (a) \ \min \left\{ \left[P_{ex,t''} - P_{ex,t''}^{set} \right]^2 \right\}, \ \forall t'' \in T' \ (b) \qquad (7)$$

Where $P_{ex,t'}^{set}/P_{ex,t''}^{set}$ is the day-ahead set-point of the tie-line power in time slot t'/t'', which is generated in the day-ahead dispatch stage; $P_{ex,t'}/P_{ex,t''}$ is the actual tie-line power in time slot t'/t'',. The constraints of the dispatch problem are the same as Eqs. (8)–(15), as shown in (Y. Mu, J. Wu, J. Ekanayake, et al) and Eq. (8).

$$P_{V2B,t''}^{lower} \le P_{V2B,t''}^{tar} \le P_{V2B,t''}^{upper} \qquad (8)$$

The upper boundary ($P_{V2B,t''}^{upper}$) and the lower boundary ($P_{V2B,t''}^{lower}$) of the $P_{V2B,t''}$ are updated in every time slot after the V2B control, which is introduced in the following text. The initial boundaries (when $t'' = 1$,) are obtained based on the V2B model without V2B control. Based on the updated upper and lower boundary of the $P_{V2B,t''}$, the energy management system solves the stage 2 dispatch problem in every time slot to generate the $P_{V2B,t''}^{tar}$ in 1 minute.

The target power output change of the V2B system is determined by Eq. (9).

$$\Delta P_{V2B,t''}^{tar} = P_{V2B,t''}^{tar} - P_{V2B,t''} \qquad (9)$$

Then, if $\Delta P_{V2B,t''}^{tar}$ is not equal to zero, the following steps are conducted to determine the updated power output of each EV in the V2B control:

Step 1) If $\Delta P_{V2B,t}^{tar}$ is higher than zero, the target power output of each EV is determined by Eq. (10). And $m_{1,t}$ is defined as the number of EVs of the V2B system which have not reached their upper limits of power output.

$$P_{j,t''}^{V,tar} = P_{j,t''}^{V} + \Delta P_{V2B,t''}^{tar} \times \frac{SOC_{j,t''}^{V} - \underline{SOC_j^V}}{\sum\limits_{j=1}^{m1,t} \left(SOC_{j,t''}^{V} - \underline{SOC_j^V} \right)} \qquad (10)$$

If $\Delta P_{V2B,t''}^{tar}$ is lower than zero, the target power output of each EV is determined by Eq. (11). And $m_{2,t''}$ is defined as the number of EVs in the V2B system which have not reached their lower limits of power output.

$$P_{j,t''}^{V,tar} = P_{j,t''}^{V} + \Delta P_{V2B,t''}^{tar} \times \frac{\overline{SOC_j^V} - SOC_{j,t''}^{V}}{\sum\limits_{j=1}^{m2,t} \left(\overline{SOC_j^V} - SOC_{j,t''}^{V} \right)} \qquad (11)$$

Step 2) Based on the obtained target power output of each EV, $m_{3,t}$ ($m_{4,t}$) is defined as the number of EVs which have violated their upper (lower) power output limits. Thus, the new revised $\Delta P_{V2B,t}^{tar}$ is determined by Eq. (12). If $\Delta P_{V2B,t''}^{tar}$ is equal to zero, the power output of EV j after V2B control is updated with $P_{j,t''}^{V,tar}$. Otherwise, Step 1) ~ Step 2) will be implemented again until the updated $\Delta P_{V2B,t''}^{tar}$ is equal to zero.

$$\begin{cases} \Delta P_{V2B,t''}^{tar} = \sum\limits_{j=1}^{m3,t} \left(P_{j,t''}^{V,tar} - \overline{P}_{j,t''}^{V} \right), & m_{3,t} > 0 \ \& \ m_{4,t} = 0; \\ \Delta P_{V2B,t''}^{tar} = \sum\limits_{j=1}^{m4,t} \left(P_{j,t''}^{V,tar} - \underline{P}_{-j,t''}^{V} \right), & m_{3,t} = 0 \ \& \ m_{4,t} > 0. \end{cases} \qquad (12)$$

4 CASE STUDIES

4.1 *Case description*

An office building Microgrid shown in Figure 1 is used to verify the effectiveness of the developed tie-line power smoothing method. The thermal parameters of the building are given in Tab. 1. The occupied hours are set to be from 8:00a.m. to 20:00p.m. In this study, the indoor temperature set-point range is set to be from 19 °C to 26 °C with VESS being dispatched. The measured actual outdoor temperature in a summer day are shown in Figure 3. The forecasted and actual solar radiation on horizontal surface are shown in Figure 4. The rated power of photovoltaic system is 100 kW. The forecasted and actual electric loads and internal heat gains of the building in a typical summer day are shown in Figure 5 and Figure 6 respectively.

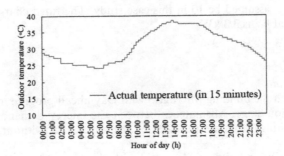

Figure 3.   Outdoor temperature.

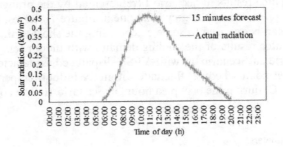

Figure 4.   Solar radiation on horizontal surface.

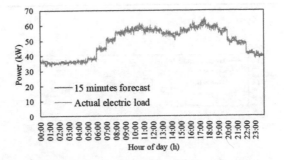

Figure 5.   Electric load of the building Microgrid.

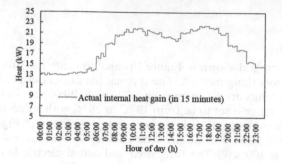

Figure 6.   Internal heat gain of the building.

The number of EVs is assumed be 10 in this case study. The rated charging and discharging power of an individual EV is 3.3 kW.

4.2   Simulation results

1) VESS dispatch results

Scenario I: Dispatch the building Microgrid with Day-ahead programming (DA-P) method at the actual operation stage. Mismatches between the energy demand and supply caused by day-ahead forecasting errors are balanced by the upper grid, without short-term dispatch of the VESS.

Scenario II: Dispatch the building Microgrid with short-term dispatch of the VESS at the actual operation stage. The VESS is dispatched to smooth the fluctuations of the tie-line power.

The day-ahead set-point of the tie-line power is shown by the black solid line in Figure 7. The comparative dispatch results in Scenario I (represented by the orange solid line in Figure 7) and Scenario II (represented by the grey solid line in Figure 7) suggest that the fluctuations of the tie-line power can be reduced to some extend using the short-term dispatch of VESS in 15 minutes. The dispatch results of the cooling demand with indoor temperature set-point of the building Microgrid in Scenario II with VESS dispatched is depicted in Figure 8. The indoor temperature set-point schedules fluctuate within the indoor temperature set-point comfort range (19 °C~26 °C) during the occupied hours in Scenario II. The cooling demand of the

Table 1.   Building parameters.

U_{wall} [W/(m². K)]	F_{wall} (m²)	U_{win} [W/(m². K)]	Window to wall ratio (%)	Long side (m)	Short side (m)	Height (m)
0.908	2400	2.750	75	40	20	30

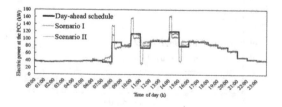

Figure 7.   Tie-line power of the building Microgrid.

595

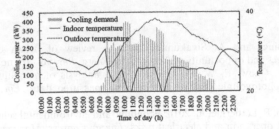

Figure 8. The cooling demand and indoor temperature of the building with VESS dispatched.

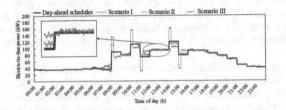

Figure 9. Tie-line power of the building Microgrid in the comparative scenarios.

building is also dispatched under Scenario II to dispatch the VESS for smoothing the fluctuations of the tie-line power.

2) V2B dispatch results

Scenario III is further developed to verify the effectiveness of the method with ultra-short term dispatch (in 1 minute) of the V2B system.

A comparison of the dispatch results in Scenario III (red solid line in Figure 9), Scenario I (orange solid line in Figure 9) and Scenario II (grey solid line in Figure 9) suggests that the electric power fluctuations at the PCC are well smoothed by the V2B system.

5 CONCLUSION

This paper proposes a two-layer tie-line power smoothing method for an office building microgrid by scheduling thermal mass of the building and EVs. To incorporate the different response characteristics and the forecasting errors of the input data at different time scales, the VESS and the V2B system are coordinated at two different time scales. Numerical studies show that the proposed method can smooth the fluctuations of the tie-line power of the office building Microgrid.

ACKNOWLEDGEMENT

Project supported by the Research on multi-energy dynamic control method and software development of regional integrated energy system(State Grid Zhejiang Power Supply Company science and technology project).

FUNDING

This work is financially supported by the research on multi-energy dynamic control method and software development of regional integrated energy system. (state grid zhejiang power supply company science and technology project).

REFERENCES

J. Keirstead, M. Jennings and A. Sivakumar. 2012. A review of urban energy system models: Approaches, challenges and opportunities. *Renew. Sust. Energ. Rev.* vol. 16, no. 6, pp. 3847–3866.

D. H. W. Li, S. K. H. Chow, E. W. M. Lee. 2013. An analysis of a medium size grid-connected building integrated photovoltaic (BIPV) system using measured data. *Energy Build.* vol. 60, no. 5, pp. 383–387.

D. Zhang, S. Evangelisti, P. Lettieri, et al. 2016. Economic and environmental scheduling of smart homes with microgrid: DER operation and electrical tasks. *Energy Convers. Manage.* vol. 110, no. Supp 14, pp. 113–124,.

Y. Zhao, Y. Lu, C. Yan, et al. 2014. MPC-Based Optimal Scheduling of Grid-Connected Low Energy Buildings with Thermal Energy Storages. *Energy Build.* vol. 86, no. 12, pp. 415–426.

JA. Duffie and WA. Beckman. 1991. Solar engineering of thermal process. *New York: Wiley.*

X. Jin, X. Wang, Y. Mu, et al. 2016. Optimal scheduling approach for a combined cooling, heating and power building microgrid considering virtual storage system. *Power and Energy Society General Meeting, IEEE.* pp. 1–5.

Y. Mu, J. Wu, J. Ekanayake, et al. 2013. Primary Frequency Response From Electric Vehicles in the Great Britain Power System. *IEEE Trans. Smart Grid.* vol. 4, no. 2, pp. 1142–1150.

EU Merge Project. 2010. Deliverable 2.1: Modelling Electric Storage devices for electric vehicles, [Online]. Available: http://www.ev-merge.eu/images stories/uploads/MERGE_WP2_D2.1.pdf.

M. Wang, Y. Mu, H. Jia, et al. 2017. Active power regulation for large-scale wind farms through an efficient power plant model of electric vehicles. *Appl. Energy.* vol. 185, no. 1, pp. 1673–1683.

X. Dong, Y. Mu, H. Jia, et al. 2016. Planning of Fast EV Charging Stations on a Round Freeway. *IEEE Trans. Sustain. Energy.* vol. 7, no. 4, pp. 1452–1461.

Z. Wang, L. Wang, A. I. Dounis, et al. 2012. Integration of plug-in hybrid electric vehicles into energy and comfort management for smart building. *Energy Build.* vol. 47, no. 47, pp. 260–266.

Emerging Developments in the Power and Energy Industry – Dufo-López, Krzywanski & Singh (eds)
© 2020 Taylor & Francis Group, London, ISBN 978-0-367-27169-5

A fault current calculation method in a neutral point ungrounded system for a single-phase ground fault

Xiao Jun Zhu & Shuang Yang
State Grid Chongqing Electric Power Research Institute, Chongqing, China

Bang Dong Yu
State Grid Beijing Electric Power Company, Beijing, China

ABSTRACT: When a single-phase ground fault in a neutral point ungrounded system occurs, the fault current includes not only a power frequency component, but also a transient component with short duration, large original amplitude, and relatively concentrated energy. Rapid and accurate calculation of the full fault current is the basis for realizing full fault current compensation. One method for full fault current calculation of a single-phase grounding fault in a non-solidly grounded system is proposed. Here a Γ equivalent circuit is applied to model the feeders, and its feasibility and rationality are analyzed. Then based on the Kirchhoff Current Law and three-phase time-domain voltage signals from bus TVs, the full fault current can be calculated via differential equations that are used to describe the Γ equivalent circuit. The calculation results can provide a basis for full fault current compensation. A simulation case indicates that the method used for calculating a full fault current based on a Γ equivalent circuit has the advantages of discernible principle, rapid and accurate calculation, less electrical quantities needed, and being independent of fault line selection. It can meet the demand of a full fault current compensation for a single-phase grounding fault in a non-solidly grounded system.

1 INTRODUCTION

In order to quench a fault arc when a single-phase grounding fault occurs in a non-solidly grounded power system, an arc suppression coil (ASC) installed on the neutral point is usually utilized to weaken the earth current of the fault spot. But the ASC can only compensate off the power frequency component rather than the full fault earth current. Generally, the remnant fault current after the ASC compensation has the characteristics of short duration, relatively concentrated energy, and wide-frequency spectrum (Wang et al. 2007; Zhang et al. 2007). It will cause fault arc reburning, and eventually lead to the failure of fault arc blowout. The continuous fault arc burning can result in the transformation from a transient single-phase grounding fault to a permanent one and even phase-phase fault, which will bring serious harm to the safe and reliable operation of a non-solidly grounded power system (Ge 2007). Therefore, study of the full fault current compensation when a single-phase grounding fault occurs in a non-solidly grounded power system has important significance.

Full fault current compensation can compensate off both the power frequency component and the high-frequency transient one. To achieve an ideal compensation result when a single-phase grounding fault occurs, calculation of the full fault earth current is necessary. At present analysis and research on full fault current calculation focus mainly on extra high voltage (EHV) transmission power systems and relatively less on distribution power system. The traveling wave method, Bergeron calculation method, and D. Alembert formula are applied into the calculation of a full fault current in an EHV transmission power system based on the two ends' voltage and current signals of the fault line (Yamaura et al. 1997; Zheng et al. 2004; Su et al. 2005). In a distribution power system, there are numerous feeders and branches. It is difficult to obtain both ends' synchronous voltage and current signals of the fault line. Therefore, finding one full fault current calculation method consistent with the characteristics of a non-solidly grounded system is required. In this article, one practical method used to calculate the full fault current of a non-solidly grounded system when a single-phase grounding fault occurs is proposed. Here the Γ equivalent circuit is applied to model distribution feeders, and the feasibility and rationality of the Γ equivalent circuit model are demonstrated. Then based on the Kirchhoff Current Law and three-phase time-domain voltage signals from the bus bar potential transformers, the full fault current is calculated via differential equations that describe a Γ equivalent circuit. Finally, calculation results are sent to the wide-frequency current generator to realize full fault current compensation. Simulation cases indicate that the method mentioned in this article can calculate the full fault current with high accuracy that can absolutely meet the requirement of full fault current compensation.

2   ANALYSIS OF FREQUENCY CHARACTERISTICS OF Π AND Γ EQUIVALENT CIRCUITS

The Π equivalent circuit is usually applied in modeling short distribution lines of a power system. In some cases, the Γ equivalent circuit is applied instead of the Π equivalent circuit for the benefit of simplifying the calculation. Figure 1 is a Γ equivalent circuit in which the whole line grounding capacitance is converted into one shunt capacitor at the line head. The parameters R and L respectively stand for single-phase equivalent resistance and inductance of distribution feeders.

The Γ and Π equivalent circuits are both dual-port networks. Within a certain frequency band, the Γ equivalent circuit equals the Π equivalent circuit in modeling the distribution feeders if their dual-port external characteristics are approximately the same. There are Z, Y, T and H parameter matrices to describe external transmission characteristics of a dual-port network and the H parameter matrix is applied most often in real projects. Generally, each matrix has four parameters that reflect the relation of both end electrical quantities of the dual-port network. Equation (1) describes the transitive relation of a dual-port network in H parameter form.

$$\begin{bmatrix} \dot{U}_1 \\ \dot{I}_2 \end{bmatrix} = \begin{bmatrix} H_{11} & H_{12} \\ H_{21} & H_{22} \end{bmatrix} \begin{bmatrix} \dot{I}_1 \\ \dot{U}_2 \end{bmatrix} = \mathbf{H} \begin{bmatrix} \dot{I}_1 \\ \dot{U}_2 \end{bmatrix} \tag{1}$$

Laplace forms of the four transitive parameters in the H parameter matrix are as follows.
  The Γ equivalent circuit:

$$H_{11} = \frac{R+Ls}{LCs^2 + RCs + 1}, H_{12} = -H_{21} = \frac{1}{LCs^2 + RCs + 1}, H_{22} = \frac{Cs}{LCs^2 + RCs + 1} \tag{2}$$

The Π equivalent circuit:

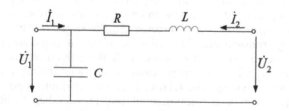

Figure 1.   Γ equivalent circuit.

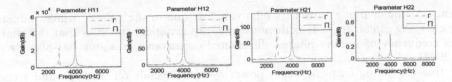

Figure 2.   Amplitude-frequency characteristic curves of Γ and Π equivalent circuits (line length is 15 km).

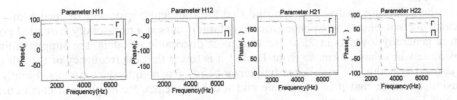

Figure 3.   Phase-frequency characteristic curves of Γ and Π equivalent circuits (line length is 15 km).

Table 1.   Applicable frequency bands under different distribution line lengths.

Line length (km)	1	5	10	15
Applicable frequency bands (Hz)	0–41,900	0–8150	0–3800	0–2600

$$H_{11} = \frac{R+Ls}{\frac{1}{2}LCs^2 + \frac{1}{2}RCs + 1}, H_{12} = -H_{21} = \frac{1}{\frac{1}{2}LCs^2 + \frac{1}{2}RCs + 1}, H_{22} = \frac{Cs}{LCs^2 + RCs + 2} \quad (3)$$

Substituting distribution line parameters (R, L, C) of a real power system into Equations (2) and (3), the amplitude-frequency and phase-frequency characteristics are analyzed. Figures 2 and 3 show respectively curves of the amplitude-frequency and phase-frequency characteristics of Γ and Π equivalent circuits according to Equations (2) and (3) when the distribution line is 15 km long.

We can conclude from Figures 2 and 3 that the applicable frequency band of two types of circuits with similar amplitude-frequency and phase-frequency characteristics is about 0–2600 Hz. In addition, the applicable frequency bands under other line lengths are shown in Table 1.

3   CHARACTERISTIC ANALYSIS AND CALCULATION OF THE FULL FAULT CURRENT

3.1   *Transient component frequency characteristic analysis of the full fault current*

When a single-phase grounding fault occurs in a non-solidly grounded system, the voltage of the fault phase drops and that of the healthy phases rises. Therefore, a transient fault current includes not only the fault phase's discharging currents but also the healthy phases' charging ones. The transient high-frequency component of the full fault current is characterized by a wide frequency band and concentrated energy, which is usually applied into fault line selection (Ma et al. 2008; Shu et al. 2008; Zhang et al. 2011). When a fault occurs, the fault phase voltage drops instantaneously and the distributed capacitance of the fault phase discharges rapidly. So the oscillating frequency is relatively high. A large amount of data indicate that the oscillating frequency of the discharging current is about 500–2500 Hz. For healthy phases, zero sequence inductance of the transformer is added to the charging circuit loop. Thus the oscillating frequency of the charging current is relatively low due to the gain of whole inductance of the charging circuit loop. It is demonstrated by empirical statistics that the charging current frequency of healthy phases' distributed capacitance is about 100–800 Hz (Xue et al. 2004).

The free oscillating frequency of a power system depends on the type and structure of distribution lines. Generally, distribution lines have larger inductance L and smaller capacitance C. So the fault current damps rapidly because of the found relation $R < < 2\sqrt{L/C}$. It is recognized that the free oscillating frequency is about 300–1500 Hz.

From the preceding analysis, we can find that the transient component frequency of a full fault earth current when single-phase grounding fault occurs in a non-solidly grounded power system is generally lower than 2500 Hz. For most distribution feeders, the supply radius is approximately less than 10 km, so from Table 1 it is found that the frequency of the full fault earth current is within the applicable frequency band of two equivalent circuits. Finally the conclusion is drawn that it is reasonable and feasible to calculate a full fault earth current based on a Γ equivalent circuit.

3.2 Calculation of the full fault earth current

A simple distributed network with two sets of distribution lines modeled by a Γ equivalent circuit is shown in Figure 4.

Here R, X, and C respectively stand for the equivalent resistance, inductance, and grounding capacitance of each line. R_g is the grounding resistor of the fault spot and i_g stands for the distributed capacitance current of each line. u_A, u_B, and u_C are the three-phase time-domain voltage signals obtained from bus bar potential transformers. The wide-frequency current generator is used to generate an objective compensation current. At the fault spot, according to the Kirchhoff Current Law, an equation can be written as follows:

$$i_k + i^I_{g.A} + i^I_{g.B} + i^I_{g.C} + i^{II}_{g.A} + i^{II}_{g.B} + i^{II}_{g.C} = 0 \tag{4}$$

Here, i_k is the full fault earth current. So

$$i_k = -[(C^I_A + C^{II}_A)\frac{du_A}{dt} + (C^I_B + C^{II}_B)\frac{du_B}{dt} + (C^I_C + C^{II}_C)\frac{du_C}{dt}] \tag{5}$$

Similarly, for a distributed network with l sets of lines, the full fault earth current can be calculated via the following equation:

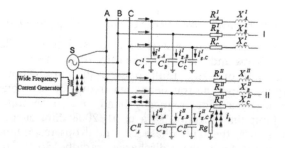

Figure 4. Fault current distribution in a non-solidly grounded system when a single-phase grounding fault occurs.

$$i_k^l = -\left(\sum_{m=1}^{l} C_A^m \cdot \frac{du_A}{dt} + \sum_{m=1}^{l} C_B^m \cdot \frac{du_B}{dt} + \sum_{m=1}^{l} C_C^m \cdot \frac{du_C}{dt}\right) \tag{6}$$

Here C_A^m, C_B^m, and $C_C^m (m = 1, 2, \ldots, l)$ respectively stand for the sum of equivalent distributed grounding capacitance of each phase. Generally, it is recognized that $C_A^m \approx C_B^m \approx C_C^m = C^m$. Thus,

$$i_k^l = -\sum_{m=1}^{l} C^m \cdot \left(\frac{du_A}{dt} + \frac{du_B}{dt} + \frac{du_C}{dt}\right) \tag{7}$$

The Laplace form of Equation (7) is

$$i_k^l(s) = -s \sum_{m=1}^{l} C^m \cdot [u_A(s) + u_B(s) + u_C(s)] + \sum_{m=1}^{l} C^m \cdot [u_A(0_-) + u_B(0_-) + u_C(0_-)] \tag{8}$$

In Equation (8), $u_A(0_-)$, $u_B(0_-)$, and $u_C(0_-)$ are instantaneous values of a three-phase voltage at the start moment of fault. Generally, their sum is about zero because of the symmetry of the power system. So calculation precision is practically immune from the second part of Equation (8). After simplification, the calculating formula becomes

$$i_k^l(s) \approx -s \sum_{m=1}^{l} C^m \cdot [u_A(s) + u_B(s) + u_C(s)] \tag{9}$$

According to bilinear Z-transformation theory, substitute $s = 2/Ts \cdot (1 - z^{-1})/(1 + z^{-1})$ into Equation (9) and the calculating equation in Z-transformation will be:

$$(1 + z^{-1})i_k^l = -\frac{2}{T_s} \cdot (1 - z^{-1}) \sum_{m=1}^{l} C^m \cdot (u_A + u_B + u_C) \tag{10}$$

Here T_s is the sample interval. Then the differential equation corresponding to (10) is

$$i_k^l(n) + i_k^l(n-1) = -\frac{2}{T_s} \cdot \sum_{m=1}^{l} C^m \cdot [u_A(n) - u_A(n-1) + u_B(n) - u_B(n-1) + u_C(n) - u_C(n-1)] \tag{11}$$

The final calculating differential equation is

$$i_k^l(n) = -\frac{2}{T_s} \cdot \sum_{m=1}^{l} C^m \cdot [u_A(n) - u_A(n-1) + u_B(n) - u_B(n-1) + u_C(n) - u_C(n-1)] - i_k^l(n-1) \tag{12}$$

Substitute three-phase voltage sampling values into Equation (12), then the full fault earth current sequence can be calculated. Simultaneously, the calculated results are sent to a wide-frequency current generator that is connected to the neutral point of distribution network, and the output current flows into the power system to compensate the fault earth current.

4 RESULTS AND CONCLUSION

4.1 *Simulation results*

In the software PSCAD/EMTDC, a 10-kV simulation model shown in Figure 5 is built. There are 13 feeders in the simulation model.

In Figure 5, C_{eq} stands for the equivalent grounding capacitance of three-phase bus bars and transformer, and the value is assumed to be 0.05 μF for each phase. Bus bar potential transformers and line current transformers are not visible. Set a single-phase grounding fault respectively at different positions of phase C of line 12. The simulation data are recorded by the recorder module and the sampling frequency is set as 6 kHz. The calculation and analysis work is completed with MATLAB®. Figures 6–8 are the calculation results, and Figure 9 shows the compensation effects of common ASC compensation versus full fault current compensation.

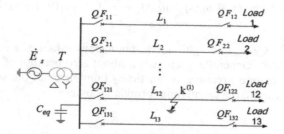

Figure 5. Simulation circuit model of 10-kV distribution power system.

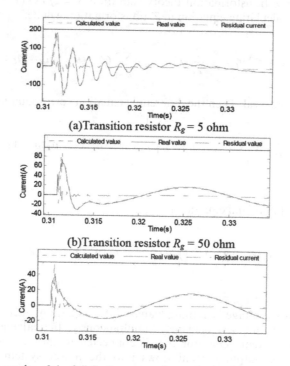

(a)Transition resistor R_g = 5 ohm

(b)Transition resistor R_g = 50 ohm

Figure 6. Calculation results of the full fault current when a single-phase grounding fault occurs at the head of the line.

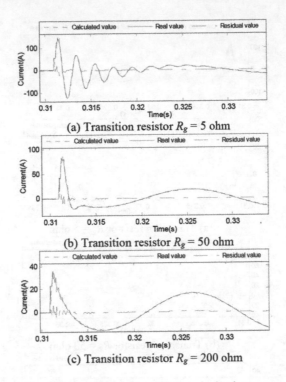

(a) Transition resistor $R_g = 5$ ohm

(b) Transition resistor $R_g = 50$ ohm

(c) Transition resistor $R_g = 200$ ohm

Figure 7.   Calculation results of the full fault current when a single-phase grounding fault occurs at the middle of the line.

4.2   *Conclusion*

By comparing calculation results and measurement values of a full fault current, the following conclusions are drawn:

1. Based on equivalent circuit and three-phase bus bar voltage sampling values, a full fault earth current can be calculated simply and rapidly with high accuracy. In addition, the method for full fault current calculation mentioned earlier has the advantages of usage of less electrical quantities and no fault line selection. This provides an accurate criterion for full fault current compensation.

2. Different fault positions cause different levels of precision in calculation because of the intrinsic characteristics of an equivalent circuit. But the calculation errors are very small, and so do the amplitude of fault spot residual currents after full fault current compensation. That is to say, full fault current compensation can achieve ideal arc extinguishing effects.

3. The larger the short circuit transition resistor is, the shorter the fault transient duration. Then transient high-frequency interference contributes less to the calculation errors and calculation results have relatively high accuracy.

4. By comparing compensation results of common ASC and full fault current, we can find that ASC cannot compensate high-frequency transient component, so in the early stage of the fault there is a residual current with large and damped amplitude. For the power frequency part of fault, the ASC compensation effect is affected by the damped DC current of ASC. On the contrary, residual current after full fault current compensation stays very small through the entire course of the fault. Above all, as far as the compensation effect is concerned, the full fault current compensation is obviously superior to the power frequency ASC compensation.

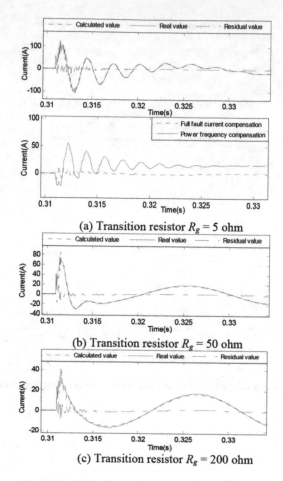

(a) Transition resistor R_g = 5 ohm

(b) Transition resistor R_g = 50 ohm

(c) Transition resistor R_g = 200 ohm

Figure 8. Calculation results of the full fault current when a single-phase grounding fault occurs at the end of the line.

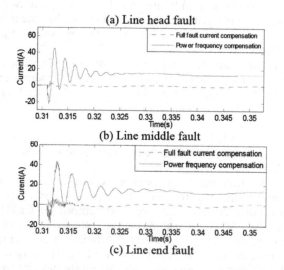

(a) Line head fault

(b) Line middle fault

(c) Line end fault

Figure 9. Compensation effects of common ASC compensation versus full fault current compensation (R_g = 5 ohm).

REFERENCES

Ge, Y. 2007. *New Principles and Techniques for Relay Protection and Fault Location*, 2nd ed. Xi'an: Xi'an Jiaotong University Press.

Ma, S., et al. 2008. An earth fault locating method in feeder automation system by examining correlation of transient zero mode currents. *Automation of Electric Power Systems* 32: 48–52.

Su, B., Dong, X., & Sun, Y. 2005. Distributed capacitive current compensation algorithm for current differential relay of UHV transmission line. *Automation of Electric Power Systems* 29: 36–40.

Shu, H., Peng, S., & Zhao, X. 2008. A new fault line detection method for hybrid transmission lines of distribution network. *Automation of Electric Power Systems* 32: 61–65.

Wang, Q., & Liu, J. 2007. Selective leakage protection based on correlation between high-frequency transient components. *Electric Power Automation Equipment* 27: 59–62.

Xue, Y., Feng, Z., & Xu, B. 2004. Analysis of transient characteristics of signals phase earth fault in non-solidly earthed network. *Journal of Xi'an Jiaotong University* 38: 195–199.

Yamaura, M., Kurosawa, Y., & Ayakawa, H. 1997. Improvement of internal changing current compensation for transmission line differential protection. In *International Conference on Developments in Power System Protection*, pp. 74–77.

Zhang, X., et al. 2007. Simulative analysis of zero modal transient characteristics for single-phase grounding fault. *Electric Power Automation Equipment* 27: 39–43.

Zhang, Z., et al. 2011. Fault line selection algorithm in small current grounding systems based on a Trous algorithm. *Automation of Electric Power Systems* 35: 66–70.

Zheng, Y., et al. 2004. Applied criterion of current differential protection based on Bergeron model. *Automation of Electric Power Systems* 28: 50–55.

Emerging Developments in the Power and Energy Industry – Dufo-López, Krzywanski & Singh (eds)
© 2020 Taylor & Francis Group, London, ISBN 978-0-367-27169-5

Compilation and calculation of prosperity index of power grid investment

Min Wang & Yuan Yuan
Business school, Renmin University of China, Beijing, China

Yanchao Lu & Dan Xu
State Power Economic Research Institute, Beijing, China

ABSTRACT: As an important part of promoting economic development, the study of investment prosperity plays an important role in monitoring the development of national industries, especially in the power industry, which is the lifeblood of the national economy. In this paper, firstly, the prosperity index of power grid investment is selected from the macro and micro levels based on the analysis of the characteristics of power grid investment and the related environment; secondly, the prosperity index system of power grid investment is determined by using time-difference correlation analysis and K-L information method, which divided power grid investment indexes into three groups; finally, the method of combining analytic hierarchy process (AHP) with composite index is used to compile the prosperity index of power grid investment, which is of practical significance for power grid enterprises to analyze the fluctuation of investment and guide their investment decision.

1 INTRODUCTION

As a vital basic energy industry in the development of the national economy, the electric power industry is a power production and consumption system composed of a series of links such as power generation, transmission, transformation, distribution and consumption. The security of electric power production, the sufficiency of electric power supply and the rationality of electric power consumption are all related to the healthy development of national economy. As the practice unit of social power supply, power grid enterprises undertake the important social responsibility of guaranteeing the rapid and steady development of national economy. Especially, with the deepening of social modernization in recent years, the demand-side power consumption pattern is increasing, which puts forward higher requirements for the security and stability of power supply. However, the establishment of a safe and stable power network requires a large number of funds and technologies. Due to the large amount of investment in the power network, the long investment cycle and the slow return of profits, grid investment requires to consider the investment demand comprehensively in many aspects, such as establishing diversified financing channels, and enhancing investment capacity so as to arrange the investment and using of funds in a reasonable way during the investment process.

There have been a lot of researches on the investment of power grid in academia. Because of the long investment cycle and slow return of the power grid, it brings a lot of uncertain risks to the power grid investment. In view of this kind of question has produced several research directions, one kind of literature constructs the power network investment risk appraisal system through the analysis power network investment risk factors (W Mianbin et al. 2006, 2010). In the other type of literature, considering the impact of investment uncertainty, the investment income is taken as the main object to analyze and study (JD Saphores et al. 2004, W Li et al. 2010, G Luo et al. 2011). In addition, due to the changes in the environment of the power industry, the reform of the power system has also had a certain impact on the investment income of the power grid. Therefore, how to face the opportunities and challenges brought about by the new system reform has also become a key consideration for the investment in the power grid. Based on this, a kind of research on the investment system of power grid (Y Ze et al. 2006, F Yongsheng 2015, L Haitao 2016) has been produced. These studies have enriched the theoretical knowledge of grid investment and provided important significance for grid investment, but grid investment is a complex issue involving many factors, and it should not be limited to considering only the influence of risk and system. Therefore, how to monitor, evaluate the investment and predict the investment benefit of power grid comprehensively has become an important topic. On the basis of synthesizing the existing literature, this paper excavates various indexes that affect the investment of power grid under the new system reform, and constructs the prosperity index of power grid investment, which provides scientific basis for the analysis of investment fluctuation and investment planning of power grid.

The method of prosperity index is an empirical research method which mainly uses statistical methods and measurement tools, which was first used by governments as an effective method for analyzing and forecasting macroeconomic conditions (P Dua & S.M. Miller 1996, R. Nilsso 2004, A. Ozyildirim 2010, D Wenquan et al. 1987). After that, it was gradually popularized and applied to all industries including automobiles, coal, steel, real estate, services, tourism, etc. (Turner et al. 1997, W Huimin & C Baoshu 1996, W Ting 2006, Y Yanbing & D Lieyun 2001, W Xiaoping & Z Shiyu, 2012, Y Wu et al. 2015). Due its validity, and also in the power industry (L Chang 2013, L Chang & G Tiemei 2011, D Haijing 2014, Z mei 2011). In this paper, we also use the prosperity index method, but different from previous studies, we mainly analyzing the indicators from micro-level, that are related industries and power grid enterprises; in addition, in the process of constructing the index model, the former research used the method of composite index to calculate more by using the form of equal weight, however, this kind of equal weight method can't reflect the actual degree of each index in the prosperity index system, therefore, in order to solve this problem, we apply the analytic hierarchy process to the composite index method to get the weight of each index, thus, we can get more scientific, more accurate and more agreeable calculation of the prosperity composite index, which has important practical significance for the analysis of the prosperity of grid investment.

2 CONSTRUCTION OF POWER GRID INVESTMENT PROSPERITY INDEX SYSTEM

2.1 *Macro factors*

In general, macro-analysis should have both long-term and overall characteristics. As a long-term behavior, grid investment is influenced by many factors, such as economic environment, related policy making, policy support, and various individual organizations, etc. Therefore, we mainly from the economic, political, social factors to consider the impact of investment in the power grid. Generally, the economic factors include the following kinds of indexes: national economy index, investment index, financial index, and fiscal index; the political factors are mainly reflected by the uncertainty of policy, and previous studies have found that policy uncertainty mainly through corporate finance affects investment (W Jinming et al. 2007), and some policies, such as tax incentives, pro-poor policies, etc., so we mainly through the level of tax preference and the power utilization rate to measure them; the social factors are mainly from the people's life and the socialized structure.

2.2 Meso factors

At the meso-level, it will analyze the market-oriented factors and relevant industry development factors that affect the development of power industry. The power market factors include market share, electrification level and market power.

Market occupancy, also known as "market share", refers to the proportion of the sales volume (amount) of commodities in the same industry. General measure market shares of specific indicators have the number of competitors and actual marketing efforts. Based on the current monopoly status of the power grid, we mainly use the actual marketing efforts to measure.

Electrification is the extensive use of electricity in all sectors of the national economy and people's lives. The electrification level is usually measured by four indicators: the amount of electricity consumed per person, the amount of electricity consumed per person, the proportion of electricity generated in the primary energy consumption, and the proportion of electricity consumed in the final energy consumption.

Market power refers to the ability of some participants in the market to take advantage of its size, or to manipulate the market by taking advantage of the defects in market structure and market rules. Market concentration ratio (CR) is usually used as a measure of market power.

2.3 Micro factors

At the microcosmic level, we will analyze the individual factors that influence the investment from the angle of the grid enterprise itself, and reflect the influence of the company's overall operating condition on the investment. The operating status of an enterprise mainly includes the operating construction level and the operating economic benefit, which can be reflected by the financial factors in the balance sheet and the income statement. There is a positive correlation between the investment ability and the business condition, because the business condition represents the economic benefit of the enterprise. The better the business condition is, the better the economic benefit is.

The impact factors of grid investment are analyzed at the macro, meso and micro levels, and then we get initial grid investment set, including more than one hundred indexes.

2.4 Selection of prosperity index of power grid investment

The effectiveness of the construction of investment prosperity index depends on the appropriateness of the selected indicators, which should reflect the overall development scale, level and speed of investment in different aspects. Based on the principles of representativeness and importance, we select the complete set of indicators mentioned above. In most cases, the selection of indicators relies on expert-made method, which is simple and easy to use widely, but this method inevitably refracts some unscientific subjective influencing factors and repeatability index, so in order to avoid the above problems, it is necessary to select the index scientifically and effectively.

In this study, the correlation analysis method is used to analyze and discuss the correlation of evaluation indexes by using the degree of mutual influence between evaluation indexes of Pearson correlation coefficient, so as to be objective and comprehensive in determining evaluation indexes and to avoid repeated interference, make sure the indicators are as scientific and operability as possible. For the indexes with high correlation degree, we further use principal component analysis to extract them and form a more representative comprehensive index system. First of all, we obtain the time series data of 2008-2017 through the National Bureau of Statistics of PRC, Wind database, World Bank database, power grid enterprise database and so on. Then we use the SPSS as the analysis tool. Through the analysis, we got the grid investment prosperity index system as shown in Table 1.

Table 1. Prosperity index system of power grid investment.

Level 1	Level 2	Level 3	Level 4
Macro Factors	Policy factors	Pro-poor Policy	Electrification rate
		Tax Policy	Level of Tax Preference
		Policy Uncertainty	Policy Uncertainty Factors
	Economics factors	General Indicators of National Economy	Comprehensive Indicators of National Economy
			Rate of Inflation
		Financial Factors	Interest Rate
			Exchange Rate
		Fiscal Factors	Fiscal Indicators
		Investment Factors	Comprehensive Indicators of Investment
	Social Factors	Indicators of The people's Livelihoods	Income and expenditure of urban and rural residents
			CPI
		Index of Socialization Level	Indicators of Socialization Level
			Number of persons with employment difficulties employed
Meso factors	Electricity Market	Market Share	Actual Level of Marketing Effort
		Electrification Level	Comprehensive Indicators of Electrification Level
		Market Power	Market Concentration
	Electricity Industry	Energy Structure	Coefficient of Elasticity of energy consumption
			Primary energy consumption in the secondary industry
			Primary energy consumption in the tertiary industry
			Primary Energy in energy intensive industries
			Proportion of raw coal consumption to total energy consumption
			Oil consumption as a proportion of total energy consumption
			Natural gas consumption as a proportion of total energy consumption
			The proportion of hydropower, nuclear power and wind power consumption in the total energy consumption
		Industrial Structure	Comprehensive Indicators of Industrial Structure
			Construction Investment of Power Supply
Microscopic Level	Operational Status	Operational Construction Level	Variable Capacity
			Line Length
			Transmission and Distribution Price
			Line Loss rate
		Economic Performance	Comprehensive Indicators of Economic Performance
			Feed-in Tariff
			Total Profit
			Yield Valve

3 THE FORMATION OF GRID INVESTMENT PROSPERITY INDICATOR GROUP

3.1 *K-L information*

K-L information, originally proposed by famous statisticians Kullback and Leibler, are used to determine the proximity of two probability distributions at the first place. As the research of prosperity index getting profounder, more and more economists use K-L information for the selection of prosperity indicators.

The practical implementation for determining the preceding, consistent, and lag indicator groups using this method are as follows:

(1) Preprocess the collected data for the purpose of consequent calculation.
(2) Select benchmark indicators. Select indicators that can sensitively reflect prosperity fluctuations in grid investments as benchmarks. The length of line construction is selected as the benchmark indicator in this case. This is partially because this indicator mainly reflects the ability of the line to deliver electrical energy, the longer the length of the line construction, the longer the electricity were able to be delivered, the more the universality of electricity consumption can be ensured, and the easier it is to meet the electricity demand of various users. On the other hand, the length of the line construction reflects the investment in the infrastructure of the power grid. More investment can strengthen the completion of the power system, and ensure the safety and stability of electricity consumption.
(3) Calculate the K-L information of all alternative indicators and choose the lagging period corresponding to the smallest value of K-L information as the lagging period of the alternative indicator.

3.2 Time-difference correlation analysis

Time-difference correlation analysis is the most used and widely accepted method of selecting the prosperity indicators.

Let the index sequence $Y = (y_1, y_2, ..., y_n)$ be the reference sequence, the index sequence $X = (x_1, x_2, ..., x_n)$ be the selected sequence, and r_l be the correlation coefficient, then the coefficient calculation formula can be expressed as equation 1:

$$r_l = \frac{\sum_{t=1}^{n} (x_{t-l} - \bar{x})(y_t - \bar{y})}{\sqrt{\sum_{t=t'}^{n} (x_{t-l} - \bar{x})^2 \sum_{t=t'}^{n} (y_t - \bar{y})^2}}, \ l = 0, \pm 1, \pm 2, ..., \pm L, \ t' = \left\{ \begin{array}{l} 1, l \geq 0 \\ 1 - l, l < 0 \end{array} \right. \tag{1}$$

The subscript l indicates the order of lag or lead, the negative value stands for the leading period and positive for the lagging period, and 0 means that the indicator is a consistent indicator. L is the maximum lagging period; n is the number of data in the lagged or leaded sequence after alignment. This research selects the indicator with the highest correlation coefficient. The specific implementation steps are as follows:

(1) Preprocess the collected data as well as mentioned above.
(2) Select the benchmark indicators. (Same as the K-L information method, the length of the line construction is used as the benchmark).
(3) Calculate the time-difference correlation, then choose the number of periods with the largest coefficient as the leading or lagging period.

3.3 The determination of grid investment prosperity indicator group

Based on the grid investment prosperity indicator system shown in Table 1, we can finally determine the grid investment prosperity indicator group by the method of K-L information and time-difference correlation analysis.

In this paper, we used the year-on-year growth rate of the sequent data, because the indicator is widely used by the most of institutions in China, such as China economic monitoring and analysis center. After that, we use time-difference correlation analysis and K-L information to calculate the correlation coefficient between the candidate index and the benchmark index, classify the grid investment prosperity indicators, determine the consistent, the leading and the lagging indicator groups, and provide a basis for the construction of the subsequent synthetic index model, as shown in Table 2.

Table 2. Grid investment prosperity indicators group.

Indicators	Lagging period	K-L information	Time-difference correlation coefficient	Group
Electrification rate	l = 9,4	1027.0685	0.416	Lag
Comprehensive Indicators of National Economy	l = 2	264.6282	0.4193	Consistent
Exchange Rate	l = 6,7	14.2873	0.3699	Lag
Comprehensive Indicators of Investment	l = 2,3	572.3173	0.4061	Consistent
CPI	l = 6,9	1.8134	0.5837	Lag
Indicators of Socialization Level	l = 2	249.2587	0.4321	Consistent
Number of persons with employment difficulties employed	l = −4,−9	9.8118	0.5744	Lead
Actual Level of Marketing Effort	l = −1	75.6698	0.4626	Consistent
Comprehensive Indicators of Electrification Level	l = 2,3	664.5534	0.4169	Consistent
Market concentration	l = 2,3	9.1606	0.4192	Consistent
Proportion of raw coal consumption to total energy consumption	l = −4, −9	1.3160086	0.559	Lead
Comprehensive Indicators of Industrial Structure	l = 2, 3	263.4418	0.4626	Consistent
Transmission and Distribution Price	l = 0, -2	1.1055	0.5667	Consistent
Total profit	l = −5, -6	135.2487	0.6939	Lead
Yield Valve	l = −5, -6	18.7263	0.7178	Lead

4 CONSTRUCTION OF GRID INVESTMENT PROSPERITY INDEX

Globally, the major methods of prosperity index preparation are diffusion index method, composite index method, and S-W prosperity index method. The Composite index method is commonly used to build up a prosperity index of particular industries because it can not only predict the turning point of economic cycle fluctuation, but also reflect the amplitude of economic cycle fluctuations, analyze the depth of each cycle contraction, the strength of recovery and the degree of expansion. Although the S-W sentiment index method can also reflect the cyclical fluctuations of the overall economy, the composite index method is easier to measure and more convenient to analyze than the S-W sentiment index. Moreover, it also supplements the shortcomings of S-W prosperity index, which is unstable and lacks of expressiveness[24]. Therefore, in order to ensure the stability of the grid investment prosperity index, the composite index method is used to analyze the growth trend and periodical fluctuation of grid investment, and provide a theoretical basis for explaining the fluctuation phenomenon of grid investment and forecasting the development trend of grid investment.

4.1 The prosperity composite index method

The complication of composite index mainly includes the following steps, detailed formulas are given in the literature[17]:

Step 1: Calculate and normalize the symmetrical rate of change for single indicators;

Step 2: Calculate the standardized average rate of change for each indicator group;

In the process, we need to calculate the weight of every index, generally, weight can be determined by the calculation of scoring system, or can also adopt equal-weight method. Here we introduce the analytic hierarchy process to calculate the weight of each index to ensure it is more scientific. The calculation results are shown in Table 3 below;

Step 3: Calculate the composite index.

Table 3. Weights of grid investment prosperity indicator system.

Indicator name	Weight	Indicator name	Weight
Electrification rate	0.017028401	Comprehensive Indicators of Electrification Level	0.093444052
Comprehensive Indicators of National Economy	0.017028401	Market concentration	0.093444052
Exchange Rate	0.00684024	Proportion of raw coal consumption to total energy consumption	0.016616026
Comprehensive Indicators of Investment	0.038101086	Comprehensive Indicators of Industrial Structure	0.016616026
CPI	0.00684024	Transmission and Distribution Price	0.091469461
Indicators of Socialization Level	0.017028401	Total profit	0.271410198
Number of persons with employment difficulties employed	0.003319296	Yield Valve	0.271410198
Actual Level of Marketing Effort	0.039403918		

4.2 Calculation of the composite prosperity index

According to the composite index method specified above and the weights of each index, the leading composite index, consistent composite index and lag composite index is calculated separately, and shown in Table 4:

The calculation results of each composite index are painted as curve shown in Figure 1, Figure 2, and Figure 3 below.

The shaded part in Figure 1 is the decline period of the leading composite index, which includes the second half of 2008 and the first half of 2009, as well as the second half of 2012, and the first half of 2013. The value of prosperity index should be between 0-200 for most of the cases. 100 is a critical value of the prosperity index: when the prosperity index is greater than 100, it indicates that the economic is prospering. When the prosperity index is less than 100, it indicates recession. As we can see in the figure, in the second half of 2008, the grid investment had been through

Table 4. Results of grid investment composite index.

	2017	2016	2015	2014	2013	2012	2011	2010	2009	2008
Leading composite index	100.05	100.00	100.05	100.12	99.83	100.63	100.27	100.00	99.61	103.48
Consistent composite index	102.91	102.42	101.34	101.34	100.58	99.87	99.17	100.00	99.15	98.49
Lag composite index	97.33	97.71	97.38	97.23	97.60	97.59	97.50	100.00	98.12	98.38

(*We mainly use 2010 as benchmark. This is mainly because that China's economy has entered the new normal period since 2010. Therefore, this composite index can be used to analyze the investment prospects of the grid under the new normal.)

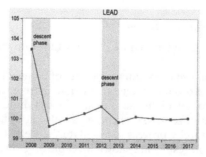

Figure 1. The lead index.

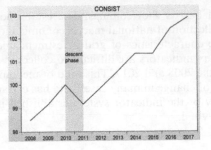

Figure 2. The consistent index.

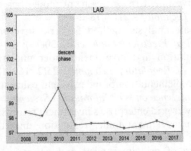

Figure 3. The lag index.

a minor recession. Until 2010, the prosperity index gradually recovered and the investment resumed the state of prosperity; in the second half of 2012, it gradually became depressed. However, during the past 10 years, the grid investment prosperity index has remained a stable value around 100, which indicates a relatively prosperous state.

Similarly, the shaded part in Figure 2 is the decline period of the consistent composite index. It can be seen from the figure that the grid investment has been in a recession until 2010, although it shows a gradual upward trend, but the index is still below 100, which mainly due to the 2008 financial crisis. In 2010, China's economy entered the New Normal, and had shown traits of recession due to the dual pressures of inflation and economic downturn. China has introduced some new macro-control policies, which have enabled the economy to rebound. Therefore, the post-2011 prosperity index is on the rise.

From the second half of 2010 to the first half of 2011 shown in Figure 3, the lag indicators of grid investment seen a period of decline, and the decline period was also one year.

It can be seen that no matter which type of index group, the value of the index is usually floated around 100. Even if rises and falls may occur, the fluctuation remains mild. It is relatively stable and basically in the prosperity range. This also illustrates the rationality and stability of the grid investment climate index we have constructed.

5 CONCLUSIONS AND DEFICIENCIES

This paper firstly analyzes the grid investment prosperity index through the three dimensions of macro, micro and meso, and then uses the index screening method, K-L information method and time difference correlation analysis method to determine the grid investment prosperity index indicator group based on the time series data. Using the combination of analytic hierarchy process and prosperity index to measure the investment prosperity index of the power grid, the results show that the prosperity index is basically in a relatively prosperous interval and relatively stable, which indicates the rationality of the constructed grid prosperity index, which is suitable to be used as a basis for grid companies to analyze the fluctuations of investment.

However, this study still has its shortcomings. The established prosperity indicators are mainly based on a large collection of national macroeconomic indicators and power industry reports, combined with the characteristics of grid investment under the new economic situation. Considering that many indicators are difficult to collect in earlier years, so the sample interval is limited between the 2008 and 2017. This kind of medium-cycle fluctuation law does not accurately reflect trend of change in many cases, and has certain limitations. In the future, the rationality and integrity of the indicator system established in this study can be further tested and improved, and revised.

REFERENCES

W Mianbin et al. 2006. Grid investment risk assessment model based on incremental method. *Journal of Electrical Engineering Technology* 21(9):18–24.

W Mianbin et al. 2010. Set pair analysis method for grid investment risk sssessment in market environment. *Chinese Journal of Electrical Engineering* 30(19): 91–99.

JD Saphores et al. 2004. Regulation and investment under uncertainty: an application to power grid interconnection. *Journal of Regulatory Economics* 25(2):169–186.

W Li et al. 2010. Research and implementation of index weight calculation model for power grid investment returns. *International Conference on Web Information Systems and Mining* 6318: 44–52.

G Luo et al. 2011. Evaluating power grid enterprise's investment returns. *Energy Procedia* 5: 224–228.

Y Ze et al. 2006. Transmission investment under regulation uncertainty and policy design. *China Industrial Economy* 7: 73–79.

F Yongsheng, 2015. The theory of grid investment and governance from the perspective of new regulation: enlightenment on China's electric power system reform. *Contemporary Finance & Economics* 10: 3–14.

L Haitao, 2016. Construction of grid investment system under government franchise mode. *Management World* 1: 178–179.

P. Dua & S.M. Miller, 1996. Forecasting and analyzing economic activity with coincident and leading indexes: the case of connecticut. *Journal of Forecasting*, 1996, 15:509–526.

R. Nilsso. 2004. OCED system of leading indicators methodology and application. *National Development and Reform Commission Workshop and Expert Hearing on Business Cycle and Indicators* 12.

A. Ozyildirim et al. 2010. Business Cycles in the Euro Area Defined with Coincident Economic Indicators and Predicted with Leading Economic Indicators. *Journal of Forecasting* 29: 6–28.

D Wenquan et al. 1987. Measurement, analysis and prediction of economic cycle in China (I)-existence and measurement of Economic Cycle. *Jilin University Journal Social Sciences Edition* 3: 1–8.

Turner L.W et al. 1997. The use of composite national indicators for tourism forecasting. *Tourism Economics* 3:309–317.

W Huimin & C Baoshu, 1996. Basic frame structure of coal industry early warning index system. *Journal of China university of coal economics*, 4: 10–13.

W Ting, 2006. Study on index system for predicting cyclical turning point in coal industry. *Beijing University of International Business and Economics*. Beijing.

Y Yanbing & D Lieyun, 2001. Research on the design of real estate early warning system. *Optimization of Capital Construction*.

W Xiaoping & Z Yuxia, 2012. Establishment and calculation of the prosperity index of service industry of China. *Finance & Trade Economics* 4: 114–120.

Y Wu et al. 2015. A research on the regional technology innovation climate index based on the quarterly data. *Science Research Management* 36(5): 55–64.

L Chang, 2013. Analysis and outlook of situation and changes of structure of China power industry in 2013. *Science & Technology for Development* 6: 91–95.

LChang & G Tiemei, 2011. Characteristics of electricity industry cycle fluctuation and influential factors of electricity demand based on business analysis and the error correction model. *Resources Science* 33(1): 169–177.

D Haijing, 2014. Construction and fluctuation analysis of China Electric Power Industry Boom Index. *Anhui University*, Hefei.

Z Mei, 2011. Research on the synthesis of provincial electricity market demand leading index-taking Chongqing as an example. *Dalian Maritime University*. Dalian.

W Jinming et al. 2007. Building SW type leading indices: an empirical study. *Chinese Journal of Management Science* 15(4):116–122.

Emerging Developments in the Power and Energy Industry – Dufo-López, Krzywanski & Singh (eds)
© 2020 Taylor & Francis Group, London, ISBN 978-0-367-27169-5

Research on overseas investment value strategy of power grid project based on social network

Haican Diao
School of Business, Renmin University of China Beijing

Yanchao Lu
State Power Economic, Research Institute, Beijing, China

ABSTRACT: Around the "One Belt, One Road" construction strategy, the investment scope of the power grid will further develop from the country and the region to the global coverage. The uncertainty, flexibility and competitiveness of investment are increasingly prominent. In order to realize the full evaluation of the value of overseas investment in the power grid and realize the scientific decision of overseas investment of power grid, this paper adopts the centrality concept of social network analysis method, and divides the overseas investment value index of power grid into strategic development, brand technology and investment return. Aspects, and identify specific investment strategy sets for each aspect of the characteristics.

1 INTRODUCTION

Under the background of economic globalization, the development environment, market structure and profit model of the power industry are constantly changing, and the position and role of the power grid in energy development are increasingly prominent. According to the strategic layout of the global energy Internet, the company's international business will continue to increase during the 13th Five-Year Plan period. The construction investment problem is not only related to economic development, but also involves issues such as political and social stability of the country. The overseas investment grid project not only meets the needs of the international development of enterprises, but also responds to the strategic requirements of the country to promote "enterprise going global". Due to the large investment amount and wide range of influences of power grid projects, the decision-making of overseas investment is more difficult than other construction projects. Therefore, reasonable investment evaluation and decision-making on the value of overseas investment is of great importance to the scientific development of power grid enterprises.

China's theoretical research on foreign direct investment started late and lags behind corporate practice. With the upgrading of China's industrial structure and the acceleration of the "going out" pace, domestic scholars began to pay more attention to the study of foreign investment. Research on the energy industry is relatively rare, and the research in the power industry or power grid enterprises is rare; and the research scope of overseas investment risk and value theory is relatively simple, and the research system needs to be improved. Most of the research focuses on the theory of Chinese enterprises' foreign direct investment system. Most scholars regard comparative advantage theory as the theoretical basis for China's foreign direct investment, and propose that foreign direct investment should make full use of comparative advantage to participate in international competition (Nie, 2001). Some scholars further pointed out that the theoretical basis of a country's foreign direct investment must be A macro and micro perspective is considered (Zhang, 2003). In addition, some people think that the comprehensive advantage theory should be adopted as the theoretical basis of foreign investment, pointing out that the comprehensive advantages are mainly reflected in the multi-polarization of investment motive, the diversification of differential advantages and the diversification of development space (Sun, 2004).

2 ESTABLISHMENT OF OVERSEAS INVESTMENT VALUE EVALUATION INDICATORS

Based on literature collation and field research, pre-process the collected overseas invest-ment data and indicators, eliminate the alienation indicators, and organize and summar-ize the indicators with higher applicability. According to the complexity, multi-level and uncertainty characteristics of overseas investment in power grid projects, fishbone dia-gram analysis method is used to analyze the formation mechanism of investment value, and the investment value indicators are screened by AHP and factor analysis. The set of evaluation indicators for establishing the value of all-factor, whole-process overseas investment is shown in Table 1 below.

After completing the construction of the full-factor and full-process evaluation system for the overseas investment value of the grid, in view of the complex factors that need to be considered in the overseas investment projects of the grid, when evaluating the value of overseas investment, it is necessary to first divide the evaluation scenarios.

3 SOCIAL NETWORK ANALYSIS METHOD TO DIVIDE INVESTMENT SCENARIOS

Nowadays, the application of social network analysis method is widely used in various fields. The research on network power now mainly has three kinds of measurement

Table 1. Full-element, full-process evaluation index system for overseas investment.

Primary indicator	Secondary indicators	Three-level indicator
Global environment (A1)	Economic development (B1)	International Market Development (C1) Economic globalization (C2) Global resource utilization (C3)
Home Country Environ-ment (A2)	Economic development (B3)	Domestic and international market demand (C8) Enterprise Internationalization Level (C9)
	Policy support (B4)	Going out strategy (C10)
Host Country Environ-ment (A3)	Economic development (B5)	Host country market size (C14) Inflation and exchange rate conditions (C16)
	Institutional quality (B6)	Degree of political corruption in the host country (C23)
	Bilateral relationship (B7)	Third country influence (C28) Investment barriers (C29)
	Difference between the two countries (B8)	Technical differences (C37)
Enterprise characteristics (A4)	Basic quality (B9)	Innovative R & D intensity (C40) Enterprise's ability to control core net-work nodes (C42)
	Management decision (B10)	Executive Education Level (C44)
Project characteristics (A5)	Profitability evaluation(B12)	Financial net present value (C50) Total investment yield (C53) Asset-liability ratio (C56)

methods: degree center degree, near center degree and feature vector center degree. For ease of analysis, a conventional node centering calculation method is used. The node's centrality is the centrality of the node's location in the social network, expressed in terms of the node's degree. The knot degree of a point is represented by the sum of the membership degrees of the lines directly connected to the node. The greater the centrality of the node, the more central the actor is in the social network, which can be expressed as:

$$C_n(n) = d(n) \tag{1}$$

where $d(n)$ is the joint strength and $C_n(n)$ is the node center degree.

When classifying the investment scenarios, using the social network analysis method, after weighted average analysis of the expert score tables, construct the adjacent matrix R of the impact value of the overseas investment value of the grid project as shown in Table 2 below, and the table is subjected to a reflexive test.

Among them, $R_{ij} = s(v_i, v_j)$, $R_{ij} = 1$, v_i and v_j represent the score of each indicator. And the R matrix satisfies reflexivity, symmetry, and transitivity, so that R is a fuzzy equivalence matrix and satisfies $R^T = R$. After scoring by experts, the R matrix was analyzed. The questionnaire data is input using Netdraw's own data table program, and the structure chart for visualizing the overseas investment value is shown in Figure 1 below.

According to the above figure, the indicators on the central node of the network are more connected to each other, and the network is denser. The network density is larger, which has a greater impact on the interrelated indicators. The connection between the indicator and other indicators is very sparse. There are more connections between each other, and the network density formed by each other is smaller, and the impact on the interrelated indicators is relatively small. The indicators with low strength associated with other nodes are now deleted, and the three key factors extracted through the structure diagram are c10 (going out strategy), c37 (technical difference), and c53 (return on investment). Based on this, the scenarios of overseas investment value evaluation are divided into three scenarios: strategic development, technological difference and return on investment.

Table 2. Overseas investment value impact relationship adjacency matrix.

	c1	c2	c3	c8	c9	c10	c14	c16	c23	c28	c29	c37	c40	c42	c44	c50	c53	c56
c1	1	1	1	0	3	5	0	0	0	1/3	0	1	9	0	0	0	5	1
c2	1	1	1/3	0	1/3	1/3	0	0	3	0	0	1/3	3	0	0	1/3	1/3	0
c3	1	3	1	0	1/3	1/5	0	0	1	0	0	1/3	0	5	5	0	1/5	0
c5	0	0	0	1	3	1	0	1/3	0	0	0	5	0	0	0	1/5	1	0
c9	1/3	3	3	1/3	1	5	0	0	1/3	0	0	1/3	0	0	0	0	1/3	1
c10	1/5	3	5	1	1/5	1	3	5	5	7	1	1	3	1/5	3	0	1	1/3
c14	0	0	0	0	0	1/3	1	3	0	0	0	1	0	0	0	0	1/3	1
c16	0	0	0	3	0	1/5	1/3	1	0	0	1/5	1/3	0	0	0	0	1/5	1/3
c23	0	1/3	1	0	3	1/5	0	0	1	0	0	1/3	0	1	1	0	1/3	0
c28	3	0	0	0	0	1/7	0	0	0	1	1/5	1/5	0	0	0	0	1/7	1/3
c29	0	0	0	0	1/5	1	0	5	0	5	1	1	0	0	0	0	1	1/7
c37	1	3	3	1/5	3	1	1	3	3	5	1	1	1/5	1/5	1/5	0	1	0
c40	1/9	1/3	0	0	0	1/3	0	0	0	0	0	5	1	1	1	0	1/3	1/5
c42	0	0	1/5	0	0	5	0	0	1	0	0	5	1	1	1	0	5	1/3
c44	0	0	1/5	0	0	1/3	0	0	1	0	0	5	1	1	1	0	1/3	1/5
c50	0	3	0	5	0	0	0	0	0	0	0	0	0	0	0	1	3	1/3
c53	1/3	1/5	5	1	3	1	3	5	3	7	1	1	3	1/5	3	1/3	1	1/7
c56	1	0	0	0	1	3	1	3	0	3	7	0	5	3	5	3	7	1

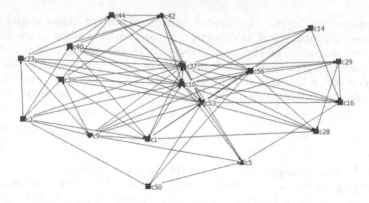

Figure 1.   Network model of overseas investment value of power grid project.

4   OVERSEAS INVESTMENT STRATEGY SET BASED ON INVESTMENT VALUE

4.1   *Strategic development perspective*

Consider the value of overseas investment based on the "going out" strategy, including two scenarios:

(1) The overall strategic synergy value and social value are high, in line with the national strategic planning of the Belt and Road Initiative, in line with the strategic layout of the company's development. After the implementation of the project, it can effectively promote the construction of the global Internet strategic layout, and drive all-round technology, standards and equipment. Going out, promoting the development of global Internet innovation and enhancing the company's voice and influence in promoting the energy revolution.

(2) The overall strategic synergy value and social value are low. Such investment projects and national industrial policies, and the national network to build a global Internet business strategy to the left, although the project can have a higher return on assets, investment strategies generally do not involve.

The set of strategies in this scenario is shown in Table 3.

4.2   *Brand technology differences*

Considering the value of overseas investment based on differences in technological development between the two countries, the following three scenarios are included:

(1) From the tactical value level

Successful overseas grid project investment is inseparable from a reasonable and effective asset assessment mechanism. All overseas investment projects of State Grid Corporation evaluate the underlying assets from the four perspectives of strategy, revenue, security and competitiveness. For each project's particularity, it is necessary to choose different business cooperation modes for different markets, such as service outsourcing, management outsourcing, leasing, franchising, BOOT and divestiture.

(2) From the level of overseas operations management team building

The investment in overseas investment projects of the power grid is large, and the operation and management cycle is long. In addition, the overseas project environment is complex, and the technical standards, investment environment, regulatory requirements, and legal environment are all different from those in China. Once the investment and operation links are wrongly managed. Directives will lead to extremely bad effects. When planning overseas investment strategies for the grid, it is necessary to component an overseas professional team.

(3) From the investment timing level

In the process of globalization, there needs to be a clear path, pay close attention to the movements of the target area or the company, have sufficient strategic patience, and be

Table 3. Set of strategies developed from a strategic development perspective.

No.	Code	Indicator	Strategy
1	c1	International market development	Formulate the development prospects of the company, adopt an active and stable overseas investment strategy, gradually increase the scale of overseas markets, increase the proportion of cross-border business operations, and actively integrate into the international market competition.
2	c5	Economic Globalization	Create a competitive business model, that is, two integrated integration models, integration of investment, construction and operation, and integration of technical equipment standards.
3	c9	Global resource utilization	Deepen international production capacity cooperation and optimize the global business layout of the State Grid.
4	c10	Enterprise internationalization level	According to the assets of the host country's power grid, choose appropriate business cooperation models such as service outsourcing, management outsourcing, leasing, franchising, BOOT and divestiture to improve the internationalization level of enterprises.
5	c29	Going out strategy	Give full play to the operational advantages of the company's group, give full play to the value of overall strategic synergy, and drive technology, standards, and equipment to go all out to promote global Internet innovation.

decisive when it is most suitable for the shot. Power grid enterprises, especially state-owned enterprise state network companies, must be profitable on the one hand. On the one hand, they must resist risks. One of the ways to resist risks is to hold assets and enjoy the reports brought by assets in the economic and business downturn.

The set of strategies in this scenario is shown in Table 4.

4.3 *Return on investment*

When considering the value of overseas investment based on the investment income of the project itself, there are two situations:

(1) The economic score is higher. The project with excellent project-level evaluation data is the main subject of investment by State Grid Corporation. In line with the project value investment concept, it can bring high return on assets. Many successful overseas investment cases of State Grid Corporation have the above characteristics. Even if the evaluation results at the micro level of the project are excellent, it is still necessary to combine the macro and meso-level situations, and work hard and steady through the various stages of the project to achieve stable investment in overseas power grids.

(2) The economic score is lower. For overseas power grid assets with low economic value, investment is generally not allowed. Further evaluation of project strategic value and location value is needed. State Grid Corporation can not blindly invest in the situation of not considering the micro level of the project, the specific economic and financial indicators of the project, and put an end to the situation of "pitching".

The set of strategies in this scenario is shown in Table 5.

5 CONCLUSION AND RECOMMENDATIONS

In view of the complexity, multi-level and uncertainty characteristics of overseas investment in power grid projects, this paper scientifically and rationally refines the evaluation factors of overseas investment in power grid projects, and establishes an evaluation index system for the whole process. The network analysis method is used to divide the evaluation scenarios, and

Table 4. Set of strategies from the perspective of brand technology.

No.	Code	Indicator	Strategy
1	c5	Domestic and international market demand	Strengthen grid load forecasting, improve the depth of planning and research in the early stage of the project, grasp the domestic network market demand, and accurately invest.
2	c10	Host country market size	Give full play to the technology and management advantages of the company's operation of the large power grid, advance planning and research, master the overall technical trends and needs of the host country's power development, and expand the international business market.
3	c14	Degree of political corruption in the host country	In-depth understanding of the host country's laws and regulations, regulatory requirements, the risk of relevant political corruption is out of scope, not in the host country to invest in the grid.
4	c29	Investment barrier	In the mature electricity market, the policies and regulations are stable and transparent, the supervision is strong, and the investment barriers are small. The investment strategies such as direct stock purchase and acquisition are selected to participate in the host country power grid operation and obtain investment returns. In the immature power market, the power market supervision policy is very unclear, the areas with high risk of capital investment and large investment barriers, and the investment strategy of selecting engineering contracting, equipment and equipment output, and green land investment. Conduct preliminary research in the market, seek to establish early-stage relationships, and cultivate the electricity market.
5	c37	Technical difference	Standards development, international energy cooperation with world-class power grid companies, occupation of technical commanding heights, cooperation and win-win.
6	c40	Innovation and research strength	Innovative R&D investment in key areas such as UHV, smart grid, and clean energy, and the launch of key innovation projects.
7	c42	Enterprise's ability to control core network nodes	Familiar with the operating rules of overseas companies, optimize the management mode of overseas companies, and improve the ability of enterprises to control the core network nodes.
8	c44	Executive education level	In selecting the overseas operation management team, we should also consider whether to serve the national strategy, firmly establish the overall situation; whether it has a prudent and pragmatic working attitude; whether it can follow the rules of the grid operation, adhere to long-term value investment; whether it is based on internal training.

three main investment scenarios are extracted, which are strategic development, brand technology and investment return. And filter out the indicator sets in the above three scenarios, and build a corresponding policy set.

The research on overseas investment strategy of grid projects based on value orientation and risk prevention is an important component of the global energy interconnection strategy research system and a hot issue in the field of overseas investment research. The overseas investment value and investment risk evaluation system of the constructed power grid project can realize the scientific measurement of value and risk; the overseas investment strategy of grid project based on value orientation and risk prevention can be widely applied to the project overseas investment decision, and the investment is lean. Management provides a reliable basis; it helps to rationally avoid investment risks, realizes scientific investment decision-

Table 5. Set of strategies from the perspective of investment return.

No.	Code	Indicator	Strategy
1	c29	Investment barrier	Ibid.
2	c42	Inflation and exchange rate conditions	Use the reverse investment strategy and other methods to grasp the timing of investment. We will prudently adopt financial leverage to conduct overseas mergers and acquisitions, reduce the risk of "foreign currency of assets, localization of liabilities", etc., and eliminate the "patch" situation, regardless of the micro level of the project, regardless of the specific economic and financial indicators of the project. Blind investment leads to increased risk.
3	c50	Financial net present value	Greater than zero, higher than the average value of domestic power grid investment. Coordinate the evaluation of economic and compliance indicators such as profitability, solvency, uncertainty analysis, investment return rate and investment recovery period.
4	c53	Total return on investment	Higher than the reference value of domestic power grid investment. Generally less than 8 years.
5	c56	Assets and liabilities	Lower than the average level of domestic power grid investment. Coordinate the evaluation of economic and compliance indicators such as profitability, solvency, uncertainty analysis, investment return rate and investment recovery period.

making, sustainable investment value, and improves investment efficiency. Due to the volatility of overseas markets, research results such as overseas investment value evaluation index system and investment strategy need to be further tested, improved and revised in practice.

ACKNOWLEDGEMENT

This research is funded by national grid science and technology project "Research on Overseas Investment Strategy of Grid Project Based on Value Orientation and Risk Prevention"(SG-TYHT/16-JS-198)

REFERENCES

Chen Yan, Zhai Ruirui, Guo Niusen. Research on Determinants of China's Foreign Direct Investment Based on Multivariate Distance Perspective [J]. Systems Engineering - Theory & Practice, 2014, 34(11): 2760–2771.

Chen Yugao, Zhang Yan. Industrial Choice of Foreign Direct Investment: Analysis Based on Industrial Status Division Method [J]. World Economy, 2007(10): 28–38.

Jiang Hui. The Impact of Host Country Infrastructure Level on China's Foreign Direct Investment——Based on the System GMM Research of the "Belt and Road" Countries [J]. Foreign Trade, 2017(3): 22–25.

Luo Weihao. Research on cost and benefit of domestic and foreign power grid investment [J]. Economic Management: Abstracts, 2016(8): 00324-00324.

Li Yan, Li Yingbo. Research on the Wealth Effect and Influencing Factors of Overseas M & A of High-end Equipment Manufacturing Enterprises——Based on the Empirical Analysis of Data from 2001 to 2014 [J]. International Business (Journal of University of International Business and Economics), 2016(1):139–149.

Ma Shuzhong, Liu Mengheng. Constraints and Strategic Choices of China's Direct Investment in the "Belt and Road" [J]. New Vision, 2017(1): 17–22.

Qu Zhi, Yang Biqin. The Influence of the System Quality of Countries along the "Belt and Road" on China's Foreign Direct Investment [J]. Journal of Economics and Management, 2017, 38(11): 15–21.

Wang Fengbin, Yang Yang. Differentiation and Integration of Foreign Direct Investment Behavior of Multinational Corporations——An Empirical Study Based on Market Value of Listed Companies [J]. Management World, 2013, 234(3): 148–171.

Wang Yi, Lu Yiqing. Discussion on Economic Benefit Analysis and Evaluation of Power Enterprise Investment Projects[J]. China International Finance (English), 2017(14): 202–203.

Wang Maojun, Xu Yongping. Basic characteristics and determinants of OFDI in China in the Americas[J]. Acta Geographica Sinica, 2017, 72(8): 1373–1391.

Xie Hui. The Impact of New International Investment Rules on China's Foreign Direct Investment [J]. Modern Trade and Industry, 2016(31): 33–34.

Yuan Haidong, Zhu Min. Research on the Impact of Overseas Chinese Network on China's Foreign Investment——Based on the Perspective of Host Country Heterogeneity [J]. International Business (Journal of University of International Business and Economics), 2017(5): 79–89.

Zhou Jing, Cai Dongqing. Micro-features of enterprises, host country factors and China's OFDI model selection [J]. International Trade Issues, 2014(2): 124–134.

Zhou Guolan, Zhou Ji, Ji Kaiwen. The Industrial Choice of China's Foreign Investment under the "Belt and Road Initiative"[J]. Enterprise Economy, 2017(9): 72–79.

Emerging Developments in the Power and Energy Industry – Dufo-López, Krzywanski & Singh (eds)
© 2020 Taylor & Francis Group, London, ISBN 978-0-367-27169-5

Study on the influence of length on frequency response characteristics of grounding pole based on CDEGS

Lin Yang & Yu Zhang
Sichuan Electric Power Research Institute, Chengdu, China

Wenyan Feng & Xiaobin Cao
Electric Engineering School of Southwest Jiaotong University, Chengdu, China

ABSTRACT: The performance of tower grounding electrode determines whether the transmission line can operate safely and steadily, so it is necessary to detect the status of tower grounding electrode. Firstly, the frequency response curves of grounding impedance at different frequencies are obtained by simulation. Then, we find out the common influence rule of length on grounding impedance, Finally, the rule is verified by experiment. It is found that the grounding electrode length has a great influence on the impedance modulus when the frequency is low. When the frequency increases to a certain extent, the impedance modulus of the grounding electrode with different length is gradually the same and the impedance angle is different. However, with the further increase of the frequency, the impedance modulus and the impedance angle of the grounding electrode with different length are the same. It is found that the length of grounding electrode can be detected by multiplying the modulus. Therefore, it is feasible to detect the length of grounding pole by frequency response characteristics

Keywords: Spectral response, CDEGS, Impedance, Tower Diagnosis, Earthing Pole

1 INTRODUCTION

The state detection of grounding device is a necessary means to discover the hidden danger of grounding device and ensure the safe operation of grounding. So far, various diagnostic methods for measuring the state of grounding conductors have been proposed (WANG Lin. et al. 2016, LIU Yugen. et al. 2015, WANG Fenghua. et al. 2016, QIAN Guochao. et al. 2014). Tower grounding has its unique characteristics (PENG Xiangyang. et al. 2011, XU Song. et al. 2014). Towers with 5 kV or more are grounded at multiple points at the foot of the tower and towers with 220 kV or more are grounded at four points. And its location is remote. The grounding pole of tower is radioactive. After long-term corrosion, the grounding pole of tower becomes shorter and thinner. Therefore, it is necessary to find a unique corrosion detection method for tower grounding. Frequency response method has been widely used in state detection. As the main detection method of transformer winding deformation (ZHOU Lin. et al. 2017), there are some similarities between transformer winding deformation detection and grounding device state detection: None of them are visible, and minor changes may occur. Therefore, it is thought to find out the change of impedance curve of grounding conductor by changing the frequency of injected power supply, and then find out a new method of diagnosis. At present, the current response of tower to different frequencies has not been studied by scholars, so it is important to study the response characteristics of grounded conductors at different frequencies. Based on CDEGS software package, two models of grounding conductor with different power supply are established. The influence of multi-frequency current on the impedance of grounding body is simulated. The

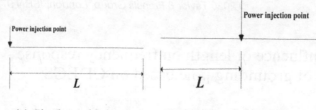

(a) Single-end injection. (b) Middle-end injection.

Figure 1. Modeling sketch map.

characteristics of grounding impedance varying with frequency under single-end injection and middle-end injection are analyzed.

2   THE ESTABLISHMENT OF SIMULATION MODELS

There are two kinds of simple grounding poles for poles and towers (LI Wei. et al. 2016, ZHANG Min. et al. 2012): single-end and middle-end injection. Horizontal grounding conductor is a typical grounding device in grounding device. The model of power grounding body is shown in Figure 1. The conductor material used is round steel. According to the actual project, its basic parameters are set up: The conductor is a round steel with length of 40 meters, depth of 0.6 meters, conductor radius of 0.005 meters, relative resistivity of 17, relative permeability of 1, and soil resistivity of 200Ω.m. In addition, the current source of 20A is injected at the injection end of the power supply, and the frequency range is swept from 2Hz to 1MHz.

3   EFFECT OF LENGTH L ON FREQUENCY RESPONSE

3.1   *Effect of the length L on impedance mode frequency response characteristics*

L varies from 40 meters to 90 meters. Simulation is made by changing *L* only. According to the simulation data, the impedance mode frequency response charts of three structures are drawn as shown in Figure 2.

It can be seen from the figure that the characteristics of impedance mode frequency response of grounded conductors under two conditions are basically similar: The impedance modes remain unchanged first. Then it began to grow rapidly. Therefore, the trend of impedance mode spectrum curves of grounding conductors with different structures is similar.

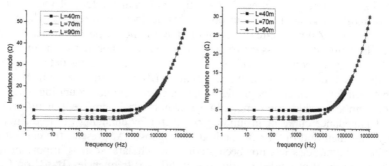

(a) Single-end injection. (b) Middle-end injection.

Figure 2.   Characteristic diagram of impedance mode frequency response.

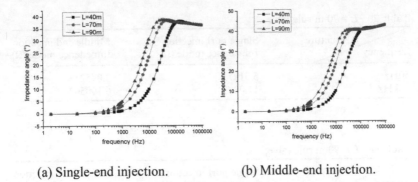

| (a) Single-end injection. | (b) Middle-end injection. |

Figure 3.   Characteristic diagram of impedance angle frequency response.

In the low frequency stage, the larger the L, the smaller the impedance modulus. Therefore, the change of L can be judged by the magnitude of impedance modes at low frequencies, and then whether the conductor has fracture corrosion or not can be judged.

3.2 *Effect of the length L on impedance angular frequency response characteristics*

According to the simulation data, the impedance angular frequency response charts of three structures are drawn as follows: Figure 3:

It can be seen from the figure that the characteristics of impedance angular frequency response of grounding conductor under two conditions are as follows: They are invariant first, then increase rapidly to reach the stationary point and then change smoothly, The larger the L, the lower the frequency of reaching the stationary change point; therefore, the frequency of reaching the stationary change point can be used to judge the change of L, and then to judge whether the conductor has fracture corrosion.

When the curve reaches the convergence point, the impedance angle curve of the central point injection is in an upward state at the end, while the single end injection is in a downward state. The trend of impedance angular frequency response characteristic curve is different with different structure.

4 ANALYSIS OF RULES OF SIMULATION DATA

From the simulation data, it can be seen that there is a certain relationship between the length L and the frequency response of the structure, as shown in Table 1, Table 2 and Table 3.

As can be seen from the table above, when $L = 40$ meters, the impedance modulus corresponding to 50 Hz is about half of the impedance modulus corresponding to 100 kHz for the three structures. when $L = 70$ meters, the impedance modulus corresponding to 50 Hz is about half of the impedance modulus corresponding to 35 kHz for the three structures. when $L = 90$ meters, the impedance modulus corresponding to 50 Hz is about half of the impedance modulus corresponding to 25 kHz for the three structures. So we can get that when the distance between the power injection point and the conductor end point L is constant, The

Table 1. $L = 40$ m rule table.

frequency〴Structure	Single port injection Impedance mode (Ω)	Middle end injection Impedance mode (Ω)
50Hz	8.650525	4.918379
100kHz	17.30225	9.9184

Table 2. $L = 70$ m rule table.

frequency Structure	Single port injection Impedance mode (Ω)	Middle end injection Impedance mode (Ω)
50Hz	5.51975	3.09525
35kHz	11.0395	6.190507

Table 3. $L = 90$ m rule table.

frequency Structure	Single port injection Impedance mode (Ω)	Middle end injection Impedance mode (Ω)
50Hz	4.517106	2.519435
25kHz	9.034212	5.03887

structure has no effect on the frequency corresponding to the change of impedance modulus multiple. That is to say, for a fixed frequency point A, such as 50 Hz, it corresponds to a fixed frequency point B, which is twice the impedance modulus. The larger L is, the smaller B is. Therefore, the change of L can be judged by the change of B.

5 EXPERIMENTAL VERIFICATION

The principle wiring diagram for measuring grounding impedance is shown in Figure 4. Because the three-pole method is simple and has high stability, it is widely used in the field at present, so the three-pole method is used in this experiment to measure the grounding resistance (ZHU Bin. et al. 2015). In order to minimize mutual inductance effect (SUN Yue. et al. 2010), let AQ⊥AP in Figure 4 (a). In Figure 4 (b), PQ is perpendicular to the ground conductor.

The grounding conductor used in this experiment is iron wire with a diameter of 4mm and is buried in soil with a depth of 0.5m. Resistance R = 5Ω, the signal generator emits a sinusoidal wave with an effective value of 20A. The power supply frequency varies from 50 Hz to 350 kHz. The actual wiring diagram is shown in Figure 5.

According to the experimental data, it is found that there are also relationships as shown in Table 4 and Table 5.

As can be seen from the table above, when $L = 40$ meters, for the power supply is injected from the single end and the power supply is injected from the middle end. The impedance modulus corresponding to 5 kHz is about half of the impedance modulus for 200 kHz. When $L = 50$ meters, the impedance modulus corresponding to 5 kHz is about half that of 120 kHz for the two structures, which is the same as the simulation data.

6 CONCLUSION

The impedance mode frequency response curves are similar for two different conductors grounded: All of them remain unchanged in the low frequency band, and then increase rapidly with the increase of frequency. In the low frequency stage, the larger the distance L from the injection point to the conductor end, the smaller the corresponding impedance modulus. Therefore, the change of L can be judged by the magnitude of impedance modes at low frequencies, and then whether the conductor has fracture corrosion or not can be judged.

Under three grounding conditions, the characteristics of the impedance angular frequency response curve are as follows: In the low frequency band, the impedance angle does not

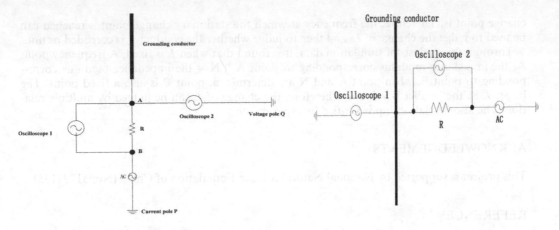

(a) Single-end injection.   (b) Middle-end injection.

Figure 4. Experimental schematic diagram.

Figure 5. Actual wiring diagram.

Table 4. L = 40m rule table.

frequency　　Structure	Single port injection Impedance mode (Ω)	Middle end injection Impedance mode (Ω)
5kHz	11.853	5.427
200kHz	23.843	10.52

Table 5. L = 50m rule table.

frequency　　Structure	Single port injection Impedance mode (Ω)	Middle end injection Impedance mode (Ω)
5kHz	10.282	4.412
1200kHz	20.444	9.022

change. When the frequency increases to a certain point, the impedance angle increases rapidly with the increase of the frequency. But unlike the impedance mode, With the further increase of frequency, the impedance angle changes slowly after reaching the stationary point. It is found that the larger the L is, the lower the frequency required to reach the stationary change point is. Therefore, the frequency at which the stationary change point is reached can be used to judge the change of L, and then to judge whether the conductor is corroded or not.

Through the analysis of simulation data, it is found that when L is fixed, A frequency point A, the impedance modulus corresponding to point A * N = the impedance modulus corresponding to point B, when point A and N are determined, point B is also a fixed point. The larger L is, the smaller B value is. Therefore, the change of L can be judged by multiple relation. The above relationship is verified by experiments.

ACKNOWLEDGEMENTS

This project is supported by National Natural Science Foundation of China (No: 51777175).

REFERENCES

WANG Lin, CHENG Feng ming, SONG Yanan. 2016, Research on the corrosion and transformation of the ground network of a certain 500kV substation [J]. electric porcelain arrester, (03): 148–151+159.

LIU Yugen, XIE Lina, SHANG Longlong. 2015, Diagnosis of grounding conductor fracture conductor based on binary genetic algorithm [J]. high voltage apparatus, 51 (01): 6–12+17.

WANG Fenghua, WANG Shaoqing, LIU Yadong. 2016, The effect of ground potential and magnetic induction on fault diagnosis of substation grounding grids is compared [J]. high voltage technology, 42 (07): 2281–2289.

QIAN Guochao, MA Yutang, LIU Yugen. 2014, Application of genetic algorithm in grounding grid corrosion fault diagnosis [J]. electric porcelain arrester, (02): 52–59+65.

PENG Xiangyang, LI Zhen, LI Zhifeng. 2011, The influence of grounding resistance on the lightning protection performance of the same tower multi circuit [J]. high voltage technology, 37 (12): 3113–3119.

XU Song, FENG Bing, HE Tiexiang. 2014, Corrosion analysis of transmission line tower grounding galvanizing line [J]. electric porcelain arrester, (04): 121–126.

ZHOU Lin, LIAO Yifan, LUO Bing. 2017, simulation study on frequency response of power transformer winding deformation based on finite element method [J]. power automation design, 37 (01): 204–211.

LI Wei, XIANG Chang Yuan, WEN Xi Shan. 2016, Simulation of the current distribution of EHV transmission line tower lightning [J]. high voltage technology, 42 (08): 2642–2650.

ZHANG Min, CAO Xiaobin, LI Ruifang. 2012, Transmission line pole to ground pole grounding resistance characteristics analysis of ground resistance [J]. electric porcelain arrester, (04): 5–9.

ZHU Bin, SIMA Wenxia, YUAN Tao. 2015, Structural optimization of needle grounding device based on electric field distribution in the earth [J]. power grid technology, 39 (10): 2907–2914.

SUN Yue, XIA Chenyang, DAI Xin. 2010, Analysis and optimization of mutual inductance coupling parameters in inductively coupled power transmission system [J]. proceedings of the Chinese electrical engineering, 30 (33): 44–50.

A method of single-phase ground fault line selection for distribution network based on correlation coefficient

Qingshan Qin, Xiuying Xie & Chongyi Tian
College of information and electrical engineering, Shandong Jianzhu University, Jinan, Shandong Province, China

Ruiqi Wang & Weijin Yang
State Grid Shandong Integrated Energy Services CO.,Ltd, Jinan, Shandong Province, China

ABSTRACT: In the resonant grounding system, the arc suppression coil only compensates for the power frequency zero-sequence current, and has almost no influence on the non-power frequency electrical quantity. When single-phase ground fault occurs, the fault lines are opposite to the zero-sequence current of the non-fault lines. In this paper, to determine the fault line, the Pearson correlation coefficient method is used to calculate the correlation coefficient of the first half-wave waveform of the zero-sequence 5th harmonic current between these distribution lines. The simulation results show that the correlation coefficient method is not affected by different factors.

Keywords: Distribution network, Single-phase ground fault, Fault line selection, Pearson correlation coefficient

1 INTRODUCTION

The construction of 6~35kV distribution network is widely used in the form of neutral point non-effective grounding, and the probability of occurrence of single-phase ground fault is 50%~80% (Shi Y.T. 2018). After a single-phase ground fault occurs in the distribution network, the fault feature quantity of each line is extracted. The fault characteristics of the fault line and the non-fault line fault are obtained by data processing to determine the fault line. This is the fault line selection technique. Fault line selection algorithm based on steady-state quantities has generated a large number of line selection devices. However, The amplitude of the steady-state signal is very small and this algorithm is only suitable for isolated neutral system (Wang, J.Y. et al. 2018, Cheng, L. et al. 2009). When the neutral point is grounded by the arc suppression coil, the zero-sequence current of the fault line and the non-fault line are not significantly different, which makes the method based on the steady-state information quantity under certain restrictions (Moldovanova, E. A. et al. 2016). Therefore, the algorithm for fault line selection by using transient information has been extensively studied.

The zero-sequence transient energy method performs the integral operation of the zero-sequence instantaneous power of the line after the line fault occurs to obtain the zero-sequence energy function (Zhuang, W. et al. 2014). The line is distinguished by judging the positive and negative values of the function. The fault function has a large amplitude and a negative polarity, which is contrary to the normal line zero sequence energy function. However, the method is essentially an active power detection method that the resistive component has a small proportion in the transient signal, thus the utilization of the transient information is insufficient, which may lead to the misjudgment. It is suggested that the fundamental components of the transient current and the transient voltage are opposite in phase of the first half of the wave after the ground fault (Hu, Z. et al. 2006). This principle is used to select the line. In the meantime, since the arc suppression coil is used for power frequency electrical, the compensation effect of the quantity causes a certain error in the method (Jin, T. et al. 2015). In this paper, the power grid system is simulated and modeled in the matlab/Simulink simulation software. The fault line selection is completed by using the amplitude and polarity information of the transient non-power frequency zero-sequence current 5th harmonic generated when the single-phase ground fault occurs. Finally the correctness of this method is verified by different fault factors.

2 FAULT CURRENT ANALYSIS

Zero-sequence current condition in single-phase ground fault is analyzed. Figure 1 shows a simple distribution network with resonant grounding, in which the AB segment is an over-head line, BC is a cable line, U is the system voltage, L is the arc suppression coil, and Z_g is the system impedance. Z_1 and Z_2 are the overhead line and the cable line respectively. Z_{load} is the equivalent impedance of the load. The equivalent composite network is shown in Figure 2.

In Figure 2, since a single-phase ground fault with a grounding resistance of Z_f occurs at point f, the fault current is I_f, it is analyzed by the composite sequence network of Figure 3.

$$\dot{I}_{Bf} = \frac{\dot{U}_s}{Z_g + Z_1 + Z_2 + Z_f // Z_{Load}} \tag{1}$$

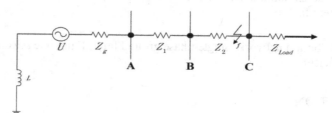

Figure 1. Simple distribution network.

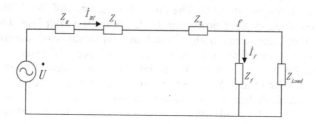

Figure 2. Equivalent composite network.

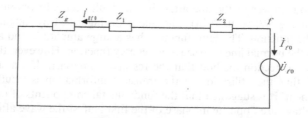

Figure 3. Equivalent zero sequence network.

And then the ground current generated by the fault point is obtained.

$$\dot{I}_f = \frac{\dot{U}_s - (Z_g + Z_1 + Z_2)\dot{I}_{Bf}}{Z_f} \qquad (2)$$

The distribution transformer at the load end is generally D/Y_n connection, i.e. the high voltage side is a triangular connection mode. The influence of system voltage and load is ignored in the zero sequence network, as shown in Figure 3. When the fault occurs, the positive sequence ground current in the equivalent load sequence network is equal to the zero sequence current in the zero sequence network.

$$\dot{I}_f = \dot{I}_{f0} \qquad (3)$$

As can be seen in Figure 3, $I_{Bf}=I_f$

$$\dot{I}_{f0} = \dot{I}_{Bf} \times \frac{Z_{Load}}{Z_{Load} + Z_f} \qquad (4)$$

(1) and (4) are combined to obtain the zero-sequence current generated by the grounding point.

$$\dot{I}_{f0} = \frac{\dot{U}_s}{Z_g + Z_1 + Z_2 + Z_f // Z_{Load}} \times \frac{Z_{load}}{Z_{load} + Z_f} \qquad (5)$$

After the above analysis, the ground current and the magnitude of the grounding resistance are nonlinear. In addition, when a single-phase earth fault occurs in a distribution network with multiple branches, the polarities of the zero-sequence current and the 5th harmonic between fault lines and non-fault lines are opposite. Therefore, when a single-phase ground fault occurs, the correlation analysis of the zero-sequence current and the fifth harmonic of each line can be performed to obtain the fault line.

3 POLARITY CORRELATION CALCULATION

Since the amplitude of the 5th harmonic current value is small and receives interference from various external factors such as the fault site, the simple method of judging the fault line from the amplitude information is more likely to cause misjudgment. Here, the correlation calculation specific to 5th harmonic current is introduced, which is the Pearson correlation coefficient calculation.

The well-known covariance represents the numerical characteristics of the relationship between the two variables X and Y. The covariance calculation formula is:

$$COV(X, Y) = \frac{1}{n-1} \sum_{i=1}^{n} (X_i - \overline{X})(Y_i - \overline{Y}) \qquad (6)$$

Although the covariance can reflect the degree of correlation between two random variables (co-variance is greater than 0 which indicates positive correlation between the two, and co-variance is less than 0 which indicates negative correlation). However, its value is greatly affected by the dimension, The judgment on the degree of correlation of the variables from the numerical value of the covariance cannot be simply made.

In order to eliminate the influence of the dimension, a correlation coefficient is introduced, i.e. the Pearson correlation coefficient, which is calculated as follows:

$$COR(X, Y) = \frac{\sum\limits_{i=1}^{n}(X_i - \overline{X})(Y_i - \overline{Y})}{\sqrt{\sum\limits_{i=1}^{n}(X_i - \overline{X})^2 \sum\limits_{i=1}^{n}(Y_i - \overline{Y})^2}} \tag{7}$$

When the correlation coefficient is 1, it becomes completely positive correlation. When the correlation coefficient is -1, it becomes completely negative correlation. The larger the absolute value of the correlation coefficient is, the stronger the correlation is. The closer the correlation coefficient is to 0, the weaker the correlation is. In this paper, the correlation coefficient row vector is established for each line under fault conditions.

$$R_1 = [r_{11}, r_{12}, r_{13}]$$
$$R_2 = [r_{21}, r_{22}, r_{23}] \tag{8}$$
$$R_3 = [r_{31}, r_{32}, r_{33}]$$

Where r_{ij} represents the correlation coefficient between line i and line j. To simplify the calculation, remove the main diagonal elements and cross-correlation elements to form a matrix of correlation coefficients

$$R = [r_{12}, r_{13}, r_{23}] \tag{9}$$

4 MODELING OF 10KV DISTRIBUTION NETWORK

4.1 *Distribution network modeling*

As shown in Figure 4, the 10kV distribution network model is established by Matlab/simulink simulation software. Among them, line L1 is an overhead line, L2 is a cable line, and L3 is a mixed line of overhead lines and cable lines. The end of each line is connected to the distribution transformer with a load of 1 MW.

Table 1 shows the circuit parameters of the cable line and the overhead line. From the parameters in Table 1, the resistance R and the inductance value L of the arc suppression coil RL in the distribution network model are obtained:

$$R = 0.03\omega L \qquad L = \frac{1}{3\omega^2(L_c C_c + L_o C_o)(p + 1)} \tag{10}$$

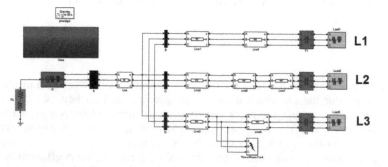

Figure 4. 10kV distribution network model.

Table 1.　Line parameter.

Type	Phase sequence	Resistance /($\Omega \cdot km^{-1}$)	Inductance /($mH \cdot km^{-1}$)	Capacitance /($\mu F \cdot km^{-1}$)
Cable	Positive	0.1800	1.2600	9.4e-3
	Zero	0.2600	5.3900	6.5e-3
Overhead	Positive	0.2710	0.2551	0.170
	Zero	2.5400	1.0193	0.153

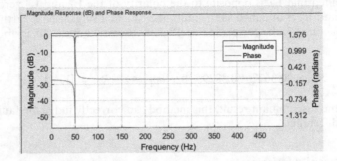

Figure 5.　Filter characteristics.

Where L_c and L_o are respectively the total lengths of the cable line and the overhead line, C_c and C_o are respectively the zero-sequence capacitance values of the cable line and the overhead line, and p is the compensation degree of the arc-suppression coil, which is 10%.

4.2　Zero sequence current extraction

In the resonant grounding system, the arc suppression coil compensates for the power frequency electrical quantity, and has almost no compensation effect on the non-power frequency electrical quantity. Therefore, the influence of the power frequency zero sequence current should be eliminated when extracting the zero sequence current. After the modeling of the distribution network is completed, the transient non-power frequency zero sequence current and its 5th harmonic needs to be effectively extracted. Here, the IIR digital filter is used to filter the power frequency current signal. The filter characteristics are shown in Figure 5. It can be seen that the filter has a good filtering effect on the power frequency 50Hz current signal, and the change of the phase information is very weak.

To get the 5th harmonic current, the 5th harmonic can be filtered out firstly by using a second-order Butter-worth band-stop filter. The current signal after filtering the power frequency signal is subtracted from the current signal after filtering out the 5th harmonic, and the non-power frequency zero sequence 5th harmonic current can be obtained.

5　SIMULATION ANALYSIS

The reliability of the above distribution network model under different fault factors is analyzed. The simulation time is 0.4s, the sampling time is set to 1MHz, and the fault phase is phase A. Due to space limitations, this paper includes simulation results in a table instead of simulation wave-forms.

5.1　Different fault distance

A phase-to-earth fault occurs at 0.12s with a transition resistance of 50Ω. There are two cases in the simulation: When a single-phase earth fault occurs at 10km on the overhead line L1, the data is brought into (7) to obtain the correlation coefficient discriminant matrix of the Zero sequence current fifth harmonic. Here is a clearer expression that an element less than 0 is equal to -1, and an element greater than 0 is equal to 1.

$$R_{L1} = [-0.9108, -0.8296, 0.6801] \quad \Rightarrow \quad R_{L1} = [-1, -1, 1]$$

When L1 fails, the correlation coefficient between L1 and L2, L1 and L3 is negative, and then it is determined that L1 is opposite to the polarity of the L2 and L3 zero sequence harmonic currents, and L1 is obtained as the fault line. When a single-phase earth fault occurs at 12km on the cable line L3, the correlation coefficient discriminant matrix is obtained.

$$R_{L3} = [0.3672, -0.3829, -0.9460] \quad \Rightarrow \quad R_{L3} = [1, -1, -1]$$

According to the discriminant result, this method can correctly judge the fault line when the fault location is different.

5.2 Different transition resistance

When 0.12s is set, a ground fault occurs at 12km on L3, and the transition resistance is set to 0.01Ω, 50Ω and 500Ω fault discrimination matrix respectively. In three cases,

$$R_{r=0.01\Omega} = [0.2249, -0.3033, -0.9527] \quad \Rightarrow \quad R = [1, -1, -1]$$

$$R_{r=50\Omega} = [0.3331, -0.2599, -0.7328] \quad \Rightarrow \quad R = [1, -1, -1]$$

$$R_{r=500\Omega} = [0.5681, -0.3807, -0.7263] \quad \Rightarrow \quad R = [1, -1, -1]$$

The polarity of the zero-order 5th harmonic current of L1 and L2 is opposite to that of L3, which proves that the line selection method is not affected by the magnitude of the transition resistance.

5.3 Different fault switching angle

A phase-to-earth fault occurs at 12km on L3, the transition resistance is 100Ω, and the fault switching angle θ is 0, 60° and 90° respectively corresponding to the A-phase voltage zero-crossing point, the phase angle is $\pi/3$ point and the peak point. The fault discrimination matrix is obtained as:

$$R_{\theta=0} = [0.4272, -0.2850, -0.6914] \quad \Rightarrow \quad R = [1, -1, -1]$$

$$R_{\theta=60°} = [0.2453, -0.2834, -0.9376] \quad \Rightarrow \quad R = [1, -1, -1]$$

$$R_{\theta=90°} = [0.2241, -0.2630, -0.9131] \quad \Rightarrow \quad R = [1, -1, -1]$$

The polarity of the zero sequence 5th harmonic current of L1 and L2 is opposite to that of L3, which proves that the line selection method can correctly identify the fault line under different fault closing angle conditions.

6 CONCLUSION

For the 10kV grid with neutral point resonant grounding mode, the arc suppression coil only compensates the fundamental frequency component of the fault current, and has almost no compensation effect on the non-power frequency current; the first half of the 5th harmonic of the zero sequence current occurs when the single phase ground fault occurs. The wave polarity is reversed, so the Pearson correlation calculation is performed on each line of the distribution network as the polarity criterion. The simulation results show that the method can correctly complete the fault line selection under different fault distances, fault closing angles and different transition resistance conditions. In the future, the research of fault line selection and location technology for multi-level branch distribution network and distribution network with distributed power supply will be focused on, and wavelet transform will be used to reduce the impact of on-site noise interference.

REFERENCES

Shi, Y.T. 2018. Study on the Method of Fault Line Selection for Single-Phase Grounding Fault in Distribution Network with DG. *College of Electrical and Information Engineering An Hui University of Science and Technology.*

Wang, J.Y. & Zhang, W. & Yang S. 2018. Research on Fault Line Selection for Distribution Networks Based on Five Times Harmonic Method and Improved Phase Locked Loop. *Journal of Northeast Electric Power University* 38(3): 1–7.

E.A. Moldovanova. & R.A. Vainshtein. & V.V. Shestakova. 2016. Detection of the faulted network element using frequency spectrum features of zero-sequence current under transient earth faults in the network with neutral grounding through Peterson-coil. *2016 Third International Conference on Electrical, Electronics, Computer Engineering and their Applications (EECEA)*: 93–97.

Zhuang, W. & Zhai, L.H. 2014. Grounding fault location of distribution network based on active component of zero sequence current. *Journal of Tongji University (Natural Science)* 42(3): 468–473.

Cheng, L. & Chen, Q.F. 2009. Overview of single-phase grounding line selection technology for small current grounding systems. *Power System Technolog* 33(18): 219–224.

Hu, Z. & Li, X.R. & Shi, J.Y. 2006. Research on Grounding Line Selection Method Based on Residual Flow and First Half Wave Synthesis. *Relay* (7):6–9+37.

Jin, T. & Zhai F.L. 2015. Grounding line selection method based on transient non-power frequency zero sequence current with DG new distribution network. *Transactions of China Electrotechnical Society* 30 (17): 96–105.

Emerging Developments in the Power and Energy Industry – Dufo-López, Krzywanski & Singh (eds)
© 2020 Taylor & Francis Group, London, ISBN 978-0-367-27169-5

Study of the modeling method of wind power time series based on correlation

Xiaoni Li

State Grid Shanxi Electric Power Company Yulin Power Supply Company, Beijing, China

ABSTRACT: Accurate prediction of wind farms' output is a vital part of research on issues of security and economy in situations in which a wind farm is connected to the power grid. Considering that there often exists an output correlation and complementarity between adjacent wind farms, a modeling method of wind power time series that is consistent with power output correlation and complementarity index is proposed. This article studies the historical data of three wind farms in northwest China. Based on the probability distribution characteristics, auto-correlation characteristics, and fluctuation characteristics of wind power output, a method of time series generation using Copula theory is proposed. A large number of simulation analyses of this method are carried out, the original wind farms in China and abroad are compared, and the accuracy of the method is verified. The method can well reflect the correlation of wind power output.

1 INTRODUCTION

Wind energy is a clean, renewable energy source. Because of the advantages of low-carbon and environmental protection, wind power is valued by all governments (Chen, 2011). With the rapid growth of the installed capacity of wind power in recent years, the characteristics of wind power have brought new challenges to the dynamic economic dispatching of power systems (Tian, 2010). In order to analyze the influence of the wind power correlation on the wind power output, it is necessary to study the time series of the wind power output and propose a model that can reflect the wind power output dependency.

There are two methods for generating wind power output time series: wind speed and direct wind power output. For the wind speed method, the literature randomly sampled the Weibull distribution according to the characteristics that the wind speed conforms to the Weibull distribution, and then converted the wind speed data to the wind power output. The method did not consider the process of the wind power output change and lacked a description of the time correlation. This method of obtaining data fluctuations in output may be very broad. In response to this problem, researchers have proposed the Vanderhoven model and the von Karman model (Lojowska, 2010), the autoregressive moving average (ARMA) series model, and the wavelet analysis model. In this article, the time series generation method of multiple wind farms with correlation is discussed. Based on the historical wind speed data, the probability distribution, autocorrelation, correlation of wind speed between wind farms, and the seasonal characteristics of wind speed, the wind speed time series is generated, and the wind power time series is converted into the wind power time series. For the first time, the spatial correlation is taken into account in the generation of a wind power time series. In terms of direct generation of wind power output, Chen et al. (2009) simulated wind power output of hours via the ARIMA model. Literature simulated the 10-minute data using the Monte Carlo Markov chain method. For the spatial correlation of wind power output, Negra et al. (2008) proposed that the Copula function can be used to model the spatial correlation of wind power output. A detailed mathematical deduction was made and the wind power data of the East Coast of the United States were used to verify the example. However, the article did not consider the seasonal characteristics of wind power output, and the results obtained cannot be verified.

Based on the probability density distribution of wind power output and the time correlation, this article proposes a time series generation method. In this article, the Copula function is introduced to model the time-dependent wind power output and a time series generation method is proposed.

2 CORRELATION ANALYSIS OF WIND POWER OUTPUT TIME

In order to describe the temporal correlation of wind power time series, this article uses the autocorrelation coefficient (autocorrelation function [ACF]). Given a time series $xt + k$, its time-shifted ACF is calculated according to Equation (1).

$$\text{ACF} = \frac{\text{cov}(xt, xt + k)}{\sqrt{\text{var}(xt) \bullet \text{var}(xt + k)}} (k = \pm 1, \pm 2, \ldots) \qquad (1)$$

where cov is convolution, var is variance; xt is based on the time series $(x1 + k, \ldots, xn)$, and $xt + k$ is based on the time series $(x1 + k, \ldots, xn)$.

Wind power output is time dependent and its contribution occurs at different times. Figure 1 shows the autocorrelation coefficient of Yulin Jingbian wind farm at different delay times. Within 24 hours, the autocorrelation coefficient decreased rapidly with the delay increasing, and dropped to about 0.2 at 24 hours. After 48 hours, the correlation coefficient is very small, below 0.1. At the same time, the autocorrelation coefficient of wind power output is cyclical. At 24-hour intervals, the decline of the correlation coefficient began to slow down and a peak appeared at every 24-hour autocorrelation coefficient because wind energy was closely related to the atmospheric phenomenon and 24 hours was one of the time periods of atmospheric phenomena. Under a short time scale, the correlation coefficient is 0.9969 when the autocorrelation coefficient is large and the delay time is 15 minutes. The autocorrelation coefficient is 0.9213 at a 1-hour delay. This shows that the short-term wind power output is highly correlated.

3 WIND POWER TIME SERIES GENERATION BASED ON WIND POWER CHARACTERISTICS AND COPULA

3.1 *Wind power output time-dependent modeling*

Through the analysis of the autocorrelation coefficient (ACF), it is confirmed that the wind power output time series has a time correlation. Before and after the time point of the output has contact. They are not completely independent random variables. This property can be described by the correlation coefficient. The original sequence and time delay sequence are expressed s two variables x^t, x^{t-1}. These two variables are in line with a beta distribution,

Autocorrelation Coefficient of Jingbian

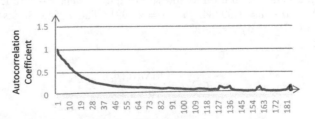

Figure 1.   Autocorrelation coefficient of Jingbian wind farm.

quantitative description of the relationship between the two variables, and one can establish a time-dependent model of wind power output.

3.2 Introduction of the Copula function

In order to model the correlation of wind power output at time points, a Copula function is introduced here. In probability statistics, the Copula function can "connect" multiple marginal distribution functions to form a multivariate joint probability distribution. The Copula function is used to model multiple variables with related relationships and it is suitable for a wide range of applications. The Copula function accurately describes the correlation measure. It will not change with the variable; at the same time, for the nonlinear correlation, asymmetric correlation, tail correlation can be described.

The definition of the binary Copula function is as follows:

The random variables x, y, Fx, and Fy are their cumulative probability distribution functions (both reversible). Fxy is the joint probability distribution function of x. If $F_X(x)$, $F_Y(y)$ are taken as random variables, and they all fit the uniform distribution:

$F_X(x) \sim \mu(0, 1) \ F_Y(y) \sim \mu(0, 1)$

The Copula theory is expressed as follows:

$$F_{XY}(x, y) = C(F_X(x), F_Y(y))$$

The multivariate cumulative probability density function C is the Copula function.

Theorem 1

$F(x1, x2, \ldots, xN)$ is the joint distribution function with edge distribution $F1(x1), F2(x2), \ldots, FN(xN)$.

$C(F1(x1), F2(x2), \ldots, FN(xN))$ is satisfactory.

$$F(x1, x2, \ldots, xN) = C(F1(x1), F2(x2), \ldots, FN(xN))$$

where $F(x1, x2, \ldots, xN)$ represents the joint distribution function of a random variable $x1, x2, \ldots, xN$.

3.3 Establishing a time correlation model based on Copula theory

A Copula function is used to establish a wind power output time correlation model. Suppose that $F(x^t, x^{t-1})$ and $f(x^t, x^{t-1})$ are the probability density joint distributions of wind farms at adjacent time points (generally 15 minutes in this article). x^t, x^{t-1} represent the output of the wind farm at time t and $t - 1$. The Copula function $F(x^t, x^{t-1})$ is recorded as $C(U, V)$, where $U = F(x_t)$, $V = F(x_{t-1})$. The time series of wind power can be described as

$$F(x_t, x_{t-1}) = C(U, V) = C[F(xt), F(x_{t-1})] \tag{2}$$

Taking the Yulin Jingbian wind farm as an example, the distributions of t and b are respectively fitted. Obtain a beta distribution as shown in Figure 2. The $t - 1$ time series beta distribution parameter is $at - 1 = 1.09784$, $\beta t - 1 = 2.4991$. The t time series beta distribution parameter is $\alpha_t = 1.09788$ and $\beta_t = 2.4983$.

By the formula:

$$F(x_t) = \frac{B(x_t; \alpha_t; \beta_t)}{B(\alpha_t; \beta_t)} \tag{3}$$

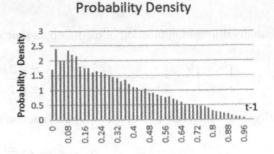

Figure 2.   Edge probability density distribution at $t - 1$.

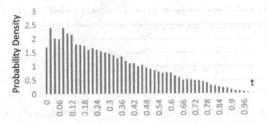

Figure 3.   Edge probability density distribution at t.

$$F(x_{t-1}) = \frac{B(x_{t-1}; \alpha_{t-1}; \beta_{t-1})}{B(\alpha_{t-1}; \beta_{t-1})} \qquad (4)$$

3.4 *Classification and selection of the Copula function*

There are many Copula functions. Their characteristics are different and need to be based on the characteristics of the variable function selection. The Copula function mainly has the following forms: normal Copula function, t-Copula function, Gumbel Copula function, Clayton Copula function and Frank Copula function. This article introduces the normal Copula function and t-Copula function.

(1) Normal Copula function

$$CN(\mu, \nu; \rho) = \int_{-\infty}^{\Phi^{-1}(\mu)} \int_{-\infty}^{\Phi^{-1}(\nu)} \frac{1}{2\pi\sqrt{1-\rho^2}} \exp\left(\frac{-(r^2 + s^2 - 2\rho r s)}{2(1-\rho^2)}\right) dr ds \qquad (3-4)$$

(2) The probability density function is

$$CN(\mu, \nu; \rho) = \frac{1}{\sqrt{1-\rho^2}} \exp\left(\frac{-\Phi^{-1}(\mu)^2 + -\Phi^{-1}(\nu)^2 - 2\rho\Phi^{-1}(\mu)\Phi^{-1}}{2(1-\rho^2)}\right) \exp\left(\frac{\Phi^{-1}(\mu)^2\Phi^{-1}(\nu)^2}{2}\right)$$
$$(5)$$

where $\Phi^{-1}(\cdot)$ is the inverse of the standard univariate normal distribution function $\Phi(\cdot)$ and $\rho \in (-1, 1)$ is the linear correlation coefficient. The normal Copula function cannot describe the tail of relevance because the tail correlation coefficient is $\lambda_L^N = \lambda_U^N = 0$.

(1) t-Copula function

The cumulative probability distribution function of the t-Copula function is:

$$CN(\mu, \nu; \rho, \eta) = \int_{-\infty}^{T_\eta^{-1}(\mu)} \int_{-\infty}^{T_\eta^{-1}(\nu)} \frac{1}{2\pi\sqrt{1-\rho^2}} \left(1 + \frac{-(r^2 + s^2 - 2\rho rs)}{\eta(1-\rho^2)}\right) dr ds \tag{6}$$

The probability density distribution function of the t-Copula function is

$$C_N(\mu, \nu; \rho, \eta) = \rho^{-\frac{1}{2}} \frac{T_\eta\left(\frac{\eta+2}{2}\right) T_\eta\left(\frac{\eta}{2}\right)}{T_\eta\left(\frac{\eta+2}{2}\right)} \frac{\left[\frac{1+\zeta_1^2+\zeta_2^2-2\rho\zeta_1\zeta_2}{\eta(1-\rho)^2}\right]^{-\frac{\eta+2}{2}}}{\prod_{i=1}^{2}\left(1 + \frac{\zeta_i^2}{\eta}\right) - \frac{\eta+2}{2}} \tag{7}$$

$\rho \in (-1, 1)$ is the linear correlation coefficient. $T_\eta^{-1}(\cdot)$ is the inverse of the t distribution function $T_\eta(\cdot)$. Its degree of freedom is η, $\zeta_1 = T_\eta^{-1}(\mu)$, $\zeta_2 = T_\eta^{-1}(\mu)$. From the definition of the tail correlation coefficient can be calculated,

$$\lambda_L^t = \lambda_U^t = 2T_{\eta+1}\left(\frac{\sqrt{\eta+1}\sqrt{1-\rho}}{\sqrt{1+\rho}}\right) \tag{8}$$

The binary t-Copula function can describe the tail correlation, but the tail correlation must be symmetrical. The Copula function's probability density distribution graph is consistent with the binary frequency histogram and frequency histogram of edge distributions $F(t)$ and $F(t-1)$ of adjacent moments. The tail of the frequency histogram is basically symmetric, indicating that the tail of the Copula function $C(\mu, \nu)$ is also symmetric. From the introduction to the Copula function above, this type of temporal correlation can be described using the normal Copula function and the t-Copula function. The two Copula functions are modeled and compared respectively.

In addition to the linear correlation coefficient, there are other indicators to measure the correlation coefficient, comparing the Kendall rank correlation coefficient and Spearman rank correlation coefficient between the original data, and finding that the normal Copula function is closer to the original data. The normal Copula function with a linear correlation coefficient of $\rho = 0.9971$ better reflects the wind power dependence at adjacent time points.

Table 1. Copula function parameters.

Parameter	Normal copula function	t-Copula function
η	0.9971	0.9986
η	—	5.6932

Table 2. Rank correlation coefficients.

Correlation coefficient types	Raw data	Normal Copula function	t- Copula function
Kendal rank correlation coefficient	0.9574	0.9523	0.9644
Spearman rank correlation coefficient	0.9972	0.9973	0.9918

Using the structure of the correlation coefficient, the simulation results of the two Copula functions are equivalent. The empirical Copula is used to analyze the accuracy of the two models.

Using the structure of the correlation coefficient, the simulation results of the two Copula functions are equivalent. The empirical Copula is used to analyze the accuracy of the two models.

Definition 2 Empirical Copula

$$\hat{C}_n(\mu, \nu) = \frac{1}{n} \sum_{i=1}^{n} I_{[F_n(x_i) \leq \mu]} I_{[G_n(y_i) \leq \nu]}, \mu, \nu \in [0, 1] \tag{9}$$

where $Fn(x)$ and $Gn(y)$ are the empirical distribution functions of X and Y. After generating the empirical Copula function A, we examine the Euclidean distance of the empirical Copula between the bivariate normal Copula and the bivariate t-Copula functions:

$$d_{Ga}^2 = \sum_{i=1}^{n} \left| \hat{C}_n(\mu_i, \nu_i) - \hat{C}^{Ga}(\mu_i, \nu_i) \right|^2 \tag{10}$$

$$d_t^2 = \sum_{i=1}^{n} \left| \hat{C}_n(\mu_i, \nu_i) - \hat{C}^t(\mu_i, \nu_i) \right|^2 \tag{11}$$

where $\mu_i = F_n(x), \nu_i = G_n(y)(i = 1, 2, \ldots, n)$. Models with smaller Euclidean distances can better fit the original data.

The calculated Euclidean distance between the bivariate normal Copula with linear correlation parameter $\rho = 0.9971$ and empirical Copula is $d_{Ga}^2 = 0.0397$. The linear correlation parameter of the binary t-Copula function is 0.9986 and the degree of freedom is 5.6830. The European distance to experience Copula is $d_t^2 = 0.0511$. Therefore, under the square Euclidean distance standard, the bivariate normal Copula function can better fit the wind farm power distribution at adjacent time points and describe the time correlation of wind power output.

4 CONCLUSION

In this article, a time-dependent model of time series is established. By comparing the time series of the Copula method and the original sequence, taking the wind field in the northwest of China as an example, the autocorrelation coefficient of the sequence generated by the Copula method is closer to the original data. The study found that wind power output is time dependent. Within a short period of time (1 hour), the correlation coefficient is very high, and with a rapid increase of time span, the correlation decreases rapidly. At the same time, the autocorrelation coefficient of wind power output has a periodicity. At the interval of 24 hours, the correlation coefficient decreases slowly, and a peak value occurs in the autocorrelation coefficient every 24 hours. This provides a theoretical basis for wind power forecasting.

REFERENCES

Bayem, H., Phulpin, Y., Dessante, P., et al. 2008. Probabilistic computation of wind farm power generation based on wind turbine dynamic modelling. In *Proceedings of the 10th International Conference on Probabilistic Methods Applied to Power Systems (PMAPS '08)*, May 25–29, pp. 1–6.

Chen, P., Berthelsen, K. K., Bak-Jensen, B., et al. 2009. Markov model of wind power time series using Bayesian inference of transition matrix. In *35th Annual Conference of IEEE on Industrial Electronics (IECON '09)*, Nov. 3–5, pp. 627–632.

Chen, P., Pedersen, T., Bak-Jensen, B., et al. 2010. ARIMA-based time series model of stochastic wind power generation. IEEE Transactions on Power Systems 25(2): 667–676.

Chen, X. 2011. Research on wind power characteristics for power system operation. Dalian University of Technology.

Global Wind Report 2014. Annual market update. Retrieved from http://www.gwec.net/index.php?id=190 (accessed December 26, 2017).

Karaki, S. H., Salim, B. A., & Chedid, R. B. 2002. Probabilistic model of a two-site wind energy conversion system. *IEEE Transactions on Energy Conversion* 17(4): 530–536.

Kitagawa, T., & Nomura, T. 2003. A wavelet-based method to generate artificial wind fluctuation data. *Journal of Wind Engineering & Industrial Aerodynamics* 91: 943–964.

Lojowska, A., Kurowicka, D., Papaefthymiou, G., et al. 2010. Advantages of ARMA-GARCH wind speed time series modelling. In *IEEE 11th International Conference on Probabilistic Methods Applied to Power Systems (PMAPS)*, June 14–17, 2010, pp. 83–88.

Louie, H. 2010. Evaluation of probabilistic models of wind plant power output characteristics. In IEEE International Conference on Probabilistic Methods Applied to Power Systems. IEEE, pp.442–447.

Negra, N. B., Holmstrom, O., Bak-Jensen, B., et al. 2008. Model of a synthetic wind speed time series generator. *Wind Energy* 11: 193–209.

Tian, X., Liu, Y., & Xia B. 2010. Research on characterization of wind power output. *Journal of the Hebei Academy of Sciences* 27(4): 43–47. DOI: 10.3969/j.issn.1001-9383.2010.04.012.

Zhang, N., Kang, C., Duan, C., et al. 2010. Simulation methodology of multiple wind farms operation considering wind speed correlation. *International Journal of Power & Energy Systems* 4).

Zhang, N., Kang, C., Xu, Q., et al. 2013. Modelling and simulating the spatio-temporal correlations of clustered wind power using Copula. *Journal of Electrical Engineering & Technology* 8(6): 1615–1625.

Emerging Developments in the Power and Energy Industry – Dufo-López, Krzywanski & Singh (eds)
© 2020 Taylor & Francis Group, London, ISBN 978-0-367-27169-5

Research on the post-evaluation of power transmission project based on GIOWA operator

Jie Zhang
State Grid Baoding Electric Power Supply Company, Baoding, China

ABSTRACT: This paper puts forward a power transmission project post-evaluation index system which investigates the power transmission project from four levels including financial benefit, environmental benefit, the project construction process and social benefit. Fifteen analysis indexes are refined from the four levels. The post-evaluation model based on fuzzy language is constructed using the GIOWA integrated operator. The feasibility of the model is verified by analysis of a power transmission project.

Keywords: GIOWA integrated operator, financial benefit, environmental benefit, social benefit

1 INTRODUCTION

In all investment of power grid enterprises, the investment in power transmission project requires large capital investment which accounts for a relatively high proportion, and affect the profitability of the whole project. For the post-evaluation method of power transmission project, Wei constructed a rational technical post-evaluation index system while the comprehensive evaluation method is applied to the technical post evaluation of power transmission and transformation project, and the before-and-after comparison method and the with-and-without comparison method are led into the comprehensive evaluation (Wei & Yang et al., 2010). Huang fully considered qualitative and quantitative factors and proposed a post-evaluation model of power transmission project based on improved analytic hierarchy process (Huang & Fu et al., 2010). Wang put forward a fuzzy comprehensive evaluation model based on the success degree methods to provide the index and method for the post-evaluation for technical renovation projects (Wang & Li et al., 2014). Ji established the index system of economic benefits evaluation of power transmission project and combined the fuzzy optimization theory with the BP neural network theory, and obtained the fuzzy optimization BP neural network weighting adjustment model with clear physical meaning (Ji, 2010). In this paper, the advantages and disadvantages of various post-evaluation methods are compared and GIOWA operator method is used to analyze the post evaluation of power transmission.

2 CONSTRUCTION OF THE POST-EVALUATION INDEX SYSTEM OF POWER TRANSMISSION PROJECT

Considering the refined safety evaluation, the detailed safety evaluation index factors and the evaluation index system of power transmission project construction are established through field research. The post-evaluation index system is shown in Figure 1.

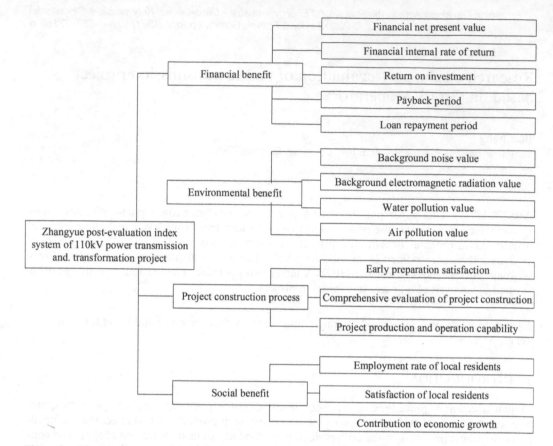

Figure 1. Zhangyue post-evaluation index system of 110kV power transmission.

3 INDEX AGGREGATION METHOD BASED ON GIOWA OPERATOR

3.1 *The mathematical model built on GIOWA operator*

We let $\hat{a} = [a^L, a^M, a^U]$, $0 < a^L \leq a^M \leq a^U$, function $\hat{a}$ is called triangular fuzzy function and its characteristic function expression is:

$$\mu_{\hat{a}} = \begin{cases} (x - a^L)/(a^M - a^L), & a^L \leq x \leq a^M \\ (x - a^U)/(a^M - a^U), & a^L \leq x \leq a^M \\ 0, & others \end{cases} . \tag{1}$$

The two algorithms for triangular fuzzy function are as follows:

$$\hat{a} + \hat{b} = [a^L, a^M, a^U] + [b^L + b^M + b^U] = [a^L + b^L, a^M + b^M, a^U + b^U] \tag{2}$$

$$\beta\hat{a} = [\beta a^L, \beta a^M, \beta a^U], \beta \geq 0 \tag{3}$$

GIOWA is called a generalized induced ordered weighted averaging operator (GIOWA operator). If $GIOWA_w(\langle \xi_1, \pi_1, \alpha_1 \rangle, \cdots, \langle \xi_n, \pi_n, \alpha_n \rangle) = \sum_{j=1}^{n} \omega_j b_j$, where $w = (\omega_1, \omega_2, \ldots, \omega_n)$ is the associated weighting vector of GIOWA, $\omega_j \in [0,1](j \in N)$, $\sum_{j=1}^{n} \omega_j = 1$, $\langle \xi_i, \pi_i, \alpha_i \rangle$ is a three-dimension data. First of all, arrange all the three-dimension data according to the first component $\xi_i(i \in N)$ from big to small. b_j will be the third component of the j-th ordered three-dimension data, so it can be called triangular fuzzy function.

3.2   Model description

We let $X = \{X_1, X_2, ..., X_i, ..., X_n\}$ be the set of power transmission post-evaluation index system, where $X_i = \{X_{i1}, X_{i2}, ..., X_{ij}, ..., X_{im_j}\}$, X_i is first grade level index, X_{ij} is second grade level index, $i = 1, 2, ..., n, j = 1, 2, ..., m_i$.

Inviting p experts to form an expert group $E = \{e_1, e_2, ..., e_k, ..., e_p\}$, $k = 1, 2, ..., p$; the linguistic assessment value of the evaluation index $X_{ij}(i = 1, 2, ..., n, j = 1, 2, ..., m_i)$ by expert $e_k(k = 1, 2, ..., p)$ is $r_{ij}^{(k)}$, and the evaluation matrix is as follows:

$$
R_i = \begin{array}{c} \\ e_1 \\ e_2 \\ \vdots \\ e_p \end{array} \begin{matrix} X_{i1} & X_{i2} & \cdots & X_{im_i} \\ \begin{bmatrix} r_{i1}^{(1)} & r_{i2}^{(1)} & \cdots & r_{im_i}^{(1)} \\ r_{i1}^{(2)} & r_{i2}^{(2)} & \cdots & r_{im_i}^{(2)} \\ \vdots & \vdots & \vdots & \vdots \\ r_{i1}^{(p)} & r_{i2}^{(p)} & \cdots & r_{im_i}^{(p)} \end{bmatrix} \end{matrix}, i = 1, 2, \ldots \ldots n \tag{4}
$$

Where $r_{ij}^{(k)} \in S, i = 1, 2, ..., n, j = 1, 2, ..., m_i, k = 1, 2, ..., p$.

3.3   First grade level index aggregation method

According to $z_i(\omega)(i = 1, 2, ..., n)$, we assemble n first grade level indexes based on GIOWA operator, and get the syntheses assessment:

$$
z(w) = \text{GIOWA}_w(\langle z_1(\omega), X_1, \hat{a}_1 \rangle, ..., \langle z_n(\omega), X_n, \hat{a}_n \rangle) = \sum_{i=1}^{n} w_i b_i \tag{5}
$$

Where $z_i(\omega) \in S$, $X_i \in X$, $\hat{a}_i$ is the triangular fuzzy number corresponding to $z_i(\omega)$. $w = (w_1, w_2, ..., w_n)$ is the weighing vector of the GIOWA operator, where $w_i \in [0,1], \sum_{i=1}^{n} w_i = 1$.

4   POST-EVALUATION OF POWER TRANSMISSION PROJECT

4.1   110kV power transmission project and post-evaluation index.

Taking Zhangyue 110kV power transmission and transformation project in Baoding, Hebei province, as an example, this paper evaluates the project from the four aspects: financial benefit, environmental benefit, project construction process and social benefit.

The evaluation index system and their weight are shown in Table 1.

Table 1. Weight of evaluation indexes at all levels.

Index	Index name	Index number	Index weight
First grade level B	Financial benefit	B_1	0.2569
	Environmental benefit	B_2	0.2864
	Project construction process	B_3	0.1915
	Social benefit	B_4	0.2652
Second grade level C_1	Financial net present value	C_{11}	0.1798
	Financial internal rate of return	C_{12}	0.2057
	Return on investment	C_{13}	0.1886
	Payback period	C_{14}	0.2220
	Loan repayment period	C_{15}	0.2028
Second grade level C_2	Background noise value	C_{21}	0.2885
	Background electromagnetic radiation value	C_{22}	0.2453
	Water pollution value	C_{23}	0.2190
	Air pollution value	C_{24}	0.2473
Second grade level C_3	Early preparation satisfaction	C_{31}	0.3501
	Comprehensive evaluation of project construction	C_{32}	0.2809
	Project production and operation capability	C_{33}	0.3690
Second grade level C_4	Employment rate of local residents	C_{41}	0.2776
	Satisfaction of local residents	C_{42}	0.3128
	Contribution to economic growth	C_{43}	0.4096

Table 2. Linguistic level and Triangular fuzzy number.

Serial number	Linguistic level	Triangular fuzzy number
s_1	best	[0.8, 0.9, 1.0]
s_2	better	[0.7, 0.8, 0.9]
s_3	good	[0.6, 0.7, 0.8]
s_4	less-good	[0.5, 0.6, 0.7]
s_5	general	[0.4, 0.5, 0.6]
s_6	less-bad	[0.3, 0.4, 0.5]
s_7	bad	[0.2, 0.3, 0.4]
s_8	worse	[0.1, 0.2, 0.3]
s_9	worst	[0, 0.1, 0.2]

Table 3. Linguistic assessment vector for first grade level index.

B_1	B_2	B_3	B_4
S_5	S_3	S_4	S_4

4.2 *Construction of linguistic assessment matrix*

In order to facilitate experts to evaluate and score the indexes, function $S = \{s_1, s_2, s_3, s_4, s_5, s_6, s_7, s_8, s_9\}$ is set as the index importance level. The specific meaning and triangular fuzzy number are shown in Table 2.

Through questionnaire survey, six experts' linguistic assessment values of indexes $B_1 \sim B_4$ are obtained.

The linguistic assessment values of first grade level indexes B_1, B_2, B_3 and B_4 are shown in Table 3.

According to the index evaluation values in Table 3, the linguistic assessment values of the four first grade level indexes B_1, B_2, B_3 and B_4 can be integrated by using the GIOWA operator.

$$z_1 = s_5, z_2 = s_3, z_3 = s_4, z_4 = s_5 \qquad (6)$$

Therefore,

$$z_2 > z_3 = z_4 > z_1 \qquad (7)$$

So,

$$b_1 = \hat{a}_2, b_2 = \hat{a}_3, b_3 = \hat{a}_4, b_4 = \hat{a}_1 \qquad (8)$$

Have to,

$$
\begin{aligned}
z(w) = GIOWA_w(&\langle z_1, B_1, \hat{a}_1 \rangle, \langle z_2, B_2, \hat{a}_2 \rangle, \langle z_3, B_3, \hat{a}_3 \rangle, \langle z_4, B_4, \hat{a}_4 \rangle) \\
= \ &0.2596 \times [0.6, 0.7, 0.8] + 0.2864 \times [0.5, 0.6, 0.7] + 0.1915 \times [0.5, 0.6, 0.7] \\
&+ 0.2652 \times [0.4, 0.5, 0.6] \\
= \ &[0.5, 0.6, 0.7] \\
= \ &s_4
\end{aligned} \qquad (9)
$$

5 CONCLUSION

According to the above evaluation process and results, the post-evaluation of the power transmission and transformation project has the best environmental benefit effect. And the project construction benefit and social benefit are better. The financial benefit of the project is modest. The post-evaluation results of the project show that the evaluation results are in line with the actual situation, proving the effectiveness and practicability of the post-evaluation.

REFERENCES

Wei G, Yang Y, et al, 2010, Technical Post-Evaluation on Power Transmission and Transformation Project Based on Comprehensive Evaluation Method. Power System Technology (3):112–116.
Huang W, Fu L and Xiao S. 2010. Post evaluation of power transmission and transformation project based on improved analytic hierarchy process. Statistics and decision-making (20):160–162.
Wang L and Li X. 2014. Post-Evaluation Index and Method of Technical Renovation Projects for Power Transmission. Power System and Clean Energy (7):25–28.
Ji D. 2010. The Study on the post evaluation of economic of the power transmission and transformation projects using fuzzy optimization neural network approach. North China electric power university.
Han S, Gao L, Liu Y, and Yang W. 2014. Post Evaluation of Wind Resource Assessment and Micro-Siting. Journal of Power and Energy Engineering 2 (4):288–296.
JA Ford, C Mackay, et al. 2013. Putting guidelines into practice: a tailored multi-modal approach to improve post-operative assessments[J]. Journal of Evaluation in Clinical Practice 19 (1):106–111.
P Kadagad, S. M. Kotrashetti. Portfolio 2013. A Comprehensive Method of Assessment for Postgraduates in Oral and Maxillofacial Surgery[J]. Journal of Maxillofacial and Oral Surgery 12 (1):80–84.
Caldwell G. 2013. An evaluation of a formative assessment process used on post take ward rounds. Acute medicine 12 (4):208.
ZS Xu and QL Da. 2010. The ordered weighted geometric averaging operators. International Journal of Intelligent Systems 17(7):709–716.

Emerging Developments in the Power and Energy Industry – Dufo-López, Krzywanski & Singh (eds)
© 2020 Taylor & Francis Group, London, ISBN 978-0-367-27169-5

Study on the effect of transition resistance on the frequency spectrum characteristics of rail impedance

K. Li
Design Institute of Guangzhou Metro Corporation, Guangzhou, China

K. Li, C. Zhu, W. Feng, R. Li & X. Cao
Electric Engineering School of Southwest Jiaotong University, Chengdu, China

ABSTRACT: Transition resistance between rail and ground is one of the main factors affecting the size of stray current. External environment can easily lead to the decline of local section transition resistance. Therefore, the research on the detection method of rail-ground transition resistance is conducive to timely discovering the problem of rail-to-ground transition resistance on metro lines and reducing the harm of stray current. In this paper, the influence of rail-to-ground transition resistance on the spectrum characteristics of rail mutual impedance is studied. The feasibility of detecting rail-to-ground transition resistance with rail mutual impedance spectrum characteristics is explored. The difference between transition resistance and other factors is found, and the spectrum characteristics of rail mutual impedance under the condition of non-uniform transition resistance are found. The research in this paper can provide reference for the detection method of rail-to-ground transition resistance.

Keywords: Transition resistance, Spectrum characteristics, Detection, Feasibility

1 INTRODUCTION

Urban rail transit (URT) plays an important role in coping with urban traffic congestion and promoting the healthy and sustainable development of the city with its advantages of large volume, fast speed, safety, high efficiency, low carbon and environmental protection.

Urban rail transit generally adopts DC traction power supply system. Traction return current flows into the earth through rails. This part of current is called stray current. The accumulated contamination layer during operation seriously affects the insulation performance of rail insulation fasteners, which reduces the transition resistance between rail and ground, increases the stray current leakage to the surrounding area (Xifeng Liu 2011), causes electrochemical corrosion of rails, structural reinforcement bars and underground metal pipelines around and along the Metro (A. Aghajani 2015), and reduces their service life (Dodds R 1996). Therefore, the transition resistance between rail and ground of newly built lines is required to be greater than 15Ω.km and that of running lines should not be less than 3Ω. km.

Nowadays, a lot of research has been done on the measurement of transition resistance at home and abroad. Goodman and others have made a detailed study of stray current and transition resistance by using finite element method (J.G.Yu 2004). In reference (Zhang Dongliang 2008), an on-line measurement method of transition resistance based on differential equation is proposed. First, the differential equation of transition resistance is derived. Then, the measured two-point track voltage and track current are brought into the general solution of differential equation to obtain transition resistance. The methods given by these studies are difficult to implement. Because the spectrum detection method can reflect some subtle changes, it has been widely studied in the detection of electrical equipment. In this paper, the feasibility of detecting the transition resistance by the spectrum characteristics of rail mutual impedance is proposed, which provides a new research idea for the detection of the transition resistance.

2 THE ESTABLISHMENT OF SIMULATION MODELS

2.1 *Corresponding relation between resistivity and transition resistance of software coatings*

In this paper, the MALZ module of CDEGS software is used to establish the simulation model of rail spectrum characteristics. Since only one layer of fixed resistivity coating can be added to the bare conductor in the MALZ module to replace the transition resistance, we need to deduce the relationship between the two. Suppose the resistivity of the coating is ρ, Conductivity is δ, The transition resistance of rail to ground is R. The radius of bare wire is r_1. The radius of the conductor after coating is r_2, r is the distance from the point of measuring potential to the center of the conductor section. Because the length of the rail is much larger than the radius of the rail cross section, we can neglect the edge effect at the end of the rail. Assuming that the current is I, the leakage current density at any point M between the two electrodes (i.e. rail heel coating) is

$$J = \frac{I}{2\pi\delta l}$$

(1)

So the electric field intensity is:

$$E = \frac{J}{\gamma} = \frac{I}{2\pi\delta l\gamma}$$

(2)

The Voltage between two electrodes is:

$$U = \int_{r_1}^{r_2} \frac{I}{2\pi\delta l\gamma} d\gamma = \frac{I}{2\pi\delta l} \ln\frac{r_2}{r_1}$$

(3)

Thus leak conductance can be obtained:

$$G = \frac{I}{U} = \frac{2\pi\delta l}{\ln\frac{r_2}{r_1}}$$

(4)

The corresponding transition resistance is

$$R = \frac{1}{G} = \frac{1}{2\pi l\delta} \ln\frac{r_2}{r_1} = \frac{\rho}{2\pi l} \ln\frac{r_2}{r_1}$$

(5)

R is the rail-to-ground transition resistance and its unit is Ω.km; ρ is the resistivity of the coating, and its unit is Ω.m. r_1 and r_2 represent the radius of bare conductor and the total radius of conductor with coating respectively, the unit is m. r is the distance from the point at which the potential is measured to the center of the conductor cross section. In this paper, r_1 is the radius of rail and $r_1 = 0.048$m, coating thickness is 0.1m, r_2 is the sum of rail radius and coating thickness and $r_2 = 0.058$m.

By substituting the data in Formula 5, the coating resistivity corresponding to a certain value of transition resistance can be calculated. The transition resistance is replaced by a fixed resistivity coating on the rail. In this paper, the radius of the rail is 0.048m and the thickness of the coating is 0.1m. The specific values are shown in Table 1.

Table 1. Resistivity of coatings corresponding to different transition resistance.

$R/\Omega \cdot km$	1	3	15	30	50	100
$\rho/k\Omega \cdot m$	31.2	99.7	510.6	1024.3	1709.2	3421.5

2.2 Establishment of simulation model

In the first part of the model, the soil layer with a certain resistivity is used to simulate the concrete layer. The resistivity is $2000\Omega.km$. In the concrete layer, the tunnel wall is simulated by 16 uniformly distributed structural reinforcing bars. The depth of the tunnel is 10 m. The resistivity of the air layer is $1*1018\Omega.km$. The radius of the structural reinforcing bars is 0.01 m. The longitudinal resistance is 320 mΩ/km. According to the relevant information, the Metro usually uses 60 kg/m and 65 kg/m rails. In this section, 60 kg/m rails are used first, and the longitudinal resistance is $0.04\Omega/km$. The equivalent radius of the cylindrical conductor is 0.048 m. The rail spacing is 1435 mm. Soil types are uniform. The resistivity of sandy clay is $100\Omega \cdot m$. The schematic diagram of the simulation model is shown in Figure 1, and the cross section of the simulation model is shown in Figure 2.

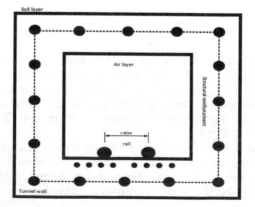

Figure 1. Simulation model diagram.

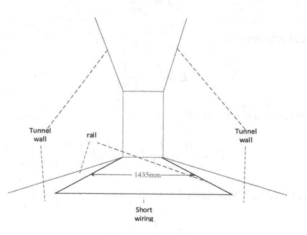

Figure 2. Cross-sectional diagram.

651

3 SPECTRUM CHARACTERISTICS OF RAIL MUTUAL IMPEDANCE WITH UNIFORM DISTRIBUTION OF TRANSITION RESISTANCE

3.1 *Simulation results of different transition resistors*

Using the model of section 2, the amplitude-frequency and phase-frequency response characteristics of rail mutual impedance under six conditions of transition resistance are simulated as shown in Figure 3 and 4 respectively. The six conditions of transition resistance is 1Ω.km, 3Ω.km, 15Ω.km, 30Ω.km, 50Ω.km, 100Ω.km.

As can be seen from Figure 3, the amplitude-frequency response characteristic curve can be divided into three sections:

In the low frequency stage (50Hz~1kHz): the curve is relatively smooth, the larger the transition resistance, the larger the mutual impedance modulus, and the smaller the variation of the curve with frequency in this frequency band.

In the intermediate frequency stage (1kHz~300kHz): the amplitude-frequency characteristic curve of the transition resistance rises faster, but the frequency required to reach the maximum modulus varies with the transition resistance.

High frequency stage (300 kHz ~ 1000kHz): Amplitude-frequency characteristic curves coincide completely under different transition resistances.

As can be seen from Figure 4, the phase response characteristic curve can also be divided into three stages:

In the low frequency stage (50Hz~1kHz), the impedance angle increases with the increase of frequency, and the smaller the transition resistance, the larger the impedance angle, the lower

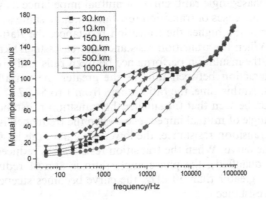

Figure 3.   Amplitude-frequency curve.

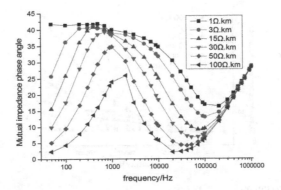

Figure 4.   Phase-frequency curve.

the frequency at which the impedance angle reaches its peak point; the larger the transition resistance, the smaller the impedance angle, the higher the frequency at which the impedance angle reaches its peak point.

In the intermediate frequency stage (1kHz~80kHz), the impedance angle decreases gradually with the increase of frequency. The larger the transition resistance, the lower the frequency required for the impedance angle to reach the valley point, the smaller the transition resistance, and the higher the frequency required for the impedance angle to reach the valley point.

In the high frequency stage (80kHz ~ 1000kHz), the phase-frequency characteristic curve almost coincides.

3.2 *Simulation results analysis*

The simulation results show that when the mutual impedance modulus is in different frequency bands, its rising speed is different, so this section will study the difference of the spectrum characteristics of different transition resistors according to this characteristic.

Considering that the mutual impedance modulus of each spectrum in the low frequency region does not change much, and the change rate of the mutual impedance frequency modulus in the middle frequency region is different, the ratio method is used to analyze this characteristic.The method divides the impedance modulus of each point by 50 Hz. Taking the simulation results of 15 Ω.km transition resistance as an example, all the data of the simulation results of amplitude-frequency characteristics are divided by 3.578 (the results at 50Hz). The mutual impedance modulus ratio curve is obtained as shown in Figure 5. Divide all the data of the simulation results of phase-frequency characteristics by 41.765 (the result at 50Hz), and draw the phase-angle ratio curve of mutual impedance as shown in Figure 6. The simulation results in other cases of transition resistance are the same as those in this case.

As shown in Figure 5, the higher the transition resistance, the flatter the cross impedance modulus ratio curve. When the transition resistance decreases to 1Ω.km (the transition resistance is not qualified), the insulation performance between rails and ground is very poor. The more obvious the interaction between rails is, the greater the variation of the rail mutual impedance modulus is. At this time, the ratio ranges from 1 to 54.37.

From Figure 6, it can be seen that the smaller the transition resistance, the smaller the variation range of phase angle of mutual impedance. When the frequency is between 50 Hz and 30 kHz, the smaller the transition resistance, the lower the curve, the larger the transition resistance and the higher the curve. When the transition resistance is reduced to 1Ω.km (the transition resistance is not qualified), the phase angle ratio is almost reduced to a straight line. When the frequency is greater than 30 kHz, the curve becomes steeper and steeper with the increase of transition resistance.

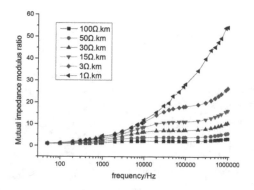

Figure 5.   Modulus ratio curve.

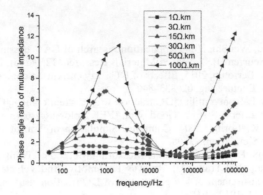

Figure 6.   Phase angle ratio curve.

From the above analysis, it can be concluded that when the impedance angular-modulus ratio diagram changes obviously with the increase of frequency, the value of transition resistance can be judged to be smaller at this time, and when the phase-angle ratio diagram changes very little with the increase of frequency, the value of transition resistance can also be judged to be smaller at this time.

4   CONCLUSION

This paper mainly studies the spectrum characteristics of rail mutual impedance, and puts forward a new solution to the detection method of transition resistance. The main conclusions are as follows:

1. The amplitude-frequency characteristics of mutual impedance with different transition resistors are studied. It is found that the curve is gentle in the low frequency stage. The bigger the transition resistance is, the bigger the mutual impedance modulus is, and the smaller the variation of the curve changes with frequency in this frequency band. In the intermediate frequency stage, the amplitude-frequency characteristic curve of the transition resistance rises faster, but the frequency required to reach the maximum modulus varies with the transition resistance. The transition resistance has no effect on the amplitude-frequency characteristic curves in the high frequency stage.
2. The phase-frequency characteristics of mutual impedance in different transition resistors are studied. It is found that the impedance angle increases with the increase of frequency in low frequency phase. The smaller the transition resistance is, the larger the impedance angle is, the lower the frequency at which the impedance angle reaches its peak point. The larger the transition resistance is, the smaller the impedance angle is, and the higher the frequency at which the impedance angle reaches its peak point. In the intermediate frequency stage, the impedance angle decreases gradually with the increase of frequency. The larger the transition resistance, the lower the frequency required for the impedance angle to reach the valley point, the smaller the transition resistance, and the higher the frequency required for the impedance angle to reach the valley point. the transition resistance has no effect on the phase-frequency characteristic curves in the high frequency stage.
3. The ratio of mutual impedance is proposed to reflect the magnitude of transition resistance. It is found that when the impedance angular-modulus ratio diagram changes obviously with the increase of frequency, the transition resistance can be judged to be smaller at this time, and when the phase-angle ratio diagram changes very little with the increase of frequency, the transition resistance can also be judged to be smaller at this time.

REFERENCES

Xifeng Liu, Chunxiang Xu, Wei Shi, 2011, Application research of CAN bussing technique in the monitor of orbit traffic stray current[J]. Advanced Materials Research, 1250(225): 1336–1340.

A. Aghajani, M. Urgen, L. Bertolini, 2015, Effects of DC stray current on concrete permeability[J]. Journal of Materials in Civil Engineering, 02: 833–840.

Dodds R, 1996, A case study from a utility [DC traction stray current control][C]//IEEE Seminar on DC Traction Stray Current –offer A Stray A Good ohm. IEEE Xplore: 6/1-6/4.

J.G. Yu, A.M.M. Khan, K. Loring. 2004, Computer Simulation of Transit Power Systems[C]. International Conference on Simulation: 93~98.

Zhang Dongliang, etc, 2008, Research on On-line Measuring Method of Transition Resistance in Power Supply Section of Urban Rail Transit [J]. Railway Locomotive and Vehicle, 2012, (6), 95–99. [8] Li Guoxin, etc. On-line measurement of key parameters of DC traction stray current [C]. 2008 National PhD Academic Forum, 11: 1912–1915.

Emerging Developments in the Power and Energy Industry – Dufo-López, Krzywanski & Singh (eds)
© 2020 Taylor & Francis Group, London, ISBN 978-0-367-27169-5

Study on the influence of AC/DC power supply on the stray current of the subway tunnel

Junhao Li & Xiaobin Cao
School of Electrical Engineering, Southwest Jiaotong University, Chengdu, Sichuan, China

Kunpeng Li & Xia Chen
Guangzhou Metro Design Research & Institute Co., Ltd., Guangzhou, Guangdong, China

ABSTRACT: There is stray current in the subway tunnel whether it is AC or DC. Reducing the influence of stray current is a problem that cannot be ignored in the process of subway design and construction. In this paper, the distribution law of stray current in subway tunnels under different power supply systems is studied. The results show that the stray current generated under the AC power supply system is much larger than the stray current under the DC system; the return lines of the two power supply modes have a certain shunting effect, so that the rail current decreases by about 1/5; In the DC power supply mode, the addition of the return line cannot effectively reduce the generation of stray current. In the AC power supply mode, the increase of the stray current after adding the return line has a certain effect.

Keywords: AC/DC, stray current, rail current, CDEGS

1 INTRODUCTION

From 2008 to 2018, the total length of subway operation lines in China was 4325.7km, an average annual increase of 472.8km (Zhao et al. 2019). For the DC-powered subway, the problem of stray current is becoming more and more prominent. Due to the influence of the track's own impedance and stray current, it will cause huge economic losses (Zhao et al. 2009, Gao et al. 1996). For example, in Hong Kong, the gas pipeline was corroded and perforated due to the stray current of the subway (Zhou et al. 1999, Lin, et al. 2002). For this reason, different scholars have studied the problem of stray current under the DC power supply system (Hu et al. 2011, Pang, et al. 2008, Dolara et al. 2012, Mei et al. 2017).

In recent years, China's subway has developed rapidly. The traditional DC power supply system is unable to meet the existing and future needs, so the demand for the AC power supply system has become urgent. Compared with the DC power supply system, the advantages of the AC system are obvious. First of all, the stray current of alternating current is mostly due to the electromagnetic coupling, and it will not corrode the metal. Secondly, it is not necessary to set the traction substation as frequently as the DC power supply (Li et al. 2015). Regardless of the power supply system, stray currents may affect the normal operation of the electrical grounding device, and even endanger human life. It is a problem that cannot be ignored in the process of designing and constructing the subway. Therefore, based on AC and DC power supply systems, this paper studies the distribution of stray current in subway tunnels.

2 ESTABLISHMENT OF SUBWAY TUNNEL MODEL

Based on the structure and related literature of a subway tunnel in China, this paper uses the grounding engineering analysis software CDEGS to establish a subway tunnel structure model composed of rail, soil structure, reinforced concrete structure and air tunnel model, and analyzes the distribution of subway stray current.

2.1 *Rail equivalent parameter*

At present, most urban rail transits use 60kg/m rails, which can be equivalent to a cylindrical section with a radius of 0.05m. The longitudinal resistance of the track is not only related to the material and size, but also to the number of joints and the welding process. According to the query literature and the actual environmental impact and measurement error, the longitudinal resistance of the rail is determined to be 0.04 Ω/km.

There is a transition resistance between the rail and the soil. During modeling, as an alternative to the transition resistance, an insulating coating having a fixed resistivity is added to the rail. The resistivity of the coating can be substituted for the transition resistance of the track by a formula.

$$R_g = \rho \cdot \int_{r1}^{r2} \frac{1}{2\pi r} \tag{1}$$

If it is a new subway line, the standard transition resistance is 15Ω•km. When the orbital radius is 0.05 m and the thickness of the insulating coating is 0.01 m, the resistivity of the coating obtained according to the above formula is 500000 Ω•m.

2.2 *Traction current and return current range*

In this paper, 15 km is selected as the traction return range. The standard current of the traction current is 2000A, which is powered by the traction substation. The injection current of the locomotive is the same as the current flowing into the traction substation.

The model consists of only one solid cylindrical conductor. The dimensions of the conductor are taken as a radius of 0.05 m, a length of 15 km, and an outer insulating layer thickness of 0.01 m. The buried depth of the track is taken as 20 m. A schematic of the model is shown below.

2.3 *Establishment of tunnel equivalent model*

Since there is an air tunnel in the subway tunnel, and the influence of the air tunnel on the overall model is unknown, it is necessary to establish an air tunnel model. According to the conventional conventions and design standards, the air tunnel length was set to 15.1 km, and the cross section was set to a square of 6 m × 6 m. If the air tunnel is regarded as a "soil" with a fixed and high resistivity, a "soil module" with a resistivity equal to the air resistivity is excavated in the earth to represent the air tunnel, and its resistivity is 1×1018 Ω•m.

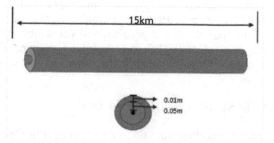

Figure 1.   Equivalent model of rail.

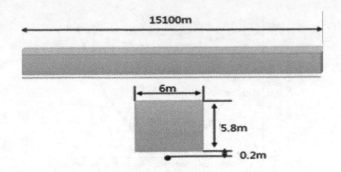

Figure 2. Equivalent model of air tunnel.

The construction methods of the subway are different in different sections of the tunnel. When the buried depth is deep, the mining method and the shield method are usually adopted; when the depth is shallow, the open cut method, the cover digging method, the shield method, and the sinking method are used. Taking the open cut method as an example, the main flow path of the stray current is composed of concrete and soil. The soil resistivity is taken as 200 $\Omega \cdot$m, and the concrete resistivity is taken as 800 $\Omega \cdot$m.

Based on the above discussion, a model is built in the CDEGS software. A perspective view of the tunnel model are as follows. This figure visually show the existence of structural reinforcement, rails, air tunnel, etc.

3 SIMULATION ANALYSIS OF STRAY CURRENT

At present, the known subway power supply systems are all DC systems. Many scholars have studied stray currents, but none of them involve the analysis of stray currents under the AC system. In order to investigate the difference between the DC system and the AC system power supply, the following simulation was performed.

3.1 *Only rail as a return channel*

The simulation results show that under two different power supply systems, the distribution of rail currents is similar. When the traction current is AC, the current drop on the rail is

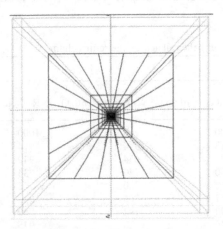

Figure 3. Subway structure model diagram.

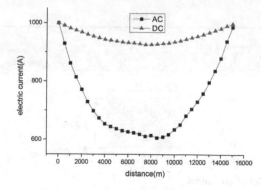

Figure 4. Rail current distribution in AC/DC power supply system.

more obvious. In the middle of the locomotive and the substation, the current is the lowest, which is 623A. The current of the rail in the AC system is much smaller than that in the DC system. The main reason is that the current in the structural reinforcement under the AC system is greater than the current in the DC system, and the mechanism is different: in the case of AC system, the traction current and the rail current will cause the induced current in the structural reinforcement in the tunnel due to the electromagnetic coupling; The structural reinforcement under the DC system do not generate induced current, but there is a leakage current that corrodes the rebar.

According to Figure 4, the single rail current near 7.5 km is the minimum, and the two rail currents are less than the traction current. According to the above figure, the stray currents at different positions under the two power supply systems are calculated by taking the rail at equal intervals. The statistics are shown in the table below.

According to the above table, the distribution of stray current under two different power supply systems can be drawn, as shown in the Figure 5. The distribution of stray current under AC system and DC system is similar, both increase first and then decrease, but the increase of stray current under AC system is larger. Taking the stray current at 7.5km from the locomotive as an example, the stray current under the AC system is about 5 times than that of the DC system.

3.2 Rails and return lines as return channels

The injection current of the locomotive is AC 2000A and DC 2000A respectively. On the basis of the simulation model of the second chapter, the return line is added together with the rail

Table 1. Stray current when only the rail acts as the return channel.

Power supply system	AC		DC	
Position(km) \ Current	Rail current (A)	Stray current(A)	Rail current (A)	Stray current(A)
0	1000	0	1000	0
1.5	813	374	974	52
3.5	673	654	949	102
5.5	630	740	934	132
7.5	610	780	925	150
9.5	617	766	932	136
11.5	704	592	950	100
13.5	836	328	975	50
15	986	28	998	4

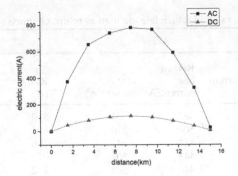

Figure 5. Distribution of stray current in subway.

as the return channels. In the model setting, the return line and the rail are connected by the suction line every 1200 meters.

As shown in the Figure 6 and 7, after the return line is added, a part of the traction current will flow back from the return line to the traction substation, and the amplitude of the current drop of the AC system will be larger than that of the DC system. Taking 7.5km from the locomotive as an example, after adding the return line, the rail current drop under the DC system is 153A, and the current drop ratio is 19.8%; the rail reduction value under the AC system is 83A, and the current drop ratio is 15.7%.

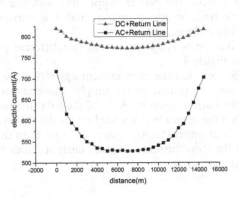

Figure 6. Rail current distribution.

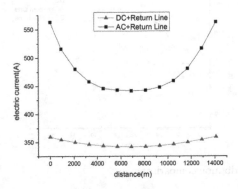

Figure 7. Current distribution of the return line.

660

Table 2. Stray current when rail and return line are used as return channels.

Power supply system	AC			DC		
Current Position(km)	Rail current (A)	Return line current(A)	Stray current(A)	Rail current (A)	Return line current(A)	Stray current(A)
0	717	562	4	819	360	2
1.5	615	500	270	801	350	48
3.5	550	453	447	786	345	83
5.5	528	443	501	775	342	108
7.5	527	440	506	772	340	116
9.5	531	450	488	775	344	106
11.5	553	475	419	786	349	79
13.5	616	530	238	801	356	42
15	710	563	17	817	359	7

According to the distribution of the current of the rail and the return line, the stray currents at different positions from the locomotive are as follows:

According to the Table 2, the current ratio of the return line at different positions can be calculated. The current flowing through the return line under AC power supply accounts for 22.0%~28.1% of the total traction current, and the current flowing through the return line under DC power supply accounts for 17.0%~18.0% of the total traction current. The return line has a certain reflow effect, and the return line's reflowing capability of the AC system is greater than the reflowing capability of the DC line. Because of the AC power supply system, the power supply line and the return line have mutual inductance due to the electromagnetic coupling, and the current of the return line is increased by the mutual inductance.

From Table 2, the distribution of stray currents in two different power supply systems can be plotted, as shown in the Figure 8.

From Figure 8, it can be seen that the stray current appears to be the maximum at 7.5 km from the locomotive. In the AC system power supply, the stray current is small when the return line is added, and the current drop is 13.7% of the total traction current; the stray current decreases by 1.7% when the return line is added under the DC system; Whether the line is added or not, the stray current under the AC system is larger than that under the DC system. At the maximum value of the stray current, the stray current under the AC system is about 5 times than that of the DC stray current.

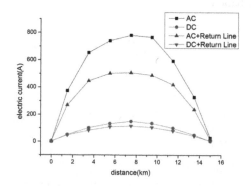

Figure 8. Stray current distribution comparison.

4 CONCLUSION

In this paper, the influence of AC/DC power supply system on the stray current of subway is studied. The main conclusions are as follows:

1) The generation of stray current under DC system is mainly due to the leakage current of the rail to the ground. The stray current under AC is mostly caused by the electromagnetic coupling between the power supply line and the rail and the structural reinforcement. Regardless of whether there is a return line, the stray current under AC is much larger than that of DC. At 7.5km from the locomotive, the stray current of the AC system is about 5 times that of the DC system;

2) The return lines of the two power supply modes have a certain shunting effect, and the ratio of the rail current decreases is about 1/5. Due to the electromagnetic coupling between the power supply line and the return line under the AC system, the current flowing through the return line during AC power supply accounts for 22.0%~28.1% of the total traction current. The current flowing through the return line during the DC system supply is 17.0%~18.0% of the total traction current;

3) In the DC power supply mode, the addition of the return line cannot effectively reduce the generation of stray current, and the stray current decreases by only 1.7%. Under the AC power supply mode, the reduction of stray current has a certain effect after adding the return line, and the drop is 13.7% of the total traction current.

REFERENCES

Zhao, X. & Gu, B.N. 2019. Statistics and analysis of China's urban rail transit operation lines in 2018. *Urban Rail Transit Research* 22 (01): 1–7.

Zhao, Y.H. & Zhou, X.J. 2009. Numerical analysis of the distribution of stray current in subway. *Urban Rail Transit* 12 (12): 42–47.

Gao, J.Y. & Yi, Y.X. 1996. Analysis of subway stray current. *Journal of Tianjin Institute of Technology* 12 (3): 52–55.

Zhou, X.J. & Gao, B. 1999. Experimental study on the corrosion of steel bars in reinforced concrete by subway turbulence. *Journal of the China Railway Society* 21(5): 99–105.

Lin, J. & Tang, H. &Yu, H.X. 2002. Subway turbulence corrosion and its protection technology. *Journal of Building Materials* 5(1): 72–76.

Hu, Y.J. & Zhong, Z. & Fang, J.P. 2011. Finite element simulation of subway stray current field. *China Railway Science* 32 (06): 129–133.

Pang, Y.B. & Li, Q.Z. & Liu, W. 2008. Research on metro stray current model based on electric field. *Urban Rail Transit Research* (02): 27–31.

Dolara, A. & Foiadelli, F. & Sonia, L. 2012. Stray current effects mitigation in subway tunnels. *IEEE Transactions on Power Delivery* Vol.27 (4).

Mei, J.W. & Lin, G.S. 2017. Analysis of subway stray current under multi-train operation. *Electrified Railway* 28 (04): 68–70.

Li, Q.Z. 2015. Urban rail transit AC traction power supply system and its key technologies. *Journal of Southwest Jiaotong University* 50(2): 199–207.

Emerging Developments in the Power and Energy Industry – Dufo-López, Krzywanski & Singh (eds)
© *2020 Taylor & Francis Group, London, ISBN 978-0-367-27169-5*

Research on transient voltage instability causes and counter-measures of multi-infeed HVDC receiving-end system

Junchuan Jia, Wuzhi Zhong, Jun Yi, Qi Wang, Chen Lv, Shigao Cheng & Shuaifei Zheng
China Electric Power Research Institute, Beijing, China

ABSTRACT: Along with the successive operation of high-capacity HVDC transmission projects, the voltage stability level of multi-HVDC-feed power system is decreasing continually. AC and DC faults can easily lead local grid or whole system voltage instability or even voltage collapse. Transient voltage stability has become a significant constraint for the stable operation of multi-HVDC-feed power grids. This paper firstly analyses various key factors related to transient voltage instability in the receiving side power grid, and points out the gradual development processes of transient voltage instability in multi-HVDC-feed power system with multi-factor-coupling effect after AC and DC faults. On this basis, the countermeasures which can effectively suppress transient voltage instability and improve the voltage stability level of the receiving end system are proposed. Finally, the effectiveness of the proposed measures is verified by simulation based on a practical grid.

1 INTRODUCTION

In order to meet the economic growth and electricity demand in Eastern China, long-distance and large-capacity UHVDC transmission has become an important means of China's West-to-East Power Transmission energy strategy, which plays the significant role in optimizing allocation of energy resources on a national scale and alleviating the pressure of air pollution. The successive commissioning of UHVDC projects has formed a whole picture of many DCs transmission out of the northwest power grid, many DCs feed into the central and eastern power grid, and strong transmission capacity of DC with weak AC capacity in China.

For the power system with multi-infeed HVDCs, on the one hand, large-capacity DC connected to the grid leads to the substitution of conventional synchronous generators in the network, resulting in the decline in the voltage support capability of the receiving system. On the other hand, due to the reactive power and voltage regulation characteristics of DC system, the DC inverter station needs to absorb more reactive power from AC system on the contrary when the AC bus voltage drops. The voltage regulation characteristics of the receiving end grid are further deteriorated. With the continuous increase of HVDC projects and their transmission capacity, the voltage stability of the receiving power grid is becoming increasingly prominent, which has become an important constraint for the stable operation of large power grid.

Voltage stability refers to the ability to maintain all buses' voltage in acceptable level with destabilization. At present, research on voltage stability of AC/DC system at home and abroad mainly focuses on static voltage stability analysis methods (Shao Yao & Tang Yong & Guo Xiaojiang & et al 2011; Lin Weifang & Tang Yong & Bu Quangquan 2008), voltage stability criteria and evaluation methods (Hou Junxian & Han Minxiao & Dong Yifeng & et al 2016; Xin Huanhai & Zhang Feng & Yu Yang & et al 2016; Yuan Zhichang & Liu Wenhua & Song Qiang 2009). However, there is few study on the causes and corresponding countermeasures of transient voltage instability in the receiving end system. For multi-infeed DC power grid, especially its key load center, three phase permanent fault of AC receiving section, DC unipolar or bipolar block fault, can easily cause the local or whole network voltage instability in the receiving side power grid, or even system voltage collapse.

Aiming at the problem and combining the characteristics of multi-feed DC power system, this paper firstly analyzes the causes of transient voltage instability from several aspects. On this basis, the control measures to deal with voltage instability of the multi-feed DC system are proposed. Finally, the effectiveness of the proposed control measures is verified by a practical grid example.

2 ANALYSIS OF TRANSIENT VOLTAGE INSTABILITY

The main reason of voltage instability in electric power system is the imbalance between reactive power supply and demand. Because of the large input total power to a multi-feed DC receiving end grid, the capacity of synchronous generators in the grid is reduced. So the dynamic reactive power reserve is insufficient and voltage support capability is poor. In the case of large disturbance of AC system such as short circuit or disconnection and so on, power flow redistribution and the dynamic characteristics of excitation system of the conventional generating units, motor load and HVDC transmission system and other factors continuously deteriorate the operating situation of the power grid, and eventually voltage instability is induced by the local induction motor instability (Liao Minchuan & Cai Guanglin & Zhang Yongjun 2009).

The reasons and development process of transient voltage instability in multi-feed HVDC power system can be roughly divided into the following stages.

The first stage: When a three-phase short circuit fault occurs at an important AC section of the grid or a DC blocking fault occurs, the active power flow is redistributed in the system, and part of the faulty line or DC power will be transferred through other tie lines, whose transmission active power and current increase. As the reactive power consumption of the line is directly proportional to the square of the current, the reactive power loss of the AC line is greatly increased, resulting in a sharp enlargement in the reactive power demand of the receiving end grid (Bai Yan & Chen Huixiang & Wang Zhonghong 2006).

The second stage: Under the circumstance of big disturbance, the conventional generator excitation system in the receiving end network is overexcited to improve the transient voltage level. However, due to the limitation of excitation winding capacity, when the excitation system of the unit reaches the maximum operating over-excitation time, the over-excitation limiter will reduce the excitation current to the rated value, resulting in the reduction of reactive power output and the further increase of reactive power shortage.

The third stage: It is known that the DC converter needs much reactive power, about 50% ~60% of the active power (Zheng Chao 2015; Li Zhaowei & Zhai Haibao & Liu Fusuo & et al 2016). In order to ensure that the reactive power exchange between DC and AC systems is very small, a large number of shunt capacitors are needed in DC converter stations under normal operation. However, because their reactive power output is proportional to the square of the voltage, the output reactive power of these shunt capacitors is greatly decreased when the converter bus voltage reduces. Hence the converter needs to absorb reactive power from AC system to maintain DC normal operation, which further makes the system voltage drop. Additionally, the voltage reduction of the inverter station is likely to lead to DC commutation failure. If the DC active power recovers quickly after the fault clearance, the converter station also needs to absorb a large number of reactive power form AC system, which is not conductive to voltage recovery and may cause DC subsequent commutation failure (Hou Jianlan & Ma Bing & Liu Yuquan & et al 2016).

The fourth stage: In order to absorb the impact energy during the fault period, the generators near the fault zone accelerate, and the relative groups of units in the grid swing to each other according to their natural oscillation frequencies. Then, the bus voltage of AC section drops, which usually locates at the oscillation center.

The fifth stage: As the most significant dynamic load in power system (Jing Yanqiang & Li Xingyuan & Guo Xiaoming & et al 2011), the electromagnetic torque of the induction motor decreases under the influence of system voltage dropping, while its rotor speed decreases and the slip increases. If the voltage dropping amplitude is increasing, the reactive power and voltage characteristics of induction motors make the terminal voltage drop continuously. Consequently the balance between mechanical torque and electromagnetic torque is destroyed, resulting in the motor speed decreasing and slip increasing further. Eventually the induction motor takes the lead in losing stability, and then deteriorate the operation state of other induction motors, and ultimately leads to the entire system voltage instability.

It should be noted that there is no definite time sequence for the five stages abovementioned. In fact, the various factors in each stage may interact with each other and interweave with each other, and eventually lead to local or whole network voltage collapse.

3 COUNTERMEASURES FOR TRANSIENT VOLTAGE INSTABILITY

From the above analysis, the fundamental cause of transient voltage instability in the receiving end system is the imbalance of reactive power supply and demand after large disturbance in the system. It is the result of combined regulating effects of multiple dynamic elements such as source, grid and load. In order to deal with the transient voltage instability of multi-feed HVDC power grid, the following measures can be taken into account.

3.1 *Increasing synchronous generator capacity and spinning reserve level in operation*

Increasing the capacity and spinning reserve level of synchronous generators in operation in the power grid, especially in the HVDC inverter station and important AC receiving sections, on the one hand can help to improve dynamic reactive power reserve, which will enhance the reactive power supply capacity during disturbance. On the other hand, the capacity increment of synchronous generators amounts to reduce the system equivalent impedance and improve system internal potential level after disturbance, which can strengthen the power system voltage support capability during disturbance. Therefore, the system voltage can be quickly restored to its initial value after the faults (Kundur P 1994; Li Mingjie 2016).

3.2 *Adding dynamic reactive power compensation devices*

Dynamic reactive power compensation device can provide reactive power support quickly when disturbance occurs, reactive power imbalanced, voltage drops, and help the grid restore reactive power balance and voltage stability. So, installing dynamic reactive power compensation devices, such as the synchronous condensers or static Var generators, in the HVDC inverter station and important AC receiving sections, can provide dynamic voltage support for the grid after faults to ensure the stability of bus voltage and improve the transient stability level of the system (Tang Yong & Zhong Wuzhi & Sun Huadong & et al 2010).

3.3 *Constraining AC and DC total received power*

With unchanged loads and spinning reserve level in the power system, constraining AC and DC total received power of the multi-infeed HVDC power grid is equivalent to increase the capacity of synchronous generators. This helps to improve dynamic reactive power reserve. In addition, reduce the active power of the feed-in HVDCs could decrease the reactive power consumption and filters capacity of the HVDCs. Consequently, the positive feedback characteristics of the reduction of filters reactive power under disturbance weakening the voltage stability level is weakened. So the bus voltage in the converter and near area is more easily recovered. Hence, under this measure, when the total input power of AC and DC to the multi-infeed HVDC power grid is limited, constraining the HVDC power is more effective than constraining the AC received power to improve the transient voltage stability level of the receiving-end power grid.

3.4 *Taking load shedding measure*

From abovementioned, it is known that the transient voltage instability is mainly caused by the reduced reactive power supply capability and increased reactive power demand of loads after large disturbance in the system. So we can take the measure of load shedding after large disturbance, especially to cut off the asynchronous induction motors, to decrease the reactive power absorbed by the induction motors. Hence the system can also achieve reactive power balance as soon as possible in the transient process after the disturbance, which will contribute to the rapid recovery of grid voltage.

4 CASE VERIFICATION

Take a practical multi-infeed HVDCs power grid of China for example to verify the effectiveness of the countermeasures proposed. Figure 1 shows the diagrammatic sketch of the multi-infeed HVDC power grid. There are 3 UHVDCs feed-into Provincial power grid D, one is from Regional power grid B and the others are from Regional power grid C. In addition, the Provincial power grid D and Regional power grid A are interconnected through 10 AC transmission lines, which constitute its AC receiving section for Provincial power grid D.

As can be seen from Figure 1, Provincial power grid D receives external power through 3 UHVDC projects and 10 AC transmission lines. In the research year, the load of the Provincial power grid D is 54 GW. The rated active powers of LG DC, ZY DC and YD DC are 10 GW, 10 GW and 4 GW respectively. Simulation results show that the maximum receiving power from 10 AC transmission lines is 18 GW, which is constrained by the thermal stability limits of AC lines XL and AC lines HB. In this case, if the foreign power occupies a relatively high proportion, like 50% and above, three-phase permanent N-1 or N-2 faults on double circuits on the same tower will lead voltage instability in Provincial power grid D. Figure 2 and Figure 3 shows the key buses voltage with the three-phase permanent N-1, N-2 faults of AC lines HQ under the 57% receiving power circumstance.

4.1 *Increasing synchronous generator and spinning reserve level capacity*

On the basis of the original mode, the system transient voltage stability is improved by increasing the unit-operating capacity of synchronous generators in the network. When the synchronous generator capacity in the network is increased to 26.2 GW and the rotary standby capacity is increased to 4.83 GW under the condition that all three DCs are fully delivered, the AC receiving section will receive 9.2 GW power in this way, and the voltage will remain stable after the AC/DC fault in the network, and the corresponding AC/DC total power receiving ratio will reach 60 % at this time. Figure 4 shows the bus voltage of the system after a three-phase permanent N-1 fault occurred on the 1000 kV HQ transmission line.

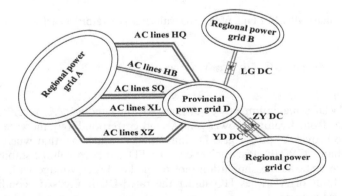

Figure 1. Multi-infeed HVDC power grid diagrammatic sketch.

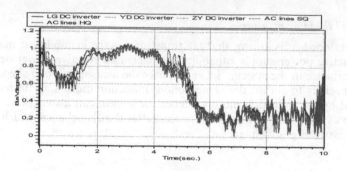

Figure 2.   Buses voltage plot under three-phase N-1 fault of AC HQ line.

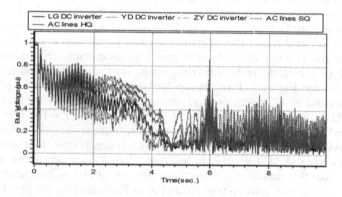

Figure 3.   Buses voltage plot under three-phase N-2 fault of AC HQ lines.

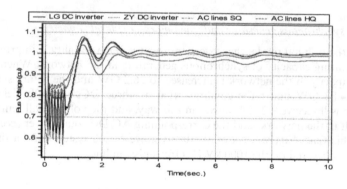

Figure 4.   Buses voltage after N-1 fault of line HQ with large generators capacity.

4.2   Constraining AC and DC total received power

4.2.1 Constraining AC power

Maintain 3 UHVDCs maximum power unchanged, only reduce AC received power of 10 transmission lines from Regional power grid A, and correspondingly increase the number of generators in the Provincial power grid D. Simulation results show that when the total input power is lower than 5 GW, the Provincial power grid D can keep voltage stability after AC or DC faults. Figure 5 shows the simulation plot of the key buses voltage with the three-phase permanent N-1 fault of AC lines HQ under the rated-DC-limited-AC condition, which is about 53% received power.

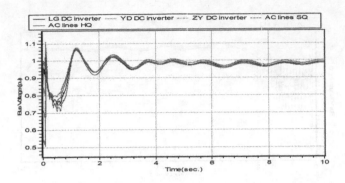

Figure 5.   Buses voltage with N-1 fault of line HQ under rated-DC-limited-AC condition.

4.2.2 *Constraining DC power*

Keep AC input power maximum, only reduce 3 UHVDCs total power, and correspondingly increase the number of generators in the Provincial power grid D. Simulation results show that when the total input power is lower than 15 GW, the Provincial power grid D can keep voltage stability after AC or DC faults. Figure 6 shows the simulation plot of the key buses voltage with the three-phase permanent N-1 fault of AC lines HQ under the max-AC-limited-DC condition, which is about 61% received power. Compared this calculated result with 4.2.1, reducing DC power is more effective for improving the transient voltage stability, which also directly validates the correctness described in chapter 3.3.

4.3 *Taking load shedding measure*

Under the basic mode, the AC/DC total power receiving ratio of the provincial power grid is about 56 %. If a three-phase permanent N-2 fault occurred on the HQ transmission line, the provincial power grid will collapse if no load shedding measures are taken. In the same mode, if a three-phase permanent N-2 fault occurred on the HQ transmission line, the system voltage can be quickly restored to the initial level by taking measures to remove 1 GW loads in the network. If the allowable load shedding is further increased after the failure, the transient stability of the system will be stronger. Figure 7 shows the buses voltage after a three-phase permanent N-2 fault occurred on the 1000 kV HQ transmission lines.

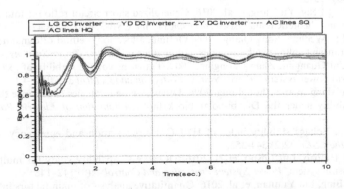

Figure 6.   Buses voltage with N-1 fault of line HQ under max-AC-limited-DC condition.

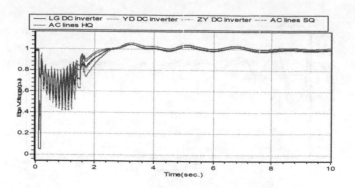

Figure 7.   Buses voltage under N-2 fault of lines HQ with 1GW loads shedding.

5   CONCLUSION

The fundamental reason of transient voltage instability in the receiving end is the imbalance of reactive power supply and demand after large disturbance occurring in the system. It is an extremely complicated dynamic process that power flow is redistributed after the large disturbance of AC system, excitation system of the conventional generating units, motor load and HVDC transmission system continuous regulating dynamically, and eventually lead to voltage instability induced by the local induction motor instability. It is the result of combined regulating effects of multiple dynamic elements such as source, grid and load.

To restrain the risk of transient voltage instability and improve the voltage stability level of the receiving power grid, the following measures can be applied: (1) increasing the capacity and spinning reserve level of synchronous generators in operation in the grid; (2) restraining the total input power of AC and DC feeding into the grid; (3) adding dynamic reactive power compensation devices near the DC inverter station and the key AC receiving sections; (4) increasing the load-shedding capacity.

REFERENCES

Shao Yao, Tang Yong, Guo Xiaojiang, et al. 2011. Transient voltage stability analysis of East China receiiing-end power grid with multi-infeed HVDC transmission lines. *Power System Technology* 35 (12): 50–55.

Lin Weifang, Tang Yong, Bu Quangquan. 2008. Study on voltage stability of multi-infeed HVDC power transmission system. *Power System Technology* 32(11):7–12.

Hou Junxian, Han Minxiao, Dong Yifeng, et al. 2016. Transient angle instability and voltage instability identification method based on oscillation center and voltage curve characteristic. *Power System Technology* 40(5):1509–1515.

Xin Huanhai, Zhang Feng, Yu Yang, et al. 2016. Generalized short circuit ratio for multi-infeed DC systems: definition and theoretical analysis. *Proceedings of the CSEE* 36(3):633–647.

Yuan Zhichang, Liu Wenhua, Song Qiang. 2009. Optimal allocation method of dynamic var compensation based on transient voltage stability index. *Automation of Electric Power Systems* 33(14):17–21.

Liao Minchuan, Cai Guanglin, Zhang Yongjun. 2009. Transient voltage stability of received power grid in AC/DC hybrid power systems. *Power System Protection and Control* 37(10):1–4 18.

Bai Yan, Chen Huixiang, Wang Zhonghong. 2006. Discussion on strategies to enhance the system transient voltage stability under the DC bi-polar block fault. *Automation of Electric Power Systems* 30 (15):93-96.

Zheng Chao. 2015. Voltage stability index of HVDC inverter station and emergency control strategy. *Porceedings of the CSEE* 32(2):344-352.

Li Zhaowei, Zhai Haibao, Liu Fusuo, et al. 2016. DC access capability study for multi-infeed HVDC power transmission system. *Power System Protection and Control* 44(8):142–148.

Hou Jianlan, Ma Bing, Liu Yuquan, et al. 2016. Quantitative analysis of main factors influencing transient voltage stability of power grid. *Power System and Clean Energy* 32(6):28–34.

Jing Yanqiang, Li Xingyuan, Guo Xiaoming, et al. 2011. A quick criterion for transient voltage stability considering induction motor load models. *Automation of Electric Power Systems* 35(5):10–14.

Kundur P. 1994. *Power system stability and control*. American:New York City.

Li Mingjie. 2016. Characteristic analysis and operational control of large-scale hybrid UHV AC/DC power grids. *Power System Technology* 40(4):985–991.

Tang Yong, Zhong Wuzhi, Sun Huadong, et al.2010. Study on mechanism of power system voltage stability. *Power System Technology*34(4):24–29.

Emerging Developments in the Power and Energy Industry – Dufo-López, Krzywanski & Singh (eds)
© 2020 Taylor & Francis Group, London, ISBN 978-0-367-27169-5

Estimation of shielding effectiveness of the lossy infinite plane with periodic circular apertures

L.L. Chen

State Key Laboratory of Advanced Power Transmission Technology, Global Energy Interconnection Research Institute, Beijing, China

W.X. Bai & C.Q. Jiao

State Key Laboratory of Alternate Electric Power System with Renewable Energy Sources, North China Electric Power University, Beijing, China

ABSTRACT: This paper focuses on the shielding effectiveness estimation of the periodic circular aperture array on a lossy infinite plane against a plane electromagnetic wave, while its wavelength is much larger than the central distance of the aperture. The circuit model and the analytical expression are proposed based on calculation for the conductor plane with periodic circular apertures and the lossy plane. The circuit model of the lossy plane with apertures is obtained by connecting the circuit of the lossy plane, which is obtained by the equivalent of the two-port network, in series with the equivalent impedance of the conductor plane with apertures. The expression of the lossy plane with apertures is obtained by doing some mathematical operation on expressions of lossy plane and conductor plane with apertures. Results shows that both the circuit model and the expression are in good agreement compared with the full-wave simulation.

1 INTRODUCTION

Due to the widespread use of electronic equipment sensitive to space electromagnetic disturbances, electromagnetic shielding as an electromagnetic disturbance suppression technology has received extensive attention (Garcia-Perez. et al. 2017, Vieira Valente. et al. 2017). Electromagnetic shielding is a practice of reducing the electromagnetic field in space by blocking the field with a metal material. The shielding effectiveness (SE) used to measure shielding performance is defined as

$$SE = 20 \log_{10}(1/|T|) \tag{1}$$

Where, T denotes the insertion loss of a shield and is calculated as the ratio of the electric (magnetic) field intensity with the shield loaded to that with the shield removed.

Electromagnetic shielding effectiveness of an infinite conductor plane with an array of apertures is a classical electromagnetic problem. However, due to ventilation, internal observation and other requirements, it is necessary to make some apertures on the plane, which are commonly circular. Previous studies have focused on the conductor plane with aperture and the lossy plane. Based on existing calculation methods for SE of two models, we study the electromagnetic shielding problem of a lossy infinite plane with periodic circular apertures, as shown in Figure 1. Wherein, d is the central distance of adjacent openings, a is the radius of circular apertures and t is the thickness of the plane. When the conductivity of the plane tends to infinity, the SE can be calculated by numerical methods and analytical methods (Chen.1971, Rubin. & Bertoni.1983, Ali. et al. 2005). Dehkhoda. et al. 2008 studied

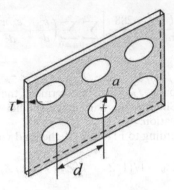

Figure 1. Conceptual diagram of a lossy infinite plane with periodic circular apertures.

the SE and the equivalent circuit of the plane with circular apertures. Similarly, there are some studies on the shielding effectiveness of the lossy plane (Dou. et al.1998, Nishizawa. & Hashimoto. 1999). This paper develops the circuit model and analytical expressions by combining the above two models to study the SE of the lossy infinite plane with circular apertures. The applicability of the circuit model and expressions to different conductivity is analyzed. This paper is organized as follows. Section II introduces the circuit model and the expression of conductor plane with circular apertures and lossy plane. The procedure to develop the circuit model of lossy plane with apertures and expressions of that are also described in this section. In section III, based on full-wave simulations (CST, Microwave studio), the applications of the circuit model and expressions are discussed. Section IV summarized the result of this paper.

2 THEORY

2.1 *The shielding effectiveness of an infinite conductor plane with periodic apertures*

An infinite conductor plane with periodic apertures against a plane wave is shown in Figure 2 (a,b) shows its equivalent circuit. The incident wave is represented by a voltage source with voltage U_s and internal impedance Z_0. Wherein Z_0 is the free-space intrinsic impedance. The plane with apertures is denoted by an admittance Y_s. According to Dehkhoda. et al. 2008, the effective of plane is equivalent to the admittance Y_s:

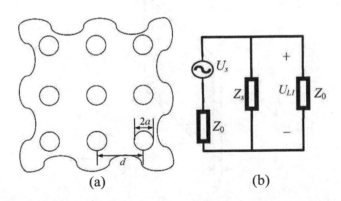

Figure 2. The infinite conductor plane with periodic circular apertures(a) and its equivalent circuit(b).

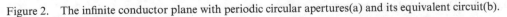

$$\frac{Y_s}{Y_0} = -j\frac{3d^2\lambda}{8\pi a^3} + j\frac{288}{4\pi\lambda a^2}\left[\sum_{m=0}^{\infty}\sum_{n=0}^{\infty}\left(\frac{\varepsilon_n^2}{d^2} + \frac{\varepsilon_n m^2}{d^2}\right)J_1^2(X)\right] \tag{2}$$

Wherein, $X = [2\pi a(m^2/d^2 + n^2/d^2)/2]^{1/2}/(m^2/d^2 + n^2/d^2)^{5/2}$. $Y_0 = 1/Z_0$. λ is the wavelength of the incident wave in free space, J_1 is the first term of the first type of Bessel function, and m, n are the mode index. Since a and d are much smaller than λ, the second term of (2) can be omitted during the calculation.

We assume that $U_S = 1$. According to Figure 2(b), the load voltage is

$$U_{L1} = 1/[Z_0 + 1/(1/Z_s + 1/Z_0)] \cdot 1/(1/Z_s + 1/Z_0) \tag{3}$$

With the conductor plane removed, the corresponding load voltage is $U_0 = 1/2$. Then, the T and SE of the plane are defined as, respectively

$$T_{PEC} = U_0/U_{L1} \tag{4}$$

$$SE_{PEC} = 20\log_{10}(1/|T_{PEC}|) \tag{5}$$

Because $Z_0 \gg Z_s$, (5) can be written as

$$SE_{PEC} = 20\log_{10}(3d^2\lambda/16\pi a^3) \tag{6}$$

2.2 *The shielding effectiveness of an infinite lossy plane*

An infinite lossy plane with thickness t against a plane wave is shown in Figure 3. The conductivity of the lossy material is σ, dielectric constant is $\varepsilon_e = \varepsilon - j\sigma/\omega$, where ω is the angular frequency. The electromagnetic field in the conductor plane is

$$\begin{cases} E_x(z) = E_x^+ e^{-\gamma z} + E_x^- e^{\gamma z} \\ H_y(z) = \frac{E_x^+}{Z_c} e^{-\gamma z} - \frac{E_x^-}{Z_c} e^{\gamma z} \end{cases} \tag{7}$$

Wherein, γ is the propagation constant of the lossy material, Z_c is the intrinsic impedance of the material, $Z_c = \sqrt{\mu/\varepsilon_e}$.

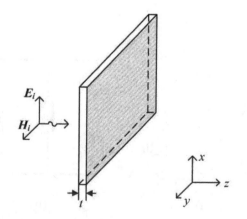

Figure 3.   The infinite lossy plane against a plane wave.

Let $E_x(0) = E_1, H_y(0) = H_1, E_x(d) = E_2, H_y(d) = H_2$. It can be deduced that

$$\begin{bmatrix} E_2 \\ H_2 \end{bmatrix} = \begin{bmatrix} \cosh \gamma t & -Z_c \sinh \gamma t \\ -\frac{\sinh \gamma t}{Z_c} & \cosh \gamma t \end{bmatrix} \begin{bmatrix} E_1 \\ H_1 \end{bmatrix} \tag{8}$$

The matrix above and relationship between two side of plane is equivalently represented by a two-port network shown as Figure 4.

The relationship between the two ports of the network is expressed as

$$\begin{bmatrix} U_2 \\ I_2 \end{bmatrix} = \begin{bmatrix} 1 + \frac{Z_1}{Z_2} & -Z_1 \\ -\frac{1}{Z_2}(2 + \frac{Z_1}{Z_2}) & 1 + \frac{Z_2}{Z_1} \end{bmatrix} \begin{bmatrix} U_1 \\ I_1 \end{bmatrix} \tag{9}$$

By comparing (9) with (8), it is deduced that

$$Z_1 = Z_c \sinh \gamma t \tag{10}$$

$$Z_2 = \frac{Z_1}{\cosh \gamma t - 1} \tag{11}$$

Therefore, the equivalent circuit of the lossy plane against a plane wave is shown as Figure 5. There is an analytical expression for solving the SE of the lossy plane (Schulz. et al, 1988)

$$SE_{lossy} = 20 \log \left\{ \left| \frac{(Z_0 + Z_c)^2}{4Z_0 Z_c} \cdot \left[1 - \left(\frac{Z_0 - Z_c}{Z_0 + Z_c} \right)^2 e^{-2\alpha t} e^{-j2\beta t} \right] \right| e^{\alpha t} \right\} \tag{12}$$

Where $\alpha + j\beta = \sqrt{j\omega\mu(\sigma + j\omega\varepsilon)}$.

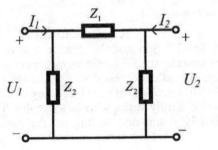

Figure 4.   The two-port network.

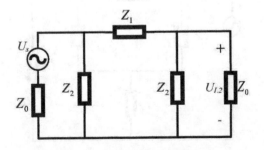

Figure 5.   The equivalent circuit of lossy plane against a plane wave.

674

2.3 The shielding effectiveness of an infinite lossy plane with periodic apertures

The effect of the lossy infinite plane with periodic apertures is equivalent to the effect of the plane with limited conductivity and the effect of the conductor plane with periodic apertures. Then the equivalent circuit of the lossy infinite plane with periodic apertures may be parallel connection in the two cases, as shown in the Figure 6.

By simplifying the circuit U_L is obtained. We assume that $U_S = 1$. With the conductor plane removed, the corresponding load voltage is $U_0 = 1/2$. Then, the $T_{l\text{-}h}$ and $SE_{l\text{-}h}$ of the plane are defined as, respectively

$$T_{l-h} = U_0/U_{L1} \tag{13}$$

$$SE_{l-h} = 20\log_{10}(1/|T_{l-h}|) \tag{14}$$

Figure 7 shows comparisons of $SE_{l\text{-}h}$, SE_{lossy}, and SE_{PEC} for different conductivities. The SE of lossy plane increases with the increase of the conductivity. For lossy plane with periodic apertures, the SE is always close to the minimum of other two cases. Let $SE_{PEC} = 20\log_{10}A_{PEC}$, $SE_{lossy} = 20\log_{10}A_{lossy}$. A_{PEC} and A_{lossy} can be obtained by (7) and (12). Thus, it is assumed that $1/A_{l-h} = 1/A_{lossy} + 1/A_{PEC}$, and the $T_{l\text{-}h}$ and $SE_{l\text{-}h}$ of the lossy plane with apertures are written as, respectively.

$$|T_{l-h}| = 1/A_{lossy} + 1/A_{PEC} = |T_{lossy}| + |T_{PEC}| \tag{15}$$

$$SE_{l-h} = 20\log_{10}(1/|T_{l-h}|) \tag{16}$$

3 RESULTS AND DISCUSSIONS

3.1 Application of the circuit model

In the following, the case of $d = 4$cm, $a = 1$cm is calculated. The circuit model is employed for $\sigma = 10$ S/m, 100 S/m, 1000 S/m, 10^4 S/m and 10^5 S/m in Figure 8, which shows the SE as a function of frequency. In general, the SE increases as the conductivity increases. When the conductivity is low, the shielding effectiveness does not change much with frequency. When the conductivity is high, the SE decreases with the frequency increases. When the conductivity is higher than 10^4 S/m, the SE hardly varies with conductivity. This is because the shielding effect in these cases is mainly dominated by the infinite conductor plane with periodic

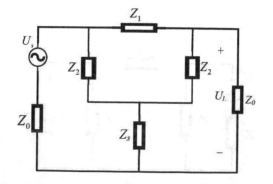

Figure 6.   The equivalent circuit of lossy plane with periodic apertures against a plane wave.

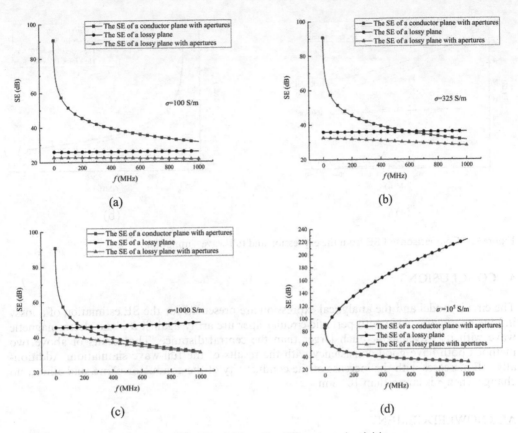

(a)

(b)

(c)

(d)

Figure 7.   Comparisons of SE_{l-h}, SE_{lossy}, and SE_{PEC} for different conductivities.

apertures. The difference between results of the circuit model and results of the full-wave simulation is within 5 dB.

The parallel model is employed for $\sigma = 10$ S/m, 100 S/m, 1000 S/m, 10^4 S/m and 10^5 S/m in Figure 9, which shows the SE as a function of frequency. A good agreement is obtained between the simulated results and the parallel model.

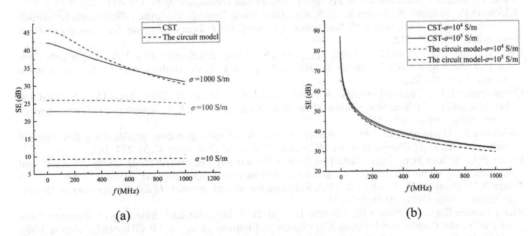

(a)

(b)

Figure 8.   Comparisons of SE from the circuit model and full-wave simulation.

676

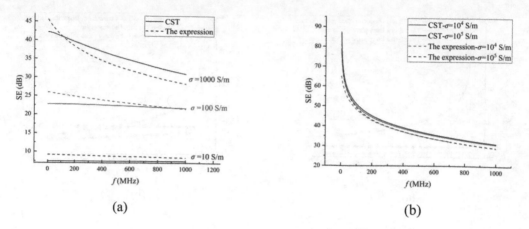

(a)   (b)

Figure 9.   Comparisons of SE from the expression and full-wave simulation.

4   CONCLUSIONS

The circuit model and the analytical expression are presented for the SE estimation of a lossy infinite conductor plane with periodic circular aperture array against a plane electromagnetic wave with its wavelength much larger than the central distance. The results of above two methods both have good consistency with the results of the full-wave simulation. Additionally, SE increases with the increase of the conductivity of the conductor plane, and almost no change when σ is larger than 10^4 S/m.

ACKNOWLEDGEMENT

This project is supported by State Key Laboratory of Advanced Power Transmission Technology (GEIRI-SKL-2018-006).

REFERENCES

Ali S., Weile D.S., Clupper T. 2005. Effect of near field radiators on the radiation leakage through perforated shields. *IEEE Transactions on Electromagnetic Compatibility* 47(2): 367–373.

Chen C.C. 1971. Diffraction of Electromagnetic Waves by a Conducting Screen Perforated Periodically with Holes. *IEEE Transactions on Microwave Theory and Techniques* 19(5): 475–481.

Dehkhoda P., Tavakoli A., Moini R. 2008. An Efficient and Reliable Shielding Effectiveness Evaluation of a Rectangular Enclosure with Numerous Apertures. *IEEE Transactions on Electromagnetic Compatibility* 50(1): 208–212.

Dou W.P., Bai T.Y., Zhang L.C. 1998. Analysis on Shielding Effectiveness of a Thick Imperfectly Conducting Plane with a Slit Using the Equivalent Magnetic Current Method. *Chinese Journal of Radio Sciense* 13(4): 398–402.

Garcia-Perez L.G., Lozano-Guerrero A.J., Blazquez-Ruiz J.M. et al. 2017. Time-Domain Shielding Effectiveness of Enclosures Against a Plane Wave Excitation. *IEEE Transactions on Electromagnetic Compatibility* 59(3): 789–7396.

Nishizawa S., Hashimoto O. 1999. Effectiveness analysis of lossy dielectric shields for a three-layered human model. *IEEE Transactions on Microwave Theory and Techniques* 47(3): 277–283.

Rubin B.J., Bertoni H.L. 1983. Reflection from a Periodically Perforated Plane Using a Subsectional Current Approximation. *IEEE Transactions on Antennas and Propagation* AP-31 (6): 829–836.

Schulz R.B., Plantz V.C., Brush D.R. 1988. Shielding theory and practice. *IEEE Transactions on Electromagnetic Compatibility* 30(3): 187–201.

Vieira Valente R.D., De Ruijter C., Vlasveld D. et al. 2017. Setup for EMI Shielding Effectiveness Tests of Electrically Conductive Polymer Composites at Frequencies up to 3.0 GHz. *IEEE Access* 5(99): 16665–16675.

Emerging Developments in the Power and Energy Industry – Dufo-López, Krzywanski & Singh (eds)
© 2020 Taylor & Francis Group, London, ISBN 978-0-367-27169-5

An energy hub based planning method for a combined heat and power based microgrid

Deshun Wang & Shanglin Zhao
New Energy Research Institute of Nanjing Branch of China Electric Power Research Institute Co., Ltd.,
Nanjing, China

Wanqing Chen
Key Laboratory of Smart Grid of Ministry of Education, Tianjin University, Tianjin, China

ABSTRACT: In recent years, distributed energy resources have been gradually applied and popularized because of their great capability of optimizing energy structure and reducing environmental pollution. For this reason, optimizing the configuration of the combined heat and power (CHP) based microgrid is of great significance. Therefore, based on the concept of the energy hub, this article proposes a planning model for determining the optimal configuration and dispatch strategy of a CHP-based microgrid.

1 INTRODUCTION

In recent years, the installed capacity of distributed energy resources (DERs) in the power system has steadily increased. DERs include photovoltaic (PV) systems, combined heat and power (CHP) units, and so on (Jing et al. 2012). But dispatching DERs with intermittent characteristics is difficult, which restricts their development in the future. The microgrid has been widely adopted in order to address this problem. The microgrid mainly includes the DERs, load, and monitoring devices. The microgrid plays a noticeable role in realizing the flexible and efficient application of DERs. The CHP-based microgrid is the most common type of microgrid in terms of protecting the environment and improving energy efficiency.

In the meantime, coordination among different energy supply systems (such as power, heat, and cold) is becoming more and more important, as the CHP-based microgrid is a complex system with multiple energy inputs and outputs (Zhang et al. 2014). The energy hub (EH) is a modeling method proposed by Geidl and Andersson (2007) from ETH, which can be used to depict the energy conversion and allocation ratio between the input and output of different types of energies.

This article develops an EH-based planning method considering the operation process for the CHP-based microgrid. A case study is used to demonstrate the effectiveness of the proposed planning method.

2 MODELING OF A CHP-BASED MICROGRID

2.1 *Objective function*

The minimal annual cost of the CHP-based microgrid (Guo et al. 2013) is used as one of the objective functions, defined as

$$f_c = C_I^{ann} + C_O^{ann} + C_M^{ann} \tag{1}$$

where f_c is the net annual value of total cost; C_I^{ann}, C_O^{ann}, and C_M^{ann} are the annual capital cost, annual operating cost, and annual maintenance cost, respectively.

The carbon dioxide emission in a life cycle is regarded as the minimum objective function with respect to environmental issues.

$$f_e = \beta_{CO_2,\, grid} W_{E,\, grid}^{ann} + \beta_{CO_2,\, fuel} W_{T,\, fuel}^{ann} \tag{2}$$

where f_e is the amount of carbon dioxide emission; $W_{E,\, grid}^{ann}$ is the annual electricity purchased from the grid; $W_{T,\, fuel}^{ann}$ is the annual natural gas consumption; $\beta_{CO_2,\, grid}$ and $\beta_{CO_2,\, fuel}$ are the emission factors of electricity from grid and natural gas, respectively.

2.2 Energy conversion technology constraints

The relationship between the input and the output of the energy conversion technologies is described by a linear equation:

$$P_{out,i} = \eta_i P_{in,i} \tag{3}$$

where $P_{in,i}$ and $P_{out,i}$ refers to the input and output of energy technology type i, and η_i refers to the conversion efficiency.

The output of conversion technologies should not excess its capacity limit, which is depicted as:

$$P_{out,i} \leq P_{out,i}^{max} \tag{4}$$

where $P_{out,i}^{max}$ is the capacity of technology i.

To make sure that the supply of different kinds of energy matches the demand, the load balance constraints is shown in Equation (5).

$$L_k = \sum_i \eta_i P_{in,i} + P_{k,D} - P_{k,C} - P_{exp,k} \tag{5}$$

where L_k refers to the demand of energy type k; $P_{k,D}$ and $P_{k,C}$ refer to the energy discharged and charged power from the energy storage systems; and $P_{exp,k}$ refers to the quantity of energy type k exported from the microgrid.

2.3 Energy storage technology constraints

The characteristic of energy storage technology is described as Equation (6):

$$W_i^1 = W_i^0 (1 - \sigma_i) + (P_{i,C} \eta_{i,C} - \frac{P_{i,D}}{\eta_{i,D}}) \Delta t \tag{6}$$

where W_i^0 and W_i^1 are the state of charge (SOC) of storage technologies i; σ_i is the loss ratio; $\eta_{i,C}$ and $\eta_{ES,D}$ are the charge and discharge efficiency, respectively; and Δt is the simulation step size.

The charge and discharge power of technology should not exceed the limits, which are depicted as follows:

$$0 \leq P_{i,C} \leq P_{i,C}^{\max} \tag{7}$$

$$0 \leq P_{i,D} \leq P_{i,D}^{\max} \tag{8}$$

The SOC of storage technologies i should not exceed its capacity limits:

$$W^{\min} \leq W_i^0 \leq W^{\max} \tag{9}$$

The initial value of SOC should be the minimal value of SOC or the same as the final value:

$$W_i^0 = W^{\min} \tag{10}$$

$$W_i^0 = W_i^N \tag{11}$$

2.4 *Other constraints*

The electricity purchased from the power grid should not exceed the capacity of the tie line between the grid and CHP-based microgrid. The constraint is shown as

$$0 \leq P_{\text{grid}} \leq P_{\text{grid}}^{\max} \tag{12}$$

where $P_{\text{grid}}^{\max}$ is the maximum capacity of the tie line.

3 SOLUTION METHOD

The mixed integer linear programming (MILP) model established in the preceding text can be solved as follows:

Step 1: Create the input data files for the model.

Step 2: Use MATLAB® to read the data and make it meet the requirements of subsequent modeling.

Step 3: Establish the optimal model of the CHP-based microgrid system. The algorithm is shown in Figure 1.

Step 4: The model generated in Step 3 is solved by a genetic algorithm (GA) and the output is presented in Excel® form.

Step 5: Visualize the output.

4 CASE STUDY

In this article, a park in the southern China is used for demonstrating the developed model and method. The microgrid consists of a gas boiler, a heat pump, a PV generator, a CHP unit, a battery, and a hot water tank. The economic and technical parameters of the equipment (Calise et al. 2017) are presented in Tables 1 and 2.

The microgrid supplies the electricity and heat to the users. The energy load profiles and the intensity of solar radiation of this park are depicted in Figures 2 and 3.

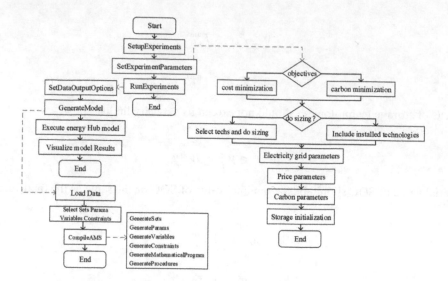

Figure 1. Flow chart of planning method for a CHP-based microgrid.

Table 1. Economic parameters of the facilities in a CHP-based microgrid.

Type of equipment	Variable initial investment, $/kW(or m^2)	Variable maintenance price, $/(kW·h)
CHP	2870	0.021
Gas boiler	730	0.01
Heat pump	2270	0.1
PV	325	0.06
Battery	430	—
Hot water tank	40	—

Table 2. Technical parameters of the facilities in CHP based microgrid.

Type of equipment	Efficiency/heat to power ratio	Largest charging and discharging ratios	Largest capacity	Life (years)
CHP	0.3/1.73:1	—	4000	20
Gas boiler	0.94	—	3000	30
Heat pump	3.2	—	5000	20
PV	0.17	—	12,500	25
Battery	0.9	0.3	100,000	20
Hot water tank	0.9	0.25	100,000	20

The electricity and natural gas prices and carbon dioxide emissions conversion factors are shown in Table 3.

To prove the effectiveness of the proposed planning method, a comparison is conducted between the planning method proposed in this article and the traditional planning method of allocating the corresponding equipment capacity according to the maximum demand of various types of energies (Wu et al. 2015). The system configuration is shown in Table 4 and the dispatch results are shown in Figures 4–7.

As shown in Figures 4–7, the capacity of the CHP unit is significantly reduced under the proposed planning method. In consideration of the economic and environmental benefits, the capacity of PV is optimized to be as large as possible. A major drawback of a heat pump is

Table 3. Electricity and natural gas prices and carbon dioxide emissions conversion factors.

Type	Price ($)	Emissions conversion factors (kg/(kW·h))
Electricity	0.24	0.13
Renewable electricity	0.2	
Natural gas	0.09	0.19

Table 4. Configurations of system.

Capacity(kW)/area (m^2)	CHP	Gas boiler	Heat pump	PV	Batteries	Hot water tank
Traditional planning method	1500	0	1500	12,500	5000	5000
Proposed planning method	400	2050	0	12,500	0	6850

Table 5. Comparison results between the two configurations.

Method	Traditional method	Proposed planning method
Annual capital cost (M$)	0.84	0.65
Annual operating cost (M$)	1.57	1.64
Annual maintenance cost (k$)	419.62	300.35
Annual electric charge (k$)	504.12	751.53
Annual income (k$)	269.19	225.71
Total cost of the planning cycle (M$)	43.08	41.23
Annual total cost (M$)	2.16	2.06
Annual carbon emission (t)	2642.05	2389.53

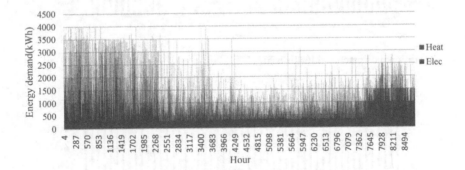

Figure 2. Energy load.

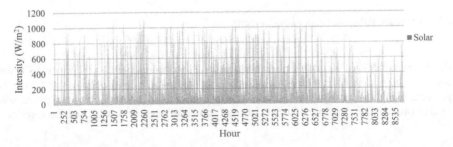

Figure 3. Intensity of solar radiation.

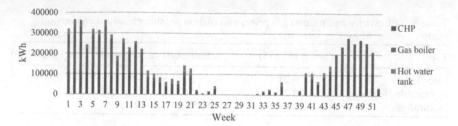

Figure 4.   Optimal dispatch results for heat.

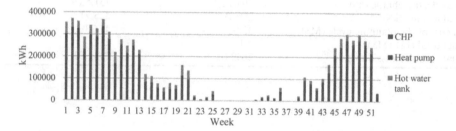

Figure 5.   Optimal dispatch results for electricity.

Figure 6.   Traditional dispatch results for heat.

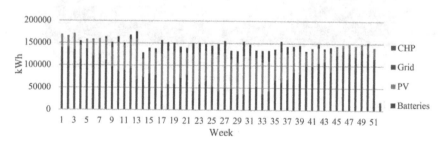

Figure 7.   Traditional dispatch results of electricity.

that the investment cost is not reasonable despite its high efficiency. In order to lower the investment cost, a gas boiler takes the place of a heat pump in load balancing. As for energy storage technology, a battery is not suitable for the microgrid under its current price, whereas a hot water tank plays an important role in peak load shifting and investment cost reduction.

5 CONCLUSIONS

In this article, an energy hub based planning method for a CHP-based microgrid is proposed. The method is proved to be correct and effective in determining the optimal configurations and dispatch strategy by a case analysis, since the cost and carbon dioxide emission is significantly reduced. The planning method, which is flexible and feasible, applies to different structures of CHP-based microgrids. Future research will focus on a planning method that takes the uncertainty of renewable energy into account.

ACKNOWLEDGMENTS

This project is supported by the Qinghai Science and Technology Department (2018-GX-A6).

REFERENCES

Calise, F., Dentice D'Accadia, M., Libertini, L., Quiriti, E., Vanoli, R., & Vicidomini, M. 2017. Optimal operating strategies of combined cooling, heating and power systems: a case study for an engine manufacturing facility. *Energy Conversion and Management* S0196890417305733.

Geidl, M., & Andersson, G. 2007. Optimal power flow of multiple energy carriers. *IEEE Transactions on Power Systems* 22(1): 145–155.

Guo, L., Liu, W., Cai, J., Hong, B. W., & Wang, C. S. 2013. A two-stage optimal planning and design method for combined cooling, heat and power microgrid system. *Energy Conversion & Management* 74(10): 433–445.

Jayasekara, S., Halgamuge, S. K., Attalage, R. A., & Rajarathne, R. 2014. Optimum sizing and tracking of combined cooling heating and power systems for bulk energy consumers. *Applied Energy* 118: 124–134.

Jing, Y., Bai, H., & Zhang, J. 2012. Multi-objective optimization design and operation strategy analysis of a solar combined cooling heating and power system. *Zhongguo Dianji Gongcheng Xuebao/Proceedings of the Chinese Society of Electrical Engineering* 32(20): 82–87.

Wu, H., Wang, D., & Liu, X. 2015. Strategies evaluation and optimal allocation of combined cooling heating and power system with solar. *Automation of Electric Power Systems.*

Zhang, X. J., Karady, G. G., & Ariaratnam, S. T. 2014. Optimal allocation of CHP-based distributed generation on urban energy distribution networks. *IEEE Transactions on Sustainable Energy* 5(1): 246–253.

Emerging Developments in the Power and Energy Industry – Dufo-López, Krzywanski & Singh (eds)
© 2020 Taylor & Francis Group, London, ISBN 978-0-367-27169-5

A scheduling method using a water pump and reservoir for renewable energy power accommodation

Qiong Tao & Bo Yang
*New Energy Research Institute of Nanjing Branch of China Electric Power Research Institute Co., Ltd.,
Nanjing, China*

Mingjun Wang & Shiheng Song
Key Laboratory of Smart Grid of Ministry of Education, Tianjin University, Tianjin, China

ABSTRACT: A scheduling method using water pumps and reservoirs in the Water Distribution System (WDS) for renewable energy power accommodation is proposed in this article. The Power Distribution System (PDS) and WDS are coupled as an energy–water nexus. The power consumption of water pumps and the water storage of a reservoir are optimized at the same time to reduce the electricity cost and accommodate renewable energy power, considering the constraints from both the PDS and WDS. A case study is carried out in this article, and the results show that the method proposed in this article can significantly improve the renewable energy power accommodation capability of PDS, as well as reduce the electricity cost of WDS in a flexible way.

Keywords: energy–water nexus, scheduling method, water storage, renewable energy power accommodation

1 INTRODUCTION

The Power Distribution System (PDS) and the Water Distribution System (WDS) are both important urban energy infrastructures. In the traditional scheduling method, these two systems are operated separately. On the one hand, the PDS regards WDS only as a load node, ignoring its flexible response capability. On the other hand, WDS also ignores the operation security and the economy of the PDS (Kurek & Ostfeld 2013; Zhang et al. 2013). The above problems lead to excessively high electricity costs in the WDS and renewable energy power accommodation issues in the PDS (Chen et al. 2016; Li et al. 2017). In order to solve this problem, the WDS and PDS are coupled into an energy–water nexus (Macknick 2011).

In recent years, many researchers have studied the energy–water nexus system. (Yoon 2018) introduced an energy–water nexus system from the perspectives of electricity consumption and water consumption, and proposed a generalized energy–water nexus system framework that includes social, economic, and political factors. Siddiqi and Anadon (2011) put forward the idea of building a photovoltaic or wind power plant together with a seawater desalination plant. Part of the renewable energy power generation is for sale, and the other part is for seawater desalination treatment. However, the aforementioned research focuses mainly on the overall planning of the two energy systems, and does not consider the WDS and PDS as a whole, ignoring the relevant constraints at the network side (Hamiche et al. 2016). Oikonomou and Parvania (2018) proposed a day-ahead optimal dispatching model based on the flexible adjustment ability of the WDS, which focuses on the coordination between the WDS and PDS. However, the coordination of water pumps and reservoirs in the WDS for renewable energy power accommodation have not been fully studied and the flexible response capability from the water pumps and reservoirs in the WDS is not well utilized.

In this article, the WDS and PDS model including renewable energy power is established, and then a scheduling method using a water pump and reservoir for renewable energy power accommodation considering the constraints from both the PDS and WDS is established, aiming at the minimum operation cost of the coupling system. Simulation results on a typical coupling PDS and WDS system show that the proposed method can effectively reduce the operation cost of the coupling system and improve the consumption level of renewable energy power.

2 MODELING OF A WDS AND PDS COUPLING SYSTEM

2.1 *The coupling of WDS and PDS*

Figure 1 shows a schematic diagram of a WDS and PDS coupling system. The Integrated Energy Service Operator (IESO) purchases electricity from the upstream power grid, dispatches the water from the reservoirs, and provides energy services of water and electricity to the users. In this system, the water pump works a coupling node between the PDS and WDS. The power consumption of water pumps will impact the operation of both the PDS and WDS.

Also, the water storage in a reservoir in a WDS is able to function as virtual energy storage for both the PDS and WDS. For example, when the electricity price is low (or the renewable energy power output is high), the water pump will increase its power for renewable energy power accommodation and pump the water into the reservoir. Conversely, the water pump will reduce its power for electricity cost reduction.

The coordinated operation of reservoir and water pump is the basis for the scheduling method in this article.

2.2 *Modeling of WDS for scheduling*

In this article, a scheduling cycle [0, T] is composed of a continuous and equal scheduling period Δt. τ is the scheduling period number during a scheduling cycle, $\tau=1,2,3,\ldots,N$. During a scheduling period Δt, the water flow velocity in the pipeline is constant. The structure diagram of the WDS is shown in Figure 2. In the WDS, the core of the hydraulic calculation is the flow continuity equation and water head equation, as shown in Equations (1) and (2).

$$\sum_{j \in i} Q_{ij,\tau} - q_{i,\tau} = 0 \tag{1}$$

$$\left(h_{i,\tau} + \overline{h_i}\right) - \Delta h_{ij,\tau} = \left(h_{j,\tau} + \overline{h_j}\right), \forall \tau \tag{2}$$

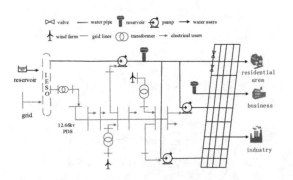

Figure 1.   Diagram of a WDS and PDS coupling system.

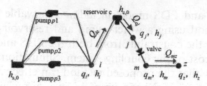

Figure 2. Sketch map of a water distribution network.

where $q_{i,\tau}$ is the water requirement at node i in the time period τ, $i = 1,2,\ldots,n_w$; $Q_{ij,\tau}$ is the flow of pipeline segment ij in the time period τ; $\overline{h}_i$ is the altitude at node i; $h_{i,\tau}$ is the water head at the node i in the time period τ; $\Delta h_{ij,\tau}$ is the water head loss of pipeline segment ij; $j \in i$ indicates that node j is connected with node i. In this article, it is assumed that the reservoir capacity for water storage is large enough, and the water head remains approximately constant.

The water head loss of pipeline segment ij in the time period τ is calculated by means of a Darcy–Weisbach formula, with details provided in Lewis and Rossman). The water head loss of valve can be calculated as Equation (3).

$$\Delta h_{jm,\tau} = m_{jm} Q_{jm,\tau}^2, \forall \tau \tag{3}$$

where $\Delta h_{jm,\tau}$ is the water head loss of valve jm in the time period τ; m_{jm} is the coefficient of the water head loss.

In order to ensure sufficient water supply, the water head should be higher than the most unfavorable water head at any node where water requirements are more than 0 except the reservoir, as shown in Equation (4) .

$$h_{i,\tau} > h_{i,\min} \tag{4}$$

The water head change of reservoir c in the time period τ is shown in Equation (5).

$$h_{c,\tau} = h_{c,\tau-1} + \left(\Delta t \sum_{i \in c} Q_{ic,\tau-1} / A_c \right) \tag{5}$$

where A_c is the average cross-sectional area of reservoir c and $Q_{ic,\tau-1}$ is water flow of the pipeline i connected with reservoir c in the time period $(\tau-1)$.

The capacity limits of the reservoir are shown in Equation (6). The inlet and outlet flow limits are shown in Equation (7) respectively. The inlet and outlet flows of reservoir c are calculated by means of Equation (8).

$$h_{c,\min} \leq h_{c,\tau} \leq h_{c,\max} \tag{6}$$

$$\begin{cases} R_{c,\tau} \leq \overline{Q_c} \\ |Z_{c,\tau}| \leq \overline{Q_c} \end{cases} \tag{7}$$

$$\begin{cases} R_{c,\tau} = \sum_{i \in c} Q_{ic,\tau-1}, \forall Q_{ic,\tau-1} \geq 0 \\ Z_{c,\tau} = \sum_{i \in c} Q_{ic,\tau-1}, \forall Q_{ic,\tau-1} < 0 \end{cases} \tag{8}$$

The total amount of water storage in the reservoir should be more than 0, as shown in Equation (9):

$$\sum_{k=1}^{n_t} \left(\Delta t \sum_{i \in c_k} Q_{ic_k, \tau-1} \right) > 0 \tag{9}$$

In addition, in order to avoid affecting the operation in the next scheduling cycle, the water storage of the reservoir at the end of this scheduling cycle should not be less than the initial water storage, as shown in Equation (10):

$$\sum_{\tau=2}^{N+1} \left(\Delta t \sum_{i \in c_k} Q_{ic_k, \tau-1} \right) \geq 0 \tag{10}$$

The power consumption $E_{p,\tau}$ of the water pump can be calculated as Equations (11) and (12).

$$E_{p,\tau} = -(\rho g \Delta h_{p,\tau} Q_{p,\tau} \Delta t) / \eta_p, \forall \tau \tag{11}$$

$$\Delta h_{p,\tau} = -(\omega_{p,\tau})^2 \cdot (h_0 - r_p (Q_{p,\tau} / \omega_{p,\tau})^{n_p}), \forall \tau \tag{12}$$

where ρ and g are the density of water and the standard gravity coefficient; η_p is the efficiency of a pump; $\Delta h_{p,\tau}$ is the water head loss of pump in the time period τ; h_0 and r_p are pump parameters; and $\omega_{p,\tau}$ is the relative speed of the pump in the time period τ.

2.3 Modeling of WDS and PDS coupling links

Water pump power consumption is the basis of coupling WDS with PDS. The power balance of PDS with a water pump can be shown as Equation (13).

$$P_{ri,\tau} - P_{i,\tau,\text{load}} - E_{pi,\tau} = V_{i,\tau} \sum_{j \in i} V_{j,\tau} [G_{ij} \cos(\theta_{ij,\tau}) + B_{ij} \sin(\theta_{ij,\tau})]$$

$$Q_{ri,\tau} - Q_{i,\tau,\text{load}} - \beta E_{pi,\tau} = V_{i,\tau} \sum_{j \in i} V_{j,\tau} [G_{ij} \sin(\theta_{ij,\tau}) - B_{ij} \cos(\theta_{ij,\tau})] \tag{13}$$

Equations (11)–(13) are the key and foundation to realizing the coupling of WDS and PDS. Based on it, the scheduling method using a water pump and reservoir can be achieved.

3 A SCHEDULING METHOD USING A WATER PUMP AND RESERVOIR

The objective of the scheduling method using a water pump and reservoir is to minimize the operating costs, as shown in Equation (14). Operating costs include PDS operating cost, pump operation cost, and the abandoned wind penalty. The renewable energy power is wind power in the article.

$$\min C = \sum_{\tau=1}^{N} \left(C_{\text{elec},\tau} + C_{\text{DER},\tau} + C_{\text{pump},\tau} \right) \tag{14}$$

where $C_{\text{elec},\tau}$ is PDS operating cost; $C_{\text{pump},\tau}$ is pump operation cost; $C_{\text{DER},\tau}$ is the abandoned wind penalty cost.

$C_{\text{elec},\tau}$ includes the purchasing power cost and line loss cost, as shown in Equations (15) and (16).

$$C_{\text{elec},\tau} = C_{\varsigma,\tau} W_\tau + C_{l,\tau} \sum_{l=1}^{n_l} \Delta S_{l,\tau} \tag{15}$$

$$\Delta S_{l,\tau} = G_{ij}^{-1} \left(\left| P_{ri,\tau} - P_{i,\tau,\text{load}} - E_{pi,\tau} \right| / \left| V_{i,\tau} \right| \right)^2 \Delta t \tag{16}$$

where $C_{\varsigma,\tau}$ is electricity price in the time period τ; W_τ is the purchasing power in the time period τ; $C_{l,\tau}$ is line loss cost per unit; and $\Delta S_{l,\tau}$ is line loss at line l in the time period τ.

$C_{\text{DER},\tau}$ is determined as Equation (17):

$$C_{\text{DER},\tau} = C_{\text{pen}} \left(W_{av,\tau} - P_{r,\tau} \right) \tag{17}$$

where C_{pen} is the abandoned wind penalty coefficient; $W_{av,\tau}$ is the wind power prediction output in the time period τ; and $P_{r,\tau}$ is the wind power actual output in the time period τ.

$C_{pump,\tau}$ is determined as Equation (18):

$$C_{\text{pump},\tau} = \sum_{k=1}^{n_v} C_\tau E_{p_k,\tau} \tag{18}$$

where C_τ is the electricity price of pump power and $E_{p_k,\tau}$ is the electricity consumption of pump in the time period τ.

The constraints of this model are shown in Equations (1)–(13).

4 CASE STUDIES

4.1 *Parameters*

A coupling case of the IEEE-33 node distribution network with wind power (Baran & Wu 1989) and the improved Richmond water distribution network (Guidoin et al. 2013) is constructed in this article, with details provided in Baran and Wu (1989) and Guidoin et al. (2013). The daily load of the PDS and the wind prediction power are respectively shown in Figures 3 and 4.

In order to study the effectiveness of the coordinated operation of reservoir and water pump, two cases are stimulated. The scheduling cycle is 24 hours, and the scheduling period Δt is 1 hour.

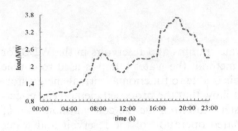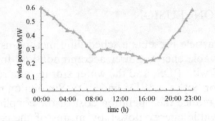

Figure 3. Load of the PDS in a day.   Figure 4. Wind power output prediction for a day.

Scenario I: WDS and PDS operate independently;

Scenario II: The scheduling method using a water pump and reservoir for renewable energy power accommodation is used. The parameters are the same as those in scenario I.

4.2 *Results*

The operation costs of the WDS and PDS in two scenarios and the wind power consumption rates of the system are shown in Table 1 and Figure 5 respectively.

Combining with Table 1 and Figure 5, we can see that: the total cost is 19,275.96 yuan in the scenario I. The WDS operation cost and the abandoned wind penalty cost are higher than that of scenario II. Therefore, the scheduling method under scenario I is not economical. For wind power consumption, wind power output is large while electricity load is small at 0:00–3:00 hours, and there is a high abandon rate under scenario I. The total cost under scenario II is 12.76% less than that of scenario I, and its economy is significantly improved. Among them, the operating cost of WDS is 628.98 yuan, which is 41.39% lower than that of scenario I. The line loss of PDS and purchasing power cost are also reduced under scenario II. For wind power consumption, the wind power accommodation rates in each period of the system reach 100%, and the abandoned wind penalty cost is 0 yuan, which effectively solves the problem of wind accommodation. The flexible response capability from the water pumps and reservoirs in the WDS is utilized, and the effectiveness of the coordinated operation of reservoir and water pump is demonstrated.

Table 1. Operation cost of two scenarios.

| Scenarios | PDS operating costs (yuan) | | WDS operating costs (yuan) | Abandoned wind penalty cost (yuan) | Total cost (yuan) |
	Line loss cost	Purchasing power cost			
I	387.75	16,828.02	1073.16	987.03	19,275.96
II	358.05	15,830.10	628.98	0.00	16,817.13

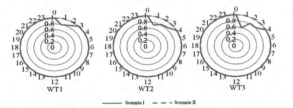

Figure 5. Wind power accommodation rates under two scenarios.

5 CONCLUSION

This article presents a scheduling method using water pumps and reservoirs in the WDS for renewable energy power accommodation. In this method, the water pump is used to couple WDS with PDS, and the water storage in a reservoir is able to function as virtual energy storage for both the WDS and PDS. Simulation results show that the proposed method can effectively reduce the operation cost of the coupling system and improve the consumption level of renewable energy power by means of the coordinated operation of the reservoir and water pump.

ACKNOWLEDGMENTS

The authors gratefully acknowledge the funding support of the State Grid Science and Technology project (NY71-18-013) and China epri self-financing project (NY83-19-008).

REFERENCES

Baran, M. E., & Wu, F. F. 1989. Network reconfiguration in distribution systems for loss reduction and load balancing. *IEEE Transactions on Power Delivery* 4(2): 1401–1407.

Chen, H. K., Yu, Y. J., & Jiang, X. 2016. Optimal scheduling of combined heat and power units with heat storage for the improvement of wind power integration. In *2016 IEEE PES Asia-Pacific Power and Energy Engineering Conference (APPEEC)*, Xi'an, pp. 1508–1512.

Guidoin, M., Kapelan, Z., & Savic, D. 2013. Using high performance techniques to accelerate demand-driven hydraulic solvers. *Journal of Hydroinformatics* 15(1): 38–54.

Hamiche, M. A., Stambouli, B. A., & Flazi, S. 2016. A review of the water-energy nexus. *Renewable and Sustainable Energy Reviews*, 65:319-331.

Kurek, W., & Ostfeld, A. 2013. Multi-objective optimization of water quality, pumps operation and storage sizing of water distribution systems. *Journal of Environmental Management* 115: 189–197.

Lewis, A., & Rossman, A. *EPANET2 Users Manual* [EB/OL]. U.S. Environmental Protection Agency. Retrieved from https://www.epa.gov/water-research/epanet.

Li, G. Q., Zhang, R. F., & Jiang, T. 2017. Optimal dispatch strategy for integrated energy systems with CCHP and wind power. *Applied Energy* 192: 408–419.

Macknick, J. 2011. A review of operational water consumption and withdrawal factors for electricity generating technologies. National Renewable Energy Laboratory.

Oikonomou, K., & Parvania, M. 2018. Optimal coordination of water distribution energy flexibility with power systems operation. *IEEE Transactions on Smart Grid* 99: 1101–1110.

Siddiqi, A., & Anadon, L. D. 2011. water-energy nexus in Middle East and North Africa. *Energy Policy* 39(8): 4529–4540.

Yoon, H. 2018. A review on water-energy nexus and directions for future studies: From supply to demand end. *Documents d'Anàlisi Geogràfica* 64(2): 365–395.

Zhang, H., Zhou, J., & Fang, N. 2013. Daily hydrothermal scheduling with economic emission using simulated annealing technique based multi-objective cultural differential evolution approach. *Energy* 50: 24–37.

Emerging Developments in the Power and Energy Industry – Dufo-López, Krzywanski & Singh (eds)
© *2020 Taylor & Francis Group, London, ISBN 978-0-367-27169-5*

Graphical and unified analysis of the unsymmetrical shunt faults

Xusheng Chen

ABSTRACT: The paper starts from forming the complete sequence network of a faulted network, from which the positive-, negative-, and zero-sequence networks are deduced by applying superposition and the system equations are obtained. Treating the fault impedances as part of the un-faulted symmetrical network makes the analysis of the shunt faults with fault impedances much simpler and clearer. The method of symmetrical components is the pillar of power system steady-state analysis. The paper presents a more straightforward, rigorous, and efficient explanation of the concepts and derivation of the equations.

Keywords: Symmetrical components, Sequence networks, Shunt faults, Single line-to-ground fault, Line-to-Line fault, Double line-to-ground fault, Substitution theorem

1 INTRODUCTION

The analysis of power systems usually implies the computation of network voltages and currents under a given set of conditions. The steady state problems are solved by phasor algebraic equations. In most cases the network is assumed to be linear, and under normal operation conditions the network is also assumed to be balanced. Thus, the network can be solved on a per phase basis. This results in a great saving of time and effort. When the system is obviously unbalanced, other methods must be used. The method generally favored is that of "symmetrical components" as proposed by Fortescue in 1918 (C. L. Fortescue, 1918). This method extends the per phase analysis to systems with unbalanced loads or with unbalanced termination of some kind, such as a short circuit or fault. Power system fault analysis has undergone nearly a century of development and almost every undergraduate and graduate textbook on power systems has a chapter on it. While it is impossible for the authors to advance the knowledge of the topic, it is possible for the authors to make the explanation of concepts and derivation of the equations more straightforward, rigorous, and efficient so as to save time for power system students and engineers to learn and grasp it. "A picture is worth a thousand words." The power of using circuits to reveal concepts are shown in the paper. The authors are confident that once a reader understands this paper, he can not only have a deep understanding of the theory of shunt faults but also remember it forever.

Figure 1 is the study network which is small enough to be solved by hand and not too small to lose generality of the analysis. E'_{abc} and E''_{abc} are balanced three-phase voltage sources and Z' and Z'_n are positive-sequence impedances under balanced operation. Z'_n and Z''_n are neutral to ground impedances. The load current $I'_{abc} = -I''_{abc}$ and fault current $I_{abc} = 0$. $Z_f = diag(Z_f\ Z_f\ Z_f)$ is the fault impedance. The short-circuit fault is applied at point F. The readers will see later that treating the fault impedance Z_f as part of the symmetrical network will make the analysis of the shunt faults with fault impedances much simpler and clearer. Section II derives the sequence networks. Section III presents the unbalanced shunt faults, and Section IV concludes the paper.

2 SEQUENCE NETWORKS

In many problems the unbalanced portion of the physical system can be isolated for study, with the rest of the system considered balanced. This is the case for an unbalanced load or fault supplied by a system of balanced or equal-phase impedances. The fault point of a system is that point to which the unbalanced connection is attached in an otherwise balanced system. A sequence network is a copy of the original balanced system to which the fault point is connected and contains the same per phase impedances as the physical balanced system, arranged in the same way, the only difference being that the value of each impedance is a value unique to each sequence (P. M. Anderson, 1973).

Before the fault at point F in Figure 1, the system is in normal operation with load current $I'_{abc} = -I''_{abc}$, fault current $I_{abc} = 0$, and pre-fault voltage V_{abcf}. When a shunt fault, for example, a single line-to-ground fault (SLG), occurs at phase a of the fault point F in Figure 2, the impedance of phase a to ground becomes zero and the impedances of the other two phases remain infinite. The three-phase voltages and currents to the ground are unbalanced: $V_a = 0$, $V_b \neq 0$, and $V_c \neq 0$, and $I_a \neq 0$, $I_b = 0$, and $I_c = 0$. Since the whole network loses its symmetry, it cannot be solved on a per phase basis in the abc variables. Although the system can be solved by general circuit analysis techniques, the approach is too complicated and provides little physical insights.

To get around the difficulty, it is reasoned that the network above the fault point F knows of the existence of the fault through fault impedances below Z_f, or voltages V_a, V_b, and V_c, or currents I_a, I_b, and I_c (A. S. Sedra & K. C. Smith, 2004). If the three unbalanced voltages are replaced by three fictitious voltage sources (although their values are not yet known and for simplicity the same notation V_a, V_b, and V_c are used), the whole network becomes symmetrical again, which is shown in Figure 3. The replacement is valid because of the substitution theorem in advanced circuits textbooks (L. O. Chua, C. A. Desoer & E. S. Kuh, 1987).

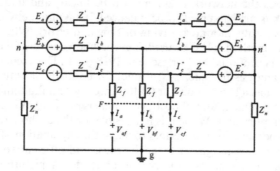

Figure 1.   The study network before the fault. V_{af}, V_{bf}, and V_{cf} are the prefault voltages at fault point F.

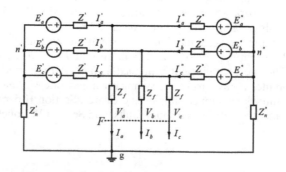

Figure 2.   Diagram of an SLG fault at F.

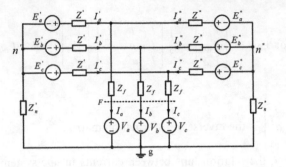

Figure 3. The network of Figure 2 with the fault point unbalanced impedances replaced by the fictitious voltage sources V_a, V_b, and V_c. The network becomes symmetrical but unbalanced.

According to C. L. Fortescue and many other practitioners (C. L. Fortescue 1918, P. M. Anderson 1973, A. S. Sedra 2004) and (John J. Grainger 1994, A. R. Bergen & V. Vittal 2000, Charles A. Gross 1986), the three unbalanced phasor voltages can be transformed into their symmetrical components:

$$V_a = V_{a0} + V_{a1} + V_{a2} = V_{a0} + V_{a1} + V_{a2} \tag{1a}$$

$$V_b = V_{b0} + V_{b1} + V_{b2} = V_{a0} + a^2 V_{a1} + a V_{a2} \tag{1b}$$

$$V_c = V_{c0} + V_{c1} + V_{c2} = V_{a0} + a V_{a1} + a^2 V_{a2} \tag{1c}$$

$$\text{or } V_{abc} = \begin{bmatrix} V_a \\ V_b \\ V_c \end{bmatrix} = \begin{bmatrix} 1 & 1 & 1 \\ 1 & a^2 & a \\ 1 & a & a^2 \end{bmatrix} \begin{bmatrix} V_{a0} \\ V_{a1} \\ V_{a2} \end{bmatrix} = A\, V_{012} \tag{2}$$

where $A = \begin{bmatrix} 1 & 1 & 1 \\ 1 & a^2 & a \\ 1 & a & a^2 \end{bmatrix}$ is the transformation matrix, $a = e^{j2\pi/3}$, $a^2 = e^{-j2\pi/3}$, $a^3 = 1$,

and $1 + a + a^2 = 0$.

Equation (1) or (2) is called the synthesis equation. Its inverse is called the analysis equation which can be derived as follows:

Eq. (1a) + Eq. (1b) + Eq. (1c):

$$V_a + V_b + V_c = 3V_{a0} + 0V_{a1} + 0V_{a2} = 3V_{a0}$$

$$V_{a0} = \frac{1}{3}(V_a + V_b + V_c) \tag{3a}$$

Eq. (1a) + a Eq. (1b) + a^2 Eq. (1c):

$$V_a + a V_b + a^2 V_c = 0V_{a0} + 3\,V_{a1} + 0V_{a2} = 3V_{a1}$$

$$V_{a1} = \frac{1}{3}(V_a + a\, V_b + a^2 V_c) \tag{3b}$$

Eq. (1a) + a^2 Eq. (1b) + a Eq. (1c):

$$V_a + a^2 V_b + a V_c = 0\,V_{a0} + 0V_{a1} + 3V_{a2} = 3V_{a2}$$

$$V_{a2} = \frac{1}{3}(V_a + a^2 V_b + a\, V_c) \tag{3c}$$

$$\text{or } V_{012} = \begin{bmatrix} V_{a0} \\ V_{a1} \\ V_{a2} \end{bmatrix} = \frac{1}{3} \begin{bmatrix} 1 & 1 & 1 \\ 1 & a & a^2 \\ 1 & a^2 & a \end{bmatrix} \begin{bmatrix} V_a \\ V_b \\ V_c \end{bmatrix} = A^{-1} V_{abc} \tag{4}$$

where

$A^{-1} = \frac{1}{3} \begin{bmatrix} 1 & 1 & 1 \\ 1 & a & a^2 \\ 1 & a^2 & a \end{bmatrix}$ is the inverse transformation matrix.

By changing V to I, the relationships between currents in abc system and 012 system are obtained.

Once the three unbalanced phase voltages, V_{abc}, are transformed into their balanced three sequence voltages, V_{012}, the resultant network, Figure 4, is symmetrical and balanced in the new zero-positive-negative sequence system or 012 system in short. The impedances, Z' and Z'', now have different values for different sequence currents.

To separate the pre-fault load currents from the fault currents, two sets of fictitious three-phase voltage sources, $E_{abcf} = V_{abcf}$, connected back-to-back are inserted into the fault point. The preparation for the derivation of the sequence networks is now complete and the complete network is shown in Figure 4. E_{abcf} is the pre-fault voltage at the fault point F. Figure 4 is really not as complicated as it looks.

The network of Figure 4 is linear and superposition can be applied. Applying the actual voltage sources E'_{abc} and E''_{abc} and the upper fictitious voltage sources $E_{abcf} = \begin{bmatrix} E_{af} & E_{bf} & E_{cf} \end{bmatrix}^t$ and setting all the other voltage sources to zero results in the pre-fault network which is shown in Figure 1. Solving Figure 1 gives the pre-fault voltages (especially $V_{abcf} = \begin{bmatrix} V_{af} & V_{bf} & V_{cf} \end{bmatrix}^t$) and the load currents of the study network.

2.1 Positive-sequence network

Applying the lower fictitious voltage source $E_{abcf} = \begin{bmatrix} E_{af} & E_{bf} & E_{cf} \end{bmatrix}^t$ and the fictitious positive-sequence voltage source $V_{abc1} = \begin{bmatrix} V_{a1} & V_{b1} & V_{c1} \end{bmatrix}^t$ and setting all the other voltage sources to zero result in the positive-sequence three-phase network of Figure 5(a). Since the positive-sequence three-phase network is symmetrical and balanced, no currents are flowing through neutral impedances

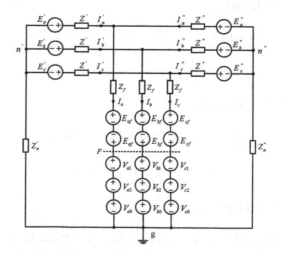

Figure 4. The study network during the fault at F. The unbalanced fault voltages are transformed into their symmetrical components and the network is now symmetrical and balanced in the 012 system. Z' and Z'' are sequence-dependent and Z_n' and Z_n'' are zero-sequence impedances.

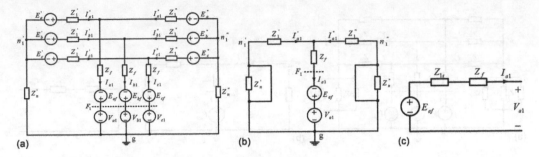

(a)   (b)   (c)

Figure 5. Positive-sequence networks. (a) Positive-sequence three-phase network. (b) The one-phase positive-sequence network. (c) Positive sequence thevenin equivalent network.

Z' and Z''_n, therefore, $V_{n'1} = V_{n''1} = V_g = 0$, and the two impedances can be short-circuited. The one-line diagram of the positive-sequence network is shown in Figure 5(b), from which the positive-sequence impedance, Z_1, which is the impedance looking back from point F_1 and g, is

$$Z_1 = Z_f + Z'_1 || Z'_1 = Z_f + Z_{1s} = Z_{1s} + Z_f \qquad (5)$$

where $Z_{1s} = Z'_1 || Z''_1$

The positive-sequence Thevenin equivalent network or positive-sequence network is shown in Figure 5(c). The Thevenin equivalent voltage in the positive-sequence network is the phase a open-circuit (pre-fault) voltage $E_{af} = V_{af}$ at the fault point.

2.2 Negative-sequence network

Applying the fictitious negative-sequence voltage source $V_{abc2} = [\, V_{a2} \quad V_{b2} \quad V_{c2}\,]^t$ and setting all the other voltage sources to zero, and following the same reasoning and procedure as the positive-sequence network, result in the negative-sequence network of Figure 6, from which the negative-sequence impedance, Z_2, which is the impedance looking back from point F_2 and g, is

$$Z_2 = Z_f + Z'_2 || Z'_2 = Z_f + Z_{2s} = Z_{2s} + Z_f \qquad (6)$$

where $Z_{2s} = Z'_2 || Z'_2$.

Note that there is no Thevenin equivalent voltage source in the negative-sequence network.

2.3 Zero-sequence network

Applying the fictitious zero-sequence voltage source $V_{abc0} = [\, V_{a0} \quad V_{b0} \quad V_{c0}\,]^t$ and setting all the other voltage sources to zero results in the zero-sequence three-phase network of Figure 7(a). Since zero-sequence currents are the single-phase currents in the three-phase system, the currents flowing through the neutral impedances Z'_n and Z''_n are $3I'_{a0}$ and $3I''_{a0}$, respectively. The voltages at the two neutral points are $V_{n'g0} = 3Z'_n I'_{a0}$ and $V_{n''g0} = 3 Z''_n I''_{a0}$, respectively. The one-line

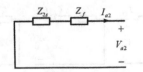

Figure 6. Negative-sequence Thevenin equivalent network.

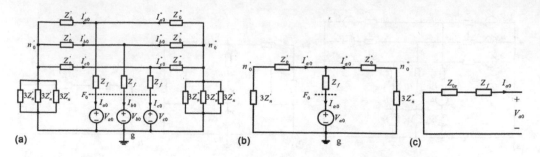

(a)   (b)   (c)

Figure 7.   Zero-sequence networks. (a) Zero-sequence three-phase network. (Down a little (b) Zero-sequence single-phase network. (c) Zero-sequence Thevenin equivalent network.

diagram of the zero-sequence network is shown in Figure 7(b). From Figure 7(b) the zero-sequence impedance, Z_0, which is the impedance looking back from point F_0 and g, is

$$Z_0 = Z_f + Z_0'||Z_0'' = Z_f + Z_{0s} = Z_{0s} + Z_f \tag{7}$$

where $Z_{0s} = Z_0'||Z_0''$

The zero-sequence Thevenin equivalent network or zero-sequence network is shown in Figure 7(c).

Note that there is no Thevenin equivalent voltage source in the zero-sequence network.

From the three sequence networks, the system equations can be written:

$$V_{a0} = 0 - Z_0 I_{a0} \tag{8}$$

$$V_{a1} = E_{af} - Z_1 I_{a1} \tag{9}$$

$$V_{a2} = 0 - Z_2 I_{a2} \tag{10}$$

The system equations have six unknowns and three additional equations are needed to solve the network, which are derived from the boundary conditions of the faults. For easy reference, the positive-, negative-, and zero-sequence networks with defined sequence quantities are shown in Figure 8.

3   THE UNBALANCED SHUNT FAULTS

Three typical unbalanced shunt faults with fault impedances are analyzed in this section.

3.1   The Single-Line-to-Ground (SLG) fault

Figure 9 shows the diagram of an SLG fault at point F. The boundary conditions in abc system are:

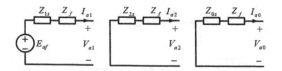

Figure 8.   Sequence network with defined sequence quantities.

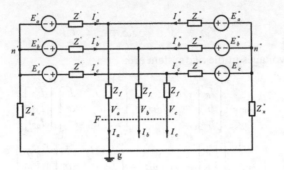

Figure 9.   Diagram of an SLG fault at F.

$$I_b = I_c = 0 \tag{11}$$

$$V_a = 0 \tag{12}$$

Using the inverse transformation matrix A^{-1} to change the constraints from abc system to 012 system results in

$$\begin{bmatrix} I_{a0} \\ I_{a1} \\ I_{a2} \end{bmatrix} = \frac{1}{3} \begin{bmatrix} 1 & 1 & 1 \\ 1 & a & a^2 \\ 1 & a^2 & a \end{bmatrix} \begin{bmatrix} I_a \\ I_b = 0 \\ I_c = 0 \end{bmatrix} = \frac{1}{3} \begin{bmatrix} I_a \\ I_a \\ I_a \end{bmatrix}$$

$$I_{a0} = I_{a1} = I_{a2} = I_a/3 \tag{13}$$

$$V_a = V_{a0} + V_{a1} + V_{a2} = 0 \tag{14}$$

Adding the system equations (8), (9), and (10) results in

$$V_{a0} + V_{a1} + V_{a2} = 0 = E_{af} - (Z_0 + Z_1 + Z_2)I_{a1}$$

$$I_{a1} = \frac{E_{af}}{Z_1 + Z_2 + Z_0} = \frac{E_{af}}{Z_{1s} + Z_{2s} + Z_{0s} + 3Z_f} \tag{15}$$

Equations (13) and (14) are the constraints in the 012 system and they suggest the sequence network of Figure 10.

Once $I_{a0} = I_{a1} = I_{a2}$ are found from (15), the sequence voltages V_{a0}, V_{a1}, and V_{a2} can be found from (8)-(10):

$$V_{a0} = 0 - Z_0 I_{a0} = - Z_0 I_{a0} \tag{16a}$$

$$V_{a1} = E_{af} - Z_1 I_{a1} \tag{16b}$$

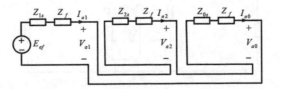

Figure 10.   Sequence network connection for an SLG fault.

$$V_{a2} = 0 - Z_2 I_{a2} = - Z_2 I_{a2} \qquad (16c)$$

The currents and voltages in the abc system are:

$$I_{abc} = \begin{bmatrix} I_a \\ I_b \\ I_c \end{bmatrix} = A I_{012} = \begin{bmatrix} 1 & 1 & 1 \\ 1 & a^2 & a \\ 1 & a & a^2 \end{bmatrix} \begin{bmatrix} I_{a0} \\ I_{a1} \\ I_{a2} \end{bmatrix} \qquad (17)$$

$$V_{abc} = \begin{bmatrix} V_a \\ V_b \\ V_c \end{bmatrix} = A V_{012} = \begin{bmatrix} 1 & 1 & 1 \\ 1 & a^2 & a \\ 1 & a & a^2 \end{bmatrix} \begin{bmatrix} V_{a0} \\ V_{a1} \\ V_{a2} \end{bmatrix} \qquad (18)$$

3.2 The Line-to-Line (LL) fault

Figure 11 shows the diagram of an LL fault at F. The boundary conditions in abc system are:

$$I_a = 0 \text{ and } I_b = -I_c \qquad (19)$$

$$V_b = V_c \qquad (20)$$

Using the inverse transformation matrix A^{-1} to change the constraints from abc system to 012 system, results in

$$\begin{bmatrix} I_{a0} \\ I_{a1} \\ I_{a2} \end{bmatrix} = \frac{1}{3} \begin{bmatrix} 1 & 1 & 1 \\ 1 & a & a^2 \\ 1 & a^2 & a \end{bmatrix} \begin{bmatrix} I_a = 0 \\ I_b \\ I_c = -I_b \end{bmatrix} = \frac{1}{3} \begin{bmatrix} I_b + I_c = 0 \\ (a - a^2)I_b \\ (a^2 - a)I_b \end{bmatrix}$$

$$I_{a0} = 0 \text{ and } I_{a1} = -I_{a2} \qquad (21)$$

$$\begin{bmatrix} V_{a0} \\ V_{a1} \\ V_{a2} \end{bmatrix} = \frac{1}{3} \begin{bmatrix} 1 & 1 & 1 \\ 1 & a & a^2 \\ 1 & a^2 & a \end{bmatrix} \begin{bmatrix} V_a \\ V_b \\ V_c = V_b \end{bmatrix} = \frac{1}{3} \begin{bmatrix} V_a + 2 V_b \\ V_a - V_b \\ V_a - V_b \end{bmatrix}$$

$$V_{a1} = V_{a2} \qquad (22)$$

Subtracting (10) from (9) and using (21) and (22) result in

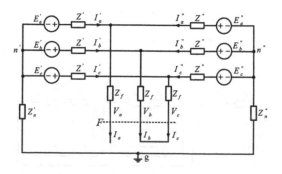

Figure 11.   Diagram of an LL fault at F.

$$V_{a1} - V_{a2} = 0 = (E_{af} - Z_1 I_{a1}) - (0 - Z_2 I_{a2})$$
$$= E_{af} - (Z_1 + Z_2)I_{a1}$$

$$I_{a1} = \frac{E_{af}}{Z_1 + Z_2} = \frac{E_{af}}{Z_{1s} + Z_{2s} + 2Z_f} \tag{23}$$

Equations (21) and (22) are the constraints in the *012* system and they suggest the sequence network of Figure 12.

Once $I_{a0} = 0$ and $I_{a2} = -I_{a1}$ are found from (23), the sequence voltages V_{a0}, V_{a1}, and V_{a2} can be found from the system equations (8)-(10):

$$V_{a0} = 0 - Z_0 I_{a0} = 0 - Z_0 \, 0 = 0 \tag{24a}$$

$$V_{a1} = E_{af} - Z_1 I_{a1} \tag{24b}$$

$$V_{a2} = 0 - Z_2 I_{a2} = -Z_2 I_{a2} \tag{24c}$$

The currents and voltages in the *abc* system are:

$$I_{abc} = \begin{bmatrix} I_a \\ I_b \\ I_c \end{bmatrix} = A \, I_{012} = \begin{bmatrix} 1 & 1 & 1 \\ 1 & a^2 & a \\ 1 & a & a^2 \end{bmatrix} \begin{bmatrix} I_{a0} = 0 \\ I_{a1} \\ I_{a2} = -I_{a1} \end{bmatrix} \tag{25}$$

$$V_{abc} = \begin{bmatrix} V_a \\ V_b \\ V_c \end{bmatrix} = A \, V_{012} = \begin{bmatrix} 1 & 1 & 1 \\ 1 & a^2 & a \\ 1 & a & a^2 \end{bmatrix} \begin{bmatrix} V_{a0} = 0 \\ V_{a1} \\ V_{a2} = V_{a1} \end{bmatrix} \tag{26}$$

3.3 The Double Line-to-Ground (LLG) fault

Figure 13 shows the diagram of an LLG fault at F.

The boundary conditions in *abc* system are:

$$I_a = 0 \tag{27}$$

$$V_b = V_c = Z_g(I_b + I_c) \tag{28}$$

Using the inverse transformation matrix A^{-1} to change the constraints from *abc* system to *012* system, results in

$$I_a = I_{a0} + I_{a1} + I_{a2} = 0 \tag{29}$$

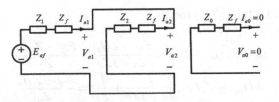

Figure 12.   Sequence network connection for an LL fault.

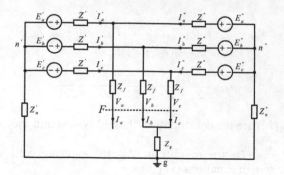

Figure 13.   Diagram of an LLG fault at F.

$$\begin{bmatrix} V_{a0} \\ V_{a1} \\ V_{a2} \end{bmatrix} = \frac{1}{3} \begin{bmatrix} 1 & 1 & 1 \\ 1 & a & a^2 \\ 1 & a^2 & a \end{bmatrix} \begin{bmatrix} V_a \\ V_b \\ V_c \end{bmatrix} = \frac{1}{3} \begin{bmatrix} V_a + 2V_b \\ V_a + (a + a^2)V_b \\ V_a + (a^2 + a)V_b \end{bmatrix}$$

$$V_{a1} = V_{a2} \tag{30}$$

$$V_b = V_c = Z_g(I_b + I_c)$$
$$= Z_g[(I_{a0} + a^2 I_{a1} + a I_{a2}) + (I_{a0} + a I_{a1} + a^2 I_{a2})]$$
$$= Z_g[2I_{a0} + (a^2 + a)I_{a1} + (a + a^2)I_{a2}]$$
$$= Z_g[2I_{a0} - I_{a1} - I_{a2}] = Z_g 3 I_{a0} = 3 Z_g I_{a0}$$

$$V_b = V_{a0} + a^2 V_{a1} + a V_{a2} = V_{a0} + (a^2 + a)V_{a1}$$
$$= V_{a0} - V_{a1}$$

$$V_{a0} - V_{a1} = 3Z_g I_{a0} \text{ or } V_{a1} = V_{a0} - 3Z_g I_{a0} \tag{31}$$

Substituting (8) and (9) into (31) results in

$$V_{a0} - V_{a1} = (0 - Z_0 I_{a0}) - (E_{af} - Z_1 I_{a1})$$
$$= -Z_0 I_{a0} - E_{af} + Z_1 I_{a1} = 3lpt\, Z_g I_{a0}$$

$$I_{a0} = -\frac{E_{af} - Z_1 I_{a1}}{Z_0 + 3Z_g} \tag{32}$$

Substituting (9) and (10) into (30) results in

$$V_{a1} = E_{af} - Z_1 I_{a1} = V_{a2} = 0 - Z_2 I_{a2}$$

$$I_{a2} = -\frac{E_{af} - Z_1 I_{a1}}{Z_2} \tag{33}$$

Substituting (32) and (33) into (29) results in

$$I_{a1} = -(I_{a0} + I_{a2}) = \frac{E_{af} - Z_1 I_{a1}}{Z_0 + 3Zg} + \frac{E_{af} - Z_1 I_{a1}}{Z_2}$$
$$= (E_{af} - Z_1 I_{a1})(\frac{1}{Z_0 + 3Zg} + \frac{1}{Z_2}) = \frac{E_{af} - Z_1 I_{a1}}{Z_\Delta}$$

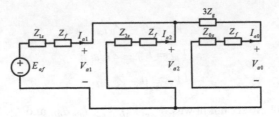

Figure 14. Sequence network connection for an LLG fault at F.

$$I_{a1} = \frac{E_{af}}{Z_1 + Z_\Delta} = \frac{E_{af}}{Z_{1s} + Z_f + Z_\Delta} \tag{34}$$

where

$$\begin{aligned} Z_\Delta &= (Z_0 + 3Z_g)//Z_2 \\ &= (Z_{0s} + Z_f + 3Z_g)//(Z_{2s} + Z_f) \end{aligned} \tag{35}$$

Equations (29), (30) and (31) are the constraints in the 012 system and they suggest the sequence network of Figure 14.

Once I_{a1} is found from (34), I_{a0} and I_{a2} can be determined by (32) and (33), respectively. Eq. (29) can be used to check the results. The sequence voltages V_{ao}, V_{a1}, and V_{a2} can be found from (8)-(10), and the currents and voltages in the abc system can be determined using (2), the synthesis equation:

$$I_{abc} = \begin{bmatrix} I_a \\ I_b \\ I_c \end{bmatrix} = A\,I_{012} = \begin{bmatrix} 1 & 1 & 1 \\ 1 & a^2 & a \\ 1 & a & a^2 \end{bmatrix} \begin{bmatrix} I_{a0} \\ I_{a1} \\ I_{a2} \end{bmatrix} \tag{36}$$

$$V_{abc} = \begin{bmatrix} V_a \\ V_b \\ V_c \end{bmatrix} = A\,V_{012} = \begin{bmatrix} 1 & 1 & 1 \\ 1 & a^2 & a \\ 1 & a & a^2 \end{bmatrix} \begin{bmatrix} V_{a0} \\ V_{a1} \\ V_{a2} \end{bmatrix} \tag{37}$$

4 CONCLUSIONS

Fig 1 is the study network which is small enough to be solved by hand and not too small to lose generality of the analysis. The network of Figure 1 is both symmetrical and balanced; therefore, per phase method can be used. The solution of Figure 1 gives the pre-fault voltages and currents of the network. When a fault occurs below point F of the network, the network becomes un-symmetrical (based on impedances) and unbalanced (based on voltages/currents), and the per phase method cannot be used. If the unbalanced fault point voltages are replaced by unbalanced voltage sources, and transformed into their symmetrical components, the network becomes symmetrical and balanced in the 012 system and can be solved one sequence at a time. To separate the pre-fault load currents from the fault currents, two sets of fictitious three-phase voltage sources, $E_{abcf} = V_{abcf}$, connected back-to-back are inserted into the fault point. The preparation for the derivation of the sequence networks is now complete and the complete network is shown in Figure 4. The network of Figure 4 is symmetrical, balanced, and linear, and superposition applies. Figures 5, 6, and 7 lead the readers to Figure 8 and from which the system equations arc written (repeated here for easy reference):

$$V_{a0} = 0 - Z_0 I_{a0} \qquad (8)$$

$$V_{a1} = E_{af} - Z_1 I_{a1} \qquad (9)$$

$$V_{a2} = 0 - Z_2 I_{a2} \qquad (10)$$

Equations (8)–(10) have six unknowns: V_{a0}, V_{a1}, V_{a2}, I_{a0}, I_{a1}, and I_{a2}. The other three independent equations are obtained from fault conditions in the *012* system.
The Single-Line-to-Ground (SLG) Fault:

$$I_{a0} = I_{a1} = I_{a2} = I_a 3 \qquad (13)$$

$$V_a = V_{a0} + V_{a1} + V_{a2} = 0 \qquad (14)$$

The Line-to-Line (LL) Fault:

$$I_{a0} = 0 \text{ and } I_{a1} = -I_{a2} \qquad (21)$$

$$V_{a1} = V_{a2} \qquad (22)$$

The Double Line-to-Ground (LLG) Fault:

$$I_a = I_{a0} + I_{a1} + I_{a2} = 0 \qquad (29)$$

$$V_{a1} = V_{a2} \qquad (30)$$

$$V_{a0} - V_{a1} = 3Z_g I_{a0} \text{ or } V_{a1} = V_{a0} - 3Z_g I_{a0} \qquad (31)$$

The positive-sequence current I_{a1} has a general form:

$$I_{a1} = \frac{E_{af}}{Z_1 + Z_\Delta} = \frac{E_{af}}{Z_{1s} + Z_f + Z_\Delta} \qquad (38)$$

The Single-Line-to-Ground (SLG) Fault:

$$Z_\Delta = Z_2 + Z_0 = Z_{2s} + Z_{0s} + 2Z_f$$

$$I_{a1} = \frac{E_{af}}{Z_1 + Z_2 + Z_0} = \frac{E_{af}}{Z_{1s} + Z_{2s} + Z_{0s} + 3Z_f} \qquad (15)$$

The Line-to-Line (LL) Fault:

$$Z_\Delta = Z_2 = Z_{2s} + Z_f$$

$$I_{a1} = \frac{E_{af}}{Z_1 + Z_2} = \frac{E_{af}}{Z_{1s} + Z_{2s} + 2Z_f} \qquad (23)$$

The Double Line-to-Ground (LLG) Fault:

$$Z_\Delta = (Z_0 + 3Zg)//Z_2$$
$$= (Z_{0s} + Z_f + 3Zg)//(Z_{2s} + Z_f) \qquad (35)$$

$$I_{a1} = \frac{E_{af}}{Z_1 + Z_\Delta} = \frac{E_{af}}{Z_{1s} + Z_f + Z_\Delta} \qquad (34)$$

Once I_{a1} is found, the other variables can be found from the system equations and the boundary conditions. The voltages and currents in the abc systems can be found using (2), the synthesis equations:

$$I_{abc} = \begin{bmatrix} I_a \\ I_b \\ I_c \end{bmatrix} = A\, I_{012} = \begin{bmatrix} 1 & 1 & 1 \\ 1 & a^2 & a \\ 1 & a & a^2 \end{bmatrix} \begin{bmatrix} I_{a0} \\ I_{a1} \\ I_{a2} \end{bmatrix} \qquad (36)$$

$$V_{abc} = \begin{bmatrix} V_a \\ V_b \\ V_c \end{bmatrix} = A\, V_{012} = \begin{bmatrix} 1 & 1 & 1 \\ 1 & a^2 & a \\ 1 & a & a^2 \end{bmatrix} \begin{bmatrix} V_{a0} \\ V_{a1} \\ V_{a2} \end{bmatrix} \qquad (37)$$

Transformation matrix A and its inverse A^{-1} play important roles in the symmetrical components method. They can be easily written out by following the procedure of (1) to (4). They are repeated below for easier reference:

$$A = \begin{bmatrix} 1 & 1 & 1 \\ 1 & a^2 & a \\ 1 & a & a^2 \end{bmatrix} \text{ and } A^{-1} = \frac{1}{3} \begin{bmatrix} 1 & 1 & 1 \\ 1 & a & a^2 \\ 1 & a^2 & a \end{bmatrix}$$

The paper convincingly shows that treating the fault impedance Z_f as part of the symmetrical network makes the analysis of the shunt faults with fault impedances much simpler and clearer.

ACKNOWLEDGEMENTS

This work was supported in part by the Sabbatical Award offered to Xusheng Chen by Seattle University and in part by Project of Shandong Province Higher Educational Science and Technology Program (J11LG30) and International Cooperation Training Program for Outstanding Teacher in Shandong Province, Shandong Education Department (SED).

REFERENCES

C. L. Fortescue, "Method of symmetrical components applied to the solution of polyphase networks," *Trans. AIEE 37, pp.* 1027–1140, 1918.

P. M. Anderson, *Analysis of Faulted Power Systems*, The Iowa State University Press/AMES, 1973, Chapter 2, pp. 19–33 and Chapter 3, pp. 36–49.

A. S. Sedra and K. C. Smith, *Microelectronic Circuits*, Fifth Edition, Oxford University Press, Inc., 2004, p. 330.

L. O. Chua, C. A. Desoer, and E. S. Kuh, *Linear and Nonlinear Circuits*, McGraw-Hill Book Company, 1987, pp. 267–269.

John J. Grainger and William D. Stevenson, Jr., *Power System Analysis*, McGraw-Hill, Inc., 1994, pp. 471–527.

A. R. Bergen and V. Vittal, *Power Systems Analysis*, 2nd ed., Prentice-Hall, INC., Upper Saddle, New Jersey 07458, 2000, pp. 445–460.

Charles A.Gross, *Power System Analysis*, 2nd ed., New York: John Wiley & Sons, 1986, pp. 333–353.

Emerging Developments in the Power and Energy Industry – Dufo-López, Krzywanski & Singh (eds)
© 2020 Taylor & Francis Group, London, ISBN 978-0-367-27169-5

Newton-Raphson discrete equivalent circuit models for the diode and their application to the photovoltaic modules

Xusheng Chen
Department of Electrical and Computer Engineering, Seattle University, Seattle, WA, USA

ABSTRACT: This paper presents the Newton-Raphson discrete equivalent circuits for the diode. The models are derived succinctly, proved theoretically, and verified computationally. The Newton-Raphson discrete equivalent circuits are the large-signal terminal models for the diode, from which simplified diode equivalent circuits and diode small-signal equivalent circuits can be easily deduced. The standard circuit model for the photovoltaic module is a current source in parallel with a diode and a shunt resistor and in series with a resistor. The proposed diode models make the solution of the photovoltaic module natural, accurate, simple, reliable, and non-software dependent.

Keywords: Diode, Newton-Raphson discrete equivalent circuit, large-signal model, nonlinear resistor, photovoltaic module, PV, Modeling

1 INTRODUCTION

The diode is the first nonlinear device that the US electrical and computer engineering students are exposed to (A. S. Sedra & K. C. Smith, 2004). Since the solution of a nonlinear circuit is qualitatively different and harder than that of a linear circuit both in concept and in computation, its analysis is a challenge to many of the students. This paper derives the Newton-Raphson (N-R) discrete equivalent circuits for the diode. The models are then proved theoretically and verified computationally. The N-R discrete equivalent circuits are the large-signal models for the diode. Since the N-R discrete circuits are linear, all the properties of a linear circuit apply (L. O. Chua, 1987).

In recent years, renewable energy sources have become more important contributors to the world's total energy consumption. Among them, photovoltaic (PV) arrays offer several advantages, such as little maintenance requirement and no environmental pollution. Nowadays, PV arrays are used in many applications such as battery chargers, solar power water pumping systems, grid-connected PV systems, solar hybrid vehicles, and satellite power systems.

The standard circuit model for the PV module is a current source in parallel with a diode and a shunt resistor and in series with a resistor. The proposed diode models make the solution of the PV module natural, accurate, simple, reliable, and *non-software dependent*.

The rest of the paper is organized as follows: Section II derives the mathematical models for the forward-bias diode, Section III verifies the new models theoretically and computationally, Section IV uses the Newton-Raphson (N-R) discrete equivalent circuits to model the PV module, and Section V concludes the paper.

2 REPRESENTATION OF DIODES

Figure 1(a) is chosen as the study circuit so as to keep the problem tractable by hand. The diode is assumed to be voltage-controlled for the time being.

The circuit equation for Figure 1(a) is

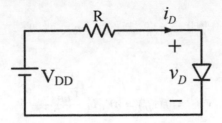

(a) Original circuit.

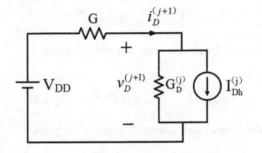

(b) Voltage-controlled diode, $i_D = g_D(v_D)$.

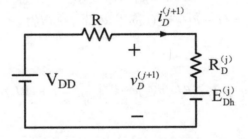

(c) Current-controlled diode, $v_D = f_D(i_D)$.

Figure 1. A diode circuit and its associated N-R discrete equivalent circuits.

$$V_{DD} = R\,i_D + v_D \tag{1}$$

with
$$i_D = g_D(v_D) = I_S\,(e^{v_D/nV_T} - 1) \approx I_S\,e^{v_D/nV_T} \tag{2}$$

where, (2) gives the *i-v* terminal characteristic of the diode in the forward region of operation when *1* in the parenthesis is omitted [1]; I_S is the saturation current, n is equal to *1* or *2*, and V_T is the thermal voltage of the diode. The default values are $I_S = 10^{-15}$ *A*, $n = 1$, and $V_T = 25$ *mV,* for this paper.

Substituting (2) into (1) results in

$$v_D = V_{DD} - R\,I_S\,e^{v_D/nV_T} \tag{3}$$

Eq. (3) is an implicit nonlinear equation for v_D, which can be solved by the Gauss iteration method:

706

$$v_D^{(j+1)} = f(v_D^{(j)}) = V_{DD} - R I_S e^{v_D^{(j)}/nV_T} \tag{4a}$$

or

$$v_D^{(j+1)} = f(v_D^{(j)}) = nV_T \ln \frac{V_{DD} - v_D^{(j)}}{R I_S} \tag{4b}$$

The convergent criterion is

$$\left| v_D^{(j+1)} - v_D^{(j)} \right| \leq TOL \tag{5}$$

Eq. (3) can also be solved by the Newton-Raphson method:
Let

$$F(v_D) = v_D + R I_S e^{v_D/nV_T} - V_{DD} = 0 \tag{6}$$

Applying the standard Newton-Raphson method to (6) gives,

$$
\begin{aligned}
v_D^{(j+1)} &= v_D^{(j)} - J^{-1} F(v_D^{(j)}) \\
&= v_D^{(j)} - (1 + R\, G_D^{(j)})^{-1}(v_D^{(j)} + RI_S\, e^{v_D^{(j)}/nV_T} - V_{DD}) \\
&= (1 + RG_D^{(j)})^{-1}(V_{DD} + RG_D^{(j)} v_D^{(j)} - Ri_D^{(j)})
\end{aligned}
\tag{7}
$$

where

$$J = (dF(v_D)/dv_D)|_{v_D^{(j)}} = 1 + (RI_S/nV_T)e^{v_D/nV_T}|_{v_D^{(j)}} = 1 + (RI_S/nV_T)e^{v_D^{(j)}/nV_T} = 1 + R\, G_D^{(j)} \tag{8}$$

with

$$G_D^{(j)} = (di_D/dv_D)|_{v_D^{(j)}} = (I_S/nV_T)\, e^{v_D^{(j)}/nV_T} \tag{9}$$

where the superscripts indicate the Newton-Raphson interaction counts; iteration j is treated as completed, while iteration j+1 is the succeeding iteration.

Equations (8) and (9) define the Jacobian and the incremental conductance at the (j+1)th Newton-Raphson iteration, respectively.

Upon convergence, $v_D^{(j+1)} = v_D^{(j)}$, (7) reduces to $F(v_D^{(j+1)}) = F(v_D^{(j)}) = 0$, that is

$$
\begin{aligned}
v_D^{(j+1)} = v_D^{(j)} &= V_{DD} - R I_s e^{v_D^{(j+1)}/nV_T} \\
&= V_{DD} - R i_D^{(j+1)} = V_{DD} - R\, g_D(v_D^{(j+1)})
\end{aligned}
\tag{10}
$$

In deriving (10), (6) and (2) are used. Thus, (2) i.e. the apparent conductance determines the solution for the diode. However, when the Newton-Raphson method is used to solve the diode circuit, incremental conductance $G_D^{(j)}$ will be used in the iteration. It is noted that as long as the iteration converges, the accuracy of the solution is determined by (2), i.e. the apparent conductance, and the accuracy of the incremental conductance determines the Jacobian and therefore affects the number of Newton-Raphson iterations, but not the accuracy of the solution.

In the above analysis, the circuit equation is formed first, which is then solved by the Newton-Raphson method. Since several software packages, such as several versions of the power system electromagnetic transients programs, EMTP, ATP, EMTDC, Microtran, etc., use nodal analysis and a fixed time-step algorithm, and especially, for small nonlinear system analysis, it is easier to individually change the diode and any other nonlinear resistors into their Newton-Raphson discrete equivalent circuits, and then form the nodal equations (L. O. Chua, 1987 & H. W. Dommel, 1986 & D. A. Calahan 1968 and 1972 & J. Vlach 1983). The derivation is shown below.

A diode can be voltage-controlled such as a tunnel or a Gunn diode and current-controlled such as a LED. A diode whose current is a "single-valued" function of the voltage is called a voltage-controlled diode. A diode whose voltage is a "single-valued" function of the current is called a current-controlled diode. If a function is one-to-one and onto, then its inverse exists (W. L. Brogan, 1974). This type of diode can be treated as either voltage- or current-controlled. In fact, the LED is current-dependent but not current-controlled in the meaning of the v-i characteristic defined above. Because the diodes are one type of nonlinear resistors, to make the derivation general, and considering that most engineers are familiar with the Thevenin rather than the Norton equivalent circuit, the current-controlled diode model is also derived here.

Figure 2(a) shows the original circuit of the diode.

2.1 Voltage-controlled diode

The branch equation for the voltage-controlled diode is

$$i_D = g_D(v_D) \tag{11}$$

Solving (11) using the most generic form of the Newton-Raphson method gives (D. A. Calahan, 1968 and 1972)

$$
\begin{aligned}
i_D^{(j+1)} &= g_D(v_D^{(j+1)}) = g_D(v_D^{(j)} + \Delta v_D) \\
&= g_D(v_D^{(j)}) + \frac{dg_D(v_D)}{dv_D}\Big|_{v_D^{(j)}} (v_D^{(j+1)} - v_D^{(j)}) + h.o.t. \\
&\approx g_D(v_D^{(j)}) + G_D^{(j)}(v_D^{(j+1)} - v_D^{(j)}) \\
&= G_D^{(j)} v_D^{(j+1)} + i_D^{(j)} - G_D^{(j)} v_D^{(j)}
\end{aligned}
$$

or

$$i_D^{(j+1)} = G_D^{(j)} v_D^{(j+1)} + I_{Dh}^{(j)} \tag{12}$$

with

$$G_D^{(j)} = (d\, g_D(v_D)/dv_D)\big|_{v_D^{(j)}} \tag{13a}$$

$$I_{Dh}^{(j)} = i_D^{(j)} - G_D^{(j)} v_D^{(j)} \tag{13b}$$

where $h.o.t.$ stands for the higher-order terms, $G_D^{(j)}$ is the incremental conductance and $I_{Dh}^{(j)}$ is a current source representing the known past history, hence the current history term at the $(j+1)$ th Newton-Raphson iteration.

Equation (12) suggests that the voltage-controlled diode can be represented by the Newton-Raphson discrete equivalent circuit of Figure 2(b) at the $(j+1)$ the Newton-Raphson iteration.

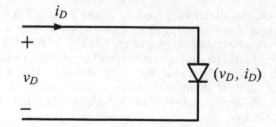

+

v_D

(v_D, i_D)

(a) Original Circuit.

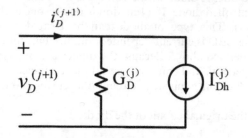

(b) Voltage-controlled diode, $i_D = g_D(v_D)$.

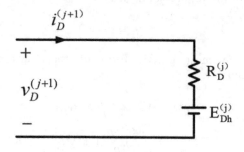

(c) Current-controlled diode, $v_D = f_D(i_D)$.

Figure 2. A diode and its associated N-R discrete equivalent circuits at the (j+1)th Newton-Raphson iteration.

2.2 *Current-controlled diode*

The branch equation for the current-controlled diode is

$$v_D = f_D(i_D) \qquad (14)$$

Solving (10) using the most generic form of the Newton-Raphson method gives [4]

$$
\begin{aligned}
v_D^{(j+1)} &= f_D(i_D^{(j+1)}) = f_D(i_D^{(j)} + \Delta i_D) \\
&= f_D(i_D^{(j)}) + \frac{d f_D(i_D)}{d i_D}\Big|_{i_D^{(j)}} (i_D^{(j+1)} - i_D^{(j)}) + h.o.t. \\
&\approx f_D(i_D^{(j)}) + R_D^{(j)}(i_D^{(j+1)} - i_D^{(j)}) \\
&= R_D^{(j)} i_D^{(j+1)} + v_D^{(j)} - R_D^{(j)} i_D^{(j)}
\end{aligned}
$$

709

or

$$v_D^{(j+1)} = R_D^{(j)} i_D^{(j+1)} + E_{Dh}^{(j)} \tag{15}$$

with

$$R_D^{(j)} = (df_D(i_D)/di_D)\big|_{i_D^{(j)}} \tag{16a}$$

$$E_{Dh}^{(j)} = v_D^{(j)} - R_D^{(j)} i_D^{(j)} \tag{16b}$$

where $R_D^{(j)}$ is the incremental resistance, and $E_{Dh}^{(j)}$ is a voltage source representing the known past history, hence the voltage history term at the (j+1)th Newton-Raphson iteration.

Equation (15) suggests that the current-controlled diode can be represented by the Newton-Raphson discrete equivalent circuit of Figure 2(c) at the (j+1)th Newton-Raphson iteration.

Note that (15) and (16) can be obtained from (12) and (13) by applying duality, and that Figure 2(c) can be obtained from Figure 2(b) through Thevenin to Norton transformation.

In summary, the mathematical model for a voltage-controlled diode is (12) and (13); and that for a current-controlled diode is (15) and (16).

3   VERIFICATION OF THE NEW MODELS

The circuit of Figures 1(b) and 1(c) are used to theoretically and computationally verify the voltage- and current-controlled diode models, respectively.

3.1   *Theoretical proof of the diode models*

1) Voltage-Controlled Diode Model: With the voltage-controlled diode represented by its Newton-Raphson discrete equivalent circuit, Figure 1(a) is changed to Figure 1(b). Solving Figure 1(b) for $v_D^{(j+1)}$ gives

$$
\begin{aligned}
v_D^{(j+1)} &= (G + G_D^{(j)})^{-1}(GV_{DD} - I_{Dh}^{(j)}) \\
&= (G + G_D^{(j)})^{-1}(GV_{DD} + G_D^{(j)} v_D^{(j)} - i_D^{(j)}) \\
&= (1 + R\,G_D^{(j)})^{-1}(V_{DD} + R\,G_D^{(j)} v_D^{(j)} - Ri_D^{(j)})
\end{aligned}
\tag{17}
$$

which is the same as given by (7).

This shows that the sequence of applying the circuit laws and the discretization of the branch equations can be swapped.

2) Current-Controlled Diode Model: Solving Figure 1(a) using the Newton-Raphson method with the diode as current-controlled, $v_D = f_D(i_D)$, gives

$$V_{DD} = R\,i_D + v_D = R\,i_D + f_D(i_D) \tag{18}$$

Let

$$F(i_D) = f_D(i_D) + R\,i_D - V_{DD} = 0 \tag{19}$$

Applying the standard Newton-Raphson method to (19) gives,

710

$$i_D^{(j+1)} = i_D^{(j)} - J^{-1}F(i_D^{(j)})$$

$$= i_D^{(j)} - (R + R_D^{(j)})^{-1}(f_D(i_D^{(j)}) + Ri_D^{(j)} - V_{DD}) \qquad (20)$$

$$= (R + R_D^{(j)})^{-1}(V_{DD} + R_D^{(j)}i_D^{(j)} - f_D(i_D^{(j)}))$$

where $\qquad\qquad J = (dF(i_D)/di_D)|_{i_D^{(j)}}$

$$= (df_D(i_D)/di_D)|_{i_D^{(j)}} + R = R + R_D^{(j)} \qquad (21)$$

With the current-controlled diode represented by its Newton-Raphson discrete equivalent circuit, Figure 1(a) is changed to Figure 1(c). Solving Figure 1(c) for $i_D^{(j+1)}$ gives

$$i_D^{(j+1)} = (R + R_D^{(j)})^{-1}(V_{DD} - E_{Dh}^{(j)})$$

$$= (R + R_D^{(j)})^{-1}(V_{DD} - f_D(i_D^{(j)}) + R_D^{(j)}i_D^{(j)}) \qquad (22)$$

which is the same as given by (20).

3.2 Numerical verification of the diode models

Assuming that the parameters for the circuit of Figure 1 are: $V_{DD} = 5\ V$, $R = 1000\ \Omega$, $I_S = 6.9144 \times 10^{-16}\ A$, $n = 1$, and $V_T = 25\ mV$ [1]. The initial guess for the voltage-controlled diode is $0.7\ V$ and that for the current-controlled diode is $1\ mA$.

Four methods are used to solve the circuit of Figure 1:

1) Gauss Iteration Method for Voltage-Controlled Diode:

$$v_D^{(j+1)} = f(v_D^{(j)}) = V_T \ln \frac{V_{DD} - v_D^{(j)}}{R\ I_S} \qquad (4b)$$

2) Gauss Iteration Method for Current-Controlled Diode:

$$i_D^{(j+1)} = f(i_D^{(j)}) = (V_{DD} - V_T \ln \frac{i_D^{(j)}}{I_S})/R \qquad (23)$$

3) N-R Method for Voltage-Controlled Diode:

$$v_D^{(j+1)} = \frac{V_{DD} + RG_D^{(j)}v_D^{(j)} - Ri_D^{(j)}}{1 + RG_D^{(j)}} \qquad (24)$$

4) N-R Method for Current-Controlled Diode:

$$i_D^{(j+1)} = \frac{V_{DD} - f_D(i_D^{(j)}) + R_D^{(j)}i_D^{(j)}}{R + R_D^{(j)}} \qquad (25)$$

The results are shown in Table 1. For this specific voltage-controlled diode and the choosing initial guess, N-R method is slower than that of the Gauss iteration method. N-R method is quadratic convergent when the initial guess is near by the solution.

Table 1.   The computed results for the diode circuit of Figure 1.

	Gauss-v (V)	Gauss-i (mA)	NR-v (V)	NR-i (mA)
1	0.7	1.0	0.7	1.0
2	0.76342	4.30000	0.80174	4.16250
3	0.76278	4.23658	0.77600	4.28150
4	0.76278	4.23722	0.76458	4.23722
5		4.23722	0.76282	4.23722
6			0.76278	
7			0.76277	

4   SIMULATION OF THE PV MODULE

Many PV models, single-diode, two-diode, and three-diode models, etc., have been proposed in the literature. For simplicity, the single-diode model of Figure 3 is studied in this paper. This PV model offers a good compromise between simplicity and accuracy. In Figure 3, I_{PV} is the photo-generated current, D is the diode, R_S is the sum of several structural resistances, and R_P is the resistance due to the leakage current of the p-n junction. A PV module is a series connection of 36 to 72 PV cells.

The diode current is given by (A. S. Sedra & K. C. Smith, 2004)

$$i_D = g_D(v_D) = I_s \left(e^{v_D/V_T} - 1 \right) \approx I_s e^{v_D/V_T} \tag{26}$$

where I_S = Reverse saturation current, in A;
V_t = kTA/q, thermal voltage of the diode, in V;
V_T = $N_S V_t$, in V;
N_S = Number of series cells in the module;
k = 1.3806503, the Boltzmann's constant, in J/K;
T = 273° + t°, the module temperature in Kelvins;
A = Diode quality factor;
q = 1.60217646 x 10^{-19}, in Coulomb.

For this study, $I_S = 0.104 \ \mu A$, $T = 273° + 25° = 298° \ K$, $A = 1.31$, and $N_S = 50$.

Replacing the diode in Figure 3(a) with its Norton N-R discrete circuit (Figure 2(b)) results in Figure 3(b).

KCL at node $v_D^{(j+1)}$ gives

$$i^{(j+1)} = I_{PV} - G_D^{(j)} v_D^{(j+1)} - I_{Dh}^{(j)} - \frac{R_S \, i^{(j+1)} + v}{R_{Sh}} \tag{27}$$

where

$$G_D^{(j)} = (dg_D(v_D)/dv_D)|_{v_D^{(j)}} = \frac{I_S}{V_T} e^{v_D^{(j)}/V_T} \tag{28a}$$

$$I_{Dh}^{(j)} = i_D^{(j)} - G_D^{(j)} v_D^{(j)} \tag{28b}$$

The parameters for the PV module are taken from (Abir Chatterjee, 2011):
$I_{PV} = 7.36 \ A$, $I_S = 0.104 \ \mu A$, $V_{oc} = 30.4 \ V$,
$R_S = 0.251 \ \Omega$, and $R_P = 1168 \ \Omega$.

A Matlab program is written to solve $v_D^{(j+1)}$ from (27). The terminal voltage V steps from 0 to $V_{OC} = 30.4 \ V$ in 200 steps. The result of the characteristic I-V and P-V curves are shown in

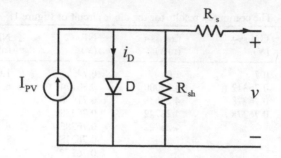

(a) Original Circuit

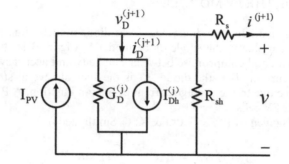

(b) The discrete circuit.

Figure 3. Single-diode circuit model of a practical PV module including the series and parallel resistances.

Figure 4. The Matlab program output gives $V_{OC} = 30.4$ V, $I_{SC} = 7.36$ A, $V_{mpp} = 24.32$ V, $I_{mpp} = 6.8764$ A, and $P_m = 167.23$ W. The parameters given by [7] are $V_{OC} = 30.4$ V, $I_{SC} = 7.36$ A, $V_{mpp} = 24.2$ V, $I_{mpp} = 6.83$ A, and $P_m = 165.29$ W. They match excellently.

5 DISCUSSIONS AND CONCLUSIONS

The large-signal models for the diode under the Newton-Raphson iteration have been derived in detail, proved theoretically, and verified computationally. Since the N-R discrete circuits are linear, all the properties of a linear circuit apply (L. O. Chua, 1987). They are accurate when exponential i-v relationship is used. When piecewise-linear i-v relationship is used and the operation does not change segment, the conductance (resistance) and history current (voltage) source of the model become constant, resulting in a linear circuit which does not require iteration to obtain a solution. This can be seen clearly by eliminating the superscripts (the iteration counters) in Figures 2(b) and 2(c) and set:

$G_D = G_D^{(j)}$ and $I_{Dh} = I_{D0}$ for Figure 2(b); and

$R_D = R_D^{(j)}$ and $E_{Dh} = E_{Dh}^j = -V_{D0}$ for Figure 2(c).

In the above equations, incremental conductance G_D and incremental resistance R_D are constant and $-V_{D0}$ and I_{D0} are the v_D axis and i_D axis intercepts of the segment, respectively; and $V_{D0} = R_D I_{D0}$.

Diode small-signal conductance or incremental conductance, $G_D = 1/R_D = g_d = 1/r_d$ is the slope of the forward i-v curve of the diode at the operating point, which is the only parameter used in the small-signal analysis circuit [1].

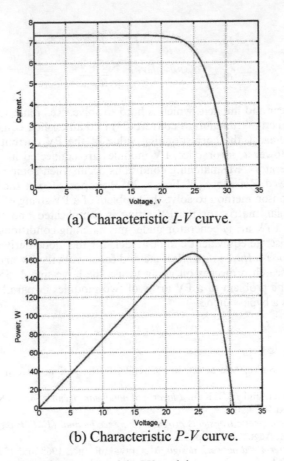

(a) Characteristic *I-V* curve.

(b) Characteristic *P-V* curve.

Figure 4.   Characteristic *I-V* and *P-V* curves of the PV module.

Diodes are used in PVs and many power electronics equipments. In (L. K. Wong, 2010), the small-signal model for an open-loop SEPIC converter is derived by applying the signal flow graph and the Mason's formula, which consists of fifty-seven (57) equations taking up five pages of the paper! The author is working on a research project to use the large-signal diode model derived in this paper and coded in Matlab, to obtain the the input to output transfer function and control to output transfer function.

It should be pointed out that although only the forward-biased diode *i-v* characteristic is used to derive the diode models, the models are general in nature and can be used in any electronic and power electronics circuits where non-ideal diode models are required and the operating frequency of the circuit is not high. In fact, (2) gives complete diode i-v characteristics: forward- and reverse-bias conditions, when *1* in the parenthesis is not omitted.

Renewable energy sources are getting more interest in recent years. Among them, PV arrays offer several advantages, such as little maintenance requirement and no environmental pollution. PV power generation is considered the ideal resource in distributed generation systems, which are located at or near by the load point. With the increasing penetration of PV generating stations in power grids, the model of a PV power station is needed (Abir Chatterjee 2011, L. K. Wong 2010, Mohamed Azab 2009, M. G. Villalva 2009, M. Abdulkadir 2012, Tarak Salmi 2012 & N. Femia 2013).

The output current of a practical PV module, Figure 3 (a), is

$$i = I_{PV} - i_D - \frac{v_D}{R_{sh}} = I_{PV} - I_s\,e^{v_D/nV_T} - \frac{v_D}{R_{sh}}$$

$$= I_{PV} - I_s\,e^{(R_s\,i+v)/nV_T} - \frac{R_s\,i + v}{R_{sh}} \qquad (29)$$

Eq. (29) is nonlinear and implicit, which is hard to solve. Reference (N. Femia, 2013) suggests to use the Lambert W function to convert (29) into an explicit equation, so that for any value of the PV voltage v the corresponding value of the PV current i can be calculated straightforwardly. However, in practice, PV modules are connected in series and parallel to form PV arrays/generators. Mismatching conditions occur when some modules in the array/generator are shadowed. Reference (N. Femia, 2013) proposes to use the Lambert W function and the Newton-Raphson method to solve the problem of a PV string of N modules. The derivation of the Jacobian matrix for the PV string is complicated and the derivation of the Jacobian matrix for a PV array/generator under mismatching conditions is much more complicated. The N-R discrete equivalent circuit models for the diode derived in this paper offer a much simpler and *non-software dependent* method for solving the PV array/generator simulation problems. A research project is on going to use the large-signal diode model derived in this paper to solve the problem of a PV array of two modules in parallel and then in series, and each module has a bypass diode.

REFERENCES

A. S. Sedra and K. C. Smith, Microelectronic Circuits, Fifth Edition, Oxford University Press, Inc., 2004, pp. 153–166.

L. O. Chua, C. A. Desoer, and E. S. Kuh, *Linear and nonlinear circuits*, McGraw-Hill Book Company, 1987, pp. 236-270 and 402–406.

H. W. Dommel, *Electromagnetic transients program reference manual (EMTP theory book)*, BPA, Portland, Oregon, U.S.A., August 1986, pp. 12–12 and I–10.

D. A. Calahan, *Computer-aided network design*, McGraw-Hill, Inc., 1968 and 1972, New York, pp. 61–73, 80-98, and 235–244.

J. Vlach and K. Singhai, *Computer methods for circuit analysis and design*, Van Nostrand Reinhold Company, New York, 1983, pp.275–280, 354–361, and 389–392.

W. L. Brogan, *Modern control theory*, Quantum Publishers, Inc, 1974, pp. 107–108.

Abir Chatterjee, Ali Keyhani, and Dhruv Kapoor, "Identification of photovoltaic source models," *IEEE Trans. Energy Conversion*, vol. 26, No.3, pp. 883–889, September 2011.

L. K. Wong, and T. K. Man, "Small signal modeling of open-loop SEPIC converters," *IET Power Electronic.*, 2010, *Vol. 3*, Iss.6, *pp.* 858–868.

Mohamed Azab, "Improved Circuit model of photovoltaic array," *International Journal of Electrical Power and Energy Systems Engineering 2:3, pp.* 185–188, 2009.

M. G. Villalva, J. R. Gazoli, and E. R. Filho, "Comprehensive approach to modeling and simulation of photovoltaic arrays," *IEEE Trans. Power Electronics*, vol. 24, No. 5, pp. 1198–1207, May 2009.

M. Abdulkadir, A. S. Samosir, and A. H. M. Yatim, "Modeling and simulation based approach of photovoltaic system in simulink model," *ARPN Journal of Engineering and Applied Sciences*, vol. 7, No. 5, May 2012.

Tarak Salmi, Mounir Bouzguenda, Adel Gastli, and Ahmed Masmoudi, "Matlab/Simulink based modeling of solar photovoltaic cell," *International Journal of Renewable Energy Research*, vol. 2, No. 2, 2012.

N. Femia, G. Petrone, G. Spagnuolo, and M. Vitelli, *Power Electronics and Control Techniques for Maximun Energy Harvesting in Photovoltaic Systems*, CRC Press, Taylor & Francis Group, Boca Raton, London, New York, 2013, pp. 22–32.

Emerging Developments in the Power and Energy Industry – Dufo-López, Krzywanski & Singh (eds)
© *2020 Taylor & Francis Group, London, ISBN 978-0-367-27169-5*

A multi-time scale optimization scheduling strategy of virtual power plant with responsive loads

F.M. Qu
China Electric Power Research Institute, Beijing, China

J. Zhao
Electric Power Research Institute of State Grid Henan Electric Power Company, Zhengzhou, Henan, China

Z. Cai
China Electric Power Research Institute, Beijing, China

C.F. Hu
State grid corporation of China, Beijing, China

S. Dai
China Electric Power Research Institute, Beijing, China

Q. Sun
Electric Power Research Institute of State Grid Henan Electric Power Company, Zhengzhou, Henan, China

ABSTRACT: From the perspective of virtual power plant (VPP) with electric vehicles, a multi-time scale scheduling strategy based on response time margin (RTM) and state of charge margin (SOCM) is proposed. Firstly, the VPP is grouped according to its output in a given scheduling period. Then, the RTM and SOCM indexes are defined on the basis of the power system scheduling target and the electric vehicles (EVs) users' travel demand, which were calculated and sorted to generate a priority queue of responsive EVs. With the progression of scheduling period and rolling iteration, the scheduling schemes of VPP for multiple time periods are determined. Finally, the VPP multi-time scale optimization scheduling strategy is validated by taking an EV aggregator containing three different traffic uses as an example.

Keywords: Electric vehicle, Virtual power plant (VPP), Multi-time scale, Scheduling strategy, Response time margin, State of charge margin

1 INTRODUCTION

With the continuous depletion of fossil energy, the world is faced with a huge challenge of energy shortage and environmental pollution issues (Chen C et al. 2016). At the end of the 20th century, the awareness of environmental protection promoted the energy technology revolution represented by new energy generation in energy production, and the development of green, clean and efficient consumption mode has also given rise to the booming of electric vehicles (EVs) in the meantime. The number of EVs in China is expected to reach 60 million by 2030 (Schuller A, et al. 2014).

Due to the small capacity of single EV battery, participating in the scheduling operation for a short time will lead to the change of State of Charge (SOC), which will have a non-negligible impact on its responsiveness at subsequent moments, and ultimately prevent VPP from formulating an effective scheduling strategy (J ZHAO, et al. 2011, Cao Y, et al. 2018).

There are two main types of research on scheduling strategy: centralized control and individual control. The centralized control formulates a unified charging and discharging strategy for VPP. Literature (J ZHENG, et al. 2013) established a charging and discharging strategy for large-scale EVs aggregator based on genetic algorithm with the goal of reducing charging load fluctuation. However, the centralized control strategy is usually unable to give a specific control scheme to EV individuals. Therefore, it is necessary to develop a series of differentiated control strategies to meet the actual needs. A multi-time-scale hierarchical control algorithm was developed in literature (Li C T, et al. 2013). Literature (J Meng, et al. 2016) proposed a frequency response control strategy of EVs aggregator considering users' travel behavior.

The above studies ignored the influence of the changes of SOC state and charge and discharge state on the response ability of VPP response system after the scheduling command, and this change of state is closely related to the scheduling strategy of VPP. In addition, with the expansion of VPP scale, the solution dimension of differential control problem increases dramatically, and the solution time of traditional nonlinear optimization algorithm cannot meet the needs of actual working conditions. In order to solve the above problems, this paper is based on a regulation strategy that comprehensively considers the response time and the SOC margin index.

2 SYSTEM MODEL OVERVIEW

The Multi-time scale scheduling model of VPP has three levels: 1) Control center level; 2) VPP level; 3) Individual EV level, as shown in the Figure 1.

In the Control Center level, the scheduling scheme is determined considering the load uncertainty, unit outage, and variation of renewable power which appears to be increasingly important as they are massively integrated into power grids. And the scheduling scheme is dispatched to the VPP within its available response capacity for regulation.

In the VPP level, the scheduling scheme issued by the control center is allocated to each EV by the strategy proposed in this paper.

In the Individual EV level, the charging/discharging state of individual EV changes correspondingly according to the control signals from the VPP.

3 MULTI-TIME SCALE SCHEDULING STRATEGY OF VPP

Before establishing the mathematical model, the response process of EV is decomposed into four combinations of response modes in Figure 2 according to the three different states of battery: discharge, charging and idle, and the following are abbreviated as modes I, II, III and IV respectively.

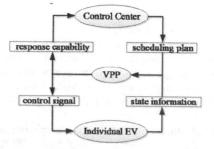

Figure 1.   Framework of VPP Multi-time scale scheduling model.

717

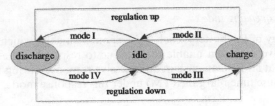

Figure 2.   Framework of VPP Multi-time scale scheduling model.

3.1   *A general model for single EV*

Taking the charging mode of slow charging for EV_i as an example, the energy storage characteristics of a single EV are shown in Figure 3. The shading part is the operation area of the EV_i in the process of V2G (M Wang, et al. 2018).

Take EV_i as an example, EV_i is connected to the power grid at $t_{i,s}$ and disconnected at the assumed $t_{i,d}$. To prevent the rapid aging of batteries, $S_{i,\max}$ and $S_{i,\min}$ are SOC upper and lower limits of EV_i respectively. EV_i starts charging at the rated power P_i^m immediately after it is connected to the grid at $t_{i,s}$, and switches to idle state when the SOC reaches $S_{i,\max}$, corresponding to the boundary *a-b-c* in Figure 3. On the contrary, boundary *a-d-e* indicates that EV_j is discharged at P_i^m since $t_{i,s}$, and stops discharging when the SOC is lower than $S_{i,\min}$. In addition, in order to ensure that the SOC of EV_i can reach the minimum SOC ($S_{i,d}$) required by the user when leaving the grid, the boundary *e-f* indicates that the battery needs to be charged compulsively before traveling.

In general, it is assumed that the dotted curve AP in Figure 3 represents the operating state of EV_i after it is connected to the power grid. Taking the operation point P as the current operating state, we can get the fastest charging process PX, idle process PY and the fastest discharge process PZ.

In the given scheduling period $[nT, (n+1)T]$, the time t_i^c when EV_i reaches the forced charging power can be obtained by Eq. (1).

$$t_i^c = \frac{\left(\frac{Q_i(S_i(nT) - S_i^d)\eta^d}{P_i^m} + nT\eta^c\eta^d + t_i^d\right)}{(\eta^c\eta^d + 1)} \tag{1}$$

where Q_i is the battery capacity and η^c, η^d respectively represent the charging and discharging efficiency.

The time t_i^l when EV_i reaches the lower bound of the controllable region can be obtained by Eq. (2). (Similarly, the time t_i^h reaching the upper bound can be calculated.)

$$t_i^l = Q_i(S_i(nT) - S_i^{\min})\frac{\eta^d}{P_i^m} + nT \tag{2}$$

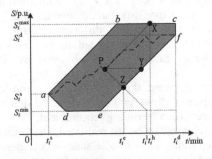

Figure 3.   Energy storage characteristic of an EV.

3.2 The definition of queueing index

Considering the energy of the battery, each EV responds quickly, which also leads to a rapid decline in its responsiveness, even unable to continue to respond. Take mode I for example, the maximum time that an EV can sustain an excessive response during the scheduling period is defined as the response time margin (RTM) under this response mode, as shown in Eq. (3).

$$RTM_i^n = \begin{cases} 0, nT \notin [t_i^s, t_i^d] \quad \text{and} \quad (n+1)T \notin [t_i^s, t_i^d] \\ T, \{t_i^s, t_i^l, t_i^c, t_i^d\} \not\subset (nT, (n+1)T] \\ \min(t_i^l, t_i^c, t_i^d, (n+1)T) - \max(t_i^s, nT), \text{otherwise} \end{cases} \tag{3}$$

From the Eq. (1), we can see that $RTM_i^n \in [0, T]$. When VPP needs to participate in optimal scheduling of power grid, the larger the RTM_i^n, the longer the response time, which means that it will be preferentially selected to participate in the response. However, in the $[nT, (n+1) T]$ period, there may be multiple EVs whose RTM_i^n is equal to T, and RTM alone will not provide clear prioritization reasons. Therefore, the state of charge margin is introduced.

The SOC margin (SOCM) index is defined as shown in Eq. (4) (Y QI, et al. 2015).

$$SOCM_i^n = \begin{cases} \dfrac{SOC_i(t) - SOC_i^{\min}}{SOC_i^{\max} - SOC_i^{\min}}, \Delta P(t) \leq 0 \\[2ex] \dfrac{SOC_i^{\max} - SOC_i(t)}{SOC_i^{\max} - SOC_i^{\min}}, \Delta P(t) > 0 \end{cases} \tag{4}$$

where $\Delta P(t)$ denotes the response demand of VPP, and the negative value means that the VPP needs to reduce the output, while the positive value means that the VPP needs to increase the output.

3.3 Multi-time scale scheduling strategy of VPP

The specific steps of multi-time scale scheduling strategy of VPP are described below:

1) VPP was divided into four groups according to four response modes. A single EV can certainly participate in multiple modes of response at the same time, which means that there is an intersection between different groups.

2) When the scheduling demand of the virtual power plant is determined, it should be allocated to all EV users in a certain way. This paper focuses on three kinds of allocation methods:

a) ID order control strategy

The EV aggregator is numbered according to the order of connecting EV charger, and the response capability is calculated in turn until the the aggregator composed of the first $No+1$ EVs can meet the response power demand. Finally, the control scheme of charging/discharging power of the No EVs is obtained.

b) prioritization control strategy

According to the four response modes of EV, the controllable EVs in VPP are divided into 4 groups. A single EV can obviously participate in multiple response modes simultaneously, i.e. there is intersection between different groups. Each controlled group is arranged in the order of RTM from large to small. For EVs with the same RTM, the corresponding priority queue of each group is obtained by referring to the descending order of SOCM.

3) According to the response demand, the EVs at the front end of the sequence is selected to adjust the charge/discharge power directly. The No can be determined by solving the minimum positive integer that meets the scheduling requirement.

When the VPP output is required to be up-regulated, the priority queue generated by the response mode group from charging state to idle state is considered first. When all EVs stop charging and still can not track the response demand curve, the "idle"→"discharge" group needs to be supplemented. Similarly, when the output of VPP is required to be down-regulated, the "discharge"→"idle" group is preferentially selected to generate a priority queue. When all EVs are charged and the response curve can not be tracked, the "idle"→"charge" group should be considered.

4 SIMULATION ANALYSIS

4.1 *Scenario setup*

Considering that EVs charging periods for different purposes have great differences, it has a critical influence on the results of calculation examples. According to statistics on registered vehicles in the UK, Private commuter vehicles, private non-commuter vehicles (generally characterized by those retired from work or who are unemployed) and corporate vehicles account for 61%, 30% and 9% of the EVs market for three different purpose, respectively (K Qian, et al. 2011). According to the above market share, this paper takes the aggregator composed of 300 EVs as an example to verify the proposed VPP multi-scale scheduling model. According to different vehicle traffic uses, the start-stop time distribution of EVs daily travel is shown in literature (M Wang, et al. 2015).

In addition, in order to realize the simulation of the control effect of VPP participating in the system scheduling, this paper set the scheduling time interval as 1h, and assumed that the scheduling demand of VPP issued by the scheduling center was shown in the dotted line of Figure 4. Finally, the change of dispatching requirement ΔP^n corresponding to the model is obtained, as shown by the real line of Figure 4.

4.2 *Results analysis*

According to the four response modes, VPP was divided into different groups and RTM and SOCM indexes were calculated respectively. Prioritization control strategy can significantly reduce the number of controlled EVs, and Figure 5 shows the number distribution of controlled EVs under different response modes. By the parameters optimal selection of RTM and SOCM, the number of EVs participating in the response mode II (i.e. "charging"→"idle" response) is significantly reduced. Even relative to the ID order control strategy, VPP do not have to discharge to the power grid in multiple periods. EVs have significant differences in RTM especially in the 19:00-22:00 period, and SOC levels were also uneven after one daytime travel in Figure 6.

The RTM and SOCM indices of each EV need to be updated and calculated continuously in each scheduling period. Different from ID order control strategy, the prioritization control strategy first ranks in descending order according to RTM. When RTM is equal, it further defines the higher priority level of the larger SOCM, participates in response preferentially on the basis of considering both scheduling and user travel demand, and finally obtains the decision-making conclusion when VPP participates in multi-time-scale scheduling. The results of EVs ranking in the above period are shown in Figure 7, and the data label in the figure is EV ID number.

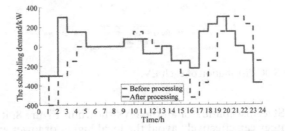

Figure 4.   The scheduling demand of VPP.

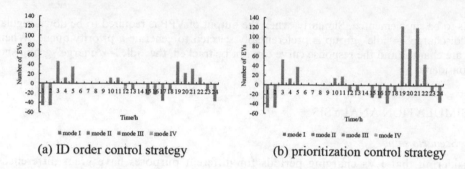

(a) ID order control strategy   (b) prioritization control strategy

Figure 5.   The number distribution of controlled EVs under different response modes.

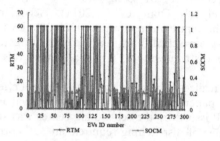

Figure 6.   Simulation results of RTM and SOCM.

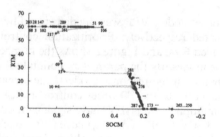

Figure 7.   The sequencing results results of each EV.

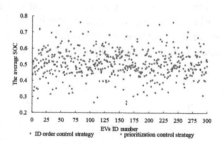

Figure 8.   The average SOC distribution of each EV.

From the average SOC distribution of all the EVs shown in Figure 8, it can be seen that the proposed control strategy can effectively avoid the local higher or lower average SOC, making the average SOC of each EV more centralized.

5 CONCLUSION

In order to develop a multi-time scale optimization scheduling strategy of VPP, the RTM and SOCM selection indexes are proposed.

1) RTM parameter is used to prioritize the EV with long duration response time, reduce the number of controlled EV and reduce the switching frequency of charging and discharging states in VPP. SOCM parameter takes the limitation and demand value of SOC as reference to realize that when VPP output is up-regulated (down-regulated), EV with a smaller difference from the lower limit of SOC (demand value of SOC) should be selected to stop charging (discharging) or even start discharging (charging) to avoid the local high or low average SOC.

2) On the basis of group division of VPP, EV aggregator is sorted according to the selected indicators, and priority queues are generated. Finally, controlled objects are determined according to scheduling demand and control instructions are issued. The VPP control strategy can meet the scheduling requirements and ensure the comfort of EV users.

ACKNOWLEDGMENT

This work was supported by The Science and Technology Project of State Grid Corporation of China (SGHADK00PJJS1700194).

REFERENCES

Chen C, Guo C, Man Z, et al. 2016. Control strategy research on frequency regulation of power system considering Electric vehicles. 2016 IEEE PES Asia-Pacific Power and Energy Engineering Conference.

Schuller A, Dietz B, Flath C M, et al. 2014. Charging strategies for battery electric vehicles: Economic benchmark and V2G potential. IEEE Transactions on Power Systems, 2014, 29(5).

ZHAO Junhua, WEN Fushuan, YANG Aimin, et al. 2011. Impacts of electric vehicles on power systems versus the dispatching and control problem. Automation of Electric Power Systems, 35(14): 2–10.

Cao Y, Tong W, Omprakash K, et al. 2018. An EV Charging Management System Concerning Drivers' Trip Duration and Mobility Uncertainty. IEEE Transactions on Systems Man & Cybernetics Systems, 48(4): 596–607.

ZHENG Jinghong, WANG Xiaoyu, MEN Kun, et al. 2013. Aggregation model-based optimization for electric vehicle charging strategy. IEEE Trans on Smart Grid, 4(2): 1058–1066.

Li C T, Ahn C, Peng H, et al. 2013. Synergistic control of plug-in vehicle charging and wind power scheduling. IEEE Transactions on Power Systems, 28(2): 1113–1121.

Meng J, Mu Y, Jia H, et al. 2016. Dynamic frequency response from electric vehicles considering travelling behavior in the Great Britain power system. Applied Energy, 162(2): 966–979.

Mingshen W, Yunfei M U, Jiang T, et al. 2018. Load curve smoothing strategy based on unified state model of different demand side resources. Journal of Modern Power Systems and Clean Energy, 6(3): 540–554.

QI Yebai, WANG Dan, JIA Hongjie, et al. 2015. Research on demand response for thermostatically controlled appliances based on normalized temperature extension margin control strategy. Proceedings of the CSEE:(21).

Qian K, Zhou C, Allan M, et al. 2011. Modeling of load demand due to EV battery charging in distribution systems. IEEE Transactions on Power Systems, 26(2): 802–810.

Mingshen W, Yunfei MU, Hongjie J I A, et al. 2015. A preventive control strategy for static voltage stability based on an efficient power plant model of electric vehicles. Journal of Modern Power Systems and Clean Energy, 3(1): 103–113.

Emerging Developments in the Power and Energy Industry – Dufo-López, Krzywanski & Singh (eds)
© 2020 Taylor & Francis Group, London, ISBN 978-0-367-27169-5

Simulation and analysis of the near-field magnetic shielding effectiveness of shielding box with aperture at low frequency

L.L. Chen
State Key Laboratory of Advanced Power Transmission Technology, Global Energy Interconnection Research Institute, Beijing, China

T.L. Li & C.Q. Jiao
State Key Laboratory of Alternate Electric Power System with Renewable Energy Sources, North China Electric Power University, Beijing, China

ABSTRACT: The transient large current during the DC circuit breaker opening process produces a high transient magnetic field, which may cause transient disturbance voltage induced at the ports of the IGBT secondary control circuit. This disturbance may threaten the normal operation of secondary control circuit. Usually, metal shielding box is used to suppress the magnetic field. In this paper, the shielding effectiveness of an aluminum shielding box used in practical engineering is simulated by CST. Based on the results, the magnetic shielding effectiveness is analyzed in both time domain and frequency domain (DC-1MHz), which is useful for guiding the design of the shielding box.

1 INTRODUCTION

With the development of flexible HVDC transmission technology, more and more DC circuit breakers are used and the voltage level is higher and higher. During the opening and closing of the DC circuit breaker, Sharp changes in voltage and current will generate high intensity transient electromagnetic field in and around the valve tower, which in turn creates spatial electromagnetic disturbances to control circuits such as IGBT control circuits placed on the valve tower. Taking the breaking process at 25kA level as Example, the peak value of the transient magnetic field near the control circuits is about 25kA/m, the falling time is about several microseconds, and the spatial distribution is relatively uniform (Zhao, Z.B. et al. 2017, Ding, X. et al. 2018). With the increasingly high level of integration, the control board is easy to be affected by electromagnetic disturbance. installing metal shielding box is necessary in practical engineering.

At present, the research on shielding effectiveness of shielding box focuses on the high-frequency band (Hu 2018, Jiao 2016), while the research on near-field shielding characteristics under low-frequency electric and magnetic field (DC-1MHz) are relatively less. Moreover, the design of the shielding box is complex and diverse, it is urgent to carry out the analysis of the shielding effectiveness of the shielding box. For low-frequency electric field, it is relatively easy to be shielded (Jiao 2015, Lee 1979). In this paper, the near-field shielding effectiveness against magnetic field is mainly illustrated.

It is relative amplitude of the magnetic field before and after the shielding box loaded that is investigated when getting the shielding effectiveness of the box. The shielding effectiveness has little correlation with the value of the magnetic field strength (for non-magnetic shielding material or magnetic material not saturate) (Bottauscio 2006). Therefore, when performing simulation calculations, it is not necessary to pay too much attention to the magnetic field strength. The advantage of simulation is that it is easier to set a uniform magnetic field to simulate the actual environment in which the shielding box is located.

2 TIME DOMAIN SIMULATION

2.1 *Time domain calculation principle*

A pulse magnetic field is created when a sharply varying current is generated when the DC circuit breaker is opening. In practical engineering, the induced voltage on the internal board is often concerned. Therefore, time domain simulation can be conducted by applying pulse current excitation, the shielding efficiency of the shielding box can be evaluated by comparing the time domain waveforms before and after shielding.

In this paper, a circle with a diameter of 5cm is used to represent the internal control board, which is regarded as a probe. According to the peak value U_{max} of the voltage waveform induced on the probe, the time domain shielding effectiveness SE_t is defined as:

$$SE_t = 20 \lg(U_{max0}/U_{maxs})(dB) \tag{1}$$

Where U_{max0} represents the peak value of the probe induced voltage without shielding, and U_{maxs} is the peak value of the probe induced voltage with shielding. SE_t is not a frequency domain magnetic field shielding effectiveness, which represents the attenuation of the induced voltage by a certain loop after shielding box loaded. This parameter can be used to evaluate the magnetic disturbance.

2.2 *Time domain simulation results and shielding effectiveness calculation*

The shielding effectiveness of the shielding box is simulated by CST. Shielding box dimensions: 225mm × 195mm × 40mm (thickness: 1mm), material: aluminum (conductivity $\sigma = 3.8 \times 10^7$S/ m, relative permeability $\mu r = 1$), and one side has 3 apertures with dimension 26mm × 9mm, one side has 2 apertures with dimension 20mm × 12mm, and one side has one aperture with dimension 24mm × 20mm.

In CST, the low-frequency module, time domain solution is used. then build a shielding box model. In order to generate a uniform pulsed magnetic field, the magnet coil model is built as a rectangular coil of 1m × 1m, which is excited by a current. The receiving probe is a circle with a diameter of 5cm and is placed in the center of the magnet coil. The induced voltage on the probe with and without shielding box loaded is recorded. Considering that the actual board is placed parallel to the larger surface of the shielding box, when the probe model is built in the software, the relative position of the box and probe should be maintained. The specific model is constructed as shown in Figure 1.

Gaussian pulse excitation current is applied, its frequency band is set between DC-1MHz, the peak value is set to 1kA, which is shown in Figure 2. For the magnet coil set in this paper, the coil factor is about 0.89, then the peak value of the magnetic field at the center of the coil is about 0.89kA/m. When the shielding box is not loaded, the induced voltage on the probe is shown in Figure 3. After shielding box loaded, the induced voltage is shown in Figure 4. It can be seen that the peak value of the induced voltage decreases obviously after shielding. If

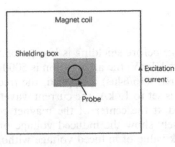

Figure 1. Time domain simulation model.

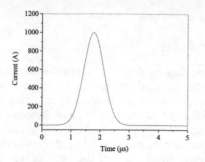

Figure 2. Gaussian pulse excitation current in the magnet coil.

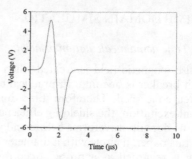

Figure 3. The induced voltage on the probe without shielding under gaussian pulse magnetic field.

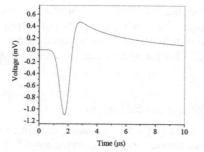

Figure 4. The induced voltage on the probe with shielding under gaussian pulse magnetic field.

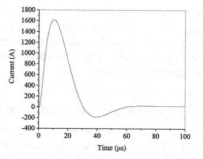

Figure 5. Standard pulse excitation current in the magnet coil.

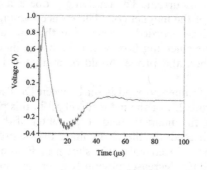

Figure 6. The induced voltage on the probe without shielding under standard pulse magnetic field.

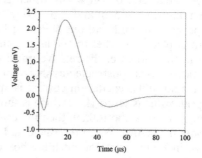

Figure 7. The induced voltage on the probe with shielding under standard pulse magnetic field.

the peak value of induced voltage before shielding is U_{0max} = 5.5V, the peak value of induced voltage after shielding is U_{smax} = 1.1mV, the attenuation is 5000 times, then SE_t = 74dB.

Standard pulse excitation current (16/64μs) is applied, the frequency is mainly concentrated around 0.1MHz, the peak value is set to 1.6kA. the current waveform is shown in Figure 5. The peak value of the magnetic field at the center of the magnet coil is approximately 1.4kA/m. Figure 6 and Figure 7 respectively show the induced voltage on the probe without and with shielding box loaded. If the peak value of induced voltage without shielding is U_{0max} = 0.88V, with shielding is U_{smax} = 2.2mv, and the attenuation is 400 times, then SE_t = 52dB.

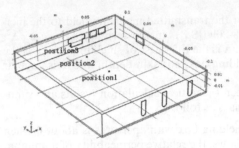

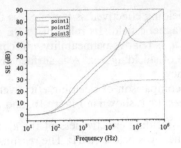

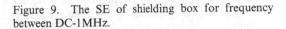

Figure 8. The position of the probe in CST.

Figure 9. The SE of shielding box for frequency between DC-1MHz.

It can be seen from the simulation results that the SE_t calculation results are different for different waveforms in different frequency band. The reason is that, within a certain range, the higher the frequency is, the higher the shielding efficiency of the shielding box will be.

Considering the actual engineering electromagnetic environment, it can be predicted that when the shielding box is installed on the secondary board, the induced voltage on the board will be greatly reduced. When the actual engineering magnetic field is 25kA/m and the board is characterized by a circle with a diameter of 5cm, the induced voltage is less than 50mV (the standard pulse result is taken).

3 FREQUENCY DOMAIN SIMULATION

Using the frequency domain solver, a uniform magnetic field is applied to the solution space, the magnetic field is still injected from the larger side, the shielding effectiveness is simulated point by point from DC to 1MHz.

The position of the observation point in the shielding box is shown in Figure 8, where point1 is located in the center of the shielding box, point2 is offset by 0.06m in the +y direction, and point3 is offset by 0.1m. The shielding performance curve is shown in Figure 9.

Piont2 can represent the position where the actual board is placed, take point2 for analysis of the characteristics of shielding effectiveness, we can have the following conclusions:

1) The magnetic shielding of aluminum shielding box with apertures in frequency domain are as follows: The shielding effectiveness is 0 when frequency is below 100Hz, then increases with the increase of frequency until reaching about 60dB at 0.1MHz, at last the shielding effectiveness remains around 60dB.
2) The closer the observation point is to the aperture, the worse the shielding effectiveness is.

4 THEORETICAL CALCULATION AND SHIELDING MECHANISM ANALYSIS

There is no accurate formula for the magnetic shielding effectiveness of the rectangular aluminum shielding box with apertures. But an analytical formula for the closed spherical and infinite long cylindrical structure is exist, which can be refer to when calculating the shielding effectiveness of closed rectangular shielding box. The approximate expression is as follows (Mager 1970, Andrieu 2012, Tesche 2009, Cooley 1968):

$$S_H = \begin{cases} 1 + \frac{j\omega\mu_0\sigma V\Delta}{A}, & \Delta \ll \delta \\ \left(\frac{1}{2} + \frac{V}{2\mu_r\Delta A}\sqrt{j\omega\tau_d}\right)e^{\sqrt{j\omega\tau_d}}, & \Delta \gg \delta \end{cases} \quad (2)$$

The shielding effectiveness factor S_H is the ratio of the transmitted magnetic field to the incident magnetic field. Its relation to SE is : $SE = -20\lg(|S_H|)$. In the formula, $\tau_d = \mu_0\mu_r\sigma\Delta^2$, $\mu_0 = 1$, μ_r is relative permeability, σ is conductivity, Δ is the thickness of shielding box, V is the volume of shielding box, A is surface area of shielding box, δ is skin depth, and ω is angular frequency.

The comparison of shielding effectiveness between theoretical calculation results and simulation results is shown in Figure 10, and the analysis is as follows:

1) The shielding effectiveness of the aluminum shielding box with apertures is about 0 when frequency is below 100Hz. The reasons are as follows: the relative permeability of aluminum is 1, at extremely low frequency, the dominant shielding mechanism is not Magnetic flux shunt, but eddy shielding. At low frequency the induced eddy current is small and the shielding efficiency is low.

2) After 100Hz, the shielding effectiveness increases with the increase of frequency, until 0.1MHZ, the shielding effectiveness reaches 60dB. At this moment, the eddy current induced on the surface of the metal shielding box plays the main role. There is a certain deviation between the simulation and the analytical results when frequency is blow 0.1MHz, which is caused by the approximate expression of the analytical solution.

3) Above 0.1MHz, the analytical results continue to rise with the frequency increasing, and the simulation results are constant above 0.1MHz, which is caused by the apertures of the shielding box. The explanation is as follows: the external magnetic field has two ways to enter the shielding box, one is through the metal plate, the other is through the apertures. The shielding effectiveness comprehensively reflects the role of the two paths. The magnetic field penetrating the metal plate decreases with increasing frequency, and the magnetic field entering the shielding box through the apertures is substantially independent of the frequency (for the frequency band under consideration). When the frequency is high, the magnetic field that penetrates the metal is small, lots enter the shielding box through the apertures. Therefore, the shielding effectiveness is maintained at a certain level instead of continuing to grow with frequency like a closed shield (Frikha 2015, Gao 2010). At this time, the shielding effectiveness is basically determined by the size and quantity of the apertures (Huang 2017).

In order to analysis the influence of the apertures, set the material as PEC (perfect electric conductor), then the magnetic field can only penetrate into the box through the apertures. Figure 10 shows the results, it indicates that the shielding effectiveness is none of relationship to the frequency, only influenced by the dimension and number of the apertures.

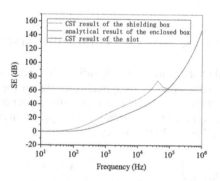

Figure 10.   Comparison of the analytical and simulation results.

5 CONCLUSION

In this paper, pulsed magnetic field is used to simulate the shielding effectiveness of aluminum shielding box used in practical engineering, and the time-domain characteristics is obtained. Then the frequency domain characteristics in the range of DC-1MHz at different position in the shielding box are obtained. The main conclusions are as follows:

1) When the DC circuit breaker opens the current of 25kA, the peak value of the transient magnetic field is about 25kA/m around the control circuit, and the falling time is about several microseconds. In this paper, Gaussian pulse and standard pulse magnetic field is used to simulate the characteristics of the magnetic disturbance.
2) It can be predicted that when the circuit breaker opens the 25kA current and the control circuit can be characterized by a circular circuit with a diameter of 5cm, the peak value of induced voltage on the control circuit will not exceed 50mV.
3) The magnetic shielding of the aluminum shielding box with apertures in frequency domain are as follows: The shielding effectiveness is 0 below 100Hz, then increases with the increase of frequency, and the shielding effectiveness reaches about 60 dB until 0.1MHz, in this frequency band, eddy current plays a major role for magnetic shielding. Then the shielding effectiveness is maintained at about 60dB. At this time, the effect of the apertures appears, and the shielding effectiveness does not continue to increase with frequency like the enclosed shielding box. Further research is needed on shielding boxes of different materials and dimension, as well as the detailed effect of apertures.

ACKNOWLEDGEMENT

This project is supported by State Key Laboratory of Advanced Power Transmission Technology (GEIRI-SKL-2018–006).

REFERENCES

Andrieu, G. & Panh, J. & Reineix, A. et al. 2012. Homogenization of Composite Panels From a Near-Field Magnetic Shielding Effectiveness Measurement[J]. IEEE Transactions on Electromagnetic Compatibility 54(3):700–703.

Bottauscio, O. & Chiampi, M. & Roccato, P.E. et al. 2006. 1–100 kHz magnetic shielding efficiency by metallic sheets: modeling and experiment by a laboratory test bed. IEEE Transactions on Magnetics, 42(10),3533–3535.

Cooley, W.W. 1968. Low-frequency shielding effectiveness of nonuniform enclosures[J]. IEEE Transactions on Electromagnetic Compatibility, 10(1):34–43.

Ding, X. & Tang, G.F. & Han, M.X. et al. 2018. Characteristic Parameters Extraction and Application of the Hybrid DC Circuit Breaker in MMC-HVDC[J]. Proceedings of the CSEE, 38 (01):309–319+369.

Frikha, A. & Bensetti, M. & Duval, F. et al. 2015. A New Methodology to Predict the Magnetic Shielding Effectiveness of Enclosures at Low Frequency in the Near Field[J]. IEEE Transactions on Magnetics, 51(3): 800404.

Gao, C. & Liu, X. & Shi L.H. et al. 2010. Measurement method and measurement equipment development for low-frequency magnetic field[J]. High Voltage Engineering, 36(9):2272–2277.

Hu, P.Y. & Zhao, Y. & Yang, J.P. et al. 2018. A Fast Approach for Calculating the Shielding Effectiveness of Rectangular Enclosures with Apertures under Oblique Plane Wave Illuminations. Transactions of China Electrotechnical Society 33(15): 3651–3660.

Huang, J. 2017. Study on Near-Field Shielding Effectiveness of an Enclosure with an Aperture[D]. Beijing Jiaotong University.

Jiao, C.Q. & Li, S.J. 2016. Shielding Effectiveness Comparison of a Rectangular Box with an Aperture Covered by Conductive Sheet. Transactions of China Electrotechnical Society, 31(1): 112–118.

Jiao, C.Q. & Niu, S. & Li, L. 2015. Experiment Study of Power Frequency Electric and Magnetic Shielding Effectiveness for composite Materials[J]. Transactions of China Electrotechnical Society, 30 (10):1–6.

Lee, K. & Bedrosian, G. 1979. Diffusive electromagnetic penetration into metallic-enclosures. *IEEE Transactions on Antennas & Propagation*, 27(2),194–198.

Mager, A. 1970. Magnetic shields. Magnetics IEEE Transactions on,6(1), 67–75.

Tesche, F.M. & Ianoz, M. & Karlsson, T. 2009. *EMC Analysis Methods and Computational Models*[M]. Beijing, China: Beijing University of Posts and Telecommunications Press, 316–338.

Zhao, Z.B. & Deng, E.P. & Zhang, P. et al. 2017. Review of the Difference between the Press Pack IGBT Using for Converter Valve and for DC Breaker. *Transactions of China Electrotechnical Society*, 32(19): 125–133.

Emerging Developments in the Power and Energy Industry – Dufo-López, Krzywanski & Singh (eds)
© 2020 Taylor & Francis Group, London, ISBN 978-0-367-27169-5

Study on corrosion fault diagnosis for grounding grid based on the law of conservation of power

Beisheng Yuan, Chaofan Li, Fujue Wang, Xinyu Feng & Yameng Fu
College of Energy and Electrical Engineering, Hohai University, Nanjing Jiangsu, China

ABSTRACT: The corrosion fault diagnosis of grounding grid requires multiple incentives and repeated measurement to get locations of the fault branches. In order to solve this problem, this paper established diagnostic equations of grounding grid fault based on the conservation between input power and consumed power, and solved the nonlinear programming problem, which indicated the approximate locations of the fault branches in the grounding grid by measuring only one time. Experiment results proved the feasibility and practicability of the diagnosis method described in this paper and indicated that it had practical value for diagnosing the locations of fault branches in small-scale grounding grid. (Wei Zhang et al. 2018, Huizhong Xu et al. 2009).

1 INTRODUCTION

Grounding grid is an important part of substation, which plays a vital role in the safe and stable operation of substation. After a certain period of operation, various branch conductors and grounding leads of the grounding grid will be corroded to varying degrees, even broken, endangering the safe operation of the grid and causing huge losses. Therefore, it is very urgent and important to detect the corrosion of grounding grid, monitor the corrosion of grounding grid in time, find out problems as soon as possible and take corresponding protective measures to complete the maintenance of grounding grid. (Jian Liu et al. 2008).

Aiming at the problem of grounding grid corrosion diagnosis, this paper first establishes the fault diagnosis equation (Lei Xu & Lin Li. 2012, Jian Liu. 2010) based on the network theory, then establishes the objective function based on the law of power conservation, solves the diagnosis equation, and obtains the network corrosion situation. Finally, a small grounding network is selected to simulate the corrosion occurrence, which verifies the effectiveness and feasibility of the proposed method.

2 EXPERIMENT PRINCIPLE

The structure of grounding network is shown in Figure 1. Assuming that there are n+1 independent nodes (one of which is a reference node), m+1 accessible nodes (one of which is a reference node) and b branches. By ignoring the effects of inductance and capacitance, the grounding grid can be equivalent to a pure resistance linear network.

According to the theory of electric network, the relation between voltage value of each accessible node and resistance value of each branch can be deduced. When the excitation current source is added externally, the voltage value of each accessible node can be measured. At the same time, we can detect the resistance of each branch by solving the equation.

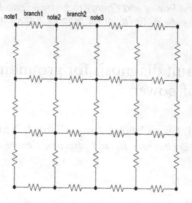

Figure 1. Equivalent diagram of grounding grid.

2.1 *Establishment of fault diagnosis equation*

For the equivalent pure resistance linear network mentioned above, it can be obtained from the electrical network theory:

$$G_n = A \bullet G \bullet A^T \tag{1}$$

$$G_n \bullet U_n = J_n \tag{2}$$

In the formula: A is the circuit network incidence matrix; G is the branch admittance matrix ($G = R^{-1}$), which is a diagonal matrix; G_n is the node admittance matrix; J_n is the current source excitation column vector of the node; U_n is the node voltage column vector.

According to formula (2):

$$U_n = G_n^{-1} \bullet J_n \tag{3}$$

Based on the above equation, the influence of resistance value changes of each branch on node voltage can be deduced. That is the partial derivative of resistance value of each branch R_i can be obtained, which is:

$$\frac{\partial U_n}{\partial R_i} = \frac{\partial G_n^{-1}}{\partial R_i} \bullet J_n + \frac{\partial J_n}{\partial R_i} \bullet G_n^{-1} \tag{4}$$

The current source excitation is constant, so

$$\frac{\partial J_n}{\partial R_i} \bullet G_n^{-1} = 0$$

Equation (4) can be changed into:

$$\frac{\partial U_n}{\partial R_i} = \frac{\partial G_n^{-1}}{\partial R_i} \bullet J_n \tag{5}$$

According to:

$$\frac{\partial (G_n^{-1} \bullet G_n)}{\partial R_i} = 0 \tag{6}$$

we can get:

$$\frac{\partial G_n^{-1}}{\partial R_i} = -G_n^{-1} \bullet \frac{\partial G_n}{\partial R_i} \bullet G_n^{-1}$$

$$= -G_n^{-1} \bullet A \bullet \frac{\partial G}{\partial R_i} \bullet A^T \bullet G_n^{-1}$$

In formula (7):

$$\frac{\partial R^{-1}}{\partial R_i} = \begin{pmatrix} 0 & & & & \\ & \ddots & & & \\ & & -\frac{1}{R_i^2} & & \\ & & & \ddots & \\ & & & & 0 \end{pmatrix} \tag{8}$$

We substitute Equation (7) into Equation (5) to obtain:

$$\frac{\partial U_n}{\partial R_i} = -G_n^{-1} \bullet A \bullet \frac{\partial G}{\partial R_i} \bullet A^T \bullet G_n^{-1} \bullet J_n \tag{9}$$

After the grounding network failure, it is assumed that the column vector composed of the voltage increment of each node is ΔU_n.
Then:

$$\Delta U_n = \sum_{i=1}^{b} \frac{\partial U_n}{\partial R_i} \bullet \Delta R_i$$

$$= -G_n^{-1} \bullet A \bullet \begin{pmatrix} -\frac{\Delta R_1}{R_1^2} & & & & \\ & \ddots & & & \\ & & -\frac{\Delta R_i}{R_i^2} & & \\ & & & \ddots & \\ & & & & -\frac{\Delta R_b}{R_b^2} \end{pmatrix} \bullet A^T \bullet G_n^{-1} \bullet J_n$$

$$= G_n^{-1} \bullet A \bullet R^{-1} \bullet \begin{pmatrix} \frac{\Delta R_1}{R_1} & & & & \\ & \ddots & & & \\ & & \frac{\Delta R_i}{R_i} & & \\ & & & \ddots & \\ & & & & \frac{\Delta R_b}{R_b} \end{pmatrix} \bullet A^T \bullet G_n^{-1} \bullet J_n$$

$$= T_1 \bullet \begin{pmatrix} x_1 & & & \\ & \ddots & & \\ & & x_i & \\ & & & \ddots \\ & & & & x_b \end{pmatrix} \bullet T_2 \tag{10}$$

In formula (10) : R_i is the resistance of the branch i. ΔR_i is the changing resistance after branch i fault. $x_i = \frac{\Delta R_i}{R_i} (i = 1, 2, \cdots, b)$ is the changing rate of resistance of each branch.

732

$$\mathbf{T_1} = \mathbf{G_n}^{-1} \bullet \mathbf{A} \bullet \mathbf{R}^{-1}.$$

$$\mathbf{T_2} = \mathbf{A}^\mathbf{T} \bullet \mathbf{G_n}^{-1} \bullet \mathbf{J_n}.$$

From the above equation, it can be seen that the voltage change of each independent node has a linear relationship with the changing rate of the resistance value of each branch. In the actual measurement, the number of accessible nodes whose voltage value can be measured is m. So we can get m equations about the changing rate of branch resistance $x_i (i = 1, 2, \cdots, b)$.

$$\begin{cases} \Delta u_1 = f_1(x_1, x_2, \cdots, x_b) \\ \Delta u_2 = f_2(x_1, x_2, \cdots, x_b) \\ \quad \cdots \cdots \\ \Delta u_m = f_m(x_1, x_2, \cdots, x_b) \end{cases} \tag{11}$$

Equation (11) is the fault diagnosis equation. In the formula: $\Delta u_i (i = 1, 2, \cdots, m)$ is the change of node voltage after failure. $f_i (i = 1, 2, \cdots, m)$ is a function of $x_i (i = 1, 2, \cdots, b)$. The change of resistance of each branch can be obtained by solving the equations.

2.2 *Objective function*

n the actual grounding grid, the number of accessible nodes (m+1) is always less than which of branches (b), so the fault diagnosis equation is an underdetermined equation without unique solution. But in the actual fault diagnosis, the solution of the diagnosis equation is required to be unique. In order to solve it, the solution of the diagnosis equation must be combined with the objective function, so that the solution can reflect the true situation.

The objective function is established based on the principle of power conservation. When the grounding grid is corroded, the sum of the power consumed by each branch should be equal to the power input in the grounding grid. Therefore, the objective function of the minimum difference between the consumed power and the input power can be established to solve the underdetermined equations of the fault diagnosis. And then, the corrosion condition of each branch will be obtained. The specific establishment process of the objective function is as follows:

$$P_{in} = U_n^T \bullet J_n \tag{12}$$

Among the above equation, $\mathbf{U_n}^\mathbf{T}$ is the node voltage column vector, and $\mathbf{J_n}$ is the node current column vector.

$$P_c = \sum_{i=1}^{b} \frac{U_i^2}{R_i} \xrightarrow{matrix} R^{-1} * \begin{pmatrix} U_1^2 \\ \vdots \\ U_i^2 \\ \vdots \\ U_b^2 \end{pmatrix} \tag{13}$$

The branch voltage column vector $\mathbf{U_b} = \mathbf{A} * \mathbf{U_n}$, that is, the branch voltage column vector $\mathbf{U_b}$ can be obtained by the product of the adjoint matrix and the node voltage column vector. The following objective function can be established from the above two equations:

$$\min P = |P_c - P_{in}| = \left| \sum_{i=1}^{b} \frac{U_i^2}{R_i} - \mathbf{U_n}^\mathbf{T} * \mathbf{J_n} \right| \qquad (14)$$

2.3 Solution process of computer

The corresponding fault diagnosis equation was established by MATLAB, and the corresponding objective function was obtained by substituting relevant data. The obtained objective function and constraint equation are solved by LINGO software for nonlinear programming, and the final result is obtained.

3 COMPUTER SIMULATION

In order to simplify the solution process, the small grounding grid model is studied first. The situation is shown in Figure 2. The grounding grid has 9 nodes and 12 branches. It is assumed that the resistance of each branch is 5Ω, and then a 10A DC current source is connected to nodes 3 and 9. Then, substitute the values of the voltages calculated by simulation into MATLAB to get the corrosion of each branch.

3.1 When single branch are corroded

When severe corrosion occurs at a certain part of the grounding grid, assume that the resistance value of the branch 11 is increased by 20 times to 100Ω, the situation is shown in Figure 3 and Table 1.

Experiment 3.1 completed the diagnosis of a single branch. When the branch 11 is corroded, the experimental results are shown in Figure 4. It can be seen that the location of the fault can be accurately determined by the method of this paper.

3.2 When two branches are corroded

When two branches are severely corroded and the corrosion conditions are different, assume that a corroded branch 4 is added to the previous single branch fault, and the resistance value is assumed to increase by 100 times to 500Ω, the situation is shown in Figure 5 and Table 2.

Experiment 3.2 completed the diagnosis of two corroded branches. As can be seen from Figure 6, the experimental model can also accurately determine the location of the fault point.

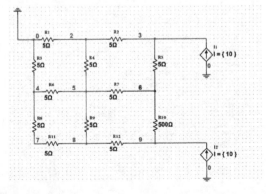

Figure 2.   2*3 Simulation grounding grid topology.

734

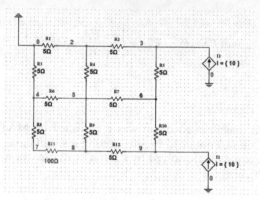

Figure 3.   Branch 11 is severely corroded.

Table 1.   The data sheet of single corroded branch test.

Branch	Change value	Branch	Change value
X1	0.4005129E-07	X7	0.4370883E-04
X2	0.1796146E-05	X8	0.3704392E-04
X3	0.1854912E-04	X9	0.2269124E-04
X4	0.4029401E-04	X10	0.4046936E-04
X5	0.1470103E-03	X11	2.904511
X6	0.1033071E-04	X12	0.2997227E-04

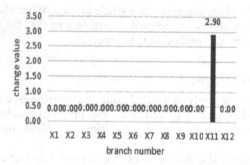

Figure 4.   The result of single branch severe corrosion.

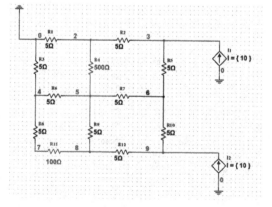

Figure 5.   Branch 4 and branch 11 are severely corroded.

Table 2. The data sheet of two corroded branch test.

Branch	Change value	Branch	Change value
X1	0.9214188E-05	X7	0.7025102E-04
X2	0.8274673E-05	X8	0.2578776E-04
X3	0.2325038E-04	X9	0.1090630E-03
X4	2.801742	X10	0.8845244E-05
X5	0.1833337E-03	X11	3.130506
X6	0.6793888E-04	X12	0.2857303E-04

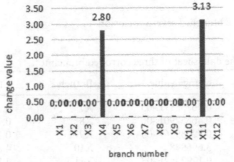

Figure 6. The result of two branched severe corrosion.

3.3 *When three branches are corroded*

On the basis of the original, an error branch is added to carry out experiments to verify the reliability of the model. Assume that the add branch 10 is a faulty branch and assume that its resistance becomes 500 Ω, the situation is shown in Figure 7, Figure 8, and Table 3.

Experiment 3.3 completed the diagnosis of severe corrosion of multiple branches. It can also be clearly seen from Figure 7 that the model proposed in this paper can accurately locate the fault point.

4 CONCLUSION

A) Fault diagnosis of grounding network based on power conservation is feasible, fault branches can be accurately diagnosed in the case of serious corrosion of multiple branches.

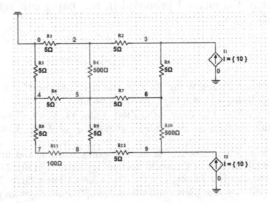

Figure 7. Branch 4, branch 10 and branch 11 are corroded.

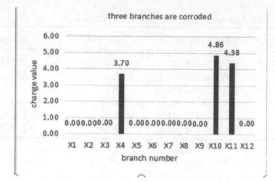

Figure 8. The result of three branched severe corrosion.

Table 3. The data sheet of three corroded branch test.

Branch	change value	Branch	Change value
X1	0.5675817E-05	X7	0.3150641E-04
X2	0.4655290E-06	X8	0.1519411E-04
X3	0.7208956E-05	X9	0.1918606E-05
X4	3.696948	X10	4.864616
X5	0.1792062E-03	X11	4.380354
X6	0.2655741E-04	X12	0.4165215E-04

B) The method adopted in this paper has the characteristics of fewer measurements and can uniquely determine the location of fault points through one measurement in small-scale fault diagnosis of grounding grids.

C) When calculating the power consumption of the whole grounding grid, it is based on the voltage values of all nodes already obtained. Therefore, the original grounding grid needs to be transformed into the intrinsic grounding grid. Therefore, the method proposed in this paper is based on the simplified cost of the actual grounding grid

D) Although the proposed method can quickly locate the fault points, the experimental results can not strictly reflect the real situation of the change of the resistance value of the fault points. Only when the fault points are seriously corroded and the resistance value increases to a certain value, such as 20 times of the value set in this paper, can it be reflected through the results. However, considering the actual situation, only when the grounding grid is seriously corroded can it be reflected through the results. Because of the need of excavation, the method proposed in this paper can achieve good practical results in practical application.

REFERENCES

Jian Liu, Shuqi Wang, Zhizhong Li, Sen Wang. 2008. Testability Study on Corrosion Fault Diagnosis of Grounding Grid [J]. High Voltage Technology, 2008(01): 64–69.

Wei Zhang, Zhuhui Zhong, Wei Li, Yang Liu, Zifei Yan, Yi Yang, Yuanran Liang, Peng Li. 2018. Research on fault location of substation grounding grid [J]. Electric ceramic arrester, 2018(03): 129–133.

Lei Xu & Lin Li. 2012. Diagnostic method of grounding grid corrosion and breakpoint of substation based on network theory [J]. Journal of Electrical Technology, 2012, 27 (10): 270–276.

Huizhong Xu, Hongzhong Ma, Zhixin Zhang, Lan Pu, Guanghe Li, Feng Zhao. 2009. Research on fault diagnosis of grounding grid in power plants and substations [J]. Power system protection and control, 2009, 37 (24): 51–54.

Jian Liu. 2010. Fault Diagnosis of Ground Grid [M]. Beijing: China Electric Power Press, 2010.

Emerging Developments in the Power and Energy Industry – Dufo-López, Krzywanski & Singh (eds)
© 2020 Taylor & Francis Group, London, ISBN 978-0-367-27169-5

A planning approach of distributed generation and energy storage based on an improved parallel P-system algorithm

Nan Wang
State Grid Shandong Electric Power Company Electric Power Research Institute, Jinan, China

Han Hao
Northeast Electric Power University, Jilin, China

Yan Cheng, Shumin Sun & Yuejiao Wang
State Grid Shandong Electric Power Company Electric Power Research Institute, Jinan, China

ABSTRACT: Aimed at distributed wind turbines integrating energy storage systems planning in the distribution network, an optimal planning approach is proposed considering the different benefit communities including distribution company and distributed generation investor. Meanwhile, in view of the excellent parallel computation ability of P-system, an improved parallel P-system algorithm based on NSGA-II is proposed to overcome the drawbacks of traditional heuristic algorithms like prematurity and easy to be trapped into local optimization. The ultimate goal is to maximize different benefit communities by sizing and locating distributed wind turbines and energy storage systems. A case study shows that with the proposed planning approach, more practical and effective schemes can be acquired.

1 INTRODUCTION

Recently, centralized wind power explodes in China, and wind curtailment is more and more strident. In this case, the distributed wind turbine (WT) with the characteristics of small capacity, tiny floor area, flexible location, quick construction period and local absorption, has received worldwide attention and been identified as the supplement of the wind power industry. Based on the distributed wind projects, the configuration of the energy storage system (ESS) can smooth the wind power's fluctuation and intermittency, and further enhance renewable energy absorption, ameliorate the power quality, and maintain the system stability.

Many studies focus on the planning modeling of the distributed generation (DG). A multi-stage coordinate planning model to minimize the present value of construction cost and operating cost over the planning horizon is presented (Zhi Wu, Yafei Liu & Wei Gu 2018). A novel planning approach in the distribution network frames based on a multi-scenario technique is proposed (Yajing Gao, Xiaobo Hu & Wenhai Yang 2017). Targeting revenue maximization of power grid, the ESS planning method is proposed based on short-term optimal power flow algorithm (M. Sedghi, Ahmadian A. & Aliakbar-Golkar M. 2015). But all of the above literature ignores that the distribution company (DISCO) and the DG owner belong to different economies, and it is not practically meaningful with viewing these two parties as a whole while planning.

The planning model is a mixed integer nonlinear programming problem, and heuristic algorithms are widespread adopted to solve such optimization problems, including non-dominated sorting genetic algorithm (NSGA) (Deb K. Pratap A. Agarwal S. et al. 2002), the strength Pareto evolutionary algorithm (Mifa Kim, Tomoyuki H. Mitsunori Miki. et al. 2004), immune algorithm (Gong M. Jiao L. Du H. et al. 2008), particle swarm optimization (Reyes-Sierra, M. & Coello Coello, C. A. 2006) and so on. Although with stable convergence, heuristic algorithms are easy to trap into local optimum, and have the low Pareto-front dispersion and the poor computational efficiency when solving iterative power flow calculation. 'Membrane computing', which is also called 'P system' proposed by Gheorghe Paun (2000), is a collateral calculation model abstracting from the structure and function of cells, tissues and organs. It is critical to guarantee the diversity and dispersion of Pareto optimal front, and enhance the convergence velocity of iteration in multi-objective optimal planning. Hence, considering the characteristic parallel capacity of P-system, we propose an improved parallel P-system algorithm based on NSGA-II aimed at the improvement of disparity and the convergence of Pareto optimal front.

Focusing on the distributed WT integrating ESS planning in the distribution network, this paper proposes a multi-objective optimal planning approach considering different benefits and presents an improved parallel P-system algorithm based on NSGA-II, in order to obtain more practical and effective schemes of the site and capacity of distributed WT and ESS.

2 MULTI-OBJECTIVE OPTIMAL PLANNING MODEL

There are different interest subjects when integrating WT and ESS in the distribution network.

- • DISCO is responsible to ensure the stability and reliability of power system. The integration of WT is beneficial for the reduction of emission cost and upgrade cost of lines and transformers. But a mass of DG power is unfavorable for the system stability. To maximize the benefit of DISCO, we choose the difference of the operation costs between the existing network and the planned network allocating WT and ESS.
- • WT investor invests WT and ESS simultaneously and concerns the return rate and the risk of the investment, so the internal rate of return (IRR) is the primary concern of the investor.

2.1 DISCO

To guarantee the profit of DISCO, maximizing the difference of the operation cost between the existing network and the planned network allocating WT and ESS is chosen as the objective of DISCO.

$$f_1 : \max\left[\sum_{y=1}^{Y} \alpha_y \left(C_{D,y} - C_{uD,y}\right)\right] \tag{1}$$

where, $C_{D,y}$ and $C_{uD,y}$ denote the operation costs in the existing network and the planned network in the yth year, respectively, which include power purchasing cost from the upstream grid, pollutant emission cost and the upgrade cost of transmission lines and transformers. In addition, $C_{D,y}$ also includes power purchasing cost from WT investor. α_y is the discount factor in the yth year. Y is the planning horizon. The formulations of $C_{D,y}$ and $C_{uD,y}$ are expressed as:

$$C_{D,y} = \sum_{s=1}^{S}\left[\lambda_s \sum_{t=1}^{T}\left(\xi_{g,t}P_{G,y,s,t}+\xi_e P_{DG,y,s,t}\right) + \left(\sum_{l=1}^{L} \alpha_{l,e}\xi_d L_{d,y,s,l} + \alpha_{l,t}T_{d,y,s}\right)\right] \tag{2}$$

$$C_{uD,y} = \sum_{s=1}^{S}\left[\lambda_s \sum_{t=1}^{T}\left(\xi_{g,t}P_{uG,y,s,t}+\xi_e P_{uDG,y,s,t}\right) + \left(\sum_{l=1}^{L} \alpha_{l,e}\xi_d L_{ud,y,s,l} + \alpha_{l,t}T_{ud,y,s}\right) + \lambda_s \sum_{t=1}^{T} \xi_{WT}P_{DG,y,s,t}\right] \tag{3}$$

$C_{\mathrm{D},y}$ and $C_{\mathrm{uD},y}$ comprise of the following parts:

- Purchasing cost for the upstream power: λ_{s} is the number of days in the sth scenario, $\xi_{\mathrm{g},t}$ is the electricity price purchasing from the upstream grid on the tth hour, $P_{\mathrm{G},y,s,t}$ and $P_{\mathrm{uG},y,s,t}$ are the purchasing power from the upstream grid in the existing network and the planned network on the tth hour of the sth scenario in the yth year, respectively.
- Pollutant emission cost: ξ_e is the emission compensation factor per unit electricity power, $P_{\mathrm{DG},y,s,t}$ and $P_{\mathrm{uDG},y,s,t}$ are the sum of the power output of WT and the discharge power of ESS in the existing network and the planned network.
- Upgrading and reconstruction cost of lines and transformers: $L_{\mathrm{d},y,s,l}$ and $L_{\mathrm{ud},y,s,l}$ are the length of the lth line needing to transform in the existing network and the planned network respectively, ξ_{d} is the upgrading cost of the lth line; $T_{\mathrm{d},y,s}$ and $T_{\mathrm{ud},y,s}$ are the reconstruction cost of transformers in the existing network and the planned network respectively, $a_{l,e}$ and $a_{l,t}$ are the conversion coefficient of uniform annual value.
- Purchasing cost for the distribution power: ξ_{WT} is the transaction price between DISCO and WT investor.

2.2 *WT investor*

IRR is chosen as the objective of WT investor, denoted with IRR_{WT}, which is the interest rate making the NPV of the project equal to 0.

$$f_2 : \max(IRR_{\mathrm{WT}}) \tag{4}$$

$$\sum_{y=1}^{Y} S_{\mathrm{WT},y} \times (IRR_{\mathrm{WT}})^t - (I_{\mathrm{WT}} + I_{\mathrm{ESS}}) = 0 \tag{5}$$

where, I_{WT} and I_{ESS} are the initial investment of WT and ESS; $S_{\mathrm{WT},y}$ is the net cash flows in the yth year, as:

$$S_{\mathrm{WT},y} = C_{\mathrm{WT},y} - C_{\mathrm{r},y} - C_{\mathrm{m},y} + B_{\mathrm{d}} \tag{6}$$

$C_{\mathrm{r},y}$ and $C_{\mathrm{m},y}$ refer to the replacement and maintenance cost respectively; B_{d} refers to the salvage value of investment, which is nonzero value only in the final year.

The model of WT and ESS could be found in other literature (L. Guo, N. Wang, H. Lu. et al. 2016). As for the operational model, the local load is supplied by distributed WT preferentially. ESS is allocated by WT investor for higher utilization of renewable energy, and charges at valley time and discharges at peak time to improve load performance and realize the revenue growth of WT investor under time-of-use electricity price mechanism. Note that, to ensure the stable operation of power system, reversed power flow from the distribution network to the upstream grid is not allowed. In the proposed model, the type and the number of WT, the capacity of ESS (including battery and PCS) and the site of WT and ESS are taken as the optimal variables.

3 IMPROVED PARALLEL P-SYSTEM ALGORITHM

3.1 *The structure of P-system*

P-system consists of objects, rules and cytomembranes. Objects denote the optimization variables in the optimal planning model. Rules include division, evolution, communication and cytolysis of cells. Cytomembranes include three types below.

- Elementary membrane: no longer containing other membrane structures.

– Nested membrane: containing other membrane structures.
– Superficial membrane: the outermost membrane.

The membrane structure can be formulated as:

$$\Pi = \{\mu, (X_1, X_2, \ldots, X_{N_0}), E, T_{\text{out}}\} \tag{7}$$

$$X_i = (x_1, x_2, x_3 \ldots x_{nvar})_i \quad i \in N_{obj} \tag{8}$$

where, μ denotes the membrane structure; $(X_1, X_2, \ldots, X_{N0})$ is the object set in the cell; $(x_1, x_2, x_3, \ldots, x_{nvar})$ is the ith object in the object set, which includes $nvar$ optimal variables, and the number of objects is N_{obj}; E denotes rule set; T_{out} is the object set deliver to the superficial membrane, which is the Pareto optimal front.

3.2 An improved parallel P-system algorithm based on NSGA-II

We established a membrane structure as follow:

$$\mu = \{(\mu_1), (\mu_2), [(\beta_1), \beta_2], \lambda\} \tag{9}$$

where, μ_1 and μ_2 are elementary membranes, within which the single-objective optimization is executed to guarantee the diversity and dispersion of Pareto optimal front generated by multi-objective optimization via communication. β_1 is an elementary membrane and β_2 is a nested membrane executing multi-objective optimization. λ is the superficial membrane.

There are three superiorities of the proposed algorithm.

– To ensure the diversity and avoid the local optimum, μ_1, μ_2 and β_1 will be dissolved with each round iteration and then start renewed division, and the objectives within cells are generated randomly.
– To improve the dispersion of Pareto optimal front, β_2 not only communicates with its inner cell β_1, but receives the excellent objects from single-objective optimization within μ_1 and μ_2.
– To approach the actually Pareto optimal front, the superficial membrane receives and stores excellent objects from all the other cells, and objects within it can be constantly updated with each round iteration.

The abridged general view is as Figure 1, and the detailed algorithm flow is as follows.

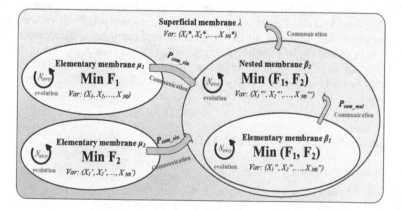

Figure 1.   The schematic diagram of proposed algorithm.

(1) Cell division

Within the superficial membrane, the cell splits into for cells as formulation (9), and objects number of N_0 within the elementary membranes and the nested membranes are generated randomly.

(2) Cell evolution

Single-objective optimization evolution: within the elementary membranes μ_1 and μ_2, crossover and mutation probability factors are introduced to execute evolution, and the parent of crossover is selected with roulette wheel selection method. New objects are ranked according to the fitness function, and the elite retention strategy is applied in evolution. The iteration number of cell evolution is N_i.

Multi-objective optimization evolution: within the elementary membrane β_1 and the nested membrane β_2, besides crossover and mutation probability factors, non-dominated sorting method and crowding distance sorting in NSGA-II are introduced to implement ranking.

Concurrent execution of evolution in all cells is allowed with characteristic parallel capacity of P-system, which can be implemented in multi-core CPU.

(3) Cell communication

After the N_{evo} times cell evolution, the nested membrane β_2 communicates with the elementary membranes μ_1, μ_2 and β_1 simultaneously to improve the dispersity and convergence of Pareto optimal front. In detail, the optimal objects in μ_1 and μ_2 are selected with the probability of P_{com_sin} according to a ranking of fitness, and the optimal objects in β_1 are selected by non-dominated sorting and crowding distance sorting with the probability of P_{com_mul}. Then the selected objects are delivered to β_2 and the original objects in β_2 will be replaced if the new objects are more excellent.

(4) Cytolysis and division again

After N_{com} times cell communication, the object set is delivered to the superficial membrane and stored for reserving excellent objects in the last iteration. Then, the elementary membranes and the nested membranes dissolve and split again, and all objects within them are recreated randomly and execute (1)–(3) again.

(5) Algorithm termination

With taking iterative N_{st} times as the end of evolution, optimum objects were outputted to the superficial membrane and viewed as the Pareto optimal front.

4 CASE STUDY

The diagram of the test case can be seen in another reference (Nan Wang, Yuejiao Wang, Yan Cheng et al. 2018), which is a typical 10kV distribution network. The network has 50 nodes and the optional sites for WT and ESS are indicated by nodes with black frames. The entire technical data for WT and ESS are shown in Table 1. Moreover, the load peak is 20.0043 MW, growing 2 percent a year, and the power factor is 0.9875. The suggested improved parallel P-system algorithm is implemented within MATLAB programming environment. Algorithm parameters are shown as in Table 2. The planning period is 5 years. The time-of-use

Table 1. Algorithm parameters.

Parameters	Value	Parameters	Value
P_{com_sin}	0.04	P_{com_mul}	0.1
N_{evo}	50	N_{com}	4
N_{st}	3	N_{obj}	50

Table 2. The economic parameters of WT and ESS.

Type	WT1	WT2	WT3	Battery	PCS
Rate capacity(MW)	1.5	2MW	2.5MW	-	-
Life (year)	15	15	15	10	10
Investment ($,$/KW)	798160	1000400	1162200	263.928	98.480
Operation cost ($,$/KW)	24100	2600	2800	2.64	-

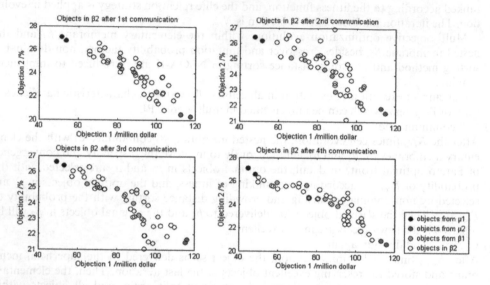

Figure 2. The optimization process in β_2.

electricity price is 0.0513$, 0.0928$ and 0.1332$ at the valley, usual and peak time respectively. The contract price between DISCO and WT investor is 0.08$.

The optimization process in β_2 can be observed visually in Figure 2, which shows the objectives of all objects in β_2 after every communication. It is obvious that the single-objective optimization improves the dispersion of Pareto optimal front, and the nested multi-objective optimization is computationally efficient and makes the solution closer to the real front.

The typical five allocation schemes from the Pareto optimal front are provided in Table 3. It can be observed that the benefits of DISCO and WT investor conflict, and more distributed power integrating the distribution network will increase the benefit of WT investor but raise the operation cost of DISCO. Hence, the Pareto optimal front considering the benefit conflict is necessary to provide a reference to the planning decision. Meanwhile, we can see the optimal location of distributed WT and ESS is the load center or the end of the feeder, which is more economic via reducing the power loss.

5 CONCLUSION

In this paper, the Pareto optimal front of the optimal type, capacity and location of WT and ESS are determined via the planning procedure, and is solved by an improved parallel P-system algorithm based on NSGA-II. The proposed multi-objective optimization model considers the benefit conflict between DISCO and WT investor, and thus more realistic schemes are obtained. The proposed improved algorithm owns the excellent ability of parallel processing, in which the single-objective optimization and the multi-objective optimization

Table 3. The typical five allocation schemes.

Schemes	Location	WT type	WT number	Battery capacity	PCS capacity	Objective 1	Objective 2
1	24	2	5	2.32	0.49	101.5956	22.6036
	39	3	5	2.60	0.75		
	32	3	4	2.25	0.65		
2	32	3	5	2.15	0.68	96.1446	23.4375
	2	2	4	1.87	0.47		
	24	3	4	2.04	0.68		
3	24	3	3	2.18	0.39	94.2028	23.1040
	32	2	5	2.63	0.72		
	39	3	5	2.95	0.75		
4	32	3	4	2.01	0.55	80.5858	24.5349
	24	3	4	1.95	0.92		
	24	3	2	1.58	0.65		
5	39	2	1	0.95	0.29	77.0822	25.7923
	24	3	5	2.25	0.54		
	39	2	5	1.96	0.41		

can be executed simultaneously, and cell division, evolution, communication and cytolysis are implemented to ensure the diversity and the dispersion of Pareto optimal front.

REFERENCES

Zhi Wu, Yafei Liu & Wei Gu. 2018. Decomposition method for coordinated planning of distributed generation and distribution network. *IET Generation, Transmission & Distribution* 12(20): 4482–4491.

Yajing Gao, Xiaobo Hu & Wenhai Yang. 2017. Multi-Objective Bilevel Coordinated Planning of Distributed Generation and Distribution Network Frame Based on Multiscenario Technique Considering Timing Characteristics. *IEEE Transactions on Sustainable Energy* 8(4): 1415–1429.

M Sedghi, Ahmadian A. & Aliakbar-Golkar M. 2015. Optimal Storage Planning in Active Distribution Network Considering Uncertainty of Wind Power Distributed Generation. *IEEE Transactions on Power Systems* 31(1):304–316.

Deb K. Pratap A. Agarwal S. et al. 2002. A fast and elitist multi. objective genetic algorithm: NSGA—II. *IEEE Transactions on Evolutionary Computation* 6(2): 182–197.

Mifa Kim, Tomoyuki H. Mitsunori Miki. et al. 2004. SPEA2+: Improving the Performance of the Strength Pareto Evolutionary Algorithm 2. International Conference on Parallel Problem Solving from Nature; PPSN 2004: Parallel Problem Solving from Nature—PPSN VIII:742–751, 18–22 September, 2004. Birmingham, UK.

Gong M. Jiao L. Du H. et al. 2008. Multi—objective immune algorithm with nondominated neighbor-based selection. *Evolutionary Computation*, 16 (2):225–255.

Reyes-Sierra, M. & Coello Coello, C. A. 2006. Multi-objective particle swarm optimizers: A survey of the state-of-the-art. *International Journal of Computational Intelligence Research* 2(3): 287–308.

Paun G. 2000. Computing with membranes. *Journal of Computer and System Science* 61(1):108–143.

L. Guo, N. Wang, H. Lu. et al. 2016. Multi-objective optimal planning of the stand-alone microgrid system based on different benefit subjects. *Energy* 116:353–363.

Nan Wang, Yuejiao Wang, Yan Cheng. et al. 2018. Distributed Wind Turbine Integrating Energy Storage System Optimal Planning Based on Trilateral Interest Subjects. The 2nd IEEE Conference on Energy Internet and Energy System Integration, 18–21 October 2018, Beijing, China.

Emerging Developments in the Power and Energy Industry – Dufo-López, Krzywanski & Singh (eds)
© 2020 Taylor & Francis Group, London, ISBN 978-0-367-27169-5

Research on three-phase unbalanced governance strategy of the microgrid based on improved PQ control

Weixiang Guan, Hailong Zhang & Enrong Wang
*School of Electric and Automation Engineering, Nanjing Normal University, Nanjing, Jiangsu Province,
China*

Liang Wang
Jiangsu Ken Engineering Power Engineering Consulting Co.LTD, Nanjing, Jiangsu Province, China

ABSTRACT: Aimed at the problem that the current deterioration and power fluctuation of
the gird-connected point exist under the three-phase unbalanced grid, the output performance
of the micro-sources is analyzed, and a grid-connected control strategy with voltage loop resonant controller applied to PQ control is proposed. The simulation results show that the
improved PQ grid-connected control strategy can realize the unbalanced governance of the
microgrid well, and also has a certain inhibitory effect on the fluctuation of microgrid output
power.

Keywords: The grid-connected point, the three-phase unbalanced grid, PQ control, resonant
controller, the unbalanced governance of the microgrid

1 INTRODUCTION

As an important auxiliary system of the grid, the microgrid has the characteristics of
environmental protection and flexible installation, and has become an indispensable part
of the grid (Meng, L. et al. 2015). However, due to grid faults, the access of nonlinear
loads and power electronic equipment in practical applications, it is difficult for the grid
to achieve three-phase balance. These circumstances will make the voltage unbalance at
the grid-connected point of the microgrid exist (Pan, H.B. et al. 2018). Traditional PQ
grid-connected control no longer guarantees good output performance of the microgrid.
Therefore, it is of realistic significance to study the microgrid control strategy under the
three-phase unbalance of the grid to ensure the power quality of the microgrid and
improve the robustness of grid-connected operation.

Chen, Y. et al. (2016) aimed at an inverter with axial pressure adjustment function, an
improved constant power decoupling control strategy is proposed, the power quality at the
end of the grid is improved by accurately designing parameters, but the inverter output performance is poor under the three-phase unbalance of the grid. Gao, J.Y. et al. (2018) proposes
a grid-connected inverter PQ control strategy which combined sagging characteristics. This
strategy can effectively adjust the active and reactive outputs to provide certain frequency and
voltage support for the grid, but it cannot achieve the three-phase unbalanced governance of
the grid-connected points, and also needs the stability of DC side voltage. Nian, H. et al.
(2015) in the three-phase unbalanced condition of the grid, the compensation term is added to
the power reference value of the PQ control, which limits the power fluctuation and reduces
the harmonic distortion rate of the grid-connected current. However, it increases the positive
and negative sequence separation module. This module increases the complexity of the control
system and parameter adjustment. Wei, Z. et al. (2016) under the unbalanced grid condition,
the resonant controller is introduced into the traditional control loop of the system to reduce

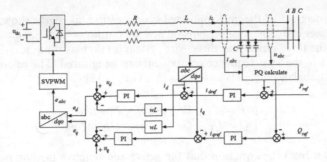

Figure 1. The PQ control structure diagram of grid-connected system of the microgrid.

the voltage positive and negative sequence separation control link, but the method is only applicable to off-grid operation mode of microgrid.

The PQ control strategy is widely used in the grid-connected system of the microgrid, however, in the actual application, due to the existence of the three-phase unbalance phenomenon of the grid, the traditional PQ control will no longer guarantee the stability of the system, resulting in poor performance of the microgrid and affecting the power quality of the grid. Aiming at the above problems of DC side voltage stability and the need of positive and negative sequence separation, this paper proposes an improved PQ control strategy that voltage loop resonant controller is added to the traditional PQ control, which effectively reduces the negative sequence component of the output voltage and realizes three-phase unbalanced governance of the microgrid.

2 RESEARCH ON GRID-CONNECTED CONTROL STRATEGY OF THE MICROGRID

When the microgrid is on-grid operation, the grid supports the voltage and frequency of the microgrid. Therefore, constant active and reactive (PQ) control is often used. Each micro-source can output power according to a given value. The PQ control structure of grid-connected system of the microgrid used in this paper is shown in Figure 1.

When the grid is unbalanced, the d-axis of the rotating coordinate system is defined to coincide with the voltage vector of the grid-connected point, that is, $u_q = 0$. If the power outer loop regulator part is ignored, the mathematical model of the PQ control is as follows:

$$L\frac{\mathrm{d}}{\mathrm{d}t}\begin{bmatrix} i_d(t) \\ i_q(t) \end{bmatrix} = \begin{bmatrix} -R & \omega L \\ -\omega L & -R \end{bmatrix}\begin{bmatrix} i_d(t) \\ i_q(t) \end{bmatrix} - \begin{bmatrix} e_d \\ e_q \end{bmatrix} + \begin{bmatrix} u_d \\ u_q \end{bmatrix} \tag{1}$$

where i_d and i_q are the d and q components of the inverter output current vector in rotating coordinate system, R and L represent the equivalent resistance and inductance, e_d and e_q stand for the d and q components of the inverter output voltage vector in rotating coordinate system, u_d and u_q present the d and q components of the grid-connected point's voltage vector in rotating coordinate system.

3 ANALYSIS OF MICRO-SOURCES' OUTPUT PERFORMANCE UNDER THE THREE-PHASE UNBALANCED GRID

In the grid-connected system of the microgrid, the three-phase unbalance of the grid will directly affect the output current quality of the microgrid inverter and affect the operating characteristics of the microgrid. In order to better grasp the output characteristics of the micro-sources under the unbalanced grid, based on th theory of instantaneous reactive power, combined with the PQ control structure diagram of the grid-connected system of the microgrid shown in Figure 1, the micro-sources' power exchange relationship and inverter output voltage quality are analyzed.

In this paper, the three-phase and three-wire system is taken as the research object, it means that the influence of the zero-sequence component can be ignored. The micro-sources' output power under the unbalanced grid is as follows (Yi, G.P. 2015):

$$\begin{cases} P = P_0 + P_{C2}\cos2\omega t + P_{s2}\sin2\omega t \\ Q = Q_0 + Q_{C2}\cos2\omega t + Q_{s2}\sin2\omega t \end{cases} \tag{2}$$

It can be known from the equation that the active and reactive instantaneous powers both contain double frequency harmonic components, and the DC side capacitor voltage also exhibits double frequency fluctuations because of the three-phase unbalanced grid. In order to analysis the influence of the DC-side capacitor voltage fluctuation on the output voltage of the grid-connected inverter, the method of the switching function is applied. The inverter output voltage is expressed as (Yi, G.P. 2015):

$$\begin{cases} e_a = AU_{dc}\sin\omega t + B\cos\omega t - B\cos3\omega t \\ e_b = AU_{dc}\sin\theta_1 + B\cos\theta_2 - B\cos(3\omega t - 2\pi/3) \\ e_c = AU_{dc}\sin\theta_2 + B\cos\theta_1 - B\cos(3\omega t + 2\pi/3) \end{cases} \tag{3}$$

where $B = AkU_{dc}/2$; $\theta_1 = \omega t - 2\pi/3$; $\theta_2 = \omega t + 2\pi/3$.

From the above equation, we can know that the output voltage of the grid-connected inverter not only contains the fundamental positive sequence component but also the fundamental negative sequence component and the triple harmonic because of the fluctuation of the DC side capacitor voltage, when the grid is three-phase unbalanced. As a result, the output current is also unbalanced, and the stable operation of the grid is affected.

4 RESEARCH ON IMPROVED PQ CONTROL STRATEGY FOR GRID-CONNECTED INVERTER UNDER THE UNBALANCED GRID

4.1 *Analysis of improved PQ control strategy*

When the grid is unbalanced, the grid voltage has positive and negative sequence components. In order to compensate the negative sequence component of the grid voltage, the inverter must emit the negative sequence current that meets the requirements. the negative sequence voltage of the grid is all reduced to the line reactance. For this problem, the positive and negative sequence separation modules are usually adopted, but it increases the complexity of the control system and parameter adjustment. So, This paper adds a voltage loop resonant controller based on traditional PQ control to achieve voltage balance, control diagram shown in Figure 2.

In Figure 2, PI is a proportional integral controller and PR is a resonant controller. Under the three-phase unbalanced condition of grid voltage, the components of the three-phase output voltage of the grid-connected inverter on the d and q axis are u_d and u_q, the error signals obtained by making the difference between u_d, u_q and the voltage reference values u_{dref}, u_{qref} are controlled by proportional integral controller and the resonance controller, superimposing the output signal with i_{dref}, i_{qref} via the active and reactive power instructions P_{ref}, Q_{ref} matrix solution are as the reference values i^*_{dref}, i^*_{qref} of the current loop, then the error signals obtained by making the difference between the inductor current feedback values i_d, i_q and the reference values i^*_{dref}, i^*_{qref} pass through the current controller.

The transfer function of an ideal resonant controller could be written as:

$$G_R(s) = \frac{2k_r s}{s^2 + \omega_0^2} \tag{4}$$

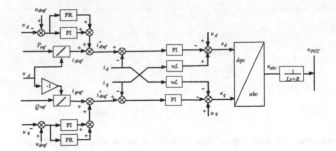

Figure 2. Improved PQ control diagram.

where k_r is the gain coefficient, ω_0 is the resonance frequency.

The resonance frequency ω_0 of the resonant controller is double frequency. Since the infinite gain of the ideal controller at the resonant frequency may make the system unstable and the gain at the non-resonant frequency is very small, the actual resonant controller with damping is applied. The transfer function of the resonant controller could be expressed as:

$$G_{PR}(s) = \frac{2k_r\omega_c s}{s^2 + 2\omega_c s + \omega_0^2} \tag{5}$$

where ω_c is cut-off frequency.

Based on the above analysis, feedforward decoupling is adopted for the improved PQ control strategy, and the control equations of e_d, e_q are:

$$\begin{cases} e_d = -\left(k_{ip} + \frac{k_{ii}}{s}\right)\left[\left(k_{up} + \frac{k_{ui}}{s} + \frac{2k_r\omega_c s}{s^2 + 2\omega_c s + \omega_0^2}\right)\left(u_{dref} - u_d\right) + i_{dref} - i_d\right] + \omega L i_q + u_d \\ e_q = -\left(k_{ip} + \frac{k_{ii}}{s}\right)\left[\left(k_{up} + \frac{k_{ui}}{s} + \frac{2k_r\omega_c s}{s^2 + 2\omega_c s + \omega_0^2}\right)\left(u_{qref} - u_q\right) + i_{qref} - i_q\right] + \omega L i_d + u_q \end{cases} \tag{6}$$

where k_{ip}, k_{ii} are parameters of current controller, respectively. k_{up}, k_{ui} are parameters of the voltage controller.

Adding the voltage loop resonance controller to the traditional PQ control increases the gain of the voltage loop at double frequency, reduces the negative sequence component of the output voltage, and can achieve the goal of governing the voltage unbalance of the grid-connected point.

4.2 Simulation analysis

The micro-source is replaced by a high-power inverter power supply, and the DC side is used to charge the capacitor through a three-phase rectifier. Based on the analysis of the micro-sources' output performance under the unbalanced grid, the conventional PQ control and the modified PQ control are simulated. In the simulation, the active and reactive power reference values are set to 100kW and 0kVar, respectively, the inductance value is 6mH, the capacitance value is 50μF, the k_{ip} and k_{ii} parameters of the current loop PI controller are 5 and 100 respectively, the voltage loop PI controller parameters are $k_{up} = 0.2$ and $k_{ui} = 0.5$, and the damped resonant controller parameters are $k_r = 8$, $\omega_c = 18$, $\omega_0 = 628$. The results are shown in Figure 3.

From (a) and (b) of Figure 3, we can see that voltage of the grid-connected point is also unbalanced by adopting traditional PQ control, but the improved PQ control results in a significant improvement in voltage unbalance, when the grid is three-phase unbalanced. In addition, the voltage loop resonance controller is added to the traditional PQ control so that the active power fluctuation of the inverter output is changed from the previous 88~120kW to 95~104kW, and the

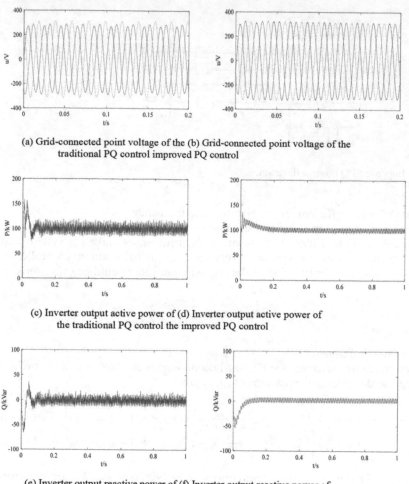

(a) Grid-connected point voltage of the (b) Grid-connected point voltage of the
traditional PQ control improved PQ control

(c) Inverter output active power of (d) Inverter output active power of
the traditional PQ control the improved PQ control

(e) Inverter output reactive power of (f) Inverter output reactive power of
the traditional PQ control the improved PQ control

Figure 3.   Traditional PQ control and improved PQ control simulation diagram.

reactive power is changed from the previous –15~12kVar to –3~7kVar, the fluctuation of active
and reactive power of the inverter output gets a certain inhibition.

5   CONCLUSION

Aiming at the shortcomings of the traditional PQ control strategy in the three-phase unbalanced condition of the grid, the output performance of the micro-sources is analyzed in detail.
In the traditional PQ control structure, the voltage loop resonance controller is added to compensate the grid unbalance. This strategy does not require positive and negative voltage
sequence separation, and the control system is relatively simple. The simulation results display
that the improved PQ control strategy effectively governs the three-phase unbalance problem
of the grid-connected point, reduces the output power fluctuations of the inverter, and
enhances the survivability of the microgrid under the unbalanced grid. The validity of the proposed method is verified.

ACKNOWLEDGMENTS

Postgraduate Research & Practice Innovation Program of Jiangsu Province (KYCX18_1226)

REFERENCES

Meng, L. & Savaghebi, M. & Andrade, F. et al. 2015. Microgrid central controller development and hierarchical control implementation in the intelligent microgrid lab of Aalborg University. *IEEE Applied Power Electronics Conference and Exposition* pp. 2585–2592.

Pan, H.B. & Wei, T.Y. & Deng, C. et al. 2018. A novel PQ control strategy for non-phase-locked loop on Hilbert transform. *IEEE Energy Conversion Congress and Exposition* pp. 2676–2682.

Chen, Y. & Zhao, J.B. & Zhu, H. et al. 2016. Research on Voltage Controlled Inverter Based on PQ Control. *Power Electronics Technology* 50(1): 28–29.

Gao, J.Y. & Zhao, J.B. & Chen, X.B. et al. 2018. Power Control and Design of Voltage Grid-Connected Inverter. *Chinese Society for Electrical Engineering* 38(6): 1788–1798.

Nian, H. & Shen, Y.B. & Yang, H.Y. et al. 2015. Flexible Grid Connection Technique of Voltage-Source Inverter Under Unbalanced Grid Conditions Based on Direct Power Control. *IEEE Transactions on Industrial Applications* 51(5): 4041–4050.

Wei, Z. & Ru, X.Q. & Shi, W. 2016. Microgrid inverter control strategy for unbalanced load conditions. *Automation of Electric Power Systems* 40(20): 76–81.

Yi, G.P. 2015. Research on Control Strategy of Microgrid Constant Power under Unbalanced Grid Voltage. *Journal of Electrical Technology* 30(14): 377–387.

Emerging Developments in the Power and Energy Industry – Dufo-López, Krzywanski & Singh (eds)
© 2020 Taylor & Francis Group, London, ISBN 978-0-367-27169-5

Study on shielding effectiveness characteristics of infinite metal sheet

A.Q. Guo & C.Q. Jiao
State Key Laboratory of Alternate Electric Power System with Renewable Energy Sources, North China Electric Power University, Beijing, China

Y.L. Li & L.J. He
China Electric Power Research Institute, Beijing, China

ABSTRACT: In order to analyze the influence of shield on low frequency electromagnetic field in valve hall, a simple electromagnetic shielding model consisting of a ring antenna and a metal sheet is presented. The programming calculation is carried out on the basis of the accurate analytical formula in the existing literature. In order to verify the reliability of the program, this paper uses CST to build a simulation model of shielding effectiveness for calculation, and the simulation results are consistent with the programming results. Change the conductivity, relative permeability, thickness and position of the metal sheet to calculate the shielding effectiveness, and the results show that the shielding effectiveness increases as the electrical conductivity, magnetic permeability and thickness of the metal sheet increase. Moreover, when the thickness of the metal sheet is much smaller than the distance from the loop antenna to the observation point, the distance from the loop antenna to the metal sheet and the distance from the loop antenna to the observation point satisfy a certain relationship. The research results of this paper are valuable for the design of low-frequency electromagnetic shielding in valve hall.

1 INTRODUCTION

With the development of modern high technology, electromagnetic interference and electromagnetic compatibility caused by electromagnetic wave are increasingly serious; after noise pollution, air pollution and water pollution, electromagnetic wave pollution has become the fourth public hazard threatening human health. Exploring efficient electromagnetic shielding materials, preventing electromagnetic radiation pollution to protect the environment and human health, and preventing electromagnetic radiation leakage to ensure information security have become urgent problems to be solved internationally. Electromagnetic shielding is to disconnect the transmission path from the electromagnetic disturbance source to the sensitive equipment to eliminate or reduce the adverse effects of the disturbance source on the sensitive equipment (Schulz R B et al. 1988). Electromagnetic shielding is the most basic and effective method for electromagnetic interference protection control. Electromagnetic shielding is usually done in two ways. One is to isolate the electromagnetic disturbance source, so that the electromagnetic field outside the shielding layer is weakened. And the other is to isolate the specific equipment so as not to be affected by the electromagnetic field outside the shielding layer. There are a lot of deductive formulas for shielding effectiveness in the existing literature (Bannister P. 2003, Bannister P. 2007, Frikha A et al. 2015, Levy S. 1936, Moser J R. 1967). In this paper, the exact analytical formula of shielding effectiveness is programmed (Ryan & Cornelius M. 1967). The effects of electrical conductivity, relative magnetic permeability, thickness and position of the thin metal sheet on the shielding effectiveness were analyzed. The shielding model was built by CST software, and the shielding effectiveness under the same conditions was calculated. The results were consistent with the experimental calculation results and the experimental results in the literature (Andrieu G et al. 2012, Manara & A. 1996).

2 ELECTROMAGNETIC SHIELDING MODEL

Figure 1 is a schematic view of an electromagnetic shielding structure, which is composed of a loop antenna, a metal sheet and an observation ring. The loop antenna has a radius of a and a current of I. The thickness of the metal sheet is t. The distance from the loop antenna to the metal sheet is b, the distance from the observation point to the metal sheet is c, and the distance from the loop antenna to the observation point is z=b+c.

Moser's formula for the shielding effectiveness of a thin, plane metal sheet of infinite extent on an electromagnetic field generated by a circular loop field source with uniform current I is shown below.

$$B_z = 2\mu_0\mu_r ai \int_0^\infty \frac{C\lambda^2\tau}{\tau_0^2} J_1(\lambda a) J_0(\lambda r) e^{-\tau_0 z - t(\tau - \tau_0)} d\lambda \tag{1}$$

$$B_{z0} = \frac{\mu_0 ai}{2}(jk_0 a + \frac{a}{\sqrt{a^2 + z^2}})\frac{e^{-jk_0\sqrt{a^2 + z^2}}}{a^2 + z^2} \tag{2}$$

$$S = 20\log_{10}\left|\frac{B_{z0}}{B_z}\right| \tag{3}$$

Where

$C = [(\tau/\tau_0 + \mu_r)^2 - (\tau/\tau_0 - \mu_r)^2 e^{-2t\tau}]^{-1}$

$\tau_0 = \sqrt{\lambda^2 - k_0^2} \ k_0^2 = \omega^2\mu_0\varepsilon_0 \ \tau = \sqrt{\lambda^2 + j\omega\mu_0\mu_r\sigma}$

$\mu_0 = 4\pi \times 10^{-7} \text{H/m}$

μ_r—the relative permeability of the shield, with respect to copper

σ—the shield conductivity

$J_1(\lambda a)$—the Bessel function of order one and argument λa

$J_0(\lambda r)$—the Bessel function of order zero and argument λr

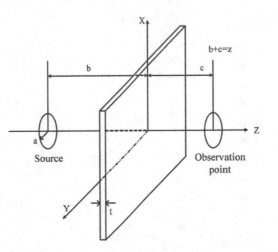

Figure 1.   Coaxial loops separated by an infinite sheet.

3 COMPARISON OF PROGRAM AND CST SIMULATION

The C/τ_0^2 in equation (1) is rewritten to the following equation, so that the denominator does not take a zero point in the integral region, which is beneficial for programming.

$$\frac{C}{\tau_0^2} = \frac{1}{(\tau + \mu_r \tau_0)^2 - (\tau - \mu_r \tau_0)^2 e^{-2t\tau}} \tag{4}$$

When the shielding sheet material is copper, select parameters: a=0.06m, b=0.04m, z=0.08m, $\mu_r = 1$, $\sigma = 5.8 \times 10^7$S/m, $t = 0.3 \times 10^{-3}$m. In the formula (1), the lower limit of the integral is rewritten as ω/c, and the upper limit of the integral is rewritten as $10/z$. Calculate the shielding effectiveness at different frequencies.

The shielding model shown in Figure 1 was built in the CST electromagnetic studio. Use a low frequency solver and replace the loop antenna with a 1A virtual current loop. Calculate the shielding effectiveness at different frequencies. The shielding effectiveness obtained by the two methods is shown in Figure 2, where the line represents the CST simulation result and the point represents the program calculation result. As can be seen from the figure, the two results are basically the same, which verifies the reliability of the program.

4 INFLUENCING FACTORS OF SHIELDING EFFECTIVENESS

The shielding effectiveness of the metal sheet is programmed to calculate when the thickness, conductivity, magnetic permeability, and relative position of the metal sheet are changed. Analyze the change law of shielding effectiveness.

4.1 Effect of conductivity of metal sheet

Set the radius of the loop antenna to a=0.06m, and the distance from the loop antenna to the observation point is z=0.075m. The thickness of the metal sheet is t=0.3×10⁻³m, and the relative magnetic permeability is $\mu_r = 100$. The shielding effectiveness under different conductivity is shown in Figure 3.

As can be seen from the figure, the higher the frequency, the greater the shielding effectiveness. When the size and relative magnetic permeability of the metal sheet are constant, the shielding effectiveness increases as the conductivity of the metal sheet increases.

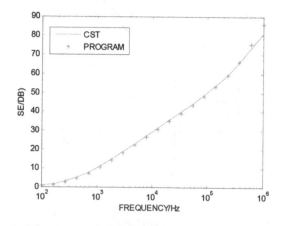

Figure 2.   Comparison of calculation results between program and simulation.

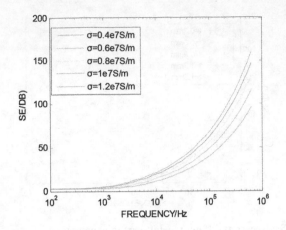

Figure 3. Shielding effectiveness at different conductivity.

4.2 *Effect of permeability of metal sheet*

When a=0.06m, z=0.075m, t=0.3×10⁻³m, $\sigma = 0.5 \times 10^7 \text{S/m}$, the shielding effectiveness under different relative magnetic permeability is calculated as shown in Figure 4.

It can be seen from the figure that when the size and conductivity of the metal sheet are constant, the higher the relative magnetic permeability, the better the shielding performance.

4.3 *Effect of permeability of metal sheet*

When a=0.06m, z=0.075m, $\sigma = 5.8 \times 10^7 \text{S/m}$, $\mu_r = 1$, the shielding effectiveness under different metal sheet thicknesses is calculated as shown in Figure 5.

It can be seen from the figure that the thicker the metal sheet, the higher the shielding effectiveness when the material and position of the metal sheet are unchanged.

4.4 *The relationship between the radius of the loop antenna and the position of the observation point*

$$S = 8.686t\tau_r + 20\log_{10}[\frac{\tau_r r'}{8.485\mu_r (z-t)} \frac{r'}{r} (\frac{r'}{r})^3] \tag{5}$$

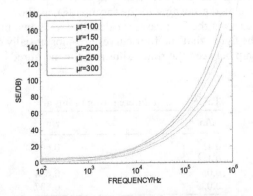

Figure 4. Shielding effectiveness at different relative magnetic permeability.

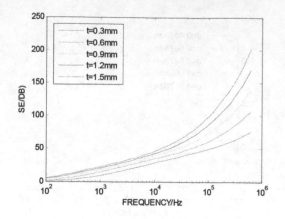

Figure 5.   Shielding effectiveness at different metal sheet thicknesses.

From the simplified formula (5) of the shielding effectiveness (Bannister P. 1969), after determining the material and thickness of the shielding sheet, the radius of the loop antenna and the distance z from the ring to the observation point satisfy the relation (6).

$$\frac{r_1'^5}{(z_1 - t)r_1^3} = \frac{r_2'^5}{(z_2 - t)r_2^3} \tag{6}$$

Where $r' = \left[a^2 + (z - t)^2\right]^{\frac{1}{2}}$, $r = \left(a^2 + z^2\right)^{\frac{1}{2}}$, then

$$\frac{\left[a_1^2 + (z_1 - t)^2\right]^{\frac{5}{2}}}{(z_1 - t)(a_1^2 + z_1^2)^{\frac{3}{2}}} = \frac{\left[a_2^2 + (z_2 - t)^2\right]^{\frac{5}{2}}}{(z_2 - t)(a_2^2 + z_2^2)^{\frac{3}{2}}} \tag{7}$$

When $z >> t$, the simplification formula can be obtained:

$$\frac{a_1^2 + z_1^2}{z_1} = \frac{a_2^2 + z_2^2}{z_2} \tag{8}$$

Select the shield material as copper. When $\mu_r = 1$, $\sigma = 5.8 \times 10^7$S/m, $t = 1 \times 10^{-3}$m, z under different a is calculated according to the formula (8), as shown in Table 1.

The shielding effectiveness in the four cases is calculated as shown in Figure 6.

It can be seen from the figure that the four curves are substantially coincident, and the correctness of the relationship between the ring radius and the measured distance z is verified.

Table 1.   z at different ring radius a.

a/m	z/m
0.15	0.601
0.1	0.622
0.05	0.634
0.025	0.637

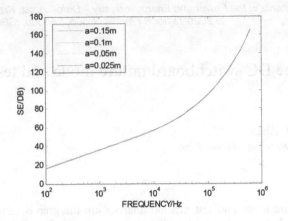

FREQUENCY/Hz

Figure 6. Shielding effectiveness of metal sheet at different positions.

5 CONCLUSIONS

1. The shielding effectiveness of infinite metal sheet increases with increasing frequency. And the higher the frequency, the greater the growth rate of the shielding performance.
2. The shielding effectiveness of the metal sheet increases as the electrical conductivity, magnetic permeability and thickness of the metal sheet increase.
3. When the thickness of the metal sheet is much smaller than the distance from the loop antenna to the observation point, the distance from the loop antenna to the metal sheet and the distance from the loop antenna to the observation point satisfy a certain relationship.

ACKNOWLEDGEMENT

This project is supported by the "Research and Application of Key Technologies for Primary and Secondary Integration and Test Detection of Distribution Equipment".

REFERENCES

Andrieu G, Panh J, Reineix A, et al. 2012. Homogenization of Composite Panels From a Near-Field Magnetic Shielding Effectiveness Measurement. IEEE Transactions on Electromagnetic Compatibility, 54(3):0-0.

Bannister P. 1969. Further Notes for Predicting Shielding Effectiveness for the Plane Shield Case. IEEE Transactions on Electromagnetic Compatibility, EMC-11(2):50-53.

Bannister P. 2003. The quasi-near fields of dipole antennas. IEEE Transactions on Antennas and Propagation, 15(5):618-626.

Bannister P. 2007. New Theoretical Expressions for Predicting Shielding Effectiveness for the Plane Shield Case. IEEE Transactions on Electromagnetic Compatibility, EMC-10(1):2-7.

Frikha A, Bensetti M, Duval F, et al. 2015. A New Methodology to Predict the Magnetic Shielding Effectiveness of Enclosures at Low Frequency in the Near Field. Magnetics IEEE Transactions on, 51(3):1-4.

Levy S. 1936. Electromagnetic Shielding Effect of an Infinite Plane Conducting Sheet Placed between Circular Coaxial Coils. Radio Engineers Proceedings of the Institute of, 24(6):923-941.

Moser J R. 1967. Low-Frequency Shielding of a Circular Loop Electromagnetic Field Source. IEEE Trans. Electromagn. Compat. 9(1):6-18.

Manara, A. 1996. Measurement of material shielding effectiveness using a dual TEM cell and vector network analyzer. IEEE Transactions on Electromagnetic Compatibility, 38(3):327-333.

Ryan, Cornelius M. 1967. Computer Expression for Predicting Shielding Effectiveness for the Low-Frequency Plane Shield Case. 9(2):83-94.

Schulz R B, Plantz V C, Brush D R.1988. Shielding theory and practice. IEEE Transactions on Electromagnetic Compatibility, 30(3):187-201.

Emerging Developments in the Power and Energy Industry – Dufo-López, Krzywanski & Singh (eds)
© 2020 Taylor & Francis Group, London, ISBN 978-0-367-27169-5

Medium voltage DC switchboard failure mode and test method analysis

W.C. Wang & W.Q. Zhan
Naval University of Engineering, Wuhan, China

ABSTRACT: Failure mode and test method analysis are the premise and foundation of testability modeling and design, and one of the important contents of testability research. Aiming at the problem of low level of testability design of existing naval power system, this paper conducts a test study on medium voltage DC switchboard of a certain naval ship by studying and analysing the potential failure mode sorting method of hierarchical definition, definition of severity category and probability level definition of fault mode method. In this paper, the set of components and parameters that need to be monitored for the type I and II fault modes that cause serious damage to distribution boards is obtained, which lays a foundation for testability design.

1 INTRODUCTION

With the increasing performance and complexity of weapon system, it is more and more difficult to detect and diagnose its faults. It is urgent to develop testability design (Chen, Qiu, & Liu 2009). In particular, the ship's integrated power system using new equipment (medium voltage DC power equipment) and new structure (DC area power distribution structure), its test design work cannot be ignored. Testability design can significantly reduce the cost of product life cycle, improve fault detection rate and shorten maintenance time (Liu & Yi 2007). Fault mode and test method analysis (Shi & Tian 2011) are the premise and basis for test modeling and test design. It is one of the important contents of test research. It is based on FMECA and combines fault related information at system level, LRU. The test level analysis is carried out in the order of SRU and SRU. The analysis mainly includes failure mode, measurement parameters or test points, and selection of test methods.

As the key equipment of medium-voltage DC distribution system, medium-voltage DC switchboard undertakes the important tasks of transporting electric energy from generator to load screen, manual closing, switching, voltage regulation, speed regulation, parallel operation, detachment and load transfer to circuit breaker. Although the United States, the United Kingdom and other countries have decades of research and development and use experience, but the development, testing and use of medium-voltage DC switchboard in China's warships is still in its infancy (Yang 2011). In this paper, aiming at the problem that the test design level of the existing ship power system is not high overall, the failure mode and test method of a medium-voltage DC switchboard of a certain type of ship are analyzed.

2 MEDIUM VOLTAGE DC SWITCHBOARD OVERVIEW

2.1 *The composition of medium voltage DC switchboard*

A medium-voltage DC switchboard of a ship is mainly composed of a gas turbine generator set control panel, a diesel generator set control panel, a bus couple control panel, a propulsion inverter control panel, a power station screen, and two chopper control panels. According to the functional division and technical requirements of the power system for the medium voltage

DC switchboard, the main components and devices of each screen of the medium voltage DC switchboard includes: medium voltage DC circuit breaker, integrated protection device, bus bar, photoelectric current mutual inductance devices, voltage transformers, grounding switches, insulation, and other secondary circuit components.

Based on the main structure and working principle of each screen of medium voltage switchboard, this paper selects the control panel of gas turbine generator set as the research object.

2.2 *Medium voltage DC switchboard fault criterion definition*

The medium voltage DC switchboard can be judged as faulty in any of the following cases:

a) The circuit breaker refuses to move (including rejection, rejection, etc.), misoperation (including after the split, no signal, etc.);
b) It is impossible to control the closing, opening, fault reset, power failure and other operations of the circuit breaker;
c) The components such as circuit breakers and bus bars are heated, and the internal temperature rise of the switchboard is too high;
d) The internal insulation of the switchboard is too low;
e) Arcing inside the switchboard;
f) Other conditions that may trigger maintenance.

3 POTENTIAL FAILURE MODE COMBING METHOD

3.1 *Analysis hierarchy definition*

3.1.1 *Initial agreement level*
The initial agreement level is the overall and complete product level for potential failure mode analysis. According to different analysis purposes, it can be either the overall system or the sub-system or equipment. The determination of the initial agreement level directly affects the description of failure impact, the division of severity level and the compensation measures adopted (Sun, Wei, & Wang 2017). In order to analyze the impact of various potential failure modes in the system on the completion of the entire system, the initial agreement level is defined as the medium voltage DC distribution board.

3.1.2 *Agreement level*
The level of agreement is based on the needs of the analysis, according to the relative complexity or functional relationship of the product, from the complex system to the product level of simple equipment division (Zhang, Ding, & Jia 2013), in the potential failure mode and test analysis, the product is being analyzed immediately. Hierarchy, the overall trade-off from performance, cost, schedule, etc. is a level of agreement, or multiple levels of agreement.

The agreed layer of this analysis includes the gas turbine generator set control panel and its component breakers, bus bars, and supporting insulators under the medium voltage DC switchboard.

3.1.3 *Minimum agreement level*
The minimum agreement level is the analysis target of the agreed-upon level that is terminated when the potential failure mode analysis is performed. The determination of the minimum agreement level of the medium-voltage DC switchboard follows the following principles:

(a) The product hierarchy in which products that are severely catastrophic (Class I) or fatal (Class II) faults are located;
(b) The minimum level of product specified or expected to be repaired, which may result in a critical (Class III) or minor (Class IV) failure.

According to this principle, the main components inside the control panels of the medium voltage DC switchboard are the minimum agreed level for the analysis of potential failure modes.

3.2 *Severity category definition*

The Severity category is a measure of the potential consequences of a failure mode at the lowest analysis level for a medium voltage DC switchboard (Lei 2014). The severity of the medium voltage DC switchboard is considered both for the success of the task and for the maintainability. According to the division of GJB1391 and the characteristics of the medium voltage DC power distribution system and power equipment, Severity is divided into four categories:

(a) Class I (disaster)—Failure will cause the switchboard to be destroyed and cannot be repaired in the short term. It needs to be docked at the home port for repair or replacement;

(b) Class II (fatal)—Failure will result in damage to the switchboard or loss of function of the switchboard, which will not meet the job requirements, and it will take longer to repair with the ship's tools and spare parts;

(c) Category III (critical)—mainly considering
The impact of service and maintainability, although the failure of the medium voltage DC switchboard function will be reduced, but the use of on-board tools and spare parts can be eliminated in a short time, has no significant impact on system function;

(d) Class IV (slight)—does not cause the consequences of the above three types of failures, but can result in unplanned repairs or minor delays in tasks.

3.3 *Failure mode occurrence probability level definition*

The failure mode occurrence probability level according to GJB1361 can be seen in Table 1.

4 ANALYSIS OF POTENTIAL FAILURE OF MEDIUM VOLTAGE DC SWITCHBOARD

Using the potential fault mode combing method studied in the previous section, the fault mode and test method are analyzed for the type I and II fault modes that can cause large damage to the medium voltage DC switchboard of a certain type of ship.

According to the level of the medium voltage DC switchboard, analyze all possible fault modes of each component from bottom to top, analyze its fault type, fault symptoms, fault

Table 1. Failure mode occurrence probability level.

Level	Definition	Characteristics of failure mode occurrence probability	Failure mode occurrence probability
A	Often occurs	The high probability	Greater than 20% of the total failure probability of the product
B	Sometimes	The medium probability	Greater than 10% of the total failure probability of the product, less than 20%
C	Occasional Occurs	Infrequent	Less than 1% of the total failure probability of the product, less than 10%
D	Seldom occurs	Unlikely to occur	Reater than 0.1% of the total failure probability of the product, less than 1%
E	Rarely occurs	Near zero	Less than 0.1% of the total failure probability of the product

Table 2. Failure mode and test analysis table of medium voltage DC switchboard (Class I severity).

Device	Failure mode	Cause of failure	Fault Phenomenon and Impact Analysis		Occurrence probability level	Test parameters/sensor types	Test Point/Sensor Layout
			Failure symptoms/precursors	Components affected by the fault			
Comprehensive and protective devices	Hardware failure; software failure	Component aging, circuit board loosening, device damage; communication problems	Rejection and misoperation make the power system lose its protection, the fault spread and spread, and damage the equipment connected with the distribution board	Switchboard	A	Volage/multimeter, Data stream/oscilloscope	Easily damaged components, circuit board connectors, digital signal output ports
DC circuit breaker	To refuse to move; mismove	Deformation and clamping of mechanical parts, loosening of core of opening and closing coil, poor contact of secondary connection, burning of opening coil, inadequate switching of auxiliary switch, faults of closing contactor and fretting switch	Circuit breakers do not operate; abnormal current/vibration signals	DC circuit breaker	C	Travel switch; Current; Vibration sensor	DC circuit breaker
Supporting insulator	Heavy contamination leads to surface discharge and leakage current may develop into arc	Insufficient daily cleaning and maintenance	Arc, Short Circuit	Supporting insulator, bus bars	D	Transient Ground Potential/TEV Sensor	Switchgear housing
Shunt	Current Burning	Material or process problems with excessive current	Large error in current measurement	Shunt	E	Current/Hall Current Sensor	Shunt
Pressure release plate	Do not act when pressure exceeds threshold	Pressure Release Plate Stick	Cabinet explosion	Cabinet body and equipment in cabinet	E	Limit switch	Cabinet body

Table 3. Failure mode and test analysis table of medium voltage DC switchboard (Class II severity).

Device	Failure mode	Cause of failure	Fault Phenomenon and Impact Analysis			Test parameters/sensor types	Test Point/ Sensor Layout
			Failure symptoms/precursors	Components affected by the fault	Occurrence probability leve		
Cable room	Cable joint virtual connection and cable insulation damage	Cable Joints Mixed with Impurities, Insulation Aging and Connection Looseness	Cable breakdown and arc generation	Cable interior cables	D	Transient Ground Potential/TEV Sensor; Current/ Current Transformer Arc/ Ultraviolet Photosensitive Diode	A cable connection; a cable chamber
Supporting insulator	Internal defects partial discharge penetration;	Bubbles caused by process defects	Arc and Short Circuit	Supporting insulator, bus bars	D	Transient Ground Potential/TEV Sensor	Switchgear housing
Hall voltage sensor	Invalid	Damage to internal electronic components	Circuit breakers refuse to act or act incorrectly; lead to misjudgment of manual or automatic control; contradict with other Hall sensor data	Controller	B	Volage/ external meter	Near the Hall Voltage Sensor
Branch bus bar	Loose, poor contact; short connection and lap connection of different mother row; oxidation failure of coatings	Vibration; Aging Failure	Excessive temperature rise, short circuit, accelerated ageing of insulators, performance degradation or burning of insulators	Associated device	D	Temperature/ Infrared Sensor; Arc/ Ultraviolet Photosensitive Diode	Near the junction of branch bus bar
Master mother row	Poor contact; Short connection of different bus bars; Oxidation failure of coatings	Vibration; Aging Failure	Excessive temperature rise, short circuit, accelerated ageing of insulators, performance degradation or burning of insulators	Circuit breaker	E	Voltage sensor, current sensor, temperature sensor	Master mother row

impact and related test information. The information required for the analysis mainly comes from the technical specifications, development plans, design drawings, reliability data and historical fault data of the medium voltage switchboard.

5 SUMMARY

In this paper, the fault modes and test methods of a certain type of medium voltage DC switchboard on warship are analyzed. The set of components and parameters that need to be monitored for the type I and II fault modes which cause great damage to the distribution board is obtained, which lays a foundation for testability design. The research methods of fault mode and test method analysis are universal and can provide reference for the test research of other power systems or equipment.

REFERENCES

Chen, X.X., Qiu, J. & Liu, G. J. 2009. Research on Test Optimal Selection Based on Hybrid Binary Particle Swarm Optimization and Genetic Algorithms. *Journal of Instruments and Instruments* 30(8): 1674–1680.

Lei, A. 2014. *Development of a series of radar automatic test system*. Harbin: Harbin University of Technology.

Liu, H.M. & Yi, X.S. 2007. Testability Modeling and Analysis of Multi-Signal Flow Diagrams. *China Testing Technology* 33 (1): 49–50.

Shi, J.Y. & Tian, Z. 2011. *Testability design analysis and verification*. Beijing: National Defense Industry Press.

Sun, P., Wei, Q.X. & Wang, K. 2017. Engineering Application Research on Testability Design Analysis and Fault Diagnosis Implementation Strategy of Aircraft Equipment. *Computer Measurement and Control* 25 (3): 11–14.

Yang, Q. 2011. Characteristics and Development Trend of Ship Medium Voltage Switchboard. *Ship Engineering* 33 (6): 45–48.

Zhang, Q., Ding, J. & Jia, A. 2013. Research and Analysis of Large Equipment Testing Technology. *Machinery Manufacturing* 4 (3): 9–12.

Emerging Developments in the Power and Energy Industry – Dufo-López, Krzywanski & Singh (eds)
© 2020 Taylor & Francis Group, London, ISBN 978-0-367-27169-5

Research on differential transformation schemes of substation secondary equipment based on improved fuzzy analytic hierarchy processes

Ling Zhou & Shuaifei Huang
College of Energy and Electrical Engineering, Hohai University, Nanjing, Jiangsu, China

Yu Cui
State Grid Jiangsu Electric Co., Ltd. Nanjing, Jiangsu, China

Dongsheng Zhu & Xiaoyue Hu
China Energy Engineering Group Jiangsu Power Design Institute Co., Ltd. Nanjing, Jiangsu, China

Jianshuang Cui*
College of Energy and Electrical Engineering, Hohai University, Nanjing, Jiangsu, China

ABSTRACT: In order to objectively and systematically evaluate the technical transformation and upgrading plans of secondary equipment in substations, a differential technical transformation model for substation is constructed to realize the differentiation and precision of technical transformation. Firstly, by analyzing the historical information record of the secondary equipment and related norms and standards, combined with the actual operation of the substation, the evaluation index system of the transformation scheme is constructed. Secondly, the evaluation rules are improved, and the calculation models and quantitative methods of each indicator are given. Thirdly, the triangular fuzzy number in analytic hierarchy process is used to solve the multi-attribute decision making model. Finally, an actual case of differential technology transformation is used to verify the validity of the improved model, and the evaluation results with better discrimination are obtained.

1 INTRODUCTION

Smart grid is the focus of the current power system development, and the number of affiliated smart substations continues to increase (Zhou, X.X. et al. 2013). During the 13th Five-Year Plan period, the State Grid Corporation continues to promote the construction of smart substations and the upgrading of traditional substations. In recent years, a large number of substations secondary equipment are close to or exceed their expected life span. Due to the various types of secondary equipment, the importance of secondary equipment varies with different voltage levels, and there are influence factors such as limited funds, capacity of maintenance crew. So it is necessary to evaluate three upgrading schemes of technical renovation: whole cabinet transformation, whole machine transformation and plug-in transformation. It should focus on the comparison from the aspects of blackout risk, transformation time, manpower cost and capital requirement, and analyze the principles of selecting renovation schemes, then form a systematic evaluation theory.

---

[1]First author. E-mail: zhoulingl@163.com
*Corresponding author. E-mail: 842507146@qq.com

In addition to the vulnerable parts and power supply modules can be replaced according to maintenance plan, the mainstream transformations are all "one-size-fits-all" schemes. There are also alterations for replacement such as the whole machine and plug-in. As for when to adopt the transformation plan, now there is no mature and universal theoretical basis.

This paper evaluates the alternative transformation schemes. Through the analysis of the operation and maintenance records and the experience of equipment technological transformation, five evaluation indicators are selected, including life after transformation, LCC, control level, operation environment and health level. On the basis of referring to various evaluation decision-making models, triangular fuzzy numbers are used to describe the content of expert evaluation, and then the relative weights of each index are determined by combining the FAHP (Zhang S. & Wang T. 2016, Xu M.M. et al. 2018). Finally, it is determined that the transformation schemes should be ranked in order to provide theoretical support for the accurate differential technical transformation. Finally, the validity of the evaluation model is verified by the analysis of specific cases.

2 EVALUATION INDICATORS OF TECHNICAL TRANSFORMATION SCHEME FOR SECONDARY EQUIPMENT

In order to accurately describe the deterioration degree of the secondary equipment, it is necessary to evaluate the criteria of the *Guidelines for Evaluation of Relay Protection Status* and to analyze the *Management Regulations for the Operation of Microcomputer Relay Protection Devices* in detail. Combined with the characteristics of power electronic components and practical engineering application experience, the principles of preventive test, live test and online monitoring score are refined (Xia Y. & Zhang X.Q. 2018, Zhou J. et al. 2019), and then the following five evaluation indicators are obtained.

2.1 *The life after transformation*

The failure rate of the secondary equipment is a key factor affecting the material life. For both the whole machine replacement and the cabinet replacement, the failure rate is determined by the set of plug-ins (Yang Y. & Yan J. 2018). After the secondary equipment fails, the protective device will not work or perform protective functions. The relationship between the plug-in failure rate and the secondary equipment failure time is expressed as follows:

$$f_1(t) = \frac{f(t)}{1 - \int_{t_0}^{t} f(t)\mathrm{d}t} = \frac{\beta}{\alpha^\beta} t^{\beta-1} \tag{1}$$

where $f(t) = \frac{\beta}{\eta}\left(\frac{t-t_0}{\eta}\right)^{\beta-1} \mathrm{e}^{-\left(\frac{t-t_0}{\eta}\right)^\beta}$ is a probability density function obeying the Weibull distribution; $f_1(t)$ is the failure rate of plug-ins; α and β are the adjustment coefficients.

The whole machine or the cabinet is composed of plug-ins, and the overall failure rate is affected by the failure rate of the component.

$$f_2(t) = \sum_{i=1}^{p} q f_1(t) \tag{2}$$

where $f_2(t)$ is the failure rate of the whole machine or the cabinet; p is the number of plug-ins that make up the whole machine or the cabinet, and q is the number of each plug-in.

From the start of a secondary equipment operation, the failure rate of all plug-ins that make up the secondary equipment begins to increase. After the plug-in replacement, only the current plug-in failure rate is reset, which can reduce the equipment failure rate compared to the before failure. After the whole machine replacement, the failure rate of the whole machine including multiple plug-ins is reset, so that the equipment failure rate is greatly reduced compared with the before failure. After the cabinet is replaced, the failure rate of all plug-ins is reset, and the failure rate of the equipment is recalculated.

2.2 *Life cycle cost (LCC)*

The economic index of secondary equipment is mainly measured by the cost generated in the whole life cycle process. In the actual power production and substation operation process, the whole LCC is divided into owner-oriented cost classification, which mainly includes initial investment cost (C_I), operation and maintenance cost (C_{OM}), failure cost (C_F) and discard cost (C_D).

It can be expressed as:

$$C = C_I + C_{OM} + C_F + C_D = (C_{I1} + C_{I2}) + (C_{O1} + C_{O2} + C_{O3}) + (C_{F1} + C_{F2}) + C_D \quad (3)$$

where C is life cycle cost; C_{I1} is procurement cost; C_{I2} is installation and commissioning cost; C_{O1} is equipment energy consumption; C_{O2} is inspection cost; C_{O3} is management cost; C_{F1} is failure maintenance cost; C_{F2} is failure loss cost.

2.3 *The control level*

The control level of secondary equipment is determined by I (equipment importance level index), and can be distinguished according to different voltage levels and the degree of impact caused by faults (Wang L.J. & Zhou K. 2018).

$$I = kP(t) \times V_i \quad (4)$$

where I is the importance of equipment; k is the fault influence coefficient; $P(t)$ is the failure rate of the equipment; V is the quantitative value of the voltage level. The value of 500kV and above is 1, 220kV is 220/500, 110kV is 110/500, below 110kV is 35/500.

2.4 *The operating environment*

The secondary equipment contains a large number of electronic components, which are affected by the environmental factors, as shown in Figure 1. Referring to the State Grid Standard Q-GDW11285–2014"*Guidelines for the Evaluation of Relay Protection Status,*" it is meaningful to evaluate the environmental status of equipment in the operating process.

When the equipment is overheated, magnetized and dusty due to temperature, humidity, electromagnetic interference, and non-standard management, the overall failure rate will increase and the life will decrease. When selecting a solution, it should be carried out in the order of the cabinet transformation, the whole machine transformation, and the plug-in transformation.

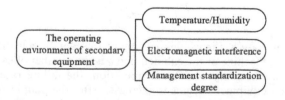

Figure 1.   Environmental factors affecting the life of secondary equipment.

2.5 The health level of secondary circuit

The health level of secondary circuit is determined by the life of secondary cable and secondary circuit wiring level.

The non-negative random variable L is used to represent the life of cable (Yuan D.H. & Zhang H.F. 2017):

$$\begin{cases} F(t) = P(L < t), t \geq 0 \\ R(t) = 1 - F(t) \end{cases} \quad (5)$$

where $R(t)$ is the reliability function of t, the reliable life of the cable is the time corresponding to the given reliability, denoted by $t(R)$. In general, reliability decreases with time, that is $t(R) = R^{-1}(t)$. And, $R^{-1}(t)$ is the inverse function of the reliability $R(t)$.

The wiring level of secondary circuit mainly includes the implementation of the requirements of strong connection, neatness of wiring specifications and labeling.

3 EVALUATION MODEL OF DIFFERENTIAL TRANSFORMATION SCHEME FOR SECONDARY EQUIPMENT

The above evaluation indicators are hierarchically processed based on the concept of AHP. Evaluate the selection of secondary equipment. Thus, the influencing factors model is established as shown in Figure 2.

In order to deal with the indices that can't be quantified accurately and have different attributes, triangular intuitionistic fuzzy numbers are introduced for the first time to improve the AHP when applying the theory of AHP to evaluate the relative importance of two elements. If there are precise values of element indexes, it can be fuzzified to facilitate uniform comparison and sorting.

3.1 Triangular intuitionistic fuzzy theory

The intuitionistic fuzzy set takes into account the information of membership degree, non-membership degree and hesitation degree at the same time (Shu M.H. et al. 2006). This paper introduces the triangular intuitionistic fuzzy theory to illustrate the principle of improving the scale of AHP.

3.1.1 Definition of triangular intuitionistic fuzzy numbers

Definition 1: triangular intuitionistic fuzzy number $\tilde{a} = \langle (\underline{a}, a, \overline{a}), w_{\tilde{a}}, u_{\tilde{a}} \rangle$ is a form of intuitionistic fuzzy number, and set up $\tilde{a} \geq 0 \cdot w_{\tilde{a}}$ and $u_{\tilde{a}}$ represent the maximum membership and non-membership degree of $\tilde{a}$, respectively. The functions are as follows:

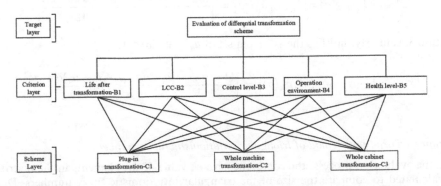

Figure 2. Hierarchical structure model for evaluation of differential transformation scheme.

$$\mu_{\tilde{a}}(x) = \begin{cases} \dfrac{w_{\tilde{a}}(x - \underline{a})}{a - \underline{a}} & ,\underline{a} \leq x < a \\ w_{\tilde{a}} & ,x = a \\ \dfrac{w_{\tilde{a}}(x - \bar{a})}{a - \bar{a}} & ,a < x \leq \bar{a} \\ 0 & ,x < \underline{a}\text{或}x > \underline{a} \end{cases} \tag{6}$$

$$v_{\tilde{a}}(x) = \begin{cases} \dfrac{a - x + u_{\tilde{a}}(x - \underline{a})}{a - \underline{a}} & ,\underline{a} \leq x < a \\ u_{\tilde{a}} & ,x = a \\ \dfrac{x - a + u_{\tilde{a}}(\bar{a} - x)}{\bar{a} - a} & ,a < x \leq \bar{a} \\ 1 & ,x < \underline{a}\text{或}x > \underline{a} \end{cases} \tag{7}$$

Meet the constraint requirements: $\begin{cases} 0 \leq w_{\tilde{a}} \leq 1 \\ 0 \leq u_{\tilde{a}} \leq 1 \\ 0 \leq w_{\tilde{a}} + u_{\tilde{a}} \leq 1 \end{cases}$, $\pi_{\tilde{a}} = 1 - w_{\tilde{a}} - u_{\tilde{a}}$ is the hesitancy degree of $\tilde{a}$, indicating the degree of confidence in the value.

3.1.2 The algorithm of triangular intuitionistic fuzzy numbers.

$$\begin{cases} \tilde{a}_1 + \tilde{a}_2 = \begin{array}{l} \langle(\underline{a}_1 + \underline{a}_2, a_1 + a_2, \bar{a}_1 + \bar{a}_2), \\ \langle\min(w_{\tilde{a}_1}, w_{\tilde{a}_2}), \max(u_{\tilde{a}_1}, u_{\tilde{a}_2})\rangle \end{array} \\ \\ \tilde{a}_1\tilde{a}_2 = \begin{array}{l} \langle(\underline{a}_1\underline{a}_2, a_1 a_2, \bar{a}_1\bar{a}_2), \\ \min(w_{\tilde{a}_1}, w_{\tilde{a}_2}), \max(u_{\tilde{a}_1}, u_{\tilde{a}_2})\rangle \end{array} \\ \\ k\tilde{a}_1 = \langle(k\underline{a}_1, ka_1, k\bar{a}_1), w_{\tilde{a}1}, u_{\tilde{a}_1}\rangle, k > 0 \\ \\ \tilde{a}_1{}^k = \langle(\underline{a}_1{}^k, a_1{}^k, \bar{a}_1^k), w_{\tilde{a}1}, u_{\tilde{a}_1}\rangle, k > 0 \end{cases} \tag{8}$$

3.1.3 Cut set of triangular intuitionistic fuzzy numbers.
Definition 2: triangular fuzzy number $\tilde{a}$ satisfies the α cut set of $0 \leq \alpha \leq w_{\tilde{a}}$:

$$\tilde{a}_{\alpha}^{+} = \{x \mid \mu_{\tilde{a}} \geq \alpha\} = [\tilde{a}_L^{+}(\alpha), \tilde{a}_R^{+}(\alpha)] = \left[\frac{(w_{\tilde{a}} - \alpha)\underline{a} + \alpha a}{w_{\tilde{a}}}, \frac{(w_{\tilde{a}} - \alpha)\bar{a} + \alpha a}{w_{\tilde{a}}}\right] \tag{9}$$

Definition 3: similarly, satisfies the $1 - \alpha$ cut set of $u_{\tilde{a}} \leq 1 - \alpha \leq 1$:

$$\tilde{a}_{\alpha}^{-} = \{x | v_{\tilde{a}} \geq 1 - \alpha\} = [\tilde{a}_L^{-}(\alpha), \tilde{a}_R^{-}(\alpha)] = \left[\frac{(1 - \alpha - u_{\tilde{a}})\underline{a} + \alpha a}{1 - u_{\tilde{a}}}, \frac{(1 - \alpha - u_{\tilde{a}})\bar{a} + \alpha a}{1 - u_{\tilde{a}}}\right] \tag{10}$$

3.2 Improved ordering method of triangular intuitionistic fuzzy number

During the evaluation process, the comprehensive 或 value of the elements in the matrix need to be calculated to compare the size of the triangular intuitionistic fuzzy numbers. Drawing on the idea of the Choquet integral aggregation operators (Zhang C.Y. et al. 2007), and, considering the weight function α of the cut set, a new score function is obtained.

$$d(\tilde{a}) = \frac{(\alpha w_{\tilde{a}}^2 + (1-\alpha)(1-u_{\tilde{a}})^2)(\underline{a} + 4a + \bar{a})}{6} \tag{11}$$

3.3 Improved evaluation model of intuitionistic fuzzy analytic hierarchy process

3.3.1 A hierarchical model for scheme evaluation
According to Figure 2, by calculating the comprehensive value of each element's triangular intuitionistic fuzzy number, the relative weights of each element are obtained by layers and recursively weighted by layers, taking B5 as an example, as shown in Figure 3.

3.3.2 Constructing a triangular intuitionistic fuzzy judgment matrix.
Language variables can be measured by index weights.

For example, when the k-th expert E_k compares the importance of element U_i and U_j of each layer, considering the former to be at least more important than the latter, at most more important than the latter and most likely more important than the latter, which can be expressed as: $\tilde{r}_{ijk} = \langle (0.75, 0.9, 0.1), 0.8, 0.15 \rangle$. 0.8 indicates the degree of support for decision makers' preference for U_i when U_j and U_j are compared, and 0.15 indicates opposition.

Then, reach the final S fuzzy judgment matrix:

$$(\tilde{r}_{ijk})_{n \times n} = \begin{array}{c|cccc} R_k & u_1 & u_2 & \cdots & u_n \\ \hline u_1 & r_{11k} & r_{12k} & \cdots & r_{1nk} \\ u_2 & r_{21k} & r_{22k} & \cdots & r_{2nk} \\ \vdots & \vdots & \vdots & \ddots & \vdots \\ u_n & r_{n1k} & r_{n2k} & \cdots & r_{nnk} \end{array} \tag{12}$$

where $\tilde{r}_{ijk} = \left\langle (\underline{r}_{ijk}, r_{ijk}, \bar{r}_{ijk}), w_{ijk}, u_{ijk} \right\rangle$ is the evaluation value of the importance of expert k to U_i compared with U_j.

3.3.3 Normalize judgment matrix
Remove the dimension of triangular intuitionistic fuzzy decision matrix, then normalize it.

For the cost attributes:

$$\tilde{f}_{ijk} = \left\langle \left(\frac{\bar{r}_{ij}^+ - \bar{r}_{ijk}}{\bar{r}_{ij}^+}, \frac{r_{ijk}}{\bar{r}_{ij}^+}, \frac{\bar{r}_{ij}^+ - \underline{r}_{ijk}}{\bar{r}_{ij}^+} \right), \frac{w_{ijk}}{w_{ij}^+}, \frac{u_{ijk}}{u_{ij}^+} \right\rangle \tag{13}$$

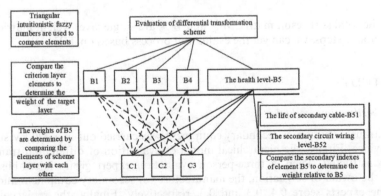

Figure 3. Hierarchical structure analysis of TIFAHP model.

For the benefit attributes:

$$\tilde{f}_{ijk} = \left\langle (\frac{r_{ijk}}{\bar{r}_{ij}^+}, \frac{r_{ijk}}{\bar{r}_{ij}^+}, \frac{\bar{r}_{ijk}}{\bar{r}_{ij}^+}), \frac{w_{ijk}}{w_{ij}^+}, \frac{u_{ijk}}{u_{ij}^+} \right\rangle \qquad (14)$$

where $\tilde{f}_{ijk}$ is the element of the normalized judgment matrix $\tilde{F}_{ijk} \cdot x_{ij}^+ = \max\limits_{k \in E_k} x_{ijk}$, and x is $\bar{r}_{ij}^+, w_{ij}^+, u_{ij}^+$.

3.3.4 *Solve the comprehensive values of the matrix elements and rank them hierarchically*

According to Section 2.1 and 2.2, the comprehensive score value $d(\tilde{f}_{ijk})$ of the i-th element of the k-th expert evaluation can be obtained by improving the score function.

Then the scores of the elements of the current layer in the index set U are ranked as:

$$W_k = \{d(u_{1k}), d(u_{2k}), \cdots, d(u_{nk})\} \qquad (15)$$

Normalize:

$$d'(u_{ik}) = \frac{d(u_{ik})}{\sum\limits_{j=1}^{n} d(u_{jk})}, (i,j = 1, 2, \cdots, n) \qquad (16)$$

The comprehensive score of layer elements is ranked as:

$$W_k' = \{d'(u_{1k}), d'(u_{2k}), \cdots, d'(u_{nk})\} \qquad (17)$$

3.3.5 *Fuzzy hierarchical total sorting with recursive weights*

Based on objective parameter data and subjective experience, the expert group gives weights $\lambda_k (k = 1, 2, \cdots, s)$ and $\sum\limits_{k=1}^{s} \lambda_k = 1$. The weighted average method is used to obtain the comprehensive weighted fuzzy set of the index set:

$$W_E = \sum\limits_{k=1}^{S} \lambda_k W_k' \qquad (18)$$

The total weight of the fuzzy hierarchy is ranked by the recursive weight of the hierarchy:

$$W = W_{Bk}' W_{Ck}' W_E \qquad (19)$$

where W is the total sort result matrix; W_{Bk}' and W_{Ck}' are single layer weight matrices.

From the above steps, we can get the evaluation process based on TIFAHP.

4 CASE STUDY

4.1 *Model verification method*

Differential transformation of secondary equipment was carried out in a 220 kV substation in Jiangsu Province. Firstly, the hierarchical model of evaluation of differential transformation scheme is constructed. Then, a three-person evaluation expert group is established, which comes from the owner of substation, the manufacturer of secondary equipment and the design institute. The experts score 0.4, 0.3 and 0.3, respectively. Finally, the replacement plan is

Table 1. Single layer triangular intuitionistic fuzzy judgment matrix. R_{A-B}.

Index/Expert	B_1	B_2	B_3	B_4	B_5
B_1	<(1,1,1),0.50,0.45>	<(3,5,7),0.65,0.20>	<(2,2,3),0.60,0.30>	<(5,7,9),0.9,0.1>	<(1,1,1),0.50,0.45>
		<(2,6,9),0.50,0.45>	<(1,1,1),0.50,0.45>	<(3,5,7),0.8,0.2>	<(1,2,4),0.50,0.40>
		<(1,3,5),0.50,0.40>	<(1,2,3),0.65,0.30>	<(4,5,8),0.7,0.3>	<(1,1,1),0.50,0.45>
B_2	<(1/7,1/5,1/3),0.65,0.20>	<(1,1,1),0.50,0.45>	<(1/9,1/7,1/6),0.8,0.2>	<(1,3,5),0.55,0.3>	<(1/8,1/7,1/6),0.8,0.1>
	<(1/9,1/6,1/2),0.50,0.45>		<(1/8,1/6,1/5),0.7,0.2>	<(2,3,6),0.6,0.3>	<(1/9,1/6,1/5),0.8,0.15>
	<(1/5,1/3,1),0.50,0.40>		<(1/9,1/8,1/7),0.85,0.1>	<(1,2,4),0.7,0.2>	<(1/9,1/8,1/7),0.9,0.1>
B_3	<(1/3,1/2,1/2),0.60,0.30>	<(6,7,9),0.80,0.20>	<(1,1,1),0.50,0.45>	<(7,8,9),0.7,0.2>	<(1/4,1/3,1),0.50,0.4>
	<(1,1,1),0.50,0.45>	<(5,6,8),0.70,0.20>		<(5,7,9),0.6,0.35>	<(1/5,1/3,1/2),0.55,0.3>
	<(1/3,1/2,1),0.65,0.3>	<(7,8,9),0.85,0.10>		<(6,8,9),0.5,0.3>	<(1/4,1/2,1),0.55,0.3>
B_4	<(1/9,1/7,1/5),0.9,0.10>	<(1/5,1/3,1),0.55,0.30>	<(1/9,1/8,1/7),0.7,0.2>	<(1,1,1),0.5,0.45>	<(1/9,1/7,1/5),0.5,0.45>
	<(1/7,1/5,1/3),0.8,0.20>	<(1/6,1/3,1/2),0.6,0.3>	<(1/9,1/7,1/5),0.6,0.35>		<(1/7,1/6,1/5),0.65,0.3>
	<(1/8,1/5,1/4),0.7,0.30>	<(1/4,1/2,1),0.70,0.20>	<(1/9,1/8,1/6),0.5,0.3>		<(1/8,1/7,1/5),0.55,0.4>
B_5	<(1,1,1),0.50,0.45>	<(6,7,8),0.80,0.10>	<(1,3,4),0.50,0.40>	<(5,7,9),0.5,0.45>	<(1,1,1),0.50,0.45>
	<(1/4,1/2,1),0.50,0.40>	<(5,6,9),0.80,0.15>	<(2,3,5),0.55,0.30>	<(5,6,7),0.65,0.3>	
	<(1,1,1),0.50,0.45>	<(7,8,9),0.90,0.10>	<(1,2,4),0.55,0.30>	<(5,7,8),0.55,0.4>	

Table 2. Global comprehensive fuzzy weight.

Target layer	Criterion level		Relative Weight of Scheme Layer			
	Index	Relevant upper-level weights	C_1	C_2	C_3	Total
Evaluation of differential transformation Scheme	B_1	0.280	0.141	0.470	0.389	1
	B_2	0.013	0.514	0.244	0.242	1
	B_3	0.301	0.423	0.309	0.268	1
	B_4	0.101	0.093	0.376	0.531	1
	B_5	0.305	0.098	0.546	0.356	1
Comprehensive weight		——	**0.213**	**0.432**	**0.355**	**1**

evaluated according to the steps of Chapter 3. In order to simplify the formulation, only the process of calculating the weight vector among the factors in the criterion layer is given. The steps of calculating the weight of the elements in the criterion layer are the same for the second-level indicators and the scheme layer.

4.2 Compute fuzzy judgment matrix

According to the hierarchical structure diagram shown in Figure 2, the expert group scores the evaluation according to its own experience and the constituent elements of the relevant object layer in Chapter 2, forming a single layer fuzzy judgment matrix R_{A-B}, as shown in Table 1:

By processing Table 2 according to equations (14–15), the standardized fuzzy judgment matrix F_{A-B} is obtained. The calculating score function is transformed into traditional AHP. The weight score ranking from criterion layer to object layer is obtained by calculating and applying equation (16–18) according to literature [15].

$$W'_{A-B} = \{f'(u_1), f'(u_2), \cdots, f'(u_5)\} = \{0.280, 0.013, 0.301, 0.101, 0.305\} \qquad (20)$$

Similarly, the single-level comprehensive fuzzy weights of the judgement matrices $R_{B-C_i}, i = 1, 2, \cdots, 5$ from the scheme level to the criterion level can be obtained.

4.3 Calculating the results of the differential transformation scheme evaluation sort

The fuzzy weights of the layers are combined to obtain the global comprehensive fuzzy weights as shown in Table 2.

4.4 Analysis of comprehensive evaluation results of transformation scheme

Through the above steps of AHP, priority is given to the replacement of the whole machine. For the substation, the equipment has been put into operation for a long time, and the plug-in transformation needs more plug-ins, whose cost is higher than that of the whole machine. At the same time, compared with the whole screen transformation, the whole machine transformation not only saves funds, improves the level of investment precision, but also reduces the load-bearing pressure of secondary teams, and ensures the reliability of power grid operation.

5 CONCLUSION

This paper puts forward the concept of differential technical transformation for substation secondary equipment, establishes the evaluation model of differential transformation, determines the reference evaluation index, and gives the corresponding evaluation criteria and calculation method. At the same time, the paper evaluates and judges the secondary equipment differential technical transformation by using triangular intuitionistic fuzzy number which evidently distinguishes the ranking results of transformer scheme; and validates the model through practical cases. It provides a theoretical support for the selection of substation technical transformation scheme.

ACKNOWLEDGEMENT

State Grid Jiangsu Electric Power Co., Ltd. Management Special Research Fund Project: SGJS0000DKWT1800676.

REFERENCES

Chang D.Y. 1996. Applications of the extent analysis method on fuzzy AHP[J]. *European Journal of Operational Research:* 1996, 95:649–655.

Cobuloglu H.I. 2015. A stochastic multi-criteria decision analysis for sustainable biomass crop, selection[J]. *Expert Systems with Applications*: 2015, 42(15–16):6065.

Gong Z.W. & Lin Y. 2013. Uncertain fuzzy preference relations and their applications[M]. *New York*: *Springer-Verlag*:45–74.

Liang T.J. et al. 2017. Comprehensive benefit evaluation of distribution network planning based on the life cycle cost theory and the cloud matter element theory [J]. *Power System Protection and Control*: 2017, 45(19):12–17.

Shu M.H. et al. 2006. Using intuitionistic fuzzy sets for fault-tree analysis on printed circuit board assembly[J]. *Micro Electronics Reliability*: 2006, 46(12): 2139–2148.

Wang L.J. & Zhou K. 2018. Risk assessment for smart substation secondary system using risk transfer network model [J]. *Power System Protection and Control*: 2018,46(06):97–105.

Xia Y. & Zhang X.Q. 2018. Research on risk assessment and maintenance decision of secondary equipment of distribution automation[J]. *Electric Power Engineering Technology*: 2018,38 (4):19–25+31.

Xu M.M. et al. 2018. Risk assessment of repeated multiple blackouts for distribution network based on fuzzy AHP[J]. *Electric Power Automation Equipment*: 2018,38 (10):19–25+31.

Xu Z & Liao H. 2014. Intuitionistic fuzzy analytic hierarchy process[J]. IEEE Transactions on Fuzzy Systems: 2014, 22(4):749–761.

Yuan D.H. & Zhang H.F. 2017. Evaluation technology of electrical cables service life based on arrhenius model[J]. *Journal of Jiamusi University (Natural Science Edition)*: 2017,35(01):114–118.

Yang Y. & Yan J.2018. Decision making method of grid sequence control reformation based on equipment failure rate curve[J]. *Smart Power*: 2018, 46 (08):1–7+24.

Zhang C.Y. et al.2007. Risk assessment system for bidding of construction projects based on Choquet fuzzy integral [J]. *China Civil Engineering Journal*: 2007(10):98–104.

Zhang S. & Wang T.2016. Synthetic evaluation of power grid operating states based on intuitionistic fuzzy analytic hierarchy process[J]. *Power System Technology:* 2016, 40(04):41–49.

Zhou J. et al.2019. Comprehensive evaluation of power failure based on improved extension analytic hierarchy process[J]. *Power System Protection and Control*: 2019, 47(03):31–38.

Zhou X.X. et al. 2013. Review and prospect for power system development and related technologies. A concept of three-generation power systems. *Journal of Chinese Electrical Engineering*: 2013,33 (22):1–11+22.

Emerging Developments in the Power and Energy Industry – Dufo-López, Krzywanski & Singh (eds)
© 2020 Taylor & Francis Group, London, ISBN 978-0-367-27169-5

Application of high precision lightning location system in distribution network and influence of topography on lightning activity

Guowei Zhao, Jie Bai, Cheng Li & Rui Zhao
Datong Power Supply Company State Grid, Shanxi Datong, Shanxi Province, China

ABSTRACT: In order to repair the transmission line which was cut off by lightning as soon as possible, it is necessary to determine the location of lightning strike quickly and accurately. The detection efficiency of wide area lightning monitoring system is about 90% and its positioning error is about 800m. It can meet the operation and maintenance requirements of transmission lines. However, for distribution network lines, its positioning accuracy is insufficient. Therefore, based on frequency domain constant false alarm rate (CFAR) detection method and wavelet analysis method, an efficient detection method for lightning impulse signal is proposed in this paper. It improves the recognition and monitoring accuracy of lightning impulse signal. A lightning monitoring device for distribution network is designed, which uses multi-channel lightning signals to locate and analyze. The device has been applied in Datong to monitor lightning activities.

The results show that the lightning monitoring system developed in this paper is more accurate in positioning. Based on the results of lightning monitoring, the influence of topography on lightning activity is analyzed.

Keywords: lightning activity, distribution network, lightning location, topography

1   INTRODUCTION

With the development of urban and rural construction, the demand for reliability of distribution network power supply is gradually improved. Because the overall insulation level of distribution network is low, it is prone to insulator flashover. The characteristics of lightning stroke fault are unclear fault features, difficult to determine fault points, and so on (A. D. Conti et al. 2010). As a result, the protective measures adopted are not well targeted. The lightning tripping rate of distribution network is difficult to be significantly reduced. For the neutral point non-effective grounding system, the system can continue to operate for two hours after the single-phase grounding fault caused by lightning strike. Although the reliability of power supply is significantly improved, the long-term operation with faults can easily enlarge the accident (S. I. Abouzeid et al. 2015, Y. Zhou et al. 2006). Therefore, it is necessary to obtain accurate lightning monitoring data after tripping accident. These data can help to locate fault points quickly. It is conducive to the repair of lines and the protection of power supply (Y. S. Fang et al. 2008). It also provides guidance for the improvement of lightning protection.

At present, the State Grid has built a wide area lightning monitoring system, mainly for large-span, high tower, long-distance transmission lines. A large number of basic lightning parameters of transmission lines are accumulated by the lightning monitoring system. It provides guidance for circuit design, technical renovation and maintenance. (X. Wang et al. 2011, G. Liu et al. 2014). The positioning principle is as follows: the lightning grounding point is determined according to the time difference of lightning electromagnetic signals arriving at different base stations. The lightning detection efficiency of the system is about 90%, and the positional error is about 800m (K. Bian et al. 2013). However, the line length of distribution network is short and the distance between towers is small. The positioning error of the system can fully meet the requirements of transmission lines, but not small-scale distribution network. (L. Li &X. J. Qi (2011)).The lightning monitoring system cannot locate fault towers in distribution network quickly. The rush repair time has increased considerably (A. Piantini & J. M. Janiszewski (2009), P. H. Larsen (2016), A. Borghetti et al. 2016).

Therefore, a high-precision lightning location device for distribution network is designed by using CFAR lightning signal detection method. A real-time high-speed data acquisition, storage and location analysis method for multi-channel lightning acoustic synchronization is proposed. Then, the high-precision lightning location device is applied in the typical distribution network area of Datong. The lightning location result of this device is compared with that of wide area lightning monitoring system. Finally, according to the monitoring results of lightning location system in distribution network, the influence of terrain on lightning activity is analyzed.

2 DETECTION PRINCIPLE OF LIGHTNING SIGNAL

2.1 *Lightning electromagnetic signal detection*

In the process of lightning occurrence, there is an inductive electric field between thundercloud and earth. When thundercloud discharges, the lightning current produces a high electromagnetic field. The electromagnetic field acts on the sensor and produces a pulse signal. Accurate detection of the time of the pulse signal is the key point of lightning location. There are three main factors for the complexity of lightning detection: 1) the starting time, spatial location and duration of lightning occurrence process are random; 2) the background change of detection environment is random; 3) the location of the receiver has an impact on the detection results.

Random noise energy distributes in the whole frequency band, and pulse signal energy concentrates in a limited frequency range. The local signal-to-noise ratio (PSNR) is obtained by Fourier transform to detect the impulse signal submerged in noise. The characteristics of background noise are as follows: poor stationarity, large fluctuation, noise and interference intensity change will cause significant changes in false alarm probability and detection probability. CFAR technology is used in background noise processing, and the effect is also analyzed. The results show that the false alarm probability can be effectively controlled under certain conditions by adaptive detection of background noise with high confidence. It also can improve the detection performance of pulse signal.

Pulse signal detection is usually expressed as binary hypothesis detection. Its observation results are either noise or signals with noise. According to Pearson criterion, it is necessary to maximize the detection probability of the pulse signal when the false alarm probability remains unchanged. The statistical characteristics of background noise are estimated by the adjacent units of the test unit. The detection threshold is adjusted adaptively according to the fluctuation of background noise. The detection performance of lightning impulse signal is improved. Background noise can be regarded as band-limited white Gauss noise. After Fourier transform, the modulus $Y_{fk}(y)$ at the same frequency point f_k can be considered to obey Rayleigh distribution. The expression of probability density function is as follows:

$$f(y) = \frac{y}{\sigma^2} \, exp\left(-y^2/\sigma^2\right) \qquad (1)$$

where σ^2 is average power of background noise.

Assuming that the transform value of noise obeys exponential distribution, the detection threshold of CFAR detection can be determined by exponential distribution. Those points below the CFAR threshold are considered as background noise, while all other points above the CFAR threshold are considered as signals plus background noise. Frequency domain CFAR detection can be described as follows:

$$Y_{f_k}^2 \underset{H0}{\overset{H1}{\gtrless}} TZ \tag{2}$$

where Z is the estimated background noise, T is the threshold weighting coefficient. The estimation methods for both are shown in the following two formulas:

$$T = (P_{fa})^{-1/N} - 1 \tag{3}$$

$$Z = \sum_{i=1}^{N} Y_{f_k}^2(y_i) \tag{4}$$

where N is the length of reference sliding window for background estimation, P_{fa} is the probability of false alarm.

The steps of detecting pulse signals in frequency domain in background noise are as follows:

1) The received data x(t) is transformed by FFT to obtain the spectrum information.

2) Search the spectrum data to determine the peak frequency and peak value.

3) Extract historical background data of peak position and calculate the estimated background noise Z of this detection.

4) The detection threshold TZ is obtained according to the principle of CFAR detection. The modulus square of peak data is compared with the threshold value TZ to determine whether the data x(t) contains a pulse signal. If no impulse signal is included, the part of the data is put into the background database.

2.2 Lightning acoustic wave signal detection

The explosive shock wave is produced by the rapid expansion of the lightning channel. This shock wave propagates outward as an acoustic signal. Lightning can be located according to the arrival time difference of lightning acoustic wave signal. In this paper, a method of constructing feature space based on wavelet transform and energy equivalence relation is proposed, which can effectively identify lightning acoustic signals. The main steps are as follows:

1) Audio signals are collected by acoustic sensors. The front-end modulator converts analog audio signal to digital signal as the original feature of lightning signal. The noise interference is eliminated by pretreatment of the signal. The time-frequency characteristics of the signal are preliminarily obtained.

2) The classical spectrum estimation based on Fourier transform is used to roughly estimate the frequency domain characteristics of signals. Firstly, the Fourier transform (spectrum) of N data is calculated, and then the product of spectrum and its conjugate is obtained.

$$P_x(\omega) = \frac{1}{N}|X_N(\omega)|^2 = \frac{1}{N}\left|\sum_{n=0}^{N-1} x(n)e^{-jn\omega}\right|^2 \tag{5}$$

3) The filter parameters are set according to the frequency response parameters, empirical values and roughly estimated signal characteristics of the acoustic sensor. And signal filtered by low pass filter. At the cost of losing some details, the computational complexity of subsequent analysis is reduced.

4) Wavelet analysis method is used to extract the characteristics of thunder signal. The structure of wavelet decomposition can be regarded as a filter bank composed of low-pass and high-pass filters. The signal is decomposed into two parts: low-frequency approximate signal and high-frequency detail signal. In the next decomposition, the low-frequency part is

decomposed into two parts: lower-frequency part and higher-frequency part. Daubechies wavelet is selected after comparing and frequency analysis of each wavelet function.

5) By normalizing the eigenvalues and doing square operation, the distance gap between the features is increased and the sensitivity is improved.

3 APPLICATION AND ANALYSIS OF LIGHTNING LOCATION SYSTEM

3.1 *Analysis of microphone array configuration*

Small sensor matrices are becoming more and more common in modern signal processing systems. The minimum number of sensors in the matrix is affected by the dimension of the direction angle to be measured. If the horizontal and vertical incidence angles need to be measured simultaneously, at least four sensors are needed. In this project, seven sensors are used to participate in the positioning. Array design is achieved by optimizing the value function. The value function is based on the following factors: 1) the number of sensors, 2) the probability of sensor failure, 3) the accuracy of estimation of different incident angles, 4) the accuracy of time difference estimation, 5) the placement of sensors, 6) the frequency components of signals. According to the quantization error, the error formula is established. Then the value function is constructed based on the error formula, which is limited by the sampling frequency f_s. The delay vector $\tau^{(m)}$ has quantization characteristics. Therefore, the propagation vector $k^{(m)}$ and incident direction obtained from the delay vector $\tau^{(m)}$ also have quantitative characteristics. θ is set as the true value, θ' is the estimated value based on the delay vector $\tau^{(m)}$, θ'' is the estimated error of incidence angle, $\theta'' = \theta - \theta'$. Estimation errors are calculated for a series of incident angles with very small intervals, and the maximum of which is taken as a value function and recorded as J_{max}.

$$J_{max} = \max\left(\theta^{(m)''}\right) = \max\left(\theta^{(m)} - \theta^{(m)'}\right) \tag{6}$$

where m is the subscript of the incident angle, $\theta^{(m)}$ is the true value, $\theta^{(m)'}$ is the estimated value.

By changing the array parameters (radius, height, sampling frequency, number of sensors), the J_{max} can be minimized to achieve the purpose of array design.

Through comprehensive analysis, a set of three-dimensional arrays composed of seven acoustic and electrical sensors is finally designed and formed as shown in Figure 1. The horizontal plane consists of four arms, one sensor on each arm. The vertical direction consists of two arms, two sensors on the upper arm and one sensor on the lower arm. The layout principles of lightning location system in distribution network are as follows: 1) Areas with higher lightning density are preferred, 2) The areas around the tower with high incidence of lightning accidents over the years are preferred, 3) Key power supply areas; 4) Towers are adjacent to ponds, reservoirs, lakes, rivers, swamps, forests or shrubs, and adjacent undulating hills, 5) Large span, and the area near the tension tower, 6) If installed in the city, try to install the

Figure 1. Sensor array and field layout.

arrays at the top of the building to avoid building occlusion around. The roof of a power supply station near the focus area is selected as the installation location of the positioning device.

3.2 *Analysis of monitoring data of lightning monitoring device in distribution network*

Several lightning monitoring devices are installed in the typical distribution network area of Datong. Contrastive Analysis of Lightning Monitoring Device and Lightning Location System in Distribution Network Based on a Direct Lightning Accident. The accident occurred in a 10kV distribution line in the southern suburb of Datong City. Distribution network lightning monitoring device monitored the process of the original signal time-domain distribution map, as shown in Figure 2. By analyzing the spectrum characteristics of this process, it can be seen that the main energy is concentrated in several hundred Hz. As shown in Figure 3, the

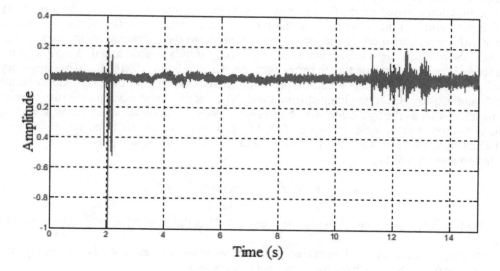

Figure 2.   Time-domain waveform of original signal.

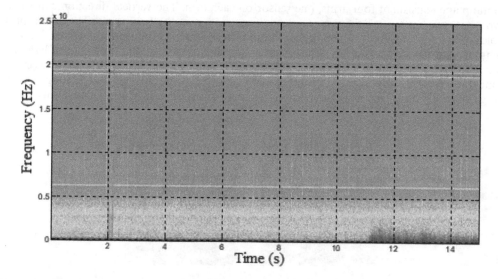

Figure 3.   Signal spectrum distribution map.

Figure 4.   Comparison of lightning location system and lightning monitoring device.

distribution network lightning monitoring device detects multiple lightning electromagnetic signals at 2s, and there are about 2s lightning acoustic signals at 6-7s. It can estimate the location of lightning occurrence and the distance from the observation point to the observation point is about 1956m. At the same time, the distribution network lightning monitoring device monitored three electromagnetic signal time points, indicating that the lightning process had three return strokes. The results of lightning location system show that this process includes one main discharge and two subsequent returns.

As shown in Figure 4, the installation location represents the position of distribution network lightning monitoring device. The fault location is the real certain lightning fault position. The location point 1 is the detection result from the distribution network lightning monitoring system. The location point 2 is given by the traditional lightning positioning system. The installation location point is 1956 meters from the fault location, and the location point 1 is 187 meters from the real lightning attack point, and so as to the location point 2 is 576 meters. From these data the distribution network lightning monitoring systems can give an optimized result for almost 389 meters. Therefore, we can declare that the distribution network lightning monitoring device has a better positioning accuracy.

4   INFLUENCE OF TOPOGRAPHY ON LIGHTNING ACTIVITY

Taking Datong City and its adjacent suburbs as an example, the topography of the area is analyzed. As shown in Figure 5, the red line identifies the separation belt between the mountainous area and plain area, and the nearby area is the piedmont. The northwest and northeast area are mountainous areas, and the southeast is flat district. The northwest side of flat district is the main city, and the southeast is the township and the paddy field. Besides, there are two main rivers and lies a reservoir in the flat district. The areas with high ground flashover density are mainly the foothill areas and the urban-rural integration areas. Although the southwestern part of Datong City has a large flash density. Some urban power lines have been converted into cable lines, and some overhead lines are also shielded by high-rise buildings. So lightning strikes faults are less. In the urban and rural contiguous areas of the mountainous area and the south of the urban area, there are many distribution power lines. The ground

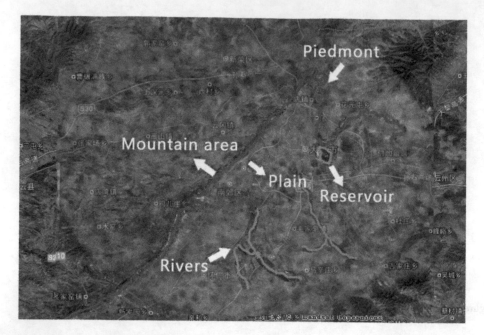

Figure 5.   Topographic of typical areas and ground flash density.

flash density is large in these areas, and the faults are concentrated. In the mountainous areas, the fault points are on the top of the mountains or in the spacious fields next to the hillside.

When the lightning current amplitude is more than 100 kA, it mostly distributes in the open suburban areas and the foothills and plains of mountainous areas, and it is less in urban areas or on the top of mountains. It is also related to rivers and reservoirs. However, the ground flashes with small current amplitude distribute more disorderly and have no specific law. The reasons are as follows: the amplitude of lightning current is mainly related to altitude and soil resistivity. Topography such as hilltop or urban high-rise buildings can easily distort the spatial electric field, and the critical breakdown field intensity is faster than that in plain areas. At this time, the amount of charges converged is also less, so the lightning current amplitude will be smaller. On the other hand, because the resistivity of soil in mountainous areas is generally high, and varies greatly with the topography. Part of the flat land at the foot of a hill is reclaimed as a field. Its resistivity is smaller than that at the top and the mountainside, and it is easier to produce larger lightning currents. The soil resistivity in suburban areas is smaller than that in urban areas. Moreover, the soil resistivity in some areas near rivers and reservoirs will also decrease. These regions are the intersecting sections of soils with different resistivity. Affected by thunderclouds, charges are more likely to accumulate in places with low soil resistivity, and this junction area will also affect the direction of lightning development. Therefore, the regions with large lightning current amplitude are also related to rivers and reservoirs, and the distribution network nearby should be strengthened protection.

5   CONCLUSIONS

1) An efficient detection technology of lightning impulse signal based on frequency domain constant false alarm detection method and wavelet analysis method is proposed. High-precision monitoring and recognition of lightning acoustic signal has been realized.

2) A real-time and high-speed data acquisition, storage and location analysis method for concurrent data of multi-channel lightning acoustic signals is proposed. An efficient lightning monitoring device for local areas of distribution network is designed.

3) Through the comparative analysis of the same lightning fault monitoring in practical engineering application, it shows that the lightning monitoring device designed in this paper is more accurate in locating distribution network lightning fault.

4) Statistical data of distribution network lightning monitoring system show that the cloud-to-ground lightning density and current amplitude in Datong area are correlated with the altitude and soil resistivity.

REFERENCES

A. D. Conti, F. H. Silveira, and S. Visacro, "Influence of a nonlinear channel resistance on lightning-induced voltages on overhead lines," IEEE Transactions on Electromagnetic Compatibility, vol. 52, pp. 676–683, 2010.

S. I. Abouzeid, G. Shabib, and A. Z. E. D. Mohamed, "Induced voltages on overhead transmission lines because of nearby included lightning channel," IET Generation, Transmission & Distribution, vol. 9, pp. 1672–1680, 2015.

Y. Zhou, A. Pahwa, and S. S. Yang, "Modeling weather-related failures of overhead distribution lines," IEEE Transactions on Power Systems, vol. 21, pp. 1683–1690, 2006.

Y. S. Fang, H. Bao, and Y. H. Yang, "Practicalization of fault location in distribution lines," Proceedings of the Csee, vol. 28, pp. 86–90, 2008.

X. Wang, S. C. Wang, and H. Liang, "Characteristics of lightning induced overvoltage of 10 kV distribution lines," High Voltage Engineering, vol. 37, pp. 599–605, 2011.

G. Liu, Y. Xi, and J. Tang, "Influence of triggered lightning of high voltage overhead transmission lines on 10 kV overhead distribution lines lightning trip characteristics," High Voltage Engineering, vol. 40, pp. 690–697, 2014.

K. Bian, W. J. Chen, and H. B. Shen, "Protective effect of erecting ground wire on distribution line to lightning induced overvoltage," High Voltage Engineering, vol. 39, pp. 993–999, 2013.

L. Li, and X. J. Qi, "Calculation of the lightning induced voltages on power distribution line," High Voltage Engineering, vol. 37, pp. 1093–1099, 2011.

A. Piantini, and J. M. Janiszewski, "Lightning-induced voltages on overhead lines—application of the extended rusck model," IEEE Transactions on Electromagnetic Compatibility, vol. 51, pp. 548–558, 2009.

P. H. Larsen, "A method to estimate the costs and benefits of undergrounding electricity transmission and distribution lines," Energy Economics, vol. 60, pp. 47–61, 2016.

A. Borghetti, F. Napolitano, and C. A. Nucci, "Influence of the return stroke current waveform on the Lightning performance of distribution lines," IEEE Transactions on Power Delivery, vol. 32, pp. 1800–1808, 2016.

Emerging Developments in the Power and Energy Industry – Dufo-López, Krzywanski & Singh (eds)
© 2020 Taylor & Francis Group, London, ISBN 978-0-367-27169-5

Non-linear parameter optimization of power system stabilizers in interconnected multi-machine power systems

A. Mahabuba
Higher College of Technology, Dubai, UAE

M. Abdullah Khan
India

ABSTRACT: This research paper presents a technique based on the non- linear optimization method for parameters of the power system stabilizers in multimachine power system. For the optimization or tuning of the parameters of PSS, the objective function is the maximization of the damping ratio of the critical modes of electro mechanical rotor oscillations. The objective of the coordinated parameter tuning is to globally optimize the overall system damping performance by maximize the damping of all both local and inter area modes of oscillations. The time domain simulation results of the multi-machine power system validate the effectiveness of the proposed approach. In this paper, the 10 machine 39bus New England system is used as the test system.

Keywords: Parameter Optimization, Tuning of Parameters, Comprehensive Damping Index (CDI), Non-Linear Optimization, Power System Stabilizer(PSS), Damping Control, Coordinated Tuning

1 INTRODUCTION

Modern Power system is characterized by the extensive system interconnection and increasing dependences on the control for optimum utilization of existing resources. Due to the rapid growth in electric power demand, and increasingly more complicated non-linear network structure, the damping in power systems deteriorates. Several unacceptable dynamic stability problems, e.g., low frequency oscillations become important to electric power utilities (Gurunath Gurrala et al. 2010).

Low frequency oscillations (0.2–2.5 Hz) restrict the steady-state power transfer limits, which therefore affect the operational system economics and security (Zhang P. et al. 2000). These oscillations are mainly due to the dynamic interactions between the various generators of a system through its transmission network. So considerable efforts have been placed on the application of Power System Stabilizers to damp low frequency oscillations and thereby improve the system stability. PSSs is well known for improving the dynamic stability of the system.

In the application of PSSs to increase the damping of certain critical modes in multi-machine power system the very first step is to determine the best location(s) for the PSS. At present, there are several methods (Madhiyeh E. et al. 2011 and Mahabuba A & Abdullah Khan M. 2013) based on right and left Eigen vectors which can accurately identify the best location of PSS. After selecting the very best location(s) of PSS, anyone of the tuning methods (Prasnti Dey. et al. 2017) can be applied for various operating condition.

In multi-machine system, in order to avoid the destabilizing interactions among the PSSs, simultaneous tuning of all the controller parameters is very essential. To improve the overall system performance many researches were made on the co-ordination between PSSs. Some of these methods are based on the complex non-linear simulation, the others are based on the linearized power system model.

In this paper, a non-linear optimization based tuning algorithm is proposed to co-ordinate among multiple PSSs simultaneously. This algorithm optimizes the total system performance by means of sequential quadratic programming. Interaction among PSSs is improved by minimizing the objective function. This will optimize the over all system damping performance.

This research paper has the following sections as follows: Section 1 is the introduction, multi-machine test system is described in section 2. Design of power system stabilizers (conventional approach) is included in section 3. In section 4, the non-linear parameter coordinated tuning method is described in detail. Section 5 comprises the simulation results and finally the brief conclusion.

2 INTERCONNECTED MULTI-MACHINE POWER SYSTEM

In the small signal stability improvement process of interconnected multi-machine system, modeling of all machines in a more detailed manner is a complex task in view of large number of synchronous machines in the system. In this research work, the New England 10-machine 39-bus power system shown in Figure 1 is considered. Each generator of the test system is described by a two-axis fourth order model. IEEE type ST1A model excitation system has been included. System data and excitation system data are extracted from (Padiyar K.R. 2002). Assumptions for the two-axis model and linearized equations used for the system modeling are described in (Anderson P.M. and Fouad A.A. 2003). The study uses the two axis model with four state variables for the synchronous machine.

The non-linear model is linearized around an equilibrium point, in order to get system model in state space form:

$$\Delta \cdot E_{d_t}' = \left\{ -\Delta E'_{di} - (x_{qi} - x'_i)\Delta I_{qi} \right\}/\tau_{q0i}'$$

$$\Delta \cdot E_{q_t}' = \left\{ -\Delta E_{FDi} - \Delta E'_{qi}(x_{di} - x'_i)\Delta I_{di} \right\}/\tau_{d0i}'$$

$$\Delta \dot{\omega}_n = \left\{ \Delta T_{mi} - \left(I_{di0} \Delta E'_{di} + I_{qi0} \Delta E'_{qi} + E'_{di0} \Delta I_{di} + E'_{qi0} \Delta I_{qi} \right) - D_i\omega_{ri} \right\}/\tau_i \quad i = 1, 2, 3, \ldots m$$

$$\Delta \delta_i = \omega_{r1} - \omega_{ri}; i = 2, 3, \ldots \ldots m \, taking \, machine \, 1 \, as \, reference$$

$$(1)$$

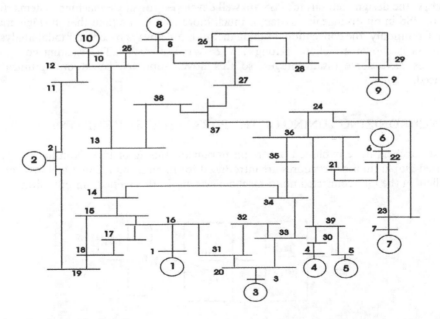

Figure 1.   New England system.

782

Where,

Ed′ – Direct axis component of voltage behind transient reactance.
Eq′ – Quadrature axis component of voltage behind transient reactance.
ω_r – Angular velocity of rotor.
δ_r – Rotor angle in electrical radians.
xd′ – Direct axis transient generator reactance.
xq′ – Quadrature axis transient generator reactance in p.u

3 TUNING THE PARAMETERS OF CONVENTIONAL POWER SYSTEM STABILIZER

PSS has a transfer function consisting of a wash–out block, a lead-lag phase compensator circuit and a stabilizer gain block (Muhammad Razali N.K. et al. 2006). PSSs are supplementary controllers in the excitation system. Main function of PSS to add damping to the generator rotor oscillations by controlling its excitation using auxiliary stabilizing signal. To provide damping, PSS must produce a component of electric torque in phase with the rotor speed deviation (Md Safiullah et al. 2018).

The stabilizer gain block determines the amount of damping introduced by the PSS in terms of gain (K_{pss}). The wash-out block serves as high pass filter with time constant T_w. The lead-lag compensator block provides the appropriate phase lead characteristics to compensate the phase lag between the exciter input and the generator electric torque with time constants T_1 and T_2. The structure of PSS used is illustrated in Figure 2.

The transfer function of the PSS is given in equation (2):

$$\Delta V_s = K_{PSS}\left(\frac{sT_\omega}{1+sT_\omega}\right)\left(\frac{1+sT_1}{1+sT_2}\right)\Delta\omega \qquad (2)$$

where KPSS is the PSS gain, Tw is the washout time constant and T1 and T2 are the compensator time constants.

The intention of PSS tuning is particularly for the damping of critical modes. Well designed PSS can provide damping over wide range of operating conditions. Based on the modern control theory, the design methods for PSS are well developed for single machine systems. But the design of PSS in multi- machine system is much more complicate than that in single machine system. Commonly, the conventional tuning methods are always based on modal analysis and the process is shown as follows (Hong Y.Y & Wu W.C. 1999).The existing conventional approaches have some disadvantages .so this novel approach based on optimization is introduced.

4 SYNCHRONIZING TUNING OF THE PARAMETERS OF THE PSS

Many researchers have published papers on parameter tuning of PSS. Non-linear optimization based global tuning procedures are introduced for minimizing the interactions among the controllers in the interconnected multi-machine system. In this section, an optimization based

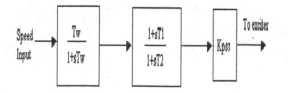

Figure 2.   PSS controller.

method for the synchronization of the co-ordinated tuning of PSSs is presented. The objective of the simultaneous parameter tuning is to globally optimize the overall system performance. This requires the simultaneous optimization and co-ordination of parameter settings of PSS controllers to maximize the damping of all modes of oscillations. Here parameters of each controller are determined by non-linear programming technique (Eduein L.M. & Percival B.A. 2018). The main procedure is as follows:

a) System linearization for analyzing the dominant modes of oscillations in large systems.
b) Allocation of controllers based on the participation factor method.
c) Using the parameter constrained non-linear optimization to optimize the global system behavior.

This paper mainly focuses on the optimization based parameter setting of PSS. This synchronizing and simultaneous tuning is done by non-linear optimization algorithm using linearized system model.

4.1 *Linearized system model of the interconnected multi-machine power system*

After the optimal allocation of the controllers is chosen, the total system linearized system model extended by PSS can be derived and represented by the following equation:

$$\Delta x = A\Delta x + B\Delta u$$
$$\Delta y = C\Delta x + D\Delta u \tag{3}$$

From (3), the Eigen values $\lambda i = \sigma_i \pm j\omega_i$ (for the mode i) of the total system can be evaluated. The proposed method is to search the best parameter sets of the of the PSS controllers, so that a comprehensive damping index (CDI) (4) can be minimized.

$$CDI = \sum_{i=1}^{n}(1 - \zeta_i) \tag{4}$$

where $\zeta_i = \frac{-\sigma}{(\sigma^2 + \omega^2)^{1/2}}$

(if $\sigma \pm j\omega$ are Dominant roots) is the damping ratio and n is the total number of dominant Eigen values which include the local modes, inter area modes. Among dominant Eigen values only those have damping ratio less than 0.4 are considered in the optimization. Maximization of damping ratio is carried out by moving the considered Eigen values to the left in the optimization technique as shown in Figure 3.

+ Eigen value before optimization
* Eigen value after optimization

In order to minimize the comprehensive damping index, the non-linear optimization technique implemented in Matlab optimization technique is used. The use of the objective function will result in PSSs that shift the lightly damped and undamped electromechanical modes to the left-hand side of a vertical line in the complex s-plane; hence, improving the damping factor. The use of the objective function will therefore guarantee that the relative stability and the time domain specifications are concurrently secured. The proposed design approach has been applied to a multi-machine power system. This is usually the case in dynamic stability where it is desired to relocate the electromechanical modes of oscillations (Sivasubramania M. et al.)

4.2 *Non-linear optimization technique*

A general-purpose nonlinear programming package is used to minimize the real part of the dominant eigenvalue in the system without implementing the eigensensitivities. The basic control theory and the root-coefficient relationship for polynomial equations are incorporated for obtaining the equality constraints. The PSS parameters bound in a set are formulated as the

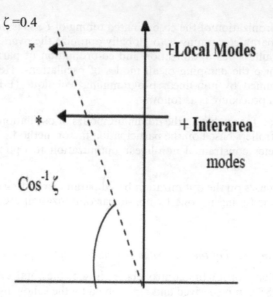

Figure 3. Objective of optimization.

inequality constraints (Ying-Yi Hong and Wen-Ching Wu 1999). A comparative result for a small-signal stability problem validates the applicability of the proposed method.

The objective of the parameter optimization can be formulated as a non-linear programming problem, expressed as follows:

Min. f (z) = CDI (from equation 3)
Subject to E (z) = 0
F (z) ≥ 0

where f (z) is the objective function defined. z is a vector, which consists of the parameters of PSSs have to be tuned. E (z) are the equality functions and F (z) are the inequality functions respectively. For this proposed method only the inequality functions which presents the parameter constraints of each controller are used (Cai L.J. & Erlich L. 2002).

The above mentioned method is a general parameter constrained non-linear optimization problem, which can be solved successfully by selecting appropriate function from the Matlab optimization tool box (Optimization Tool Box 1999). In order to select the function the objective function is derived in terms of PSS parameters (Yao –nan Yu). The flowchart of the optimization based co-ordinated tuning algorithm is shown in Figure 4.

The optimization begin with the pre-selected initial values of the PSS parameter z0.Then the non-linear algorithm is applied to tune the parameters of the controller iteratively until the objective function is minimized. These so determined parameters are the optimal settings of the PSS controllers.The proposed method permits considering several operating points of the system simultaneously. In this case, the CDI is calculated for each state successively and added to global CDI provided for the optimization algorithm.

5 SIMULATION RESULTS

To verify the performance of the proposed tuning method, the algorithm is tested in a multi-machine test system. Firstly the test system is evaluated without PSS controllers. The system is unstable as shown in Figure 5. [rotor angle (radians) δ vs. time (sec)].

Conventional sequential tuning of PSSs is done in order to make the system stable. But damping of some critical modes of oscillations is not satisfactory.

Co-ordinated tuning based on optimization is done and the result is shown in Figure 6.

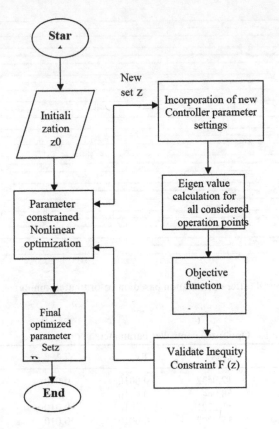

Figure 4. Flowchart of optimization based tuning.

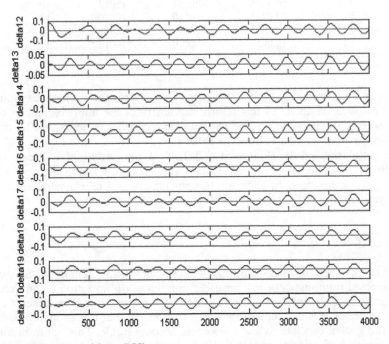

Figure 5. Unstable system (without PSS).

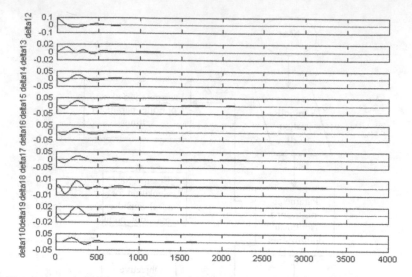

Figure 6.   Simulation result (after optimization based on co-ordinated tuning).

Table 1.   Optimized controller parameters.

Gen	kpss	T1	T2	Tw
1	12.0432	0.0010	0.010	10
2	38.064	0.0010	0.010	10
3	11.2791	0.0010	0.010	10
4	11.2792	0.0010	0.010	10
5	9.0056	0.0010	0.010	10
6	11.2792	0.0010	0.010	10
7	10.6075	0.0010	0.010	10
8	12.0432	0.0010	0.010	10
9	12.0432	0.0010	0.010	10
10	12.0432	0.0010	0.010	10

After co-ordinated tuning, as shown in Figure 6, local modes and inter-area modes of oscil-lations are now well damped. So this results demonstrates the improvement in damping of overall power oscillations in the system. The detailed controller parameters after optimization based on co-ordinated tuning are shown in Table 1.

From the simulation results (Figure 6) and Table 1 the optimized PSS can improve the small-disturbance stability of the power system and efficiently damp the power oscillation caused by large disturbances. In addition, the practicality and effectiveness of the proposed method are verified on IEEE 10-generator New England test systems.

6   CONCLUSION

This is a novel approach for the synchronizing parameter optimal tuning of PSS controllers in the interconnected multi-machine system using non-linear optimization technique. The opti-mization algorithm is based on linearized power system model and parameter constrained non-linear optimization technique. Simulation results reveal that this proposed method effect-ively damp the system oscillations and the system become stable. This optimization based on co-ordinted tuning is simple and easy to realize in large power system.

REFERENCES

Gurunath Gurrala and Indraneeth Sel. 2010, Power system Stabilizers Design for Interconnected Power System, IEEE Transactions on Power Systems, 25(2): 1042–1051.

Zhang P. and Coonick H. 2000, Co-ordinated Synthesis of PSS Parameters in multiachine powersystems using the method of inequalities applied to genetic algorithms, IEEE Trans.on Powersystems, 5(2).

Madhiyeh Eslami, Hussain Shareef, Azar Mohamed. 2000, Optimal location of PSS using improved PSO with chaoatic sequence, International Conference on Electrical, Control and Computer Applications InECCE.

Mahabuba A. and Abdullah Khan M. 2013, Identification of Optimum locations of Multi machine Power system using Second order eigen sensitivity Analysis, Smart Grid Rnewable Energy 4(01): 35–42.

Prasanti Dey, Aniruddha Bhattarchaya and Priyanth Das 2017, Tuning of power system Stabilizer for small signal stability improvement of interconnected power system, Applied Computing and Informatics.

Padiyar K.R., Power System Dynamics and Control. Second Edition, BS Publication, Hyderabad, AP, 2002.

Anderson P.M and Fouad A.A. 2003, Power System Control and Stability, The IOWA State University Press, Ames, IOWA, USA, Galgotia Publication.

Muhammed Razali N.K., Ramachandra Murthy V.K and Mukherjee R.N. 2006, Power System Stabilizer Placement and Tuning Methods for Inter-area Oscillation Damping, IEEE International Power and Energy Conference, 28–29 November 2006, Malaysia.

Md Safiullah, Md Jual Rena Md. Shafiul Alam and Abido M.A. 2018, Online tuning of Power System Stabilizers employing genetic programming for stability enhancement, Journal of Electrical Systems and Information Technology, 5(3): 287–299.

Hong Y.Y. and Wu W.C. 1999, A new approach using optimization for tuning the parameters of power system stabilizers, IEEE trans. Energy conversion, 14(3): 780–786.

Eduein Luiz Miotto, Percival Bueno de Araujo, Coordinated Tuning of the parameters of PSS and POD Controllers using Bioinspired Algorithms. 2018, IEEE Transactions on Industry Applications, 54(4): 3845–3847.

Sivasubramanina M., Mustafa P. and Sudheer K., Simultaneous Tuning of Power System Stabilizers parameters for Multi Machine Power System, International Journal of Modern Engineering Research (IJMER), 1(1): 229–235.

Cai L.J. and Erlich L. 2002, Simultaneous co-ordinated tuning of PSS and Facts Controller for damping power system oscillations in multi machine systems, IEEE Transactions on Power systems, 5(1).

Optimization tool box, 1999 user's guide, and version 2.

Yao-nan Yu, Electric power system dynamics, Academic press publications, New York.

Ying-Yi Hong and Wen-Ching Wu. 1999, A new approach using optimization for tuning parameters of Power System Stabilizers, IEEE Transactions on Energy Conversion 14(3): 780–786.

Emerging Developments in the Power and Energy Industry – Dufo-López, Krzywanski & Singh (eds)
© 2020 Taylor & Francis Group, London, ISBN 978-0-367-27169-5

Study of energy generation through biogas from landfill waste

K. Kanimozhi
Professor/Electrical and Electronics Engineering, Sethu Institute of Technology, Kariapatti, Tamil Nadu, India

B. Raja Mohamed Rabi
Professor/Mechanical Engineering, Sethu Institute of Technology, Kariapatti, Tamil Nadu, India

ABSTRACT: The intensification of human and industrial activities in the last few decades has generated increase in the production of municipal solid wastes (MSW), becoming a serious problem for the society. Furthermore, the uses of large landfills in great urban centers are still common, which causes sanitary and ambient problems. Gramacho's landfill was chosen as study case, for technical and economical feasibility analysis of energy generation though the biogas from waste. The more important environmental contribution associated to this paper is the reduction of greenhouse gases emissions (GHG), by means of the conversion of methane into carbon dioxide. Studies and comparative analysis were presented demonstrating when gas turbine, internal combustion engines (Otto or Diesel cycles) or other technologies of energy conversion have technical and economical feasibility for implantation of the thermoelectrical plant. The results are based on limited factors of contingency enclosed in the estimates of capital and the operation and maintenance costs.

1 INTRODUCTION

Waste disposal in landfills can generate environmental problems such as water pollution by unpleasant odors, risks of explosion and combustion, risk of asphyxiation, vegetation damage, and greenhouse gas emissions (Kanimozhi 2017). According to prediction of the United Nations Organization, the world-wide population must grow until 2050 about 40%, reaching 8.9 billion people. Agenda 21 from the ECO-92 Conference foresees the duplication of the amount of residue produced in the world until 2010, based on values of 1990 and they will quadruplicate until 2025.

The amounts of wastes generated by the societies are increasing in the whole world, either due to population increase or due to increment of the per capita production of residues. Additionally, current production and consumption models prioritize the use of disposable materials and products, not taking in account the necessity of maintenance of a sustainable ambient (Abreu 2011). Landfill gas is generated under both aerobic and anaerobic conditions. Aerobic conditions occur immediately after waste disposal due to entrapped atmospheric air. The initial aerobic phase is short-lived and produces a gas mostly composed of carbon dioxide.

Since oxygen is rapidly depleted, a long-term degradation continues under anaerobic conditions, thus producing a gas with a significant energy value that is 55% methane and 45% carbon dioxide with traces of a number of volatile organic compounds (Booma 2017). The biogas generated in landfills is basically composed of methane (CH_4, 55 to 65%), carbon dioxide (CO_2, 35 to 45%), nitrogen (N_2, 0 to 1%), hydrogen (H_2, 0 to 1%) and sulfuric gas (H_2S, 0 to 1%). The anaerobic process begins after the waste has been in the landfill for 10 to 50 days. Although the majority of CH_4 and CO_2 are generated within 20 years of landfill completion, emissions can continue for 50 years or more (Ben 2009).

Table 1. MSW disposal in Brazil.

Region	Total (tones/day)	Open dump (%)	Control Landfill (%)	Landfill (%)	Others (%)
North	*11.06*	*56.7*	*28.3*	*13.3*	*1.7*
Northeast	41.55	48.2	14.6	36.2	1.0
Southeast	141.7	9.7	46.5	37.1	6.7
South	19.87	25.7	24.3	40.5	9.5
Center-west	14.2	21.9	32.8	38.8	6.5
Brazil	228.4	21.2	37.0	36.2	5.6

Table 1 shows MSW disposal in Brazilian geographical regions. Brazilian Energy Matrix is compound of approximately 48.4% from renewable energy sources and 51.6% from non-renewable.[6] Electricity generation in Brazil reached 463.1 TWh in 2008, or 4.2% higher than 2007 total. Main contributors are public utilities, with 89.0% of shares. From those, hydro utility plants remain as main source, even with a reduction of 1.4% in comparison to 2007. Thermal generation increased in 63.2%, especially from natural gas (116.6%) and nuclear (13.1%) (Comlurb 2015).

Landfill gas (LFG) recovery and utilization have not been significantly evaluated in Brazil. A number of reasons might have contributed for this scenario, including: public regulation uncertainties, lack of financial incentives, absence of public and private investments, operational conditions of landfills, and low level of technical support.(Lora 2015) The only full scale LFG power plant started its operation in the beginning of 2004 with an installed capacity of 20 MW (Maciel 2005). Bandeirantes and Sao Joao landfills were disabled in 2007 and 2009, respectively, and thermoelectric power plants were installed to burn LFG produced by the decaying waste. Table 2 shows potentials of methane recovery and electricity generation in main Brazilian landfills Although the proposed strategy to include social costs is quite reasonable, and it increases the feasibility of the thermo power facility, difficulties in accounting add to lack of precise data do not allow that social costs were included in the present study (Zamorano 2007).

The proposed work aims at presenting a technical and economical evaluation of energy generation from MSW at Gramacho's landfill in Brazil. Waste-to-energy (WTE) technologies, which combust municipal solid waste to produce energy.

2 METHODOLOGY

2.1 *Gramacho's landfill*

Gramacho's landfill was chosen as study case because its importance for the city of Rio of Janeiro and its metropolitan region. Gramacho's landfill is located at the following

Table 2. Potentials of methane recovery and electricity generation in main Brazilian landfills.

Municipality	Unit of treatment	Waste disposal (tones/years)	Methane recuperation (MM m³/day)	Power generation (MW average)
Duque de Caxias/RJ	*Gramacho Landfill*	*2.258.429*	*484*	*53.8*
Rio de Janeiro/RJ	CTR Gericinó	1.081.848	232	25.8
Caucaia/CE	ASMOC Landfill	1.038.670	223	24.8
Jaboatão	Muribeca Landfill	955.746	205	22.8
Belo Horizonte/MG	CTRs BR040	909.520	195	21.7
Brasília/DF Joquei	Landfill	846.669	182	20.2
Manaus/AM KM 19	Landfill	709.696	152	16.9
Curitiba/PR	Caxima Landfill	670.790	144	16.0

coordinates: 22°44′46″ South and 43°15′37″ West, Gramacho's landfill operations started as an open dump in a mangrove swamp in 1978. Initial filling was performed by pushing waste into the swamp area to fill it to a point where it was above high sea level. Subsequent fill activities consisted of haphazard dumping, waste burning, and uncontrolled scavenging.

Since the beginning of the decade of 1990 it has started to receive some cares to minimize its environmental impact. In the early 1990s, the landfill operator, Companhia de Limpeza Urbana (COMLURB), began converting the open dump into a sanitary landfill. By 1996, most of the attributes of a modern sanitary landfill were in place, including controlled access, a recycling facility, well-maintained access roads, waste compaction by bulldozers, and the application of daily and intermediate cover soils. Table3 shows solid waste disposal evolution in Gramacho's landfill. All waste deposited prior to 1993, during the open dump operations, were not included in the present study. Excluding waste disposed in the open dump, it is estimated that there was more than 40 million tonnes of waste in place at the Landfill as of the end of 2010.

By 1996, most of the attributes of a modern sanitary landfill were in place, including controlled access, a recycling facility, well-maintained access roads, waste compaction by bulldozers, and the application of daily and intermediate cover soils. Table 3 shows solid waste disposal evolution in Gramacho's landfill. All waste deposited prior to 1993, during the open dump operations, were not included in the present study. Excluding waste disposed in the open dump, it is estimated that there was more than 40 million tonnes of waste in place at the Landfill as of the end of 2010. The site currently receives approximately 250,000 tonnes per month (3,000,000 tonnes per year).

2.2 Technical solutions for energy generation in landfills

Most suitable conventional technologies for direct electric energy conversion from biogas are gas turbines and internal combustion engines, since steam turbines require a furnace e for steam generation. From small to medium power generation capacities, internal combustion engines are more appropriated because of its lower cost and greater efficiency in this range. Only for higher capacities, gas turbines are competitive, and their yielding is improved when they are used in combined cycles. Internal combustion engines are more efficient within the operation range of this project. Diesel cycle engines work on higher compression rates,

Table 3.   Solid waste disposal in Gramacho's landfill.

Year	Waste disposed Tonnes	Waste in place Tonnes
1993	1.646.374	1.646.374
1994	1.669.443	3.315.817
1995	1.800.209	5.116.026
1996	2.325.161	7.441.187
1997	2.414.508	9.855.695
1998	2.390.021	12.245.716
1999	2.403.311	14.649.027
2000	2.454.563	17.103.590
2001	2.417.409	19.520.999
2002	2.473.918	21.994.917
2003	2.359.715	24.354.632
2004	2.400.000	26.754.632
2005	2.400.000	29.154.632
2006	2.568.000	31.722.632
2007	2.747.760	34.290.632
2008	2.920.000	37.210.632
2009	3.000.000	40.210.632

requiring that biogas is fed mixed with diesel or biodiesel, which would represent an additional input to the energy facility. Moreover, in the Brazilian internal market, Otto cycle engines can be more easily adapted to operate with biogas.

2.3 *Economical analysis*

The following assumptions have been considered:

(i) The economical analysis was carried out through a 15-year period.
(ii) Two financing options have been evaluated: one without financing of capital expenditures and another with a 75% financing of the initial capital expenditures.
(iii) Recipes from RECs have been included, with the selling price of US$ 17 per ton of CO2 equivalent.
(iv) The same 8% interest tax has been adopted for the liquid present value (LPV) determination and for the financing of the loan.
(v) The loan's payment period for the initial investment is 15 years.
(vi) The payment of approximately 20% of REC recipes to the landfill proprietor for the biogas use has been considered, representing a tax of $0.43/MMBtu;
(vii) The value of biogas has a 3% annual readjustment.

For biogas generation potential calculation, it has used the model

$$Q_m = \sum_{i=1}^{n} 2KL_0 M_i e^{-kti}$$

where Q_m= methane generation (m³/years); L_o = potential methane generation capacity (m³/tonnes); M_i = annual waste disposal in year i (tonnes); k = methane generation (decay) rate constant (1/years); t = time elapsed (years); i = time increment in one year).

The methane recovery potential (Lo) is the total amount of methane that a unit mass of refuse will produce given enough time and is a function of the organic content of the waste. For Gramacho's Landfill, started with a default Lo value based on 1,140 mm of annual precipitation, and then adjusted this value based on the ratios of organic and moisture contained in U.S. waste and waste at the Landfill. The methane recovery potential for Gramacho's Landfill is 84.8 m³/Mg. Table 4 summarizes TEP schedule, proposed by SCS Engineers [12].

3 RESULTS AND DISCUSSIONS

The costs of capital for the development of a biogas recovery project and those related to the operation, maintenance and regular expansion of the biogas collection system were estimated, including recurrent costs for capacity expansion of the ventilation and burning station. Figure 1 shows the energy efficiency in function of the thermoelectric plant (TEP) capacity, for gas turbines, internal combustion engines (Otto and Diesel cycles) and combined cycles.

Table 4. Summarizes TEP schedule, proposed by SCS Engineers (2005).

Years	Planning of TEP – Biogás
1	System of collection of gas and burning in construction
2	Beginning of the collection system and burns. Plant in construction
3	Beginning of the functioning of the energy plant; System to operate the capacity of 10 MW
4 to the 8	System with capacity of 10 MW
9 and 10	System with capacity of 7,2 MW
11 to the 15	System with capacity of 4,3 MW

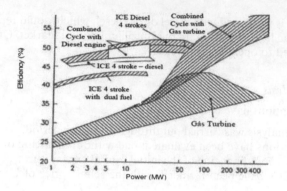

Figure 1. Efficiency comparison among diverse energy conversion technologies.

Table 5. Costs of the thermoelectrial plant (TEP).

Detail	Estimated total cost ($)[1]
Plant of Energy of 10MW supplied with biogas	$9,910,875
Interconnection of 3 km	$617,500
Construction of the Plant/work in the place (including tubing)	$214,890
Measurement of biogas and equipment of register	$61,750
Engineering/contigency (10% of other costs)	$1,080,625
Total costs	$11,885,640

Since Gramacho's potential power generation has been estimated at 10 MW, internal combustion engines present better performance than gas turbines form this application. The initial cost for accomplishment of the 10 MW (bulk) TEP has been estimated as US$ 11,885,640 (shown in Table 5) The generation is made using internal combustion engines, fed with biogas, intended to attain all landfill and its own energy consumption and to sell the exceeding energy to the electrical grid.

The costs of the biogas collection and burning system were added (cost of 7,164,086 US$). It was assumed that the plant will start to operate in first day of the third year of the project and will continue to operate until 15th year (in this case until 2024). So, the value of investment is US$ 19,049,726. Table 6 shows the other costs of Thermoelectrical Plant. The typical payback for Thermoelectrical Plant is nine years, in this scenario actual. Then, 2014 is the year of payback of this project. Table 7 shows a summary of the results of the economic evaluation in the scenario without taking account recipes from RECs or carbon credits. Table 8 shows sensibility analysis, scenario with carbon credits ($17 tCO2 eq. – Gramacho's adopted tax). The economic projections of the TEP are presented attractive for financing scenarios. On the other hand, the scenario without carbon credits is not attractive.

4 CONCLUSIONS

Biogas energy is one of the important options which might gradually replace oil, which is facing increasing demand and may be exhausted early in this century. In the meantime, the biogas energy can help to save exhausting the oil wealth. Based on results, the landfill biogas energy exploitation of Gramacho's Landfill is viable taking as reference the value of CER in $17 of ton.CO2eq and any of the financing options analyzed. The results are based on limited factors of contingency enclosed in the estimates of capital

Table 6. Other costs of thermoelectrical plant.

Year	Annual cost O&M thermoelectrial plant	Annual O&M of the collection system and gas of control and ampliation of costs	CDM register and annual verification	Comlurb Recipe	Payment of Garbage's participation deep
2005	-	-	-	-	-
2006	-	-	-	-	-
2007	-	-	-	-	-
2008	-	-			
2009	-	-	-	-	-
2010	-			$741,000	$1,482,000
2011	-	$435,023	$58,986	$770,640	$1,541,280
2012	$2,010,809	$448,073	$60,755	$801,465	$1,602,931
2013	$2,071,133	$461,516	$62,578	$833,524	$1,667,048
2014	$2,133,267	$475,361	$64,455	$866,865	$1,733,730
2015	$2,197,265	$489,622	$66,389	$901,539	$1,803,079
2016	$2,263,183	$504,311	$68,381	$937,601	$1,875,202
2017	$2,331,079	$519,440	$70,432	$975,105	$1,950,210
2018	$1,715,031	$535,023	$72,545	$1,014,109	$2,028,219
2019	$1,766,482	$551,074	$74,721	$1,054,674	$2,109,348
2020	$1,819,476	$567,606	$76,963	$1,096,861	$2,193,722
2021	$1,874,061	$584,634	$79,272	$1,140,735	$2,281,470
2022	$1,930,283	$602,173	$81,650	$1,186,364	$2,372,729
2023	$1,988,191	$620,238	$84,100	$1,233,819	$2,467,638
2024	$2,047,837	$638,846	$86,623	$1,283,172	$2,566,344

Table 7. Investment analysis (scenario with carbon credits).

Value of initial investment	Percentual value of the initial investment of capital(%)	LPV	RIT*
19.160.877	100	$33.833.352	24.95
4.790.219	25	$32.768.859	35.40

and the operation and maintenance costs. Improvements to be added in some of the used estimates in the economic evaluation, mainly the electricity sale price, can positively modify the results of this analysis.

Energy generation facilities shall be included in future landfill projects. A methodology for evaluation of social and environmental costs shall be added in economical evaluation of WTE. Energy generation from landfills does not impact Brazilian Energy Matrix. The main advantage of implementing WTE facility is waste volume reduction, lengthening landfill useful life and technical servicing. The following measures are being suggested to promote the growing of energy production through biogas from the waste:

(i) Simplification of the environmental licensing procedures for landfills.
(ii) Adoption of fiscal favorable instruments
(iii) Dissemination of technical and economical data on construction and operation of landfill with exploitation of biogas, as well as the achieved benefits.
(iv) Establishment of special credit lines by development banks (as BNDES) with favored taxes and dedicated calls in official researching support agencies to promote the scientific initiation and technological innovation for energy exploitation from biogas in landfill.

REFERENCES

Abreu FV, Costa Filho MAF, Souza MCL. 2011. Technical and economical feasibility analysis of energy generation though the biogas from waste in landfill—an alternative of renewable energy generation. In: 20th International Congress of Mechanical Engineering (COBEM), Gramado: pp. 50–60.

Ben 2009. National Energy Balance 2009 Booma, Mahadevan, Kanimozhi.2017. Perspectives of Hydro Power Plant and Pumped Storage System in Tamil Nadu International Journal of Chemtech Research, 11(4):145–152.

Comlurb 2015. Transforming Ambient Liabilities into an Energy Resource: The garbage as power plant, Rio de Janeiro.

Kanimozhi K., Raja Mohamed Rabi B., 2017. Hydrogen synthesis from Solar reactor using Reverse Photo synthesis," International Journal of Chemtech Research, 10(9):212–219 https://ben.epe.gov.br/downloads/Relatorio_Final_BEN_2009.pdf - Access in: August of 2014.

Lora EES, Nascimento 2014. Thermoelectrical Generation: plan, project and operation. São Paulo: 457–461.

Maciel FJ, JUCÁ JFT 2005. Gas Recovery Investigation in a Brazilian Landfill. 3:45–52.

Zamorano M, Pérez JL, Aguilar I, Ramos A 2007. Study of the energy potential of the biogas produced by an urban waste landfill in Southern Spain, Renew. Sustain. Energy Rev., 11(5): 909–922.

Emerging Developments in the Power and Energy Industry – Dufo-López, Krzywanski & Singh (eds)
© 2020 Taylor & Francis Group, London, ISBN 978-0-367-27169-5

Comparing the wind and solar power uncertainty in virtual power plants using IGDT method

M.J. Moradof
Electrical Engineering Department, Islamic Azad University Research and Science branch, Tehran, Iran

A. Sheikhi
Sharif Energy Research Institute (SERI), Tehran, Iran

M. Rayati & Ranjbar
Electrical Engineering Department, Sharif University of Technology, Tehran, Iran

ABSTRACT: To decrease dependency on fossil fuels, almost all countries plan to invest more in renewable energy sources (RESs), e.g., wind and solar power plants. RESs provide substantial benefits for our climate and economy. However, there are still some technical and financial challenges to generate large amount of power through RESs. Increasing RESs capacities in a power system escalates the risk of instability. In this paper, RESs are deployed in the framework of virtual power plants (VPPs). The VPPs, as a new concept in power systems and electricity markets, have a substantial impact on increasing efficiency, engagement, and synergy effects of RESs in power systems. Optimal amounts of power generation of different generating units in a sample VPP are calculated and financial indicators are studied in this paper. To achieve more reasonable results with respect to stochastic nature of wind and solar generations, we apply Information Gap Decision Theory (IGDT) method to evaluate the system uncertainty. The results are achieved by GAMS software. Moreover, the impacts of different sources of uncertainties are compared with each other.

1 INTRODUCTION

Traditionally, fossil energy sources were the single origin of energy. However, these finite sources could not be proper choices for going on human's life with respect to growing population. Therefore, scientists propounded renewable energy sources (RESs), e.g., wind or solar power plants. RESs can produce energy for unlimited time without contamination; hence, these sources are recognized as clean energy. RESs are considered as a key solution to pave the way of sustainable development, mitigate greenhouse gas emission, and decrease global warming (Panwar, N. L. et al. 2011). Although the pros definitely outweigh the cons of applying these kinds of sources, RESs have some disadvantages that should be dealt with their adverse effects thoughtfully. For instance, instability issues due to the presence of high penetration of RESs made some hesitations in deploying these resources (Shim, Jae Woong et al. 2013) or uncertain nature of them is considered as a serious hurdle in their applications (Hadjipaschalis, Ioannis et al. 2009), (Mehrabankhomartash, Mahmoud et al. 2017), (Roustai, Mahmoud et al. 2018). With respect to these facts, these resources are not perfectly reliable to supply power in the long-term and in high penetration. To tackle these issues, many solutions were proposed categorizing in different threads.

Energy storage systems (ESSs) are used in one of the important method for addressing and solving RESs problem. This method has been studied in many publications. In (Evans, Annette et al. 2012) ESS technologies are reviewed and a comparative summary of performances of the major ESS options are given. Despite the fact that integration of electricity from

RESs in a power supply system is very important, the dependency of integration procedure on different ESSs parameters is unknown yet. To address this deficiency, a method is proposed in (Weitemeyer, Stefan et al. 2015) to utilize data series of long-term power productions of solar and wind energy to investigate ESS size effects and its efficiency. The results show that a precise compromising between both of installation of additional RESs generation capacities and storage capacities is needed. Another study about ESS field was done in (Zheng, Yu et al. 2014) where the operation of distribution companies (DISCOs) is considered and a new battery energy storage system (BESS) is presented based on energy purchasing model. This model uses ESSs optimally and declines operational risks due to the electricity price fluctuations. Moreover, a cost-benefit analysis method was suggested to obtain an optimal size and operation of BESS.

Another method to tackle RESs problem is proposed in some recent studies. In (Kusakana, Kanzumba et al. 2014) a method for optimal operation of a hybrid multi-source system is presented. A hybrid multi-source system includes various types of energy sources, e.g., solar, wind, hydro, diesel generator, and ESS. The proposed model is based on taking into account different power sources, losses, and their impacts on the operation of power system. An optimal design for a hybrid renewable energy system (HRES) is studied in (Sharafi, Masoud et al. 2014) which uses a concurrent method to minimize the total costs of the system, unmet loads, and greenhouse gas emissions. Results of (Sharafi, Masoud et al. 2014) show that at the same emission level and loss of load probability (LLP), by applying the proposed approach, depreciation in total cost is obtained in comparison with the previous methods. A robust energy and reserve dispatch (RERD) model is presented in (Wei, Wei et al. 2015) that divides operation decisions into pre-dispatch and re-dispatch steps. Results of (Wei, Wei et al. 2015) show that RERD model can decrease both the required reserve and the total costs of the power system operation. It is worth mentioning that in (Wei, Wei et al. 2015), the operational reliability of power system in the presence of uncertainties is also improved while probability distribution functions of uncertainties are considered unknown. In (Tazvinga, Henerica et al. 2014), an energy dispatch model to satisfy the load demand is presented in (Tazvinga, Henerica et al. 2014), in which a hybrid power supply system is modelled including solar photovoltaic, wind, diesel, and ESS units. The outcomes demonstrate that the proposed method in (Tazvinga, Henerica et al. 2014) has the ability to diminish uncertainties and resist against external disturbances.

Recently, virtual power plant (VPP) concept has also been introduced as one of the effective tools for proposing methods of mitigating and managing RESs barriers. The VPP is a way for gathering different kinds of RESs while they are not located at the same geographical situation for better management targets. The VPP concept was defined for the first time in (Dielmann, K. et al. 2003) and then VPP concept has been used in many studies.

There are different definitions for VPP concept. In FENIX project, VPP is defined as aggregation of different distributed energy resources (DERs) capacities (Pudjianto, Danny et al. 2007). In other study, VPP is defined as combinations of different technologies with different availabilities and utilization patterns that can be connected to various parts of distribution networks (Saboori, H. et al. 2011). Finally, in (El Bakari, Khalil et al. 2010), VPP is defined as an information and communication system with centralized control over the aggregation of DERs, controllable loads, and ESS devices.

While these studies are applied to define VPP concept, there are other groups of papers dealing with other aspects of VPPs. In (Tascikaraoglu, A. et al. 2014), a method is presented for dispatching and forecasting an adaptive load for a VPP which consists of RESs. In (Holjevac, Ninoslav et al. 2015), a two-stage stochastic mixed integer linear programming (MILP) model is investigated. This model maximizes the expected profit of VPP, which consists of an alternative source, an ESS, and a power plant. In (Holjevac, Ninoslav et al. 2015), numbers of scenarios are presented for considering uncertainties based upon historical data as well. Another significant study is (Peik-Herfeh, Malahat et al. 2013), in which the optimal bidding strategy of a commercial VPP (CVPP) in day ahead (DA) market is investigated. The CVPP consists of DERs, BESSs, and electrical consumers. In (Peik-Herfeh, Malahat et al. 2013), the uncertainty of produced power by

DERs in DA and also the uncertainty of load consumption are formulated by a three-stage stochastic optimization model.

According to the previous studies carried out about VPPs, most of them investigated an objective function for a VPP and then proposed a new method, model, or a strategy to optimize VPP parameters. While, none of them are modelled with various type of components, i.e., wind and solar power plants, micro turbines, controllable loads, and ESSs. This study proposes an optimal dispatch approach for a VPP that consists of micro turbines, wind and solar power generation, controllable loads, and ESSs. Simultaneously, wind and solar outputs fluctuations are mitigated by using information gap decision theory (IGDT) method. Moreover, three different strategies are proposed for analyzing the impacts of RESs uncertainties on the performance of the proposed VPP. In the first and second strategies, only wind and solar output variations are considered respectively, and in the third one, both wind and solar uncertainties are considered at the same time. The proposed model is simulated in GAMS software and results are compared with each other.

The remainder of this paper is organized as follows.

In Section 2, the model of power dispatch in the proposed VPP is formulated. In Section 3, the proposed method based on IGDT is developed to calculate the best strategies of VPP operation of to mitigate and manage the uncertainties of RESs. In Section 4, simulation results are presented for three different scenarios of solar and wind generation. Finally, concluding remarks are given in Section 5.

2 POWER DISPATCH IN THE PROPOSED VPP

Both managing and dispatching of produced power of a VPP are the most important issues in operation of a VPP. In this paper, a VPP is defined as combination of micro turbines, wind and solar power generation, controllable loads, and ESSs in an electrical grid. The coordinated operation of these units in the proposed VPP is imperative to gain more profit in the electricity market. Moreover, by managing and dispatching of the proposed VPP, the impacts of RESs uncertainties are mitigated.

In this section, an optimal power dispatch problem for the proposed VPP is formulated. The objective function and constraints of the optimization problem for calculating the optimal power dispatch of VPP are given in following.

2.1 Objective function

In this paper, the objective function is defined as the difference between income and cost of the proposed VPP. This objective function is presented in (1).

$$PROFIT = INCOME - COST, \tag{1}$$

where, INCOME and COST are the total income and cost of proposed VPP, respectively. In following equations, i.e., (2) and (3), they are formulated.

$$INCOME = \sum_{t=1}^{T} \left(P^{VPP}(t) \cdot \pi_r(t) \right), \tag{2}$$

$$COST = F^{ES} + F^G, \tag{3}$$

here, $P^{VPP}(t)$ is total output power of VPP, $\pi_r(t)$ is the price at time t, F^{ES} is the cost function of using ESS, and F^G is the cost function of energy generation via micro turbine units. The cost functions, i.e., F^{ES} and F^G, are as (4) and (5).

$$F^G = \sum_{t=1}^{T} \sum_{i=1}^{N} \left(aP_i^2(t) + bP_i(t) + c \right), \tag{4}$$

$$F^{ES} = \sum_{t=1}^{T} \left(\frac{1}{2}.e.\left| P^{ES}(t) \right| \right), \tag{5}$$

in which, $P_i(t)$ is the active power generated through each micro turbine. Parameters a, b, and c are the fuel cost coefficients of the unit (Wang, Yao et al. 2016). Variable $\left| P^{ES}(t) \right|$ is the abstract of injected or supplied power of ESS and e is the operational cost of ESS that represents its efficiency. Note that if $P^{ES}(t)$ is negative or positive, ESS imports or exports energy, respectively.

2.2 Constraints

The constraints of the optimization problem of proposed VPP are divided into three categories; power balance and required reserves, systems limitations, and controllable load constraints.

The first thread is entailed as constraints related to power balance and required reserves of the system, which are formulated in (6)-(10). Repeat this procedure until all affiliations have been typed.

$$P^{VPP}(t) \leq \sum_{i \in N} P_i(t) + P^{WT}(t) + P^{PV}(t) + P^{ES}(t), \tag{6}$$

$$R^{VPP}(t) = r_1.D^{VPP}(t) + r_2.P^{WT}(t) + r_3.P^{PV}(t), \tag{7}$$

$$P^{VPP}(t) = D^{VPP}(t), \tag{8}$$

$$P_{max}^{VPP}(t) - D^{VPP} \geq R^{VPP}(t), \tag{9}$$

$$P_{i_{min}} \leq P_i(t) \leq P_{i_{max}}, \tag{10}$$

where, $D^{VPP}(t)$, $R^{VPP}(t)$ and $P_{max}^{VPP}(t)$ are the total load, required reserve, and maximum active output power of the VPP, respectively. Parameters r_1, r_2 and r_3 are the coefficients of impacts of load, wind, and photovoltaic system on required reserve, correspondingly and equal to 0.05, 0.2 and 0.15. Parameters P_{imin} and P_{imax} are the minimum and maximum values of output power of unit i.

The second group of constraints is related to ESS. These constraints are given in (11)-(13).

$$\sum_{t=1}^{T} P^{ES}(t).\Delta t = 0, \tag{11}$$

$$E_{min} \leq E_0 - \sum_{\tau=1}^{t} P^{ES}(\tau).\Delta t \leq E_{max}, \quad \forall t = 1, \ldots, T \tag{12}$$

$$-P_{max}^{charge} \leq P^{ES}(t) \leq P_{max}^{discharge}, \tag{13}$$

here, E_{min} and E_{max} are the minimum and maximum storage capacity of ESS, E_0 is the initial storage capacity of batteries, P_{max}^{charge} and $P_{max}^{discharge}$ are the maximum charge and discharge power. Finally, Δt is the dispatch period, e.g., 1 hour.

Third group consists of constraints related to controllable loads in the VPP. These constraints are presented in (14)-(20).

$$D^{VPP}(t) = d(t) + d^{up}(t) - d^{do}(t), \tag{14}$$

$$d^{up}(t).d^{do}(t) = 0, \tag{15}$$

$$\sum_{t=1}^{T} d^{up}(t) = \sum_{t=1}^{T} d^{do}(t), \tag{16}$$

$$B^{up}(t).d(t) \geq d^{up}(t) \geq 0, \tag{17}$$

$$B^{do}(t).d(t) \geq d^{do}(t) \geq 0, \tag{18}$$

$$d^{up}(t) \geq \varepsilon^{up}.d(t).\left(1 - \frac{\pi_r(t)}{\pi_r^{ref}}\right), \tag{19}$$

$$d^{do}(t) \geq \varepsilon^{do}.d(t).\left(\frac{\pi_r(t)}{\pi_r^{ref}} - 1\right), \tag{20}$$

in which, $d(t)$ is must-run part of load consumption during time t, $d^{up}(t)$ and $d^{do}(t)$ represent increasing and decreasing parts of the flexible load during time t. Parameters $B^{up}(t)$ and $B^{do}(t)$ are the maximum limitations of upward and downward parts of flexible demand during time t. Parameters ε^{up} and ε^{do} are defined as load elasticity coefficients. As shown in (19) and (20), both of ε^{up} and ε^{do} are used to determine the lower limitations for flexible loads. Finally, π_r^{ref} is a reference tariff and defines according to (Wang, Yao et al. 2016) as the average of electricity price in the market.

3 FINDING OPTIMAL STRATEGY OF VPP BY USING IGDT METHOD

As stated before, in this paper, an IGDT method is used for modeling both wind and solar power plants uncertainties. To find optimal strategy by using IGDT method, three basic components are required:
- System model: It shows inputs and outputs structures of the system. Moreover, the income and cost functions are also required in the system model. Here, the system model is denoted by $R(q, u) = INCOME(q, u) - COST(q, u)$. Vector q is used to denote the decision variables and vector u is used to model uncertain parameters. In addition, Q is the set of feasible decision variables.
- Uncertainty model: It is used to model the gap between the known parameters and what needs to be known. There are different types of uncertainty model, one of which is envelope-bound model that is presented in (Mohammadi-Ivatloo et al. 2013) and in this paper is used as follow.

$$U(\alpha, \tilde{u}) = \{u : |u(t) - \tilde{u}(t)| \leq \alpha\varphi(t)\}, \tag{21}$$

where, $U(\alpha, \tilde{u})$ is the uncertainty model, $\tilde{u}(t)$ is the known, and $u(t)$ is what needs to be known. Function $\varphi(t)$ determines the shape of envelope and can have different forms depending on the applications, e.g., $\varphi(t) = 1$ and $\varphi(t) = \tilde{u}(t)$. Finally, variable α shows the range of uncertain parameters.
- Performance requirement: It is a function for decision variable q and uncertain parameter u. Fluctuations of uncertain parameter can be either increase or decrease the net profit of the proposed VPP. In each problem, decision makers can chose two different behavioral approach, being risk-averse or risk-seeker. In this paper, we chose to be risk-averse; hence, we need to define robustness function and if we opt risk seeker approach, we have to define opportunity function.

With respect to the fact that in this paper we seek to protect consumers against fluctuation of uncertainty parameters, we define robustness function as (22).

$$\hat{\alpha}(q, r_c) = max_\alpha \left\{ \alpha : \min_{u \in U(\alpha, \tilde{u})} R(q, u) \geq r_c \right\}, \tag{22}$$

in which, r_c is referred to a critical value of performance function that is guaranteed to be met or exceeded by the VPP. In other word, robustness function guarantees that minimum value of function which is defined as r_c is satisfied. This value can be defined as $(1 - \sigma)R_0$, in which R_0 is the expected profit of the VPP when expected values are used for generation of wind power and photovoltaic units, and σ is an indicator to model the maximum deviation of the net profit from the expected scenario.

In this paper, the decision variables are $P_i(t)$. Moreover, the uncertain parameters are $P^{PV}(t)$ and $P^{WT}(t)$. With respect to this fact, the IGDT optimization model is as (23)-(27).

$$\max_{\alpha \geq 0} \alpha \tag{23}$$

Subject to:

$$\min_{u \in U(\alpha, \tilde{u})} \sum_{t=1}^{T} P^{VPP}(t).\pi_r(t) - F^{ES} - F^G \geq (1 - \sigma)R_0, \tag{24}$$

$$P^{VPP}(t) \leq (1 - \alpha)\left(P^{WT}(t) + P^{PV}(t)\right) + P^{ES}(t) + \sum_{i \in N} P_i(t) \tag{25}$$

$$R^{VPP}(t) = r_1.D^{VPP}(t) + r_2.(1 - \alpha)P^{WT}(t) + r_3.(1 - \alpha)P^{PV}(t), \tag{26}$$

$$(8)\text{-}(20), \tag{27}$$

As it is observable in (23) and (24), α is maximized which was defined as the range of uncertain parameters when a minimum amount of PROFIT function is greater than $(1 - \sigma)R_0$. In order to prove equations (25) and (26), the uncertainty model in equation (21) is changed as follow:

$$U\left(\alpha, \widetilde{P^{WT}}\right) = \left\{ u : \left| P^{WT}(t) - \widetilde{P^{WT}}(t) \right| \leq \alpha \widetilde{P^{WT}}(t) \right\}, \tag{28}$$

where, only wind uncertainty is considered. Expanding a rewriting this relation as:

$$P^{WT}(t) \leq (1 + \alpha)\widetilde{P^{WT}}(t), \tag{29}$$

$$P^{WT}(t) \geq (1 - \alpha)\widetilde{P^{WT}}(t), \tag{30}$$

Inasmuch as in one hand optimization model maximizes α and on the other hand with respect to the amounts of $\widetilde{P^{WT}}(t)$, minimum values of wind power are achieved when we consider below equation:

$$P^{WT}(t) = (1 - \alpha)\widetilde{P^{WT}}(t), \tag{31}$$

This is the worse situation in this problem and under this situation, optimal model guarantees profit is always greater than r_c. Finally, there is the same process for solar power.

4 CASE STUDY AND SIMULATION

To verify the proposed method, the IEEE-9 buses network is considered as a case study and the selected part in the Figure 1. is defined as a VPP. The readers can refer to (Wang, Yao et al. 2016) for the detail information of this standard IEEE network. The proposed VPP includes six identical micro-turbine units and the characteristics of these units are presented in Table 1. Here, parameters a, b, and c of all units are considered equal to 0.004, 1.475 and 9.986, respectively.

For the produced power by wind and photovoltaic units in the VPP, the generations are according to data of (Wang, Yao et al. 2016). The produced power of wind and photovoltaic units are shown in Figure 2. Moreover, must-run part of load, i.e., $d(t)$, is also depicted in this figure as well.

The characteristics of ESS of proposed VPP are given in Table 2. In this paper, parameter e of ESS is considered equal to 0.08 In addition, the numerical constants of the controllable load constraints are expressed in Table 3.

The procedure applying in this paper for finding the optimal strategy of proposed VPP is as follow. First, the optimal power dispatch of the VPP is calculated while our aim is maximization of the proposed VPP. Then, three different cases are defined to evaluate the impacts of photovoltaic and wind power uncertainties. Case I and II associate with wind and photovoltaicpower uncertainties, respectively. Unlike these two cases which include either wind or photovoltaicuncertainties, case III will be defined to express both wind and photovoltaicuncertainties. Moreover, related parameters in each case are calculated and eventually, we compare all results. This procedure is depicted in Figure 3.

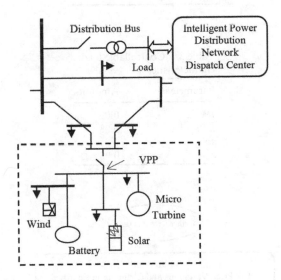

Figure 1. VPP schematic.

Table 1. The number of officially reported plague cases in the world.

Type	Capacity (KW)	Amounts	Upper Limit Of Output (KW)	Lower Limit Of Output (KW)
micro-turbine	30	6	180	–

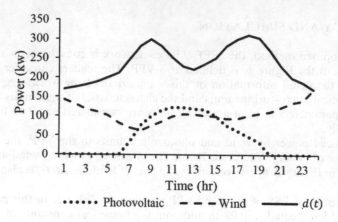

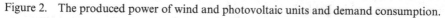

Figure 2. The produced power of wind and photovoltaic units and demand consumption.

Table 2. ESS characteristics.

Parameter	Amounts
$E_{\min}(kw)$	0
$E_{\max}(kw)$	1.2
$P_{\max}^{charge}(kw)$	15
$P_{\max}^{discharge}(kw)$	15

Table 3. Controllable loads characteristics.

Parameter	Amounts
B^{up}	0.09
B^{do}	0.08
ε^{up}	0.04
ε^{do}	0.03

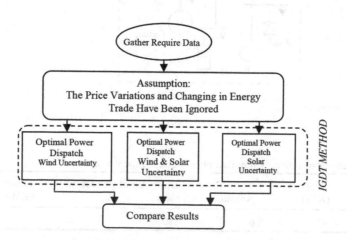

Figure 3. The procedure used in this paper.

Table 4. Specifications of hourly prices.

Type	Peak	Flat	Valley
Time division (hr)	8-10,16-20	6,7,11-15,21	1-5,22-24
Price	0.8789	0.6171	0.3819

For the sake of simplicity, we assume that the impacts of price variations and changing in energy trade in power market are negligible. Therefore, to compute the VPP's income, the time of use (TOU) price regime is used. These hourly prices are shown in Table 4.

From the above table it is obvious that the prices are categorized into three different time zones, i.e., peak, flat, and valley. Note that the reference price π_r^{ref} is equal to 0.56 according to (Wang, Yao et al. 2016).

The results of simulations in different cases after running the program in GAMS software are presented in following three subsections.

4.1 Case I, Considering wind uncertainty

In this case, uncertainty of wind power is considered. Moreover, the uncertainty of photovoltaic units is neglected. In Figure 4, the total amount of $\sum_{i \in N} P_i(t)$ are shown. The power fluctuations in each hour is in accordance with the load variations. Noteworthy, as all micro-turbine units have same characteristics, the amounts of generated power by each unit is equal. Figure 5 shows the total and predicted loads, i.e. $d^{up}(t)$ and $d^{do}(t)$. Again, it shows here the values of total load are almost equal to predicted load values and fluctuations of $d^{up}(t)$ and $d^{do}(t)$ have compatibility to total load variations. Finally, total cost, income and profit in VPP are shown in Table 5.

4.2 Case II, Considering photovoltaic units uncertainty

In this case, the uncertainty of photovoltaic units in the proposed VPP is considered. Figure 6 shows values of produced and consumed power in case II. Values of $d^{up}(t)$, $d^{up}(t)$, $d(t)$ and $D^{VPP}(t)$ in case II are illustrated in Figure 7 as well. This figure shows that the amounts of $D^{VPP}(t)$ and $d(t)$ are so close to each other. All results are summarized in Table 7.

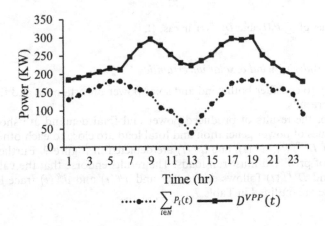

Figure 4. The values of $\sum_{i \in N} P_i(t)$ and $D^{VPP}(t)$ in case I.

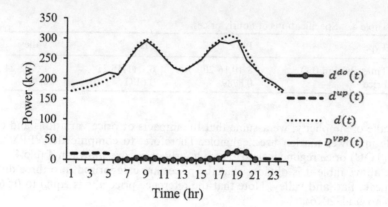

Figure 5. The values of $d^{up}(t)$, $d^{do}(t)$, $d(t)$ and $D^{VPP}(t)$ in case I.

Table 5. Total results in case I.

Parameter	Value
COST	7770.246
INCOME	37845.644
PROFIT	30075.398
α	51%

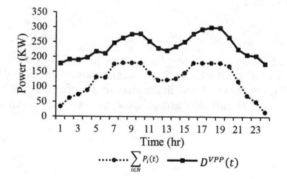

Figure 6. The values of $\sum_{i \in N} P_i(t)$ and $D^{VPP}(t)$ in case II.

4.3 Case III, Considering wind & solar uncertainties

Here, we are going to consider both wind and solar power uncertainties and find their impacts on the simulation results.

In the first step, the results of produced power and final demand are shown in Figure 8. Obviously the values of power generation and total load are close to each other. In the second step, the values of $D^{VPP}(t)$, $d^{up}(t)$, and $d^{do}(t)$ are depicted in Figure 9. Furthermore, Figure 9 shows the values of predicted must-run loads. The results endorse that the values of predicted must-run loads and $D^{VPP}(t)$ follows each other and $d^{up}(t)$ and $d^{do}(t)$ trace loads variations. Results in this case are outlined in Table 7.

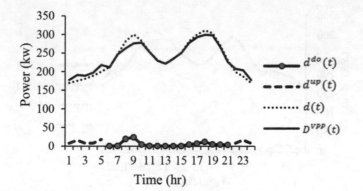

Figure 7.   The values of $d^{up}(t)$, $d^{do}(t)$, $d(t)$ and $D^{VPP}(t)$ in case II.

Table 6.   Total results in case II.

Parameter	Value
COST	6810.383
INCOME	37885.724
PROFIT	31075.341
α	92%

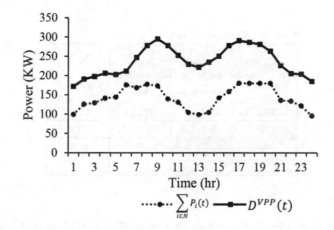

Figure 8.   The values of $\sum\limits_{i \in N} P_i(t)$ and $D^{VPP}(t)$ in case III.

4.4   *Compare three cases*

In this section, the outputs of three cases will be compared with each other. A new parameter, which is called ΔA, is defined to describe the area between $d(t)$ and $P^{VPP}(t)$ curves. A lesser ΔA in each case gives more optimal solution for the problem. Figure 10 shows ΔA with green color in three cases. All results are presented briefly in Table 8.

The comparison is divided into three various categories. In other words, wewill investigate the results from three different point of views which they are economic, uncertainty parameter, and ΔA. The economic division consists of COST, INCOME, and PROFIT parameters. Although minimum COST and maximum INCOME belongs to case II, there is 92% of

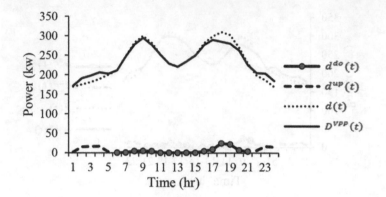

Figure 9. The values of $d^{up}(t)$, $d^{do}(t)$, $d(t)$ and $D^{VPP}(t)$ in case III.

Table 7. Total results in case III.

Parameter	Value
COST	7783.194
INCOME	37879.095
Profit	30095.9
α	38%

Table 8. Compare all results.

ΔA			
8.705641	15.255	15.255	
Wind & Solar	Photovoltaic	Wind	Uncertainty Parameter
7783.194	6810.383	7770.246	COST
37879.095	37885.724	37845.644	INCOME
30095.9	31075.341	30075.398	PROFIT
38%	92%	51%	α

uncertainty in this case. The outcomes indicate that minimum amount of α is equal to 38% in case III, however VPP profit in this case is not too much lower than case II. Eventually, with respect to ΔA values in all three cases, it is understandable that the area between $d(t)$ and $P^{VPP}(t)$ in case III is minimum. Therefore, case III releases better consequences in both α and ΔA comparisons and from economic point of view.

5 CONCLUSION

In this paper, we worked on an optimal power dispatch problem in a VPP and considered three different cases. Case I and II includes uncertainty for wind and solar, respectively. In case III, the uncertainty for wind and solar were considered and the optimal power dispatch problem was calculated in the VPP afterward. We used IGDT method for modeling uncertainties and calculating α as uncertainty parameter rate. Economical parameters such as COST, INCOME, and PROFIT were computed in those three cases. To compare the outputs better,

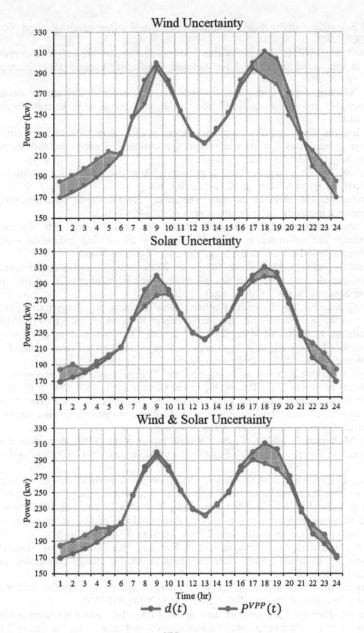

Figure 10.   Difference area between $d(t)$ and $P^{VPP}(t)$ curves in all three cases.

we defined a new parameter, ΔA, to represent difference area between the graph of $d(t)$ and $P^{VPP}(t)$ in all cases. Three different comparisons demonstrated that when both wind and solar uncertainties were modeled in case III, we had minimum values for α and ΔA simultaneously. In addition, PROFIT is closer to its maximum amount.

REFERENCES

Dielmann, K., and Alwin van der Velden. "Virtual power plants (VPP)-a new perspective for energy generation?" In Modern Techniques and Technologies, 2003. MTT 2003. Proceedings of the 9th

International Scientific and Practical Conference of Students, Post-graduates and Young Scientists, pp. 18–20. IEEE, 2003.

El Bakari, Khalil, and Wil L. Kling. "Virtual power plants: An answer to increasing distributed generation." In Innovative Smart Grid Technologies Conference Europe (ISGT Europe), 2010 IEEE PES, pp. 1–6. IEEE, 2010.

Evans, Annette, Vladimir Strezov, and Tim J. Evans. "Assessment of utility energy storage options for increased renewable energy penetration." Renewable and Sustainable Energy Reviews 16, no. 6 (2012): 4141–4147.

Hadjipaschalis, Ioannis, Andreas Poullikkas, and Venizelos Efthimiou. "Overview of current and future energy storage technologies for electric power applications." Renewable and sustainable energy reviews 13, no. 6-7 (2009): 1513–1522.

Holjevac, Ninoslav, Tomislav Capuder, and Igor Kuzle. "Adaptive control for evaluation of flexibility benefits in microgrid systems." Energy 92 (2015): 487–504.

Kusakana, Kanzumba, and Herman Jacobus Vermaak. "Hybrid diesel generator/renewable energy system performance modeling." Renewable energy 67 (2014): 97–102.

Mehrabankhomartash, Mahmoud, Mohammad Rayati, Aras Sheikhi, and Ali Mohammad Ranjbar. "Practical battery size optimization of a PV system by considering individual customer damage function." Renewable and Sustainable Energy Reviews 67 (2017): 36–50.

Mohammadi-Ivatloo, Behnam, Hamidreza Zareipour, Nima Amjady, and Mehdi Ehsan. "Application of information-gap decision theory to risk-constrained self-scheduling of GenCos." IEEE Transactions on Power Systems 28, no. 2 (2013): 1093–1102.

Panwar, N.L., S.C. Kaushik, and Surendra Kothari. "Role of renewable energy sources in environmental protection: a review." Renewable and Sustainable Energy Reviews 15, no. 3 (2011): 1513–1524.

Peik-Herfeh, Malahat, H. Seifi, and M.K. Sheikh-El-Eslami. "Decision making of a virtual power plant under uncertainties for bidding in a day-ahead market using point estimate method." International Journal of Electrical Power & Energy Systems 44, no. 1 (2013): 88–98.

Pudjianto, Danny, Charlotte Ramsay, and Goran Strbac. "Virtual power plant and system integration of distributed energy resources." IET Renewable Power Generation 1, no. 1 (2007): 10–16.

Roustai, Mahmoud, Mohammad Rayati, Aras Sheikhi, and AliMohammad Ranjbar. "A scenario-based optimization of Smart Energy Hub operation in a stochastic environment using conditional-value-at-risk." Sustainable Cities and Society 39 (2018): 309–316.

Saboori, H., M. Mohammadi, and R. Taghe. "Virtual power plant (VPP), definition, concept, components and types." In Power and Energy Engineering Conference (APPEEC), 2011 Asia-Pacific, pp. 1–4. IEEE, 2011.

Sharafi, Masoud, and Tarek Y. ELMekkawy. "Multi-objective optimal design of hybrid renewable energy systems using PSO-simulation based approach." Renewable Energy 68 (2014): 67–79.

Shim, Jae Woong, Youngho Cho, Seog-Joo Kim, Sang Won Min, and Kyeon Hur. "Synergistic control of SMES and battery energy storage for enabling dispatchability of renewable energy sources." IEEE Transactions on Applied Superconductivity 23, no. 3 (2013): 5701205–5701205.

Tascikaraoglu, A., O. Erdinc, M. Uzunoglu, and A. Karakas. "An adaptive load dispatching and forecasting strategy for a virtual power plant including renewable energy conversion units." Applied Energy 119 (2014): 445–453.

Tazvinga, Henerica, Bing Zhu, and Xiaohua Xia. "Energy dispatch strategy for a photovoltaic–wind–diesel–battery hybrid power system." Solar Energy 108 (2014): 412–420.

Wang, Yao, Xin Ai, Zhongfu Tan, Lei Yan, and Shuting Liu. "Interactive dispatch modes and bidding strategy of multiple virtual power plants based on demand response and game theory." IEEE Transactions on Smart Grid 7, no. 1 (2016): 510–519.

Wei, Wei, Feng Liu, Shengwei Mei, and Yunhe Hou. "Robust energy and reserve dispatch under variable renewable generation." IEEE Transactions on Smart Grid 6, no. 1 (2015): 369–380.

Weitemeyer, Stefan, David Kleinhans, Thomas Vogt, and Carsten Agert. "Integration of Renewable Energy Sources in future power systems: The role of storage." Renewable Energy 75 (2015): 14–20.

Zheng, Yu, Zhao Yang Dong, Feng Ji Luo, Ke Meng, Jing Qiu, and Kit Po Wong. "Optimal allocation of energy storage system for risk mitigation of DISCOs with high renewable penetrations." IEEE Transactions on Power Systems 29, no. 1 (2014): 212–220.

Evaluation bioethanol production from fusant of *S. cerevisiae* and *P. stipitis*

F.J. Shalsh
Industrial Microbiology Department, Directorate of Agricultural Research/Ministry of Science and Technology, Baghdad, Iraq

N.A. Ibrahim & M. Arifullah
Faculty of Agro Based Industry, Universiti Malaysia Kelantan, Malaysia

A.S. Meor Hussin
Faculty of Food Science and Technology, Universiti Putra Malaysia, Malaysia

ABSTRACT: A protoplast fusion between *S. cerevisiae* and xylose-fermenting yeast *Pichia stipitis* ATCC 58785 was conducted to produce the fusant which could improve bioethanol production. Among the five selected fusants, the fusant F24 (MH429783.1) was able to attain maximum bioethanol production from the glucose-xylose mixture. The fusant F24 produced 30.24 g/L bioethanol. Bioethanol has been gaining attention as a clean and renewable fuel because of its immense environmental benefits. Worldwide bioethanol demand has been continuously increasing due to the rapid growth in population and industrialization. Co-fermentation was achieved by P.stipitis ATCC 58785 in the medium containing the glucose and xylose mixture. P.stipitis ATCC 58785 fermented both sugars and produced 28.30 g/L bioethanol. The co-culture fermentation of *S. cerevisiae* and P. stipitis ATCC 58785 produced 30.12 g/L bioethanol. The fermentation using fusant gives the advantage to the production of bioethanol by reducing the duration of fermentation and cost of production.

1   INTRODUCTION

Fusants from protoplast fusion may provide a potential solution to integrate the traits of different species into one strain (Peris *et al.*, 2018). *S. cerevisiae* and *P. stipitis* provide a cost-effective challenge in bioethanol production. On the other hand, the concurrent fermentation of glucose and xylose mixture by co-cultures have been linked with a number of challenges which include the diauxic behavior of xylose-fermenting microbial organisms which are unable to utilize xylose in the presence of glucose (Karagoz & Ozkan, 2014). The intensity of oxygen intensity is a critical factor that influences P. stipitis in the fermentation of glucose and xylose to bioethanol. High aeration boosts biomass production but reduces the production of bioethanol. Under strict anaerobic conditions, *P. stipitis* cells will not be able to produce bioethanol and they will not be able to survive beyond one generation (Papini et al., 2012). The micro-aerophilic condition as well as the glucose affinity with P. stipitis reduces its xylose fermentation capacity. On the contrary, *S. cerevisiae* does not require oxygen on glucose fermentation (Rastogi & Shrivastava, 2017).

2   METHOLODGY

2.1   *Yeast strain*

This study used two yeast strains *S. cerevisiae* and *P. stipitis* ATCC 58785. *P. stipitis* ATCC 58785 was bought from the American Type Culture Collection (ATCC). The fusants F12,

F18, F22 and F24 obtained previously from yeast by protoplast fusion with the Genbank NCBI accession no. as MH429780.1, MH429781.1, MH429782.1 and MH429783.1 respectively

2.2 Preparation of fermentation media

The synthetic fermentation medium containing of 5.0 g/L yeast extract, 5.0 g/L peptone, 5.0 g/L KH_2PO_4, 0.2 g/L $(NH_4)_2SO_4$ and 0.4 g/L $MgSO_4$.$7H_2O$, xylose and glucose were employed as a carbon sources either as single carbon source or mixture of carbon source for the production of bioethanol. Both 60 g/L glucose and xylose used as single carbon source. Meanwhile the mixture of glucose and xylose were used 60 g/L and 20 g/L respectively. The pH of the media was adjusted to pH of 5 and sterilization of the medium components was performed at 121°C for 15 min (Okuda et al., 2008).

2.3 Co-culture fermentation using synthetic media by S.cerevisiae and P.stipitis ATCC58785

5% (v/v) of inoculum size of S. cerevisiae was transferred into the synthetic fermentation media containing a mixture of glucose and xylose glucose as 60 g/L and 20 g/L respectively. Fermentation experiments was performed in bioreactor with a total working volume of 0.75 L at 30 °C for 72 h. For glucose fermentation, S. cerevisiae was subjected to anaerobic incubation at 30 °C for 24 h with agitation at 150 rpm on first phase of fermentation time. Subsequently, 5% (v/v) of inoculum size of P. stipitis ATCC 58785 was transferred into the fermentation culture in the bioreactor with the help of a syringe. On second phase of fermentation time xylose fermentation with agitation at 300 rpm, 30 °C and aeration conditions at 0.25 vvm (Kordowska & Targon, 2002; Silva et al., 2011). Samples were obtained at regular intervals 12 h from sampling system of bioreactor for sugar and bioethanol analysis.

3 RUSTLE AND DISCUSSION

The summary of bioethanol production by parent and fusants performance in a medium containing 60.0 g/L glucose and 20 g/L xylose at 30 °C with initial pH 5 during 48 h fermentation time in triplicate as presented in Table 1. S. cerevisiae, P. stipitis ATCC 58785 and the fusants F4, F12, F18, F22 and F24 were found to have high sugar consumption rate, even though there was variations in the amount of bioethanol produced. Although fusants F4, F18 and F22 able to consume mixed sugar glucose and xylose at high concentration 73.52 g/L, 73.05 g/L and 73.14 g/L respectively, bioethanol production less than fusant F24, the reduction of bioethanol production could be attributed to the fact that fusants F4, F18 and F22 produce another byproduct on pathway for bioethanol production.

Protoplast fusion between S. cerevisiae and P. stipitis ATCC 58785 gave a stable and good performing fusant F24 with maximum utilization of sugars in the media more than any of its parents. The sugar consumed were 71.90 g/L, 59.30 g/L and 70.30 g/L for fusant F24, S. cerevisiae and P. stipitis ATCC 58785 receptively. Amazingly, fusant F24 produced more bioethanol compared to any of its parents using mixed sugar. The bioethanol produced were 30.24 g/L, 24.64 g/L and 28.30 g/L for fusant F24, S.cerevisiae and P. stipitis ATCC 58785 receptively and the difference was significant according to Tukey's test (P > 0.05). Fusant F24 showed significant increase in bioethanol production compared with fusants F4 which showed lowest bioethanol production with significant decrease in production of bioethanol at 23.20 g/L, hence F4 was eliminated in the subsequent experimental (Table 1). Bioethanol productivity of fusant F24 more compared to any of its parents using mixed sugar at 48 h. The bioethanol productivity were 0.63 g/Lh, 0.51 g/Lh and 0.58 g/Lh for fusant F24, S. cerevisiae and P. stipitis ATCC 58785 receptively. F24 and co-culture fermentation by S. cerevisiae and P. stipitis ATCC 58785 had shown comparable bioethanol productivity of 0.63 g/Lh and 0.62 g/Lh respectively. F24 is able to consume mixed sugar glucose and xylose at concentration of 71.90 g/L compared

Table 1. Bioethanol production by *S. cerevisiae*, *P. stipitis* ATCC 58785 and the fusants using synthetic media contains mixture of glucose and xylose.

Strain	Sugar consumed (g/L)	Bioethanol production (g/L)	Bioethanol productivity (g/Lh)
S. cerevisiae	59.30 ± 1.550^b	24.64 ± 1.102^c	0.51 ± 0.014^c
P. stipitis	70.30 ± 1.550^a	28.30 ± 0.149^b	0.58 ± 0.006^b
F 4	73.52 ± 0.835^a	23.20 ± 0.020^d	0.48 ± 0.003^d
F12	70.81 ± 1.556^a	28.29 ± 0.310^b	0.58 ± 0.004^b
F18	73.05 ± 0.416^a	29.95 ± 0.062^{ab}	0.62 ± 0.003^{ab}
F22	73.14 ± 1.400^a	29.20 ± 0.364^{ab}	0.60 ± 0.004^{ab}
F24	71.90 ± 1.090^a	30.24 ± 0.519^a	0.63 ± 0.002^a
Co - culture	70.01 ± 2.000^a	30.12 ± 0.120^a	0.62 ± 0.009^a

Note: Mean values with the same letter in the same column do not differ significantly according to the Tukey's test ($p > 0.05$).

with 70.01 g/L by co culture fermentation with bioethanol produce were 30.24 g/L by F24 at 48h of fermentation time and 30.12 g/L by co-culture fermentation at 60 h of fermentation time.

These results of ours are in accord with the results presented in an earlier study during which protoplast fusion of *S. cerevisiae* 2.0251 with *P. tannophilus* ATCC 2.1662 produced a fusant (Fusant 1) that was generate more bioethanol (11.76 g/L) during fermentation of mixed sugars (30 g/L glucose and 20 g/L xylose) compared to either the *S. cerevisiae* 11.2 g/L or P. tannophilus 8.8 g/L produced by the parental strains (Yan et al., 2009).

Based on these findings, it can be concluded that the genetic improvement of yeast strains through protoplast fusion could give rise to novel strains with a higher bioethanol tolerance level and high bioethanol production at the rate of 30.24 g/L, as seen in the case of fused F24 strain. While the production of bioethanol using co-culture fermentation and fusant F24 are comparable. F24 has the advantage of reducing production cost and time as it involve the use of only single isolate as against the use of two isolates in co-culture fermentation that able to growth together, at compatible condition there are usually compatibility issues with the fermentation parameters of the two microbial organism combinations. For example, there is compatibility issues in the temperature, agitation and pH for *S. cerevisiae* to use and convert glucose to bioethanol with those of the xylose-fermenting yeast. Yoon et al (1996) isolated strains with improved mixed sugar (xylose and sucrose) fermentation by protoplast fusion S. cerevisiae STV 89 and *Scheffersomyces* (*Pichia*) *stipitis* CBS 5776. One fusant (F5) based on its fermentation performance of 100 g/L xylose medium supplement with 5 g/L yeast extract, 5 g/L peptone and mineral salts produced more ethanol with a higher yield 0.44 g/g compared to the parental strain *S. cerevisiae* STV 89 and S. stipitis ethanol yield was 0.39 g/g, 0.436 respectively. Spent & Yeas (2003) investigate the possibility of the use of interspecific hybrids of yeast in ethanol fermentation of media containing xylose. Distilling yeast *S.cerevisiae* D43, xylose fermenting yeast Yamadazyma stipitis ATCC 58376. The parental strain Yamadazyma stipitis ATCC 58376 produced ethanol 0.389 g/g from xylose. The hybrids YD43-6 and YD43-11 did not statistically differ from Yamadazyma stipitis in respect to efficiency of ethanol from xylose, but produced less xylitol. In the model medium, with the ratio of glucose to xylose (7:3) .the fusant YD43-6 produced the most ethanol (0.377 g/g). From protoplast fusion of S. cerevisiae 2.0251 with *P. tannophilus* ATCC 2.1662 the best isolate (Fusant 1) produced more ethanol (11.76 g/L) during fermentation of mixed sugars (30 g/L glucose and 20 g/L xylose), compared to either the *S.cerevisiae* 11.2 g/L or *P. tannophilus* 8.8 g/L parental strains (Yan, 2009). Kumari and Pramanik (2012) fused *S. cerevisiae* NCIM-3090 with different xylose -fermenting yeasts *C.shehatae* NCIM-3500, *P. tannophilus* NCIM-3502 and *P. stipitis* NCIM-3507. The fermentation performance of the fusants in a glucose 150 g/L and xylose 50 g/L mixture and supplement with 2 g/L yeast extract and minerals. Three fusant strains produced more ethanol with a greater yield than the S. cerevisiae parental

Table 3. The evaluation of fusants bioethanol production stability.

Subculturing times	Bioethanol production (g/L)			
	7/F12	F18	F22	F24
4	27.28 ± 0.31^b	29.95 ± 0.06^a	29.20 ± 0.36^a	30.28 ± 0.51^a
8	28.32 ± 0.13^a	27.54 ± 0.24^b	26.43 ± 0.55^b	29.98 ± 0.12^a
12	26.43 ± 0.43^c	27.11 ± 0.21^b	23.33 ± 0.43^c	30.24 ± 0.31^a

Note: Duration of subculturing is 6 months; Mean values with the same letter in the same column do not differ significantly according to the Turkey's test (p > 0.05). Fusant F4 eliminated due to low performance in bioethanol yield.

strain 0.436 g/g. *S. cerevisiae* and *P.tannophilus* fusant (RPR39) was found to be the most efficient strain giving high ethanol yield (0.458 g/g)

Evaluation of Fusants Bioethanol Production Stability

Stability feature is most necessary for the utilization of fusants for commercial production of bioethanol. In this study, fusants bioethanol production stability was evaluated as presented in Table 3. Fusants were sub-cultured up to 12 times within 6 months and tested for bioethanol production in order to evaluate their stability. From this stability evolution, F24 was found to be stable, producing almost the same yield of bioethanol (30.28 g/L, 29.98 g/L and 30.24 g/L) as no significant difference was found in bioethanol productions within this period according to the Turkey's test (p > 0.05). These findings are in agreement with those reported by Krishnamoorthy et al. (2010), who found that 60% of fusants in their study have stability in terms of bioethanol production over a period of weeks. The other fusant strains F12, F18 and F 22 were found loose stability on bioethanol production after 8 times of subculturing within 2 months, showing significant difference in bioethanol production according to the Turkey's test (p > 0.05). These differences in terms of stability of bioethanol production by fusant over a period of time as observed in this study corroborate with the findings of Krishnamoorthy et al. (2010), who assessed the stability of fusant strains from protoplast fusion of *S. cerevisiae* and *Kluyveromyces marxianus* during 2, 4, 6, 8, 10, 15 and 20 weeks post protoplast fusion for bioethanol production and found that, almost 40% of the fused cultures lost their features received from parental strains in the 4th week following protoplast fusion.

Fusant F12 showed initial bioethanol production rate of 27.28 g/L which subsequently declined to 26.43 g/L while the initial bioethanol production for fusant F18 was 29.95g/L which later declined to 27.11 g/L and F22 had initial bioethanol production rate of 29.20 g/L which also declined to 23.33 g/L at end sub-culturing duration. In an earlier study the stability of 3 fused cultures over a period of 12 month was examined and the authors reported that, only one fused culture was stable following 12 month while the 2 other cultures lost their stability feature during that period. Hence, F24 being the fusant with the highest efficiency and stability was selected for production of bioethanol using rice straw hydrolysate. Fusant F4 was eliminated due to low performance in bioethanol yield (Pasha et al., 2007).

The study was evaluated the performance of *S. cerevisiae* and *P. stipitis* ATCC 58785 in the production of bioethanol using synthetic media glucose-xylose mixture by batch culture fermentation in which *S. cerevisiae* and *P. stipitis* ATCC 58785 co-culture showed enhanced bioethanol production, reduced fermentation time and lower process costs even though fusant F24 showed greater bioethanol production at 30.24 g/L compared to the parent. Based on the findings of this study, the following recommends are made. Future studies should subject fusants to a whole scheme of genetic analysis studies in order to improve the potentials of the yeast strains towards having the desired fermentation features.

REFERENCES

Karagoz, P., & Ozkan, M. (2014). Ethanol production from wheat straw by Saccharomyces cerevisiae and Scheffersomyces stipitis co-culture in batch and continuous system. Bioresource Technology, 158, 286–293.

Kordowska, W. M., & Targon, Z. (2002). Ethanol fermentation on glucose/xylose mixture by co-cultivation of restricted glucose catabolite repressed mutants of Pichia stipitis with respiratory deficient mutants of Saccharomyces cerevisiae. Acta Microbiolgica Polonica, 51(4), 345–352.

Krishnamoorthy, N., Vijila, K., & Kumutha, K. (2010). Intergeneric protoplast fusion of yeast for high ethanol production from cheese industry waste Whey. Journal of Yeast and Fungal Research, 1 (5),81–87.

Kumari, R., & Pramanik, K. (2012). Improved bioethanol production using fusants of Saccharomyces cerevisiae and xylose-fermenting yeasts. Applied Biochemistry and Biotechnology, 167(4), 873–884.

Okuda, N., Ninomiya, K., Katakura, Y., & Shioya, S. (2008). Strategies for reducing supplemental medium cost in bioethanol production from waste house wood hydrolysate by ethanologenic Escherichia coli: inoculum size increase and coculture with Saccharomyces cerevisiae. Biochemical Engineering Journal, 105(2),90–96.

Papini, M., Nookaew, I., Uhlen, M., & Nielsen, J. (2012). Scheffersomyces stipitis: a comparative systems biology study with Crabtree positive yeast Saccharomyces cerevisiae. Microbial Cell Factories, 11(1), 136–136.

Pasha, C., Kuhad, R., & Rao, L. V. (2007). Strain improvement of thermotolerant Saccharomyces cerevisiae VS3 strain for better utilization of lignocellulosic substrates. Journal of Applied Microbiology, 103(5), 1480–1489.

Peris, D., Perez-Torrado, R., Hittinger, C. T., Barrio, E., & Querol, A. (2018). On the origins and industrial applications of Saccharomyces cerevisiae × Saccharomyces kudriavzevii hybrids. Yeast, 35(1), 51–69.

Rastogi, M., & Shrivastava, S. (2017). Recent advances in second generation bioethanol production: An insight to pretreatment, saccharification and fermentation processes. Renewable and Sustainable Energy Reviews, 80, 330–340.

Silva, J. P. A., Mussatto, S. I., Roberto, I. C., & Teixeira, J. (2011). Ethanol production from xylose by Pichia stipitis NRRL Y-7124 in a stirred tank bioreactor. Brazilian Journal of Chemical Engineering, 28(1), 151–156.

Spent, N. F., & Yeast, B. (2003). Selected biotechnological features of hybrids of Saccharomyces cerevisiae and Yamadazyma stipitis. Biotechnology, 6(1), 1–8.

Yan, F., Bai, F., Tian, S., Zhang, J., Zhang, Z., & Yang, X. (2009). Strain construction for ethanol production from dilute-acid lignocellulosic hydrolysate. Applied Biochemistry and Biotechnology, 157(3), 473–482.

Yoon, G. S., Lee, T., Ryu, Y. W., Kim, C., & Seo, J. (1996). Characterization of alcohol fermentation and segregation of protoplast fusant of Saccharomyces cerevisiae and Pichia stipitis. Journal of Microbiology and Biotechnology, 6(4), 286–291.

Emerging Developments in the Power and Energy Industry – Dufo-López, Krzywanski & Singh (eds)
© 2020 Taylor & Francis Group, London, ISBN 978-0-367-27169-5

Methodology of analysis, modeling and supervision of SCADA systems in thermal power plants

M.F. Karoui & M.N. Lakhoua
Research Laboratory Smart Electricity & ICT, University of Carthage, Tunisia

ABSTRACT: A thermal power plant is a power station in which heat energy is converted to electric power. This complex system necessitates a methodology of analysis, modeling and supervision. The objective of this paper is to present initially the architecture of SCADA systems (Supervisory Control and Data Acquisition) in thermal power plants. Then, we present two structured analysis methods: SADT (Structured Analysis and Design Technique) and SA-RT (Structured Analysis for Real Time). Next, we present the diverse steps and tools of a general model for analysis and supervision of control - command systems. Finally, these functional analysis techniques are discussed.

Keywords: SCADA systems, SADT method, SA-RT method, Thermal power plants

1   INTRODUCTION

SCADA systems (Supervisory Control and Data Acquisition) are commonly used in industry for supervisory control and data acquisition of industrial processes. The process can be industrial, infrastructure or facility (Kumar, 2018)

The SCADA system generally consists of the next subsystems (Babunski, 2018):

– A Man-Machine Interface (MMI) is the equipment which presents process data to a human operator, and through this, the human operator, monitors and controls the process.
– A supervisory system, acquiring data on the process and sending commands to the process.
– Remote Terminal Units (RTU) connecting to sensors in the process, converting sensor signals to digital data and sending digital data to the supervisory system.
– Communication infrastructure connecting the supervisory system to the RTU.

In fact, the majority control actions are performed automatically by RTU or by PLC (programmable logic controllers). Multitude control functions are frequently restricted to basic overriding or supervisory level intervention. For example, a PLC may control the flow of cooling water through part of an industrial process, however the SCADA system may allow operators to change the set points for the flow, and enable alarm conditions, for example loss of flow and high temperature, to be displayed and recorded (Irmak, 2018), (Jyotsna, 2017), (Moshko, 2018). The feedback control loop passes through the RTU or PLC, while the SCADA system monitors the overall performance of the loop (Ben Hammouda et al., 2015).

2   ARCHITECTURE OF SCADA SYSTEMS

SCADA system is one of the effective tools in power network automation (Islamova, 2016). In fact, the diverse equipment's, including SCADA system, have been installed in many companies over the world to modernize their power system distribution networks (Wiles, 2008), (Chan, 1992), (Patel, 2004), (Lakhoua, 2009).

In this part we present an example of the architecture of a SCADA system used in a thermal power plant. In fact, the architecture of the SCADA system of a thermal power plant (Lakhoua, 2010), (Lakhoua, 2012), (Glaa R. et al., 2016) is constituted of next elements (Figure 1):

- A plate of bornier;
- FBM (Field Bus Modules);
- FCM (Field Bus Communication Module);
- CP60 (Control Process 60);
- DNBT (Dual Node Bus base_T interface);
- AW (Work Station Processor);
- WP (Application Work Station).

Among software of the SCADA system, we cite:

- System monitor (SYS MON) that supervises the good working of all the facilities of the system.
- Interfacing operator (FOXVIEW) to visualize the tabular with a slim rod to activate the main functions of the SCADA system.
- Creator of tabular FOX DRAW.
- Software FOX SELECT that permits to make the different elements of the hierarchy of the data base of the CP60.
- Software Integrated Control Configuration (ICC) that permits to create and to configure programs residing in the CP60.
- Software AIM HISTORIAN that permits to collect, to organize and to protect data for storage, it also permits to configure features of points to archive, as messages partners to events.

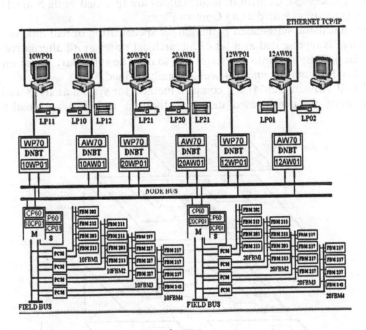

Figure 1.   Architecture of the SCADA system of a TPP.

3 PRESENTATION OF STRUCTURED ANALYSIS METHODS

In this part, the two structured analysis methods SADT and SA-RT are presented.

3.1 *SADT method*

The SADT represents an image of the system. It is a method of analysis to understanding why a system exists, or must be designed, what functions it must complete and finally, how they are realized, and whatever of the complexity (Jaulent, 1992), (Strohmeier, 1996). The method is based on a graphical model, proceeds by down approach in the sense that are going from general to more detailed, by focusing on system activity (Figure 2).

The SADT method seems adapted to the modeling such systems for at least one motive: this method applies perfectly to the multi-technological systems; it adapts to mechanical, electronic and software systems. But it does not consider the dynamic aspect of system (Zennir, 2015).

The boxes called ICOM's input-control-output-mechanisms are hierarchically decomposed. At the top of the hierarchy, the overall principle of the system is shown, which is then decomposed into components-subactivities. The decomposition process continues until there is sufficient detail to serve the principle of the model builder. SADT/IDEF0 models ensure constancy of the overall modelled system at each level of the decomposition (Lakhoua et Khanchel, 2011), (Mahdi, 2015), (Lakhoua et al., 2016).

3.2 *SA-RT method*

Among the graphical methods most commonly used in industry, two of the leading methods are SA-RT and Statecharts.

SA-RT is a short name for Structured Analysis Methods with extensions for Real Time. The model is represented as a hierarchical set of diagrams that includes data and control transformations (processes). Control transformations are specified using State Transition diagrams, and events are represented using Control Flows.

The other graphical and state-based paradigm for specification of real time systems is Statecharts. The system is represented as a set of hierarchical states as an alternative of processes. Each state can be decomposed into sub states and so on. The statecharts notation is additional dense than the SA-RT notation and has been formally defined.

Therefore, SA-RT (Naoui, 2014) is a complex method for system analysis and design. This is one of the most frequently used design method in technical and real-time oriented

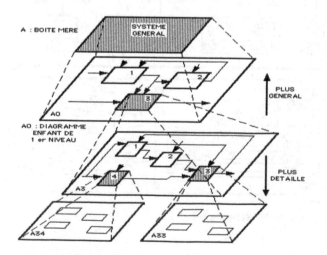

Figure 2. Structure of an SADT model.

applications adopted by various Case-Tools. It is a graphical, hierarchical and implementation independent method for top-down development (Figure 3).

SA-RT method enables us to identify an entrance and an exit of data in an algorithm or a computer program. It is divided in three modules: Diagram of Context, Data Flows Diagram and Control Flows Diagram. Every module includes in its graphic interpretation different symbols.

Certainly, the Context Diagram in the SA-RT method is going to enables us to identify a process in a program in relation to the entered and exits of data. This process can have diverse units. This process will be able to be identified per seconds, in term of constant or variable but as this process will be able to be material type (Process interfacing) (Lakhoua, 2010).

The diverse symbols used in the Context Diagram of the SA-RT method are (Jaulent, 1992), (Strohmeier, 1996):

- The terminator is the element in end, final element that encloses the action.
- The data flow is the final element that opens up on a last action.
- The control flow is generally a tie back of the process toward the terminator. It can be a main element of the process.

The termination is generally a direct tie between a terminator and the process.

4   MODEL FOR ANALYSIS AND SUPERVISION OF AN INDUSTRIAL PROCESS

In this paragraph, we present a model for analysis and supervision of an industrial process.

4.1   *Phase 1: Functional analysis of an industrial process*

The first step is to carry out a functional modeling of industrial process using the SADT method. The structure of this method to master the complexity of the process thanks to its modular and top-down analysis. The steps one such approach to functional analysis by the SADT of the industrial process method are:

Step 1: prepare the model SADT: creating a SADT model begins with the definition of two concepts that are the goal which sets out the objectives of the model and the point of view that sets to what audience the model is created. These two determined concepts to guide the decomposition of each box;

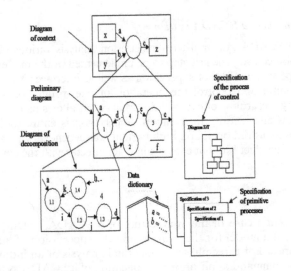

Figure 3.   Different diagrams of an SA-RT model.

Step 2: create the diagram A-0 representative General to analyze in an actigramme activity;

Step 3: create the A0 activity chart. The name inside boxes drawn on a new form (3 to 6 boxes) of activities and we track the arrows of interface between different boxes;

Step 4: from the A0 diagram, select the less clear box and by decomposition giving more information to create the diagram I $(1 < i < 6)$ with i corresponding to the number of the box in A0;

Step 5: repeat the same principle of decomposition for the other diagrams;

Step 6: represent all the diagrams of the model on the SADT forms.

This first phase of the model allows us the decomposition of the process one way prioritized to bring it back to basic situations and develop a static model describing the activities of the process. This first phase of the model allows us the decomposition of the process one way prioritized to bring it back to basic situations and develop a static model describing the activities of the process.

4.2 *Phase 2: Identification of control – command applications*

In order to identify applications of control once the functions and activities of every function have been identified, the next step is to analyze an application of control using the SA-RT method. The steps one such approach functional increases by the SA-RT one application of the control method are:

Step 1: Set up the context diagram representing the different endings as well as incoming data flows and outgoing;

Step 2: Set up the preliminary diagram (diagram of data flows) that represents the list of functional processes necessary to the application with the flood of data matching.

Step 3: Set up the streams of decomposition diagram associated with one of the identified functional processes;

Step 4: Set up the diagram of water control representative aspect control of the SA-RT method;

Step 5: Set up the diagram State/Transition representing behavioral aspect or real-time application;

Step 6: Specification of business processes (data dictionary) based on a procedural specification.

This second phase of the model allows us the decomposition of one-way hierarchical studied application to bring it back to basic situations and lead to preparation one dynamic model describing the various processes of this application as well as the flow of data and control.

4.3 *Phase 3: Application of a SCADA system*

The application on a SCADA system after a supervision analysis various processes and streams of data and application control, the last step to be implemented in the model of analysis and one industrial process supervision a tool of supervision as well as interfaces Man - Machine. Supervision is aimed to monitor and control the operation of one installation for remaining in the normal operating range regardless of external disturbances. It allows us to detect in real time the faults from the alarms and identify their causes. Supervision is entire tools and methods that allow driving facilities industrial operation both normal interdependency presence of failures. It is the tool of operator of conduct him but can also interact directly with the control system.

5 CONCLUSION

In this paper we present a case of the architecture of a SCADA system in a thermal power plant. Then, we present a model for the analysis and the supervision of an industrial process. This model is decomposed of three phases: Functional analysis of an industrial process; Identification of control - command applications; Application of a SCADA system.

Consequently, we present on the one hand, the diverse steps of programming in a SCADA environment and the configuration of the tabular of the control-command application.

Starting from this study of the methodology of analysis and supervision of SCADA systems used in thermal plants presented in this paper, we will extend this approach on Smart Grids.

REFERENCES

Babunski, D. & Zaev, E. & Tuneski, A. & Bozovic, A. 2018. Optimization methods for water supply SCADA system, 7th Mediterranean Conference on Embedded Computing (MECO), pp. 1–14.

Ben Hammouda M. & Lakhoua M.N. & El Amraoui L., 2015, Dependability Evaluation and Supervision in Thermal Power Plants, International Journal of Electrical and Computer Engineering, Vol. 5, N°5.

Chan, E.K. & Ebenhon, H. 1992. The implementation and Evolution of a SCADA System for a Large Distribution Network, IEEE Transactions on Power systems, Vol.7, No.1, 1992, pp.320–326.

Glaa R. & Lakhoua M.N. & El Amraoui L., 2016, Using SA-RT method and SCADA for the analysis and the supervision of an hydrogen circuit, Journal of Electrical Engineering, Vol.16, N°3.

Irmak, E. & Erkek, I. 2018. An overview of cyber-attack vectors on SCADA systems, 6th International Symposium on Digital Forensic and Security (ISDFS), pp. 1–5.

Islamova, O.V. & Zhilyaev, A.A. & Bozieva, A.M. 2016. SADT technology as a tool to improve efficiency in the use of process approach in management of engineering enterprise, IEEE Conference on Quality Management, Transport and Information Security, Information Technologies (IT & MQ & IS), pp. 65–668.

Jaulent, P. 1992. Génie logiciel les méthodes: SADT, SA, E-A, SA-RT, SYS-P-O, OOD, HOOD, Armand Colin.

Jyotsna, K. & Sharma, A. & Kapadia, H. 2017. Data acquisition in wind power plant using SCADA, International Conference on Intelligent Computing, Instrumentation and Control Technologies (ICICICT), pp. 296-2301.

Kumar, S. & Gaur, N. & Kumar, A. 2018. Developing a Secure Cyber Ecosystem for SCADA Architecture, 2018 Second International Conference on Computing Methodologies and Communication (ICCMC), pp 559–562.

Lakhoua, M.N. & Khanchel, F. & Laifi S. & Khazemi S., 2016, System analysis of medical equipment for healthcare management, Annals of the Faculty of Engineering Hunedoara 14 (4), 17.

Lakhoua, M.N. & Khanchel, F., 2011, Overview of the methods of modeling and analyzing for the medical framework, Scientific Research and Essays, Academic Journals, ISSN: 1992–2248, Vol. 6(19), pp. 3942–3948.

Lakhoua, M.N. 2009. Application of Functional Analysis on a SCADA system of a Thermal Power Plant, AECE Journal, Issue No2/vol. 9.

Lakhoua, M.N. 2010. SCADA applications in thermal power plants, IJPS, vol.5, N°7, 2010, pp 1175–1182.

Lakhoua, M.N. 2012. Application of Functional Analysis Techniques and Supervision of Thermal Power Plants, Thermal Power Plants, Mohammad Rasul (Ed.).

Mahdi, I. & Nadji, B. 2015. Contribution to the functional and dysfunctional analysis of the photovoltaic systems, 3rd International Renewable and Sustainable Energy Conference (IRSEC), pp. 1–14.

Moshko, S.V. & Stotckaia, A.D. 2018. Principles of SCADA-system development, IEEE Conference of Russian Young Researchers in Electrical and Electronic Engineering (EIConRus), pp. 937–9940.

Naoui, A. & Bel Hadj Ali, S & Afilal, L.E. & Abdelkrim, M.N. 2014. Application of functional specification and operational safety conventional methods for a networked control system suitable qualitative analysis, 15th International Conference on Sciences and Techniques of Automatic Control and Computer Engineering (STA), pp. 44-452.

Patel, M. & Cole, G.R. & Pryor, T.L. & Wilmota, N.A. 2004. Development of a novel SCADA system for laboratory testing, ISA Transactions 43, pp 477–490.

Strohmeier, A. & Buchs, D. 1996. Génie logiciel: principes, méthodes et techniques, Presses polytechniques et universitaires Romandes, Lausanne.

Wiles, J. 2008. Techno Security's Guide to Securing SCADA: A Comprehensive Handbook On Protecting The Critical Infrastructure, Elsevier.

Zennir, Y. & Bendib, R. 2015. Modeling and dependability analysis of an industrial plant: Case study, International Conference on Industrial Engineering and Systems Management (IESM), pp. 1012–11018.

Emerging Developments in the Power and Energy Industry – Dufo-López, Krzywanski & Singh (eds)
© 2020 Taylor & Francis Group, London, ISBN 978-0-367-27169-5

Energy harvesting from the railroad bed: Choice of the optimum frequencies

Valery Tsaplev, Sergey Konovalov, Roman Konovalov & Ivan Pavlov
Department of Electroacoustics and Ultrasonic Engineering, Saint-Petersburg State Electrotechnical University (LETI), Saint-Petersburg, Russia

ABSTRACT: The railway bed is a source of powerful vibrations. These vibrations are a source of waste energy suitable for conversion into electrical energy. To build the adequate generator-harvester, it is necessary to choose the most effective oscillation frequencies within the spectrum of the railway track. Two types of piezoelectric transducers are compared in the paper: flexural beam-type transducer and disk bimorph transducer of the umbrella-type. Using these transdusers, and the vibrometer-phonendoscope, the frequency spectrum of a real railway embankment is obtained. Examples of spectra are given, and also the most effective frequency is estimated from 10 different measurement results.

Keywords: bimorph piezoelectric element, harvester, phonendoscope, railway track, vibration spectrum

1 INTRODUCTION

The usage of the so-called waste energy waste energy (that is, the energy dissipated by different mechanisms) is now quite relevant (Tsaplev, V. et al. 2016). Vibrations are present everywhere, and in some cases they have a very high level of energy (P. Fiala, et al. 2017, L. G. Kurzweil, 2017, J.G.S. da Silva, 2004). The usage of this waste energy can solve two problems at the same time. One task is to reduce the harmful effects on the environment and on humans. The second, equally important, is the generation of electricity, especially in areas where the supply of electricity is difficult. The railway track, which is heavily trafficked (especially freight), is one such source of energy, the use of which can bring significant economic benefits. To solve both of these problems, it is necessary to study the spectrum of oscillations of the railway track along which the train passes. This spectrum, of course, depends on the type of train (passenger or freight), as well as its weight and speed. In general, however, we can say that the oscillations of the railway track are in the low frequencies range (from some Hertz up to several tens of Hertz). This means that low-frequency bending piezoelectric transducers must be used to extract energy and/or to actively suppress vibrations.

2 BENDING-TYPE TRANSDUCERS

Different types of bending transducers are considered in the reviews (Sodano H., et al. 2004, Erturk A., Inman D. J., 2011). Bending transducers are usually made in the form of beam-type bimorph elements. However, for the usage on the railway track or on the highway, much more acceptable are the transducers of the piezoelectric cymbal type, which were also considered by different authors (Hyeoung Woo Kim, et al. 2004, Ochoa P., et al. 2002, Dogan A., et al. 1996, Sun CL,, 2005, A. Moure et al. 2016), or the umbrella-type transducers (V.M. Tsaplev et al. 2014, 2015). A detailed study of bending transducers is out of the scope of our study. There is a lot of works in the scientific literature on this subject (Erturk A., Inman

D. J., 2011, S.A Rybak, B.D. Tartakovsky, 1996, R.W. Trail-Nash, A.R. Collar., 1953, Yu.I. Bobrovnitsky, 1975, Antonyak Yu.T., Vassergisser M.E., 1982, V.M. Sharapov, 2013). Here it should be noted only two important features inherent in both the bending piezoelectric elements of the beam type and the bending piezoelectric elements of the umbrella type. Figure 1 illustrates these features. The piezoelectric element consists of a flexible body *1* (beam or disc) attached to the oscillating base *B*, and two or more piezoelectric plates *2* glued to both (or one) surfaces of the flexible body. Both piezoelectric plates may be poled (*P* shows the direction of the polarization) in the same direction (a), or in opposite directions (b).

Coaxially or counteraxially polarized piezoelectric plates are connected to an electrical load *R* as shown in Figure 1. It is clear that the electric load is actually a fairly complex electrical circuit designed to collect, accumulate and then use the electric charge, but here we do not consider this circuit. The theory and applications of the corresponding circuits one can find in numerous papers (e.g. A. Badel et al. 2005, Hui Shen et al. 2010). The natural frequency of the flexible element can be adjusted by changing the mass *M*. The mass *M* is not a necessary element of the system, and is used only to adjust the oscillatory system to the desired frequency. In the case of an umbrella piezoelectric element, the mass *M* is a metal ring attached to the outer edge of the flexible disc

Piezoelectric plates do not necessarily have to cover the entire area of the flexible body. Moreover, piezoelectric elements located in the region of low mechanical stresses absorb energy produced by the active elements located in the region of high mechanical stresses, and thereby reduce the useful energy at the output of the system.

This can be shown with the help of Figure 3, which shows the distribution of mechanical stresses across the bended bimorph piezoelement. More or less significant mechanical stresses occur only near the clamped base of the bimorph and near its surface, and the farther from the neutral surface, the more is the stress. In the average cross-section mechanical stresses are zero, as expected, and as follows from the analytical calculations. Consequently, the density of the electrical energy produced by the transducers of such a type, is extremely low, and it is unlikely that such transducers can be any suitable for generating energy on railway tracks.

The amount of electrical energy produced can be slightly increased by increasing the thickness of the flexible body, as well as by increasing the thickness of the piezoelectric plates, but it reduces the flexibility of the elastic system and, as a consequence, increases the resonant frequency of the piezoelectric transducer. In this case, the main resonant frequency of the transducer goes out of the spectrum range of natural frequencies of vibrations of the railway track.

Finally, it can be concluded that bending transducers of beam type are unsuitable for harvesting electricity from the vibrations of the railway track.

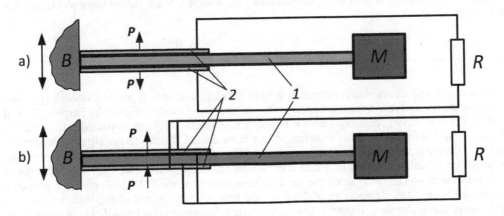

Figure 1. The piezoelectric element with the inverted polarizations *P* (a) and piezoelectric element with the equally oriented polarization *P* (b).

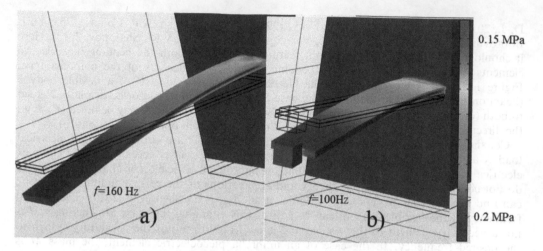

Figure 2. Distribution of mechanical stresses along the bended bimorph piezoelement.

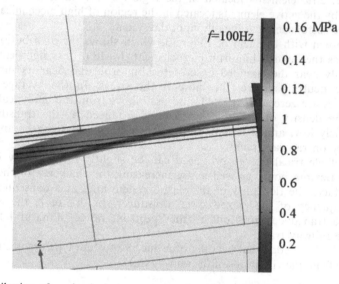

Figure 3. Distribution of mechanical stresses across the thickness of the curved bimorph element near the clamped base.

3   UMBRELLA-TYPE BIMORPH TRANSDUCERS

To overcome the above disadvantages, we have developed several types of bimorph-type disk transducers, or umbrella-type transducers. One of them is shown in Figure 4 (Tsaplev V. et al. 2015). The transducer consists of a disk bimorph piezoelectric element 1, pasted into a massive bronze ring 2, and the leg 3 for attachment to the oscillating object. The massive bronze ring 2 reduces the natural frequency of the transducer. The advantage of this type of transducer over a beam-type transducer described above is the significant larger specific electrical energy produced. The transducer was modeled using the finite element method in the COMSOL package, manufactured and tested. The dependence of the electrical power at the output of the transducer on the load resistance is shown in Figure 5. One can see, that the experimental results are in good agreement with the theoretical ones. The transducer with the vibration sensor fixed on it is shown in

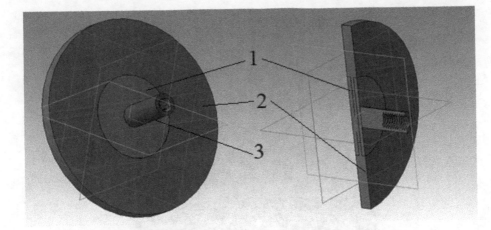

Figure 4.   Umbrella-type bimorph piezoelectric transducer for energy harvesting.

Figure 6. The volume of the piezoceramic element was equal to 1 cm^3, while the power produced by the transducer was equal to 9 mWt/sm^3, that is, the specific power was equal to 9 mWt/cm^3. The whole transducer was calibrated with the help of the miniature accelerometer PCB Piezotronics Inc.

Further development of the bimorph disc transducers is the transducer shown in Figure 7. One may name it "partitioned" disk bimorph umbrella transducer (Tsaplev V. et al. 2015). It contained some (6 in our experiment) pairs of piezoceramic plates 1, glued to a flexible diaphragm 2, forming thus some bimorph piezoelements. These piezoelements may have different resonance frequencies and may be connected in parallel, or in series, as necessary, forming thus rather wide frequency response. The massive brass ring 3 serves to reduce the resonance frequency. The whole system is placed on the leg 4 and is mounted on the vibrating surface. The cross-section of the oscillatory system is shown in Figure 8.

Figure 9 shows the simulation results of a six-section disk transducer in the COMSOL package. The distribution of deformations over the entire volume of the oscillatory system is shown here. One can see, that the distribution of deformations is much more uniform than that in the case of the previous disk transducer.

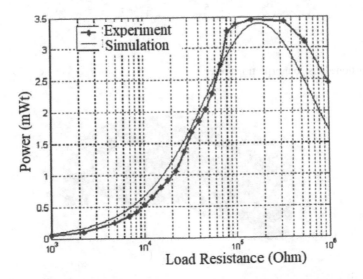

Figure 5.   The electrical power at the output of the transducer from the load resistance.

Figure 6. The disk-type bimorph transducer with the vibration sensor KS92 fixed on it.

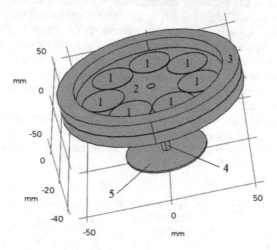

Figure 7. Partitioned disk bimorph transducer.

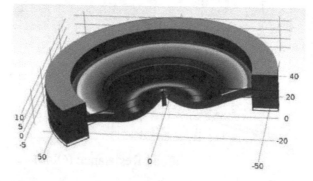

Figure 8. The cross-section of the "partitioned" disk bimorph umbrella transducer (Tsaplev V. et al. 2015).

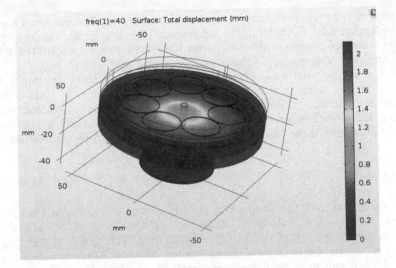
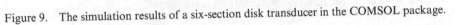

Figure 9.   The simulation results of a six-section disk transducer in the COMSOL package.

4   EXPERIMENTAL RESULTS

All three described types of piezoelectric transducers (beam-type and disk-type) were used for experimental study of the vibration spectrum of the railway bed. The oscillation spectrum has a wide frequency band with discrete peaks. Therefore, a specially developed broadband vibrometer-phonendoscope was also used for the study. This vibrometer-phonendoscope can work both in vibration measurement mode and in the mode of measuring the level of sound vibrations. All three transducers and the vibrometer-phonendoscope were accurately calibrated with the help of the miniature accelerometer PCB Pi-ezotronics Inc. The bandwidth of the vibrometer-phonendoscope was 40 - 5000 Hz.

Measurements were carried out on the stretch between stations to ensure stability in the speed of trains. The transducer (or vibrometer-phonendoscope) was installed on a massive structure (concrete pedestal), dug into the railway bed.

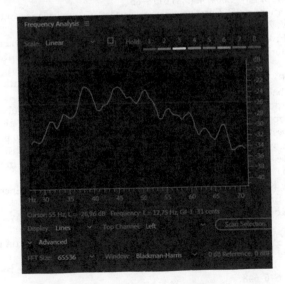

Figure 10.   Example of vibration spectrum of railway bed.

An example of spectral characteristics is shown in Figure 10. The spectrum was obtained using Adobe Audition CC 2018. Using the right mouse button, scale the resulting spectrum. If necessary, you can set a larger number of points for fast Fourier transform (FFT Size in the Frequency Analysis window). In this study, this parameter is 65536.

The given spectral characteristic is a typical one. Similar measurements were performed for different modes of freight trains. In each mode, measurements were performed 10-15 times, followed by the subsequent statistical processing. The general pattern, however, can be seen in Figure 10. The most intensive vibrations are within the frequency range 30-70 Hz with a difference in intensity of 22 – 34 dB. These frequencies were selected to build disk energy piezoharvesters.

5   CONCLUSION

Thus, the possibility of using the waste energy of vibrations of the railway bed to harvest electricity is shown. It is shown that the most suitable type of electroacoustic transducers for power generation are bimorph piezotransducers of umbrella type. The design of this type of transducer, for which a patent is obtained, is demonstrated. The simulation of such a transducer in the COMSOL environment by the finite element method is performed.

REFERENCES

Tsaplev V. M., Abbakumov K. E., Kononalov R. S. Nonlinear piezoceramic materials and small-sized harvesters. Saint-Petersburg State Electrotechnical University «LETI», – 2016. – 215 p. (Rus)

P. Fiala, G. Degrandeb, F. Augusztinovicza. Numerical modelling of ground-borne noise and vbration in buildings due to surface rail traffic //J. of Sound and Vibration. – 2007. – V. 301. – P. 718–738.

L. G. Kurzweil. Ground-borne Noise and Vibration from Underground Rail Systems //J. of Sound and Vibration. 1979. V. 66. №3. P. 363-370.

J.G.S. da Silva. Dynamical performance of highway bridge decks with irregular pavement surface surface, Computers and Structures, 2004, VOLUME 82, p. 871–881

Sodano H., Park G., Inman D.J. A Review of Power harvesting from Vibration Using Piezoelectric Materials //The Shock and Vibration Digest. 2004. V. 36. 7 №3. P. 197–205.

Erturk A., Inman D. J. Piezoelectric Energy Harvesting. N.-Y.: Wiley, 2011, 392 p.

Hyeoung Woo Kim, Amit Batra, Shashank Priya, Kenji Uchino, Douglas Markley, Robert E. Newnham and Heath F. Hofmann, Energy Harvesting Using a Piezoelectric "Cymbal" Tranducer in Dynamic Environment, Jpn J Appl Phys 2004; 43(9A):6178–83.

Ochoa P, Villegas M, Fernández JF. Effective piezoelectric coefficient calculation of cymbal piezocomposite. Ferroelectrics. 2002. V. 273. p.p. 315–320.

Dogan A, Fernandez JF, Uchino K, Newnham RE. Cymbal electromechanical actuator. In: Proceedings of the tenth IEEE international symposium on applications of ferroelectrics. ISAF 1996, vol. 1; 1996. p. 213–6.

Sun CL, Guo SS, Li WP, Xing ZB, Liu GC, Zhao X-Z. Displacement amplification and resonance characteristics of the cymbal transducers. Sens Actuators, A 2005;121:213–220.

A. Moure, M.A. Izquierdo Rodríguez, S. Hernández Rueda, A. Gonzalo, F. Rubio-Marcos, D. Urquiza Cuadros, A. Pérez-Lepe, J.F. Fernández, Feasible integration in asphalt of piezoelectric cymbals for vibration energy harvesting, Energy Conversion and Management 112 (2016), p. 246–253.

V.M. Tsaplev, Abbakumov K. E., Konovalov R. S. Experimental study of the disk bimorph piezoelectric energy harvester. Izvestia LETI, Saint-Petersburg State Electrotechnical University «LETI», № 9, 2014, c. 59 – 63 (Rus).

Tsaplev V., Abbakumov K., Konovalov R. Disk Bimorph-Type Piezoelectric Energy Harvester. Journal of Power and Energy Engineering, Vol. 3, №4, Apr. 2015, p.63-68.

S.A Rybak, B.D. Tartakovsky, On the oscillations of slim plates, Acoustical Physics, 1963, v.9, №1, p.p. 66–71 (Rus.).

R.W. Trail-Nash, A.R. Collar. The effect of shear flexibility and rotatory inertia on the bending vibrations of beams. Quart.j. mech. and appl. math., 1953, v. 6, p.2, p.p. 186–222.

Yu.I. Bobrovnitsky. On the acoustical model of the bending oscillations of the rod. Acoustical Physics. 1975. V. 21. № 4. p.p. 646–648.

Antonyak Yu.T., Vassergisser M.E. Calculation of the characteristics of the flexural membrane type piezoelectric transducer, Acoustical Physics, 1982, V. 28, № 3, p.p. 294-302

V.M. Sharapov, I.G. Minaev, J.V. Sotula, L.G. Kunitskaya, Electroacoustical Transdusers 2013. M. Technosphera. 296 p.

A. Badel, D. Guyomar, E. Lefeuvre and C. Richard, Efficiency Enhancement of a Piezoelectric Energy Harvesting Device in Pulsed Operation by Synchronous Charge Inversion. Journal of Intelligent Material Systems and Structures. 2005. V.16. № 10. p.p. 889–901

Hui Shen, Jinhao Qiu, Hongli Ji, Kongjun Zhu and Marco Balsi, Enhanced synchronized switch harvesting: a new energy harvesting scheme for efficient energy extraction. Smart Mater. Struct. 2010. V.19, 115017 (14pp)

Tsaplev V., Abbakumov K., Konovalov S., Ronovalov R., Disk-bimorph piezoelectric harvester. Rus. Patent №154688 from 07.08.2015

Emerging Developments in the Power and Energy Industry – Dufo-López, Krzywanski & Singh (eds)
© 2020 Taylor & Francis Group, London, ISBN 978-0-367-27169-5

Improved DFIG control with decoupled control to enhance LVRT response

Jignesh Patel & S.K. Joshi
The Maharaja Sayajirao University of Baroda, Vadodara, Gujarat, India

ABSTRACT: Increasing wind power penetration in the existing power system may challenge the controllability aspects of the power system. Virtually the renewable power system has no or very low inertia. Any change in the generation or load will directly impact the change in the system parameter. The control structure plays important role in keeping Wind Turbine in sync with the rest of the power system. In this work the decouple Control of Rotor Side Converter (RSC) and Grid Side Converter (GSC) is discussed. The DFIG response to the grid fault is simulated in ATP program is also discussed.

Keywords: Rotor Side Converter (RSC), Grid Side Converter (GSC), Grid Fault, DFIG, Power System, Fault Ride Through (FRT)

1 INTRODUCTION

The trend of rate of rise of wind power share in existing power system has given rise to several challenges. Frequency control is one of them. As more and more wind power is being introduced in the existing power system, the load of control is being increased on the conventional power system, which would not be feasible beyond certain level of wind power penetration. Several failures of coal-based steam turbines have been reported in recent years in the state of Gujarat, India. Although this is correlated with the increasing wind power in the state, but it is still difficult to prove in absence of proper data. However, the probable problems which could arise in the future can be well predicted. Frequency control is one of them.

In order to maintain the frequency and rate of change of frequency (ROCOF), the storage devices of any kind is inevitable. Battery-based storage devices mostly are being used. But due to its unrenewable nature and the cost of the batteries has imposed limitation of its use. Three security indices, Rate of Change of Frequency, Frequency Nadir, and Steady State Frequency Deviation. The frequency control can be improved by using the self-inertia of turbine and the sub optimal control of MPPT device.

Wind power may adversely impact system frequency if wind power variation is larger than that of system load [2–3]. In [4], author has presented the power and frequency fluctuations occurring due to inclusion on large wind farms in existing power system. [5–6] have studied the effect of increased share of wind power on system frequency and voltage profile. Reference [7] takes further ahead to this work and studied the impact of a much large share of wind power (more than 50%) on system frequency and voltage profile during a contingency situation. Reference [7] has also discussed the possible control measures to be implemented in wind generating plants to counter the frequency sag arising from a sudden loss of generation or increase in the system load. The stability of operation of power system with increasing number of wind energy is also dependent on the Ramp rate of the conventional energy sources. Ramp rate is defined as the change in the mean value of power from one period to the next [8]. The negative ramp rate means the decreasing wind power, which requires positive ramping of other power plants. Fluctuations in Wind power causing frequency deviation of utility grid. As the grid frequency is regulated, the deviation can limit high penetration of

Wind Power [9]. In this paper author has concluded that 5% wind power in a system can cause frequency deviation of 1%.

The present share of wind power in the state of Gujarat, India, is only about 5% maximum up to the starting of year 2018. Though it is expected to rise in near future due to lucrative government policies, but still a vast portion of the wind power potential is not tapped yet. With increase in penetration of wind power, some of the issues due to uncertainty of wind power is well expected to arise. One of them is the integration of wind power. The effective integration of wind power with existing system needs to be check with reference to two major aspects, one is the technical and second is regulatory [10–11].

Curtailment of conventional generation with increase in wind power is inherently linked with the reduction of the system inertia, which increases the magnitude of frequency excursion with load – generation imbalance occurs [12]. Change in reserve policy and modification of wind turbine inertial response characteristics will be essential with increased level of wind generation. In this work, the author has simulated the frequency response at different penetration level of wind power coming from different proportion of fixed WTG and DFIG. The frequency control of interconnected system is still manageable due to the frequency support from conventional generators, but it is difficult with isolated wind power system [13]. In this, the frequency control of autonomous power system with high wind power penetration is discussed. According [14–15], maximum around 30% of wind power penetration is predicted. However, the careful examination of dynamic security is also advised at higher penetration.

Different Controllers is proposed [16–20] in the literature for variable speed wind turbines (VSWTs) in order to improve their frequency response and achieve active participation in frequency control. In [16] and [17], controllers reestablishing inertia response in doubly fed induction generator (DFIG) WTs are proposed and in [18] the concept of droop control is added to the proposed control scheme. In [21], pitch control is additionally used to reduce the steady state frequency error with DFIG WTs. Various relevant studies are done, but these are based on simple models for either the power system or the WTs and the response is studied mostly for power system disturbances under constant wind conditions [20–21]. The most of such studies focus primarily on the DFIG scheme. Further, these studies is carried out considering large power systems, rather than on smaller, isolated island grids. The fundamental issue of controller parameter selection and its implications are usually ignored, as is the matter of the required primary reserves to be maintained in order for the WFs to provide the desired response.

The frequency control after a turbulence or imbalance in large system is carried through in different phase and time steps. Immediately after the disturbance, the inherent action of inertial frequency response, which consumes part of the kinetic energy in rotating mass of WTG to counter the frequency deviation from the scheduled or specified frequency. This is known as the primary control (Primary ALFC). The primary control only an automatic response to the transient situation. Still after the primary control frequency settles to a new value, which may not be the desired or target value. This results in steady state frequency error and needs to be corrected. This is done in the next phase of frequency control known as the secondary control. In this phase automatic governing system is activated to keep the frequency error in acceptable range. This also restore the used reserves by increasing the kinetic energy along with the reduction of steady state error.

The majority of WTGs currently employed are either of Type 3 or 4. Type -3 is a Full Converter WTG and type-4 is DFIG based WTG. Both of this type of WTG employs power electronic converters. In type-3, the WTG is fully isolated from the system, so, the system frequency doesn't affect WTG directly and making WTG insensitive to change in system frequency and consequently unable to provide any inertial response. But in Type-4, the WTG remains directly connected to the system and thus system frequency affects the speed of the WTG. To handle the problem of frequency insensitivity with Type-3 WTG, some researchers has proposed the synthetic inertia as a potential solution to it. This is nothing but the automatic controller embedded in wind turbine acting in the time frame of few seconds.

Though VSC (Voltage Source Converter) makes the inertial system poor, but DFIG is exception in this case, as the back to back converter used in DFIG is of only about one-third capacity linking grid and rotor circuit of IG. The stator of IG remains connected to the supply and this provides the higher frequency sensitivity and better inertial response as compared to type-3 WTG.

In [22], the author has studied the frequency response in the power system of California and also the impact of the current projections of high integration of wind power on the system. In this different performance metric factors to measure the frequency performance by wind turbines has been proposed. A detailed discussion on the inertial frequency response and the impact of wind farms on the system frequency response can be found in [23–27].

In a large power system with interconnection of different areas with tie lines, the wind power density on any particular area will cause flow of asynchronous power in the tie line due to frequency difference in two areas. Area with high wind power may cause power fluctuation and hence the frequency. So, this makes different areas oscillate with respect to each other. This is not but swinging of power system and it may result in to the collapse of power system. The frequency of oscillation depends on the regulation of the governors of the generating system. Lower the regulation, better is the frequency quality.

2 WIND POWER MODEL

Wind power follows the Weibull distribution and the probability distribution can be estimated by 2- or 3-parameter Weibull function. The captured aerodynamic power is given by equation (1). It depends on air density, swept area of wind turbine, the power coefficient, and wind speed.

$$P_w = \frac{\rho}{2} A_r U_w^3 C_P(\lambda, \beta) \tag{1}$$

Wind power is a non-linear function, depends on few parameters. The wind turbine can be run either in constant speed mode or in constant power mode. The motivation behind the constant power assumption is that remains constant for any changes in shaft speed. This is achieved by varying the blade pitch. The effect on the output of the WTG models can be examined for the both the assumption of constant power and constant torque.

3 WIND TURBINE MODEL

There are various literatures are available on the modelling of the Induction Generator (IG). Starting from a third-order to seventh-order model of DFIG is available. According to the requirement of accuracy different order of DFIG is used in the study. Here, a fifth-order d-q DFIG model is used for the study purpose as it have very good accuracy in simulating the rotor speed, electromagnetic torque, both active and reactive power, and stator currents. Control of electromagnetic torque allows for the control of the speed, the basis of the variable speed nature of the DFIG wind turbine.

Independent control of the electromagnetic torque through the control of a single variable: the rotor current i_{dr} is possible by the selection of an appropriate reference frame in the controller of the back-to-back converter. The equation for the electromagnetic torque T_{em} is given in terms of the rotor current:

$$T_{em} = \frac{3PL_m}{4L_s} \left(\psi_{qs} i_{dr} - \psi_{ds} i_{qr} \right) \tag{2}$$

Using the appropriate reference frame, the torque equation can be decoupled:

$$T_{em} = \frac{3PL_m}{4L_s} \left(\psi_{qs} i_{dr} \right) \tag{3}$$

The control of i_{dr} may also be simplified considerably through the careful design of the d-axis and q-axis voltages, such that

$$v_{dr} = v'_{dr} - (\omega - \omega_r)\frac{L_m}{L_s}(\psi_{qs}) + (\omega - \omega_r)\frac{L_a}{L_s}(i_{qr}) \qquad (4)$$

and

$$v_{qr} = v'_{qr} + \frac{L_m}{L_s}(p\psi_{qs}) - (\omega - \omega_r)\frac{L_a}{L_s}(i_{dr}) \qquad (5)$$

where v'_{dr} and v'_{qr} are auxiliary signals in the controller reference frame and are the outputs from the d-axis and q-axis proportional integral current controllers:

$$v'_{qr} = \left(i^*_{qr} - i_{qr}\right)\left(K_{P2} + \frac{K_{I2}}{p}\right) \qquad (6)$$

$$v'_{dr} = \left(i^*_{dr} - i_{dr}\right)\left(K_{P1} + \frac{K_{I1}}{p}\right) \qquad (7)$$

4 WIND TURBINE INERTIAL RESPONSE

The inertial response control loop is added to the DFIG active power control loop, as shown in Figure 1. The intent is that when system frequency changes, not only do the synchronous generators respond to this change, the wind plant also quickly changes its output active power to share in the frequency restoration process during such short term frequency fluctuations. The output active power reference is predefined for different wind speeds. With the given DFIG rotor speed, optimal power signal $P_{optimal}$ is obtained and compared with measured output electrical power. The error signal is regulated by the rotor side converter controller to obtain the required rotor current I_{qr}. The inertial control loop is responsible for sending additional power regulation signal $\Delta Pref$ to the rotor side converter controller. When the system load increases leading to a frequency drop, the inertial control loop sends additional active power $\Delta Pref$ to the DFIG active power reference $Pref$.

The reactive power is directly related with the voltage control. The rotor side controller supports the grid voltage and also fulfils the power factor control requirement.

The rotor controller for d-axis and q-axis are given in Figure 3.

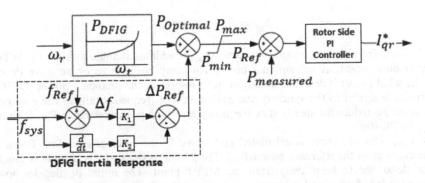

Figure 1.   Primary frequency control.

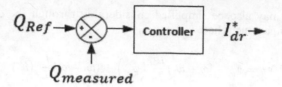

Figure 2. Voltage control.

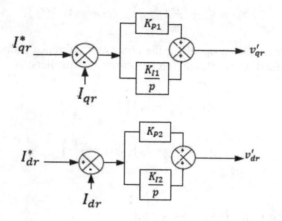

Figure 3. PI-Controller for *d*-axis and *q*-axis.

5 PITCH ANGLE CONTROL

Pitch angle control of the wind turbine mimics governor control of synchronous machine. Pitch angle control is sluggish in response and contributes in the steady state frequency error control. The DFIG, in this case, operate on the sub-optimal active power operation to reserve the some power for dynamic frequency control. If WTG is run on the MPPT point, it will run on one speed only. But if, it is operated at a sub-optimal point, there will be two speeds for a given power point, it can be seen and verified from the graph in Figure 4. So, there is a trade-off between the maximum extracted energy and the dynamic response capability of the WTG. It has advantage that, if WTG is operated sub-optimal point at the higher speed, there will be additional power available. This additional power comes from two stream, one from the stored kinetic energy of the rotor of WTG and second is the difference between optimal and suboptimal power point, i.e. (PMPPT– $P_{suboptimal}$). The pitch angle control loop diagram is given below in Figure 4 .

Pitch angle is regulated to follow the β_{Ref}. It is given by equation (8).

$$\beta_{ref} = \beta_0 + \Delta\beta + \beta_{Bias} \qquad (8)$$

where $\Delta\beta$ comes from the β – Frequency curve. β_{Bias} is additional angle to keep the WTG away from the optimal point, at sub-optimal point. This may be positive or negative depends on whether the wind power is deficient or surplus with respect to the predicted power. DFIG droop characteristics is similar to the synchronous generator. The frequency falls as the load on generator increases. To reduce the steady state frequency error, the power reference P_{Ref} is changed by changing the PL-Ref.

The β- P_{Ref} reference curve is calculated and stored in the form of equation. For a given β value, the curve gives the reference power P_{Ref}. The β value is decided based on how much WTG should be de-loaded to keep away from the MPPT point. The initial βo decides how much power is de-loaded. As the load increases, pitch angle gets adjusted to reach at the MPPT point.

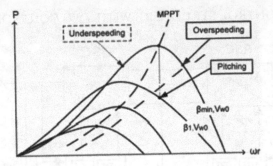

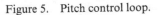

Figure 4.   MPPT power control of wind Turbine.

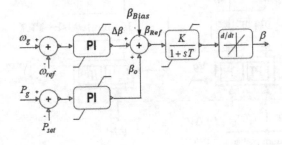

Figure 5.   Pitch control loop.

6   DECOUPLED CONTROL STRUCTURE WITH TWO CONTROL LOOPS

In DFIG, the stator is directly connected to the grid and the slip ring rotor is connected to the grid through back to back AC-DC-AC converter as shown in Figure 6. The converter consists of two converter Rotor Side Converter (RSC), and Grid Side Converter (GSC) connected by a DC Link. A harmonic filter and line inductor is connected between the Grid and Grid Side Converter to reduce the harmonics and fault current. A craw bar is used to limit and control the overcurrent and over voltages in the rotor circuit. A DC chopper is used in the DC Link to limit the DC link overvoltage.

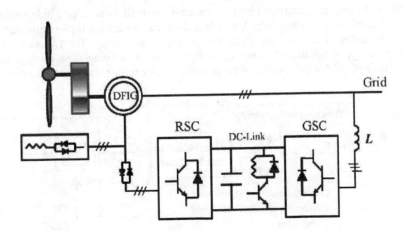

Figure 6.   DFIG schematic.

7 DECOUPLED CONTROL STRUCTURE WITH TWO CONTROL LOOPS

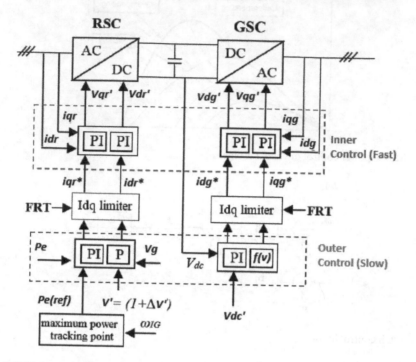

Figure 7. DFIG decoupled control structure.

8 CONTROL PROBLEM

The DFIG Control scheme is illustrated in Figures 7 and 8. The i_{dr} and i_{qr} are the d – axis and q – axis currents of the RSC. Similarly, i_{dg} and i_{qg} are the d- axis and q – axis currents of the GSC. V_{dc} is the DC link voltage of the converter. P_e is the electrical active power. V_g is the positive sequence voltage of the grid at the MV terminal. The rotor current controls the active power and the grid terminal voltage, whereas the grid side current i_{dg} controls the DC voltage and i_{qg} used to support the reactive power on event of faults.

Both RSC and GSC are controlled by two separate controllers with two independent loops, as shown in Figure 7. The outer control, which is slow, calculates the dq – frame current reference and faster inner controller generates the reference ac voltage for two converters. The active power $Pe_{(ref)}$ is calculated from the MPPT Control scheme. The change in positive sequence voltage of grid is calculated using reactive power requirement as shown in Figure 8.

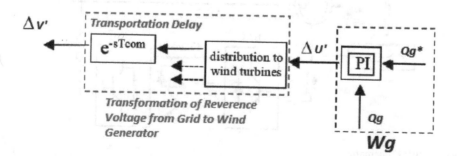

Figure 8. Grid voltage deviation control.

9   TEST SYSTEM

A 4-Bus test system is selected for the simulation purpose. The Wind Farm, with 45 DFIG Wind Generator of 1.5 MW capacity each, is connected at Bus-1. Whereas the Conventional Generator and load is placed at Bus-2. The load (30 MW, 15 MVAR) is connected to Bus-3 and Bus-4. The fault is created at the Wind Farm at Bus-1. The node voltages at different buses are captured along with various control signals and is present here in the next section.

10   SIMULATION RESULTS

The simulation results are given in Figures 10 to 14.

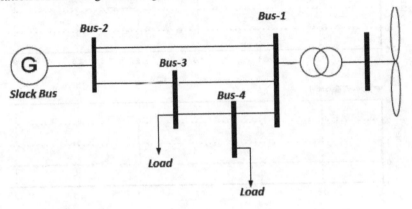

Figure 9.   4-bus test system.

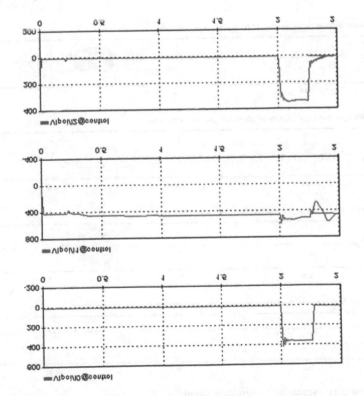

Figure 10.   Positive, negative and zero sequence control current.

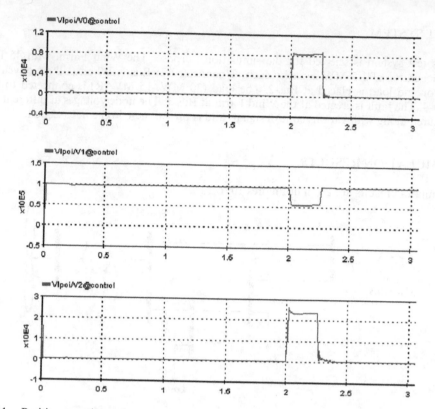

Figure 11.   Positive, negative and zero sequence control voltage.

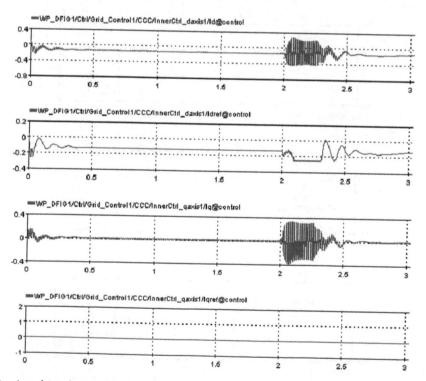

Figure 12. i_d and i_q axis actual and reference current.

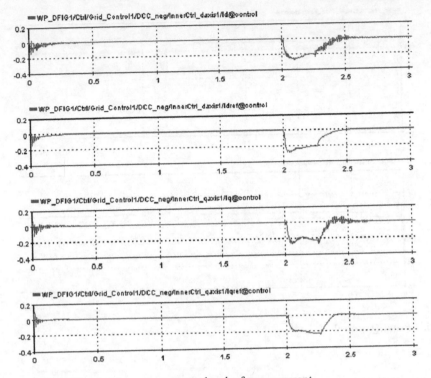

Figure 13. i_d and i_q axis negative sequence actual and reference current.

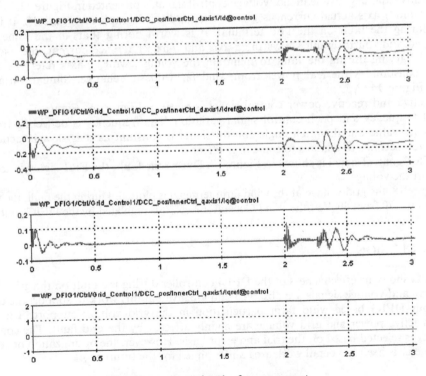

Figure 14. i_d and i_q axis positive sequence actual and reference current.

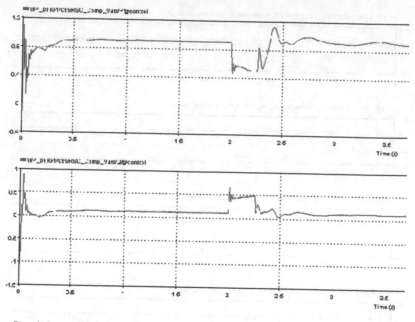

Figure 15. *P* and *Q* control.

In Figure 10, the zero, positive and negative sequence control current is given. The change zero and negative sequence control current is higher than the positive sequence. Similarly, the zero, positive and negative sequence voltage control signal is presented in Figure 11.

The *d* and *q* axis actual current is give in first and third graph of Figure 12. It is oscillating during the fault on the grid terminal. It is worth noting that, in the same figure, the second and fourth graph shows the *d*-axis and *q*-axis reference current, the *d*-axis reference is varying during the fault to maintain the voltage, while *q*-axis reference current remains constant as the real power from wind farm doesn't change. Similar variation is seen in Figure 14.

The active and reactive power control signal is given in Figure 15. Figure 16 depicts the result of frequency and pitch control signal variation. The frequency is deflecting from the steady state value but the pitch angle control signal is constant as it is slow in action and requires few seconds to minutes to come in to effect.

The FRT control signal is shown in Figure 17. During the fault, the FRT signal is activated to support the voltage.

In Figure 18, the grid voltage at the wind farm terminal is plotted. During the fault, the voltage is reduced to 270 from 400V at the Wind Generator level. It is regulated by the DFIG controller.

11 CONCLUSION

The results show the effectiveness of the DFIG controller during the fault on the grid. Different control and voltage signals are observed and are given in this work to support the conclusion about DFIG based wind farm participation in the grid voltage support. The active power, reactive power and grid voltage are mainly affected by the grid fault. The controller parameter is selected based on the trial and error basis. However, the optimization of controller parameter is itself is a detail work area and is aimed for the future work.

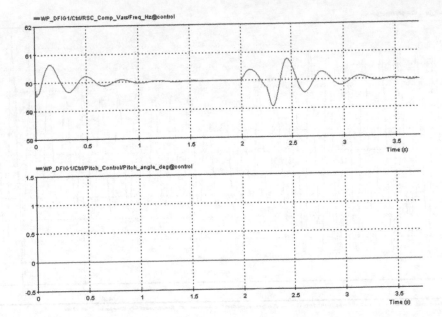

Figure 16.   *Frequency* and *pitch* control.

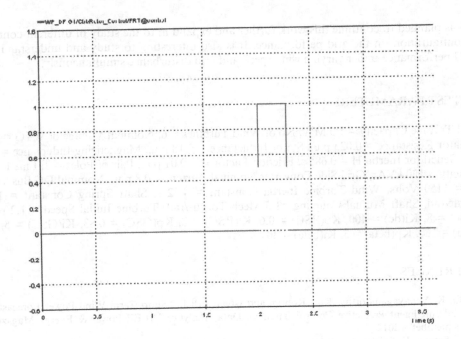

Figure 17.   *FRT* control signal.

12 FUTURE WORK

There are numerous literature available on the different control strategies for optimal operation of wind farm is available. But the optimization of DFIG controller parameter to obtain
the balance performance under different contingencies is still to be investigated. The future

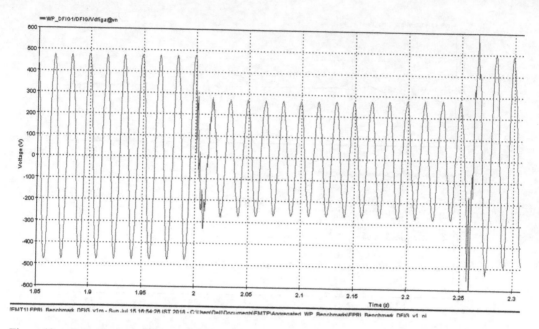

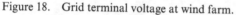

Figure 18.   Grid terminal voltage at wind farm.

work is planned to continue this work further and expand in to the study of different controller configuration on the grid performance. It is also interesting to study and understand the WTG performance under varying wind speed and grid disturbance simultaneously.

13   TEST PARAMETERS

Wind Turbine Capacity = 1.5 MW, No of Wind Turbines = 6, Nominal Voltage of WTG = 575 V, Stator Resistance = 0.023 p.u., Stator Inductance = 0.18 p.u., Magnetizing Inductance = 2.9 p.u., Generator Inertia H = 0.685 s, Friction Factor F = 0.01 p.u., Pair of Poles = 3, Line Filter Capacity = 120 kVAr, Grid Side Converter Nominal Current = 0.8 p.u., Nominal DC Bus Voltage = 1150 Volts, Wind Turbine Inertia Constant = 4.32 s, Shaft Spring Constant = 1.11 Torque/rad, Shaft Mutual Coupling =1.5 Mech Torque/rad, Turbine Initial Speed = 1.2 p.u., Kp(dc) = 8, Ki(dc) = 400, Kp(RSC) = 0.6, Ki(RSC) = 8, Kp(GSC) = 0.83, Ki(GSC) = 5, K-(Pitch) = 160, Kp(beta) = 3, Ki(beta) = 30.

REFERENCES

[1]   G. K. Venayagamoorthy, Kurt Rohrig, and Istvan Erlich, "Short Term Wind Power Forecasting and Intelligent Predictive Control Based on Data Analytics," IEEE Power & Energy Magazine, September – 2012.

[2]   T. Senjyu, R. Sakamoto, N. Urasaki, T. Funabashi, H. Fujita, and H. Sekine, "Output Power Leveling of Wind Turbine Generator for all operating regions by Pitch angle Control," IEEE Transactions on Energy Conversion, Vol-21, No-2, June-2006.

[3]   T. Kaneko, T. Senjyu, A. Yona, M. Datta, T. Funabashi, and C. H. Kim, "Output Power Co-ordination Contol for Wind Farm in Small Power System," Proc. 14th Int. Conf. Intelligent Systems Applied to Power Systems (ISAP-07), November 2007.

[4]   Poul Sorensen, Nicolaos Antonio Cutululis, Antonio Vigueraz-Rodrigues, Leo E. Jensen, Jesper Hjerrild, Martyn Heyman Donovan, and Henrik Madsen, "Power Fluctuations From Large Wind Farms," IEEE Transactions on Power Systems, Vol 22, No-3, February-2007.

[5] Seon-Ju Ahn, Jin-Woo Park, Il-Yop Chung, Seung Il-Moon, Sang Hee-Kang and Syun Ryul Nam, "Power Sharing Method of Multiple Distributed Generators Considering Control Modes and Configurations of Microgrids," IEEE Transactions on PD, Vol 25, 2010.

[6] F. Koch, I. Erlich, "Simulation of Dynamic Interaction of Large Offshore Wind Farms with the Electric Power System," OWEMES-2003.

[7] F. Koch, I. Erlich, "Dynamic Simulation of Large Wind Farms Integrated in a Multi Machine Network," IEEE PES General Meeting, 2003.

[8] Poul Sorensen, Nicolaos Antonio, Antonio Vegureras-Rodriguez, Leo E. Jensen, Jesper Hjerrild, Martin Heyman Donovan, and Henrik Madsen, "Power Fluctuations from Large wind Farms," IEEE Transactions on Power Systems, Vol-22, No-3, August-2007.

[9] Changling Luo, and Boon-Tek Ooi, "Frequency Deviation of Thermal Plants due to Wind Farm," IEEE Transactions on Energy Conversions, Vol-21, No-3, September-2006.

[10] Jignesh Patel, and Satish Joshi, "Power System Operation with Increased Wind Power-A Gujarat State Case Study," submitted to NPSC-2018, India.

[11] T. Ackermann, "Wind Power in Power System," New York, Wiley Publication, 2005.

[12] Gillian Lalor, Student Member, IEEE, Alan Mullane, Member, IEEE, and Mark O'Malley, Senior Member, IEEE, "Frequency Control and Wind Turbine Technologies," IEEE Transactions On Power Systems, Vol. 20, No. 4, November 2005.

[13] Ioannis D. Margaris, Stavros A. Papathanassiou, Senior Member, IEEE, Nikos D. Hatziargyriou, Fellow, IEEE, Anca D. Hansen, and Poul Sørensen, Senior Member, IEEE, "Frequency Control in Autonomous Power Systems With High Wind Power Penetration," IEEE Transactions On Sustainable Energy, Vol. 3, No. 2, April 2012.

[14] N. D. Hatziargyriou, E. S. Karapidakis, and D. Hatzifotis, "Dynamic behaviour of power systems in large islands with high wind power penetration," in Proc. Bulk Power Systems Dynamics and Control Symp.—IV Restructuring, Santorini, Greece, Aug. 23–28, 1998.

[15] N. D. Hatziargyriou and E. S. Karapidakis, "Online preventive dynamic security of isolated power systems using decision trees," IEEE Trans. Power Syst., vol. 17, no. 2, pp. 297–304, May 2002.

[16] G. Lalor, A. Mullane, and M. J. O'Malley, "Frequency control and wind turbine technologies," IEEE Trans. Power Syst., vol. 20, no. 4, pp. 1905–1913, Nov. 2005.

[17] J. Morren, J. Pierik, and S. W. H. de Haan, "Inertial response of variable wind turbines," Elect. Power Syst. Res., vol. 76, pp. 980–987, 2006, Elsevier, Wind Energy.

[18] G. Ramtharan, J. B. Ekanayake, and N. Jenkins, "Frequency support From doubly fed induction generatorwind turbines," IET Renew. Power Gener., vol. 1, no. 1, pp. 3–9, 2007.

[19] J. Ekanayake, L. Holdsworth, and N. Jenkins, "Control of DFIG wind turbines," Power Eng. J., vol. 17, no. 1, pp. 28–32, 2003.

[20] J. Ekanayake and N. Jenkins, "Comparison of the response of doubly fed and fixed-speed induction generator wind turbines to changes in network frequency," IEEE Trans. Energy Convers., vol. 19, no. 4, pp.800–802, Dec. 2004.

[21] L. Holdsworth, J. B. Ekanayake, and N. Jenkins, "Power system frequencyresponse from fixed speed and doubly fed induction generator- based wind turbines," Wind Energy, vol. 7, pp. 21–35, 2004, DOI: 10.1002/we.105.

[22] N.W. Miller, M. Shao, S. Venkataraman, C. Loutan, M. Rothleder, "Frequency Response of California and WECC Under High Wind and Solar Conditions," Power and Energy Society General Meeting, 22–26 July 2012.

[23] H. Shun-Hsien, D. Maggio, K. McIntyre, V. Betanabhatla, J. Dumas and J. Adams, "Impact of Wind Generation on System Operations in the Deregulated Environment: ERCOT Experience," Power & Energy Society General Meeting, 26–30 July 2009.

[24] E. Muljadi, V. Gevorgian, M. Singh, S. Santoso, "Understanding Inertial and Frequency Response of Wind Power Plants," Power Electronics and Machines in Wind Applications, July 2012.

[25] L. Meegahapola, and D. Flynn, "Impact on Transient and Frequency Stability for a Power System at Very High Wind Penetration," Power and Energy Society General Meeting, 25–29 July 2010.

[26] J. Morren, S. W. De Haan, and J.A. Ferreira, "Primary Power/Frequency control with Wind Turbines and Fuel Cells," Power Engineering Society General Meeting, 2006.

[27] H.Chavez, R.Baldick, and S.Sharma, "Regulation Adequacy Analysis Under High Wind Penetration Scenarios in ERCOT Nodal," IEEE Transactions on Sustainable Energy, vol.3, no.4, pp.743,750, Oct. 2012.

Emerging Developments in the Power and Energy Industry – Dufo-López, Krzywanski & Singh (eds)
© 2020 Taylor & Francis Group, London, ISBN 978-0-367-27169-5

Smart energy measurement using Ethernet

Deepa Jose
Head Research, Associate Professor, Department of Electronics and Communication, KCG College of Technology, Chennai, Tamil Nadu, India

B. Balaji, A. Arun Raj, S. Abdullah & T. Anbumani
Department of Electronics and Communication, KCG College of Technology, Chennai, Tamil Nadu, India

ABSTRACT: The days are gone where we used analog energy meters; digital energy meters are used extensively used today. It has recently been uncovered that these meters are not completely 100% reliable. As with any other device they are subject to mechanical and human errors. Human error rating in these meters is about 0.5% which accounts to about 78,426 cases just in Tamil Nadu alone. The proposed paper was designed to overcome this human error through automation. Through this design users will be able to read their usage data within the comforts of their home with just a click of a button. This is possible through a proprietary application that is developed for both iOS and android. The design will not only be beneficial for the users but also be useful for the body that provides the energy used by the users, as it reduces the errors and also reduces the number of personnel employed to be physically present at the meter to read the data.

Keywords: Smart grid, smart energy meter, home automation, automation

1 INTRODUCTION

According to the census data the number of houses with electricity; keep increasing every cycle which is 10 years. Now, according to the 2011 census almost 92% of the homes in Tamil Nadu have electricity and 90% of these homes have installed a digital energy meter. Although these digital meters seem like an upgrade from the old analog meter, they still have defects. For example, the analog meters have an instrumental error rate of 5% and digital meters have a rate of 2.5%. Along with this there is also a human error rate as well, which is 0.5%. Overall the total error rate of the digital meter is 3%. Considering only the human error, there are about 78,246 (theoretically) errors or misreads every two months. Although this number may seem small with just a change of one number 100 units could become 700 units and the bill for these falls on the consumers. Therefore, the idea proposed here reduces the human error rate and also introduces a new module into the home automation field. With this module in our homes there will be a possibility to view our usage in real time through an application that will also be developed.

1.1 *Abbreviations*

Global System for Mobile communication (GSM), Kilobits per second (Kbps), Megabits per second (Mbps), General Packet Radio Service (GPRS), Internet of Things (IoT), Wireless Fidelity (Wi-Fi), Common Meter Reading Instrument (CMRI), Integrated Circuit (IC), Static Random Access Memory (SRAM), Electronically Erasable Programmable Read Only Memory (EEPROM).

1.2 *Units*

Kilobyte (Kb), Voltage (V), Mega Hertz (MHz), Direct Current (DC), Input/Output (i/o) Mille Ampere (mA), Mille Meter (mm), Grams (g),

2 LITERATURE SURVEY

Claudio De Capua, Gianluca Lipari, Mariacarla Lugarà, Rosario Morello, et al. (2014) presents the design and implementation of smart energy meters using a ZigBee and GSM module. A ZigBee controller is a high protocol transmission device and also is expensive. It doesn't seem feasible to be using Zigbee when Arduino is comparatively cheaper and more versatile. Its range is about is 10–20 meters and that to only in line of sight so in practicality it is even lesser. The data rate is about 256 Kbps as opposed to the 10–100 Mbps on the Ethernet shield used in Arduino. Therefore, Arduino is the better choice. Coupled with this, the GSM module is also used in this paper. GSM module and the technology are outdated. The majority of the services used today by the people are 3G and 4G services and GSM still uses 2G services. Another factor that influences the using of internet is using a GSM module requires the user to have a sim card from a telecom service provider and do frequent top-ups in order for its messaging service to work.

The system in Himshekhar Das, L.C. Saikia, et al. (2015) provides a method for controlling energy theft. The payment method for the energy consumption is through prepaid payment and this uses GSM technology to be able to pay the bill online. Reducing the theft of energy from other homes needed to be controlled and this system gave an answer for this. This was a necessity in the previously used analog energy meters but with the currently used digital meters this is not the case. They come with an array of security features and hence once these meters are installed it is possible to take the readings from them but energy theft is close to impossible. As stated earlier, the GSM technology is an outdated one with regards to the current communication bands and also is in the long run expensive. The author has proposed an idea (L. C. Saikia, Himshekhar Das, N.B. Dev Choudhury, T. Malakar, et al. 2016) which can help in making the current meter into a smart meter with the help of a GPRS module and also displaying the consumption on an in home display. Similar to the previous paper discussed the GPRS technologies relies on the 2G and 3G bands which are slower in comparison with the 4G bands and also the internet speeds provided by various internet providers. Therefore, with a transmission rate of 10–100 Mbps the Ethernet coupled with the Arduino microcontroller seems to be the perfect choice. An additional aspect of this is in home display. These are also very expensive and bulky; the same can be accomplished with a liquid crystal display. This idea (M. S. Khandare and A. Mahajan, et.al. 2010) also deals with converting the current energy meter into a smart meter by adding a GSM module but also deals with home automation or automation of home appliances. GSM-based home automation is a thing of the past as today the era has shifted from GSM to IoT. IoT, or Internet of Things, is a concept that comes out the home automation field by connecting the appliances to the internet through a single controller through Wi-Fi (Nabil Mohammad, Anomadarshi Barua and Muhammad Abdullah Arafat, et al. 2013) or Ethernet. This way the user would be able to access the appliances from anywhere around the world if need be.

3 HARDWARE AND COMMUNICATION

Common meter reading instruments, more formally known as CMRIs, are used not only to read these data from the optical port of the meter but also to test and troubleshoot the meters (O. Homa Kesav and B. Abdul Rahim, et al. 2012). This would also require the presence of a human to read the data and hence the error would still exist. The only way to reduce the human error is by eliminating the human factor from the equation, which could not only be beneficial for the consumer but it could also help the government to use them elsewhere; where they are needed. Automation can be achieved by developing a module that read the data from the meter in real time and transfer the data to cloud storage from where the user would be able to download it through the application provided.

3.1 *Microcontroller*

In order to do this, the hardware chosen is a microcontroller (as seen in Figure 1) instead of a microprocessor. The reasons for this are:

1. To maintain cost efficiency such as to develop it into a product.
2. Microprocessor would not be required as a microprocessor can perform multiple functions but the only function here is to read the data and transfer it to the cloud storage.

The type of microcontroller that is used here is the Arduino Mega board based on the Atmega 2560. This particular IC consists of a 128Kb flash memory of which 4Kb will be used by the bootloader (P. Prudhvi, D.Bhalodi, M Manohar, V.Padidela and S.Adapa, et al. 2012). Along with the flash memory it also comes with an SRAM and an EEPROM. The SRAM capacity is about 8kb and that of EEPROM is 4kb. Apart from the IC the board consists of 54 digital I/O ports and 16 analog i/o ports. This gives us the ease to use send or receive data in digital or analog format. The operating voltage of this board is in the range of 7V to 12V and the output voltage is 5V. The clock speed for this board is 16MHz and the DC per I/O pin is 20mA and that of the 3.3 V pin is 50mA. It is also small and compact; the length of the board is 101.6mm and its width 53.34mm. It is also lightweight, just 37g.

3.2 *Communication*

From Figure 1 we can see that an Ethernet module would be to transfer the data from the microcontroller to the cloud storage.

The reasons for choosing Ethernet when there are other ways to transfer the data, including Wi-Fi, GSM or Bluetooth are:

1. When using a Wi-Fi module, it has to be connected to the router in the home by searching for the signal and also by typing in the password. This increases the cost and also the connection would not be very secure. Ethernet is a wired connection so it is very secure and there is no need to input passwords.
2. GSM is pretty much an outdated technology. There are a number of disadvantages when it comes to applying it in this project. The first is that is expensive, that is, if the user were to use data through GSM, the user would have to constantly recharge the SIM used in it. The second factor is that it does not work without a SIM card which needs to be bought from a telecom carrier. The third reason is that the data would be just a series of messages and hence would be able to be stored properly. Ethernet has none of these problems and hence Ethernet is preferred over GSM.
3. Bluetooth data transfer is short range and very slow and hence it is very hard to see the data while away from your house. Since Ethernet uses the internet to transfer the data, this data can be viewed from anywhere in the world with the right credentials.

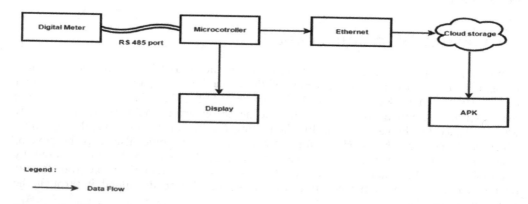

Figure 1.   Block diagram representation of the methodology.

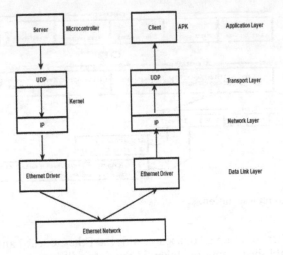

Figure 2. Ethernet communication structure.

3.3 *Energy meter*

The type of meter that will be used is one that works on the MODBUS protocol. The reasons for this are:

1. According to the inputs from various industries like L&T, HPL, etc., in the future meters running on the MODBUS protocol are going to be preferred and hence in order to be future ready this type of meter has been chosen.
2. MODBUS protocol is user-friendly and also add-ons like this module can work easily with a meter running on the MODBUS protocol.

These are the blocks which shown in the block diagram and a small gist of how they work also the reasons for choosing these particular parts.

Ethernet is used in the transmitting of data from the microcontroller to the server where the data is going to be stored. In order to this, a few other aspects of internet communication need to be looked into. These aspects are further explained below.

3.4 *MAC address*

A MAC address, that is, a media access control address, is a unique identifier assigned to the data link layer in the OSI reference model used for communication (S. S.Ali, M. Maroof, S. Hanif, et al. 2010). It is used as a network address in most IEEE 802 network technologies including Ethernet and Wi-Fi. The reason for the discussion on the Mac address is because this project requires the use of an Ethernet shield to transfer the data from the controller to the database from where the data can again be retrieved through any other device such as an android app (V. Preethi; G. Harish, et al. 2016).

The MAC address is generally assigned by the manufacturer to the network interface controller and is stored in its read-only memory. The MAC address encoded by the manufacturer is generally known as burned in address or Ethernet hardware address. This address would be formatted according to the rules of IEEE 802. The original IEEE MAC address comes from the Xerox Ethernet addressing scheme which is MAC-48. It has about 2^{48} possible MAC addresses which are contained in a 48-bit address space.

Although the MAC-48 represented in Figure 3 was mainly used for the addressing scheme, IEEE started using the EUI-48 for which it owned a patent. This in turn realized the MAC-48 as an obsolete term. This was the case when IPv4 was being used, now that IPv6 is one of the most predominant protocols users use the modified version of the EUI-64. This Mac address

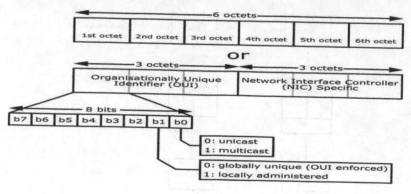

Figure 3.   MAC address representation.

is used by the controller to set up a connection with the Ethernet card and the database, once the connection is established, using the token id the data will be sent to the particular server for storage. Once the data is stored in its database, using the API given the user can retrieve the data through the Android app.

3.5   *Token identifier and API*

A token identifier is nothing but an integer that tells what type of data has been read in the input stream.

API is nothing but a set of instructions or procedures that allows the creation of an application that can read the data from operating system, application or any other service. The token id is a form of variable of id that is used to send the data from the Ethernet card and the API is a set of procedures used to retrieve the data from the database to be displayed on the application.

3.6   *IP (Internet protocol) address*

The IP address, or internet protocol address, is assigned to each device participating in a computer network such as the internet. This IP address is the identifier of the device in a global scale. The IP address is of two versions

1. IPv4
2. IPv6

From Figure 4 it is seen that the IPv4 version is a 32-bit address. This limits it to about 2^{32} unique addresses where some of them are reserved by the protocol in case of any private

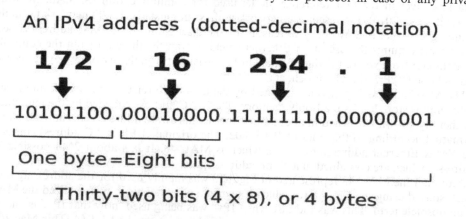

Figure 4.   IPv4 representation.

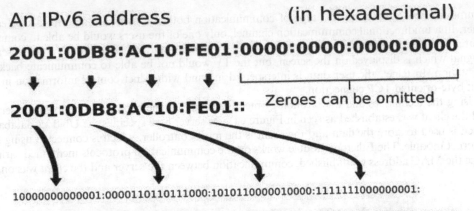

An IPv6 address   (in hexadecimal)

2001:0DB8:AC10:FE01:0000:0000:0000:0000

2001:0DB8:AC10:FE01::   Zeroes can be omitted

10000000000001:0000110110111000:1010110000010000:1111111000000001:

0000000000000000:0000000000000000:0000000000000000:0000000000000000

Figure 5.   IPv6 representation.

networks. Although the protocol consisted of 2^{32} unique addresses, they were exhausted quickly due to the high demand. This led to the development of the IPv6. In 1995 the IPv6 was coined and seemed sufficient for the foreseeable future. The IPv6 which is represented in Figure 5 is a 128-bit address, which means that it has 2^{128} unique addresses.

The main intent of the design was not only to provide sufficient quantity addresses but also to redesign routing in the internet by more efficient sub-network routing prefixes.

3.7   Sub-network

A sub-network or a subnet is a logical division of a single network into two or more networks. This way numerous computers can be connected to the same network at the same time. Most often a computer connected in a sub-network will be addressed with a common, identical, most significant bit in its IP address. Let us see this with an example. The internet is a network of computers all over the world, but these computers are not connected on the same network directly (Deepa Jose, Roshini Tamil Selvan, et al. 2016). Let us assume that there is a main network and to this network only a limited number of computers can connect at any given moment, therefore in order to accommodate the large volume of devices all over the world, this network is divided into networks and these networks are further divided into more networks. Hence at the end it finally reaches "US" the end user. In order for the message to pass through the subnet, it has to pass through a gateway.

3.8   Gateway

A gateway in basic terms is a node or a connection between the internet and the user. It acts as the protocol converter. To explain it more it is like your login page on Facebook or Gmail. Unless the data is authorized it cannot pass through this gateway. Now, as cloud storage is being used, let us see about cloud storage gateway. The cloud storage gateways allow companies to integrate their private cloud storages into application without moving the application into a public cloud, this way data security is maintained. The cloud storage gateway converts the cloud storage APIs to block-based storage protocols.

3.9   Telnet

Telnet is a protocol used on the internet to provide bidirectional communication. This way the server and the client can communicate with each other. The difference between a bidirectional and unidirectional communication is, let is the example of a telephone; this is a bidirectional communication channel. In this type of communication both the users are able to talk to each other. In a unidirectional communication channel, only one of the users would be able to communicate. The perfect example for this would be a TV and its station. Here the TV station sends a signal which is displayed on the screen, but the TV would not be able to communicate back to the station. In telnet, the user data is interspersed in band with telnet control information in an 8-bit byte-oriented TCP connection.

Using the techniques and the technology mentioned earlier, the connection between the server and the client was established as seen in Figure 6a. The server here specifies the Ubidocs database, which is used to store the data and the client is the microcontroller, which is connected using the Ethernet module. The Ethernet module works on the communication protocols mentioned earlier. Once the MAC address is established, communication between the server and the client was on its

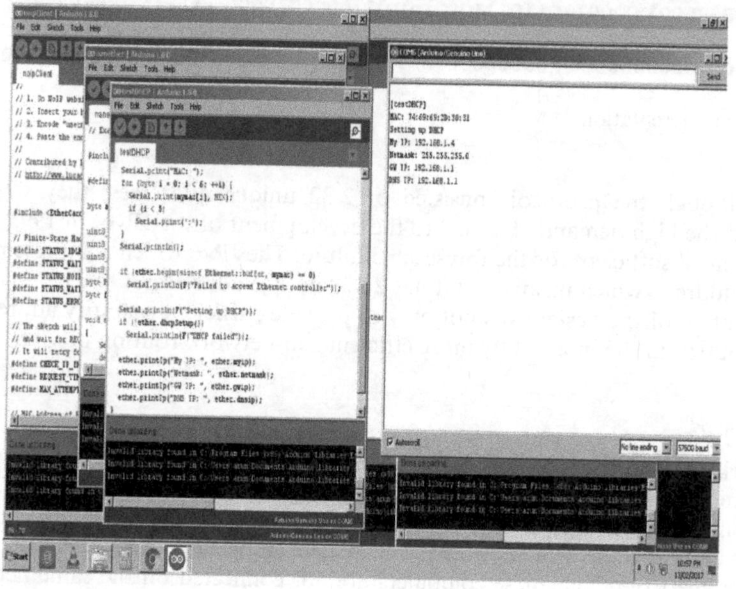

Figure 6a.   Establishment of connection.

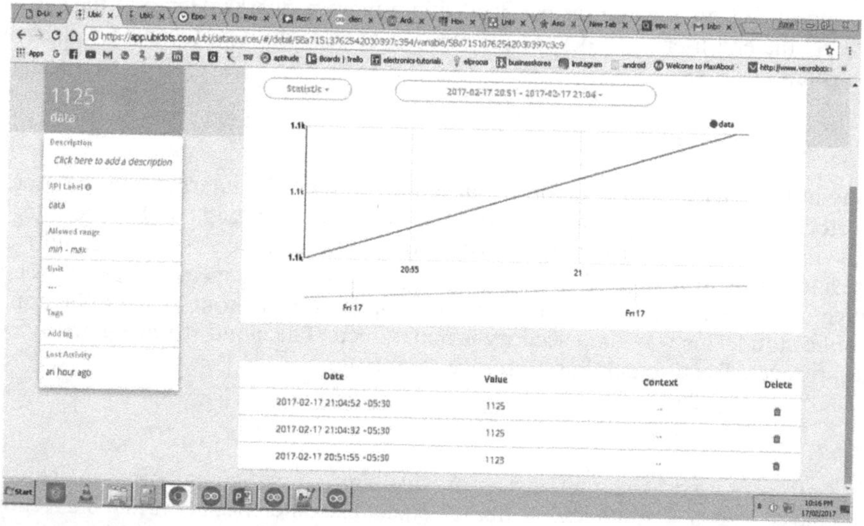

Figure 6b.   A graph representing the storage used in the server.

way. This what the first image is showing. The establishment of the connection is shown through the serial monitor of the Arduino IDE software. This software is used because this is the software used for coding the microcontroller which is an Arduino Uno board. Once the server was established, a series of data was sent to the server to be stored in the database. A graph of the storage used through these is shown in Figure 6b. This graph shows the data storage used on each of the day's the data was sent to the server. The database is also gives and API code which is used to download the data from the server to the application to be viewed by the user. This way the user is able to view the data stored in the server and database on a real-time basis.

4 IMPLEMENTATION

The plan is to implement this in two ways:

1. To provide the government a better way of reading the energy meter through automation which would reduce losses and also reduce the number of personnel employed to physically take these readings. This way the government would be able to employ them in other sectors where they are necessary.
2. This module can be developed further such it not only performs as a standalone module but also is able to combine with the existing home automation systems. This way home automation can develop in a field where it hasn't yet entered into before.

5 CONCLUSION

From the results it is possible to gather that such a system is possible and can also be cost-efficient. Being cost-efficient enables more users to be able to install it in their homes. As stated earlier this not only benefits the user by also benefits the government body providing this energy to the consumers, as it reduces the cost of employment and also reduces the cost to read the data each cycle.

REFERENCES

Claudio De Capua, Gianluca Lipari, Mariacarla Lugarà, Rosario Morello, 2014. A Smart Energy Meter for Power Grids: In International Instrumentation and Measurement Technology Conference (I2MTC) Proceedings.

Himshekhar Das, L.C. Saikia, 2015. GSM Enabled Smart Energy Meter and Automation of Home Appliances: In International Conference of Energy, Power and Environment.

L. C. Saikia, Himshekhar Das, N.B. Dev Choudhury, T. Malakar, 2016. GPRS Enabled Smart Energy Meter with In-Home Display and Application of Time of Use Pricing: In Annual India Conference.

M. S. Khandare and A. Mahajan, 2010. Mobile Monitoring System for Smart Home: In 3rd International Conference on Emerging Trends in Engineering and Technology.

Nabil Mohammad, Anomadarshi Barua and Muhammad Abdullah Arafat, 2013. A Smart Prepaid Energy Metering System to Control Electricity Theft: In International Conference on Power, Energy and Control (ICPEC).

O. Homa Kesav and B. Abdul Rahim, 2012. Automated Wireless Meter Reading System for Monitoring and Controlling Power Consumption: In International Journal of Recent Technology and Engineering 2(1).

P. Prudhvi, D. Bhalodi, M. Manohar, V. Padidela and S. Adapa, 2012. A Smart Meter Architecture in Indian Context: In 11th International Conference on Environmental and Electrical Engineering.

S. S. Ali, M. Maroof, S. Hanif, 2010. Smart Energy Meters for Energy Conservation & Minimizing Errors: In PEDES, Joint International Conference on Power Electronics, Drives and Energy Systems.

V. Preethi; G. Harish, 2016. Design and Implementation of Smart Energy Meter: In International Conference on Inventive Computational Technologies (ICICT).

Deepa Jose, Roshini Tamil Selvan, 2016. Fault Tolerant and Energy Efficient Signal Processing on FPGA Using Evolutionary Techniques, Advances: In Intelligent Systems and Computing (AISC) Computational Intelligence, Cyber Security and Computational Models, 412:155–164.

A novel low power BIST-TPG with reduced testing time

V.M. Thoulath Begam

Department of ECE, A.R. College of Engineering and technology, Tirunelveli, Tamil Nadu, India

ABSTRACT: Power dissipation is the major problem in testing VLSI circuits. The application of random test input patterns causes high switching activity which increases dynamic power dissipation. The proposed Test Pattern Generator (TPG) generates test patterns with high correlation and reduces power dissipation in Circuit Under Test (CUT) with reduced test time. It identifies the consecutive test patterns that have more transitions and inserts test vectors in between them. Test vectors are formed using random bits which have low transitions. The results on ISCAS bench marks prove that the proposed method reduces the 72 % of dynamic power dissipation.

Keywords: low power TPG, random bit insertion, reduced test time

1   INTRODUCTION

Circuits are often designed to operate in two modes: normal and test modes. The test mode usually dissipates more power than the normal mode. Power dissipation during testing is one of the most important issues. Integrated circuits and systems have been tested using a pseudorandom Built-in Self Test (BIST) generator which generates test inputs. For circuits with hard-to-detect faults, a large number of random patterns have to be generated before high fault coverage is achieved. This increases test time. These random test patterns are less correlative which increases the dynamic power dissipation also. Reducing power consumption in test mode with less test time is becoming an important objective in circuit design.

2   LITERATURE REVIEW

That the power dissipation is 200% more in test mode than normal mode is reported in (Zoria. 1993). High switching activity is one of reasons for power increases during test mode (Girard. 2002, Pomeranz I. 2014). Test vector insertion to reduce vertical switching activity between consecutive test patterns (Nourani M. et al. 2008) is the popular low power TPG method. Clock frequency also contributes to dynamic power dissipation. A modified clock scheme (Girard et al. 2001) has been proposed for low power BIST which concentrates on modifying the clock.

Sometimes altering LFSR structure (Li R. 2001) also can reduce power dissipation in test mode. Many types of low power ATPGs were proposed in (Zoria. 1993). Twisted Ring Counter (TRC) is used (Lien W. C. et al. 2013) as TPG due to their small area over head. TRC also achieves reduced test time. The new TRC scheme results in significant reductions on test application cycles and test data volume. Test sequence generation by ordering the pre computed test vectors has been presented in (Pomeranz. I. 2010). The advantage of this approach is that its computational complexity is limited by limiting the search space to a given set of input vectors and test sequence length.

Many VLSI circuits have accumulators. During test mode accumulators can be utilized as TPG (Paschalis A. et al. 2012, Keane J. et al. 2011). The accumulator based TPG scheme generates set of patterns with weights 0, 0.5, and 1. Since accumulators are commonly found in current VLSI chips, this scheme can be efficiently utilized to drive down the hardware of BIST pattern generation. Reducing power dissipation at the input of scan cells has been presented in (Atoofian et al. 2005). TPG for both CUT and scan cells are given in (Wang H. S. 2007), which has high fault coverage and low power dissipation. The patterns generated using LFSR and NFSR are analyzed and the obtained results are compared in (Zhang J. M. et al. 2015). Test patterns generated using LFSR have high fault coverage.

A High-Speed Parallel Architecture was presented in (Ayinala, Parh. 2011) where the LFSR circuits transform into equivalent state space formulations. This transformation achieves a full speed-up compared to the serial architecture at the cost of an increase in hardware overhead. The possibility of reducing the complexity of deterministic sequential test pattern generation by using subsets of primary input vectors of limited sizes during test generation for target faults were studied in (Chakrabarty, Swaminathan. 2000). In the method proposed in (Dobai. 2011) a certain number of populations had been generated and evaluated and the system was assumed to be in a non-converging state.

Seed and Polynomial selection play an important role in improving fault coverage. A method for seed generation using LFSR as TPG is proposed in (Moriyasu, Ohtake. 2013). A new polynomial selection procedure is proposed in (Haoqi Ren, Sch & Zhenya Xiong. 2015), with the minimum degree that can encode all deterministic test vectors for the hard faults. But this method performs better for the n-detection of single stuck-at -faults only. The method presented in (Irith. Pomeranz 2014) addresses the problem in computing a polynomial of small degree directly from the given test set without having to solve multiple non-linear systems and fixing a priori the polynomial degree. The polynomial computation method (Agaris. May 2015) uses an adaptation of the Berlekamp- Massey algorithm and the Sidorenko-Bossert theorem. It performs the computation of small degree polynomial.

A new feedback LFSR reseeding technique method is reported in (Lien W. C. et al. 2013). This method employs the internal net responses of the circuit itself as the control signals to change the states of the LFSR. The proposed TPG reduces power dissipation without affecting the fault coverage. It also reduces testing time. Test vector insertion increases fault coverage also.

The paper is organized as follows: Section 3 explains the proposed low power TPG design in detail. The proposed TPG for two small combinational circuits are implemented and test patterns required for 100% FC is shown in Section 4. This section clearly explains the function of TC and shows how the testing time is reduced. Section 5 shows the experimental results. Section 6 concludes the paper.

3 PROPOSED LOW POWER BIST-TPG

In the proposed TPG, generated test patterns are first applied to Transition Counter (TC) which determines the total number of vertical switching transitions between consecutive test

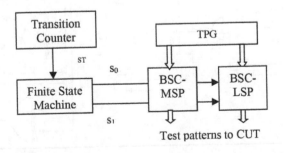

Figure 1. Proposed low power BIST-TPG.

patterns. If the number of transition is more, the output of TC ("ST") is '1' otherwise it is '0'. Depending on the value of "ST", Finite State Machine (FSM) changes its state. At each state 2 outputs are generated and applied as inputs to Bit Selector Circuit (BSC). BSC either generates and sends test vector or allows the test pattern to CUT. The test pattern is equally divided in to two parts as Most Significant Part (MSP) and Least Significant Part (LSP).For an 8-bit test pattern the MSP and LSP are having 4 bits each. Each part is processed separately.

3.1 Transition counter

TC is designed to identify more transition patterns. It is constructed using XOR gates, adders and an OR gate. Transitions between consecutive TPs are identified by using 'N' number of XOR gates, where N is the number of bits in TP. The output of XOR gates are applied to adders and their outputs are given to an OR gate which generates the TC's final output "ST".

The Figure 2 shows TC circuit for an 8-bit LFSR. Present and next test pattern bits are applied as inputs to each XOR gate. If the test bits are not same, then the XOR gate will produce the output as 1. These 1's are added using the proposed adder. More numbers of 1's are indicated by sum and carry output of adder. The last bits of sum and carry outputs only indicate more number of 1's. Therefore they are only given as OR gate inputs. Depending on the requirement, the OR gate inputs can be changed. For an example, an eight bit LFSR needs 8 XOR gates, one 8-bit adder and one OR gate. By changing the inputs of the OR gate, test patterns with different number of transition can be identified. Here S (Paschalis A. et al. 2012) and C (Paschalis A. et al. 2012) are given to OR gate to identify the patterns that have more than 3 transitions.

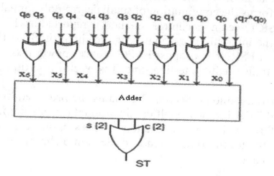

Figure 2. Transition counter.

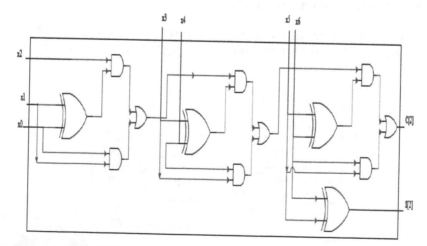

Figure 3. Proposed adder circuit.

There are many types of adders like Carry Look Ahead adder (CLA), Ripple Carry Adder (RCA) and Carry Skip and Carry Save Adder (CSA). CLA and RCA have more combinational delay. CS adders have lesser delay than RCA but with increased area. Here adders are used only to identify the number of ones at XOR outputs. Therefore initial sum bits can be neglected. In an 8-bit adder, it is found that sum (Paschalis A. et al. 2012) and carry (Paschalis A. et al. 2012) bits reflect the more of number ones.

All carry generator circuits with required sum generator are enough to identify more number of transitions. The required sum and carry bits generator circuits only included in the proposed adder. Thus TC has less delay than other adders. It identifies less correlative test patterns using fewer components.

3.2 *FSM*

FSM has 4 states and each with 3 outputs (s_0, s_1 and en). The enable line "en" is used to disable test pattern generation of LFSR during test vector insertion. s_0 and s_1 are the select lines to BSC. Figure 4 shows the state diagram of FSM.

The input of FSM is the output of TC. If the input is zero, the outputs $s_0 s_1 = 00$. Then the generated test patterns are applied to the CUT without any change. If the input is one FSM changes the states as A-B-C-D and the outputs $s_0 s_1$ is 11-10-01-00. Test vectors are inserted between the consecutive TPs. "en" is '0' and it is applied to LFSR to hold TP generation during this time.

Table 1 show the operation performed at each state. The outputs s_0 and s_1 decide which bit to be sent to the CUT. When the transitions between two test patterns are more, three test

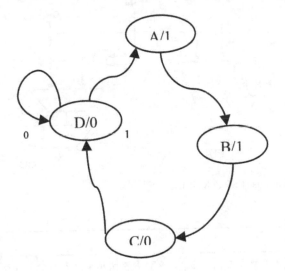

Figure 4. FSM - state diagram.

Table 1. Low power TPs generated by the proposed TPG.

FSM I/P (from TC circuit)	O/Ps States- $S_1 S_0$	TP applied to CUT	
		MSP	LSP
ST = 0	D-01	q_{n+1}	q_{n+1}
	A-11	q_n	R
ST =1	B-10	q_n	q_{n+1}
	C-00	R	q_{n+1}
	D-01	q_{n+1}	q_{n+1}

vectors are inserted between them. Test vectors are formed using random bits which reduce switching transitions.

3.3 *Bit selector circuits*

Two different types of MUX networks are used for MSP and LSP.These structures select any one of the bits among present value,q_n next value q_{n+1} or random bit (R).

The total number of test pattern bits are divided in to two parts MSP and LSP. Each part is separrately applied to BSC- MSP and BSC- LSP circuits (see Figure 5). Outputs (of FSM) s_0 & s_1 are used as select lines multiplexers. In BSC-MSP, the ouput is q_n or R when the switching transition is more and the output is q_{n+1} when the number of transition is less. Similarly, in BSC-LSP the output is R or q_{n+1} depending on the number of transition between consecutive test patterns.Outputs of FSM s_0 and s_1 are used to select BSC outputs.

3.4 *Implementation of the proposed TPG*

Figure 6 shows the implementation of the proposed method for 8-bit CUT. The polynomial of the LFSR implementation is x^7+x+1. The XOR output of first and last flip flop is given as input to the first flip flop. Inputs to BSC are the select line from FSM, present and next test pattern bits. Test pattern is divided into MS and LS parts and applied to the respective BSC circuits. Outputs of BSC are applied to CUT as test patterns.

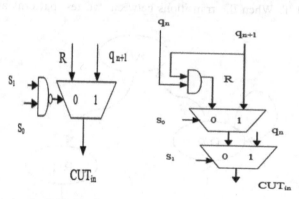

Figure 5.   Bit selector circuits (LSP & MSP).

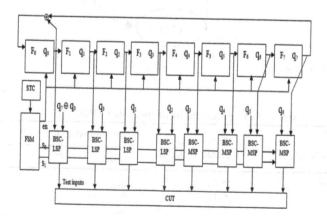

Figure 6.   Proposed method implementation (8-bit CUT).

4 PERFORMANCE OF THE PROPOSED APPROACH FOR SMALL CIRCUITS

To analyze the use of TC circuit in detail, two small combinational circuits, Magnitude Comparator and C-17 bench mark, were taken and the results were shown in Table 2. Both circuits shown in Figure 7 were tested using all three ATPG methods (Conventional LFSR, LT-LFSR and the proposed TPG) and it is proved that the proposed TPG identifies less correlative test patterns using TC circuit and inserting test vectors only in between them. Therefore the proposed TPG reduces the number of sequences required to cover all faults and reduces dynamic power dissipation.

Both circuits shown in Figure 7 were tested using all three ATPG methods (Conventional LFSR, LT-LFSR and the proposed TPG) and it is proved that the proposed TPG identifies less correlative test patterns using TC circuit and inserting test vectors only in between them. Therefore the proposed TPG reduces the number of sequences required to cover all faults and reduces dynamic power dissipation.

LT-LFSR needs more number of sequences to achieve 100% fault coverage. But the proposed TPG achieves FC with less number of sequences. LFSR has very high power dissipation. But the proposed TPG has low power dissipation (see Table 4).

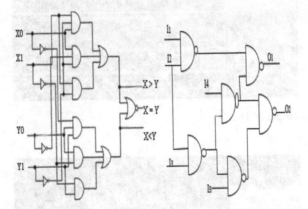

Figure 7.   a) Magnitude comparator b) C17- benchmark.

Table 2.   Comparison of existing methods with the proposed TPG.

CUT	Seed applied	Number of sequences required for 100% FC		
		LFSR	LT-LFSR	Proposed TPG
Magnitude	0101	7	11	19
Comparator	1000	9	30	18
	1101	9	18	17
	1110	10	23	13
	0100	11	37	25
C-17 bench	00001	11	38	19
mark	00011	11	35	18
	10001	15	57	27
	01111	9	26	16
	01011	10	48	42

5 EXPERIMENTAL RESULTS AND DISCUSSION

The proposed approach was verified using Xilinx 13.2 software and Verilog language. Results proved that the vertical switching transitions between consecutive test patterns are reduced using an external circuit without modifying the test patterns. Therefore fault coverage is not affected.

In the simulation output of TC (see Figure 8), the generated test patterns are indicated by 'q' and the output of TC is indicated by 'st'. Output 'st' is high for high transition test patterns and it is low when the transitions between two consecutive test patterns are less.

The simulation output of FSM is shown in Figure 9. FSM inputs are 'clk' and 'st' and 's' indicates the 2-bit outputs at each state. For low value of 'st', FSM stays at a single state D (D=00). When 'st' is high FSM travels through four states A-B-C-D (as shown in Table 1).

In the simulation output of BSC (see Figure 10), the input 'q' indicates the test patterns generated by LFSR. Select lines s_0 s_1 are the FSM outputs. Output 'out' indicates the reduced

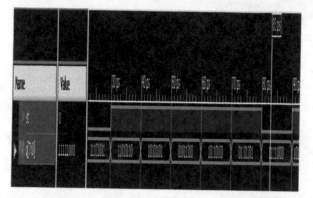

Figure 8. Simulation output of TC.

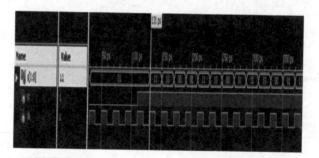

Figure 9. Simulation output of FSM.

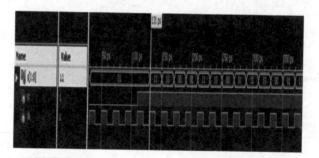

Figure 10. Simulation output of BSC.

Table 3. Application of the proposed TPG to ISCAS benchmarks.

| CUT | FC % | Number of sequences required | | | TT reduction (%) |
		LFSR	LT-LFSR	Proposed TPG	
c1908	95.9	996	863	734	14.9
c3540	97.8	1164	1052	894	15
c5315	99.7	1129	1111	944	14
s15850	98.8	96640	96413	86772	9.9
s38417	97.5	115882	116208	104587	10
s38584	99.3	79712	79360	71424	10

* LT-LFSR (Nourani M. et al. 2008)

Table 4. Peak power dissipation using different LFSRs.

| Circuit | P_{peak} (in mW) | |
	LFSR	Proposed TPG
c1908	1455.3	396.9
c3540	2205	573.3
s38417	1675.8	441
s38584	1234.8	308.7

transitions test patterns. If the number of transitions is more, BSC inserts 3 test vectors in between the consecutive test patterns.

In Table 3, FC and TT denote Fault Coverage and Test Time respectively. Number of sequences required by different TPG methods to achieve the FC % (column 2) has been determined. It is proved that using the proposed method 10 to 15% of testing time is reduced. Random pattern generation is done here by using a simple AND gate. But LT-LFSR generates random bit using 3 AND gates and 1 OR gate. Sometimes this logic increases number of transitions instead of reducing it. The proposed method does not increase unnecessary transitions like LT-LFSR. The generated test patterns are not modified by the proposed approach. Therefore fault coverage is not affected. Required number of test sequences for particular fault coverage is lesser than other methods (see Tables 2 and 3).

Table 4 shows the peak power dissipation measured for different test bench marks. The voltage applied is 1.2V, load capacitance value is 2pF and the cycle frequency is 100MHz. Minimum 72% of peak power is reduced in all circuits. The insertion of test vectors also improves fault coverage. Using TC circuit, test vector insertion is done whenever it is needed. This reduces testing time.

6 CONCLUSION

In the proposed TPG, less correlative TPs were identified by TC circuit and test vectors were inserted only in between them. These reduced transition patterns are applied to the CUT. Insertion of test vectors increases the correlation between test patterns by reducing the number of switching transitions. This reduces dynamic power dissipation and test time. But the existing low power TPG methods insert test vectors in between all consecutive test patterns.

The proposed approach uses simple circuit with fewer components. Therefore the combinational delay is lesser than LT-LFSR. Thus the proposed test pattern generation method reduces vertical transitions and test time without affecting the fault coverage.

The future development includes the design of horizontal transition reduction circuit with the proposed TPG which is used to test registers in sequential circuits.

REFERENCES

Zoria. 1993. A Distributed BIST Control Scheme for Complex VLSI Devices, IEEE VLSI Test Symposium, pp. 4–9.

Girard. 2002. Survey of Low Power Testing VLSI Circuits. IEEE Design and Test of Computers, 19/3, pp.80–90.

Irith. Pomeranz 2014. Unknown Output Values of Faulty Circuits and Output Response Compaction, Volume 33, Number 2, pp. 323–327.

Mehrad Nourani, Mohammad Tehranipoor, and Nisar Ahmed. 2008. Low Transition test pattern generator for BIST-based applications, IEEE transactions on computers, Vol.57.

Girard, Guiller, Landrault. 2001. A Modified clock scheme for a low power BIST test pattern generator, Proceedings VLSI test symposium, pp.306–311.

Rui Li. 2001. A new low power BIST methodology by altering the structure of linear feedback shift registers, Proceedings IEE International Conference on ASIC.

Wei-Cheng Lien, Kuen-Jong Lee, Tong-Yu Hsieh, and Wee-Lung Ang. 2013. An Efficient On-Chip Test Generation Scheme Based on Programmable and Multiple Twisted-Ring Counters, IEEE Transactions on Computer-Aided Design of Integrated Circuits and Systems, Vol. 32, No. 8, August 2013, pp.1254–1264.

Irith Pomeranz. 2010. TOV: Sequential Test Generation by Ordering of Test Vectors, IEEE Transactions On Computer-Aided Design of Integrated Circuits and Systems, Vol. 29, No. 3, pp.454–465.

Antonis Paschalis, Ioannis Voyiatzis, and Dimitris Gizopoulos. 2012. Accumulator Based 3-Weight Pattern Generation IEEE Transactions on Very Large Scale Integration (VLSI) Systems, Vol. 20, No. 2, pp. 357–361.

John Keane, Shrinivas Venkatraman, Paulo Butzen and Kim. 2011. An Array-Based Test Circuit for Fully Automated Gate Dielectric Breakdown Characterization, IEEE Transactions on Very Large Scale Integration (VLSI) Systems, Vol. 19, No. 5, pp. 787–795.

Atoofian, Hatami, Navabi, Alisaface, and Afzali-Kusha. 2005. A New Low-Power Scan-Path Architecture, IEEE International Symposium, Vol.5, pp.5278–55281.

He Seongmoon Wang. 2007. A BIST TPG for low power dissipation and high fault coverage, IEEE transactions on VLSI systems, Vol.15, No.7.

Jia-Min Zhang, Wen-Feng Qi, Tian Tian, Zhong-Xiao Wang. 2015. Further Results on the Decomposition of an NFSR Into the Cascade Connection of an NFSR Into an LFSR, IEEE Transactions on Information Theory Vol.61, Issue.1, pp. 645–654.

Ayinala, Parh. 2011. High-Speed Parallel Architectures for Linear Feedback Shift Registers, IEEE Transactions on Signal Processing, 59/9, 9, pp. 4459.

Chakrabarty, Swaminathan. 2000. Built In Self Testing Of High Performance Circuits Using Twisted Ring Counters, Proceedings of the IEEE Internaitonal Symposium on Circuits and Systems, pp. 72.

Dobai. 2011. Test Generation for Asynchronous Sequential Digital Circuits, Information Sciences and Technologies Bulletin of the ACM Slovakia, vol. 3, no. 1, pp. 73–83.

Moriyasu, Ohtake. 2013. A Method of LFSR Seed Generation for Scan-Based BIST Using Constrained ATPG, Proceedings of The Seventh International Conference on Complex, Intelligent, and Software Intensive Systems (CISIS), IEEE Computer Society 2013, pp. 755–759.

Haoqi Ren, Sch & Zhenya Xiong. 2015. A Multi-polynomial LFSR Based BIST Pattern Generator for Pseudorandom Testing Information Science and Control Engineering (ICISCE), 2nd International Conference, pp .568.

Irith. Pomeranz 2014. Unknown Output Values of Faulty Circuits and Output Response Compaction, Volume 33, Number2, pp. 323–327.

Agaris. May 2015. On The Computation of LFSR Characteristic Polynomials for Built-in Deterministic Test Pattern Generation, IEEE Transactions on Computers Vol. PP, Issue. 99.

Wei-Cheng Lien, Kuen-Jong Lee, Tong-Yu Hsieh. 2013. A New LFSR Reseeding Scheme via Internal Response Feedback 22nd Asian Test Symposium.

Mehrad Nourani, Mohammad Tehranipoor, and Nisar Ahmed. 2008. Low Transition test pattern generator for BIST-based applications, IEEE transactions on computers, Vol.57.

Moriyasu, Ohtake. 2013. A Method of LFSR Seed Generation for Scan-Based BIST Using Constrained ATPG, Complex, Intelligent, and Software Intensive Systems (CISIS), Seventh International Conference in 2013, pp. 755–759.

Emerging Developments in the Power and Energy Industry – Dufo-López, Krzywanski & Singh (eds)
© 2020 Taylor & Francis Group, London, ISBN 978-0-367-27169-5

Straightforward derivation of the wye-delta transformations based solely on Kirchhoff's law

Xu-Sheng Chen
The Department of Electrical and Computer Engineering, Seattle University, Seattle, WA, USA

Kirstin Schauble, Noah Weller & Mirka Mandich
Electrical and computer engineering students of Seattle University, Seattle, WA, USA

ABSTRACT: Wye-delta and delta-wye transformations are derived based solely on Kirchhoff's current and voltage laws and by using the nodal and mesh analysis by inspection. The duality between the wye and delta connections is explored.

Keywords: Wye-delta transformations, Kirchhoff's laws, Nodal analysis, Mesh analysis, Duality

1 INTRODUCTION

Situations often arise in circuit analysis when the resistors (or impedances) are neither in parallel nor in series. They occur in electrical filters, matching networks, and especially in three-phase power systems. The wye-delta circuit transformation theory was published by A. E. Kennelly in 1899 (A. E. Kennelly, 1899). It is widely used in the analysis of three-phase electrical power circuits (W. H. Hayt, 1997). Today, wye-delta transformations appear in every introductory circuit textbooks and many of the electrical machinery and power system analysis textbooks. It is amazing that none of the current derivations of the wye-delta transformations are based solely on Kirchhoff's voltage and current laws (KVL and KCL). This paper fills in the gap by presenting straightforward derivations of both the wye-delta and delta-wye transformations.

Together with parallel and series reduction, the three techniques can be used to simplify any complicated planar networks. Because of the computer, the computational importance of wye-delta transformations is decaying, however, its theoretical and educational value remains.

2 WYE-DELTA TRANSFORMATION

2.1 Figure 1(a) shows the wye-connected circuit powered by three DC current sources. Although for simplicity, DC sources and resistors are used in the circuit, AC sources and impedances can be used as well. Nodal analysis by inspection gives,

$$G_{11}V_{an} + G_{12}V_{bn} + G_{13}V_{cn} + G_{14}V_{on} = I_I$$
$$G_{21}V_{an} + G_{22}V_{bn} + G_{23}V_{cn} + G_{24}V_{on} = I_{II}$$
$$G_{31}V_{an} + G_{32}V_{cn} + G_{33}V_{cn} + G_{34}V_{on} = I_{III}$$
$$G_{41}V_{an} + G_{42}V_{cn} + G_{43}V_{cn} + G_{44}V_{on} = I_{IV}$$

Where

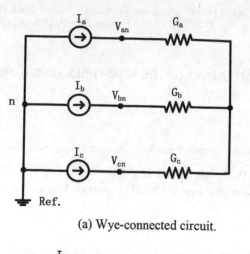

(a) Wye-connected circuit.

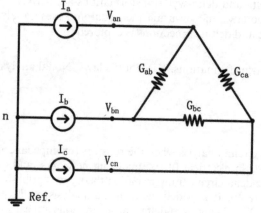

(b) Delta-connected circuit.

Figure 1.   The wye- and delta-connected circuits. (a) Wye-connected circuit (b) Delta-connected circuit.

$$G_{11} V_{an} + G_{12} V_{bn} + G_{13} V_{cn} + G_{14} V_{on} = I_I$$
$$G_{21} V_{an} + G_{22} V_{bn} + G_{23} V_{cn} + G_{24} V_{on} = I_{II}$$
$$G_{31} V_{an} + G_{32} V_{cn} + G_{33} V_{cn} + G_{34} V_{on} = I_{III}$$
$$G_{41} V_{an} + G_{42} V_{cn} + G_{43} V_{cn} + G_{44} V_{on} = I_{IV}$$

Where:

$$G_{11} = G_a, \quad G_{12} = 0, \quad G_{13} = 0, \quad G_{14} = -G_a, \quad I_I = I_a$$
$$G_{21} = 0, \quad G_{22} = G_b, \quad G_{23} = 0, \quad G_{24} = -G_b, \quad I_{II} = I_b$$
$$G_{31} = 0, \quad G_{32} = 0, \quad G_{33} = G_c, \quad G_{34} = -G_c, \quad I_{III} = I_c$$
$$G_{41} = -G_a, \quad G_{42} = -G_b, \quad G_{43} = -G_c,$$
$$G_{44} = G_a + G_b + G_c = G, \quad I_{IV} = 0$$

Therefore:

861

$$G_a V_{an} - G_a V_{on} = I_a \tag{1}$$

$$G_b V_{bn} - G_b V_{on} = I_b \tag{2}$$

$$G_c V_{cn} - G_c V_{on} = I_c \tag{3}$$

$$-G_a V_{an} - G_b V_{bn} - G_c V_{cn} + G V_{on} = 0 \tag{4a}$$

To verify the equations obtained by inspection, KCL at V_{an}, V_{bn}, and V_{cn} is performed, respectively:

$$G_a(V_{an} - V_{on}) = I_a$$
$$G_b(V_{bn} - V_{on}) = I_b$$
$$G_c(V_{cn} - V_{on}) = I_c$$
$$G_a(V_{on} - V_{an}) + G_b(V_{on} - V_{bn}) + G_c(V_{on} - V_{cn}) = 0$$

The above equations are identical to Eqs. (1)-(4a).
Solving V_{on} from Eq. (4a) results in,

$$V_{on} = (1/G)(G_a V_{an} + G_b V_{bn} + G_c V_{cn}) \tag{4b}$$

Substituting Eq. (4) into Eq. (1) and (2) results in:

$$\frac{G_a G_b + G_c G_a}{G} V_{an} - \frac{G_a G_b}{G} V_{bn} - \frac{G_c G_a}{G} V_{cn} = I_a \tag{5}$$

$$-\frac{G_a G_b}{G} V_{an} + \frac{G_a G_b + G_b G_c}{G} V_{bn} - \frac{G_b G_c}{G} V_{cn} = I_b \tag{6}$$

$$-\frac{G_c G_a}{G} V_{an} - \frac{G_b G_c}{G} V_{bn} - \frac{G_c G_a + G_b G_c}{G} V_{cn} = I_c \tag{7}$$

2.2 Figure 1(b) shows the delta-connected circuit powered by three DC current sources. Nodal analysis by inspection gives,

$$G_{11} V_{an} + G_{12} V_{bn} + G_{13} V_{cn} = I_I$$
$$G_{21} V_{an} + G_{22} V_{bn} + G_{23} V_{cn} = I_{II}$$
$$G_{31} V_{an} + G_{32} V_{bn} + G_{33} V_{cn} = I_{III}$$

Where:

$$G_{11} = G_{ab} + G_{ca}, \quad G_{12} = -G_{ab}, \quad G_{13} = -G_{ca}, \quad I_I = I_a$$
$$G_{21} = -G_{ab}, \quad G_{22} = G_{ab} + G_{bc}, \quad G_{23} = -G_{bc}, \quad I_{II} = I_b$$
$$G_{31} = -G_{ca}, \quad G_{32} = -G_{bc}, \quad G_{33} = G_{ca} + G_{bc}, \quad I_{III} = I_c$$

Therefore:

$$(G_{ab} + G_{ca}) V_{an} - G_{ab} V_{bn} - G_{ca} V_{cn} = I_a \tag{8}$$

$$-G_{ab} V_{an} + (G_{ab} + G_{bc}) V_{bn} - G_{bc} V_{cn} = I_b \tag{9}$$

$$-G_{ca}V_{an} - G_{bc}V_{bn} + (G_{ca} + G_{bc})V_{cn} = I_c \qquad (10)$$

Again, the equations attained above can be verified using KCL at nodes V_{an}, V_{bn}, V_{cn} respectively:

$$G_{ab}(V_{an} - V_{bn}) + G_{ca}(V_{an} - V_{cn}) = I_a$$
$$G_{bc}(V_{bn} - V_{cn}) + G_{ab}(V_{bn} - V_{an}) = I_b$$
$$G_{ca}(V_{cn} - V_{an}) + G_{bc}(V_{cn} - V_{bn}) = I_c$$

The above equations are identical to Eqs. (8)-(10).

2.3 The wye and delta circuits are equivalent if the same three current sources, I_a, I_b, and I_c, generate the same terminal voltages, V_{an}, V_{bn}, and V_{cn} in the circuits shown in Figures 1(a) and 1(b). This requirement can be satisfied if the coefficients in Eqs. (5) and (8), and in Eqs. (6) and (9) are equal, respectively.

$$\frac{G_a G_b + G_c G_a}{G} V_{an} - \frac{G_a G_b}{G} V_{bn} - \frac{G_c G_a}{G} V_{cn} = I_a \qquad (5)$$

$$-\frac{G_a G_b}{G} V_{an} + \frac{G_a G_b + G_b G_c}{G} V_{bn} - \frac{G_b G_c}{G} V_{cn} = I_b \qquad (6)$$

$$-\frac{G_c G_a}{G} V_{an} - \frac{G_b G_c}{G} V_{bn} - \frac{G_c G_a + G_b G_c}{G} V_{cn} = I_c \qquad (7)$$

$$(G_{ab} + G_{ca})V_{an} - G_{ab}V_{bn} - G_{ca}V_{cn} = I_a \qquad (8)$$

$$-G_{ab}V_{an} + (G_{ab} + G_{bc})V_{bn} - G_{bc}V_{cn} = I_b \qquad (9)$$

$$-G_{ca}V_{an} - G_{bc}V_{bn} + (G_{ca} + G_{bc})V_{cn} = I_c \qquad (10)$$

Setting the coefficients equal results in wye to delta transformation:

$$G_{ab} = \frac{G_a G_b}{G} \quad (11a) \qquad\qquad R_{ab} = R_a + R_b + \frac{R_a R_b}{R_c} \quad (11d)$$

$$G_{bc} = \frac{G_b G_c}{G} \quad (11b) \quad \text{or} \quad R_{bc} = R_b + R_c + \frac{R_b R_c}{R_a} \quad (11e)$$

$$G_{ca} = \frac{G_c G_a}{G} \quad (11c) \qquad\qquad R_{ca} = R_c + R_a + \frac{R_c R_a}{R_b} \quad (11f)$$

3 DELTA-WYE TRANSFORMATION

Now that the wye to delta transformation is complete, similar techniques can be used to derive a delta to wye transformation. For this derivation, the delta-connected circuit is shown in Figure 2(a) and the wye-connected circuit is shown in Figure 2(b).

3.1 Figure 2(a) shows the delta-connected circuit powered by three DC voltage sources. Nodal analysis by inspection gives,

$$R_{11}I_1 + R_{12}I_2 + R_{13}I_3 = V_I$$
$$R_{21}I_1 + R_{22}I_2 + R_{23}I_3 = V_{II}$$
$$R_{31}I_1 + R_{32}I_2 + R_{33}I_3 = V_{III}$$

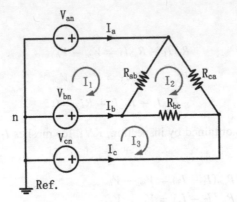

(a) Delta-connected circuit.

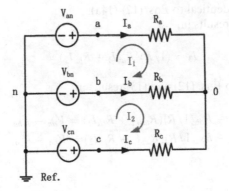

(b) Wye-connected circuit.

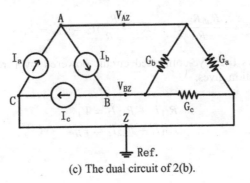

(c) The dual circuit of 2(b).

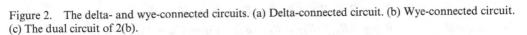

Figure 2.   The delta- and wye-connected circuits. (a) Delta-connected circuit. (b) Wye-connected circuit. (c) The dual circuit of 2(b).

Where:

$$R_{11} = R_{ab}, \quad R_{12} = 0, \quad R_{13} = -R_{ab}, \quad V_I = V_{an} - V_{bn}$$
$$R_{21} = 0, \quad R_{22} = R_{bc}, \quad R_{23} = -R_{bc}, \quad V_{II} = V_{bn} - V_{cn}$$
$$R_{31} = -R_{ab}, \quad R_{32} = -R_{bc}, \quad R_{33} = R_{ab} + R_{bc} + R_{ca} = R, \quad V_{III} = 0$$

864

Therefore:

$$R_{ab}I_1 - R_{ab}I_3 = V_{an} - V_{bn} \tag{12}$$

$$R_{bc}I_2 - R_{bc}I_3 = V_{bn} - V_{cn} \tag{13}$$

$$-R_{ab}I_1 - R_{bc}I_2 + RI_3 = 0 \tag{14a}$$

To verify the equations obtained by inspection, KVL for meshes I_1, I_2, and I_3 is performed, respectively:

$$R_{ab}(I_1 - I_3) = V_{an} - V_{bn}$$
$$R_{bc}(I_2 - I_3) = V_{bn} - V_{cn}$$
$$R_{ab}(-I_1 + I_3) + R_{bc}(-I_2 + I_3) + R_{ca}I_3 = 0$$

The above equations are identical to Eqs. (12)-(14a).
Solving I_3 from Eq. (14a) results in,

$$I_3 = (1/R)(R_{ab}I_1 + R_{bc}I_2) \tag{14b}$$

Substituting Eq. (14b) into Eqs. (12) and (13) results in:

$$R_{ab}I_1 - R_{ab}(1/R)(R_{ab}I_1 + R_{ca}I_2) = V_{an} - V_{bn}$$
$$R_{bc}I_2 - R_{bc}(1/R)(R_{ab}I_1 + R_{bc}I_2) = V_{bn} - V_{cn}$$

which simplifies to:

$$(R_{ab} - \frac{R_{ab}^2}{R})I_1 - \frac{R_{ab}R_{bc}}{R}I_2 = V_{an} - V_{bn} \tag{15}$$

$$-\frac{R_{ab}R_{bc}}{R}I_1 + (R_{bc} - \frac{R_{bc}^2}{R})I_2 = V_{bn} - V_{cn} \tag{16}$$

3.2 Figure 2(b) shows the wye-connected circuit powered by three DC voltage sources. Mesh analysis by inspection gives,

$$R_{11}I_1 + R_{12}I_2 = V_I$$
$$R_{21}I_1 + R_{22}I_2 = V_{II}$$

Where:

$$R_{11} = R_a + R_b, \quad R_{12} = -R_b, \quad V_I = V_{an} - V_{bn}$$
$$R_{21} = -R_b, \quad R_{22} = R_b + R_c, \quad V_{II} = V_{bn} - V_{cn}$$

Therefore:

$$(R_a + R_b)I_1 - R_bI_2 = V_{an} - V_{bn} \tag{17}$$

$$-R_bI_1 - (R_b + R_c)I_2 = V_{bn} - V_{cn} \tag{18}$$

Again, Eqs. (17), (18) and (19) can be verified using KVL for meshes I_1, I_2, and I_3:

865

$$R_a I_1 + R_b (I_1 - I_2) = V_{an} - V_{bn} \tag{19}$$

$$R_b (I_2 - I_1) - R_c I_2 = V_{bn} - V_{cn} \tag{20}$$

The above equations are identical to Eqs. (17) and (18).

The delta and wye circuits are equivalent if the same three voltage sources, V_{an}, V_{bn}, and V_{cn}, generate the same mesh currents, I_1 and I_2, in Figure 2(a) and Figure 2(b). This requirement can be satisfied if the coefficients in Eqs. (15) and (17) and in Eqs. (16) and (18) are equal respectively.

$$(R_{ab} - \frac{R_{ab}^2}{R})I_1 - \frac{R_{ab} R_{bc}}{R} I_2 = V_{an} - V_{bn} \tag{15}$$

$$-\frac{R_{ab} R_{bc}}{R} I_1 + (R_{bc} - \frac{R_{bc}^2}{R})I_2 = V_{bn} - V_{cn} \tag{16}$$

$$(R_a + R_b)I_1 - R_b I_2 = V_{an} - V_{bn} \tag{17}$$

$$-R_b I_1 - (R_b + R_c)I_2 = V_{bn} - V_{cn} \tag{18}$$

Setting the coefficients equal results in delta to wye to transformation:

$$R_b = \frac{R_{bc} R_{ab}}{R} \tag{19b}$$

$$R_a + R_b = \frac{R_{ab} R_{bc} + R_{bc} R_{ca}}{R} \tag{21}$$

$$R_c + R_a = \frac{R_{ca} R_{ab} + R_{bc} R_{ca}}{R} \tag{22}$$

Substituting Eq. (19b) into Eqs. (21) and (22) results in,

$$R_a = \frac{R_{ab} R_{ca}}{R} \tag{19a}$$

$$R_b = \frac{R_{bc} R_{ab}}{R} \tag{19b}$$

$$R_c = \frac{R_{ca} R_{bc}}{R} \tag{19c}$$

Where $R = R_{ab} + R_{bc} + R_{ca}$
or

$$G_a = G_{ab} + G_{ca} + \frac{G_{ab} G_{ca}}{G_{bc}} \tag{19d}$$

$$G_b = G_{bc} + G_{ab} + \frac{G_{bc} G_{ab}}{G_{ca}} \tag{19e}$$

$$G_c = G_{ca} + G_{bc} + \frac{G_{ca} G_{bc}}{G_{ab}} \tag{19f}$$

where $G = G_a + G_b + G_c$.

4 DISCUSSION AND CONCLUSION

In summary, wye-to-delta transformation is given by,

$$R_{ab} = R_a + R_b + \frac{R_a R_b}{R_c} \tag{11d}$$

$$R_{bc} = R_b + R_c + \frac{R_b R_c}{R_a} \tag{11e}$$

$$R_{ca} = R_c + R_a + \frac{R_c R_a}{R_b} \tag{11f}$$

and delta-to-wye transformation is given by

$$R_a = \frac{R_{ab} R_{ca}}{R} \tag{19a}$$

$$R_b = \frac{R_{bc} R_{ab}}{R} \tag{19b}$$

$$R_c = \frac{R_{ca} R_{bc}}{R} \tag{19c}$$

where $R = R_{ab} + R_{bc} + R_{ca}$

Figure 3 shows the wye-delta transformations. For wye to delta transformation, R_{ab} is equal to the sum of the two resistors connected to point a, R_a, and point b, R_b, plus $R_a R_b/R_c$. For delta to wye transformation, R_a is equal to the product of the two resistors connected to point a, divided by the sum of the three delta-connected resistors. Because of symmetry, the other terms of the wye-delta transformations can be readily written out.

W. H. Hayt states, "Two circuits are *duals* if the mesh equations that characterize one of them have the same mathematical form as the nodal equations that characterize the other. They are said to be *exact duals* if each mesh equation of the one circuit is numerically identical with the corresponding nodal equation of the other." (C. K. Alexander & M. N. O. Sadiku, 2013)

Comparing Eqs. (11d), (11e), and (11f) with Eqs. (19d), (19e), and (19f), and Eqs. (19a), (19b), and (19c) with Eqs. (11a), (11b), and (11c), it is clear that wye and delta connections

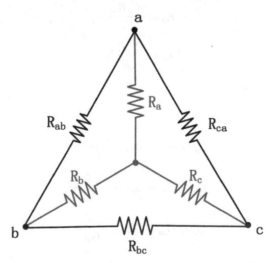

Figure 3. Wye-Delta transformations.

are duals, but not exact duals. It is important to point out that the circuits of Figures 1(a) and 1(b), and Figures 2(a) and 2(b) are not dual circuits. Following the procedure outlined by C. K. Alexander, the dual circuit for Figure 2(b) is not Figure 2(a), but Figure 2(c).

REFERENCES

A. E. Kennelly, "The Equivalence of Triangles and Three-Pointed Stars in Conducting Networks." Elcc. World and Engr.; Vol. 34, 1899, pp. 413–414.

W. H. Hayt, Engineering Circuit Analysis, 7th Ed, p.0875 Prentice-Hall, 1997, pp. 96–105.

C. K. Alexander and M. N. O. Sadiku, Fundamentals of Electerica Circuits, McGraw Hill Companies, 5th Ed, 2013, pp. 351–353.

Emerging Developments in the Power and Energy Industry – Dufo-López, Krzywanski & Singh (eds)
© 2020 Taylor & Francis Group, London, ISBN 978-0-367-27169-5

Author Index